ENCYCLOPEDIA OF

science
AND technology
ethics

EDITORS AND CONSULTANTS

ENCYCLOPEDIA OF
science
technology
AND ethics

EDITED BY

CARL MITCHAM

volume

3

l–r

MACMILLAN REFERENCE USA

An imprint of Thomson Gale, a part of The Thomson Corporation

Detroit • New York • San Francisco • San Diego • New Haven, Conn. • Waterville, Maine • London • Munich

Encyclopedia of Science, Technology, and Ethics
Carl Mitcham, Editor in Chief

LIBRARY OF CONGRESS CATALOGING-IN-PUBLICATION DATA

Encyclopedia of science, technology, and ethics / edited by Carl Mitcham.
p. cm.
Includes bibliographical references and index.
ISBN 0-02-865831-0 (set, hardcover : alk. paper)—ISBN 0-02-865832-9 (v. 1) — ISBN 0-02-865833-7 (v. 2)—ISBN 0-02-865834-5 (v. 3)—ISBN 0-02-865901-5 (v. 4)
 1. Science—Moral and ethical aspects—Encyclopedias.
 2. Technology—Moral and ethical aspects–Encyclopedias.
 I. Mitcham, Carl. Q175.35.E53 2005
503—dc22   005006968

While every effort has been made to ensure the reliability of the information presented in this publication, Thomson Gale does not guarantee the accuracy of the data contained herein. Thomson Gale accepts no payment for listing; and inclusion in the publication of any organization, agency, institution, publication, service, or individual does not imply endorsement of the editors or publisher. Errors brought to the attention of the publisher and verified to the satisfaction of the publisher will be corrected in future editions.

This title is also available as an e-book.
ISBN 0-02-865991-0
Contact your Thomson Gale representative for ordering information.

Printed in the United States of America
10 9 8 7 6 5 4 3 2 1

L

LASSWELL, HAROLD D.

• • •

Born in Donnellson, Illinois, Harold D. Lasswell (1902–1978) was an innovator in a number of scientific disciplines and the major figure in developing the policy sciences. The son of a teacher and a Presbyterian minister, he was educated at the University of Chicago, earning a doctorate in political science and then joining the faculty in 1926. In 1938 Lasswell moved to Washington, DC, to serve as a researcher and policy adviser. After the war, as a professor at Yale, Lasswell collaborated with the lawyer legal scholar Myres S. McDougal (1906–1998) and others on law, science, and policy. His broad interests and travels brought him into direct contact with many of the major intellectual and political figures of his time.

Lasswell wrote that "it is growth of insight, not simply of the capacity of the observer to predict the future operation of an automatic compulsion, or of a non-personal factor, that represents the major contribution of the scientific study of interpersonal relations to policy" (1951, p. 524). Insight brings those factors into conscious awareness, leaving the individual free to take them into account in making choices. Freedom through insight often modifies interpersonal relationships; hence, all propositions about those relationships are subject to new insight. Lasswell took the lead in developing the intellectual tools of the policy sciences to integrate and apply natural and social science insight to the fuller realization of human dignity for all, including freedom.

In his presidential address to the American Political Science Association, Lasswell chose "to inquire into the possible reconciliation of man's mastery over Nature [through science-based technologies] with freedom, the overriding goal of policy in our body politic" (1956, p. 961). At the outset he considered atomic weapons in order to entertain the proposition that "our intellectual tools have been sufficiently sharp to enable political scientists to make a largely correct appraisal of the consequences of unconventional weapons for world politics." After using those tools to sketch the kind of analysis that could have been done before the use of atomic weapons in 1945, he concluded that the profession had not institutionalized procedures to anticipate technical developments that had been reported publicly before the war and clarify in advance the main policy alternatives open to decision makers: "As political scientists we should have anticipated fully both the bomb and the significant problems of policy that came with it" (Lasswell 1956, p. 965).

Lasswell qualified this statement of professional responsibility, however: "I do not want to create the impression that all would have been well if we had been better political scientists, and that we must bear upon our puny shoulders the burden of culpability for the state of the world today. We are not so grandiose as to magnify our role or our responsibility beyond all proportion. Yet I cannot refrain from acknowledging ... that we left the minds of our decision makers flagrantly unprepared to meet the crisis precipitated by the bomb" (1956, p. 965). Moreover, the profession was not responsible for information on the bomb withheld by officials. "We must however assume responsibility for any limitation of theory or procedure that prevented us from making full use of every opportunity open to us" (Lasswell 1956, p. 964).

Turning to the future, Lasswell asserted, "It is our responsibility to flagellate our minds toward creativity, toward bringing into the stream of emerging events conceptions of future strategy that, if adopted, will increase the probability that ideal aspirations will be more approximately realized" (1956, p. 966). Lasswell accepted that responsibility when he applied the intellectual tools of the policy sciences to potential applications of science in production of material goods and evolution of intelligent organisms (including humans) and machines as well as weapons. Particularly creative and prescient were certain remarks on the implications of genetics, embryology, and intelligent machines for evolution (Lasswell 1956, pp. 975–977):

- Because new species already had been created or re-created experimentally, "A garrison police regime fully cognizant of science and technology can, in all probability, eventually aspire to biologize the class and caste system by selective breeding and training."

- Because machines already had solved complex problems, "at what point do we accept the incorporation of relatively self-perpetuating and mutually influencing 'super-machines' or 'ex-robots' as being entitled to the policies expressed in the Universal Declaration [of Human Rights]?"

- Perhaps most disturbing was "the possibility that super-gifted men, or even new species possessing superior talent, will emerge as a result of research and development ... introducing a biological elite capable of treating us [as] imperial powers have so often treated the weak."

Lasswell concluded by outlining a program of contextual and problem-oriented research using the tools of the policy sciences to address the aggregate effects of any specific innovation: "Our first professional contribution ... is to project a comprehensive image of the future for the purpose of indicating how our overriding goal values are likely to be affected if current policies continue" (1956, pp. 977–978). The concluding task is "inventive and evaluative. It consists in originating policy alternatives by means of which goal values can be maximized. In estimating the likely occurrence of an event (or event category), it is essential to take into account the historical trends and the scientifically ascertained predispositions in the world arena or any pertinent part thereof."

Lasswell later noted discrepancies between the earlier promises of science-based technology and current reality: "If the promise was that knowledge would make men free, the contemporary reality seems to be that more men are manipulated without their consent for more purposes by more techniques by fewer men than at any time in history" (1970, p. 119). After a diagnosis of such discrepancies, he observed that their potential effects on science are not trivial, "for science has grown strong enough to acquire visibility, and therefore to become eligible as a scapegoat for whatever disenchantment there may be with the earlier promises of a science-based technology." The proposal again called for the perfecting of institutions to apply the intellectual tools of the policy sciences (Lasswell 1971, Lasswell and McDougal 1992) on a continuous basis toward policies to advance human dignity for all.

Relatively few scientists have answered the call despite the continuing relevance of Lasswell's proposal. This may be partly the result of a specialized vocabulary that critics claim is a barrier to the policy sciences. Nevertheless, if more scientists do not come forward, humankind's growing mastery of nature will jeopardize human dignity and the privileged position of science in society.

RONALD D. BRUNNER

SEE ALSO *Freedom; Governance of Science; Political Economy; Political Risk Assessment; Science Policy; Soft Systems Methology.*

BIBLIOGRAPHY

Lasswell, Harold D. (1951). "Democratic Character." In his *The Political Writings of Harold D. Lasswell.* Glencoe, IL: Free Press.

Lasswell, Harold D. (1956). "The Political Science of Science: An Inquiry into the Possible Reconciliation of Mastery and Freedom." *American Political Science Review* 50, no. 4 (December): 961–979.

Lasswell, Harold D. (1970). "Must Science Serve Political Power?" *American Psychologist* 25(2) (February): 117–123.

Lasswell, Harold D. (1971). *A Pre-View of Policy Sciences.* New York: Elsevier.

Lasswell, Harold D., and Myres S. McDougal. (1992). *Jurisprudence for Free Society: Studies in Law, Science and Policy.* New Haven, CT: New Haven Press; Dordrecht, The Netherlands: Martinus Nijhoff Publishers.

LEIBNIZ, G. W.

• • •

Diplomat and court councilor to the house of Brunswick in Hanover, Gottfried Wilhelm Leibniz (1646–1716) was born in Leipzig on July 1. By the age of twenty-one he had earned a doctorate of law and written a *Dissertation on the Art of Combination*, which allowed him to lecture in philosophy. Though he never formally held an academic position (he had jobs as a jurist, librarian, mining engineer, and historian), his duties in Hanover enabled him to travel and meet many well-known thinkers of his time, such as mathematician Christian Huygens (1629–1695), who tutored Leibniz in mathematics during the latter's visit to Paris from 1672 through 1676. While he published several scholarly articles and only one book during his lifetime, the *Theodicy*, his large body of posthumously published work reveals Leibniz's contributions to mathematics, logic, science, law, philosophy, and ethics.

A rationalist, Leibniz exhibited a characteristically modern ambition with an ambitious scientific attempt to create a universal science of all human knowledge, which consisted of a universal, simple (i.e., numerical) language and a formalized calculus for reasoning. Though he eventually acknowledged the impossibility of completing the task because of the perspectivity of human knowing, he pursued this project until the end of his life. Leibniz's crowning achievement was his discovery of the infinitesimal calculus. Although Isaac Newton (1643–1727) discovered the infinitesimal calculus several years earlier, their achievements were independent and Leibniz's system of notation (published before Newton's) continues to be used in the early twenty-first century.

To understand Leibniz, one must acknowledge the fundamental premise behind his thought: God created the best of all possible universes by achieving the maximum amount of diversity consonant with unity. This cannot be proven but must be accepted as true for rational inquiry to be possible. From this premise Leibniz identified five basic *a priori* metaphysical principles to guide inquiry: the principle of sufficient reason (for every event or thing there is a reason for its being what it is rather than otherwise) the principle of non-contradiction (that an essence cannot contain opposite properties in the same way at the same time) the principle of perfection (that God always creates by choosing the maximum amount of perfection) the principle of the identity of indiscernibles (that no two things can be identical in all respects save spatial location) finally, the principle of continuity (that there are no "gaps" in the perfection of the created order). In revised version, these premises may still be argued to underlie even empirical scientific research.

Leibniz's scientific method, "the conjectural method a priori," assumes certain hypotheses to demonstrate that natural occurrences follow from them. It is a

G. W. Leibniz, 1646–1716. Leibniz was a German mathematician and philosopher. Known as a statesman to the general public of his own times and as a mathematician to his scholarly contemporaries, he was subsequently thought of primarily as a philosopher. (*The Library of Congress.*)

priori because it relies on his five basic metaphysical principles. Leibniz used it to improve the mechanics of philosopher René Descartes (1596–1650) by distinguishing between speed and velocity, and to criticize Newton's description of force. Moreover, this method was not meant merely for demonstration, but also for technological invention (which motivated Leibniz: for example, he invented a calculator). Most of his technologies nevertheless failed, but many of his proposals foreshadowed later technological developments. For example, he attempted to use windmills to remove water from mines and proposed a system of ball bearings to improve the efficiency of carriage rides.

Leibniz rejected Descartes's metaphysical dualism of mind and matter, and its major scientific presupposition, namely that the physical universe is a *res extensa*, whose causality is exclusively mechanistic. One reason for rejecting matter as the basic element of the universe is its infinite divisibility. This leads to an infinite regress when trying to explain matter, thereby constituting a violation of the principle of sufficient reason. Instead, Leibniz argued for the monad as the most basic element of reality.

Monads are immaterial, "windowless" (that is, there is no causal interaction between monads), microcosms of the universe, the basic activity of which is perception. God harmonizes each monad (which contain all of their predicates analytically) according to his supremely perfect divine plan. Moreover, each person, as a unified collection of monads, has a unique perspective on the universe and, consequently, gets at some degree of truth. Hence, Leibniz insisted that rational inquiry must take place within an intersubjective community.

Leibniz's emphasis on intersubjectivity is reflected in his ethics, which focuses on three concepts: wisdom, virtue, and justice. Wisdom leads to happiness because all moral action must be guided by thought. Happiness is a durable state of pleasure (i.e., understood as perfection). Virtue is the habit of acting according to wisdom, and justice is the charity of the wise person, who pursues the good of others. These are assumed to be the motivations of all technology.

Leibniz's impact cannot be adequately measured. In addition to influencing such thinkers as Immanuel Kant, Edmund Husserl, and the quantum physicist David Bohm, Leibniz's aspirations continue to be a resource for those seeking to reconcile modern science, technology, and ethical responsibilities.

CHRISTOPHER ARROYO

SEE ALSO *Husserl, Edmund; Kant, Immanuel; Theodicy.*

BIBLIOGRAPHY

Collins, James. (1954). "Leibniz." In his *A History of Modern European Philosophy.* Milwaukee, WI: Bruce.

Hostler, John. (1975). *Leibniz's Moral Philosophy.* London: Gerald Duckworth.

Leibniz, Gottfried Wilhelm. (1989). *Philosophical Essays,* trans. Roger Ariew and Daniel Garber. Indianapolis, IN: Hackett.

Leibniz, Gottfried Wilhelm. (1991). *Discourse on Metaphysics and Other Essays,* trans. Daniel Garber and Roger Ariew. Indianapolis, IN: Hackett.

Rescher, Nicholas. (1979). *Leibniz: An Introduction to His Philosophy.* Oxford: Basil Blackwell.

Ross, G. MacDonald. (1984). *Leibniz.* Oxford and New York: Oxford University Press.

LEOPOLD, ALDO

• • •

Aldo Leopold (1887–1948), who was born in Burlington, Iowa, on January 11, was a pioneer of the American environmental movement. His essay "The Land Ethic," published in *A Sand County Almanac* (1966 [1949]), has become a foundational text of American environmental ethics. Leopold challenges his readers to reevaluate their relationship to the land they inhabit and act in accordance with a "land ethic" that "enlarges the boundaries of the community to include soils, waters, plants, and animals, or collectively: the land" (Leopold 1966, p. 239). In his work the land and the biotic community become more than symbolic or abstract entities; they become beings with an intrinsic right to exist. Extending ethics and rights to the land, according to Leopold, necessarily "changes the role of *Homo sapiens* from conqueror of the land-community to plain member and citizen of it" (Leopold 1966, p. 240). Leopold died in Baraboo, Wisconsin, on April 21.

Leopold's love of the land began when as a young naturalist he hunted and fished in his native Iowa. He took his interest in the natural world to Yale's School of Forestry in 1904. During his four years at the school founded by Gifford Pinchot (1865–1946), the first director of the U.S. Forest Service, Leopold absorbed the utilitarian philosophy of the early conservationists (Nash 1989). He served in the Forest Service from 1909 to 1928, working in Apache National Forest in Arizona and then managing the Carson National Forest in New Mexico. By 1928 his earlier studies in ecology and practice of game and forest management had taught him to see the world as a web of interrelated systems. He also came to understand the lasting consequences of individual action on the landscape. In "The Land Ethic" Leopold uses the term *biotic pyramid* to describe the dynamic relationships that exist among organisms and their environments. "Land," he argues, "is not merely soil; it is a fountain of energy flowing through a circuit of soils, plants, and animals" (Leopold 1966, p. 253). In 1933 Leopold accepted an appointment in wildlife management at the University of Wisconsin.

The year 1935 was an important one for Leopold: His concern for vanishing American primitive areas led him to cofound the preservationist group the Wilderness Society. Leopold also purchased an abandoned, 120-acre farm in Sauk County, Wisconsin. It was in that setting that Leopold tried to articulate what it means to have an ethical relationship to the land. *A Sand County Almanac,* the record Leopold created of his years on the farm and his maturing environmental philosophy, was published in 1949, a year after he died fighting a fire on a neighbor's farm.

In his short piece "Axe in Hand" from *Almanac* Leopold provides an illuminating vignette on bias, showing how he imagines his relationship to the plants and animals that coinhabit his space and how he executes, sometimes literally, his decisions involving land management. The context for Leopold's dilemma is the felling of a tree; the decision he must make is between the white pine and the red birch, two species that crowd each other in those woods. Leopold examines the biases that influence a conservationist, which he defines as the axe wielder "who is humbly aware that with each stroke he is writing his signature on the face of the land." He is specifically intent on examining the "logic, if any" behind his own biases (Leopold 1966, p. 73). Leopold understands that his biases are a filter through which he passes the details of the landscape, making his world and the objects in it comprehensible.

The examination of individual biases—in this case Leopold's inquiry into his preference for the pine over the birch—forms the first stage in the development of an ethical relationship to the land. What Leopold describes is land as a system with an integrity of its own. The boll weevil, for instance, will or will not attack the pine if certain relations with the birch exist or do not exist. Some plants will thrive and others will not, depending on whether the birch or the pine is there to give them shelter. When the axe wielder enters the scene, he has the potential to disrupt that system. His examination of bias enables Leopold to see all the possible consequences of his actions and act in a thoughtful manner.

In this essay Leopold paints a portrait of a community in which he is as much a part of the environment as are the trees, insects, and birds; he, like them, has a role to play. In "Axe in Hand" Leopold demonstrates what he calls in "The Land Ethic" the "ecological conscience"; that conscience, he writes, "reflects a conviction of individual responsibility for the health of the land" (Leopold 1966, p. 258). Leopold summarized the principle behind the land ethic as follows: "A thing is right when it tends to preserve the integrity, stability, and beauty of the biotic community. It is wrong when it tends otherwise" (Leopold 1966, p. 262). Leopold's land ethic forces a reevaluation of the "value" of land broadly conceived and requires that limits be placed on the individual in favor of the health of the biotic community.

TINA GIANQUITTO

SEE ALSO *Environmental Ethics; Multiple Use; Wildlife Management.*

Aldo Leopold, 1886–1948. Leopold was an early environmentalist who laid the groundwork for many of the conservation laws and policies in place today. (*AP/Wide World Photos.*)

BIBLIOGRAPHY

Callicott, J. Baird. (1989). *In Defense of the Land Ethic: Essays in Environmental Philosophy.* Albany: State University of New York Press. See the essay, "Leopold's Land Aesthetic."

Leopold, Aldo. (1966 [1949]). *A Sand County Almanac.* New York: Oxford University Press. Classic text of American environmental ethics; collection of essays detailing Leopold's experiences on his farm and his environmental philosophy.

Meine, Curt. (1991). *Aldo Leopold: His Life and Work.* Madison: University of Wisconsin Press. Authoritative biography of Leopold.

Nash, Roderick Frazier. (1989). *The Rights of Nature: A History of Environmental Ethics.* Madison: University of Wisconsin Press. Traces the history of environmental ethics.

Scheese, Don. (1996). *Nature Writing: The Pastoral Impulse in America.* New York: Twayne.

LEVI, PRIMO

• • •

Primo Levi (1919–1987) was born to an assimilated Jewish family in Turin, Italy. In 1944, after training as a chemist, Levi joined a group of antifascist partisans, was captured, and was deported to the concentration camp at Auschwitz. He survived and returned to Turin in

Primo Levi, 1919–1987. An Italian author and chemist, Levi was considered one of the foremost writers of concentration camp literature. (*The Library of Congress.*)

1945, at which point he embarked on joint careers as an industrial chemist and an author, publishing the account of his experiences titled *Se questo e un uomo* (If this is a man) in 1947. The book, published in the United States as *Survival in Auschwitz*, is considered to be among the finest accounts of the death camps.

Levi retired from his work as a chemist in 1978 and fell to his death in his Turin apartment building on April 11, 1987. Debate continues about whether Levi, who experienced repeated bouts of depression, killed himself or fell by accident.

Throughout his work Levi stressed the connections between science, literature, and ethics. His use of chemistry as an inspiration for storytelling in *The Periodic Table* (1984) made scientists more attuned to literature and readers of literature more appreciative of science.

One theme unifying Levi's diverse essays and short stories is his belief in the importance and value of work. Levi believed that human beings are naturally constituted to need to work, to strive toward a goal and solve problems encountered in doing so. He emphasized the importance of practice and effort and saw science as a particularly important forum for the struggle to survive and grow.

Levi argued that technology does not necessarily alienate humanity from nature but can enhance the rapport between them. At the same time he emphasized the capacity of humanity for self-transformation, which necessarily means defying and altering nature. He believed that through its inventions humankind has turned its back on nature, damaging both people and the natural world but also improving the lot, and raising the stature, of individuals. Levi argued that one must learn from nature but that one also learns from struggling against it.

Levi eschewed both triumphalism and despair regarding humanity's prospects and the contributions to them made by science. He emphasized that progress will always be noisy, dangerous, and limited. However, because people are adaptable and capable of courage, reason, and strength, progress is possible. Levi celebrated the "cheerful strength" and "sober joy" connected with thought and invention, which allow human beings to endure and learn. He spoke of himself as a man sustained by curiosity about the world and emphasized the value of the inquiry that human curiosity fuels. However, he also acknowledged that the struggle to unlock the secrets of nature through measurement and categorization can be monstrous as well as heroic.

Levi, who was particularly worried by the proliferation of nuclear weapons, called on his fellow scientists and technicians to "return to conscience," to become aware of their immense and potentially sinister power. He insisted that science is not neutral; it either helps or harms human beings. Scientists should not stop doing research for fear of the possible negative consequences of their work, but they should concern themselves with the results of their work and avoid research that leads to immoral results. Scientists should resist the temptation of material rewards and intellectual stimulation, engage in work that will benefit and not harm their fellow human beings, and speak out against the misuse of science by others.

Levi's short stories often satirize the arrogance, ambition, and desire for control or enrichment that can lead scientists to ignore or abandon moral scruples in pursuing and applying knowledge. He warned against submissiveness to power and urged that "a precise moral consciousness" be instilled in scientists as part of their training; he also recommended that scientists take a sort of Hippocratic oath to do no harm (Levi 2001, pp. 71, 89–90).

Levi's reflections on the ethical dimension of science emphasize potential benefits as well as limitations, hope as well as danger, and the joys of discovery as well as moral responsibility. He believed that human beings are alone in a universe not made for their well-being and warned that although science gradually reveals the secrets of the cosmos, those secrets do not provide answers to "big questions" regarding the aims of human life; those answers can come only from within human beings. People's reason for being, he concluded, rests on their nature as, in the words Levi quoted from Pascal, "thinking reeds" who seek knowledge and excellence, and this quest is the source of human dignity.

JOSHUA L. CHERNISS

SEE ALSO *Holocaust; Science, Technology, and Literature; Work; Scientific Ethics.*

BIBLIOGRAPHY

Levi, Primo. (Italian edition 1947; English edition 1961). *Se questo e un uomo* [Survival in Auschwitz: The Nazi assault on humanity], trans. Stuart Woolf. New York: Collier. Reprint, New York: Simon & Schuster, 1996. Levi's first book, an account of, and meditation on, his experiences at Auschwitz.

Levi, Primo. (1966). *Storie naturali.* Selections published in English in *The Sixth Day,* trans. Raymond Rosenthal. Turin: Einaudi; London: Michael Joseph, 1990

Levi, Primo. (1971). *Vizio di forma.* Selections Published in English in *The Sixth Day,* trans. Raymond Rosenthal. Turin: Einaudi; London: Michael Joseph, 1990. A collection of short stories, many of them fantastic parables addressing the moral dimensions of science and technology.

Levi, Primo. (Italian edition 1975; English edition 1984). *Il sistema periodico* [The periodic table], trans. Raymond Rosenthal. Turin: Eidaudi; New York: Schocken. A collection of autobiographical tales and reflections, each one associated with and inspired by an element from the periodic table.

Levi, Primo. (Italian edition 1986; English edition 1989). *Racconti e saggi* [The mirror maker], trans. Raymond Rosenthal. New York: Schocken. A collection of short stories and essays, many of them addressing the ethics of scientific work.

Levi, Primo. (2001). *The Voice of Memory: Interviews, 1961–87,* ed. Marco Belpoliti and Robert Gordon. New York: New Press.

Thomson, Ian. (2002). *Primo Levi.* London: Hitchison.

LEVINAS, EMMANUEL

• • •

Emmanuel Levinas (1906–1996), who was born in Lithuania of Jewish parents, studied the Hebrew Bible along with the works of the Russian authors Aleksandr Pushkin (1799–1837), Fyodor Dostoyevsky (1821–1881), and Lev Tolstoy (1828–1910). In 1928 and 1929

Emmanuel Levinas, 1906–1995. Levinas was a major philosopher of the 20th century who attempted to proceed philosophically beyond phenomenology and ontology and to engage in a more immediate and irreducible consideration of the nature and meaning of other persons. (© *Bassouls Sophie/Corbis Sygma.*)

he attended the philosopher Edmund Husserl's (1859–1938) lectures in Freiburg, Germany, and started writing a dissertation on Husserl's theory of intuition. He also attended lectures given by the philosopher Martin Heidegger (1889–1976). Levinas was largely responsible for introducing Husserl and Heidegger to French philosophers, most notably Jean-Paul Sartre (1905–1980).

Levinas's first major work, *Totality and Infinity*, was published in 1961. It was only in the 1980s that a wider audience acknowledged Levinas's work, and his thought eventually became central to postmodern ethics. A number of authors, including philosophers and theorists such as Jacques Derrida (b. 1930), Zygmunt Bauman (b. 1925), John D. Caputo (b. 1940), Robert Bernasconi (b. 1950), and Simon Critchley (b. 1960) adopted his ideas, so that any discussion of ethics outside the analytical tradition would be incomplete without reference to Levinas. This is also true with regard to ethics in science and technology.

Ethics: Not Theory but Happening

For Levinas ethics is not a theory, a rule, an idea, or knowledge of how people ought to act or live. In this sense it can be said that his work falls outside the traditional field of ethical theory. For Levinas ethics is a profound and disruptive event in which the Other disrupts and shatters the self-certain I. Levinas uses the term *Other* (with a capital letter) to refer to the absolute singularity of each human being. Ethics is a disruptive event in which a person's claims to rights and deserts is questioned radically in the face of the infinitely singular person before that individual here and now—the "widow and the orphan." If such persons call on an individual for help or support, that act recalls the individual's guilt, pointing out that that individual has from his or her very beginning taken the place in the sun of the person who has asked for assistance. Levinas would argue that an individual's particular existence has its origin in and through a terrible and violent act seizing the place of the Other who is calling on that individual. This primitive primacy of the individual's guilt, the birth of the ethical question, is Levinas's most profound insight, elaborated in all his works.

Why is the individual already guilty? In taking up his or her personal existential project (to be that particular person), the individual has taken the "place in the sun" of the Other. Further, in making sense of the world and those who cross his or her path, the individual continues to *reduce* the Other to the themes and categories (mother, criminal, politician, manager, man, black, etc) of his or her comprehension. Others become "domesticated" as themes or categories "for-me" through and by the individual's ongoing comprehension of them. This domestication prolongs and extends the violence that began at the birth of a person's individual existential project. Thus, that person has been guilty from the start. For Levinas ethics becomes possible when a person acknowledges that the Other—the particular singular person facing the individual—is infinitely more than any idea (theme, category, attribute) that the individual can use in his or her ongoing comprehension. How, then, can a relationship with the Other be anything but comprehension, how can one encounter the other as Other? Working this out is Levinas's task.

Levinas claims that ethics happens in the "saying" or speaking of language. When the particular Other faces a person and speaks or makes a nonlinguistic gesture, there is more in the words than the message: There is a residue, a trace, of the Other that disturbs the hearer. Levinas uses the familiar event of a doorbell ringing and disturbing one's work and thoughts, but when one opens the door, there is nobody there. Was there nobody there? Did the hearer imagine it? The hearer cannot recall anything but the disturbance. Just when the hearer settles back into his or her thoughts, the doorbell rings again, but there is never somebody there. In the recalling of ethics people are affected without the source of the affection becoming something they can think about as such. It is this relationship of incessantly there but never present that Levinas calls proximity: the disturbing face before the individual that is (re)calling that individual's responsibility. The only recourse in this moment of ethics is to respond, to take up the responsibility for one's original and ongoing violence. For Levinas one is a particular person because one has these particular responsibilities. This is the only possibility for ethics. As he expresses it: "In her face the Other appears to me not as an obstacle, nor as a menace I evaluate, but as what measures me. For me to feel myself unjust I must measure myself against infinity" (Levinas 1996b, p. 58).

Is the individual not also a face? Who will look out for that person? These questions lead to the issue of justice. The radical asymmetrical ethics of Levinas must be reinserted into the symmetrical relationship of justice in which all people are equal before the law. Thus, Levinas claims that it is necessary to add "the third" (all other people) to the relationship of the self and the Other. This is the moment of justice. It involves the need to compare what is never comparable, the dilemma a judge faces in the courtroom every day: to treat all people as equal even though they are absolutely different ("singulars" in Levinas's terminology). Nevertheless, for Levinas the urgency of justice stems from the radical asymmetry of the original ethical relationship. Without such a radical asymmetry—the ethical relationship—the claim of the Other always can be subject to codes, rules, and regulations. Then justice becomes mere calculation and (re)distribution. Thus, justice has its standard, its force, in the proximity of the face of the Other: "The equality of all is born by my inequality, the surplus of my duties over my rights. The forgetting of self moves justice" (Levinas 1991 (1974), p. 159).

Implications for Science, Technology, and Ethics

Levinas's ethics is important in thinking about ethics more generally. One could say that it is a call to rescue ethics from theory. Nevertheless, Levinas's work is particularly important to science and technology. In the epistemological categories of science and the mechanisms and algorithms of technology the absolute singular (the individual particular person) does not fit well. One could see how the singular person becomes a subject, subjected to the logic of the method. In the mechanisms and algorithms of technology the individual person can become an exception (perhaps an error) to be discarded in favor of the categories those technologies rely on for their smooth operation.

Given this seemingly obvious conclusion one could draw from Levinas' ethics above, it is surprising to find that Levinas (1990) takes a very positive view of science and technology. In discussing the space program he argues that science and technology strips nature of its divine pretentions, thereby allowing humans to harness it in the service of humanity. Nevertheless, such a view that posits science and technology as neutral 'tools' that can and ought to be applied in the service of humanity denies the value ladenness of science and technology as well as the political structures within which these human endeavours function. Thus, as Peperzak (1997) argues: "the inherent violence of technology cannot be overcome by technological practice. The micro-ethical practice of persons who are well disposed to others, nature, and art, notwithstanding the distorting networks in which these people function, can point the way towards a better disposed constellations of justice, technological utility, and natural beauty" (p. 143).

Thus, ethically minded designers of technology must ask which categories they assume when they are designing. What about those who do not fit? Moreover, as people apply science and technology in the ordering of society, many singular faces may suffer as they fall through the cracks of method and machine. Does that mean that science and technology are inherently violent? Levinas (1990) would argue that this is necessary violence in the service of freedom and justice. Nevertheless, in its service of justice the ultimate measure should be the proximity of the face of the Other; without this standard it would pursue its path as pure violence.

One could say that Levinas's ethics leaves humankind with plenty and with nothing. The call of the Other is powerful, but how can it be worked out in every instance? Ethical theories such as utilitarianism and consequentialism provide resources to decide what one ought to do in a particular case. However, according to Levinas, all people are guilty and must respond, yet when they respond, they may perpetuate violence. Derrida (1992) claims that Levinasian ethics is impossible because it provides no clear answer or procedure for deciding what to do. This, paradoxically, is an answer. It is the impossibility of ethics that provides the urgency of ethics and interrogates every decision. If making ethical decisions were possible through the use of a rule or procedure, people might forget the plight of the particular individual, the Other. Impossibility is what keeps people open to the possibility of encountering the other

as Other in every situation. For Derrida and Levinas it is impossibility that makes ethics possible.

Is Levinas's ethical system anthropocentric? Can other animals and other things have a face? Are they also absolute singulars? Does Levinas deny a responsibility toward nonhuman others? A number of authors have argued against Levinas's ethics on these grounds. Feminist authors have stated that his work is based on the predominant view of the male ego of autonomy and competition as opposed to the female ego of affiliation, empathy, and nurturing (Chanter 1988). Deep ecologists have argued against his exclusion of nature from the realm of morality (Gottlieb 1994). Levinas scholars such as John Llewelyn (1991) and Adriaan Peperzak (1997) have responded to these criticisms. In contrast to these critical comments, Benso (2000), with the help of Heidegger, uses Levinas to make a powerful argument for an "ethics of things." Such an approach points toward the application of Levinas's thought to science, technology, and ethics.

LUCAS D. INTRONA

SEE ALSO *French Perspectives; Heidegger, Martin; Phenomenology.*

BIBLIOGRAPHY

Benso, Silvia. (2000). *The Face of Things.* Albany: Sate University of New York. Presents an argument for the "enlargement" of Levinas's ethics to include objects.

Chanter, Tina. (1988). "Feminism and the Other." In *The Provocation of Levinas: Rethinking the Other,* ed. Robert Bernasconi and David Wood. London: Routledge. A feminist response to the ethics of Levinas.

Derrida, Jacques. (1992). "Force of the Law: The 'Mystical Foundation of Authority.'" In *Deconstruction and the Possibility of Justice,* ed. Drusilla Cornell, Michel Rosenfeld, and David Gray Carlson. London: Routledge. An important essay that draws on Levinas's ethics to give an account of the way ethics interacts with the question of justice.

Gottlieb, Roger S. (1994). "Ethics and Trauma: Levinas, Feminism, and Deep Ecology." *Cross Currents* 44(2): 222–240.

Levinas, Emmanuel. (1969 [1961]). *Totality and Infinity,* trans. Alphonso Lingis. Pittsburgh: Duquesne University Press. This book is Levinas's first major account of his ethics.

Levinas, Emmanuel. (1990) "Heidegger, Gagarin and Us." In *Difficult Freedom: Essays on Judaism,* trans. S. Hand. London: The Athlone Press, 231–234. This is the only essay where Levinas explicitly addresses the question of science and technology.

Levinas, Emmanuel. (1991 [1974]). *Otherwise Than Being or Beyond Essence,* trans. Alphonso Lingis. Dordrecht, Netherlands: Kluwer Academic. This book is the later, and more subtle account of Levinas's ethics.

Levinas, Emmanuel. (1996a). "Ethics as First Philosophy." In *The Levinas Reader*, ed. Sean Hand. London: Blackwell. A very good, succinct account of why Levinas's ethics is so radically different from traditional ethical theory.

Levinas, Emmanuel. (1996b). "Philosophy and the Idea of Infinity." In *Collected Philosophical Papers: Emmanual Levinas*, ed. Alphonso Linis. Bloomington: Indiana University Press.

Llewelyn, John. (1991). "Am I Obsessed by Bobby? (Humanism of the Other Animal)." In *Re-Reading Levinas*, ed. Robert Bernasconi and Simon Critchley. Bloomington: Indiana University Press. Addresses the question of the applicability of Levinas's ethics for other animals.

Peperzak, Adriaan Theodor. (1997). *Beyond: The Philosophy of Emmanuel Levinas*. Evanston, IL: Northwestern University Press. Comprises of a number of essays on Levinas's philosophy. It serves as a very useful introduction written in an accessible style.

LEWIS, C. S.

• • •

Novelist, critic, poet, essayist, and Christian apologist Clive Staples Lewis (1898–1963) was born in Belfast on November 29, served in France, and was wounded during World War I. He completed his undergraduate studies at University College, Oxford, in 1922, and from 1925 until 1954 was a Fellow of Magdalen College, Oxford, and tutor in English. From 1954 until just before his death he was Professor of Medieval and Renaissance Literature at Cambridge.

Lewis once wrote that although he was a rationalist who had *scientific impulses*, he could have never been a scientist. He considered the role and direction of science for nearly three decades and mentioned and alluded to it in many of his works. He was aware of its limitations and methodology, and was respectful of its status as a type of knowledge that could be used for the benefit of humanity. Lewis praised genuine scientific accomplishment and said that scientific reason, if accurate, was valid, although it was not the only kind of reasoning. Truth, value, meaning, and other ideals were necessary presuppositions to the scientific method but were not themselves scientific phenomena.

Lewis was sometimes accused of being unscientific and discrediting, or even attacking, scientific thinking. In reality he criticized what he called *scientism*, a reductionist outlook on the world that popularized the sciences. Scientism (*science deified*) occurred when a naturalistic worldview was linked to the empirical method of experimentation. Scientism as radical empiricism rejected the truth of a nonquantifiable reality such as God.

C. S. Lewis, 1898–1963. An author and scholar, Lewis is known for his work on medieval literature and for his Christian apologetics and fiction, especially *The Chronicles of Narnia*. (AP/Wide World Photos.)

Lewis saw the Genesis creation accounts as non-literal folk tales or myths. In *The Problem of Pain* (1940), he presented a modified view of creation and the Fall because scientific evidence that "carnivorousness was older than humanity" had led him to believe that evil had manifested itself long before Adam (Lewis 1940, p. 121). He had a theistic view of evolution but resisted attempts to draw broad philosophical implications from various scientific theories of it. He was never directly opposed to science, but believed many scientific theories were tentative and dependent on changing presuppositions and *climates of opinion*. Early evidence from his letters indicate that he denied that biological evolution was incompatible with Christianity; in later letters he became increasingly pessimistic about evolutionism as a progressive philosophy. Earlier he felt that the theory of evolution was often held because of dogmatic, not scientific reasons, but he never gave up his long-held view that biological evolution was compatible with Christian accounts of creation. He opposed evolutionism as a philosophical theory, not evolution as a biological theory.

In many of his writings Lewis tried to redefine the role of science and its proper role in society. He believed that scientism was in error in that it reduced life to abstractions and denied the possibility that physical events and human experiences had God behind them. He observed that since scientism was only concerned with how things behave, it was not qualified or capable of *looking behind things*, particularly the power behind the universe.

In his much-praised defense of natural law, *The Abolition of Man* (1943), Lewis discussed the possibility of a world that no longer believed in objective truth and value. He saw this as possibly leading to a power struggle in which societal elites tried to control and recondition society. "Man's conquest of Nature, if the dreams of some scientific planners are realized, means the rule of a few hundreds of men over billions and billions of men ... Each new power won *by* man is a power *over* man as well" (Lewis 1955, p. 70).

Many of Lewis's ideas in *The Abolition of Man* were expressed dramatically in his space novel *That Hideous Strength* (1945). In the story, the degeneration of humanity nearly occurs as a result of a gross scientific materialism controlled by bureaucrats that is devoid of all idealistic, ethical, and religious values. Lewis satirized materialistic scientists in *That Hideous Strength* by showing them as ignoring metaphysical reason and refusing to submit their claims to any kind of moral or religious authority.

He wrote his trilogy of space novels (the others being *Out of the Silent Planet* [1938], and *Perelandra* [1943]) as a result of reading Olaf Stapledon's (1886–1950) *Last and First Men* (1930) and the Cambridge biochemist J. B. S. Haldane's (1892–1964) essay "Man's Destiny" (1927), both of which took interplanetary travel seriously but contained an immoral outlook that denied God. He was openly critical of Stapledon's fictional universes, in which science represented the greatest good and Christian ideals played no essential role. After reading Stapledon's *Star Maker* (1937), Lewis said that the race Stapledon described was concerned primarily for the increase of its own power by technology, a technology that was indifferent to ethics, and a *cancer in the universe*.

PERRY C. BRAMLETT

SEE ALSO *Anglo-Catholic Cultural Criticism; Christian Perspectives.*

BIBLIOGRAPHY

Aeschliman, Michael D. (1998). *The Restitution of Man: C. S. Lewis and the Case against Scientism*. Grand Rapids, MI: Eerdmans. A succinct examination of Lewis's role as spokesman for the classical Christian philosophical tradition that has opposed scientific materialism since the seventeenth century.

Lewis, C. S. (1940). *The Problem of Pain*. London: Geoffrey Bles.

Lewis, C. S. (1955). *The Abolition of Man*. New York: Collier Books.

Lewis, C. S. (1996). *That Hideous Strength*. New York: Scribner.

Philmus, Robert M. (1972). "C. S. Lewis and the Fictions of Scientism." *Extrapolation* 13(2)(May): 92–101.

Sammons, Martha D. (1976). "C. S. Lewis's View of Science." *CSL: The Bulletin of the New York C. S. Lewis Society* 7(10)(August): 1–6.

Schultz, Jeffrey D., and John G. West Jr., eds. (1998). *The C. S. Lewis Readers' Encyclopedia*. Grand Rapids, MI: Eerdmans. See particularly entries on M. D. Aeschliman, "science," Perry C. Bramlett, "the fall," Thomas Howard, "that hideous strength," Thomas T. Talbott, "the problem of pain," John G. West, Jr., and "the abolition of man."

LIBERALISM

• • •

Liberalism as a theory about politics and society upholds freedoms of belief, inquiry, expression, action, association and elections. In liberalism, freedom coalesces with value-commitments to equality, individualism, toleration, pluralism, and rationality. All of these commitments have interacted with science and technology in multiple ways.

Classical Liberalism

Liberals differ over determining the nature of freedom. Isaiah Berlin's distinction between negative and positive freedoms (freedom *from* as against freedom *to*) is useful in explaining the difference between classical and modern forms of liberalism.

In classical liberalism, freedom is interpreted in terms of a *private* sphere of non-interference that is supported by the rule of law. Free agents are protected from arbitrary interference, being left to enjoy their possessions, to retain personal beliefs, and to act in preferred ways on the condition that they respect the freedom of others to do the same. Support for private property and free markets goes hand in hand, in classical liberalism, with a prescription that power (economic as well as political) be divided so as to alleviate the risk of its being abused.

John Locke (1632–1704), whose *Second Treatise of Civil Government* (1690) started the tradition of liberal thought, encapsulated classical liberalism in stating that "Liberty is to be free from restraint and violence from others which cannot be, where there is no Law" (1960, p. 324). Locke's form of liberalism supported parliamentary government and the rule of law in England against absolute monarchy.

Among French thinkers, Charles de Secondat, Baron de Montesquieu, in *The Spirit of the Laws* (1748), praised the English constitution for its separation of the powers of government and reflected adversely on the absolutism of the French monarchy. Tolerance, aversion to fanaticism, and advocacy of freedom of discussion and of the press characterize the writings of the eighteenth-century *philosophes*, including Marquis de Concordet and Francois-Marie Voltaire. After the turmoil of the French Revolution and of Napoleon Bonaparte's rule, Benjamin Constant (1767–1830) and Francois Guizot (1787–1874) conceived of a liberalism that was conservative and admiring of English political institutions, while Alexis de Tocqueville (1805–1859) warned that democracy gives no guarantee of freedom and might end in tyranny.

Pre-eminent among German liberals, Immanuel Kant (1724–1804) conceived of liberty as the will determining itself according to its rational law, converting pure reason into practical reason. Kant's state is a legal organization, *limited* in its role of ordering legal rights, reconciling the free will of each individual with that of all others. The sphere of morality, for Kant, consists in individual conscience as the judge of the righteousness of acts. In 1927 Guido de Ruggiero contended that Kant's liberalism served to constrain the exercise of power in Germany through the nineteenth century because, "even in periods of the strictest absolutism," governments were checked "by a profound consciousness of" being restricted to the sphere of rights (de Ruggiero 1927, p. 220). In his *Essay on the Limits of the Action of the State* (1851), Wilhelm von Humboldt argued that the worthy faculties and qualities of individuals only develop in an environment that a minimalist state protects as free and pluralist. In Germany, as in France, liberals were nowhere near as committed to the market economy as were their English counterparts.

Among the sources of nascent U.S. liberalism was Locke with his ideas of natural rights, government by consent, and the entitlement of subjects to revolt against a government that betrays their trust. French *philosophes* could envision human perfectibility, but the liberals who contributed to the formation of the U.S. republic were skeptical. Their understanding of human nature derived from the Scottish Enlightenment: Adam Ferguson, David Hume and, particularly, Adam Smith who, in *The Wealth of Nations* (1776), argued for capitalist economics (the price mechanism as a beneficent *invisible hand*), the rule of law in a constitutional order, and equal freedom. Smith believed that a strong presumption exists against governmental activity, but his advocacy of laissez faire was not doctrinaire. A rule of thumb with Smith was that government should arrange social conditions in ways that would assist the market to provide public services; Jeremy Bentham and his circle embraced this theory. Adopting the principle of utility as his axiom, policies being calculated to advance the greatest happiness of the greatest number in society, Bentham inferred that joint stock companies should bid for government contracts to operate public institutions (prisons and poor houses).

Modern Liberalism

The emphasis in *modern* liberalism is placed on freedom as empowerment (freedom *to*). There has been no closer approximation to the ideal type of classical liberal society than nineteenth-century England in the era of William Gladstone and Richard Cobden. Nevertheless, after the reform of Parliament in 1832, governments in England—partly from the impetus received from Benthamite utilitarianism—became more active: reforming the administration of the poor law and of public health; regulating working hours, the police, and inspection of factories; and overhauling the civil service and local government.

Liberal thought in England also underwent a major revision with Lionel Hobhouse in 1911 describing the *liberal socialism* of John Stuart Mill (1806–1873) as the link *between the old and the new liberalism*. The *new liberalism* of the Hegelians Thomas Hill Green (1836–1882) and Bernard Bosanquet (1848–1923), appreciated the value of freedom as a positive power and recommended a more constructive mode of government. Agreeing with Mill that the core of liberalism consists in the "liberation of ... [the] spiritual energy" of agents (Hobhouse 1911, p. 137), Hobhouse proposed that the state should act so as to secure the economic conditions that would enable individuals to develop their faculties and to fully participate in the life of the community.

Two world wars and the intervening Great Depression led governments to assume a greater role in European and North American societies. John Maynard Keynes's *General Theory of Employment* (1936) explained how governments should use their fiscal powers of taxing and spending to regulate economic activity and control money supply as a means of mitigating the business cycle and unemployment.

In 1935, in the United States, John Dewey (1859–1952) expressed hostility to the free market order and its disparities in wealth. The Humboldt-Mill ideal of individual development as grounded in freedom that had impressed Green and Hobhouse was assimilated by Dewey and by many other liberal philosophers through the twentieth century. Dewey saw the ends of liberalism—"liberty and the opportunity of individuals" to fully realize "their potentialities"—as requiring governmental planning of "industry and finance" (1963, p. 51, 55).

The ideal of individual development is discernible in the most important work of liberalism to appear in the second half of the twentieth century, John Rawls's *A Theory of Justice* (1971). Arguing for redistribution and the welfare state, Rawls relied on principles of liberal justice. One of Rawls's tenets attributes freedoms of conscience, conduct, and religion to citizens; his other basic belief dictates that a redistribution of resources may only take place on the condition that the least well-off members of society will benefit from it. As a corollary, inequalities determined by an agent's social circumstances, and by that person's talents and abilities, are deemed to be illegitimate.

Prominent among the responses to Rawls's Kantian liberalism is *communitarianism*. Michael Sandel in *Liberalism and the Limits of Justice* (1982) demurred to Rawls's use of an abstracted individual to reason about justice, envisaging the self as being socially formed, and the individual as exercising reason only within the community.

The term *modern* liberalism does not mean that classical liberalism is an anachronism. The writings of neoliberals—Ludwig von Mises, Friedrich Hayek, Ayn Rand, and Milton Friedman—that influenced the governments of Margaret Thatcher and Ronald Reagan, confirm the durability of the classical liberal position. Neoliberals argued that the meliorist activity of democratic governments must be kept to a minimum if liberal societies are to avoid what Hayek sign-posted as *the road to serfdom*.

The distinction between classical and modern liberalisms is not a sharp one, the positions shading off into each other. Walter Lippmann (1889–1974), for example, was convinced that many services in modern society can only be provided by large governmental enterprises and he defended a redistribution of wealth as socially stabilizing. Lippmann held with the ideals of Smith, however, which turned him against Franklin Roosevelt's New Deal and other forms of *collectivism*. The political thought of Karl Popper (1902–1994) can be located with Lippmann's near the middle of the continuum between classical and modern liberalisms.

Science and Technology as Supporting the Achievement of Liberal Ideals

Liberalism and science have commonly been seen as buttressing each other. While recognizing that scientific research needed governmental funding, liberals argued that because scientists are experts in research they should be free to select their topics of, and methods for, research. In the 1940s, Michael Polanyi defended the autonomy of science against Soviet-style planned research, and Popper supported free inquiry by showing that knowledge advances in an unpredictable manner. Like Polanyi and Popper, Robert Merton depicted science as an exemplary liberal community, highlighting norms of universalism, communalism, disinterestedness, and organized skepticism.

Since the detonation of the atomic bomb, with the proliferation of weapons of mass slaughter and with the deterioration of the environment, even liberals have become ambivalent toward science and technology, although most remain sure that science and technology are conducive to liberal values. Without science and technology, liberals argue, freedoms of modern society—of the press and of the airwaves, for example—would be attenuated. Freedoms of election and association benefit from electronic communications and rapid transport. The technology of publishing serves the marketplace of ideas, and media technology helps in checking the power of government. Travel and the mass media expose more people to foreign cultures, encouraging tolerance of ethnic and cultural diversity. Dissemination of information by way of the Internet assists people in making free choices on matters of health, religion, education, and politics. In contributing to the material conditions of life that underlie the enjoyment of all liberties, science and technology have helped people, particularly in Europe and North America, to live longer, suffer less pain, and enjoy better health and greater comfort.

Science and Technology as Impeding the Attainment of Liberal Ideals

Much of the liberal image of science is out of date. In the early twenty-first century most scientists are a part of *big science*. Typically research is conducted by large teams, is capital-intensive, and is shrouded in secrecy because most scientists aim at producing innovations for industrial and governmental sponsors. While liberals are correct in claiming that science ails when governments and corporations instruct scientists on how to conduct their research, the fact remains that governmental controls on scientific research have become more stringent.

Science and technology may support the liberal values of freedom and tolerance, but in a number of ways they also *standardize* culture and social practices, as James Scott has argued. Paul Feyerabend (1924–1994) examined the idolization of science and technology—scientism and the cult of the expert—that so often takes responsibility away from laypeople and leads to the denigration of non-scientific beliefs and practices. In the first half of the nineteenth century, liberals (Tocqueville, Humboldt, and Mill) worried that newspapers and railways were creating a social *mass* that was hostile to individuality, diversity, and freedom. Concern about technology and the masses was also voiced by Max Weber (1864–1920) and José Ortega y Gasset (1883–1955). In the twentieth century, assembly line mass production and deskilling of the workforce in accordance with the precepts of Frederick Winslow Taylor's *scientific management* gave further impetus to standardization.

Social elites of scientists and technologists have privileged access to government policy makers and to funding agencies. They promote and benefit from scientism and standardization, having a major say over the curriculum and attracting the lion's share of resources for research in their fields.

In the hands of governments and corporations, modern science and technology have intruded deeply into the *private realm*. Although totalitarianism provided the most graphic evidence of mental regimentation by the electronic mass media, the mass media in democracies have been accused of *manufacturing consent*, indoctrinating consumers, and promoting irrationality. Computers and other information handling systems, security cameras, wire taps, and interception of on-line communications represent technologies that subject a citizenry to electronic *surveillance*.

STRUAN JACOBS

SEE ALSO *Civil Society; Communitarianism; Conservatism; Democracy; Lock, John; Merton, Robert; Mill, John Stuart; Neoliberalism; Polanyi, Michael; Popper, Karl; Rawls, John.*

BIBLIOGRAPHY

Arblaster, Anthony. (1984). *The Rise and Decline of Western Liberalism*. Oxford: Basil Blackwell. A major critique of the tradition.

Bellamy, Richard. (1992). *Liberalism and Modern Society*. Cambridge, UK: Polity Press.

de Ruggiero, Guido. (1927). *The History of European Liberalism*, trans. R. G. Collingwood. London: Oxford University Press. A classic study of liberal theory and practice up to the early twentieth century.

Cranston, Maurice. (1967). *Freedom*, 3rd edition. London: Longmans.

Dewey, John. (1963). *Liberalism and Social Action*. New York: G. P. Putnam's Sons. Diagnosis of, and prescription for, the ills of liberalism during the Great Depression.

Feyerabend, Paul (1975). *Against Method*. London: New Left Books. How to liberalize science.

Gray, John. (1986). *Liberalism*. Minneapolis: University of Minnesota Press.

Greenleaf, W. H. (2003). *A Much Governed Nation*. London: Routledge. Part of his magisterial history of liberalism in England.

Hart, David. (1998). *Forged Consensus: Science, Technology, and Economic Policy in the United States, 1921–1953*. Princeton, NJ: Princeton University Press.

Hobhouse, Lionel. (1911). *Liberalism*. London: Thornton Butterworth. The evolution of liberal thought in the second half of the nineteenth century.

Locke, John. (1960). *Two Treatises of Government*. Cambridge, England: Cambridge University Press. The great source of liberal ideas.

Manent, Pierre. (1994). *An Intellectual History of Liberalism*, trans. Rebecca Balinski. Princeton, NJ: Princeton University Press. Instructive on the development of French liberalism.

Manning, D. J. (1976). *Liberalism*. London: J. M. Dent & Sons.

Massimo, Salvadori. (1977). *The Liberal Heresy*. London: The Macmillan Press. Wide-ranging; somewhat uncritical.

Mulhall, Stephen, and Adam Swift. (1996). *Liberals and Communitarians*. Oxford: Blackwell. Careful examination of recent theories.

Ryan, Alan. (1995). *John Dewey and the High Tide of American Liberalism*. New York: W. W. Norton. A scholarly account of the career of Dewey.

Scott, James. (1998). *Seeing Like a State: How Certain Schemes to Improve the Human Condition Have Failed*. New Haven, CT: Yale University Press. The case against sweeping radicalism.

LIBERTARIANISM

• • •

Libertarianism is the belief that one has the right to dominion over one's own person, including the fruits of one's labor. Adults are entitled to make their own decisions and agreements. Coercion, particularly by the government, is wrong.

In contemporary American politics libertarians side with the far left in favoring personal freedom and side with the far right in favoring economic freedom. Thus, libertarians argue for the decriminalization of recreational drug use on the grounds that adults should have the right to make choices about their bodies. Libertarians oppose a national health care system as coercive and inevitably interfering with the rights of individuals to make their own choices about health care.

Libertarians view other ideologies as overly paternalistic. Politicians routinely begin a sentence with "We must," as in " We must reduce our dependence on foreign oil" or "We must spend more on education." A libertarian asks, "Who is this 'we'?" Libertarians argue that individuals can decide for themselves how much to spend on their own education. Moreover, people who want to see others obtain more education are free to donate funds for that cause. To libertarians "We must spend more on education" translates into "The government is going to coerce individuals into paying for their own or others' education."

For many people economic freedom is justified on utilitarian grounds. Those individuals endorse free markets because markets deliver economic growth and a high average standard of living. For libertarians economic freedom is justified on first principles. Even when government regulation is intended to make people better off, libertarians oppose such regulation as coercive. Thus, libertarians would not endorse most regulation carried out in the name of protecting consumers, preferring instead that consumers be expected to protect themselves.

Libertarianism faces a number of challenges. First, libertarians must establish the boundaries between freedom and coercion. In theory, one person's freedom can negate another's. The libertarian solution to this problem is to focus on property rights. If a person's property is clearly defined, no one may take that property without that person's consent. The libertarian's ideal role for government is to enforce property rights and nothing else.

Second, libertarianism is criticized for taking social institutions and cultural norms for granted. That is, libertarians speak as if society could function with only markets as institutions. However, markets operate in a context of cultural values and government protections, and chaos would result if those protections were taken away.

On the left critics of libertarianism argue that without social welfare programs the poor might turn to crime or armed insurrection. Without public education people might not acquire the basic tools needed to function in and maintain their society. On the right critics of libertarianism argue that individual morality is too fragile to

prevail in the noncoercive environment favored by libertarians. Without the restraints imposed by religion, social opprobrium, and legal sanction people's behavior would degenerate, ultimately reaching the point where they no longer were capable of respecting themselves or one another.

Third, libertarianism is criticized as an ideology that ignores inequality and scorns the disadvantaged. This line of criticism is embedded in lines such as "The rich man and the poor man have equal freedom to sleep in the gutter" and "Freedom of the press exists only if you own one" (the second quote is attributed to the journalist A. J. Liebling).

These critics argue that property rights are not sufficient to make everyone free. They suggest that those born without sufficient endowments of land, capital, and aptitude are at the mercy of the powerful even in the absence of coercion. In response libertarians argue that government programs enacted for the benefit of the disadvantaged often are counterproductive, circumscribing freedom without aiding the intended beneficiaries.

History of Libertarianism

Libertarianism has its roots in Enlightenment philosophy, particularly the writings of the philosopher John Locke (1632–1704). Locke argued that dominion over one's own body and one's own property is a natural right. Locke viewed government as legitimate only if it has the consent of the governed. In Chapter 8 of the *Second Treatise on Government* Locke wrote, "The only way whereby any one divests himself of his natural liberty, and puts on the bonds of civil society, is by agreeing with other men to join and unite into a community for their comfortable, safe, and peaceable living one amongst another, in a secure enjoyment of their properties, and a greater security against any, that are not of it." Locke was a major influence on the founders of the United States, who embodied the contractual theory of government in the U.S. Constitution. The U.S. Bill of Rights also reinforced libertarian ideas of natural rights.

Another major libertarian work is *On Liberty* by the philosopher John Stuart Mill (1806–1873). Mill argued that social condemnation could be as oppressive as government coercion.

In the twentieth century one of the most important libertarian thinkers was Friedrich Hayek (1899–1992), who argued against the dominant view that a modern economy requires central planning and a welfare state. Hayek believed that the price system, fed by local information in markets, is more efficient than any central planner. For him the coercion required to implement the welfare state would undermine freedom and thus was *The Road to Serfdom* (1944).

The Internet and Libertarianism

In 1996, John Perry Barlow, a writer and activist in the Electronic Freedom Foundation (EFF), composed "A Declaration of the Independence of Cyberspace," which argued that government should adopt a hands-off approach with respect to the Internet. Barlow's declaration exemplifies the symbiotic relationship between the Internet and libertarian thinking. Barlow's words contain echoes from Locke ("We are forming our own Social Contract."), Mill ("We are creating a world where anyone, anywhere may express his or her beliefs, no matter how singular, without fear of being coerced into silence or conformity."), and Hayek ("our culture, our ethics, or the unwritten codes that already provide our society more order than could be obtained by any of your impositions") (quoted in Barlow 1996, Internet site).

The Internet is, like the U.S. Constitution, designed as an agreement among consenting individuals. It is a set of communication protocols that allow data to be transmitted from one computer to another. Any communication that uses Internet Protocols (IP) can be sent over the Internet. The protocols impose only minimal constraints on the information that can be transmitted. Video, telephony, text, and data all can be sent via IP.

The Internet is also decentralized. No single computer acts as a hub or main distribution point. Instead, like Hayek's spontaneous order, the Internet relies on local information, contained in routing tables, to pass data from any computer on the network to another. Also, the Internet is configured to facilitate anonymity. This tends to shift the balance of power away from government officials and toward individuals. As a result it has proved all but impossible to regulate pornography and junk mail on the Internet.

The Internet was designed to have multiple routes between endpoints, which makes it more difficult both to attack militarily and to regulate. John Gilmore, a libertarian Internet activist, famously said, "The Internet interprets censorship and damage, and routes around it."

Personal computers and the Internet have changed the relationship between individuals and large organizations. One does not need to own a printing press to publish ideas that can reach the masses. One does not need

to lease stores to sell goods to people all over the world. One does not need a mainframe computer costing millions of dollars to write a piece of software.

Because individuals are now better able to bypass large organizations, the rationale for government intervention as a check against corporate power has lost its appeal to many people who make a living using computers and the Internet. In *Cyberselfish*, a critical survey of libertarianism in the technology community, the journalist Paulina Borsook wrote that "with geeks, the attitude, mind-set, and philosophy is libertarianism" and "libertarians are the most vocal political thinkers and talkers in high tech" (Borsook 2000, pp. 3 and 7).

Intellectual Property

The low cost of distributing and copying content on the Internet has opened a schism within the libertarian community concerning the issue of intellectual property. Some libertarians argue that intellectual property rights are legitimate, based on Locke's principle that one has a natural right to property created by one's labor. According to this view, if one composes a song or another creative work, one has a property right that should be protected.

Other libertarians, including Barlow, believe that ideas should not to be regarded as property. One person can use an idea without infringing on another person's ability to use that idea. Barlow argues in the tradition of Thomas Jefferson, who wrote, "He who receives an idea from me, receives instruction himself without lessening mine; as he who lights his taper at mine, receives light without darkening me" (Quoted in Barlow 1996, Internet site).

A potential libertarian approach to the issue of copyright is Digital Rights Management (DRM). The idea behind DRM is that the composer of a creative work would embed in its digital representation a digital "lock" that could be opened only by a consumer who agreed to purchase and use the work within the limitations intended by the author.

However, there are those who doubt that DRM can be effective. Those critics say that the ability of individuals to circumvent DRM will make it impossible to rely on the private sector alone to protect intellectual property. Instead, DRM will require government involvement in the design and enforcement of restrictions on the specifications of equipment. For example, the Digital Millenium Copyright Act (DMCA) criminalized the production of technology that could be used to circumvent copyright restrictions. Many libertarians were troubled by the DMCA.

Biotechnology

The libertarian position on biotechnology, nanotechnology, and other potentially revolutionary scientific developments is one of laissez-faire. The libertarian view is that individuals are capable of addressing the ethical issues raised by new technologies without government interference.

Libertarians tend to dismiss concerns such as those raised by the President's Commission on Bioethics. In *Beyond Therapy* (President's Council of Bioethics 2004) the commission argues that biotechnology poses ethical problems by potentially enhancing human capabilities, eliminating death, and giving parents control over the characteristics of their children. Libertarians believe that individuals are capable of dealing with these issues as they arise. Moreover, libertarians argue that the sort of regulatory regime that would be needed to enforce controls over such technologies would be draconian.

Privacy

Libertarians are mindful of the effect of technology on privacy. Some technologies, such as miniature cameras, radio identification tags, and powerful storage and processing for large databases, seem to threaten privacy. Other technologies, such as the decentralized Internet and cryptography, seem to enhance privacy.

David D. Friedman has painted one scenario for the way these technologies could play out. In Chapter 1 of his draft *Future Imperfect* he writes, "Put all of these technologies together and we may end up with a world where your realspace identity is entirely public, with everything about you known and readily accessible, while your cyberspace activities, and information about them, are entirely private—with you in control of the link between your cyberspace persona and your realspace identity."

The last point—that the individual will control the link between electronic identity and physical identity—is crucial. If the opposite scenario were to emerge, in which the government always would have the ability to trace electronic communications to an individual person, the potential for totalitarian control would appear to be high.

In *The Transparent Society* (1998) David Brin has suggested that the inevitable improvement in surveillance technology is going to cause privacy to be replaced by transparency. Cameras are certain to become smaller, digital radio tracking devices will become more powerful, and all forms of surveillance will become cheaper. In light of this outlook Brin argues that freedom and autonomy can best be preserved by ensuring that individuals have as much access to information about government and large corporations as those organizations have access to information about individuals.

The Future of Libertarianism

In the late industrial age libertarianism went into eclipse. For most of the twentieth century it appeared that the future belonged to powerful manufacturing enterprises and the large government that was thought necessary to regulate and plan the industrial economy. In the Internet age many people are seeing the potential for unplanned order emerging from the decisions of individuals. This has revived libertarianism as an important philosophy.

Libertarianism may have reawakened, but it is far from triumphant. Libertarian approaches to government policy on recreational drugs, education, and health care remain far from the mainstream, where paternalism remains entrenched. Moreover, technology poses problems for which libertarianism, typically absolutist and unabashed, lacks clear answers. Intellectual property poses a conflict between the natural right to own the product of one's labor and the right to engage in free expression and activities that do not infringe directly on another person. New technologies also provide surveillance potential in ways that require libertarians to reconsider the fundamental basis for privacy.

ARNOLD KLING

SEE ALSO *Communitarianism; Democracy; Human Rights; Locke, John; Market Theory; Natural Law; Skepticism; Smith, Adam.*

BIBLIOGRAPHY

Boaz, David. (1994). *Libertarianism: A Primer*. New York. Free Press.

Borsook, Paula. (2000). *Cyberselfish: A Critical Romp through the Terribly Libertarian Culture of High Tech*. New York: PublicAffairs.

Brin, David. (1998). *The Transparent Society: Will Technology Force Us to Choose between Privacy and Freedom?* Reading, MA: Perseus. A remarkably prescient book on the challenges that surveillance technology will pose for privacy and liberty.

Caldwell, Bruce. (2004). *Hayek's Challenge: An Intellectual Biography of F.A. Hayek*. Chicago: University of Chicago Press.

Hayek, F. A. (1944). *The Road to Serfdom*. London: Routledge.

Lessig, Lawrence. (1999). *Code and Other Laws of Cyberspace*. New York: Basic Books.

Postrel, Virginia. (1998). *The Future and Its Enemies: The Growing Conflict over Creativity, Enterprise, and Progress.* New York: Free Press.

President's Council on Bioethics. (2004). *Beyond Therapy: Biotechnology and the Pursuit of Happiness.* Washington, DC: President's Council on Bioethics.

Stock, Gregory. (2003). *Redesigning Humans: Choosing Our Genes, Changing Our Future.* Boston: Houghton Mifflin.

Thierer, Adam, and Clyde Wayne Crews, Jr., eds. (2002). *Copy Fights: The Future of Intellectual Property in the Information Age.* Washington, DC: Cato Institute. Exemplifies the differences among libertarians on intellectual property issues.

INTERNET RESOURCES

Barlow, John Perry. (1996). *A Declaration of the Independence of Cyberspace.* Available at http//:www.eff.org/~barlow/Declaration-Final.html.

Friedman, David D. (2003). *Future Imperfect.* Available at http//:www.patrifriedman.com/prose-others/fi/commented/Future_Imperfect.html.

Locke, John. *Second Treatise of Civil Government*, Chapter 8. Available at http://www.constitution.org/jl/2ndtr08.htm.

Mill, John Stuart. *On Liberty.* Available at http://www.bartleby.com/130/.

LIBERTY

SEE *Freedom.*

LIFE

• • •

In consideration of the ethical uses of science and technology the phenomenon of life, especially human life, has repeatedly played significant roles in both progressive and conservative arguments. In modern philosophy notions of life have also made repeated appearances, from Thomas Hobbes's claim that the fundamental aim of politics is to replace the insecurity of life in the state of nature with a more secure life by means, in part, of technology, to Friedrich Nietzsche's appeal to a life ideal that transcends concerns of personal security. Contemporary debates about the limits of biomedical interventions in terms of whether or not human life begins at conception and feminist criticisms of cultural tendencies to disembody life thus reflect and advance long-standing concerns. Indeed, at the beginning of philosophy in Europe, one of Socrates's fundamental theses was that "The unexamined life [*bios*] is not worth living for humans" (*Apology* 38a); and as a manifestation of his divinity, the Christian scriptures record Jesus's claim to being "the way, the truth, and the life [*zoe*]" (John 14:6).

Life Sciences

Science has from its earliest forms distinguished two fundamental realms in nature: the nonliving and the living. Aristotle (384–322 B.C.E.) was among the first systematic investigators of nature and for centuries provided an authoritative orientation that took its bearings from the living. For Aristotle, living entities reveal the workings of nature better than the nonliving; life provides the key to explain the nonliving—in contrast to modern natural science, which seeks an explanation of life in terms of nonlife. Certainly life more clearly displays the dynamism and purposefulness that Aristotle sees as central to reality as a whole. Purposefulness, final causation, and teleology conceptualize that by which entities seek natural states or places proper to their kind. The acorn matures in order to become an oak tree because that is its inner nature; the oak tree maintains its state through metabolism because this inner nature has been achieved. Living things have an internal principle of motion and rest, which can be grasped by reason, whereas the nonliving are moved by external forces, the rationality of which is more difficult to comprehend.

For modern natural science, however, it is the external forces moving nonliving entities that are most readily calculable, thus giving rise to physics in a new sense. René Descartes (1596–1650), for instance, proposed that animals are simply complex machines, and that all life functions (except human thinking) could be explained in terms of mechanical interactions. From the beginning, however, the adequacy of this view has been contested, and the reduction of life to physics and chemistry challenged. The vitalism of Hans Driesch (1867–1941) and Henri Bergson (1859–1941), who argued that life involved some nonphysical element or is governed by special principles, was but one of the more pronounced examples.

Traditional explanations for the variety of life—namely, that either species are eternal or divinely created—and how organisms change over time had long been scrutinized before Charles Darwin (1809–1882) published *On the Origin of Species.* It was Darwin's theory of evolution by natural selection, however, that produced the first comprehensive account of the changing diversity of life that appeared to go beyond simple mechanism without rejecting it. Fused with the model of biological inheritance developed by Gregor Mendel (1822–1884), the synthesis of evolution by natural selection operating on the gene became the cornerstone of modern biology.

In the early 1940s the Austrian physicist Erwin Schrödinger (1887–1961) proposed that genes functioned by means of a "molecular code-script" present in chromosomes. This pointed toward the idea of molecular biology. A decade later, in 1953, James D. Watson (b. 1928) and Francis Crick (1916–2004) discovered the double-helix molecular structure of DNA. Analyses of DNA eventually elucidated the connection between genetic information and the traits of living organisms, which describes the transcription and translation of genetic information into proteins.

Redefining Life

Difficulties nevertheless remain for developing a post-Aristotelian definition of life as a biological phenomenon. One common approach has been to consider an entity living if it exhibits the following characteristics at least once during its existence: growth, metabolism, reproduction, and response to stimuli. Yet in some sense fire meets all these criteria. Moreover, some entities are not clearly either living or nonliving. Chief among these are viruses, which contain protein and nucleic acid molecules that make up living cells but require the assistance of those cells to replicate. In response, life can be further described as cellular and homeostatic—even though this would continue to classify viruses as anomalous.

Systems theorists such as Ilya Prigogine, Fritjof Capra, and Francisco Varela, however, have preferred to define life as a complex, autopoeitic (self-creating), dissipative feedback system. This conception gave rise to the Gaia hypothesis of James Lovelock and Lynn Margulis, which conceives of the entire biosphere as living insofar as it maintains conditions favorable to its continued existence.

What about the possibility of human-made, artificial life? This term can refer to a number of different research programs. Genetic engineering (and even animal breeding) creates forms of life that might not otherwise occur in nature. For Christopher G. Langton computer programs that model life processes by means of complex algorithms constitute artificial life or "a-life." Some theorists go even further to argue that beyond modeling, life is a process that can be abstracted away from any particular medium and need not necessarily depend on carbon-based chemical solutions.

Precisely when human life begins, whether at conception or some point further along in embryonic development, is also a highly contested issue. The premodern view that human life begins at the "quickening"—that is, when a woman experiences the first movements of a new child in her womb—has been altered by the very biological science that often proposes to treat embryos as no different than many other rudimentary organisms.

Life Philosophies

All such modern definitions have difficulty accounting for life as having any intrinsic ethical significance. The purposelessness of natural selection and the lowered status of humans in a hierarchy of being challenge traditional moral and theological beliefs. When life is conceived as an assemblage of adaptations to random and constantly changing circumstances, there remain no forms or essential types to imitate, and no harmonious order or basic good to maintain. Yet despite the most sophisticated explanations, purposefulness does appear to be an aspect of the living.

One response has been the development of a life philosophy (German *Lebensphilosophie*) that arose as a reaction against Enlightenment rationalism. Life is prioritized over mere understanding, and life philosophy has had many variants, including artistic movements in which life is used as a concept to assess and critique modern society. Certainly over the course of the nineteenth and twentieth centuries life as "vitality" or vividness, a sense of both spiritual striving and joyous experiencing, played an important role in literature, art, and music as a touchstone of criticism of the scientific and technological. Among the most important representations of this view are attempts made by Arthur Schopenhauer (1788–1860) and Friedrich Nietzsche (1844–1900) to grasp life as an all-encompassing metaphysical category or first philosophy.

Nietzsche's life philosophy differed from the thought of Schopenhauer in its naturalism. In his genealogical work, he traced the development of the life-denying ascetic ideal that he saw as dominant in Western (and most Eastern) philosophy and religion. Value comes to being always in support of life, but ascetic philosophies give vital ideals a life-devaluing interpretation. Anything that is part of the natural, changing, lifeworld is interpreted as wrong and sinful, and ideals of truth and virtue are rooted in otherworldly, changeless realms. The ascetic ideal removes all source of value from nature, whereas modern natural science removed any faith in a realm outside of nature. One interpretation of the "death of God" is the extinction of this transcendent, nonhuman, and ahistorical realm to ground human values. There is nothing but life on which to base values, including truth. Whether Nietzsche successfully distinguished this revaluation of values from nihilism remains a subject of dispute.

During the mid-twentieth century life philosophy made a new appearance in the forms of phenomenology and existentialism. Phenomenology especially criticized science as separating itself from the human lifeworld or as disembodying experience. Related arguments have been carried forward in feminist criticisms such as those of Barbara Duden and Donna Haraway. In her studies of women's medicine and experiences such as pregnancy, Duden (1993) defends the primacy of lived experience over its conceptual analysis. In her notion of "companion species," Haraway (2003) criticizes the primacy of conceptual oppositions in favor of mutuality of living relationships, which harks back to the work of Pytor Kropotkin (1842–1921) and his notion of "mutual aid" among organisms.

Whether molecular biology can account for what is apparently goal-directed behavior in organisms likewise continues to spark controversy (see, e.g., Allen, Bekoff, and Lauder 1998). Finally, given the difficulties of understanding the ethical significance of biological life in the modern sense, philosophers such as Hans Jonas (1966) and Leon R. Kass (1985) have even attempted to revive an Aristotelian approach that would understand the most elementary forms of life in terms of higher forms of life rather than vice versa.

The Human Condition, Bioethics, and Biotechnology

According to Hannah Arendt (1958) the life of human activity, or *vita activa*, may be distinguished into labor, work, and action. Labor pertains to the biological processes of the human body, work to the world of artifice, and action to politics. Political action is so central to the human condition that the Romans used the same term (*inter homines esse*) to signify both "to live" and "to be among men." But as Arendt also notes, "life" takes distinct forms in each level of the *vita activa*. In the first instance life is related to the futile, biological labors of the body in which there is a kind of "deathless everlastingness of the human as of all other animal species" (p. 97). In the second instance life takes on the worldliness of work with distinct beginnings and ends and can be told as a story.

The first notion of life corresponds to the Greek *zoe*, from which English derives *zoology*; the second corresponds to the Greek *bios*, from which comes *biography* and a sense of the historical. For Arendt the modern world may be characterized by an effulgence of *zoe* as labor moved from the most-despised to the most-esteemed position with a productivity that outstripped all traditional work and overwhelmed action. But action and speech, beyond the necessary but lower forms of the *animal laborans* (labor) and *homo faber* (work), is the highest form of human life. The measure of all things, she claims, "can be neither the

driving necessity of biological life and labor nor the utilitarian instrumentalism of fabrication and usage" (p. 174).

The term *bioethics* was initially coined by the biologist Van Rensselaer Potter (1911–2001) to refer to an ethics grounded on the science of life, rather than on religion or philosophy. It has since come to signify the field that studies the intersection of biology and biography, or the science of life studied scientifically and life lived experientially (Kass 2002). The focus on biography and the good life, rather than mere biological life, has taken on more importance as new biomedical technologies expand the capacities of human biology, or what Arendt would call the labor of human bodies. This is best illustrated by advances in life-extending techniques used in palliative care. In many instances, one's biological life is extended well beyond the duration of one's biographical life among the world of things and within the plural realm of action and speech. This raises ethical questions about what it means to die a dignified death and who should make such decisions in various circumstances.

Advances in biotechnology offer new powers to alter and to some degree control the phenomena of life. This has brought both reward and risk. In agricultural uses, biotechnology has raised concerns about risks, especially involving uncertain ecological interactions and health effects. In biomedical uses, similar health risk issues occur along with questions of informed consent and privacy. Additionally, the controversial techniques of abortion, cloning, and stem cell research sustain heated debates about when human life begins. New reproductive techniques have stimulated questions about how much control the present generation ought to have over future generations.

This last issue highlights the fact that both in agricultural and medical biotechnology, traditional ethical issues are complemented by deeper concerns about the proper limits to the human activity of "remaking Eden" and "relieving man's estate." How ought humankind responsibly exercise its power over life and where should limits be drawn? For example, even though biomedical technologies offer obvious rewards in terms of satisfying deep human desires, they can also serve (intentionally or not) to diminish human life. As the President's Council on Bioethics remarked in *Beyond Therapy* (2003), "To a society armed with biotechnology, the activities of human life may come to be seen in purely technical terms, and more amenable to improvement than they really are" (p. xvii). Promoting the genuine flourishing of human life is foremost a matter of understanding the good life rather than commanding the tools to manipulate life processes.

CARL MITCHAM
ADAM BRIGGLE

SEE ALSO *Bioethics; Environmental Ethics; Medical Ethics.*

BIBLIOGRAPHY

Allen, Colin; Marc Bekoff; and George Lauder, eds. (1998). *Nature's Purposes: Analyses of Function and Design in Biology.* Cambridge, MA: MIT Press.

Arendt, Hannah. (1958). *The Human Condition.* Chicago: University of Chicago Press.

Duden, Barbara. (1993). *Disembodying Women: Perspectives on Pregnancy and the Unborn,* trans. Lee Hoinacki. Cambridge, MA: Harvard University Press.

Haraway, Donna. (2003). *The Companion Species Manifesto: Dogs, People, and Significant Otherness.* Chicago: Prickly Paradigm Press.

Jonas, Hans. (1966). *The Phenomenon of Life: Toward a Philosophical Biology.* New York: Harper and Row.

Kass, Leon R. (1985). *Toward a More Natural Science: Biology and Human Affairs.* New York: Free Press.

Kass, Leon R. (2002). *Life, Liberty and the Defense of Dignity: The Challenge for Bioethics*San Francisco: Encounter Books.

Schrödinger, Erwin. (1944). *What Is Life? The Physical Aspect of the Living Cell.* Cambridge, UK: Cambridge University Press. Based on lectures delivered at Trinity College, Dublin, 1943.

U.S. President's Council on Bioethics. (2003). *Beyond Therapy: Biotechnology and the Pursuit of Happiness.* Washington, DC: President's Council on Bioethics.

LIMITED NUCLEAR TEST BAN TREATY

• • •

The Limited Nuclear Test Ban Treaty (LTBT) was signed by the United States, Great Britain, and the Soviet Union in Moscow on August 5, 1963. Ending more than eight years of negotiations, the LTBT prohibits nuclear weapons tests or other explosions in the atmosphere, outer space, or underwater. While the treaty does not ban underground nuclear explosions, it does prohibit tests if they would cause "radioactive debris to be present outside the territorial limits of the State under whose jurisdiction or control" the explosions were conducted. In addition, by signing on to the treaty the countries agreed to the goal of "the discontinuance of all test explosions of nuclear weapons for all time."

Emergent History

After the end of World War II, Great Britain and the Soviet Union joined the United States in the nuclear club and the United States and the Soviet Union tested their first hydrogen bombs in 1952 and 1953 respectively. Public concern about nuclear testing began to grow, especially after the March 1954 test of a thermonuclear device by the United States at Bikini atoll. This test was expected to have a yield equivalent to approximately eight million tons of trinitrotoluene (TNT), but in actuality was about fifteen megatons, or almost double the predictions. The fallout from the explosion greatly exceeded geographical expectations, contaminating a Japanese fishing vessel, the Lucky Dragon, as well as Bikini atoll.

This incident, as well as others, increased the awareness of the effects of fallout and the issue of continued nuclear tests garnered greater public scrutiny. Organizations such as Women Strike for Peace and Physicians for Social Responsibility were formed to increase public pressure on western governments for signing a treaty, as well as informing the public of the dangers of nuclear testing. For instance, Women Strike for Peace originated from an international protest of women against atmospheric testing. Physicians for Social Responsibility documented the presence of strontium-90—a highly radioactive waste product of atmospheric nuclear testing—in children's teeth across the country. As it became apparent that no region of the world was untouched by radioactive fallout, there was increasing apprehension about the possibility of global environmental contamination and the resulting genetic effects. It was in this atmosphere that efforts to negotiate an end to nuclear tests began in May 1955 in the Subcommittee of Five of the United Nations Disarmament Commission.

International interest in the course of the negotiations was intense and sustained. The issue was brought up in statements and proposals at international meetings and the United Nations General Assembly addressed the issue in a dozen resolutions, repeatedly pressing for an agreement to be reached. While the United States, Great Britain, and the Soviet Union engaged in a tripartite effort—The Conference on the Discontinuance of Nuclear Weapons Tests—almost continuously from October 31, 1958 to January 29, 1962, no treaty could be drafted due to differences on a number of issues.

Basic Treaty Issues

The issue of a control and enforcement mechanism to verify compliance to a comprehensive test ban was the primary point of disagreement between the parties. Western European and U.S. powers, especially, were concerned that it would be more dangerous to accept pledges without the means to verify that they were being complied with than to not have a treaty at all. The Soviets, for their part, felt that because, "in the present state of scientific knowledge" (Premier Bulganin writing to President Eisenhower on October 17, 1956, from U.S. Department of State Bureau of Arms Control) no explosion could be produced without being detected, then there could be an immediate agreement to prohibit tests without an international control mechanism at all.

To resolve the issue of how compliance could best be verified, the Geneva Conference of Experts met in July and August 1958 and was attended by representatives from the United States, Great Britain, Canada, France, the Soviet Union, Poland, Czechoslovakia, and Romania. The group of experts developed and agreed on the technical aspects of a verification system to monitor a ban on atmospheric, underwater, and underground tests. This control system included an elaborate network of more than 150 land control posts, ten ship-borne posts, and special aircraft flights. In addition it allowed for on-site inspections to determine whether seismic events were caused by earthquakes or by explosions. While the United States and Great Britain said they would be willing to negotiate an agreement based on the establishment of an international control system, the Soviet Union responded by linking the test ban to other arms control issues and resumed testing. The other nuclear powers refrained from testing until 1961, after France tested its first nuclear weapon in 1960, and in 1962, the four nuclear powers conducted a record 178 nuclear tests.

Disagreement on a control system was focused on four main areas:

(a) The Veto. The Soviet Union wanted all operations to be subject to a veto while the United States maintained that the inspection process should be automatic in order to be effective.

(b) On-Site Inspections. The Soviet Union capped on-site inspections at three per year while the United States and Great Britain insisted that the number should be determined by detection capability and necessity. Eventually the United States said it would accept a minimum of seven inspections, which was rejected by the Soviet Union.

(c) Control Posts. Neither side could agree on the number and location of posts or of the automatic seismic observation stations that would supplement nationally owned control posts. The argument of the Soviet Union that these national posts and observation stations would make inspections unnecessary was rejected by the United States and Great Britain.

(d) The Organization and Control Commission. The Soviet Union proposed a *troika* of administrators for the Control Commission, including one neutral, one Western European or North American, and one Communist member. The Western European and North American countries argued that this would make the Control Commission powerless and unable to take action. The Soviet Union eventually acquiesced to opposition concerns and abandoned this position.

Treaty Creation and Ratification

After the Cuban Missile Crisis in October 1962, both sides were anxious to alleviate public fears about nuclear weapons and therefore restarted the three-power conference on a test ban treaty in July 1963. While the Soviet Union would not agree to a treaty that prohibited underground testing, the three powers were able to agree on a partial ban on atmospheric, outer space, and underwater testing, which were all easily verifiable without intrusive inspections. In just ten days, the three parties had developed and signed the LTBT. The U.S. Senate ratified the agreement on September 24, and President John F. Kennedy signed the LTBT into law on October 7, 1963. The LTBT formally entered into force on October 10, and it is of unlimited duration.

Although the LTBT was touted by all parties as a success, and indeed it was so as it greatly reduced dangerous atmospheric fallout and deadly radiation, including strontium-90, secondary results were mixed. Because neither France nor China signed the LTBT, they continued to test intermittently until the early 1980s. India, Pakistan, and Israel, all signatories of the treaty, were able to join the nuclear club despite the limited ban. And in the United States and the Soviet Union, nuclear weapons development and testing continued unabated, although all tests were moved underground. Additionally there was less international public pressure to develop a comprehensive test ban treaty as the most visible sign of the arms race, atmospheric testing, was eliminated. However despite these failings, the LTBT was an important and symbolic first step and served as a precedent for future arms control treaties.

JESSICA L. COX
MARGARET COSENTINO

SEE ALSO *Baruch Plan; International Relations; Nuclear Ethics; Nuclear Non-Proliferation Treaty; Weapons of Mass Destruction.*

BIBLIOGRAPHY

"Treaty Banning Nuclear Weapons Tests in the Atmosphere, in Outer Space and Under Water (Partial Test Ban Treaty—PTBT)." (August 2000). *Inventory of International Nonproliferation Organizations and Regimes.* Monterrey, CA: Monterrey Institute For International Studies.

INTERNET RESOURCES

Burr, William, and Hector L. Montford, eds. (2003). "The Making of the Limited Test Ban Treaty, 1958–1963." George Washington University. Available from http://www.gwu.edu/~nsarchiv/NSAEBB/NSAEBB94/.

Kimball, Daryl, and Wade Boese. (2003). "Limited Test Ban Treaty Turns 40." *Arms Control Today* 33, no. 8 (October). Available from http://www.armscontrol.org/act/2003_10/LTBT.asp

"Treaty Banning Nuclear Weapon Tests in the Atmosphere, in Outer Space and Under Water." U.S. Department of State. Bureau of Arms Control. Available from http://www.state.gov/error_404.html

LIMITS

• • •

The question of human limits, both cognitive and moral, is a persistent theme in the history of religion and philosophy. Both Siddhartha Gautama (Buddha, c. 563–c. 483 B.C.E.) and Socrates (469–399 B.C.E.) argued, in quite different ways, for the human acceptance of limits. Indeed, in general premodern traditions in human culture widely acknowledged both theoretical and practical limits on human knowledge and action.

Thus ever since the founding of modernity, with its appeals to transcend many traditional limits in the development of science and technology—and even certain aspects of the human condition—the question of whether and to what extent there might be new limits to the modern project has been a recurring theme. Late eighteenth and early nineteenth century poets such as Johann Wolfgang von Goethe (1749–1832) and William Blake (1757–1827) called for recognition of cognitive limits in modern science; nineteenth and early twentieth century novelists such as Charles Dickens (1812–1870) and John Steinbeck (1902–1968) argued for placing social and political limits on industrial technological practices; and philosophers of limits such as Karl Marx (1818–1883), Friedrich Nietzsche (1844–1900), and Oswald Spengler (1880–1936) proposed the existence of historical and cultural limits to modern development as a whole.

Limits to Growth

Such general discussions were given a new, specialized form with the 1972 publication of *The Limits to Growth* by the team of Donella Meadows, Dennis Meadows, and Jørgen Randers, which brought the environmental predicament of industrial progress to the attention of a world audience. On the basis of a computerized world model, the celebrated but controversial study claimed that continuing high rates of growth would lead to (a) a depletion of vital global resources, (b) increasing pollution, and (c) population outrunning the world's potential food supplies. The study suggested that, unless swift action was taken, absolute limits to growth would appear in the course of the twenty-first century, causing population size and industrial capacity to drop rapidly. This message was instantly seen as a blow against the creed of economic growth dominating at the time, both in the Western and the Communist world. Subsequently, the rift between growth advocates and growth skeptics has continued to divide the contemporary world of science and of politics; in fact, this division reaches deeper than conventional distinctions such as conservative/progressive or right/left.

Do Limits Exist?

The debate on limits carries on where classical economics had left off. Thomas Malthus (1766–1834), for example, still had the implicit vision of the Earth as a closed space, with limits to the size of population and level of human achievement it could sustain. He argued that lack of food supply would ultimately constrain population growth, throwing into doubt the idea of the inevitability of progress. However, he underestimated both the variability of growth and the capacity of technology to overcome natural limits. In contrast, neo-classical economics, operating on the background assumption of the infinite power of science and technology, had subsequently ignored the dependence of economic systems on natural systems completely. This shortcoming had left economic science blind to the impending environmental crisis in the twentieth century.

The attempt of Meadows, Meadows, and Randers to expose this failure set off a replay of the controversy between the "closed space" and "infinite ingenuity" schools of thought. While the former insists on the finiteness of both resource inputs and waste sinks, the latter emphasizes the practically infinite substitutability of natural resources by technology and organizational innovation (Simon 1981). What matters to the biosphere is the scale of resource flows, not just their efficient allocation (Daly 1996). Markets may reduce the volume of resource use through substitution of natural inputs, but continuing growth will eventually cancel out these efficiency gains, increasing volumes again. It is the overall scale of resource flows with respect to both input sources and waste sinks that determines the relationship between the economy and the biosphere.

Scientific findings suggest that for the first time in history, human-induced material flows are presently outgrowing nature-induced flows. In other words, the technosphere eclipses the biosphere. Some well-known facts are symptoms for this imbalance: Humankind has already exhausted 40 percent of known oil reserves, transformed nearly 50 percent of the land surface, appropriates more than half of all accessible freshwater, increases greenhouse gases in the atmosphere over and above natural variability, and causes extinction rates to increase sharply in marine and terrestrial ecosystems (Steffen et al. 2004, p. 6). In general terms, human impacts on the Earth are approaching or exceeding in magnitude the impact of some of the great forces of nature. In addition, they operate on much faster time scales than rates of natural variability. Estimates following the ecological footprint methodology imply that human activities presently exceed the Earth's capacity by 15 to 20 percent—without taking the needs of other living beings into account (Wackernagel et al. 2002). Ecological overshoot has become the distinguishing mark of human history.

What To Do about Limits?

The way "limits" are understood has consequences for politics and ethics. One metaphor for conceptualizing limits is that of a cliff face: The concept implies a fixed line beyond which collapse looms. It insinuates that crucial changes happen in an abrupt as well as catastrophic fashion, making everyone suffer equally. However, changes may also occur in a gradual as well as insidious fashion, and may burden some more than others. A metaphor based on a tapestry—each act of destruction is like pulling a thread from the tapestry—would emphasize linear and not just non-linear processes, multiple smaller losses and not just overall collapse. In particular, it would highlight the presence of political choices along the gradient of degradation (Davidson 2000). The tapestry metaphor, more than the cliff metaphor, encourages one to judge wreckage not only as prelude to the collapse, allows one to trace the differential impact of losses on social groups, and stimulates the

politically and ethically essential question: What thresholds are considered tolerable/intolerable for whom and on what grounds?

Thresholds of ecosystem changes represent "limits" only for humans; any definition of limits is therefore a political act. Moreover, limits are rarely scientifically knowable; their definition is therefore an ethical act as well. As a consequence, any definition implies choices in terms of human welfare, equity, and the common good. A first approach centers on risks, putting the spotlight on possible physical, technical, and economic losses resulting from the technology or economic policy in question. Emphasis is placed on the precautionary principle of preventing the worst from happening. Guardrails, for instance, are suggested in order to avoid abrupt and irreversible changes from which human societies would find it difficult or impossible to recover (German Advisory Council on Global Change 1997).

A second approach focuses on institutions, because the rise of external limits is brought about by structures of growth and accumulation that are internally insatiable and limitless. This approach highlights the constellation of social and economic factors driving perilous developments (Harvey 1996). Proposals range from the reform of price structures to the containment of the profit motive, from the reallocation of research funds to the phase-out of certain technologies.

Finally, a third approach calls for a reconsideration of values, bringing into sharp relief the civilizational losses incurred by the predominance of the logic of growth. Natural limits are often preceded by the appearance of social and cultural limits; before growth causes physical perturbations, collective and individual well being has suffered (Illich 1973, Hirsch 1976). Recognizing limits, therefore, implies the emergence of fresh opportunities by restoring a balance. In this approach, limits acquire a positive connotation, making a more accomplished life possible. They turn out to be productive for a civilization that regards economic power and growth only marginally important.

WOLFGANG SACHS

SEE ALSO *Ecological Footprint; Precautionary Principle.*

BIBLIOGRAPHY

Daly, Herman E. (1996). *Beyond Growth: The Economics of Sustainable Development.* Boston: Beacon Press.

Davidson, Carlos. (2000). "Economic Growth and the Environment: Alternatives to the Limits Paradigm." *BioScience* 50: 433–440.

German Advisory Council on Global Change. (1997). *World In Transition: The Research Challenge*. Berlin: Springer

Harvey, David. (1996). *Justice, Nature, and the Geography of Difference*. Cambridge, MA: Blackwell.

Hirsch, Fred. (1976). *Social Limits to Growth*. Cambridge, MA: Harvard University Press.

Illich, Ivan. (1973). *Tools for Conviviality*. New York: Harper & Row.

Meadows, Donella, Dennis Meadows, and Jørgen Randers. (1972). *The Limits to Growth*. New York: Universe Books.

Meadows, Donella, Dennis Meadows, and Jørgen Randers. *Beyond the Limits: Confronting Global Collapse, Envisioning a Sustainable Future*. White River Junction, VT: Chelsea Green.

Meadows, Donella, Dennis Meadows, and Jørgen Randers. (2004). *Limits to Growth: The 30-year Update*. White River Junction, VT: Chelsea Green.

Simon, Julian L. (1981). *The Ultimate Resource*. Princeton, NJ: Princeton University Press.

Steffen, Will, et al. (2004). *Global Change and the Earth System: A Planet Under Pressure*. Berlin: Springer.

Wackernagel, Mathis; Niels B. Schulz; Diana Deumling; et al. (2002). "Tracking the Ecological Overshoot of the Human Economy." *Proceedings of the National Academy of Sciences* 99: 9266–9271

John Locke, 1632–1704. An English philosopher and political theorist, Locke began the empiricist tradition and thus initiated the greatest age of British philosophy. He attempted to center philosophy on an analysis of the extent and capabilities of the human mind. (*Rutgers University Library.*)

LOCKE, JOHN

• • •

John Locke (1632–1704), was an English philosopher, Oxford academic, and occasional bureaucrat. He was born at Wrington, Somerset, on August 29 and died at Oates, Essex on October 28. Locke's fame as a philosopher rests chiefly on two works: *An Essay Concerning Human Understanding* (1689) and *Two Treatises of Government* (1689). The former became a chief textbook of the European enlightenment and subsequent philosophy. The latter deeply influenced both the Declaration of Independence (1776) and the Constitution of the United States (1787), a document that made promoting the "progress of science and useful arts" one of its distinguishing features (Article I, section 8). These facts establish his reputation as one of the most influential modern philosophers and signal his importance in issues related to science, technology, and ethics.

Locke's strategy in his two most influential works is characteristic of early modern thought. First he sets out to clear away errors and conceits left over from classical and medieval science. Next he reduces the subject to its most basic natural constituents, as yet unmodified by culture. Only then does he set about reconstructing new systems of epistemology and political philosophy.

The *Essay*

Part One of the *Essay* is devoted to a refutation of the doctrine of innate ideas, according to which all human beings are born with certain principles already stamped upon their minds. It might seem doubtful that the importance of this doctrine justifies the attention that Locke devotes to it; however, its demolition whets the appetite for a more satisfactory account of the mind.

Locke holds to the view that all human ideas are reducible to experiences, a doctrine known as *empiricism*. An *idea* here means anything in the contents of the mind that is definite enough to have a name. Impressions, such as *hot* and *red*, received from the external world are the primary source of ideas. But unlike more uncompromising empiricists, such as David Hume, Locke admits of a second source of ideas: reflection upon the operations of human minds. One may observe what the mind does with the material provided by sensation and so acquire ideas of *thinking*, *willing*, and the like. So, though there are no innate ideas, there are innate sources of information.

Some ideas, such as *hard* or *perception* are indivisible. These are received passively by the mind. But the mind can also act on elementary ideas in three ways: by combining several into one complex idea; by comparing one with another; and by abstracting some idea from the setting in which it actually occurs. By such operations the mind can furnish itself with a potentially unlimited stock of complex ideas. These in turn fall into three categories: *relations* between ideas, *substances* that may exist on their own; and *modes* that exist only in something else. Thus the sun is a substance; it is bright in relation to terrestrial fire; and its brilliance is one of its modes.

Though all complex ideas are products of the mind, they can be anchored in the real world. A substance is known only by its qualities, which are the impressions it makes on the senses. Its primary qualities belong to it independently of observation, so a stone has weight and shape whether anyone perceives it or not. Secondary qualities depend on an observer. The stone is brown only in the right light, and in the eyes of some beholder. One cannot conceive but that these qualities subsist in some underlying thing, but has no idea of what that thing is. Locke subscribes, however, to the corpuscular or atomic theory of matter and supposes that the substratum consists of invisibly small particles.

Locke's philosophy of mind narrows the distance between speculation and technology. Chemistry, once it has purged itself of any alchemical conceits and has arrived at knowledge of the elements, not only understands the world better but provides human beings with means to manipulate it. Similarly Locke offers both a better account of human knowing and a set of useful instruments both for scientific and philosophical investigation.

This raises the question of the rank of philosophy with respect to science and technology. In one respect Locke's view of this matter seems closer to the medieval than to the classical conception. For the Greeks, philosophy was more elevated and more complete than any science, if indeed it did not incorporate all the sciences. In medieval scholarship, philosophy is famously regarded as the *handmaiden of theology,* usually in so far as it supports and clarifies faith. For Locke, philosophy seems to become the handmaiden of the sciences.

In the Epistle to the *Essay* Locke distinguishes between the *Master-Builders* and the *Under-Labourers* of the sciences. Among the former are Robert Boyle, Thomas Sydenham, Christiaan Huygens, and Isaac Newton, whose works stand as monuments to posterity. Locke counts himself among the latter, whose job it is merely to clear the ground and remove the rubbish that obstructs the advance of science. If this is Locke's view, he has reduced philosophy to a preparatory exercise, much of which is necessary only because of the abuses of language committed by psuedophilosophers. Locke's *Essay* is certainly similar to contemporary academic philosophy, which understands itself as clarifying questions up to the point that science can get a grip on them.

The scientists named by Locke are conspicuous for both theoretical and technological achievements. Boyle constructed an air pump; Newton and Huygens built advanced telescopes; Sydenham pioneered new medical treatments. But it is clear that for Locke their greatness lay more in their theoretical work than in any useful devices they may have contrived. He shows no inclination to subordinate the sciences to technology. A few lines after mentioning Newton, he identifies philosophy as "nothing but a true knowledge of the nature of things" (Locke 1975, p. 10). Whatever Locke's view of his business in the "Essay," he had a view of philosophy broad enough to encompass the sciences. It is closer to the classical view than is often supposed.

The *Two Treatises*

In his *First Treatise,* Locke demolishes Robert Filmer's argument in favor of the divine right of kings. This sets the stage for the *Second Treatise*: If political authority does not originate in God's appointment of Adam, then its origin must be sought in human nature.

Typically Locke identifies and isolates the elementary building block of political societies: This is the human being in the *state of nature.* The latter indicates a condition of perfect freedom and equality, with no one having any authority over another. But it is not, as Thomas Hobbes (1588–1679) supposed, a state of license. For there is a natural law available to all human beings, directing them to respect one another's life, liberty, and property.

Oddly enough, it is not viciousness that requires the formation of governments, but the human capacity for righteous indignation. In the state of nature, each person is entitled to punish any transgression of rights. But as each person judges primarily in his or her own favor, one person's enforcement of natural law is another's transgression of the same. Thus the universal distribution of the executive power can lead to endless cycles of revenge. The way to avoid this is for all to surrender their portions of the executive power to some common judge, to whom appeal may be made in case of conflict.

Human beings thus leave the state of nature in order to more securely enjoy those rights that they possessed while still in it. Universal consent is the foundation of political authority, which may be invested in such forms (for example, kings and parliaments) as the subjects think fit. However that grant of authority is always conditional rather than absolute. When the government forfeits the consent of its subjects, or by aggression or neglect fails to protect their liberties, it effectively abdicates. The people are then entitled to abolish it and form a new one.

Property Rights

Locke's theory of property, set forth in Chapter 5 of the *Second Treatise*, is among the greatest achievements of seventeenth-century political and economic thought. Here Locke cuts to the original position immediately: In the beginning all things belonged equally to all human beings, and each had leave to take from the earth whatever he or she needed. What then is the origin of any private rights to property?

Each person has ownership of his or her own body and labor. In order for some external good such as food to be enjoyed it must sooner or later be appropriated. After an apple is consumed it joins with the perfect privacy of the flesh. Locke argues that the moment of appropriation comes when someone's labor is mixed with the bounty of nature. When acorns are first gathered from the wild, they become private property. The right of appropriation is universal, the only limit is that one may gather only what one can use.

Locke weds this account with a theory of economic progress, which includes in turn a labor theory of value and a theory of money. The greater part of the value of any product originates in the labor required to produce it. Invested in a loaf of bread, for example, is a plowed and cultivated field, harvested and milled wheat, a bricked and furnished bakery. All this labor represents a vast increase in the wealth available to humankind over what unimproved nature provides.

But how is it possible to encourage people to labor beyond what their needs require or the durability of their produce allows? The answer lies in money, the exchange of the products of one's labor for some durable medium of nominal rather than real value. When someone settles and improves a piece of land, it is taken out of the common stock; however, in return for money, the settler gives back more value than he or she took away. Locke understood that this process, repeated across a wide range of industries, was an engine of unprecedented economic growth. For that reason, one of the most important ends of government was the protection of private property.

Locke's theory of property may be set comfortably in the context of a fundamental modern project: the conquest of nature. The natural world is not charitable to human beings. It provides little of what they need in advance of their labor. But the potential wealth that exists in nature is vast beyond calculation. Thus the aboriginal inhabitants of America who, Locke says, "are rich in land but poor in the comforts of life" exemplify the situation of human beings in the state of nature (Locke 1988, p. 296). By encouraging labor, a system of money and property rights will result in the most thorough cultivation of nature, for the comfort of all humankind.

It is clear that Locke's approach to all three topics elevates the products of human invention far above the natural materials from which they are fashioned. Complex ideas are more interesting and useful than simple ones. There is both more security and more freedom under government than out from under it. If a government acts to protect property rights, human beings will then make whatever they need to relieve the poverty into which the species was born. Nature will be reduced to a storehouse of useful materials.

KENNETH C. BLANCHARD JR.

SEE ALSO *Hume, David; Liberalism; Libertarianism; Mill, John Stuart; Skepticism.*

BIBLIOGRAPHY

Goldwin, Robert. (1987). "John Locke." In *History of Political Philosophy*, 3rd edition, ed. Leo Strauss and Joseph Cropsey. Chicago: University of Chicago Press. 1st edition, 1963; 2nd edition, 1972. One of the most thorough essays on Locke's place within the larger context of modern thought.

Locke, John. (1975). *An Essay Concerning Human Understanding*, ed. Peter H. Nidditch. Oxford: Oxford University Press.

Locke, John. (1988). *Two Treatises of Government*, ed. Peter Laslett. Cambridge, UK: Cambridge University Press. A very useful edition of Locke's most important political work, including a very extensive introductory essay on the historical context of the works.

Locke, John. (1997). *Political Essays*, ed. Mark Goldie. Cambridge, UK: Cambridge University Press. A thorough collection of Locke's major and minor essays, including his essays on the Law of Nature.

Mackie, John L. (1976). *Problems from Locke*. Oxford: Oxford University Press. A consideration of Locke's essay in light of contemporary science and analytical philosophy.

Walmsley, Jonathan. (2004). "Locke's Natural Philosophy in Draft A of the Essay." *Journal of the History of Ideas* 65(1): 15–37. A treatment of Locke's early views and of his collaboration with Sydenham.

LOGICAL EMPIRICISM

• • •

Logical empiricism (LE) is a term that was coined by the Austrian sociologist and economist Otto Neurath (1880–1945) to name the philosophical work of the Vienna Circle and related work being pursued by the physicist and philosopher Hans Reichenbach (1891–1953) and his associates. Related terms include *logical positivism*, *neopositivism*, and *scientific empiricism*. The basic intention of LE was to formulate a scientific philosophy for understanding the relationship between science and society. In historico-philosophical terms the aim was to combine the empiricist legacy of philosopher-scientists such as Hermann von Helmholtz (1821–1894), Ernst Mach (1838–1916), Henri Poincaré (1854–1912), and Pierre Duhem (1861–1916), with the new logic developed by Gottlob Frege (1848–1925), David Hilbert (1862–1943), and Bertrand Russell (1872–1970). The intended synthesis was not simply a theoretical project. Logical empiricists considered themselves part of a progressist movement for a more rational and enlightened society. As stated in the so-called Manifesto of the Vienna Circle, LE aimed to foster a "scientific world-conception" ("wissenschaftliche Weltauffassung") that would help create a better world for all people.

The Scientific World-Conception

The characteristic method of LE was logical analysis, which used mathematical logic to clarify the logical structure and meaning of assertions. In this way LE aimed for a logical analysis of scientific and philosophical language that would distinguish clearly between meaningful and meaningless sentences; fight against metaphysics, which was considered as a hotbed of meaningless "pseudo-sentences"; and provide a "unified science" (*Einheitswissenschaft*) that would be formulated in a logically analysed language cleansed of metaphysical elements.

LE claimed that logical analysis demonstrated that there are only two kinds of meaningful propositions, the analytic a priori propositions of logic and mathematics and the synthetic a posteriori propositions of empirical sciences. All other assertions were to be considered cognitively meaningless. This holds in particular for all metaphysical propositions. The most famous argument to this effect is found in "Overcoming Metaphysics by Logical Analysis of Language" 1932 by Rudolf Carnap (1891–1970). Moreover, "overcoming metaphysics" was not simply an internal philosophical issue because logical empiricists considered metaphysics to be a medium for propagating politically and morally pernicious ideologies that had to be fought not only in the academic sphere but also in the political arena.

Politically, most logical empiricists were democratic socialists or unorthodox Marxists and thus were partisans of an "engaged scientific philosophy." A few, such as Moritz Schlick (1882–1936) and Friedrich Waismann (1896–1959), were less political but shared a progressive, liberal outlook.

For all logical empiricists scientific philosophy was a collective enterprise that had to contribute to the construction of a modern, enlightened society. That task was to be carried out in close collaboration with the sciences and other progressive cultural forces, such as the artists and architects belonging to the *Neue Sachlichkeit* movement or the Bauhaus. When LE was at its peak in the late 1920s and early 1930s, the more radical logical empiricists of the Vienna Circle, such as Neurath and Carnap, regarded themselves as "social engineers" engaged in the task of forging the philosophical and scientific tools for building a new socialist society. This is expressed emphatically in the concluding lines of the *Manifesto of the Vienna Circle*: "We witness the spirit of the scientific world-conception penetrating in growing measure the forms of personal and public life, in education, upbringing, architecture, and the shaping of economic and social life according to rational principles. The scientific world-conception serves life, and life receives it" (Sarkar 1996, Vol. I, pp. 329–330).

LE included a multifaceted and variegated group of philosophers and scientists. Its internal diversity often is underestimated. LE was less a school with a common doctrine than a movement whose members shared vaguely progressist convictions. Even closely related thinkers such as Carnap and Neurath disagreed on many basic philosophical issues. Here the focus is on few leading figures of the Vienna Circle: Schlick, its founder; Carnap and Neurath; and Carl Gustav Hempel (1905–1997), the most influential representative of LE in the United States.

In the early 1930s the LE movement in Europe gradually dissipated as a result of disastrous, political developments and indivdual events. The mathematician Hans Hahn (1879–1934), considered by some to be the "real" founder of the Vienna Circle, died in 1934, and Schlick was murdered by a demented student in 1936. In 1934 Carnap left Vienna and moved to the German university in Prague. After the rise of National Socialism in Germany (1933) and clerical fascism in Austria (1934) most logical empiricists emigrated. The majority went to the United States, including Carnap, Reichenbach, and Hempel. The history of LE thus divided into two periods: a European period ending in the mid-1930s and an Anglo-American period from the 1930s until its dissipation in the 1960s.

Major Figures and Their Ideas

The founder and official leader of the Vienna Circle was Schlick, who studied physics under Max Planck (1858–1947). Later Schlick turned to philosophy, and in 1922 he was appointed to the chair of natural philosophy at the University of Vienna as the sucessor to Ludwig Boltzmann (1844–1906) and Ernst Mach (1838–1916). Beginning in 1923, he and his assistants Herbert Feigl (1902–1988) and Friedrich Waismann organized a discussion group (first called the "Schlick circle") that soon became known as the "Vienna Circle."

Schlick had begun as a "critical realist", and later was influenced by Ludwig Wittgenstein (1889–1951). In *The Turning-Point in Philosophy* (1930) Schlick emphatically endorsed Wittgenstein's thesis that the philosophy of science is not to be considered a system of knowledge but instead a system of acts: "[P]hilosophy . . . is that activity whereby the meaning of statements is established or discovered. Philosophy elucidates propositions, science verifies them" (Sakar 1996, vol. II, p. 5). This entailed the idea that only propositions that are meaningful can be verified. Philosophy, as philosophy of science, thus is left with the task of explaining what is meant by verification. Following Wittgenstein, Schlick proposed that the meaning of a proposition is established by its method of verification, that is, method for determining whether it is true or false. Formulated negatively, a proposition for which no verification procedure can be imagined is a meaningless pseudo-sentence.

The principle of verifiability initially appears to be quite plausible. However, it turns, out to be impossible to construct a definition that would classify all the statements of empirical science as meaningful while disqualifying all metaphysical assertions as meaningless. Even if it was easy to formulate criteria that rendered meaningful observational statements such as "it is cold outside now," it turned out to be extremely difficult to distinguish in a principled manner meaningful scientific statements such as "all electrons have the same charge" or "f = ma" from meaningless metaphysical pseudo-statements such as "the absolute is perfect".

Probably the best-known representative of LE is Carnap; there is even a misleading tendency to identify LE with Carnap's philosophy. Carnap began his philosophical career as a neo-Kantian with *The Logical Structure of the World* (*Der Logische Aufbau der Welt*) (1928), which proposed constitutional theory as a scientific successor to traditional epistemology and philosophy of science. Constitutional theory was to be a general theory of rational reconstruction of scientific knowledge in the logico-mathematical framework of Alfred North Whitehead (1861–1947) and Bertrand Russell's (1872–1970) *Principia Mathematica*. In informal terms the constitution of a concept provides coordinates that determine its logical place in a conceptual system.

Subsequently, Carnap replaced constitutional systems with more empiricist constitutional languages and pursued the philosophy of science as the study of the structure of the languages of science. According to Carnap, the task of philosophy is to construct linguistic and ontological frameworks that can be used in the ongoing progress of scientific knowledge. In *Testability and Meaning* (1937) he argued that philosophy should not formulate its principles as assertions such as "All knowledge is empirical" or "All synthetic sentences that we can know are based on experiences" or the like—but rather in the form of a proposal or requirement. By such a formulation, he maintained, "greater clarity will be gained both for carrying on discussion between empiricists and anti-empiricists as well as for the reflections of empiricists" (Sakar 1996, Vol. II, p. 258). Throughout his philosophical career Carnap saw the task of logical empiricist philosophy of science as formulating a general theory of linguistic frameworks to provide conceptual tools for the enhancement of science and philosophy, as already had been done implicitly in the 1929 manifesto.

The sociologist, economist, and philosopher Neurath was the most radically "engaged philosopher" in the Vienna Circle. He was the driving force behind the rapid change from an academic discussion group to an international philosophical movement that eventually was to dominate the philosophy of science in the mid-twentieth century. A pitiless fighter against traditional metaphysics, Neurath made his most important positive contribution to the scientific world-conception in the form of the project of "unified science."

In contrast to the essentially negative program of eliminating metaphysics, the project for a unified science is the great constructive paradigm of LE. According to Neurath, scientific knowledge does not have the form of an all-embracing deductive system but constitutes an encyclopedia. According to encyclopedism, as he termed his account, scientific knowledge has the following five characteristics: It is fallible, pluralistic, holistic, and locally but not globally systematizable, and it is not an image of the real world. Neurath conceived the encyclopedistic project as a large-scale politico-scientific and philosophical program aimed at the highest possible level of the integration of the sciences without succumbing to the temptation of an exaggerated rationalism that would force the sciences into the straihtjacket of a metaphysical system.

The foundation for Neurath's encyclopedism was a robust physicalism according to which all concepts can be defined ultimately and entirely in terms of physicalist concepts and/or the concepts of logic and mathematics. Physicalist concepts are not simply the concepts of physics but instead are the concepts of everyday language dealing with middle-sized spatio-temporally located things and processes. Physicalist language, cleansed of metaphysical phrases and enriched by scientific concepts, was conceived as a mixed language containing precise and vague terms side by side. Depending essentially on the concrete practices of everyday life, Neurath's encyclopedism turned scientific knowledge into historically and socially situated knowledge. This had strong implications for its form. Instead of the "pseudorationalist" conception of a timeless objective "system" of knowledge that would create a picture of the world "as it really is," Neurath put forth a more flexible, nonhierarchical encyclopedia as the appropriate model for human knowledge.

Although Neurath's account of LE is the version most congenial to science, technology, and social studies, this has not been recognized widely. One reason for this misunderstanding is Neurath's death in 1945, which made it impossible to promote his version of LE in the Anglo-American world. Since the 1980s, however, Neurath's vision has received a considerable reconsideration in both the United States and Europe.

Carl Gustav Hempel was Reichenbach's student in Berlin but also spent time in Vienna. After emigrating to the United States via Belgium he became Carnap's assistant in 1937. He began his philosophical career with a dissertation on the logical analysis of the concept of probability. In the 1950s and 1960s he became the most influential logical empiricist in the English-speaking philosophical community. His papers set a standard for the logical analysis of concepts. For instance, his contributions to the theory of scientific confirmation and explanation, especially the covering-law model, determined the agenda of analytic philosophy of science for decades. His "Fundamentals of Concepts Formation in Empirical Science" (1952) served as an introduction to philosophy of science for generations of students.

Hempel was particularly engaged in pointing out difficulties and paradoxical features in many core concepts of the philosophy of science, arguing for the necessity of a thoroughgoing logical analysis. The "raven paradox" is a famous example: If it is a law of nature that all ravens are black, the observation of a black raven may count as a (partial) confirmation of this law. Moreover, it is reasonable to assume that laws of nature should be independent of their logical formulation. Thus, the law that all ravens are black has the logical form "All R are B," which is logically equivalent to "All non-B are non-R." With this conceded, a green frog, as something that is not black and not a raven, counts as a (partial) confirmation of the original law. However, this is absurd. Hence, something in the conception of natural law and confirmation seems to be wrong. The raven paradox shows that philosophers do not understand even the most basic concepts in the philosophy of science fully.

Hempel's philosophical work was characterized by a careful and circumspect application of modern logic that made the achievements of logical analysis attractive even for those who were not professional logicians and philosophers. For instance, *The Function of General Laws in History* (1942) exerted influence far beyond the confines of philosophy. It is one of the few LE analyses that has had an impact in the humanities. In *Problems and Changes in the Empiricist Criterion of Meaning* (1950) Hempel further criticized the various logical empiricist attempts to formulate a waterproof criterion for distinguishing meaningful and meaningless assertions. In later years Hempel was influenced by Thomas Kuhn (1922–1996), belying the claim that LE and historical accounts of science are necessarily opposed.

Assessment

A special problem in LE is the transformation of the movement when the intellectual exodus from Europe to the United States took place in the 1930s. The transplantation of LE did not leave its philosophical content unaffected. Although a comprehensive history of LE has not been written, important differences between the two versions can be noted easily. European LE was politically much more radical than its U.S. successor. Although the Vienna Circle showed a vigorous interest in political and social issues such as education, technology, architecture, and art, in the United States the political dimension of LE became less visible. For instance, Carnap was a dedidated supporter of the civil rights movement until the end of his life.

One factor in this change from a radically "engaged scientific philosophy" to an academically confined "philosophy of science" is surely the fact that logica empiricists had to adapt to a different political and societal context in which the application of their traditional political categories was difficult. Another reason may have been that to survive in exile it was expedient to use a language that was more cautious than that which was acceptable in the "Red Vienna" of the late 1920s. After all, LE started in the United States among a rather obscure philosophical group of emigrants without much of a reputation. Only gradually did it become the mainstream in Anglo-American philosophy of science and epistemology in the 1940s and 1950s.

The dominance of LE did not last long, however. First, many of the internal problems of the movement, such as the issue of distinguishing neatly between meaningful and meaningless statements, stubbornly resisted a satisfying solution. Second, analytic philosophers such as Willard van Orman Quine (1908–2000) and Hilary Putnam (b. 1926) attacked the very basis of the logical empiricist philosophy of science, that is, the distinction between the synthetic/analytic and the observational/theoretical levels of empirical knowledge. Third, authors such as Norwood Russell Hanson (1924–1967) and Thomas Kuhn (1922–1996) shifted the emphasis from the strictly logical toward the historical and sociological aspects of scientific theorizing, thus challenging the autonomy of a logical philosophy of science in the style of Carnap.

In a sense these and related developments were welcomed as liberations from the straitjacket of the so-called "received view." For instance, one immediate consequence of the logical empiricist thesis that meaningful statements are either analytic or empirical was that all value judgments are cognitively meaningless. Value statements are not analytic because they say nothing about the world and are not empirical because they cannot be verified. Hence, they are meaningless. The dichotomy between analytic and empirical statements led logical empiricists to a strictly noncognitivist (emotivist) ethics according to which there can be no knowledge of values in a proper sense. This stance is not to be considered as necessarily leading to a loss of interest in moral and political problems. All members of the Vienna Circle took a strong interest in the political and social events they were living through. These problems, however, were considered as practical problems, to be strictly separated from the theoretical problems science and philosophy were dealing with.

This emotivist account of ethics, which leaves only a small niche for "theoretical meta-ethics," that is, the logical analysis of moral statements, is insufficient. In a world in which science and technology present increasing numbers of ethical questions and difficulties, it does not provide reasoned arguments formorally relevant actions.

At the same time the complete dismissal of LE by the self-proclaimed "revolutionary" postpositivist philosophy of science might have been a bit hasty, especially if one takes into account its lesser-known European variants. Indeed, the differences between LE and postpositivist philosophy of science might have been unfairly exaggerated. With regard to Neurath's and Hempel's versions of LE, it does not seem far-fetched to suggest that to some extent the allegedly unbridgeable gap between LE and its successors has been an interest-guided social construction. As usual, the critics of LE were unaware of how much they had absorbed of the belief system they so eagerly berated.

In summary, one may propose that LE was a rich philosophical movement that set the stage for a large part of the philosophy of science and epistemology during the twentieth century. However, despite this general claim, a balanced assessment of the movement has not been formulated. In particular, the relationships between LE and its successor disciplines, such as the various currents of "postpositivist" philosophy of science, cultural studies of science, and science, technology, and society studies (STS), are not yet fully appreciated.

ANDONI IBARRA
THOMAS MORMANN

SEE ALSO *German Perspectives; Science, Technology, and Society Studies; Wittgenstein, Ludwig.*

BIBLIOGRAPHY

Carnap, Rudolf. (1967 [1928]). *The Logical Structure of the World.* Berkeley and Los Angeles: University of California Press. Basic work of logical empiricism of the Vienna Circle.

Carnap, Rudolf. (1937). *The Logical Syntax of Language.* London: Routledge and Kegan Paul. Basic work of logical empiricism.

Friedman, Michael. (1999). *Reconsidering Logical Positivism.* Cambridge, UK: Cambridge University Press. Collection of trail-blazing essays for the recent re-evaluation of logical empiricism, particularly Carnap's philosophy.

Giere, Ronald N., and Alan W. Richardson, eds. (1996). *Origins of Logical Empiricism.* Minnesota Studies in the Philosophy of Science XVI. Minneapolis and London: University of Minnesota Press.

Hempel, Carl Gustav. (2000). *Selected Philosophical Essays,* ed. Richard Jeffrey. Cambridge, UK: Cambridge University Press.

Kuhn, Thomas S. (1970). *The Structure of Scientific Revolutions.* Chicago: Chicago University Press. Basic work of post-empiricist philosophy of science.

Neurath, Otto. (1981). *Gesammelte philosophische und methodologische Schriften,* 2 vols., ed. Rudolf Haller and Heiner Rutte. Vienna: Hölder-Pichler-Tempsky. Collected philosophical and methodological works of Neurath.

Sakar, Sahotra. (1996). *Science and Philosophy in the Twentieth Century: Basic Works of Logical Empiricism,* 6 vols. New York and London: Garland. Six volumes of reprints of the classic papers of logical empiricism and related currents of philosophy.

Stadler, Friedrich. (2001). *The Vienna Circle: Studies in the Origins, Developments, and Influence of Logical Empiricism.* Vienna and New York: Springer. Provides the most complete account of logical empiricism in Europe.

Depiction of the Luddite Rebellion. The rebellion began in 1811 when organized bands of men in England s Midlands began breaking into hosiery factories and smashing looms used to weave stockings. Claiming allegiance to "General Ludd," the Luddites were skilled craftsmen driven to despair by changes in weaving technology that cost them wages and worsened the effects of the already ongoing economic crisis. (© Mary Evans/Thomas Philip Morgan.)

LUDDITES AND LUDDISM

• • •

Luddite and *Luddism* are terms of both derision and praise. Depending on context, they have been used to indicate either mindless opposition to or critical assessment of technology and science.

Origins

The first Luddites were English textile workers who in 1811 and 1812, during the Industrial Revolution, resisted and rebelled against the use of wide-frame knitting machines, shearing machines, and other machines of mass production. The term is based on a mythical Ned Ludd who supposedly led the workers in their resistance. The Luddites, however, were not one unified political group. They reflected their regions and local trade organizations, hence the more appropriate use of the terms Manchester, Yorkshire, and Midland Luddites.

Much of the knitting of stockings and other apparel was done in cottages and small shops by knitters (stockingers) who sometimes owned their own frames but usually rented them from the hosiers (the knitting-frame was invented by William Lee in 1589 and introduced in the Midlands in the mid-1600s). The knitting-frame, operated by an individual at home, could make 600 stitches per minute as opposed to about 100 stitches by hand-knitters. Frame-knitting in cottages sustained a way of life for more than a century.

The rebellion began in March 1811 in the Midland shire of Nottingham (home of the legendary Robin Hood) and then spread north to Manchester and Yorkshire. At the height of the rebellion, knitters, croppers, and other textile workers smashed textile machinery almost on a daily basis. The Midland Luddites were particularly well organized and led a sustained campaign of focused machine breaking without resorting to the more general violence evident in their northern counterparts. The open rebellion ended in 1812 with arrests and subsequent hangings.

The original Luddite rebellion grew out of intolerable economic and political conditions that threatened the livelihoods of the textile workers and eventually destroyed their cottage industry and their way of life. Economic factors included a depressed market resulting in part from Napoleon's economic blockade of British trade and Britain's counter-blockade of European ports. Wages decreased substantially at a time when a number of poor harvests in 1809 nearly doubled the price of bread.

Political conditions also fueled the rebellion. Fearful the French Revolution would spread to the working class, the Parliament passed the Combination Acts of 1799 and 1800 to outlaw trade unions and muzzle workers, making it a criminal offense for workers to join together to petition employers for fair wages and better working conditions. Furthermore the government's policy of non-intervention in industrial relations abandoned the working class to the captains of capitalist industry. In addition the Midland Luddites believed the acts of Parliament contravened the charter from King Charles II that founded the Framework Knitters' Company. In rebelling, the Midland frame-knitters upheld the principles of their charter to regulate their trade.

Historically Luddism may thus be described as an assertion of the right of organized trade to protect its way of life from the unfair introduction of technology, from technology that reduces the quality of the product, and from political measures that would change the trade without the consent of the trade workers.

Developments

Although the Romantic poet George Gordon Lord Byron (1788–1824) defended Luddites against their critics, by the mid-1800s the term had largely disappeared from use. Then in 1959 the novelist C. P. Snow in his famous lecture defending "The Two Cultures and the Scientific Revolution" revived it to stigmatize *literary intellectuals* such as T. S. Eliot and William Butler Yeats as *natural Luddites*. Following Snow, the term became a common way to disparage critics of the cultural influence of modern science as simply uninformed antitechnologists.

In the late-twentieth century, however, critics attempted to turn the tables on those who would dismiss them as technophobes by adopting the term *neo-Luddite* and *neo-Luddism* as a badge of honor for those who refuse to uncritically accept virtually everything that techno-economic momentum throws up. As Langdon Winner (1986) argued, technology critics are no more antitechnology than art and literature critics are anti-art and anti-literature. The most influential defense of this critical stance was perhaps Chellis Glendinning's "Notes Toward a Neo-Luddite Manifesto" (1990), which argued that technology and technological systems may be beneficial to global capitalism but are not necessarily beneficial to human beings, the environment, and the common good. Although neo-Luddism is not a well-defined creed, it commonly includes critiques of consumer culture, television, and high-energy use automobiles while promoting enhanced participation in technological design, social and economic equity, and respect for nature. Some representatives draw inspiration from religious traditions, especially Quakers, Mennonites, Amish, and Shakers. Others argue an inherent will to power in modern technology that threatens human dignity rather than enhancing it.

FRANK H. W. EDLER

SEE ALSO *Industrial Revolution; Modernization.*

BIBLIOGRAPHY

Bailey, Brian. (1998). *The Luddite Rebellion.* New York: New York University Press.

Binfield, Kevin, ed. (2004). *Writings of the Luddites.* Baltimore, MD: Johns Hopkins University Press.

Fox, Nichols. (2002). *Against the Machine: The Hidden Luddite Tradition in Literature, Art, and Individual Lives.* Washington, DC: Island Press/Shearwater Books.

Glendinning, Chellis. (1990). "Notes Toward a New-Luddite Manifesto." *Utne Reader* 38(March/April): 50–53.

Sale, Kirkpatrick. (1995). *Rebels Against the Future: The Luddites and Their War on the Industrial Revolution: Lessons for the Computer Age.* Reading, MA: Addison-Wesley.

Winner, Langdon. (1986). *The Whale and the Reactor: A Search for Limits in an Age of High Technology.* Chicago: University of Chicago Press.

LUHMANN, NIKLAS
• • •

German sociologist Niklas Luhmann (1927–1998) was born in Lüneburg on December 6. In more than seventy books and 450 papers, he developed what is perhaps the most comprehensive theory of modern society, in which ethics plays an important, but secondary, role. Educated in legal science, Luhmann was inspired by the phenomenology of Edmund Husserl, the systems theory of Talcott Parsons, the theory of autopoiesis of Humberto Maturana, the second order cybernetics of Heinz von Foerster, and the form calculus of G. Spencer-Brown. He synthesized these elements into a systems theory of impressive scope and radicalism, representing what he saw as a paradigm shift in the social sciences. He died on November 6 in Bielefeld, Germany.

A Universal Systems Theory

Luhmann distinguished between physical, biological, mental, and social systems, but his main focus was on social systems, which he subdivided into interactions, organizations, and society as a whole. His main theoretical tool was the *distinction*. In order to observe social systems, the observer must use a *guiding distinction*. Luhmann chose the distinction between system and environment, but admitted that others were possible.

A radical tenet of Luhmann's systems theory is the thesis that social systems consist only of communication—not of persons, of artifacts, or even of actions. Communication is defined as the unity of three selections: information, utterance, and understanding, to which is added the acceptance or rejection of the receiver to continue the communication. Because communications are transient events, the system must generate linguistic structures and themes to create and combine new communications. Social systems are autopoietic systems, creating their own elements within their network of elements. Even though human beings, as information-processing units, are necessary for communication, they are not part of the communication, but of its environment. The physical world is likewise not part of the communication, but is only its object, and it is not the function of communication to mirror the physical world. By using the theory of autopoiesis, Luhmann made systems theory dynamic, with time and change at its center. Everything in a social system is contingent, meaning that alternatives are always possible.

According to Luhmann, social systems cannot be understood in terms of *rationality, norms,* or *human beings.* Change must be seen as evolution, a choice among existing alternatives. There is no one point of view from which society can be correctly observed and described. With the cultural death of God, and the attendant loss of the only ostensibly *right* worldview, a poly-centered world remains. In his late-twentieth-century analysis, Luhmann claims that the most fruitful way of imagining society is as a world community with no center, no purpose, and no overarching rationality.

Luhmann analyzes society as a unity of functional subsystems, each having is own symbolic generalized medium and its own guiding distinction. Society can be observed from many points of view, economic (where the medium is money), political (power), scientific (truth), intimate (love), and more. The number of functional subsystems is an empirical question. In addition to his two principal works, *Soziale Systeme* (1984) and *Die Gesellschaft der Gesellschaft* (1997), Luhmann wrote a series of monographs dealing with the various social subsystems.

Functional subsystems make communication more effective. By using symbolic generalized media, it is possible to communicate on a world scale because the simple binary form allows for simplification, motivation, and measurement of success or failure. An observer can quickly decide whether or not he will take over the point of view inherent in the medium. Symbolic generalized media can differ—in operation mode and time relations, among others—but all share a common structure. Though the most effective communications in modern society are oriented towards functional subsystems, Luhmann acknowledged that what is good for a functional subsystem is not necessarily good for society as a whole because proponents of each subsystem have biased and narrow views.

Technology can also be seen as a functional subsystem, operating in the medium of effectiveness. Its code is functioning or broken, its programs are blueprints, its institutions are organizations and universities, and its contribution to society is maintenance of regular processes. Technology has its own internal dynamics and thus it might clash with or be helpful to other functional subsystems.

Functional subsystems are not action systems. They *do* nothing, but can be conceived as semantic discourses. The action systems of twenty-first-century society are organizations; specialized organizations define themselves as agents of a particular functional subsystem, such as technology, religion, or law.

Morals and Ethics in Functional Subsystems

In real life, subsystems must cooperate. Because their respective criteria for success and failure are not the same, conflicts arise with no objective solution, thus creating a need for normative or ethical solutions. As a consequence, many functional subsystems develop special professional ethics criteria to deal with the integration of highly specialized products and methods in society.

It should be noted that no functional subsystem uses the moral distinction between right and wrong. One reason for this is empirical: A moral distinction is not precise enough to facilitate communication. It has too many dimensions. A moral evaluation might focus on motives or on consequences, and be dependent on religious or subcultural assumptions. Moralizing creates conflict, not consensus. Instead Luhmann views morality as a tool for distributing *esteem,* which depends not on professional skills but on the qualities of a person as a whole.

Morals have important social functions and Luhmann wrote extensively on moral issues though he flatly rejected any attempt to understand society in moral or purposive terms. Luhmann conceded that moral distinctions are used with the same spontaneity as empirical distinctions in daily life. Using the distinction between moral and ethics, he argued that ethics is a theoretical reflection of the social phenomenon of morals, and concluded that the most important task of ethics is to warn against morals. He had no illusions as to the effectiveness of ethics to control technological development. Because there is no ethical consensus in modern society, no ethical control is possible or desirable.

Each functional subsystem has its own criteria for success or failure, but it also has a tendency to exaggerate its own importance and blind itself to other criteria. Economy focuses on money, politics on power, and science on truth. When criteria clash, no super rationality can create a rational solution. Luhmann had a lifelong debate with the German philosopher Jürgen Habermas regarding this issue. Habermas stresses the possibility of rational consensus, while Luhmann argues that conflict is not only inevitable, but also fruitful. Consensus is only a transient phase in the ongoing communication of social systems.

Luhmann accepted that functional subsystems have evolved as centers for solving specific tasks, however, he argued the need for *criteria for criteria* or second order criteria. But such criteria, which might be called ethical criteria, are not socially binding. There is no universally accepted viewpoint from which the social and moral implications of technology or pollution, for example, can be observed and judged right or wrong.

Luhmann described each functional subsystem as having its own complexity and society as a whole as a hypercomplex entity composed of many functional subsystems. However Luhmann posited no solutions to the problems he presented. With no rationality, there is only evolution to rely on: Something will happen, perhaps better, perhaps worse, perhaps catastrophic. When nations, organizations and persons try to control technology, they are controlled by the technology they want to control and are unable to control all the other actors trying to control. Technology, like life, will find its way.

OLE THYSSEN

SEE ALSO *German Perspectives; Habermas, Jürgen.*

BIBLIOGRAPHY

Luhmann, Niklas. (1984). *Soziale Systeme*. Frankfurt am Main, Germany, Suhrkamp Verlag. English translation, *Social Systems*. Sanford, CA: Stanford University Press (1995).

Luhmann, Niklas. (1997). *Die Gesellschaft der Gesellschaft*. Frankfurt am Main, Germany: Suhrkamp Verlag

LYOTARD, JEAN-FRANÇOIS

• • •

French philosopher Jean-François Lyotard (1924–1998), who was born in Vincennes, France, on August 10, was an originator of what became known as postmodernism.

After teaching philosophy in secondary schools in France and Algeria, Lyotard was awarded a position at the University of Paris VII, where he also served as a council member of the Collège international de philosophie. Toward the end of his life he also held visiting professorships in the United States. Lyotard died of leukemia in Paris on April 21.

Lyotard's work is marked by a persistent interest in the relations between science, technology, ethics, and politics, as can be seen in the work for which he is most well known, *The Postmodern Condition: A Report on Knowledge* (1984), which focuses on the state of knowledge in highly developed countries. According to Lyotard, the sciences and late twentieth-century societies were in the midst of a legitimation crisis because of the inability to provide a justification in the form of an overarching explanation of the relations between science, technology, and society.

Lyotard explains the crisis using Ludwig Wittgenstein's (1889–1951) notion of language games. A language game is a field of discourse defined by a set of internal rules that establish the types of allowable statements. Different discourse practices, such as science and ethics, have become distinct language games, adhering to different sets of rules. Because disparate language games prohibit statements that fail to conform to their rules, it is impossible to give a single, overarching account that would guarantee the legitimacy of all possible discourse practices. For this reason, Lyotard states that the postmodern situation is marked by an "incredulity toward meta-narratives" (Lyotard 1984, p. xxiv).

If Lyotard is correct and it is no longer permissible to give an overarching account for the diversity of discourse practices, then the postmodern condition demands a new response to the problem of legitimation. Lyotard claims that the appropriate response to the problem in a society marked by the postmodern condition is "paralogy." In the practice of paralogy, the goal of producing an overarching legitimation narrative is replaced by an attempt to increase the possible language moves in a particular language game. Hence, paralogy champions the diversity of discourse practices by prohibiting the hegemony of a single discourse over all others. Paralogy thus resists the tendency to treat ethics and politics as forms of scientific knowledge or technology.

The Postmodern Condition has implications for ethics that are further developed in *The Differend: Phrases in Dispute* (1988). A *différend* is Lyotard's label for an irresolvable conflict between two phrases or parties. The *différend* as a conflict between phrases was implied in Lyotard's earlier work as the inability to unify diverse language games. In this work, however, rather than being concerned with the legitimation of knowledge, Lyotard develops the notion of the *différend* to include a certain type of injustice that occurs to differing language games (or genres), specifically the cognitive and ethical.

The ethical genre, according to Lyotard, is concerned with prescriptive statements of the form "you ought," whereas the cognitive genre consists of descriptive statements. Ethics, with its prescriptive statements, is a discourse of obligation. As such, ethics takes the form of phrases marked by an asymmetry between the addressor and the person addressed. The person who says "You shall not lie" commands interlocutors and places obligations upon them, but the statement "Lying is wrong" leaves out the relation between persons that is characteristic of ethical discourse. Consequently for Lyotard, the nature of ethics is covered over in attempts to transform the prescriptive into the descriptive.

In response to this threat, the task of philosophy, according to Lyotard, is to champion and protect the diversity of discourse and practice. While not providing a unifying account of the relations between genres, philosophy is marked by an obligation to bear witness to the *différend*. Although primarily focused on discourse, this responsibility extends to the sociopolitical world, in which there is the continuous threat of one social entity (individual persons or cultures) being overpowered by another.

Lyotard's thinking continues to be a powerful, cautionary note for the relations between science, technology, and ethics. Rather than subsume distinct discourses under a unifying account, his work argues for maintaining that which marks each as different.

KEM D. CRIMMINS

SEE ALSO *French Perspectives; Postmodernism.*

BIBLIOGRAPHY

Bennington, Geoffrey. (1988). *Writing the Event.* New York: Columbia University Press. Critical exposition of Lyotard's philosophy.

Lyotard, Jean-François. (1984). *The Postmodern Condition: A Report on Knowledge,* trans. Geoffrey Bennington and Brian Massumi. Minneapolis: University of Minnesota Press. Originally published as *La condition postmoderne: Rapport sur le savoir* (Paris: Minuit, 1979).

Lyotard, Jean-François. (1988). *The Differend: Phrases in Dispute,* trans. Georges Van Den Abbeele. Minneapolis: University of Minnesota Press. Originally published as *Le différend* (Paris: Minuit, 1983).

Lyotard, Jean-François. (1988). *Peregrinations: Law, Form, Event.* New York: Columbia University Press.

Lyotard, Jean-François. (1997). *Postmodern Fables,* trans. Georges Van Den Abbeele. Minneapolis: University of Minnesota Press. Originally published as *Moralités postmodernes* (Paris: Galilée, 1993).

Rojek, Chris, and Bryan S. Turner, eds. (1998). *The Politics of Jean-François Lyotard: Justice and Political Theory.* New York: Routledge. Collection of eight articles.

Silverman, Hugh J., ed. (2002). *Lyotard: Philosophy, Politics, and the Sublime.* New York: Routledge. Collects sixteen articles.

LYSENKO CASE

• • •

The debate on the relative influence of heredity and environment took a distinctive form in the Soviet Union in the turbulent years between the 1920s and the 1960s. There was among many committed communists a sense that the socialist revolution should transform everything, including the foundations of knowledge. There was intense debate about what constituted a Marxist approach to every discipline, including biology.

Lysenko's Practice and Theory

Into this context came Trofim Denisovich Lysenko (1898–1976), a young agronomist from the Ukraine, who emerged into the limelight in 1927 in connection with an experiment in the winter planting of peas to precede the cotton crop in the Transcaucasus. The results he achieved in a remote station in Azerbaijan were sensationalized in the national Communist Party newspaper *Pravda*. The article projected an image of him as a sullen, barefoot scientist close to his peasant roots. Lysenko subsequently became famous for *vernalization*, an agricultural technique that allowed winter crops to be obtained from summer planting by soaking and chilling the germinated seed for a determinate period of time. Lysenko then began to advance a theory to explain his technique. The underlying theme was the plasticity of the life cycle. Lysenko came to believe that the crucial factor in determining the length of the vegetation period in a plant was not its genetic constitution, but its interaction with its environment. By the mid-1930s he rejected the existence of genes and held that heredity was based on the interaction between the organism and its environment, through the internalization of external conditions. He recognized no distinction between genotype and phenotype.

Lysenko's theory was an intuitive rationalization of agronomic practice and a reflection of the ideological environment surrounding it rather than a response to a problem formulated in the scientific community and pursued according to rigorous scientific methods. Lysenko seemed to achieve results at a time when there was a great demand for immediate solutions and a growing impatience with the protracted and complicated methods employed by established scientists. This brought a sympathetic predisposition to whatever theoretical views Lysenko chose to express, no matter how vague or unsubstantiated.

Even Lysenko's practical achievements were extremely difficult to assess. His methods were lacking in rigor. His habit was to report only successes. His results were based on extremely small samples, inaccurate records, and the almost total absence of control groups. An early mistake in calculation, which caused comment among other specialists, made him extremely negative regarding the use of mathematics in science.

But Lysenko was the man of the hour, one who had come from humble origins under the revolution and who directed all his energies into the great tasks of socialist construction. He was pictured as the model scientist for the new era, and was credited with conscientiously bringing a massive increase in grain yield to the Soviet state, while geneticists idly speculated on eye color in fruit flies.

Genetics on the Defensive

Catching the ideological demagoguery that was beginning to flourish among a certain section of the young intelligentsia, some denounced the science of genetics as reactionary, bourgeois, idealist, and formalist, and contrary to the Marxist philosophy of dialectical materialism. Its stress on the relative stability of the gene was supposedly a denial of dialectical development as well as an assault on materialism. Its emphasis on internality was thought to be a rejection of the interconnectedness of every aspect of nature. Its notion of the randomness and indirectness of mutation was held to undercut both the determinism of natural processes and human abilities to shape nature in a purposeful way.

The new biology, with its emphasis on the inheritance of acquired characteristics and the consequent alterability of organisms through directed environmental change, was well suited to the extreme voluntarism that accompanied the accelerated efforts to industrialize and collectivize. The idea that the same sort of willfulness could be applied to nature itself was appealing to the mentality of those who believed that Soviet man could transform the world. Lysenko's voluntarist approach to experimental results and to the transformation of

Trofim Lysenko, kneeling in a field, measuring the growth of wheat. During the Soviet famines of the 1930s, Lysenko proposed techniques for the enhancement of crop yields, rejecting orthodox Mendelian genetics on the basis of unconfirmed experiments, and gained a large popular following. But in 1964 his doctrines were officially discredited, and intensive efforts were made toward reestablishing orthodox genetics in the Soviet Union. (© Hulton-Deutsch Collection/Corbis.)

agriculture was the counterpart of Joseph Stalin's voluntarist approach to social processes, undoubtedly a factor in Stalin's enthusiastic support of Lysenko during this period.

Other political leaders and scientific administrators were not so easily swayed. Geneticists defended their work and had very influential support. There was strong resistance within the Academy of Sciences. The debate reached a climactic point at a special session of the Lenin Academy of the Agricultural Sciences in 1936, devoted to a discussion of the two trends in Soviet biology. The official goal was to achieve a reconciliation of the two schools, some kind of accommodation for genetics within the framework of Lysenko's agrobiology. The outcome was the opposite. The open confrontation of the two trends resulted in drawing the lines more sharply than ever and in highlighting the irreconcilability of the two contrasting approaches.

The sharpest speech in the defense of genetics came from the American geneticist Hermann J. Muller, a foreign member of the Academy of Sciences, who had come to work in the Soviet Union out of a belief in the possibilities of science under socialism. Muller was also inclined to philosophical reflection on science and had definite views as to the place of genetics within the framework of a dialectical materialist philosophy of science. He turned the charge of idealism against the Lysenkoites and accused them of hiding behind the screen of a falsely interpreted dialectical materialism.

The growing ascendancy of Lysenko coincided with the purges that reached into virtually every Soviet institution from 1936 to 1939. The campaign against geneticists became more and more vicious and slanderous. Scientific and philosophical arguments gave way to political ones. The pursuit of genetics was branded as racism and fascism. Geneticists were named and accused of sabotage, espionage, and terrorism. Many were arrested. Of these some were shot, while others died in prison. Still others were witch-hunted, lost their jobs, and were forced into other areas of work. Institutes were closed down. Journals ceased to publish. Books were removed from library shelves. Texts were revised. Names became unmentionable. The 7th International Congress of Genetics, which was scheduled to be held in Moscow in August 1937, was cancelled. When the congress did take place in Edinburgh in 1939, no Soviet scientists were present, not even the internationally respected geneticist N. I. Vavilov, who had been elected its president.

By 1938 Lysenko had been elected to the Academy of Science and replaced Vavilov as president of the Lenin Academy of Agricultural Sciences. In 1940 Vavilov was arrested and Lysenko replaced him as director of the Institute of Genetics of the Academy of Sciences. In 1941 Vavilov stood trial and was found guilty of sabotage in agriculture. After several months of incarceration, Vavilov's death sentence was commuted, but he died in prison in 1943 of malnutrition. Although some of the more outspoken and defiant survived, many gave way under the pressure, engaged in abasing self-criticism, and acknowledged the superior wisdom of Lysenko. The degree of demoralization was overwhelming.

Assessment

Lysenkoism reached its peak in 1948 with official Communist Party endorsement. But almost immediately after Stalin's death in 1953 it went into decline. Vavilov, for instance, was posthumously rehabilitated in 1955. However Lysenkoism continued to be a force in Maoist China, where a promotional congress was held in 1956. The case was thus a protracted episode in the history of science under Communism, and has been the subject of many commentaries.

These analyze the scientific, political and philosophical issues in quite divergent ways. Soyfer and others represent it as a story of personal opportunism and political terror, as a cautionary tale against the dangers of ideological distortion of science. This position tends to see philosophy and politics as alien impositions upon science. Joravsky, Graham and Lecourt put more emphasis on the complexity of the philosophical issues,

although with varying degrees of hostility or sympathy with Marxism. Medvedev's account is of historical significance as a critique coming from someone within the world of Soviet science. Some searching and sophisticated explorations of the issues have come from within Marxism, most notably by Lewontin, Levins, and Young. This position is marked by an insistence that science is inextricably tied to philosophy and politics, even to ideology, opening up a more nuanced investigation of the varying modes of interaction and a more complex critique of Lysenkoism.

HELENA SHEEHAN

SEE ALSO *Communism; Russian Perspectives.*

BIBLIOGRAPHY

Graham, Loren. (1973). *Science and Philosophy in the Soviet Union.* London: Allen Lane.

Huxley, Julian. (1949). *Heredity, East and West: Lysenko and World Science.* New York: H. Schuman.

Joravsky, David. (1970). *The Lysenko Affair.* Cambridge, MA: Harvard University Press. Reprint, Chicago: University of Chicago Press, 1986.

Lecourt, Dominique. (1977). *Proletarian Science: The Case of Lysenko,* trans. Ben Brewster. Atlantic Highlands, NJ: Humanities Press. Introduction by Louis Althusser.

Levins, Richard, and Richard Lewontin. (1985). *The Dialectical Biologist.* Boston: Harvard.

Lysenko, Trofim Denisovich. (1946). *Heredity and Its Variability,* trans. Theodosium Dobzhansky. New York: King's Crown Press.

Lysenko, Trofim Denisovich. (1948). *The Science of Biology Today.* New York: International Publishers. The Presidential address at the V.I. Lenin Academy of Agricultural Sciences, July 31, 1948.

Manevich, Eleanor D. (1990). *Such Were the Times: A Personal View of the Lysenko Era in the USSR.* Northampton, MA: Pittenbruach Press. Foreword by Eric Ashby.

Medvedev, Zhores A. (1969). *The Rise and Fall of T. D. Lysenko,* trans. Il Michael Lerner, with Lucy G. Lawrence. New York: Columbia University Press.

Roll-Hansen, Nils. (2004). *The Lysenko Effect: The Politics of Science.* Amherst, NY: Prometheus Books.

Sheehan, Helena. (1993). *Marxism and the Philosophy of Science: A Critical History.* Atlantic Highlands, NJ: Humanities Press.

Soyfer, Valerii. (1994). *Lysenko and the Tragedy of Soviet Science,* trans. Leo Gruliow, and Rebecca Gruliow. New Brunswick, NJ: Rutgers University Press.

Young, Robert. (1978). "Getting Started on Lysenkoism." *Radical Science Journal* 6/7: 81–105.

M

MACHIAVELLI, NICCOLÒ

• • •

Niccolò Machiavelli (1469–1527), in Florence on born May 3, was a Florentine statesman and Renaissance Italy's greatest political philosopher; he died in Florence on June 21. He is often regarded as the first to take a scientific approach to politics.

Major Contributions to Political Thought

Machiavelli is known chiefly as the author of two books, *The Prince* and *The Discourses on Livy* (both c. 1517). The former concerns the acquisition of principalities, a form of government in which the state belongs to an individual or a family. The latter is a meditation on republics, in which the state is public rather than private property. The notoriety of these books is largely due to the absolute ruthlessness advocated by Machiavelli. In *The Prince*, he recommends acting against faith, charity, humanity, and religion. In *The Discourses*, he criticizes Giovampagolo Baglioni because that tyrant had the opportunity, but not the courage, to murder the Pope.

Despite their practical orientation, *The Prince* and *The Discourses* are works of political science. Machiavelli asks theoretical questions: how states are born and what sustains them. But his work marks a fundamental break with premodern political thought. Classical and medieval thinkers were concerned above all with the difference between good and bad forms of government; Machiavelli ignores that distinction in favor of hard realism. In the first chapter of *The Prince*, he classifies states solely according to how they are acquired. In chapter fifteen, he dismisses those who

Niccolò Machiavelli, 1469–1527. Machiavelli was an Italian political philosopher during the Renaissance. His most famous book, *Il Principe*, was a work intended to be an instruction book for rulers. Published after his death, the book advocated the theory that whatever was expedient was necessary—an early example of utilitarianism and realpolitik. (*Corbis. Archivo Iconografico, S.A./Corbis.*)

dream of imagined principalities; perhaps referring to heaven, or Plato's *Republic*. Machiavelli thus narrows the horizon of political science; the question is not what kind of government is best, but how do people get the kind they want.

To answer this question, Machiavelli first explains the origin of states. He observes that hereditary principalities are established based on habit: People accept the regime because they are accustomed to it. But every established government was once new. How does a new state survive long enough to *become* hereditary? Machiavelli ignores the traditional answers: God's blessing or natural development. Perhaps just dumb luck? But fortune is fickle by definition, and does not sustain any one thing for long. Because all states originated from some source, Machiavelli proposes that certain people have, within themselves, the power to conquer fortune, to create armies, and to establish and maintain states.

He calls this power *virtue*, a word suggesting the premodern idea of moral excellence. But in fact, Machiavelli's definition of virtue supports the ruthlessness he advocates. Morality and justice as commonly understood exist only as the products of established states. Machiavellian virtue must exist before the state is founded, and is therefore beyond ordinary right and wrong. It does, however, require that certain temptations be resisted: The prince must never rely on fortune or the grace of others, or put off until tomorrow a murder he needs to commit today.

Whereas ancient philosophers were conservative, more concerned with preserving decent governments than with creating new ones, Machiavelli encourages innovators. He especially admires those who create principalities and republics from scratch, or rejuvenate existing ones. In all cases, he insists that the innovator must rely on his own virtue, and have *arms of his own*. By this, Machiavelli means soldiers, loyal to the prince alone. He severely criticized Italian states for their reliance on mercenary and *auxiliary* arms. Paid soldiers, or those borrowed from another prince, have no connection to the innovator's virtue, and so cannot be a secure foundation for the state.

Pertinence to Modern Political Thinking

Machiavelli is regarded by some as the founder of value-free political science. He describes politics as it is, not as it might be, and shows how this knowledge can be exploited to bring greater order into human affairs. But Machiavelli's science is anything but value-free: He prefers glory to security, and admires innovators more than conservatives. Though he writes both for republics and tyrants, many have argued that he favors one over the other. In fact, he clearly has a preference for republics, but believes that the founding father of every republic needs to possess unrestrained power.

Machiavelli's writing has never gone out of fashion. Perhaps this is because he had the courage to face certain hard truths about modern thought. In order to conquer chance and nature, the early moderns were willing to reject the authority both of divine and natural right, thus imposing no moral restraints on the technological power unleashed by their new sciences. Machiavelli's political science vividly illustrates the consequences of their boldness.

Machiavelli paid relatively little attention to the rise of modern science and technology, concentrating much more on the topic of political reform. It was left to Francis Bacon and others to apply Machiavellian principles to the conquest of nature as a whole. But Machiavelli's thought did at least hint at the Baconian project. He speculates that it was natural famine that drove large populations of barbarians out of their homelands in the east to inundate the Roman empire. He likens the movement of such peoples to floods, and speaks of strong political institutions as dams and dikes that can restrain such floods. Machiavelli is thus developing a science of politics that is technological in the modern sense.

KENNETH C. BLANCHARD, JR.

SEEALSO *Modernization; Scientific Revolution.*

BIBLIOGRAPHY

Machiavelli, Niccolò. (1996). *The Discourses on Livy*, trans. Harvey C. Mansfield and Nathan Tarcov. Chicago: University of Chicago Press.

Machiavelli, Niccolò. (1998). *The Prince*, trans. Harvey C. Mansfield. Chicago: University of Chicago Press. Mansfield's translations are by far the best available. They are readable and precise, and include useful introductions and a glossary of Machiavelli's terms.

Skinner, Quentin. (2001). *Machiavelli: A Very Short Introduction.* Oxford: Oxford University Press. A very brief introduction to Machiavelli aimed at the general reader.

Strauss, Leo. (1995). *Thoughts on Machiavelli.* Chicago: University of Chicago Press. A difficult but penetrating study of Machiavelli. Strauss argues that Machiavelli was the founder of modern political thought.

MANAGEMENT

• • •

Overview
Models of

OVERVIEW

The term *management* can name both an activity and persons in charge of the activity. As activity, the term derives from the Italian *maneggiare*, meaning to handle or control a horse, which is itself rooted in the Latin *manus*, or hand. In the late 1500s the word was applied to the governing body of a theater and from there to other business activities, including those involved with industrial manufacture. Shifts in the ownership of large-scale manufacturing companies led to what has been termed a managerial revolution, in which direct control and decision-making became invested in neither owning capitalists nor wage-earning workers but in salaried managers (Burnham 1941, Chandler 1977). This shift has influenced both science and technology, with "big science" and "technoscience" increasingly managed by neither science nor engineering workers—a development that poses questions of ethical responsibility for both technical professionals and managers. Attempts to systematize informal management techniques into either a science or a technology of management further highlight ethical issues.

Historical Background

Humans have always collaborated to reach shared goals. Distributed tasks for common ends require coordination, planning, control, and organization—all of which are as subject to ethical assessment along with the ends to which they are subordinate. For example, in Plato (c. 428–347 B.C.E.) one can find both praise for the division of labor that engenders expertise in specialized workers (*Republic*) and criticisms of the pretensions of technical specialization (*Apology* and *Gorgias*). Thus, although the term did not exist as such, "management" has often been read back into such preindustrial orders of household, tribe, city-state, military, or church. What distinguishes modern management from traditional political or religious organization and leadership is its greater emphasis on the systematic coordination of means.

Management did not take on its contemporary connotations until the technological, economic, political, and social changes of the Industrial Revolution (c. 1750–1850). Specifically, certain organizational problems arose in the embryonic factory system that led to the genesis of modern management practices and eventually the formalization of management study (Wren 2005). It was also during this era that attitudes to work began to change, although slowly, from ceaseless, futile labor to opportunities for personal wealth and social progress. Central to this transformation were the Renaissance revival of science and reason and the Protestant work ethic with its notion of a worldly "calling" that Max Weber (1930) argued paved the way for market-based capitalist economies.

The modern understanding of management in terms of leading an organization toward a goal through the deployment and manipulation of resources (material, human, financial, and intellectual) was further shaped by classical and nineteenth-century economic theory and the development of technical production elements such as standardization, specialization, and work planning. The emergence of modern technologies and the market economy challenged managers to develop a body of knowledge on how best to administer and utilize human and technological resources. By the middle of the nineteenth century, Robert Owen (1771–1858) and others were developing theories pertaining to the human element of management including worker training, organizational structure, span of control, and the effects of fatigue on performance. By the 1880s, university courses in management were being offered, based in part on the work of Andrew Ure (1778–1857), who developed training programs for managers in the early factory system.

The first comprehensive theories of management appeared around 1920 in the work of scholars such as Henri Fayol (1841–1925), who outlined five functions for managers and synthesized fourteen principles for organizational design and effective administration. Some theorists such as Ordway Tead (1891–1973) applied principles of psychology to management, whereas Elton Mayo (1880–1949) and others approached it from a sociological perspective. In *The Practice of Management* (1954), Peter F. Drucker (b. 1909) presents a contrast to the Fayolian process texts by introducing the notion of "management by objectives," which replaces control from above with self-control and greater worker empowerment in the goal of reaching well-defined objectives.

In *The Managerial Revolution* (1941), James Burnham (1905–1987) sets management theory within a broad historical narrative of political economy and technological change. Burnham saw industrial production coming to be controlled neither by the owners (capitalism) nor the working class (socialism). Rather, a new managerial class was replacing the bourgeois capitalist as a dominant social force, as ever more complex systems of production separated control from ownership. For Burnham, technological progress necessitates a hierarchy of managers among whom direction and coordination of production becomes a highly specialized skill.

In *The Visible Hand* (1977), Alfred D. Chandler Jr. (b. 1918) presents a similar argument but one less oriented toward prophecy. Chandler claims that neither the traditional family firm nor market mechanisms are able to coordinate the increasingly swift and complex flows of goods made possible by technological innovation. Managers of large, multiunit businesses fill this need for coordination, and in so doing assume strong economic and social power, giving rise to managerial capitalism: "In many sectors of the economy the visible hand of management replaced what Adam Smith referred to as the invisible hand of market forces" (p. 1). But while acknowledging the centrality of technology in bringing about increased managerial control, Chandler fails to explore fully the role of scientists and engineers.

The managerial revolution may have held true in heavy industry, but it seems less valid for service and information economies, where bigger and more complex is not always better. Indeed the continual evolution of technological, political, and economic contexts ensures that management theories are constantly being revised. Some of the more recent developments in management thought include operations research, the theory of constraints, reengineering, complexity theory, and information technology–driven theories. A general trend in management thought is toward systems-based, adaptive processes capable of integrating several categories (e.g. human resources, marketing, and production) into a complex, flexible web of organizational administration.

Management as Science

The conceptualizing and ordering of management as a science did not begin in earnest until the nineteenth century. And although Charles Babbage (1792–1871) made significant contributions to management science, Frederick Winslow Taylor (1856–1915) is viewed as the founder of the field. In 1895 Taylor wrote a seminal paper titled "A Piece-Rate System" that developed a set of management techniques designed to stimulate maximum worker productivity and efficiency. This helped fuel the rising emphasis on efficiency and rationality in decision-making that sought the "one best way." Theodore Roosevelt, Gifford Pinchot, and other conservationists spearheaded this movement by preaching a "gospel of efficiency" in natural resource management, which was "an attempt to supplant conflict with a 'scientific' approach to social and economic questions" (Hays 1959, pp. 266–267).

In *The Principles of Scientific Management* (1985 [1911]), Taylor acknowledged the inefficiencies in natural resource use, but argued that wasteful practices in human resource management were just as damaging to the goals of efficiency, productivity, and prosperity. The Industrial Revolution had vastly increased resources and

capital and improved technologies, but crude ways of organizing and administering these resources hampered productivity. Taylor set out to prove that the best management is a true science, resting upon a clearly defined foundation of laws, rules, and principles. Furthermore, he sought to show that the fundamental principles of scientific management are applicable to all kinds of human activities, from the simplest individual acts to the work of huge corporations.

Among other organizational techniques, this "true science" involved standardizing measures of productivity and quality; developing time, motion, and method studies; and improving the relationship between mangers and workers. In one instance, Taylor was able to reduce the number of people shoveling coal at Bethlehem Steel Works from 500 to 140 by designing more ergonomic shovels. Taylor believed the credo of rational efficiency would lead to prosperity for all, thus abolishing class hatred, but many labor leaders felt that scientific management meant autocracy in the workplace. In fact, Taylor was questioned at length by Congress in 1911 and 1912 on the grounds that some of his methods treated workers like machines.

Frank Gilbreth (1868–1924) and Lillian Gilbreth (1878–1972) were associates of Taylor, and their studies culminated in laws of human motion from which evolved principles of motion economy. The Gilbreths coined the term *motion study* and used cameras to record motions and improve efficiencies even in domestic chores. Other important pioneers in scientific management included Henry Gantt (1861–1919) and Charles Bedaux (1886–1944). After World War II, scientific management played a key role in boosting economic productivity. Statistical and mathematical techniques were applied to planning and decision analyses. Physics Nobel laureate Patrick Blackett (1897–1974) combined these techniques with microeconomic theory to produce the science of operations research, which has been greatly enhanced by the use of computers.

The work of social scientists such as Elton Mayo uncovered many aspects of human interaction in the workplace that had been ignored by other theorists. Specifically, he noted that worker motivations (e.g., feelings, multiple needs, personal goals) are often outside the bounds of the logical, rational human being posited by scientific management, and that workers think and act not as individuals but as members of formal or informal groups (see also McGregor 1960). This type of work led to the rise of human relations management. The period between 1950 and 1970 witnessed a sevenfold increase in managerial employment. It was during this time that behavioral science became widely applied to management practices by theorists such as Rensis Likert (1903–1981). There is a wide range of contemporary scientific theories of management, and it is clear that the best fit for improving performance depends in part on contextual contingencies.

Indeed in many areas alternatives and complements to scientific management stress the importance of building flexibility into systems in order to accommodate the surprises generated by nature, cognitive limitations, and the pace of global commerce. One example is adaptive management (e.g., Brunner et al. 2005), which is a diverse field developed in the 1970s and based on the incorporation of multiple stakeholders in decision-making processes in order to shift to bottom-up, open-ended management structures. In natural resource management, the underlying realization is that the politics of most problems (even many highly technical ones) cannot be elided by focusing solely on scientific expertise and efficiency. In the business world, the driving factors in the shift away from overly rigid forms of scientific management are the need for flexibility to maintain competitiveness and the realization that many valued outcomes are not readily captured by quantification.

Thus scientific management has from its beginnings been a diverse field that has given rise to equally diverse criticisms. It has been both praised and stigmatized as technocratic, insofar as technocracy can be conceived as an ideological-free pursuit of efficient production and a form of production that excludes the consideration of human values. In natural resource policies, technical management has been argued to impede common-interest solutions (Brunner et al. 2005). In business, although it can lead to greater competitiveness via increased efficiency, scientific management can also rigidify an organization, robbing it of flexibility and creativity.

More generally, Alasdair MacIntyre (1984) criticizes the notion of managerial expertise that derives from the dominant conception of the social sciences as somehow mimicking the natural sciences. For MacIntyre, "What managerial expertise requires for its vindication is a justified conception of social science as providing a stock of law-like generalizations with strong predictive power" (p. 88). He then identifies four sources of systematic unpredictability in human affairs, which he claims undermine the very notion of managerial expertise. He concludes that the concept of managerial expertise, or the idea that anyone can consciously manipulate the social order, is a moral fiction: "Our social order is in a very literal sense out of our, and indeed anyone's, control" (p. 107). What appears to be pragmatic, scientifically managed social control is but the skillful imitation of such control. This does not deny the enormous power exercised by bureaucratic managers, it is just that "the most effective bureaucrat is the best actor" (p. 107).

Nevertheless, regardless of outcomes and the fact that the term has fallen out of use, "'scientific management,' as well as its near synonym, 'Taylorism,' have been absorbed into the living tissue of American life" (Kanigel 1997, p. 6). Indeed, the history of scientific management mirrors the development of science more broadly, having evolved from the ideal of disclosing a single right answer to the reality of uncovering an imbroglio of human values intertwined with artifacts and systems, in which uncertainty and ambiguity are multiplied along with the importance of context and values.

Management as Technology

Parallel with attempts to develop management as a science—and as a science with applications—have been attempts to conceptualize management as a technology. Here the leading theorist has been Peter Drucker, who argues for an identification between management and modern technology. Just as in premodern technology work was more important than the tools with which work was performed—that is, work is the context from which tools receive their meaning—so in modern technology management or the organization of activity is the whole that unifies material resources, human labor, financial capital, and machines. Central to any wealth production is the process of ordering, interrelating, or managing the parts in order to assemble a productive business enterprise, which Drucker identifies as a "*system* of the highest order" (1970, p. 55).

For Drucker, management as technology may also be understood as an extension of biological evolution. Management is an adaptive process that orders (and reorders) different aspects of the world (through productive work); as such management is the most general contemporary expression of the human capacity for purposeful, nonorganic evolution. Tools and technologies are not just givens for management but, like the materials and human beings who make up a productive enterprise, are able to be transformed by management—and then transformed again in response to the changed context that the original transformation produces. Management involves a recursive process in which it takes its own successes and failures into account. "The organization of work, in other words, is . . . the major means of that purposeful and nonorganic evolution which is specifically human" (pp. 48–49).

Related to Drucker's view of management as technology is an argument by intellectual historian Bruce Mazlish (1993) regarding the relation between humans and machines. For Mazlish modern history is characterized by the rejection of four discontinuities: between Earth and the rest of the cosmos (Newtonian mechanics, which used the same laws to explain terrestrial and planetary phenomena), between animals and humans (Darwinian evolution, which argued for a natural development from animals to humans), between the unconscious and rationality (Freudian psychology, which presented reason as tied to the unconscious), and between machines and humans (through the integration of computers and humans). By arguing that human beings are defined by their coevolution with machines, a coevolution they must learn to manage, Mazlish likewise presents management (without using the term) as the fulfillment of technology.

Insofar as this is the case, of course, the science and technology of management must also be brought to bear on science and technology, especially big science or technoscience, which has become a complex enterprise. As first identified by the historian of science Derek J. de Solla Price (1963) and scientist-science administrator Alvin M. Weinberg (1967), science that depends on large-scale funding and coordinates many disciplines to achieve a common goal (such as the Manhattan Project to create the atomic bomb) requires increasingly sophisticated techniques of management. The same goes for macroengineering projects such as the U.S. interstate highway system or the European Channel Tunnel (or Chunnel). When this is the case it can reasonably be argued that the science and technology involved have become manifestations of management.

Management Ethics and Policy

In an influential analysis of how theories of human nature influence managerial practice, Douglas McGregor observed that "the more professional the manager becomes in his use of scientific knowledge, the more professional he must become in his sensitivity to ethical values" (1960, p. 12). Indeed, professionals can expect to be granted professional autonomy by the societies in which they operate only "to the extent that human values are preserved and protected" (p. 14). As the prominence of scientific and technological management has increased, so has the question of the relation between management and ethics—both ethics in management and the management of ethics.

In many instances management ethics is not strongly distinguished from business ethics. As in business ethics, key issues in management ethics include standards of communication, conflict of interest, responsibilities to stockholders, treatment of employees, social and environmental responsibilities, leadership obligations, and more. But because of their managerial roles, managers more than businesspersons or entrepreneurs also have to deal with the ethics of introducing ethics into business operations. One of the central issues in management ethics is thus how to introduce and manage ethics in a corporation or other enterprise that is also being managed for shareholder profit and/or stakeholder interests. One of the key questions for management ethics is thus: What is the proper role for ethics in management? Given the practical orientation of management, this includes: How is ethics best managed?

With regard to managing science and technology, the distinctive forms of scientific research and technological development organizations and processes must also be taken into account. Claude Gelès and colleagues (2000), for instance, argue that because most management texts assume a context of traditional business organizations using repetitive tasks and mass production to make a profit, they are not relevant to the management of scientific laboratories that use exploratory research and creativity to produce new knowledge and technical innovation. To achieve their aim of managing innovation to produce more innovation, science and technology managers need to be aware of the special characters of scientists and engineers, and of institutional resistances to new knowledge and technical innovation. They also need to be aware of the special ethical challenges involved in the scientific production of knowledge associated with temptations to scientific misconduct and the need to promote best practices in the responsible conduct of research.

Finally, because management takes place largely by means of establishing policies, the management of science is intimately related to science policy, especially that type of science policy known as policy for science. Here the work of Weinberg, as a reflective scientist manager of a big science and technology organization (Oak Ridge National Laboratory), provides basic orientation. For Weinberg, it is useful to distinguish internal and external criteria for decision-making in the management of science. Internal criteria focus on whether a particular research program is ripe for pursuit and on the competencies of the scientists involved. External criteria are of three types: scientific merit, technological merit, and social merit. Finally, Weinberg argues that especially in big science, which depends for its existence on financial support from the larger non-scientific community, and because science cannot be presumed to be the summum bonum (supreme good) of a society, "the most valid criteria for assessing scientific fields come from without rather than from within the scientific discipline" (1967, p. 82).

CARL MITCHAM
ADAM BRIGGLE

SEE ALSO Business Ethics; Science Policy; Science, Technology, and Society Studies; Work.

BIBLIOGRAPHY

Bowie, Norman E., and Patricia H. Werhane. (2005). Management Ethics. Malden, MA: Blackwell. A business ethics text with a "management ethics" title.

Brunner, Ronald D.; Toddi A. Steelman; Lindy Coe-Juell; et al. (2005). Adaptive Governance: Integrating Science, Policy, and Decision Making. New York: Columbia University Press. A comprehensive critique of scientific management with the preferred alternative of adaptive governance demonstrated through several case studies.

Burnham, James. (1941). The Managerial Revolution: What Is Happening in the World. New York: John Day.

Chandler, Alfred D., Jr. (1977). The Visible Hand: The Managerial Revolution in American Business. Cambridge, MA: Harvard University Press, Belknap Press.

Drucker, Peter F. (1954). The Practice of Management. New York: Harper.

Drucker, Peter F. (1970). Technology, Management, and Society. New York: Harper and Row. See especially the chapter "Work and Tools."

Fayol, Henri. (1949). General and Industrial Management, trans. Constance Storrs. London: Pitman. Originally published, 1916.

Gelès, Claude; Gilles Lindecker; Mel Month; and Christian Roche. (2000). Managing Science: Management for R&D Laboratories. New York: Wiley. An examination of the most appropriate principles and techniques for the management of research organizations.

Hays, Samuel P. (1959). Conservation and the Gospel of Efficiency: The Progressive Conservation Movement, 1890–1920. Cambridge, MA: Harvard University Press.

Hosmer, LaRue Tone. (2003). The Ethics of Management, 4th edition. Chicago: McGraw-Hill/Irwin. 1st edition, 1987.

Kanigel, Robert. (1997). The One Best Way: Frederick Winslow Taylor and the Enigma of Efficiency. New York: Viking. A 675-page treatise on the implications of Taylor's life and work for modern society.

MacIntyre, Alasdair. (1984). After Virtue, 2nd edition. Notre Dame, IN: University of Notre Dame Press. An account of the decline of virtue ethics into modern emotivism and a defense of virtue ethics as a viable moral framework for the modern world.

Mayo, Elton. (1933). The Human Problems of an Industrial Civilization. New York: Macmillan.

Mazlish, Bruce. (1993). The Fourth Discontinuity: The Co-evolution of Humans and Machines. New Haven, CT: Yale University Press.

McGregor, Douglas. (1960). The Human Side of Enterprise. New York: McGraw-Hill. Distinguishes two theories of human nature (Theory X, in which human beings dislike and avoid work, and Theory Y, in which work is as natural as play or rest), and argues that the second is more consistent with social scientific research and has better implications for management practice.

Price, Derek J. de Solla. (1963). Little Science, Big Science. New York: Columbia University Press. 2nd edition published as Little Science, Big Science—and Beyond (New York: Columbia University Press, 1986).

Spender, J.-C., and Hugo J. Kigne, eds. (1996). Scientific Management: Frederick Winslow Taylor's Gift to the World? Boston: Kluwer Academic. A collection of seven essays on Taylor's contributions and their impacts around the world.

Taylor, Frederick Winslow. (1985 [1911]). The Principles of Scientific Management. Easton, PA: Hive Publishing.

Van de Ven, Andrew H.; Harold L. Angle; and Marshall Scott Poole, eds. (2000). Research on the Management of Innovation: The Minnesota Studies. New York: Oxford University Press. Originally published, New York: Harper and Row, 1989. A series of longitudinal case studies undertaken by faculty at the University of Minnesota.

Weber, Max. (1930). The Protestant Ethic and the Spirit of Capitalism, trans. Talcott Parsons. New York: Scribner. Also translated by Stephen Kalberg (Los Angeles: Roxbury, 2002), with various printings of each version. Originally published as journal articles in 1904 and 1905, then later in book form as Die Protestantische Ethik und der Geist des Kapitalismus.

Weinberg, Alvin M. (1967). Reflections on Big Science. Cambridge, MA: MIT Press. Influential, critical studies by a leading scientist and manager of science.

Wren, Daniel A. (2005). The History of Management Thought, 5th edition. Hoboken, NJ: Wiley. Previous editions were titled The Evolution of Management Thought.

MODELS OF

Management is the process of reaching individual and collective goals by working with and through human and nonhuman resources to improve the world. Managerial values include performance effectiveness (achieving goals), operational efficiency (not wasting resources in the process), sustainable innovation (continually improving outputs and processes), and adding value (as measured by stakeholder responsiveness). Good managers demonstrate sound judgment by balancing these four competing but complementary values.

The four values inherent to some degree at all levels of management are embodied in four management mod-

FIGURE 1

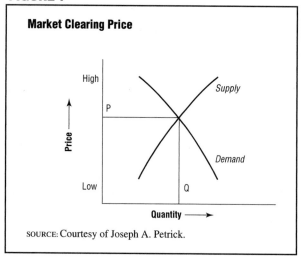

Market Clearing Price

SOURCE: Courtesy of Joseph A. Petrick.

els. Those models focus on rational goals, internal process, human relations, and open systems (Quinn et al. 1995), each of which involves ethical issues that have relevance for the management of science and technology.

Rational Goal Model

The rational goal model, which Frederick Taylor (1856–1915) introduced at the beginning of the twentieth century, stresses the importance of managerial external control that results from the exercise of director and producer role responsibilities in order to employ humans and other tools to engineer optimal productivity (Taylor 1911). Performance effectiveness is achieved through setting goals, speeding productivity, and increasing profits faster than external competitors can and by using time-and-motion studies, financial incentives, and technological power to maximize output.

Three of Taylor's followers—Henry Gantt (1861–1919) and Frank (1868–1924) and Lillian (1878–1972) Gilbreth—expanded the rational goal approach by using new engineering techniques (time and motion studies) that enhanced the ability of technological experts to expand productivity. Time and motion studies provided detailed information about job activities such as grasping, searching, transporting, or assembling and the time it took to complete them in order to measure normal and superior productivity standards.

The strength of this model is that it accounts for managers' providing structure and initiating action. The exclusive and extreme emphasis on the rational goal model, however, imposes fast-paced, robotlike movements on people that were impossible to sustain, and this neglect of individual psychosocial needs in the pur-

suit of economic returns tends to result in offended individuals and destroy cohesion at the organizational level.

At the microeconomic and geopolitical levels the rational goal model of management was advanced indirectly by Alfred Marshall (1842—1924) and James Burnham (1905—1987), respectively. Marshall was a neoclassical economist who explained how the price and output of a good are determined by both supply and demand curves, such as the price and output of new automobiles that are determined by the demand of the buyers and the supply from the manufacturers, that are like scissor blades that intersect at an optimal point of equilibrium. It is at this point of equilibrium that buyers, sellers, and/or managers could and should rationally optimize their utility values by clearing the external market (see Figure 1).

Burnham's later neoconservative geopolitical works argue that because of the unceasing desire for power among an oligarchy of managerial elites from the three major global "super-states," the struggle for external political control of the world requires a decisive victory by strong-willed U.S. political leadership that exercises an aggressive geopolitical strategy by using all the offensive resources at its disposal. The perceived overreliance on the rational goal model at the microeconomic and geopolitical levels to secure external global control has led to the expected results of offended stakeholders and has destroyed cohesion at those extraorganizational levels as well.

Internal Process Model

The internal process model introduced by Henri Fayol (1841–1925) in the first quarter of the twentieth century stresses the importance of managerial internal control that results from the exercise of the monitor and coordinator role responsibilities in order to exert authority over humans to maintain the stability of hierarchic administration. Operational efficiency is achieved through information management, documentation control, and consolidated continuity and by emphasizing process measurement, smooth functioning of organizational operations, and the maintenance of structural order (Fayol 1916). Fayol described the five functions of management as planning, organizing, commanding, coordinating, and controlling and laid down fourteen principles of good administration, with the most important elements being specialization of labor, unity and chain of command, and the routine exercise of authority to ensure internal control.

Another key exponent of operational efficiency in managing large groups was the sociologist Max Weber (1864–1920), who described and advocated the indispensability of bureaucracy. Weber's ideal bureaucracy included authority, hierarchy, formal rules and regulations, and impersonality in rule application. His ideal bureaucrat neutrally and efficiently manages by the book and follows orders from above even if they go against his or her personal convictions.

When the internal process model is applied to politico-economic control, socialist and communist regulatory infrastructures constrain the negative externalities of the free market but create the risk of stifling technological and politico-economic innovations through overregulation. The strength of this model is that it accounts for managers' maintaining structure and collecting information. The exclusive and extreme emphasis on the internal process model, however, results in stifled progress and neglected possibilities at the organizational and extraorganizational levels.

Human Relations Model

The human relations model, which Elton Mayo (1880–1949) popularized in the second quarter of the twentieth century, stresses the importance of the managerial internal flexibility that results from the exercise of facilitator and mentor role responsibilities in order to improve human relations at work and enhance extraorganizational stakeholder responsiveness. Stakeholder responsiveness is achieved by showing managerial consideration for employees' psychosocial needs to belong, fostering informal group collaboration, and providing recognition at work as well as promoting managerial social responsibility and humane community building in society (Mayo 1933). Mayo's research at the Hawthorne Works demonstrated that management consideration, employee group affiliation, and special recognition motivated can increase productivity.

Peter Drucker (b. 1909), although critical of Mayo's perceived psychological manipulation of employee loyalty, promotes the value of the socially responsible use of managerial power and humane community building. He argues that in a global knowledge society managerial power can and should be applied to the nonprofit sector because that appears to be the primary sector that is focusing on creating socially responsible citizens and giving knowledge workers a sphere in which they can make a positive difference and re-create meaningful communities.

The strength of this model is that it accounts for managers' showing consideration and facilitating supportive interaction with intraorganizational and extraorganizational stakeholders. The exclusive and extreme emphasis on the human relations model, however, creates the risk of slowing production at work and abdicating decision-making authority in society.

Open Systems Model

The open systems model introduced by Paul Lawrence (b. 1933) and Jay Lorsch (b. 1934) in the third quarter of the twentieth century stresses the importance of the managerial external flexibility that results from exercising the innovator and broker role responsibilities in order to adapt continually to changing environmental forces (Lawrence and Lorsch 1967). Sustainable innovation is achieved by cultivating organizational learning cultures, developing cross-functional organizational competencies for continuous creativity, and respecting quality and ecological system limits while negotiating for external resource acquisition, building sustainable entrepreneurial networks, and enabling creative system improvement.

W. Edwards Deming (1900–1993) used statistical quality control to separate special and common causes of variation, fixing the former and accepting the latter to improve production systems continually by narrowing the range of acceptable performance variation over time. Deming's message to managers was that because most performance variations are the result of common causes, that is, fall within a normal range of statistical variation, managers should focus on improving the production system instead of overcontrolling employees.

Paul Shrivastava (b. 1939) focuses on entrepreneurial ecocentric management of sustainable development systems that technologically prevent and/or control pollution of nature and corruption of sociopolitical systems over time. The strength of this model is that it accounts for managers' envisioning improvements and acquiring resources for sustainable system development. The exclusive and extreme emphasis on the open systems model, however, results in disrupted operational continuity and energy wasted on unrealistic change projects.

Ethics of Management

The four management models for handling behavioral complexity have management ethics parallels in handling moral complexity, that is, inclusively balancing the competing moral values of achieving good results, following the right rules, cultivating a virtuous character, and creating supportive contexts (Petrick and Quinn 1997). In effect, the way people manage—make managerial judgments—implicitly and/or explicitly discloses their moral value priorities: the relative emphases they place on results, rules, character, or context in their moral choices. Rational goal "bottom line" managers are naturally disposed to emphasize results-oriented teleological ethics theories; internal process "by the book" managers are naturally disposed to emphasize rule-oriented deontological ethics theories; human relations "bleeding heart" managers are naturally disposed to emphasize character-oriented virtue ethics theories; and open systems "change agent" managers are naturally disposed to emphasize context-oriented situation ethics theories. Nevertheless, just as the balance and inclusiveness of the four management models determine the quality of managerial behavioral complexity judgment, the balance and inclusiveness of the four ethics theories determine the quality of managerial moral complexity judgment as well.

Especially in bringing these ethical issues to bear in the management of science and technology, the economist Adam Smith's (1723–1790) social calculus of adding individual selfish motives to the greater good must be supplemented by the insight that managers often are faced with ethical responsibilities that run counter to their actual or perceived self-interest. Otherwise, management ethics would be synonymous with corporate profit or self-promotion. A case in point would be the uncritical scientific endorsement of genetically modified human foods for global profit without morally considering the harmful effects of genetically modified foods on the health of current and future human generations.

Management ethics involves a complex and inclusive balancing of multiple stakeholder interests, internal and external to organizations, domestically and globally. For example, business managers that focus only on advancing the financial interests of investors while neglecting other stakeholders' interests, such as those of employees, society, and nature, are increasingly criticized for an unduly narrow and short-term managerial ethics perspective. The ability to simultaneously and/or sequentially optimize moral results, rules, character, and context in a sustained way for multiple stakeholders at intraorganizational and extraorganizational levels is becoming the touchstone of sound management ethics and the basis of hope for moral progress in the future.

JOSEPH A. PETRICK

SEE ALSO *Bureaucracy; Engineering Ethics: Overview; Entrepreneurism; Stakeholders; Work.*

BIBLIOGRAPHY

Deming, W. Edwards (1993). *The New Economics for Industry, Government, Education.* Cambridge, MA: MIT Center for Advanced Engineering Study.

Drucker, Peter F. (1973). *Management: Tasks, Responsibilities and Practices.* New York: Harper and Row.

Evans, William A. (1981). *Management Ethics: An Intercultural Perspective.* Boston: Martinus Nijhoff. Covers several topics and focuses on the theme of individual responsibility. Develops a guide to management ethics.

Fayol, Henri. (1916). *General and Industrial Management.* London: Pittman.

Lawrence, Paul, and Jay Lorsch. (1967). *Organization and Environment.* Cambridge, MA: Harvard University Press.

Mayo, Elton. (1933). *The Human Problems of an Industrial Civilization.* New York: Macmillan.

Petrick, Joseph, and John Quinn. (1997). *Management Ethics: Integrity at Work.* Thousand Oaks, CA: Sage.

Quinn, Robert E.; Sue R. Faerman; Michael P. Thompson; and Michael R. McGrath. (1995). *Becoming a Master Manager: A Competency Framework,* 2nd edition. New York: Wiley.

Shrivastava, Paul. (1996). *Greening Business: Profiting the Corporation and the Environment.* Cincinnati, OH: Thomson Executive Press.

Taylor, Frederick. (1911). *The Principles of Scientific Management.* New York: Harper and Brothers.

MARCUSE, HERBERT

• • •

Herbert Marcuse (1898–1979) was born in Berlin on July 19. After earning a doctorate in literature in 1922, he studied philosophy with Martin Heidegger (1889–1976) in Freiburg from 1928 to 1933. Troubled by Heidegger's affiliation with the National Socialist party, Marcuse joined the philosophers Max Horkheimer (1895–1973) and Theodore Adorno (1903–1969) at the Institute for Social Research in Frankfurt before fleeing to New York in 1934. Marcuse remained for the rest of his life in the United States, where he continued the institute's interdisciplinary work in critical social theory. He died on July 29 in Starnberg, after having suffered a stroke on a trip to Germany. Marcuse synthesized the works of Heidegger, Karl Marx (1818–1883), and Sigmund Freud (1856–1939) into a unique philosophical perspective from which he analyzed the nature of social control and the prospects for liberation in advanced industrial capitalist and communist societies.

Among Marcuse's contributions to critical social theory was his analysis of science and technology as instruments of social and political domination. Echoing Heidegger, Marcuse spoke of the "technological a priori" of scientific-technical rationality that projects nature as potential instrumentality. Technological rationality homogenizes people and nature into neutral objects of manipulation. That rationality is easily co-opted by economic and political power. However, science and technology merely function in the service of social control; they could be transformed to serve different ends, such as freedom, individuality, and creativity.

Marcuse's 1941 article "Some Social Implications of Modern Technology" argued that technological rationality undermines traditional "individual rationality" (autonomy) by employing efficiency as the single standard of judgment. Industrialized societies take advantage of the notion of efficiency to induce people to accept mass production, mechanization, standardization, and bureaucracy. Consequently, Marcuse argued, appeals to enlightened self-interest and autonomy appear progressively quaint and irrational in the face of a technological rationality that makes conformity seem reasonable and protest seem unreasonable.

In the mid-twentieth century political power—including state capitalism, fascism, and state socialism—developed seemingly rational, even pleasurable, means of social control that integrated individuals into a homogeneous society. The result was a "one-dimensional" society that eroded the capacity for individuality, critical thinking, and practical resistance. However, Marcuse maintained that the same impersonal rationality that made individualism unnecessary could be harnessed to realize rather than repress human capacities. Technological rationality could be used as an instrument to foster democracy, autonomy, and individuality. Marcuse was pessimistic about the prospects for that transformation because the technological apparatus tends to incorporate and subsume all opposition. However, despite Marcuse's pessimism regarding the achievement of such a transformation, he maintained that it was in principle possible.

In his most influential book, *One-Dimensional Man* (1964), Marcuse continued to argue that advanced industrialized societies employ science and technology to serve existing systems of production and consumption but claimed that technological rationality itself required transformation; it could not remain value-neutral if it were to lead to real human liberation. Marcuse also extended his analysis of the role of science and technology in manipulating human needs through advertising, marketing, and mass media. The scientific and technical aspects of a society are used to increase productivity and dominate humans and nature. The result is a carefully managed society that creates a one-dimensional person who willingly conforms to a society that limits freedom, imposes false needs, stifles creativity, and co-opts all resistance.

At the end of *One-Dimensional Man* Marcuse expresses the hope that humans one day will develop technologies for the "pacification of the struggle for

Herbert Marcuse, 1898–1979. Marcuse was a leading 20th-century New Left philosopher in the United States and a follower of Karl Marx. His writing reflected a discontent with modern society and technology and their "destructive" influences, as well as the necessity of revolution. He was considered by some to be a philosopher of the sexual revolution. (© UPI/Corbis-Bettmann.)

existence" that will reduce misery and suffering and promote peace and happiness. Developing those technologies would require a political reversal, not simply more technological advances. A radical break from existing capitalist modes of production is needed to generate a new science and new technology. Science and technology then would become the instruments of liberation, not domination. New technologies would lead to new modes of cooperative production, energy sources, management, and communities; a new science of liberation would serve the interests of freedom and help satisfy genuine human needs. In his later work Marcuse considered the contributions that utopianism, student revolts, feminism, and aesthetic interests might make to the emergence of a new science and technology.

Marcuse was enormously popular in the 1960s and 1970s, and although his fame has been eclipsed since that time by that of Jürgen Habermas (b. 1929) and

French postmodern thinkers, he left an enduring legacy in critical social theory. He created a widely influential framework for analyzing the connections among political economy, science, technology, mass media, and culture in a way that not only identifies social domination and oppression but also attempts to identify the potential for social transformation leading to human liberation.

DAVID M. KAPLAN

SEE ALSO *Critical Social Theory; Habermas, Jürgen.*

BIBLIOGRAPHY

Alford, C. Fred. (1985). *Science and the Revenge of Nature: Marcuse and Habermas.* Gainesville: University of Florida Press.

Held, David. (1980). *Introduction to Critical Theory: Horkheimer to Habermas.* Berkeley: University of California Press.

Kellner, Douglas. (1984). *Herbert Marcuse and the Crisis of Marxism.* Berkeley: University of California Press.

Marcuse, Herbert. (1941). "Some Social Implications of Modern Technology." *Studies in Philosophy and Social Science* 9, no. 3 (1941): 414–439.

Marcuse, Herbert. (1964). *One Dimensional Man.* Boston: Beacon Press.

Marcuse, Herbert. (1969). *An Essay on Liberation.* Boston: Beacon Press.

Marcuse, Herbert. (1972). *Counterrevolution and Revolt.* Boston: Beacon Press.

Marcuse, Herbert. (1978). *The Aesthetic Dimension.* Boston: Beacon Press.

Pippen, Robert, and Andrew Feenberg, eds. (1988). *Marcuse: Critical Theory and the Promise of Utopia.* South Hadley, MA: Bergin and Garvey.

MARKETING

SEE Advertising, Marketing, and Public Relations.

MARKET THEORY

• • •

The market system allows individuals to exchange goods and services voluntarily, based on prices, without knowing one another. For instance, the cup of coffee a person drinks in the morning was brought to that person by thousands of strangers, who cultivated, harvested, pro-cessed, manufactured, packaged, shipped, stocked, and sold goods at various stages of production along the way.

One way to appreciate the distinctiveness of market-mediated trade among strangers is to contrast it with other ways in which people transact with one another. The anthropologist Alan Fiske (2004) suggests that all interpersonal transactions can be sorted into four relational models:

- In a communal sharing transaction, such as a family dinner, every member in the relationship is entitled to share in what is available.

- In an authority ranking transaction, such as a decision made in a traditional military unit or a corporation, there is a clear hierarchy, with people lower in the hierarchy deferring to those who are higher up.

- In an equality matching transaction, such as taking turns going through a four-way stop, people operate according to an intuitive sense of balance and fairness.

- In a market pricing transaction, such as buying a used car, people make decisions on the basis of their calculations of the costs and benefits.

The cognitive psychologist Steven Pinker, author of *The Blank Slate* (2002), argues that among these four modes of transactions market pricing is a relatively new phenomenon in the development of the human species:

> Market Pricing is absent in hunter-gatherer societies, and we know it played no role in our evolutionary history because it relies on technologies like writing, money, and formal mathematics, which appeared only recently (Pinker 2002, p. 234).

An important aspect of hunter-gatherer societies is that people belonged to tribes or bands of fewer than 150 people. Everyone knew everyone else, and people expected to interact with one another repeatedly. Small groups with repeated interactions are conducive to establishing trust and confidence in reciprocity, which are requirements for communal sharing and equality matching. When societies become larger and people must interact with strangers, something must replace trust and confidence. Only authority ranking or market pricing can "scale up" to large groups.

Economic historians see the modern market system as having arisen only within the last 300 years. Two features of the modern market system were largely absent until that time. One was flexibility of prices in response to supply and demand. In contrast, ancient and feudal trade took place at prices fixed by custom, authority, and tradition. A second feature of modern markets is that they enable people to work for money and trade for food. Before modern times markets did not have sufficient depth and breadth to allow for specialization and cash crops.

Before 1500 almost all people existed at a subsistence level, living on what they could cultivate. Feudal lords took any excess production and in return provided some public goods, notably protection. As late as 1700 the practice of raising a crop for cash and buying goods and services for money was relatively unknown. Even under late feudalism trade was relatively unimportant, and the terms of exchange were fixed by tradition rather than adjusting to supply and demand. The feedback loop between prices and production did not operate.

Between 1700 and 1850 the market system arose in Western Europe and North America. Better farming techniques allowed people to produce surplus food, giving them something to trade and releasing labor to work in manufacturing. Improvements in transportation, particularly railroads, facilitated specialization and trade. Increasingly, people moved from subsistence farming to a money economy in which they obtained cash for either a crop or physical labor. They then exchanged money for goods and services. Land, labor, and capital became responsive to market conditions.

Adam Smith was the first philosopher to articulate the virtues of the market system fully. In *The Wealth of Nations* (1776) Smith argued that trade was more efficient than self-sufficiency. With trade people can enjoy a wide variety of goods and services while specializing in their labor. In addition, Smith pointed out that the self-interest of producers worked to the benefit of consumers. When consumer demand increases for a good, the price goes up, attracting more producers.

The fact that higher prices induce more production is known as the law of supply. Similarly, a higher price for one good induces consumers to buy less of that good. This is known as the law of demand. Together, the laws of supply and demand determine an equilibrium price and level of output for each good. This impersonal, self-adjusting process is what distinguishes a market economy. In contrast, in a planned economy a bureaucrat determines prices and output levels. In a feudal economy prices are set by custom.

The concept of a market remains counterintuitive in the early twenty-first century. This can be seen in discussions of energy policy, in which it is suggested that the United States could become independent of foreign oil by reducing its domestic consumption and increasing the production of alternative energy. In fact, the world energy market is highly integrated. If the United States

reduced its demand for oil, the world oil price would be reduced. However, Americans still would be affected by a disruption in the world supply of oil because such a disruption still would cause the price to rise.

The Ethics of the Market

The market system has ethical virtues in the view of libertarians and utilitarians. The libertarian view is that voluntary exchange among consenting adults is preferable to coercive allocation of resources by government. The utilitarian case for markets, which goes back to Smith, is that market exchanges make people better off.

Markets improve living standards in two ways. First, for any state of knowledge and technology markets achieve an efficient allocation of resources. Flexible prices and competition send signals that accomplish this. Consumers choose the goods and services that satisfy their wants most effectively. Firms choose the inputs and outputs that maximize the value of what is produced. Workers choose the occupations that best apply their talents and interests to social needs.

The second way in which markets improve living standards is through a Darwinian selection of innovative products and processes. Entrepreneurs attempt new techniques, with successful methods surviving and achieving widespread adoption. As unprofitable firms go out of business, failed innovations and obsolete methods fall by the wayside.

The support that markets give to innovation accounts for the high standard of living in the contemporary developed world relative to the past or to the underdeveloped world. The difference is large. Whereas the poorest people in the early 2000s and people who lived 500 ago lived on the equivalent of less than a dollar per day, the average American consumes more than $30,000 in goods and services each year. Market-driven South Korea has a standard of living more than ten times that of communist North Korea.

Feedback between Technological Innovation and Markets

Technological innovation and markets reinforce each other. Markets promote innovation by rewarding success and punishing failure. Technological change broadens markets and makes them more efficient.

Every innovation faces resistance. Scientists may doubt the validity of the theory behind an innovation. Firms are reluctant to discard tried-and-true production methods. Workers in existing industries find their livelihoods threatened by new competition. Consumers may be afraid of new products.

Interest groups that are threatened by new technology attempt to mobilize social institutions to retard innovation. Governments are asked to intervene. For example, some countries in Europe have banned genetically modified food. In the United States opposition to Wal-Mart stores often is driven by store owners and labor unions seeking to stifle competition.

Markets overcome resistance to innovation. The impersonal price system gives its approval to innovations that increase productivity and consumer well-being as firms that adopt the innovations earn profits. Simultaneously, the demise of unprofitable businesses frees resources to be used in more productive ways.

In addition to the ability of markets to foster innovation there is positive feedback from technological innovation to markets. Each improvement in transportation, communication, and trading technology serves to strengthen the market system, increasing the scope of transactions occurring in markets.

The revolution in oceangoing shipping that took place in the fifteenth century helped spur trade, which in turn fostered the transition from feudalism to a market economy. The invention of the steam engine and the railroad lowered shipping costs, enabling cash crops to replace subsistence farming. The internal combustion engine increased the mobility of labor and goods, leading to an increased share of economic activity taking place in the market. Electric motors and labor-saving devices helped release women from household labor and move into market-paid work. In modern times the Internet has increased the breadth of markets, including new possibilities for international trade in white-collar services.

Ethical Concerns with the Market System

There is a long-standing set of ethical concerns with markets. Major problems include inequality, failure to provide public goods, and erosion of cultural traditions.

Markets provide different rewards to different individuals. Those with talent, capital, entrepreneurial instincts, and luck do well. Those who lack valuable talents and/or encounter bad luck do poorly.

Critics of the market system believe that goods and services should be distributed more equally. The socialist thinker Karl Marx (1818–1883) described capitalism not as a neutral system of market pricing but a hierarchical system, with the ruthless capital-owning class exploiting the helpless working class. "From each according to his abilities, to each according to his needs" was Marx's slogan, promising the alternative of communal sharing. However, as anti-Marxists such as

Max Weber (1864–1920) and Friedrich Hayek (1899–1992) predicted, large economies could not be made to operate efficiently without markets. Hayek in particular emphasized that the information developed by the price system and individual incentives is much more effective than is central planning.

Critics of inequality tend to view the economy as a zero-sum game, with the success of some individuals necessarily coming at the expense of others. Supporters of the market system view it as a positive-sum game, making it possible for nearly all people to raise their standard of living.

Another area where critics see a zero-sum game is in terms of resource constraints. The argument is that the earth's resources are finite and will be "used up." Economists counter by pointing out that human ingenuity seems boundless. As a result, Jerry Muller comments, "the history of capitalism, as Schumpeter observed, is of finding new ways to make use of formerly insignificant resources. Coal ... petroleum ... uranium ... sand for silicon chips. We may well be at the beginning of the fourth wave of capitalist industrial innovation, the biotechnology revolution" (Muller 2002, p. 391).

Federal Reserve Chairman Alan Greenspan is fond of pointing out that the physical weight of the American gross national product (GDP) is declining, an indication of reduced pressure from economic growth on physical resources. This trend may continue as nanotechnology allows products to be built from raw atoms. Rodney Brooks of the Massachusetts Institute of Technology talks about the possibility of not having to cut down trees and carve wood to make a table but instead simply growing a table with genetic engineering. The technology futurist Ray Kurzweil has suggested in *The Age of Spiritual Machines* (1999) that the information component of GDP is asymptotically approaching 100 percent, which would imply that physical scarcity will never constrain growth.

Another criticism of markets is that they give choices to individuals at the expense of collective purpose. It is argued that there is no overall direction or goal for a market economy. Those who want society to have a common objective see the market as too anarchic. A related criticism of markets is that they fail to pursue cultural ideals: The market may not reward fine art, classical music, or religion.

One strength of the market is that it promotes innovation. However, the market may fail to preserve cultural values and institutions. Occupations made obsolete by market forces represent ways of life that are no longer sustainable. Unique cultural identity may be replaced by homogeneous, anonymous market forces.

Market Imperfections

Economists have found a number of flaws in the market system. The most important are externalities and imperfect information. An externality is a cost or benefit that is not internalized by the market. Pollution is the classic example. The pollution caused by an automobile does not cost its owner anything but the total pollution caused by all automobiles is costly to society. Even though laissez-faire leads to too much pollution, economists still favor market-oriented approaches, including taxes on pollution and tradable pollution "permits." These solutions preserve the flexibility and efficiency of the market while forcing the market to internalize the cost of pollution. Consumers' lack of information provides a rationale for a number of government interventions in the market. For example, government meat inspection helps ensure the safety of meat and regulation of medicines helps protect consumers from harmful or ineffective drugs.

Modern Challenges for the Market System

The market system faces a number of challenges from modern technology. The increased importance of health care and education, the increased role of research and development, the issue of network externalities, and the increased importance of information goods all raise issues for the market.

As human capital increases in importance relative to material resources, health care and education are accounting for an increasing share of the economy. These sectors traditionally have been ones in which government involvement has been extensive.

Health care expenses can soar for the people least able to afford them. Someone who is sick often cannot work. The elderly, who are most likely to have illnesses, are on fixed incomes. Private health insurance may be prohibitively expensive for those with the highest likelihood of needing costly health care. All these issues provide a rationale for government provision of health-care coverage, at least for some segment of the population.

The question is where to draw the line between the market and government involvement. At one extreme are national health-care systems that attempt to put the entire sector under government control. However, this leads to bureaucratic rationing of care and, as is the case any time market forces are suppressed, to slow adoption of new technology and lack of innovation. The United States, which has the most market-oriented health-care system in the industrialized world, also does the most to advance the state of the art through pharmaceutical development, diagnostic equipment, and innovative medical procedures.

Education is another area where the individuals with the greatest needs may be least able to afford the best service. As with health care there is a long tradition of government involvement. Critics argue that this has meant slow innovation and the persistence of ineffective schools. Some economists believe that a more market-oriented approach of giving parents vouchers and letting entrepreneurs supply schooling would be more effective.

The inequality that characterizes market outcomes may be a more significant issue as education and health care increase in importance. One may be able to shrug at the differences between what the rich and the poor can afford in terms of cars or wine, but it is more difficult to feel comfortable when the rich are able to obtain better medical care and education.

Economic growth depends on research and development. In the future the fields of computer science, biotechnology, and nanotechnology will be particularly important to the economy. As a theoretical matter, "basic research," which is generally applicable but yields no immediate profits, will be undersupplied by markets and will have to be supported by the government. By the same token "applied research," which is specific and provides immediate rewards, is best done by private firms so that unprofitable ideas are discarded quickly.

In practice the distinction between basic research and applied research is not as easy to draw. In any event the questions of how much the government should invest in research and where it should invest are very important. People's future standard of living will depend to a large extent on how well those decisions are made.

Modern technology gives rise to networks, in which the size of the network is a source of value. For example, the value of a fax machine is low if no one else has one. When everyone else has a fax machine, the value is much higher. The same is true for e-mail accounts, instant messaging services, CD burners, and popular word-processing file formats.

People may choose a word-processing program for compatibility with their colleagues even though they would prefer the features in a different program. In theory everybody could choose to use an inferior program because it is the program others are using. In that way the market gravitates toward an inferior standard. This possibility is called a network externality.

Another aspect of the economy that has changed in recent years is the increased importance of information goods relative to physical goods. Information goods pose a challenge to the market system.

With physical goods the price system is effective at allocating resources. The price of a bicycle or an apple reflects the marginal cost of producing and distributing those goods. Moreover, there is rivalry in consumption: The bicycle that one person rides is one that another person cannot ride; the apple that a person eats is an apple that nobody else can eat.

With information goods the marginal cost of production and distribution approaches zero. Once an essay or a song is stored as information (bits) on a computer, it costs very little to copy those bits or send them to another computer halfway around the world. Furthermore, an author's ability to read an essay on his or her computer does not interfere with another person's ability to read that essay.

The dilemma caused by information goods is that the marginal cost of production and distribution is zero but the up-front development costs may be substantial. For example, consider the case of a new pharmaceutical to treat diabetes or AIDS. That drug may cost hundreds of millions of dollars to develop. However, the pills can be manufactured for pennies apiece. What should be the price? On the one hand, the price should be low enough not to discourage use, which at the margin costs very little. On the other hand, the price should be high so that companies recover their up-front costs and have an incentive to continue to innovate.

There are a variety of possible pricing mechanisms for information goods, none of which is perfect. In the case of pharmaceuticals the government grants a temporary monopoly in the form of a patent. This allows drug companies to set prices above marginal cost so that they can recover the cost of research. However, at the margin this discourages the use of medications because the price is higher than the marginal cost of production.

The challenge with research-intensive goods is to come up with a way to cover fixed costs while leaving the marginal price as low as possible to encourage broad use. Price discrimination—charging higher prices to the consumers most willing to pay—can be not only profitable but also socially optimal. Alternatively, it may be desirable for many consumers to combine to cover up-front costs through a subscription model or a membership model. It may be desirable for taxpayers to cover some up-front costs through a subsidy or prize offered by the government.

Doomsday Scenarios

There is a long-standing tension between economic growth and cultural stability. Markets, which facilitate the former, undermine the latter. Many futurists project an acceleration of technological change in the twenty-first century. This has the potential to raise the standard of living dramatically, but it also has the potential to cause great culture discontinuity. There are many examples:

- In computer science, Kurzweil (1999) argues that Moore's law, which roughly states that the power of computers doubles about every eighteen months, implies that there will be a computer with the intelligence of a human brain by about 2030. Moreover, once computers catch up with humans, they will surpass humans rapidly. Thus, the long-term future is one in which humans and machines will be integrated and coevolve, with the human species becoming inferior or extinct.

- In nanotechnology Eric Drexler (1986) and Bill Joy (2000) warn of the possibility of chemical production processes expanding uncontrollably. In the worst case, dubbed the "gray goo scenario," a substance could reproduce indefinitely until it swallowed the planet.

- In biotechnology the President's Commission on Bioethics (2003) emphasized a number of possible dystopian scenarios, including one in which human beings are designed and created to serve the purposes of their masters. The commission also pointed to issues raised by medicines that enhance performance or might prolong life indefinitely.

If these doomsday scenarios are possible technologically, markets are unlikely to prevent them. Accordingly, fear of doomsday scenarios could lead people to favor strong, worldwide government action to intervene in markets. Opposition in Europe to genetically modified food and opposition in the United States to embryonic stem-cell research could be symptoms of antimarket regulation to come.

The Future

Markets are conducive to technological innovation, and vice versa. People who place a high value on the benefits of technological innovation tend to want to expand the scope of the market. People who are more concerned with the risks of technological innovation are more inclined to favor government intervention.

The chief benefit of technological innovation is that it raises people's standard of living. People's labor, capital, and natural resources become more productive as they use science and engineering to develop more efficient techniques for satisfying human wants.

The combination of markets and technological innovation creates economic inequality. Successful entrepreneurs, business leaders, and others earn outstanding rewards. Unskilled workers have a higher standard of living than was the case a century ago, but they are significantly less wealthy than those at the top of the income distribution.

Markets and innovation also cause cultural dislocation. Old ways of life disappear, and people must adapt to new circumstances. The possibility appears to exist for dramatic, discontinuous change.

People are close to having capabilities that may undermine their identity as human beings. Will people merge with machines? Will pharmacology or genetic engineering give people control over their emotions, memories, aging process, and physical and cognitive skills? Will scientific discoveries serve primarily to enhance the lives of the rich, or will they also give new opportunities to the poor?

The market offers only one way to answer these types of questions: with trial and error. Individual responses to opportunities and incentives will cumulate to an overall social result. Those who want the outcome to be arrived at by a different process, such as the deliberations of moral philosophers and experts, will seek to find a way to disrupt the decentralized, experimental market mechanism and replace it with something more planned and controlled.

ARNOLD KLING

SEE ALSO Capitalism; Environmental Economics; Libertarianism; Smith, Adam.

BIBLIOGRAPHY

Blinder, Alan S. (1987). Hard Heads, Soft Hearts: Tough-Minded Economics for a Just Society. Reading, MA: Addison-Wesley.

Drexler, K. Eric. (1986). Engines of Creation. Garden City, NY : Anchor Press/Doubleday.

Kurzweil, Ray. (1999). The Age of Spiritual Machines: When Computers Exceed Human Intelligence. New York: Viking.

Mokyr, Joel. (2002). The Gifts of Athena: Historical Origins of the Knowledge Economy. Princeton, NJ: Princeton University Press.

Muller, Jerry Z. (2002). The Mind and the Market. New York: Knopf. Muller provides an intellectual history of pro-market and anti-market philosophers from Adam Smith through the contemporary age.

North, Douglass C. (1981). Structure and Change in Economic History. New York: Norton. Emphasizes the role of political and social institutions in the development of modern markets.

Pinker, Steven. (2002). The Blank Slate. New York: Viking.

President's Council on Bioethics. (2003). Beyond Therapy: Biotechnology and the Pursuit of Happiness. Washington, DC: Author.

Rosenberg, Nathan, and L. E. Birdzell, Jr. (1986). How the West Grew Rich: The Economic Transformation of the Industrial World. New York: Basic Books. The authors explain how market institutions evolved from feudalism in Western Europe.

Shapiro, Carl, and Hal. D. Varian. (1999). Information Rules: A Strategic Guide to the Network Economy. Boston: Harvard Business School Press.

Stock, Gregory. (2003). Redesigning Humans: Choosing Our Genes, Choosing Our Future. Boston: Houghton Mifflin.

Smith, Adam. 1998 (1776). An Inquiry into the Nature and Causes of the Wealth of Nations. Washington, DC: Regency.

INTERNET RESOURCES

Brooks, Rodney. (2002). "Beyond Computation: A Talk With Rodney Brooks." Available from http://www.edge.org/3rd_culture/brooks_beyond/beyond_index.html.

Fiske, Alan Page. (2004). "Human Sociality." Available from http://www.sscnet.ucla.edu/anthro/faculty/fiske/relmodov.htm.

Joy, Bill. (2000). "Why the Future Doesn't Need Us." Wired 8.04. Available from http://www.wired.com/wired/archive/8.04/joy.html.

MARX, KARL

• • •

Karl Marx (1818–1883) was born in Trier, Prussia on May 5 and died in London on March 14. He was educated in Trier and at the universities of Bonn and Berlin, thus coming under the influence of Georg Wilhelm Friedrich Hegel (who he later radically criticized) before receiving his doctorate in philosophy from the University of Jena in 1841. Throughout most of his adult life, he was assisted both financially and intellectually by Friedrich Engels (1820–1895), with whom he coauthored such works as The German Ideology (1845–1846) and "The Communist Manifesto" (1848).

Marx wrote mainly on capitalism as an economic system, and is most closely identified with the multivolume Capital (Vol 1 [1867]; Vol. 2 [1885]; Vol.3 [1894], Vols. 2 and 3 published by Engels after Marx's death). This massive 2,500 page work explores the capitalist system in terms of the logic of its functioning, its historical progression, and its fate. Marx's writings on science are scattered and fragmentary, and his discussions of technology, though more detailed, are largely unsystematic. Therefore this entry will concentrate more on his views on ethics and morality, the implications of which are enormous.

Technology and Science

Technology and science played an important role in Marx's thought. His general theory of human history, *historical materialism,* gave technology a major role in forming the foundation of society and in the process of historical change. Every society rests on an economic base or mode of production, which includes both forces and relations of production. The forces of production consist mainly of the level of technological development a society has achieved and of the features of the natural environment in which it is located. Relations of production are the social and economic relations people enter in the process of production and involve the ownership of the productive forces. The productive forces might be owned and controlled by the entire society, or, more commonly, by a relatively small segment of society. Those who own the productive forces dictate their operation and often subject the mass of the population to conditions of severe exploitation and oppression. The other major part of every society is the superstructure, which consists of politics, law, family life, religion, and the mode of consciousness, or collective forms of thought and feeling. The superstructure rests on the economic base and is largely determined by it.

Marx regarded the earliest societies as constituting forms of primitive communism. Here people lived by using simple technologies of hunting, fishing, agriculture, and animal husbandry. Because of the communal nature of such societies and the absence of class divisions and exploitation, they would have been idyllic except for their low level of technological development, which prevented people from adequately satisfying basic needs. Gradually, however, progress in technology enhanced human power to manipulate the environment, but in ways that led to the formation of private property and class divisions. European society passed through a slave mode of production in ancient times and then a feudal stage. Capitalism succeeded feudalism.

Despite his savage criticisms, Marx appreciated the great achievements of capitalism, the foremost being its enormous capacity for the development of technology in the form of modern industry. In his general theory of history, Marx saw capitalism as a prerequisite for the development of socialism because the latter, in order to meet basic human needs and allow for everyone's self-realization and self-fulfillment, requires material abundance. Capitalism developed technology to a level sufficient for the creation of this abundance. But socialism would develop technology even further, thus allowing for the elimination, or at least the reduction, of the most unpleasant and burdensome forms of work.

Karl Marx, 1818–1883. This German philosopher, radical economist, and revolutionary leader founded modern "scientific" socialism. His basic ideas—known as Marxism—form the foundation of socialist and communist movements throughout the world. *(The Library of Congress.)*

Marx had much less to say about science than he did about technology, but he was a major proponent of science, both because of its ability to produce intellectual knowledge and its capacity for the development of industry. In the section of the "Economic and Philosophical Manuscripts" (1844) devoted to private property and capitalism, Marx writes that "natural science has invaded and transformed human life all the more *practically* through the medium of industry; and has prepared human emancipation" (Marx 1978b, p. 90). Also "Natural science will in time subsume under itself the science of man, just as the science of man will subsume under itself natural science: there will be *one* science" (Marx 1978b, p. 91).

Indeed Marx regarded historical materialism as a scientific theory that could be empirically verified (Husami 1980). He was also a great admirer of Charles Darwin and highly commended *Origin of Species* (1859) to Engels, saying that it served as a basis in nature for their theory of history. Later, in his speech at Marx's grave, Engels was to say, "Just as Darwin discovered the law of development of organic nature, Marx discovered the law of development of human history" (Engels 1978, p. 681).

Ethical Perspective

Marx did not have an ethical theory, or a theory of justice, in the sense of such great moral philosophers as Immanuel Kant or John Rawls. In fact Marx explicitly disavowed all talk of justice and rights, in part because they belong to the juridical superstructure rather than the technoeconomic base. In capitalist society, juridical notions are part of the way in which the capitalist mode of production and its ruling class are maintained. In "Critique of the Gotha Programme" (1875) he argues that, in discussions of socialism, notions of justice and rights are *obsolete verbal rubbish* and *ideological nonsense*. Under socialism there will be no need for rights and liberties, their *raison d'etre* having disappeared. The rights and liberties found in capitalist society only exist because capitalism is a highly inadequate mode of production from a human point of view (Buchanan 1982).

In his famous essay "On the Jewish Question" (1843), Marx drew an important distinction between political freedom and human freedom. Political freedom consists of the constitutional liberties that people have in capitalist society: the right to property, speech, and assembly, equal treatment before the law, and so on. Political rights are a cover for an absence of human rights. Human freedom involves the opportunity of all individuals not only to have the full satisfaction of their basic needs, but also the opportunity to realize their essential nature as human beings through creative and self-fulfilling work. In capitalist society, everyone has political freedom but only a few can achieve true human freedom. Only in socialist society can human freedom become commonly achieved. This vision of freedom is intimately tied to Marx's views on technology, because true human freedom requires a very advanced level of technology, which a fully realized socialist society will have.

Nevertheless although Marx did not develop an ethical theory and rejected its need or desirability, he did have moral or evaluative notions that guided his critique of capitalism and his advocacy of socialism. Marx was a moralist who had no moral theory, that is, he "advocates principles that are supposed to guide present-day social and political choice in the same way as a political morality" (Miller 1984, p. 51). In various writings, Marx refers to the misery and sufferings of the working class under capitalism, of the deadening and degrading nature of work created by the capitalist division of labor (and thus of the alienation and dehumanization of the worker), and of how capitalism "enforces on the laborer abstinence from all life's enjoyments" (Husami 1980, p. 43). The capitalist class receives all the material and intellectual benefits of society while the proletariat assumes all its burdens. Capitalism exploits the worker, and exploitation is variously described as robbery, embezzlement, plunder, and theft. Husami argues that these evaluative notions are tantamount to a conception of justice despite the fact that Marx formally rejected all talk of justice.

Marx also seemed to have a theory of distributive justice (Husami 1980). As set forth in *Critique of the Gotha Programme* (1875), the first phase of the new socialist society will be guided by the principle *to each according to his abilities*. Workers receive from society payment in accordance with the labor contribution they make. Individuals differ in their mental and physical endowments and some contribute more labor than others; those who contribute more receive more in return. But inequalities never become significant because society provides for every person's social needs (healthcare, education, and so on). Whatever inequalities do exist are not the result of power and class differences because private ownership of the means of production has been abolished.

But this first phase of socialist society, having just emerged from capitalist society, is still stamped with defects. There will emerge a higher phase of socialist or communist society, and "only then can the narrow horizon of bourgeois right be crossed in its entirety and society inscribe on its banner: From each according to his ability, to each according to his needs" (Marx 1978b, p. 531). In this phase, society takes into consideration the fact that individuals differ not only in their talents and abilities, but also in their needs. Because some individuals have greater needs than others, they should be rewarded accordingly. This highest form of socialist society is guided by the principle of full individual self-development, and as such must provide each person with the resources necessary for that development. Inequalities therefore remain. Again, however, these inequalities do not arise from class position (because there are no classes) and do not involve any exploitation. Moreover the inequalities are not great and do not affect the satisfaction of basic needs related to physical well-being and education, because these are automatically provided to everyone. (See Wood [1980] for a very different interpretation of Marx on justice. For an interpretation partway between Husami's and Wood's, see Brenkert [1980].)

Historical Failures and Legacy

The implications of Marx's thinking on science and technology are relatively minor, but his thought has enormous implications for an ethical assessment of society. Marx's predictions concerning future socialist revolutions and the content and nature of socialist society have been overwhelmingly repudiated by the past 100 years of history. Socialist revolutions occurred where Marx did not expect them, and utterly failed to occur in those places where he thought they would. And the so-called socialist societies that did develop were for the most part a grotesque deformation of what he expected. These failures lie both in a flawed theory of history—Marx badly misunderstood the historical trajectory of capitalism—and in a failure to appreciate the importance of a theory of justice and morality. Marx's view that political rights and liberties are merely expressions of a defective bourgeois mode of production, and as such will be irrelevant and unnecessary in a socialist mode of production, opened the way for, and gave license to, some of the most brutal dictatorial regimes in human history. Marx did not foresee this outcome, and certainly would have vehemently rejected it. The ideals may have been noble, but their actual implementation proved to be an entirely different matter.

Many different kinds of Marxism have developed since Marx's time, including the critical theory of the Frankfurt School (Adorno, Horkheimer, Marcuse, Habermas), the Italian Marxism of Antonio Gramsci, French existentialist Marxism (Sartre), Wallerstein's world-system theory, and anticolonialist theory. Some of these are as different from one another, and from classical Marxism, as they are similar. Critical theory, for example, is highly critical of modern science and technology in a way that would have been inconceivable to Marx. In terms of ethics, a wide range of complex positions can be found.

STEPHEN K. SANDERSON

SEE ALSO Alienation; Capitalism; Communism; Critical Social Theory; Freedom; Hegel, Georg Wilhelm Friedrich; Marxism; Political Economy; Socialism; Sociological Ethics; Work.

BIBLIOGRAPHY

Brenkert, George G. (1980). "Freedom and Private Property in Marx." In Marx, Justice, and History, eds. Marshall Cohen; Thomas Nagel; and Thomas Scanlon. Princeton, NJ: Princeton University Press. An attempt to map out a position on Marx and justice part way between the positions of Husami and wood.

Buchanan, Allen E. (1982). Marx and Justice: The Radical Critique of Liberalism. Totowa, NJ: Rowman & Allanheld. Takes the position that Marx had no formal theory of justice or rights and disdained any discussion of such with respect to the virtues of socialism.

Cohen, Marshall; Thomas Nagel; and Thomas Scanlon, eds. (1980). Marx, Justice, and History. Princeton, NJ: Princeton University Press. An important collection of essays on the role of justice in Marx's thinking.

Engels, Friedrich. (1978). "Speech at the Graveside of Karl Marx." In The Marx-Engels Reader, ed. Robert C. Tucker. Text originally published in 1883.

Husami, Ziyad I. (1980). "Marx on Distributive Justice." In Marx, Justice, and History, eds. Marshall Cohen; Thomas Nagel; and Thomas Scanlon. Princeton, NJ: Princeton University Press. Sets forth the position that Marx had a theory of justice (including two principles of distributive justice) and judged capitalism to be unjust.

Marx, Karl. (1967 [1867]). Capital, Vol. 1, trans. Samuel Moore and Edward Aveling. New York: International Publishers. The most important volume of Marx's great work in which he sets forth the laws of functioning of capitalist society and its evolution.

Marx, Karl. (1967 [1885]). Capital, Vol. 2, ed. Friedrich Engels. New York: International Publishers.

Marx, Karl. (1967 [1894]). Capital, Vol. 3, ed. Friedrich Engels. New York: International Publishers.

Marx, Karl. (1978a). "Critique of the Gotha Programme." In The Marx-Engels Reader, ed. Robert C. Tucker. Text originally published in 1875. An important discussion by Marx of the place of human abilities and needs in the future socialist society.

Marx, Karl. (1978b). "Economic and Philosophical Manuscripts of 1844" In The Marx-Engels Reader, ed. Robert C. Tucker. Text originally published in 1844. Youthful essays written by Marx on such topics as alienation, private property and communism, and Hegelian philosophy.

Marx, Karl. (1978c). "On the Jewish Question." In The Marx-Engels Reader, ed. Robert C. Tucker. Text originally published in 1843. A famous essay in which Marx draws an important distinction between political freedom and human freedom. The famous early essay on the coming of communist society, including a brief sketch of its principles of organization.

Marx, Karl, and Friedrich Engels. (1970 [1845–1846]). The German Ideology, ed. C. J. Arthur. New York: International Publishers. The work in which Marx and Engels lay out their general theory of society and history, historical materialism.

Marx, Karl, and Friedrich Engels. (1978). "Manifesto of the Communist Party." In The Marx-Engels Reader, ed. Robert C. Tucker. Text originally published in 1848.

Miller, Richard W. (1984). Analyzing Marx: Morality, Power, and History. Princeton, NJ: Princeton University Press. An extremely important book that critically examines the nature of historical materialism and the extent to which the concept of justice played an important role in Marx's critique of capitalism.

Tucker, Robert C., ed. (1978). The Marx-Engels Reader, rev. edition. New York: Norton. An excellent collection of selections from all of the important works of Marx and Engels.

Wood, Allen W. (1980). "The Marxian Critique of Justice." In Marx, Justice, and History, ed. Marshall Cohen; Thomas

Nagel; and Thomas Scanlon. Princeton, NJ: Princeton University Press. Argues that Marx had no formal theory of justice and did not condemn capitalism as unjust.

MARXISM

• • •

An intellectual tradition and political movement initiated by Karl Marx (1818–1883) and Friedrich Engels (1820–1895), Marxism has devoted much attention and debate on matters of science, technology, and ethics. Marx and Engels themselves were particularly influenced by Darwinism and saw themselves as extending an understanding of organic evolution into human history. They believed that developments in the natural sciences of their times required elaboration of the philosophical and sociological consequences in the direction of a dialectical and historicist form of materialism. But they were critical of existing materialist currents as undialectical and existing dialectical positions as idealist. In the intellectual division of labor between Marx and Engels, Marx devoted his efforts to economics, while Engels wrote on philosophy, science, culture, morality, and gender, and entered into polemics with critics. His *Dialectics of Nature,* published posthumously in 1927, explores the philosophical implications of the natural sciences.

Marxism held that capitalism has played a crucial part in developing science and technology, but that only socialism could fulfill their potential and organize an equitable distribution of their benefits. For Marxism, capitalism was an inherently contradictory mode of production. It was a system based on the primacy of market forces and private ownership of the means of social production, generating a basic class division between those who own the means of production and those who own only their labor power. Although capitalism led to an unprecedented development of productive forces, rising standards of living, and advances in science and technology, it also created massive inequality, parasitism, and alienation. Capitalism was a historically necessary stage in human development, but socialism was a necessary next step. A socialist system based on the social ownership of the means of social production would create a social order based on the principle "from each according to his or her abilities, to each according to his or her needs."

Marxism pioneered the field of sociology of knowledge, including the sociology of science and technology. It has insisted that science and technology are not iso-

lated, self-contained activities, but develop in complex interaction with a whole range of other processes: philosophical, cultural, political, and economic forces. Within this interaction, the mode of production is decisive. All existing scientific theories, technological developments, economic structures, political institutions, philosophical positions, legal codes, moral norms, sexual roles, cultural trends, aesthetic tastes, and even common sense are inextricably interrelated and determinately shaped by the dominant mode of production. Marxism thus made extraordinarily strong claims regarding the philosophical assumptions and sociohistorical basis of scientific knowledge. At the same time it put considerable emphasis on ideology, arguing against the view that science itself is neutral and that only the use or abuse of science is ideological. Yet Marxism perceived recognition of these aspects as enhancing science and not being in conflict with the rationality and credibility of science.

Developments in the USSR

There have been many twists and turns in the history of Marxism due to the impact of new scientific discoveries, technological developments, philosophical trends, and political formations. Marxists of subsequent generations got caught up in many controversies. Along with political conflicts over evolutionary versus revolutionary paths to socialism, those of the second generation took various positions on the epistemological implications of the natural sciences. Vladimir Ilyich Lenin's (1870–1924) *Materialism and Empirio-Criticism* (1909) is a product of the philosophical debates of that period.

After the October revolution of 1917 that gave rise to the Union of the Soviet Socialist Republics (USSR), Marxism came to power as the official ideology of the new Soviet state, meaning that its visionary ideas could be tested in social practice. There were fiery debates about how to do so in virtually every sphere: from strategies for industrialization and agriculture to nationalities policy about the fate of different nationalities/national cultures within the USSR, socialist morality, science policy, free love, and the future of the family.

In the early years of the revolution, the movement for proletarian culture, *proletkult*, led by Alexander Alexandrovich Bogdanov (1873–1928), a doctor who advocated a collectivist subjectivism in the philosophy of science, argued that the culture of the bourgeoisie—from art and literature and morality to science and technology—was saturated with class ideology and could not serve the needs of the proletariat. *Proletkult* required a specifically proletarian culture, including proletarian science, because science had been shaped by the capitalist mode of production and needed to be collectivized and revolutionalized, putting an end to the fragmentation of scientific knowledge and the competitive drive of capitalist production. For *Proletkult* socialism was impossible without science, but it was also impossible with bourgeois science. Lenin and others took issue with this argument, contending that it was premature and sectarian to sweep aside the existing intelligentsia and existing knowledge. Lenin insisted that it was necessary to embrace bourgeois science and knowledge while critically reconstructing it. Bogdanov's movement dissipated within a few years, especially after he, as director of the Institute for Research in Blood Transfusion, died in an experiment on himself.

Nevertheless the USSR put much emphasis on working out a distinctive approach to science and technology under the banner of Marxism. Many political and philosophical debates flourished through the 1920s. The relationship of philosophy to the empirical sciences was very much in play through the prolonged debate between those who were grounded in the empirical sciences and emphasized the materialist aspect of dialectical materialism and those who were more grounded in the history of philosophy, particularly Hegel, and emphasized the dialectical dimension of dialectical materialism. It has been an ongoing tension in the history of Marxism, playing itself out in the intellectual ferment and institutional transformation of a socialist revolution. Philosophy was considered to be integral to the social order. Political leaders, particularly Lenin and Nikolai Ivanovich Bukharin (1888–1938), participated in philosophical debates as if these issues were matters of life and death, of light and darkness. Even while preoccupied with urgent affairs of state, they polemicized passionately on questions of epistemology, ontology, ethics, and aesthetics.

Bukharin was an advocate of the new economic policy aimed at achieving agricultural productivity and steady industrialization, but was outmaneuvered and defeated by Joseph Vissarionovich Stalin (1879–1953). Although he had fallen from the heights of political power, he continued to work as constructively as possible and devoted himself particularly to the application of science to economic planning during the first five-year plan. Bukharin believed that Marxists should study the most advanced work in the natural and social sciences and cleanse their thinking of the lingering idealism inherent in quasimystical Hegelian formulations. In *Historical Materialism* (1921), used as a basic text in educational institutions, he interpreted dialectics in terms of conflict and equilibrium. Other Marxists, such as the Italian Antonio Gramsci (1891–1937) and the Hungarian Georg Lukacs (1885–1971), saw Bukharin as the personification of a positivist tendency in Marxism. Lukacs's book *History and Class Consciousness*, rejecting Engels's concept of the dialectics of nature, drew a storm of controversy.

In 1931 Bukharin led a Soviet delegation to the Second International Congress of History of Science in London, projecting enormous enthusiasm for the role of science in a socialist society. Boris Mikhailovich Hessen (1883–c. 1937) delivered one of the most influential papers ever in the historiography of science, giving an ideological analysis of Newton's *Principia*, setting it firmly within the social, political, and economic struggles of the seventeenth century.

Both Hessen and Bukharin perished in the purges. Bukharin was the most prominent defendant in the spectacular Moscow trials and was executed. Even during his imprisonment he continued to write of how Marxism forged the most progressive path for science and technology, as affirmed in his posthumous work *Philosophical Arabesques* (2005), which was discovered decades after his death.

Another Marxist intellectual who espoused ideas relevant to science and technology was Leon Trotsky (1879–1940). He was inclined to the mechanist position in the debates of the 1920s and saw the role of philosophy as systematizing the conclusions of all the positive sciences. After Lenin's death in 1924 Stalin also outmaneuvered Trotsky, rejecting his pursuit of a worldwide socialist revolution in favor of developing socialism in the Soviet Union. Dismissing him from the government and expelling him from the party, in 1929 Stalin forced Trotsky into exile where he was assassinated.

Beyond and Within the USSR

The intellectual energy and social purpose of the Soviet philosophers and scientists had great impact on their international audience, especially in Britain, where influential scientists, such as J. D. Bernal (1901–1971), J. B. S. Haldane (1892–1964), and Joseph Needham (1900–1995) took up the challenge of a sociohistorical analysis of science and put their energies into a movement for social responsibility in science.

Marxism captured the imagination of many intellectuals in the west in the 1930s. Some of the most brilliant, such as David Guest (1911–1938) and Christopher Caudwell (1907–1937), died in the Spanish Civil War. In *The Crisis in Physics* (1939), Caudwell extended his ideological analysis of all spheres of thought into physics, seemingly the area most remote from ideological involvement. Caudwell saw a causal connection between the crisis in physics and those in biology, psychology, economics, morality, politics, art, and, indeed, life as a whole. The cause of the crisis in physics was not only the discrepancy between macroscopic or relativity physics and quantum or subatomic physics, but the deeper problem was the metaphysics of physics. What it came down to was the lack of an integrated worldview that could encompass all the sciences with their dramatically expanding experimental results. Science was decomposing into a chaos of highly specialized, mutually repellent sciences, whose growing separation increasingly impoverished each of them and contributed to the overall fragmentation of human thought. Ironically the very development of each of the sciences in this situation accentuated the general disorientation and resulted in scientists falling back on eclecticism, reductionism, positivism, and even mysticism.

Back in the USSR, a number of those who were fervent advocates of the new social order being created there were accused of undermining it and perished. All the debates of the 1920s took a sharp turn from 1929 on with the frenzy of the first five-year plan and the intensified pressure to bolshevize every institution and discipline. The intelligentsia was told that the time for ideological neutrality was over. They had to declare themselves for Marxism and for the dialectical materialist reconstruction of their disciplines or evacuate the territory. All controversies, whether between Marxism and other intellectual trends or between different trends within Marxism, were sharply closed down through the 1930s. There was to be one correct line on every question. Any deviation was considered to be not only mistaken but treacherous.

There was resistance in many areas. Geneticists fought back against attempts by brash bolshevizers to override the process of scientific discovery. The protracted struggle over the theories of Trofim Denisovich Lysenko (1898–1976) took the debate over proletarian science into difficult and dangerous territory, making legitimate issues such as hereditarianism versus environmentalism into a struggle for power where all intellectual and ethical criteria were at times abandoned. Nikolai Ivanovich Vavilov (1887–1943), an internationally prominent geneticist and ardent advocate of the unity of science and socialism, defended genetics and resisted the onslaught of Lysenkoism. He was accused of sabotage of agriculture and died in a prison camp.

These developments in Soviet intellectual life were inextricably tied to the rhythms of Soviet political and economic life. The way forward with the first five-year plan was far from smooth and uncomplicated. There was violent resistance to the collectivization of agriculture and peasants were burning crops and slaughtering livestock rather than surrender. There was one disaster after another in the push to industrialization. There was a fundamental contradiction between the advanced goals that were to be achieved and the level of expertise in science, engineering, agronomy, and economics, indeed a general cultural level, needed to achieve them. There was panic and confusion and desperation. There was reckless scapegoating. Breakdowns, fires, famine, and unfulfilled targets were attributed to sabotage and espionage. There was a blurring of the lines between bungling and wrecking, between association with defeated positions and treason, between contact with foreign colleagues and conspiracy with foreign powers.

After the death of Stalin, subsequent Soviet leaders, particularly Nikita Sergeyevich Khrushchev (1894–1971), in the critique of Stalinism after the Twentieth Party Congress (1956), and Mikhail Sergeyevich Gorbachev (b. 1931), in the period of glasnost and perestroika (1985–1991), attempted to put Soviet life, including its science, on a new basis, but, some contend, the traumas of the period prevented such changes.

Outside the USSR: New Left Marxism

From the 1940s on, Marxism came into the ascendancy in the academies of much of Eastern Europe and parts of Asia, Africa, and Latin America following the succession of communist or socialist parties to power in such countries as Czechoslovakia, Yugoslavia, China, Mozambique, and Cuba. The academicians of the German Democratic Republic were particularly devoted to developing a philosophy of science in the sense of elucidating the philosophical implications of the natural sciences.

Marxism also played a special role in French intellectual life. Some Marxist scientists, such as the physicist Paul Langevin (1872–1946) and biologist Marcel Prenant (1893–1983) saw dialectical materialism as illuminating their sciences and looked to the Soviet Union as developing science in a way that would liberate human society. Georges Freidmann (1902–1977), however, who made original contributions to industrial sociology, came to think that Soviet science was drowning in facile formulas and sterile polemics. Later many French Marxists, such as Jean Paul Sartre (1905–1980) and Maurice Merleau-Ponty (1908–1961) adapted their Marxism to existentialism or phenomenology. Others such as Louis Althusser (1918–1990) took Marxism in the direction of structuralism. It emphasized scientificity, but did not engage meaningfully with actual science.

In the 1960s and 1970s the influence of Marxism again became a formidable force, not only in countries defining themselves as socialist, but in the most prototypically capitalist ones as well. Although it never took state power in these milieus, Marxism did seize the intellectual and moral initiative for a time.

During this period a new left arose, posing new questions to the old left, as well as to the old right and the ever shifting center. Eurocommunism represented a merging of old and new left currents, which promised much at the time. The most vibrant debates of the day were conducted within the arena of Marxism. There were many journals such as *Science and Society* (1936–), *Marxism Today* (1953–1991), Socialist Register (1964–), and *New Left Review* (1960–) in which the discussion flourished.

On all matters touching on science, technology, and ethics, there was a new left challenge. The new left view of science represented a sharp break from the old left, for example the older radical science movement in Britain, exemplified by such figures as Bernal and Haldane. Science, as the older left saw it, was a progressive force. It was essential to socialism and socialism was essential to science. The *Radical Science Journal* (1974–1983) took the Marxist emphasis on the ideological nature of science in the direction of a radical social constructivism that sometimes tended to reject the cognitive and liberating potential of science. A long-standing leftist position, characterized by a blending of neo-Kantian, neo-Hegelian, and, more recently, postmodernist ideas with Marxist ideas, is represented by the Frankfurt School's (1923–) critical social theory, which identifies science with bourgeois ideology, counterposes scientific with humanistic values, and tends to hostility toward the whole sphere of the natural sciences. The divisions of the left on the question of science flared up in the science wars of the 1990s and were dramatized by the controversy that arose between the journal *Social Text* and Alan Sokal in 1996.

From the mid-nineteenth century and continuing into the early twenty-first century, Marxism made major contributions to intellectual history. It may at times seem to be a discarded theory, but one would be mistaken in believing that Marxism might not surge again.

HELENA SHEEHAN

SEE ALSO *Class; Communism; Conservatism; Critical Social Theory; Marx, Karl; Socialism; Weil, Simone.*

BIBLIOGRAPHY

Bernal, J. D. (1954, 1957, 1965, 1969). *Science in History*, 4 vols. London: Watts, Penguin.

Bukharin, Nikolai I. (2005). *Philosophical Arabesques*. New York: New York University Press.

Caudwell, Christopher. (1939). *The Crisis in Physics*. London: John Lane.

Engels, Friedrich. (1940). *Dialectics of Nature*. London: Lawrence and Wishart.

Graham, Loren R. (1987). *Science, Philosophy, and Human Behavior in the Soviet Union*. New York: Columbia University Press.

Haldane, J. B. S. (1938). *The Marxist Philosophy and the Sciences*. London: University of Birmingham.

Jay, Martin. (1984). *Marxism and Totality: The Adventures of a Concept from Lukács to Habermas*. Oxford: Polity Press.

Joravsky, David. (1961). *Soviet Marxism and the Natural Sciences, 1917–1932*. New York: Columbia University Press.

Kolakowski, Leszek. (1978). *Main Currents of Marxism Its Rise, Growth, and Dissolution*, trans. P. S. Falla, 3 vols. Oxford: Clarendon.

Lenin, Vladimir I. (1948). *Materialism and Empirio-Criticism*. London: Lawrence and Wishart.

Lewis, William. (2005). *Louis Althusser and the Traditions of French Marxism*. Lanham, MD: University Press of America.

Rose, Hilary, and Steven Rose, eds. (1976). *The Radicalisation of Science: Ideology of/in the Natural Sciences*. London: Macmillan.

Sheehan, Helena. (1993 [1985]). *Marxism and the Philosophy of Science: A Critical History*. Atlantic Highlands, NJ: Humanities Press.

Timpanaro, Sebastiano. (1975). *On Materialism*, trans. Lawrence Garner. London: New Left Books.

MATERIAL CULTURE

• • •

Material culture may be defined as the human significance of the totality of tangible artifacts that humans have produced. These artifacts range from the mundane and perishable to the monumental and enduring, and have been linked together in distinctive ways across place and time. Scholarly attention to material culture beyond technical analyses is divided among mainstream disciplines such as history and anthropology and specializations such as art history, archaeology, history of technology, cultural geography, and philosophy of technology. In all instances, questions of the ethical implications of material culture call for reflective consideration.

Basic Transformations

Despite the manifold plurality of material cultures across places and times, the Industrial Revolution of late-eighteenth-century England introduced a watershed into human history that began a radical transformation in the general character of material culture across all of its permutations. The steam engine for the first time in human history provided a tireless, ubiquitous, and powerful prime mover. Coal became a seemingly limitless energy source, and iron and steel constituted a material for structures that were both large and finely articulated.

Already in the nineteenth century, this transformation exhibited creative and destructive aspects, both noted by Karl Marx (1818–1883) and Friedrich Engels (1820–1895) in *The Communist Manifesto* (1848). About the creative side they said: "The need of a constantly expanding market chases the bourgeoisie over the whole surface of the globe. It must nestle everywhere, settle everywhere, establish connections everywhere" (Marx and Engels 1955 [1848], p. 13). This creative process has continued over the past century and a half and is much discussed in the early 2000s under the term *globalization*.

The destructive side Marx and Engels described as follows: "All that is solid melts into the air, all that is holy is profaned, and man is at last compelled to face with sober senses his real conditions of life and his relations with his kind" (Marx and Engels 1955 [1848], p. 13). They described the dissolution more specifically in their description of how the labor power of workers was being torn out of its traditional context of personal relations, social bonds, and ownership of stores and tools and converted into a commodity whose price was being more and more depressed. Marx's *Capital* (1867) extended this analysis to all those things that used to be rooted in the production and consumption of the household and were pulled into the market by industry and commerce. This process too is still being discussed vigorously, and Anglo-American scholars have coined the term *commodification* as a covering concept.

Both creation and destruction are pervaded by a third process, a dematerialization and refinement of production and consumption. John Kenneth Galbraith (1967) noted how the basis of economic power had shifted since the eighteenth century from land via capital to expertise. Daniel Bell (1973) described a similar shift from extraction via fabrication to processing. Remarkably, Thomas J. Schlereth (1982) observed a broadly analogous process of sophistication in the scholarly concern with material culture. He distinguished

the "The Age of Collecting (1876–1948)" from the "The Age of Description (1948–1965)" and the "The Age of Analysis (1965–)." The current end phase of this development is also much considered and contested in the early twenty-first century under such headings as *the computer era* or *the information age*.

Modern technology began as a widespread activity of inspired tinkering and ingenious inventing in the last third of the eighteenth century. It was well underway before the natural sciences in the nineteenth century caught up with technology and, through the explanation of heat, pressure, electricity, and materials, became an engine of innovation. Technological devices, in turn, began to open up deeper dimensions of familiar phenomena and entirely new areas of investigation. Research and development have to this day been the major sources of productivity growth and thus of an exploding material culture. By now technology and science have so fulsomely embraced one another that it has become fashionable to see them as one creature—technoscience (Ihde and Selinger 2003). It is an undeniable fact, to be sure, that much of science is undertaken for technological gain and that technology has stimulated science and made it more effective; yet technology and science remain distinguishable and, from the moral point of view, need to be distinguished.

Ethical Assessment

When it comes to its ethical examination, Marx may again be considered a founding figure in his ambivalence about the moral quality of the newly emerging material culture. Under the surface, Marx regretted the loss of traditional things and relations. Overtly, however, he considered the world of the past as one of oppression, exploitation, and even idiocy, and he embraced the Industrial Revolution and its fruits. What he emphatically found objectionable and doomed was not the quality of the new material culture, but maldistribution in the power over production and in the blessings of consumption.

Because it does not examine or question the internal moral structure and properties of the artifacts modern technology has produced, Marx's moral judgment of the material culture is an extrinsic one. It has in fact become the received wisdom of social theory that there are no morally significant internal structures or properties and that tangible technology is thus morally neutral. Accordingly, when considering how standard ethical theories and more popular moral positions bear on contemporary material culture, all those bearings turn out to be extrinsic.

This does not mean they are unimportant. Consider the two leading contemporary ethical theories. The first is the ethics of equality and liberty, masterfully represented by John Rawls (1999) and technically known as deontology. It contends that inequalities in power and prosperity are warranted only if everyone has an opportunity to become powerful and prosperous, and if inequalities are to the benefit of the poor and powerless. This implies a significant and well-warranted critique of how prosperity and the material objects of which it consists are distributed nationally and globally. At the same time Rawls makes the debatable claim that prosperity and opportunity in themselves can be defined in a morally thin or neutral sense.

The other leading contemporary moral theory is utilitarianism, which is concerned with maximizing the happiness of a given population (Sidgwick 1981 [1907]). The animating principle of utilitarianism is as intuitively simple and attractive as it is technically difficult and forbidding. Finding a measure for happiness, establishing the maximizing procedure, and defining the relevant population have turned out to be endlessly complicated and controversial problems that at every turn threaten implementation with paralysis. Utilitarianism becomes a feasible program if one substitutes prosperity for happiness and agrees to measure prosperity with money. The resulting moral theory—what may be termed monetary utilitarianism—dominates public policy decision-making in the advanced industrial countries and retains some of the affirmative and forward-looking spirit of the original conception. Maximizing becomes equated with increasing the gross domestic product by all available means, a person's happiness is measured by income and prosperity, and the relevant population is the citizenry of a nation. All this is animated by a spirit of optimism and tolerance. But utilitarianism, monetary or not, remains neutral when it comes to the moral quality of the goods that, along with the services, compose prosperity or lead to happiness. This is how utilitarians understand tolerance.

Environmentalism and Religion

The two more popular moral positions that bear on the material culture are environmentalism and religion. Environmentalists, broadly speaking, regard contemporary material culture as hypertrophic (growing excessively) and ruinous. Hence they counsel a reduction of material possession and consumption. This too is a moral injunction on the material culture—and one that is important and would be beneficial if heeded. But as practiced, environmentalism would not require a deeper

understanding and a transformation of the moral quality of material culture. One might continue to enjoy the same tangible and consumable objects, albeit in environmentally sustainable versions—sitting on natural-fiber couches, drinking beer brewed from organically grown barley and hops, eating chips made from genetically unmodified corn, staring at a television set that, at the end of its useful life, the producer has to take back and recycle in its entirety. All of this would make the material culture simpler in quality and reduced in quantity, but not essentially different in character.

The most pointed and the best-known critique of the material culture comes from religious ethics. It condemns materialism—the excessive concern with material goods. Pope John Paul II has been a vocal proponent of this criticism, and his voice may seem a lonely one because, at least in the United States, Christianity and materialism seem to be anything but antagonistic. When questioned, however, Americans profess to be worried about materialism (Wuthnow 1996, Schor 1998). These worries surface in movements that range from Luddism to voluntary simplicity (Elgin 1981).

Materialism is an ill-defined phenomenon. The concern with material objects covers such disparate things—television sets and sport-utility vehicles (SUVs) are material objects, but so are musical instruments and bicycles. Can't one at least say that, no matter the kinds of material objects, there are simply too many? Aren't humans consuming too much and thus running out of raw materials, food, timber, and energy? And in the process, aren't the industrialized countries of the northern hemisphere exploiting those of the globe's southern half? According to Mark Sagoff (1997), however, these apprehensions turn out to rest on misconceptions.

Two conclusions appear to follow. First, the religious objection to materialism stands no matter how materialism is defined. Excessive concern with any kind of material object is a distraction from spiritual matters or the afterlife. Second, secular worries about materialism are unfounded, and a secular outlook on life cannot have objections in principle to the current way of taking up with material culture. Both conclusions leave one uneasy, however. As to the first, excessive concern with tangible stuff is morally objectionable by definition. But what about appreciation and enjoyment of the visible world? Some religious traditions at least think of the tangible world as created by God and therefore as fundamentally good. Secular folks who worry about materialism have something specific in mind, namely, consumerism (Wuthnow 1996, Schor 1998). Materialism in this sense is a preoccupation with a particular kind of material object, consumable objects, presumably. There is a need, then, for an intrinsic analysis of material goods and for a determination of whether their internal structure is ethically potent.

Material Goods Themselves

One school of thought has it that material goods are used to mark and enforce class distinctions (Veblen 1992, Douglas and Isherwood 1979, Schor 1998). Though this is certainly true and morally troubling, it reveals little about the specific quality of goods produced by modern technology. Horses, servants, and mansions were used to signal high status prior to the Industrial Revolution, and sumptuary laws were used to enforce class distinctions more rigorously than even Ferraris do in the early 2000s. Here again a cue may be taken from Marx or at least from his progeny. Like Marx, more recent left-liberal theorists have examined the transformation things undergo when they are drawn into market. Commodification is the term used to name this phenomenon, and the term carries connotations of disapproval, unlike the coreferential term that conservatives prefer, namely, privatization, or the term of mixed connotations, namely, commercialization.

Commodification has a clean and crisp economic definition: the process of moving something into the market—from either the intimate sphere or the public sphere—so that it becomes available for sale and purchase. In the case of a good from the public sphere, a public good is converted into a commodity, and, speaking more precisely, privatization is commodification in this latter sense only. Some of the public goods, such as justice and elementary education, are not material, of course, but others, such as transportation or a healthy environment, clearly are. The same distinction applies to intimate goods. Friendship and freedom are not material goods, but food and clothing are.

Commodification of intangible goods is morally objectionable because in this case a good commodified becomes a good corrupted. Justice bought is no longer justice, and friendship paid for is not real friendship. But no such opprobrium seems to taint tangible goods. Railroads are managed as public goods by governments in some countries, whereas in others they are private enterprises run for profit. Food and clothing have left the intimate sphere of the household so long ago that people no longer notice their peculiarities as commodities. Accordingly, Michael Walzer (1983), who has thought deeply about commodification (though he does not use the word), has drawn up a list of never-to-be-commodified goods, all of which are intangible.

Is there a way of capturing the apprehensions about consumerism, the suspicion that commodification of material goods is a process whereby "all that is holy is profaned" or that at least some holy things are profaned? The sacredness of food is certainly lost when it is shelved in a supermarket. The sacredness of nature is gone when it becomes an engineered setting for the wilderness lodge in Disney World. The holiness of things, or, more prosaically, their power to engage people deeply, is lost when things are stripped of their spatial, temporal, and social contexts, when those contexts are reconstituted and concealed technological means, and when the resulting commodities are made available for sale.

Commodification, then, is a cultural as well as an economic process. These two processes largely overlap, but not entirely. The food in a supermarket is commodified both economically and culturally. A typical farmers' market is a scene of economic commodification. The food, after all, is for sale. But significant contexts are there to be experienced directly. The local market reflects its special context in the fruits and vegetables that the local soil and climate can produce. It reflects the season with the hardy stuff appearing early in the year and the more tender things not until summer. Sellers are known for their expertise in growing this or that, and they establish ties of expectations and pleasure with their customers.

Conversely, tourists whose only concern is to capture the sights and scenes with their cameras deracinate treasures, trees, and towers and make them available as videos that can be shown anywhere and any time. They commodify their travels culturally though rarely economically. The things on those videos are severed from their here and now, but few would pay to see those desiccated things.

What is driving commodification? In its economic aspect it is certainly propelled by the pursuit of prosperity. This is a creditable desire, and many are grateful beneficiaries of at least some important parts of this affluence. The less noticed kinetic force of commodification is the desire for liberty—less noticed because one tends to think of liberty exclusively as political, the freedom from the oppression by persons. But, prior to the Industrial Revolution, there were also burdens and claims of material reality: the need to shear, card, and spin wool, and knit it into sweaters; the need to plant, water, weed, harvest, clean, prepare, and cook beans; and so on. Commodification, taken culturally, disburdens people of these requirements, and consumption can be taken in a culturally corresponding sense as the unencumbered enjoyment of commodities. Dematerialization turns out to be a consistent tendency of commodification. The less materially heavy and imposing commodities are, the more variously and easily they will be available and consumable. Technologically perfect virtual realities are the endpoint of this process.

Disburdenment too has its undeniable moral benefits, certainly when it comes to such basic parts of the material culture as water, warmth, and light. But disburdenment can hypertrophy from liberation to disengagement and lead to the physical and mental shapelessness that plagues the most advanced industrial societies. There is then a need to save or selectively reintroduce those material things that rightfully claim people's engagement and exertion, things such as musical instruments, gourmet kitchens, running trails, urbane cities, and more.

Morally debilitating commodification is not a problem for most people on the globe, namely, those who suffer from hunger, disease, illiteracy, and confinement. Appropriate globalizing of commodification is morally desirable. But finding a measure for appropriate globalization and for the readjustment of the material culture requires understanding the cultural and moral aspects of commodification. It is hard, however, to meet this task when science and technology are conceptually fused or rather confused into technoscience. Consider genetics. There are things to be found out about how genes and proteins relate to one another and how genes cooperate with one another and with environmental conditions to help produce brains, dispositions, and behavior. To come to understand these things is progress, and once clearly understood, the resulting knowledge compels assent. But there is nothing obviously progressive or compelling in the application of such knowledge. The eradication of aging and a massive deferral of dying may not be progress at all, and nothing compels one to think of those goals as desirable. These are moral issues that call for wisdom and persuasion.

ALBERT BORGMANN

SEE ALSO *Consumerism; Distance; Place.*

BIBLIOGRAPHY

Bell, Daniel. (1973). *The Coming of Post-industrial Society.* New York: Basic.

Borgmann, Albert. (1984). *Technology and the Character of Contemporary Life: A Philosophical Inquiry.* Chicago: University of Chicago Press.

Borgmann, Albert. (1995). "The Moral Significance of Material Culture." In *Technology and the Politics of Knowledge,* ed. Andrew Feenberg and Alastair Hannay. Bloomington: Indiana University Press.

Douglas, Mary, and Baron Isherwood. (1979). *The World of Goods.* New York: Basic.

Elgin, Duane. (1981). *Voluntary Simplicity.* New York: Morrow.

Galbraith, John Kenneth. (1967). *The New Industrial State.* Boston: Houghton Mifflin.

Ihde, Don, and Evan Selinger, eds. (2003). *Chasing Technoscience: Matrix for Materiality.* Bloomington: Indiana University Press. Includes work by Donna Haraway, Bruno Latour, Andrew Pickering, and the editors.

Marx, Karl. (2003 [1867]). *Das Kapital.* Cologne, Germany: Parkland.

Marx, Karl, and Friedrich Engels. (1955 [1848]). *The Communist Manifesto,* ed. Samuel H. Beer. New York: Appleton-Century-Crofts.

Rawls, John. (1999). *A Theory of Justice,* rev. edition. Cambridge, MA: Harvard University Press, Belknap Press.

Sagoff, Mark. (1997). "Do We Consume Too Much?" *Atlantic Monthly* 279(6): 80–96.

Schlereth, Thomas J. (1982). "Material Culture Studies in America, 1876–1976." In *Material Culture Studies in America,* ed. Thomas J. Schlereth. Nashville, TN: American Association for State and Local History.

Schor, Juliet B. (1998). *The Overspent American.* New York: Basic.

Sidgwick, Henry. (1981 [1907]). *The Methods of Ethics.* Indianapolis, IN: Hackett.

Veblen, Thorstein. (1992 [1899]). *The Theory of the Leisure Class.* New Brunswick, NJ: Transaction Publishers.

Walzer, Michael. (1983). *Spheres of Justice.* New York: Basic.

Wuthnow, Robert. (1996). *Poor Richard's Principle: Recovering the American Dream through the Moral Dimension of Work, Business, and Money.* Princeton, NJ: Princeton University Press.

MATERIALISM

• • •

Materialism is a term with both metaphysical and social meanings. As a metaphysical position materialism regards matter (Latin *materia*) as the primary or most real substance. In modern times materialism also has taken practical forms. Because science studies empirical objects and because material entities are more perceptible than are immaterial ones, the scientific worldview tends to assume materialism at least for heuristic purposes or on provisional grounds. Moreover, modern technological progress, especially in its early phases, provided mostly material improvements. Thus, one effect that technology seems to have had on culture is the creation of social forms of materialism such as consumerism.

Metaphysical Materialism

As a form of metaphysical monism, materialism attempts to reduce all phenomena to a single basic substance: matter. Thus, the opposites of metaphysical materialism are doctrines such as spiritualism, which holds that spirit is the ultimate reality; idealism, which sees the phenomenal world and matter as creations of the mind; and immaterialism, which rejects the reality of matter itself.

The idea of materialism was present when ancient Greek philosophy originated with Ionian natural philosophers who began to explain phenomena by referring to natural causes instead of religious myths in the sixth century B.C.E. The first systematically materialistic philosophers were the atomists Democritus and Leucippus of Abdera in the fifth century B.C.E. Among the major schools of philosophy in antiquity, Epicureanism professed materialism. In the modern period important materialists have included Pierre Gassendi (1592–1655), Thomas Hobbes (1588–1672), Heinrich Dietrich d'Holbach (1723–1789), Karl Marx (1818–1883), and Friedrich Engels (1820–1895).

One important difference between premodern and modern materialism is that the former tended to promote acceptance of the state of affairs in the world, whereas the latter is used to promote human action to change the world. Marxist materialism strongly illustrates the modern version of materialism. Indeed, Marx and Engels's philosophy developed in the former socialist countries into what was called dialectical materialism. It was materialism in the sense that it strictly denied the existence of immaterial entities, arguing that, for example, religious beliefs were part of a false ideology. The word *dialectical* referred to the quality of the laws that govern transformations in nature, history, and the human mind. Dialectical materialism saw these laws as based on the interplay of opposites.

Science, Materialism, and Ethics

Because science in principle does not make metaphysical commitments, science is not materialistic in the strict sense of the word. In fact, a more proper term for describing the way science perceives reality is *naturalistic*. The progress of modern natural science, however, has made materialism a more creditable stance than it was previously. Science studies phenomena that can be experimented on or otherwise brought to the impartial attention of the community of scientists. Clearly imma-

terial things such as the soul, supernatural events, values, ideals, and meanings are difficult or impossible to research scientifically. Thus, it seems from a scientific perspective that things one cannot examine scientifically are not real.

In practical life and in the adaptation of science the tendency toward materialism is manifested, for instance, in measuring. Measuring is essential in all science-related activities because exact scientific research is based on calculating measured quantities. An object of science must be measurable in some sense. Hence, it is difficult to do scientific research on phenomena in their qualitative aspects. For example, a scientist easily can determine the weight, size, and age of an ancient Chinese vase, but it is impossible to specify scientifically its degree of beauty. In consequence, quantity appears to be "more real" category than quality.

In ethics the success of natural science has had both implicit and explicit consequences. The most explicit consequence was the logical positivist argument in the 1920s that ethics is a merely emotional use of language that lacks empirical content. Although this extreme view soon softened, ethics nevertheless struggled throughout much of the twentieth century against the tendency in a culture dominated by science to perceive reality as being defined by the possible objects of science. For instance, medicine can study whether smoking harms health, but it is a value question whether harming health is wrong. The only scientific approaches to value in this sense appear to consist of empirical research on expressed preferences or arguments for the evolutionary development of certain behaviors. Because values, norms, and ideals in the normative sense—moral sociology is another question—are not objects of scientific inquiry, ethics as a rational pursuit has had a credibility problem.

Technology and Materialistic Culture

Until recently technological advancement has contributed mainly to the improvement of the material conditions of life. This has meant highly increased material well-being for the majority of the people in industrialized societies.

According to some cultural critics, however, this development has not been free of malaise. It appears to those critics that human life has lost some of its dignity in the course of material success. This lack of dignity has been pointed out in consumerism, the loss of traditional skills, the sacrifice of ideals in the search for economic profit and quick satisfaction, and so on. Culture itself has been turned into a commodity to be mass produced and marketed industrially. The rule of quantity over quality in social and political life often is expressed in attitudes that make money and financial success the final arbiters of the good.

Some analyses of contemporary culture have suggested that classical Western ethics is incapable of addressing current issues because it does not pay sufficient attention to the material culture, that is, the production and use of material goods. At the heart of such criticisms is the notion of alienation. Cultural critics are afraid that the materialistic mass culture estranges human beings from themselves, other people, and nature. When it comes to nature, ecological problems are the most pressing issues related to materialistic consumerism.

Immateriality in Science and Technology

However, science and technology also have crucial immaterial aspects. Mathematics is indispensable for science, and mathematical abstractions are clearly immaterial. Moreover, science attempts to find regular patterns in reality and to form lawlike theories to describe those patterns. The structures, laws, and theories that science develops while investigating material reality are all immaterial. In this sense the object of science is material phenomena but the results of research are immaterial concepts that give new meanings to material reality. This is especially true in the most recently developed fields in science, such as computer science, genome studies, and neurological research.

Science can ask the question "What is matter?" but its answers are extremely complex and theoretical. Matter appears to consist mostly of empty space between elementary particles. Modern physics thus challenges any idea of matter "in itself" because what can be known about matter in the early twenty-first century is eminently theoretical and experiment-dependent.

In the realm technology information technologies and nanotechnology, which are highly theory-based forms of technology, deal mostly with immaterial phenomena. Generally speaking, technology can be interpreted as making matter less significant for human beings. For instance, communication and transportation technologies have made the globe "smaller" and reduced the role of time and place, which form the ultimate framework for matter, in human life. In this sense technology has made matter "serve" humankind.

Some essential immaterial aspects can be found in production as well. The emphasis of the economic struc-

ture in advanced societies has moved increasingly toward the production of immaterial services and information processing. Furthermore, in designing and marketing material commodities, aesthetic values, symbols, concepts, and myths form something that is now called a "brand." More and more companies do not sell only a material product but market an idea and a lifestyle. One does not buy a cell phone, one buys a successful person's phone.

These transformations in the economic structure and the style of production have been referred to as dematerialization. This term denotes the reduction of material used to produce specific goods and services. Dematerialization has raised hopes that economic growth and ecological sustainability may be reconciled so that consumers characteristically will purchase functions rather than material objects.

These reflections indicate how materialism is an ambivalent issue for science, technology, and ethics. Techno-scientific development has passed through a phase of studying and molding material reality, but currently the most important fronts in science and technology involve work on largely immaterial phenomena.

TOPI HEIKKERÖ

SEE ALSO *Consumerism; Dematerialization; Material Culture; Two Cultures.*

BIBLIOGRAPHY

Cornforth, Maurice. (1971). *Dialectical Materialism.* New York: International Publishers. A basic exposition of the Soviet Marxist version of materialism.

Moser, Paul K., and J. D. Trout. (1995). *Contemporary Materialism: A Reader.* New York: Routledge. Sixteen essays defending science-based materialism or "physicalism," with the final three essays addressing issues of materialism and ethics.

Rosenberg, Bernard, and David Manning White, eds. (1971). *Mass Culture Revisited.* New York: Van Nostrand Reinhold. A good collection of articles criticizing the materialistic culture industry; first published in 1957.

Trungpa, Chögyam. (1973). *Cutting through Spiritual Materialism,* ed. John Baker and Marvin Casper. Boston: Shambhala. Finds materialism a temptation even in religious practice.

Twitchell, James B. (2000). *Lead Us into Temptation: The Triumph of American Materialism.* New York: Columbia University Press. A witty defense of cultural materialism that reviews and rejects many of the criticisms.

McCLINTOCK, BARBARA

• • •

Nobel Prize-winning geneticist Barbara McClintock (1902–1992) was born in Hartford, Connecticut on June 16, and earned a doctorate in botany at Cornell University in 1927. Her early work on maize cytogenetics in R. A. Emerson's group at Cornell University in the 1920s and 1930s (where she worked with Marcus Rhoades, George Beadle, Harriet Creighton, Charles Burnham, and others) provided crucial evidence for the chromosomal basis of genetic crossover. Later, McClintock moved to the Cold Spring Harbor Laboratory in New York where she continued her groundbreaking research in genetics. But of her many achievements, her work on genetic transposition stands out as the most revolutionary. This work, establishing the mobility of genetic elements, defied conventional assumptions of the fixity of genes on the chromosomes and went unheeded for many years by most geneticists. But in 1983, thirty-two years after her first definitive paper on the subject, she was awarded the Nobel Prize for Physiology and Medicine, and her vindication was complete. After a lifetime pattern of relative obscurity and isolation, this prize ushered in a period of widespread public recognition—recognition not only for the quality of her work, but also for the model of scientific research she both advocated and exemplified. In her own words, good scientific research needed to be premised on "a feeling for the organism." She died near Cold Spring Harbor on September 2.

McClintock is of particular interest to historians of biology for her success in breaking with tradition on a number of fronts: as a geneticist whose understanding of genes was shaped by her interests in development; as a woman who refused to be constrained by conventional notions of gender; as a scientist who dared to affirm the importance of cultivating an intimate relation to the object of one's study in the rational construction of knowledge. For her, understanding a plant requires following it from its beginning: "I don't feel I really know the story if I don't watch the plant all the way along. So I know every plant in the field. I know them intimately, and I find it a great pleasure to know them" (Keller 1983, p. 198). But McClintock has also become a controversial figure, largely owing to differences in perspective between the two biographies that have been published (Keller 1983, Comfort 2001). Controversy centers largely on two issues: first, the extent to which her early work on transposition was in fact neglected; and second, on whether or not her particular methodological

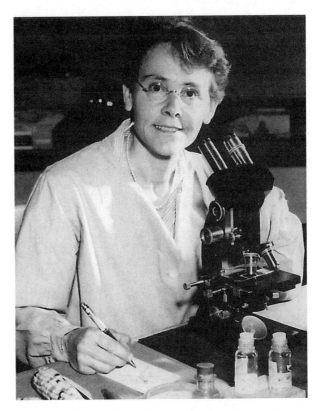

Barbara McClintock, 1902–1992. American geneticist McClintock received the Nobel Prize in Physiology for her discovery that genes could move from place to place on a chromosome. (AP/Wide World Photos.)

style can be taken as representative of either a "feminine" or a "feminist" approach to science.

Perceptions of neglect and recognition are inevitably at least partly subjective. Certainly, McClintock felt her work to be neglected, or at best, misunderstood. Equally certainly, many colleagues held her in enormously high regard. Nevertheless, prior to her Nobel Prize, and even after the rediscovery of transposition in the mid-1970s (under the name "jumping genes"), the phenomenon was widely regarded as of marginal significance to the general processes of genetics and development. Furthermore, interviews conducted prior to 1983 provide strong support for a fairly widespread tendency, perhaps especially among molecular biologists, to regard her and her work as eccentric curiosities. After 1983, however, a sea change could be seen to take place.

As a Nobel Laureate, McClintock suddenly became a heroine with whom virtually everyone wished to be identified, including feminists and mainstream scientists. Indeed, it was only at this point that McClintock began to be perceived as a feminist heroine, and that Keller's book (published some months before the prize) began to be read as a feminist manifest. Both readings fly in the face of the evidence—evidence provided both by McClintock's life and by Keller's biography. Comfort's biography goes some way toward correcting the record, and in deflating the "McClintock myth." Unfortunately, in the process he may have unwittingly contributed to the creation of a new myth, making of McClintock too much a practitioner of "normal science," and one who now appears to have been more fully embraced by the community around her than the historical record suggests. However, the scientific community's celebration of McClintock after 1983 is evident, and attested to by numerous publications (such as, for example, the excellent overview of her work by Federoff and Botstein 1992).

EVELYN FOX KELLER

SEE ALSO Genetic Research and Technology; Sex and Gender.

BIBLIOGRAPHY

Comfort, Nathaniel. (2001). The Tangled Field: Barbara McClintock's Search for the Patterns of Genetic Control. Cambridge, MA: Harvard University Press.

Federoff, Nina V., and David Botstein, eds. (1992). The Dynamic Genome: Barbara McClintock's Ideas in the Century of Genetics. Cold Spring Harbor, NY: Cold Spring Harbor Laboratory Press.

Keller, Evelyn Fox. (1983). A Feeling for the Organism: The Life and Work of Barbara McClintock. New York: W. H. Freeman.

McLUHAN, MARSHALL

• • •

Herbert Marshall McLuhan (1911–1980) spent nearly all of his life in Canada. Born in Edmonton on July 21, he was raised in Winnipeg and developed an early interest in engineering. There, he earned an M.A. in English, then went to Cambridge University and received additional B.A. and M.A. degrees, and also a Ph.D. (English). A widely published author of more than thirty books, one of which has been translated into more than twenty-five languages, McLuhan taught for three decades at the University of Toronto and died in Toronto on December 31.

McLuhan virtually invented the field of media studies and its relation to culture and society. McLuhan argued that the initial content of any new medium is always a preexisting medium (so radio, for example, takes over from the music hall and the newspaper; TV subsumes radio drama and film; and so on), so that the study of how a medium is used reveals little or nothing about its formal character or effects. Content study invariably leads to moral declaration and away from knowledge of the new form. Each major new medium means a new culture, and often a new war (McLuhan and Fiore 1968). For McLuhan the usual "moralistic" approach to media matters was incapable of producing real insight into the working of media as potent cultural forms.

Works and Insights

His groundbreaking Understanding Media: The Extensions of Man (1964) was the first to examine the effects of technologies of communication on shaping the culture and sensibility of the users. Ralph Waldo Emerson (1803–1882) had observed, "The human body is the magazine of inventions, the patent-office, where are the models from which every hint was taken. All the tools and engines on earth are only extensions of its limbs and senses" (1870). This was a key to McLuhan's insight into human artifacts. McLuhan thus pioneered the study of the human senses as they are extended and modified by old and new media alike. The Gutenberg Galaxy (1962) details the impact of the printing press on late-medieval European sensibility and how it brought about the Renaissance. Later works traced the effects of electric technologies, beginning with the telegraph, in dissolving print culture and literacy and instituting a new kind of tribal mentality that extends worldwide. Although he approached the study of media by observation and analysis, the major criticism leveled at his work was that it was "not scientific."

In posthumous works such as Laws of Media: The New Science (with Eric McLuhan; 1988) and The Global Village (with Bruce R. Powers; 1989), McLuhan synthesized his major discoveries and identified four scientific laws that govern the action of all human artifacts: amplification, obsolescence, reversal, and retrieval. He explored how his work integrated and updated the work of Francis Bacon (Novem Organum) and Giambattista Vico (The New Science).

McLuhan had a facility for aphorism, encapsulating a complex process in a memorable phrase such as "The medium is the message." He went to great lengths to point out that each medium, independent of the content it mediates, has its own intrinsic effects that are its unique message.

The message of any medium or technology is the change of scale or pace or pattern that it introduces into human affairs. The railway did not introduce movement or transportation or wheel or road into human society, but it accelerated and enlarged the scale of previous human functions, creating totally new kinds of cities and new kinds of work and leisure. This happened whether the railway functioned in a tropical or northern environment, and is quite independent of the freight or content of the railway medium (McLuhan 1964, p. 8).

What he writes about the railroad applies with equal validity to the media of print, television, computers, and now the Internet. "The medium is the message" because it is the "medium that shapes and controls the scale and form of human association and action" (p. 9).

Another McLuhan term that has entered common usage is "the global village." In *Understanding Media* he wrote, "since the inception of the telegraph and radio, the globe has contracted, spatially, into a single large village. Tribalism is our only resource since the electromagnetic discovery. Moving from print to electronic media we have given up an eye for an ear" (pp. xii–xiii). The "global village," which many now see forming as a result of the Internet, was a side effect of the telegraph and of radio.

Influences On and From

McLuhan's work absorbed influences from prior work on the social and cultural impact of communications technology by Harold Innis (1894–1952) and others in the arts. In integrating and extending such perspectives, McLuhan created a distinctive approach to media studies often erroneously described as emphasizing a kind of technological determinism with rhetorical excess. In reality, however, McLuhan was simply pointing out how certain technologies influence the world so that their users could learn to control them.

After a decline in reputation during his later years and soon after his death, McLuhan was rediscovered in the 1990s, and his insights into media found new application in interpreting twenty-first-century global communications developments. Among those who have taken up the study of technologies and culture, McLuhan offers one of the more comprehensive and consistent explanations for the welter of changes that accompany science and technology—changes that include new challenges for ethics and politics. Although some scholars continue to dismiss him as a maverick, he has been welcomed by pioneers in digital communications

Marshall McLuhan, 1911–1980. A Canadian professor of literature and culture, McLuhan developed a theory of media and human development claiming that "the medium is the message." (© Bettmann/Corbis.)

such as those associated with *Wired* magazine (founded 1993). Moreover, philosopher and media theorist Paul Levinson (1997) has drawn connections between McLuhan and the evolutionary epistemologies of Karl Popper (1902–1994) and Donald T. Campbell (1916–1996), both of which have ethical dimensions.

ERIC McLUHAN

SEE ALSO *Internet; Science, Technology, and Society Studies; Television.*

BIBLIOGRAPHY

Carpenter, Edmund, and Marshall McLuhan, eds. (1960). *Explorations in Communication: An Anthology.* Boston: Beacon Press.

Gordon, W. Terrence. (1997). *McLuhan for Beginners.* New York: Writers and Readers Publishing.

Levinson, Paul. (1997). *The Soft Edge: A Natural History and Future of the Information Revolution.* New York: Routledge.

McLuhan, Marshall. (1962). *The Gutenberg Galaxy: The Making of Typographic Man.* Toronto: University of Toronto Press.

McLuhan, Marshall. (1964). *Understanding Media: The Extensions of Man.* New York: McGraw-Hill.

McLuhan, Marshall. (1967). *Verbi-Voco-Visual Explorations.* New York: Something Else Press.

McLuhan, Marshall. (1995). *Essential McLuhan,* ed. Frank Zingrone and Eric McLuhan. Concord, Ontario: House of Anansi Press.

McLuhan, Marshall, and Barrington Nevitt. (1972). *Take Today: The Executive as Drop-Out.* Don Mills, Ontario: Longman Canada.

McLuhan, Marshall, and Bruce R. Powers. (1989). *The Global Village: Transformations in World Life and Media in the Twenty-First Century.* New York: Oxford University Press.

McLuhan, Marshall, and David Carson. (2003). *The Book of Probes,* ed. Eric McLuhan and William Kuhns. Corte Madera, CA: Gingko Press.

McLuhan, Marshall, and Eric McLuhan. (1988). *Laws of Media: The New Science.* Toronto: University of Toronto Press.

McLuhan, Marshall, and Quentin Fiore. (1967). *The Medium Is the Massage.* New York: Random House.

McLuhan, Marshall, and Quentin Fiore. (1968). *War and Peace in the Global Village: An Inventory of Some of the Current Spastic Situations that Could Be Eliminated by More Feedforward.* New York: McGraw-Hill.

Sanderson, George, and Frank Macdonald, eds. (1989). *Marshall McLuhan: The Man and His Message.* Golden, CO: Fulcrum.

MEAD, MARGARET

• • •

The most celebrated anthropologist of the twentieth century, Margaret Mead (1901–1978) was born in Philadelphia, Pennsylvania on December 16, and died in New York City on November 15. Her career began with a shift from psychology when Ruth Benedict (1887–1948) and Franz Boas (1858–1942), two of her teachers at Columbia, attracted her with Benedict's challenge that they had "nothing to offer but an opportunity to do work that matters." Bridging these two fields, Mead became a founder of the culture and personality school of anthropology; she was deeply committed to making anthropological knowledge matter—especially in a world of rapid scientific and technological change.

Mead's career took off when she went to Samoa at age twenty-three to study adolescent girls and to explore whether the emotional strains of adolescence were uniform across cultures or varied depending on socialization and experience. This led to her first book, *Coming of Age in Samoa* (1928), a bestseller that gave many readers their first awareness that their assumptions about human behavior might not always apply. Although this book was caricatured and attacked by the anthropologist Derek Freeman in 1983, twenty years of debate has affirmed her descriptions, showing that Freeman's insistence on the biological determination of variations observed fifty years after Mead's work in other areas of Samoa supplemented but could not refute Mead's basic emphasis on learned—and therefore potentially variable—behavior.

Mead's subsequent fieldwork up until World War II took her to four different New Guinea societies and to the Omaha tribe of Nebraska with her second husband, Reo Fortune, and then to Bali and another New Guinea society, the Iatmul, with her third husband, the anthropologist and ecological thinker Gregory Bateson. During this period, she focused primarily on child rearing and personality development and secondarily on gender differences, where she pioneered the comparative study of gender roles. Her work appeared both in further trade books such as *Sex and Temperament in Three Primitive Societies* (1935) and in detailed technical monographs such as *The Mountain Arapesh* (published in three parts, 1938–1949), establishing the pattern of applying her findings in the field to the dilemmas of industrialized society, and writing in several genres for different audiences. She also innovated in methodology, beginning the use of projective tests in fieldwork and, with Bateson, invented a new technique of visual anthropology exemplified in *Balinese Character* (1942). Her fieldwork archives are available at the Library of Congress.

World War II led Mead and other social scientists to focus on industrialized nations as part of the war effort. Mead collaborated with Benedict in developing the application of anthropology to contemporary cultures made inaccessible by war and political conflict, primarily through the Columbia University Research in Contemporary Cultures project. This methodology, described in *The Study of Culture at a Distance* (1953), which led to multiple publications by many authors, involved the creation of interdisciplinary and intercultural teams not unlike contemporary focus groups, and the analysis of literary and artistic materials in ways that anticipated contemporary cultural studies. Mead founded the Institute for Intercultural Studies in New York in 1944 to house these projects and a variety of later activities.

The war had precipitated rapid and often devastating culture change, and Mead's postwar focus was on change, particularly the possibilities of purposive culture change. In 1953 she returned to Pere, a Manus village in the Admiralty Islands (now part of Papua New

Guinea) she had studied with Fortune, to analyze the effects of the war on a community with little previous outside contact. In Manus, she found that a charismatic leader had promoted the choice of integration into the outside world and the villagers were positive about change rather than demoralized by it; that rapid change is sometimes preferable to gradual change; and that children could play a key transformative role (Mead 1956). Mead was one of those who introduced the concept of "culture" into the thinking of readers, with profound intellectual and ethical results, but her emphasis on purposive culture change reaffirmed ethical issues avoided by some cultural relativists, and she insisted that many human institutions, such as those of warfare and racism, be seen as human inventions that could be modified or replaced, rather than as "natural" and unavoidable. Her understanding of the role of individuals and groups in the remaking of Manus society was key to her book *Continuities in Cultural Evolution* (1964), best summarized in her often quoted phrase, "Never doubt that a small group of thoughtful committed citizens can change the world."

Mead believed that the understanding of cultural diversity offered a new kind of freedom to human societies, and she worked tirelessly and skillfully to disseminate anthropological ideas, lectured widely, published profusely, and was quick to understand the possibilities of new media. Unlike many academics, she saw communicating to the public as a professional obligation of comparable intellectual integrity to her more narrow professional writing. She also taught for many years at Columbia University and the New School for Social Research. At the same time, Mead worked with colleagues in other fields who kept her close to new developments in biology and neurology. She was an active member of the Macy Conferences on Cybernetics and on Group Process in the postwar period and of the World Federation for Mental Health. She was associated for more than fifty years with the American Museum of Natural History, serving in her later years as its Curator of Ethnology. She served as president of the American Association for the Advancement of Science and the American Anthropological Association, and was a founder of the Scientists' Institute for Public Information. She received twenty-eight honorary degrees, more than forty academic and scientific awards, and was awarded the Presidential Medal of Freedom following her death in 1978.

MARY CATHERINE BATESON

SEE ALSO *Cultural Lag; Modernization.*

Margaret Mead, 1901–1978. An American anthropologist, Mead developed the field of culture and personality research and was a dominant influence in introducing the concept of culture into education, medicine, and public policy. *(AP/Wide World Photos.)*

BIBLIOGRAPHY

Banner, Lois W. (2003). *Intertwined Lives: Margaret Mead, Ruth Benedict, and Their Circle.* New York: Knopf. A scholarly examination of Mead's early personal and professional relationships.

Bateson, Mary Catherine. (1984). *With a Daughter's Eye: A Memoir of Margaret Mead and Gregory Bateson.* New York: Harper Collins.

Côté, James E. (1994). *Adolescent Storm and Stress: An Evaluation of the Mead/Freeman Controversy.* Hillsdale, NJ: L. Erlbaum. One of many scholarly refutations of the Freeman attacks.

Freeman, Derek (1983). *Margaret Mead and Samoa: The Making and Unmaking of an Anthropological Myth.* Cambridge, MA: Harvard University Press.

Mead, Margaret. (1928 [2001]). *Coming of Age in Samoa: A Psychological Study of Primitive Youth for Western Civilization.* New York: Perennial Classics.

Mead, Margaret. (1934 [2001]). *Kinship in the Admiralty Islands.* New Brunswick, NJ: Transaction Publishers.

Mead, Margaret. (1935 [2001]). *Sex and Temperament in Three Primitive Societies.* New York: HarperCollins.

Mead, Margaret. (1938 [2002]). *The Mountain Arapesh,* 2 Vol. New Brunswick, NJ: Transaction Publishers.

Mead, Margaret. (1942 [2000]). *And Keep Your Powder Dry: An Anthropologist Looks at America.* New York: Berghahn Books. Written as a contribution to the war effort.

Mead, Margaret. (1956 [2001]). *New Lives for Old: Cultural Transformation—Manus, 1928–1953.* New York: Perennial.

Mead, Margaret. (1964 [1999]). *Continuities in Cultural Evolution.* New Brunswick, NJ: Transaction Publishers. Includes Mead's theoretical discussion of the role of small groups in cultural change.

Mead, Margaret. (1972 [1995]). *Blackberry Winter: My Earlier Years.* New York: Kodansha. Mead's partial autobiography.

Mead, Margaret. (2004). *The World Ahead: An Anthropologist Anticipates the Future,* ed. Robert B. Textor. New York: Berghahn Books. A selection of Mead's writings about the future.

Mead, Margaret, and Gregory Bateson. (1942). *Balinese Character: A Photographic Analysis.* New York: New York Academy of Sciences.

Mead, Margaret, and Rhoda Metraux, eds. (1953 [2000]). *The Study of Culture at a Distance.* New York: Berghahn Books. A manual developed from research in contemporary cultures.

MEDICAL ETHICS

• • •

Medical ethics is the most prominent branch of the broader field of bioethics. In general, medical ethics concerns itself with issues arising in the relationship between a health care professional, primarily a physician, and a specific patient. To a lesser extent medical ethics is concerned with issues of justice and equity in the delivery of and access to medical care.

Three sets of issues have dominated the discussion of medical ethics as a discipline since the 1960s. Each of these sets of questions has been decisively influenced by the development of modern medical science and technology. In fact, it can be argued that if not for the advances in medical technology between the early decades of the nineteenth century and the first half of the twentieth century, medical ethics as the discipline it currently is simply would not exist.

Doctor and Patient

The first set of issues of decisive interest in medical ethics are those having to do directly with the relationship between the physician (or other professional) and the patient. The most important of these concerns is that having to do with the informed consent of the patient to medical interventions. In the discussion of medical ethics since World War II the principle of informed consent has achieved universal, canonical status. One may not provide any care to otherwise competent patients without first explaining the situation and the options and securing patient agreement to proceed. This principle was first enshrined in medicine after World War II when the abuses of Nazi doctors in so called "experiments" came to light. The Nuremberg Code, formulated at the famous war crimes trial, enunciated the principle for researchers clearly: *The voluntary consent of the human subject is essential.* Later, in the 1960s, when it was discovered that some American physicians were ignoring this principle, renewed emphasis was placed on it in law and medicine.

The very emergence of this bedrock principle has been decisively shaped by technology. Prior to the twentieth century little could be done to actually treat most forms of illness and disease. What could be done required the active involvement of patients both in telling the doctor their symptoms, and their stories (travel, diet, lifestyle, etc.), and in following a therapeutic regimen of diet, rest, fluids, or other recommendations. An unconscious or unwilling patient would not reveal much nor cooperate in therapy. Pedro Laín Entralgo (1908–2001), the great medical historian of the premodern period, aptly called Greco-Roman medicine the "therapy of the word" in which the spoken word was crucial to diagnosis and treatment.

Consider now a patient who is brought unconscious into a twenty-first-century emergency room. The stethoscope can alert the physician to heart or lung problems, and scanning technology can reveal the presence of various brain injuries such as blood clots or strokes. Further scannings and laboratory techniques can reveal the precise source of problems from heart infections and heart attacks to pneumonia or drug abuse. Broken ankles and sprained ankles can be differentiated with technology, as can warts and melanoma.

Treatment can likewise be provided even if the patient refuses. Surgery, which is dominated by technologies, is performed on an anesthetized patient, not a conscious one. Prisoners can be treated for infectious diseases whether they wish it or not in the name of prison safety. Intravenous medication and hydration can be provided to the unconscious or the unwilling. Thus, the modern insistence on informed consent as a moral principle makes sense only in a world in which technology has decisively objectified the patient in the physician's hands and made possible medical care without patient involvement.

The same may be said of the importance of patient competency in contemporary medical care. The concept of competency is highly complex. The general idea is that in order for informed consent to be required patients must be capable of comprehending their medical situation and making choices about it. But competency is crucial only if one can, with technology, offer plausibly beneficial therapy to patients who are not competent. Competency and the associated questions regarding who should make decisions when patients cannot (for example, physicians, families, courts, committees) becomes a serious issue only when treatment is possible without the interpersonal word passing between doctor and patient. When therapies of the word are the only therapies possible, then any therapy presumes that the patient is plausibly competent. It is only when therapies of impersonal technology have surpassed therapies of the word that competency becomes an essential focus.

Technology has also profoundly altered the context in which one of the oldest principles of medical ethics, confidentiality, is viewed and defended. Though enshrined as early as the fourth century B.C.E. in the celebrated Hippocratic oath, and in the latest code from the American Medical Association, this concept has been decisively pressured if not altered by modern technology in three important ways. First, the early-twentieth-century growth of complicated technology such as clinical laboratories and X-ray machines caused a centralization of medical services in the modern hospital. Doctor's offices became appendages of the hospital often physically connected by tunnels or walkways. Records once kept confidential in a physician's office became centralized in the hospital and available for many more to see.

Second, technology led to increasing specialization both in medicine itself and in allied fields such as nursing, laboratory technology, physical and respiratory therapy, and more. Each of these specialists, from cardiac surgeons to cardiac rehabilitation technologists, has a legitimate need for access to a medical record both to document their care and to see what other care has been given. Thus, gone are the days of a specific private communication between two and only two persons: physician and patient. Now anonymous lab technologists who have just met a patient will know, and arguably need to know, that the patient from whom they are drawing blood has, for example, a bloodborne disease such as AIDS.

Third, advanced information technologies have become a standard way for storing information. They provide the most efficient means of data storage and retrieval both in hospital and out. The idea is that if a patient is brought to an emergency room thousands of miles from home the emergency room staff can have nearly instantaneous access to a patient's medical history, which they need to know to provide adequate care. But the very promise of easy access to sensitive information for professionals also suggests easy access for those with no need to know: reporters, hackers, angry relatives, titillated billing clerks, and nosy neighbors.

Life and Death

The second great set of issues in medical ethics are those having to do with the beginning and ending of life: abortion and the variety of issues dealing with euthanasia. Abortion has been an issue within medicine since Greco-Roman times. The Greek physician Soranus (second century C.E.), author of the first gynecological textbook, describes methods of producing abortion and then proceeds to criticize abortion for "cosmetic" reasons (for what he regarded as reasons of personal comfort or vanity, such as the fact that pregnancy altered one's looks or figure).

Though abortion has been a staple of medical ethics since, modern science and technology have decisively altered debates about the morality of abortion. It is often said that "science" believes or has "proven" that life starts at conception. Though technically correct now, the beginning of life at conception was discovered only in the early nineteenth century. Furthermore, contrary to the wishes of those who often make this claim, the claim itself does not lead to a moral conclusion unless one adds a moral principle such as "all human life of whatever sort should be preserved." Whether such a principle is sound is widely debated, but something like it must be added to the embryological claim to lead to a moral conclusion about abortion.

Furthermore, some of the most contested issues about selective abortion and partial-birth abortion exist only because of advances in medical technology. It was only in the late 1960s that the first process of prenatal diagnosis, amniocentesis, was developed to diagnose fetuses with chromosomal abnormalities such as trisomy 21 (Down syndrome), trisomy 18 or 13, fragile X syndrome, or broken chromosomes. In the early 1980s sonography (ultrasound) and blood screening technology advanced to the point that it could reliably diagnose in utero the second most common birth defect, spina bifida. In the future scientists hope to move beyond analysis of chromosomal abnormalities to genotyping of specific genetic abnormalities such as those that cause a variety of ills from blindness and Huntington's chorea to vaguer conditions such as tendencies to substance abuse and depression. To a limited extent this is already done in fertility clinics with preimplantation genetic diagnosis. Whatever the outcome, the issue of abortion for reasons of parental deselection of undesirable characteristics would not exist except for the technology that allows for the identification of such characteristics and the safe abortion of second-trimester fetuses or the existence of fertility clinics.

The same influence of technology is evident in the much contested situation of "partial-birth" abortions and/or very late term abortions. It is only because of advanced medical technology that late-term abortions are relatively safe, so the morality of taking the life of those fast approaching birth becomes an issue. Before the relatively recent past, abortion of any sort was preformed only infrequently because it was simply medically too dangerous for the woman.

The second cluster of life and death issues, those having to do with end-of-life care, have been even more decisively shaped by technological change. The first of these issues, that concerning the concept of death, would not exist but for the advancement of technology. Before the middle decades of the twentieth century the legal and moral definition of death was simple: complete and irreversible cessation of vital signs, specifically heartbeat and respiration. Physicians routinely called a person dead when the vital signs had ceased for a period of time that made them irreversible. In the 1950s technology decisively altered this framework. Respirator technology could pump air into a patient's lungs, forcing them out and weakly pumping blood to the heart and the body. Vital signs might never stop, and even the permanently unconscious might never "die." Technology seemed to promise longevity even to those whose conscious life had ended.

In this context, medicine and society were compelled to develop new understandings of death. Thus came the well-known concept of "brain death" in which persons could be considered brain dead if certain brain activities had ceased, even though other vital signs were artificially maintained. A debate has followed over different conceptions of brain death—centering on a cautionary "whole brain" formulation versus a broader "higher brain" formulation—but the important point here is that such a debate would not exist were it not for respirators, feeding tubes, and intravenous hydration and antibiotics that allow persistently unconscious human bodies to be kept alive indefinitely.

The same technological revolution brought out the importance of many other issues surrounding end-of-life care. How aggressive of an approach should be taken in keeping individuals alive who are gravely or terminally ill or severely brain damaged? The question of whether to go to extraordinary lengths to keep persons alive with advanced Alzheimer's disease or other brain deterioration is different from whether a doctor can just declare them dead. These questions have become crucial questions of end-of-life care. They become questions, however, only if there is a possibility of aggressive treatment of those who are gravely ill or severely handicapped. The morality of prolonging the life of the critically ill with technology becomes an issue only when the technology exists, such as respirators or dialysis machines, that will help preserve life.

A similar question involves when to resuscitate or not to resuscitate a patient who goes into cardiac arrest. Of course, one resuscitates in the emergency room and in cases of simple cardiac arrest in otherwise healthy persons. Furthermore, if patients have stated their wishes to be resuscitated, one honors them. But most hospitalized patients have never let their wishes be known. Should gravely ill persons in the intensive care unit routinely be revived even though data shows that such patients have very poor outcomes? The issue is widely debated, but the debate follows only from the existence of resuscitation technology such as defibrillators and heart-stimulating drugs.

So also does the agonizing debate since the 1970s about treatment for critically ill newborns follow from the advance of technology. Critically ill newborns may be saved with extensive interventions. But they may be left with severe handicaps as a result of many deficits. Parents and physicians are left with serious questions about when to intervene to save the life of such infants. Questions about the sanctity of all life, the quality of life, and when if ever life itself is not worth living swirl around these cases. Agonizing moral and legal debates both at the individual and policy levels have been involved. The debates, however, follow only from the dramatic advances in medical technologies that allow evermore fragile newborns to be saved.

Though issues of euthanasia and physician-assisted suicide have been around for millennia, as witnessed by the condemnation of euthanasia in the Hippocratic oath, they have taken decisive new turns in the modern period with the development of pharmacological means of causing death relatively painlessly and with a high degree of certainty. When suicide was limited to guns, knives, and poisons, the pain of the act was a deterrent. But when an injection of morphine and potassium chloride from a physician will end life quickly and painlessly, the issue takes on new dimensions. The same is true of the contested issues of physician-assisted suicide where doctors provide the means and patients take the action. This is hardly an issue when a person can buy a gun or rat poison at a hardware store. But with the advent of modern pharmacology, physicians can provide their terminally ill patients with strong painkillers such as Demerol and verbal instructions to enhance the power and speed with whiskey. Patients will then go unconscious and die without much suffering. Technology enhances the question: Should doctors ever do this?

Justice and Distribution

The third and final set of issues that has dominated the field of medical ethics in the last generation has been those related to access to and distribution of health care. One subset of issues here has to do with access to scarce lifesaving technology. In the early 1960s it was the development of dialysis, in the early twenty-first century it is organs for transplant. In the future it is likely to be new genetic technologies. Technologies change but issues of equitable access continue.

Basically there have been two broad contenders: (a) a merit-based selection or deselection scheme or (b) some form of randomization. Merit schemes are intuitively appealing but notoriously difficult to practice. Who is not moved by the plight of a mother of young children who needs a liver? Better she get it than a fifty-year-old who has grown children. Or who is not adversely affected by the thought of giving a liver transplant to someone, even as famous as the New York Yankees baseball star Mickey Mantle, who needed a new liver because drinking destroyed his original one? Though appealing, criteria of social worth are notoriously slippery. Perhaps Mantle stopped drinking years ago. Is he now to be thought of as less meritorious because of what he did as a younger person? Perhaps the fifty-year-old has a handicapped grandchild and her child care is much needed. Once carefully thought through, it seems that most people have merits and demerits in their lives. No one is so stellar that their case for new organs or other technologies shines clearly above the rest. Nor is anyone so completely unworthy that they can make no reasonable claim on a scarce medical resource.

Such considerations have led many to support some kind of randomization as a means of selection. The most common, especially in the case of transplants, is first come first served. In the case of transplants, patients are first screened medically to see if they are candidates for surgery—for example: Do they need a transplant? Could they survive such major surgery? And so on. Then they are broadly ranked according to medical need: How soon would they die without surgery? Finally, they wait their turn. When an organ becomes available that is tissue compatible, the person at the top of the list goes first. Though common this is not the only random method discussed in the literature. For example, though not often used, a lottery would be just as random and may have other advantages such as giving every needy person an equal opportunity to be served.

Finally the discussion of medical ethics has focused intensely on the question of whether there is a "right to health care" and if so how best to provide access to health care to those without it. It is now widely held that a society should provide basic medical care to all. Once this is granted two problems remain. First, how should the range of services to be provided be determined? Should services for some or all citizens be cut to free resources for those who do not have access? Plastic surgery might be an obvious cut, but what about expensive surgery that has very limited chances of success, such as treatment for some forms of cancer? For the person who needs the treatment as their only hope of survival, the question is answered one way. For the rest of society trying to find resources to provide prenatal care for poor women, the question might be answered differently.

Though this problem is difficult, a second sort of discussion has centered around how to provide access to basic services. Two broad approaches have dominated the discussion. The first is a government-run system in which doctors are paid by the government and tax revenues are used to provide health care for everyone. The second is to use tax revenues to move those without care into private health care plans. Each approach has its own difficulties. Government-run plans are often overused for minor problems and can result in long waiting lines for needed care. Private insurers can become bankrupt when they enroll too many sick persons at low rates. The problems are only compounded by the development of new technologies that increase the cost of health care in general.

For present purposes the most important points concern questions of access that follow advances in medical technology. One cannot talk about rationing access to dialysis or organ transplants until there are dialysis machines or transplant capabilities. Like other issues, this one too has been decisively shaped by the advances of medical technology.

Assessment

Medical ethics is representative of a larger field of professional and applied ethics in two important ways. First medical ethics involves the application of generally recognized principles in specific social, economic, and cultural settings. All cultures place a very high value on human life. But how that is balanced against quality of life and the use of scarce resources may vary in different settings. In a wealthy country such as the United States keeping someone alive at great expense may look very different than the use of scarce resources on a single life may look in a poor country with many public health needs. High technology may be afforded in one country but where even low technology is socially expensive the choices are much different.

Secondly, medical ethics combines both universal moral principles such as honesty and integrity with intra-profession principles or norms that are unique to that profession. Empathy is a highly valued virtue in medicine and less so in other professions such as engineering. Empathy is also a decisive virtue in modern times when technology can so easily separate doctor and patient. At other times such a virtue may require less effort.

Medical ethics, like medicine itself, has been profoundly shaped by modern science and technology. Without technology, the moral choices will look very different. However, without guidance from general principles such as respect for life and liberty technology may challenge the profession in uncharted ways.

RICHARD SHERLOCK

SEE ALSO *Abortion; Acupuncture; Aging and Regenerative Medicine; Bioengineering Ethics; Bioethics; Brain Death; Cancer; Complementary and Alternative Medicine; Death and Dying; DES (Diethylstilbestrol) Children; Drugs; Embryonic Stem Cells; Emergent Infectious Diseases; Genethics; Health and Disease; HIV/AIDS; Persistent Vegetative State.*

BIBLIOGRAPHY

Battin, Margaret P.; Rosamond Rhodes; and Anita Silvers, eds. (1998). *Physician Assisted Suicide.* New York: Routledge.

Bayertz, Kurt, ed. (1996). *Sanctity of Life and Human Dignity.* Dordrecht, Netherlands: Kluwer Academic.

Beauchamp, Tom L., and Robert M. Veatch, eds. (1996). *Ethical Issues in Death and Dying,* 2nd edition. Upper Saddle River, NJ: Prentice Hall.

Berg, Jessica W.; Paul S. Appelbaum; Charles W. Lidz; and Lisa S. Parker. (2001). *Informed Consent: Legal Theory and Clinical Practice,* 2nd edition. New York: Oxford University Press.

Brody, Baruch A., ed. (1989). *Suicide and Euthanasia.* Dordrecht, Netherlands: Kluwer Academic.

Buchanan, Allen E., and Dan W. Brock. (1989). *Deciding for Others: The Ethics of Surrogate Decisionmaking.* New York: Cambridge University Press.

Daniels, Norman. (1985). *Just Health Care.* New York: Cambridge University Press.

Daniels, Norman, and James E. Sabin. (2002). *Setting Limits Fairly: Can We Learn to Share Medical Resources?* New York: Oxford University Press.

Faden, Ruth R.; Tom L. Beauchamp; and Nancy M. P. King. (1986). *A History and Theory of Informed Consent.* New York: Oxford University Press.

Hursthouse, Rosalind. (1987). *Beginning Lives.* Oxford: Basil Blackwell.

Kamm, F. M. (1992). *Creation and Abortion.* New York: Oxford University Press.

Ramsey, Paul. (1978). *Ethics at the Edges of Life.* New Haven, CT: Yale University Press.

Reiser, Stanley Joel. (1978). *Medicine and the Reign of Technology.* New York: Cambridge University Press.

Steinbock, Bonnie, ed. (2002). *Legal and Ethical Issues in Human Reproduction.* Aldershot, UK: Ashgate/Dartmouth.

Weir, Robert F. (1989). *Abating Treatment with Critically Ill Patients.* New York: Oxford University Press.

MERTON, ROBERT

• • •

American sociologist considered to be the father of the sociology of science, Robert King Merton (1910–2003) was born in Philadelphia, Pennsylvania, on July 4, and died in New York City on February 23. His scholarly career spanned more than seven decades. Merton's contribution to ethics in science and technology was his elaboration of the social, and human, nature of scientific research.

After undergraduate study at Temple University, Merton attended Harvard University. He began his doctoral thesis in 1933 and completed it two years later with the title "Sociological Aspects of Scientific Development in Seventeenth Century England." In 1938 Merton's revised thesis was published in *Osiris: Studies on the History and Philosophy of Science, and on the History of Learning and Culture* as *Science, Technology and Society in Seventeenth Century England (STS).* In this work, Merton explored the reciprocal relationships between the development of science and the religious beliefs associated with Puritanism. He concluded that cultural attributes, religious beliefs, and economic influences made it possible for science and its technical applications to flourish.

Merton later indicated that when *STS* was first published, it was generally ignored by sociologists (see Cohen 1990 and Chapter 20 by I. Bernard Cohen in Clark, Modgil, and Modgil 1990). More than three decades later, Merton's *STS* was published by a commercial publisher. By then, his reputation in sociology generally and in the sociology of science particularly was so broad that *STS* was widely studied and was considered a classic. It was both criticized and praised by historians, sociologists, and others.

After completing his doctorate, Merton taught at Harvard and published his most famous paper, "Social Structure and Anomie" (see Stephen Cole in Coser 1975). Merton's theory asserted that in the United States, people are taught to pursue the goal of economic success regardless of their location in the social structure. Yet the means to achieve success are not always available, resulting in a social condition conducive to deviant behavior.

After Harvard, Merton taught for two years at Tulane University. In 1941 he was invited to join the faculty at Columbia University; he remained affiliated with that university for the rest of his career. Soon after joining the faculty, he began to serve as associate director of the Bureau of Applied Social Research.

Merton published several articles from his thesis analyzing the social contexts of scientific advancement. In 1942, he described the normative structure of science in "Science and Technology in a Democratic Order" (reprinted in Merton 1973). He explains how the social institution of science involves a normative structure that works to support the goal of science—the extension of certified knowledge. Modern science has at least four norms or behavioral constraints that constitute its unique ethos.

Organized skepticism requires that any claim to new knowledge stand up to the same scrutiny, regardless of its source, before it becomes part of the accepted body of certified knowledge. *Universalism* requires that age, sex, race, or creed should not influence a decision about the acceptance or rejection of scientific information. Only the logical structure of the argument and the quality of the data are relevant. *Communism* (or *communality*) requires that once scientific information has been created or discovered and made public, the originator has no future intellectual claims to it. All scientists are free to use it in their work (with appropriate attribution). *Disinterestedness* requires scientists to be motivated to extend knowledge, not to seek personal gain.

This 1942 paper had a passing reference to a remark by Sir Isaac Newton stating, in effect, that if he had seen

Robert Merton, 1910–2003. Merton was a sociologist, educator, and internationally regarded academic statesman for sociology in contemporary research and social policy. He is considered the founder of the sociology of science. (*Archive Photos, Inc.*)

farther (in his work), it was by standing on the shoulders of giants. In the two decades that followed, Merton traced backward (and forward) the twelfth century origins of that phrase. *On the Shoulders of Giants* (1965) became a classic for its bibliographic erudition and style, and is recognized as a literary masterpiece.

During the 1940s and 1950s, the Bureau of Applied Social Research provided unusual opportunities to collect data and conduct sociological analyses, and Merton developed a large body of theory that established his sociological talents. His new ways of seeing social realities invaded popular and official language. His work included such concepts as manifest and latent functions, self-fulfilling prophecy, goal displacement, local and cosmopolitan influentials, accumulation of advantage, the Matthew effect, theories of the middle range, sociological ambivalence, and obliteration by incorporation (Clark et al. 1990).

For two decades after Merton's 1938 contribution to the historical sociology of science, research by others in the sociology of science was largely dormant. In 1952, Merton explained why social aspects of science would be neglected by sociologists (Merton 1973). Most sociological research focuses on social problems such as deterioration of the family, political unrest, urban congestion, race relations, the media, and so on. Consequently, until either scientific knowledge or science as an institution is defined as a problem for society, scholarly investigators likely would not select science as the subject of social analysis.

In 1957, Merton's American Sociological Association presidential paper "Priorities in Scientific Discovery" continued his exploration of the developing sociology of science (reprinted in Merton 1973). That paper eventually became the most cited publication in the sociology of science (see Cole and Zuckerman's chapter in Coser 1975). It was full of ideas for further research, and provided a broad foundation for a growing interest in the sociology of science. During the 1970s, as science became to be perceived as a social problem, the number of scholars specializing in the sociology of science increased much faster than the growth of the field of sociology in general.

By the 1980s, Merton's influence was evident in the United States and in Europe. Colleges established courses and degree programs, and research centers focusing on social studies of science were created. Sociologists successfully organized specialty scholarly groups nationally and internationally. Although Merton was recruited to organize these societies, he mostly encouraged others and provided moral support.

During the last twenty years of the twentieth century, many competing ideas about the social nature of science developed. Controversies flourished about the foci of inquiries, research methodologies, and the validity of Merton's and other theories. These issues were debated internationally among historians, philosophers, sociologists, and others.

The Mertonian view of science based on the institution's normative structure was criticized as empirically invalid, especially by scholars outside sociology. Because social norms are not absolute, and compliance is rarely total, some deviance among community members is expected. Deviance among scientists, however, provided the basis for scholars to question Merton's perspective.

Merton was arguably the most influential sociologist in the twentieth century. Even scholars who did not see his scholarship as the final word on a subject nevertheless studied his work to create their own interpretations of the nature of society and the reciprocal relationships between science and society.

JERRY GASTON

SEE ALSO *Science, Technology, and Society Studies; Skepticism; Sociological Ethics.*

BIBLIOGRAPHY

Clark, Jon; Celia Modgil; and Sohan Modgil, eds. (1990). *Robert K. Merton: Consensus and Controversy.* London: Falmer Press. A critique of Merton's wide-ranging contributions across the entire field of sociology.

Cohen, I. Bernard, ed. (1990). *Puritanism and the Rise of Modern Science: The Merton Thesis.* New Brunswick, NJ: Rutgers University Press. An extensive discussion of the ideas developed in Merton's 1935 doctoral thesis.

Coser, Lewis A., ed. (1975). *The Idea of Social Structure: Papers in Honor of Robert K. Merton.* New York: Harcourt Brace Jovanich. Former students and colleagues discuss the breadth and depth of Merton's first forty years of scholarship and his career as teacher, collaborator, and colleague.

Merton, Robert K. (1965). *On the Shoulders of Giants: A Shandean Postscript.* New York: The Free Press. A literary exposition of the origin and persistence of the humble notion asserting that scientists are able to be creative because of the foundation laid by predecessors.

Merton, Robert K. (1970). *Science, Technology and Society in Seventeenth Century England.* New York: Howard Fertig. The 1935 doctoral thesis that first established comprehensively the sociological perspective of scientific development.

Merton, Robert K. (1973). *The Sociology of Science: Theoretical and Empirical Investigations.* Edited by Norman W. Storer. Chicago: University of Chicago Press. Major collection of Merton's papers on science as a social institution.

META-ANALYSIS

• • •

Meta-analysis is the quantitative review of the results of a number of individual studies in order to integrate their findings. The term (from the Greek *meta* meaning after) refers to analysis of the conclusions of the original analyses. The methodology can in principle be applied to quantitative studies in any area of investigation, but it has become a basic tool in healthcare research. It is part of the broader approach of *research synthesis*, which also includes qualitative aspects.

Evolution of Meta-Analysis

Gaining an overview of the outcomes of different experiments is the constant aim of science, and statisticians have been concerned with the combination of results since the emergence of formal statistical inference in the early twentieth century. The basic principles were established by the 1950s (Cochran 1954), and the need became clear with the subsequent rapid increase in research publications. The procedure was first developed in the social sciences, and the term meta-analysis introduced in the educational literature in 1976. The 1980s saw mounting interest in the combination of results of clinical trials, and since the early 1990s meta-analysis has experienced explosive growth in medical applications.

Although there seems little doubt that meta-analysis is here to stay, it has been fraught with controversy. There is the problem of the *quality* of individual studies, with their own biases, often small clinical trials with poor design and execution. There is the problem of *heterogeneity*, studies that measured different effects, used different populations, had different aims. A further problem is that of *publication bias*, the fact that studies with positive results are more likely to get published than those with negative outcomes, leading to an inflation of the effect estimate. Related to this is *Tower of Babel bias*, meaning that most meta-analyses identify only reports published in English.

An international conference on meta-analysis was held in Germany in 1994, to review problems and progress (Spitzer 1995). A strong opponent present called the method "statistical alchemy for the 21st century" (Feinstein 1995). But work has continued, with the development of guidelines for doing meta-analyses, emphasizing the need to identify unpublished studies, eliminate incomplete reports and those of flawed research designs, and include only quality studies that appear to address the same well-defined question. The gold standard is that of *Individual Patient Data* (IPD), where the original data are available for reanalysis in the combined context. *Cumulative meta-analysis* is the systematic updating of the analysis as new results become available. There is also extensive research on meta-analysis for observational studies.

The Cochrane Collaboration

An important, promising development is the vigorous Cochrane Collaboration, "an international nonprofit and independent organization, dedicated to making up-to-date, accurate information about the effects of health care readily available worldwide. It produces and disseminates systematic reviews of health care interventions and promotes the search for evidence in the form of clinical trials and other studies of interventions" (Cochrane Collaboration). The movement was inspired by Archibald Cochrane (1909–1988), the British epidemiologist best known for his 1972 work *Effectiveness and Efficiency: Random Reflections on Health Services*. Cochrane urged equitable provision of those modes of healthcare that had been shown effective in properly designed studies, preferably randomized clinical trials. He considered the latter among the most ethical forms of treatment, and he emphasized the need for systematic critical summaries, with periodic update by specialty, of all relevant randomized clinical trials.

The first Cochrane Center opened in the United Kingdom in 1992, followed by the founding of the Cochrane Collaboration in 1993. In November 2004 its web site listed twelve Cochrane centers worldwide (using six languages) that serve as reference centers for 192 nations and coordinate the work of thousands of investigators. The main output of the Cochrane Collaboration is the *Cochrane Library* (CLIB), published and updated quarterly by Wiley InterScience and available by subscription via the Internet and on CD-ROM. Its contents include the Cochrane Database of Systematic Reviews (CDSRs), over 3,000 reviews prepared by fifty Collaborative Review Groups (CRGs), the Cochrane Central Register of Controlled Trials, bibliographic data on hundreds of thousands of controlled trials, as well as methodologic information on the rapidly developing field of research synthesis, and critical assessment of systematic reviews carried out by others.

The Ethics of Evidence

Meta-analysis, an attempt to integrate the information already on hand from past studies, enhanced by guidelines that it be done on the highest professional level, fits into the framework of the *Ethics of Evidence*, a multidisciplinary approach proposed for dealing with the uncertainties of medicine (Miké 1999). The Ethics of Evidence calls for the development, dissemination, and use of the best possible evidence for decisions in healthcare. As a complementary precept, it points to the need to accept that there will always be uncertainty.

To explore the quality of evidence from meta-analyses, a 1997 study compared the results of twelve large randomized clinical trials (RCTs) published in four leading medical journals with the conclusions of nineteen previously published meta-analyses addressing the same questions, for a total of forty primary and secondary outcomes (LeLorier et al. 1997). The agreement between the meta-analyses and the subsequent large RCTs was only somewhat better than chance. A third of the meta-analyses failed to correctly predict the outcome of the RCTs, and would have led to adoption of an ineffective treatment or the rejection of a useful one. (The actual differences between effect estimates were not large, but that did not count in this adopt/reject type of analysis.) Then in 2002 the long-held belief that menopausal hormone replacement therapy offered protection against heart disease, a medical consensus supported by meta-analyses, was shockingly reversed by RCT evidence (Wenger 2003).

The Cochrane Collaboration, as a worldwide, integrated movement, has the great potential to promote cooperation on high-quality, controlled clinical trials. Systematic reviews of these, with regular update and dissemination, should help improve the evidence available for the practice of medicine. But it is important to keep in mind that even the best meta-analysis cannot take the place of original research. *Evidence-based medicine*, which makes heavy use of the results of meta-analyses, cannot apply evidence that does not exist. Scientists need to stay close to the primary literature, with an open mind, to get new ideas, seek new insights, and generate new hypotheses.

The public needs to have a cautious view of meta-analysis, judging each case in its proper context. For example, the meta-analysis showing that more than 100,000 Americans die each year from the side effects of legally prescribed drugs (Lazarou et al. 1998) merits serious concern, even if the estimate is not quite accurate. There is no substitute for being informed, getting involved, and taking personal responsibility.

VALERIE MIKÉ

SEE ALSO *Biostatistics; Statistics.*

BIBLIOGRAPHY

Bailar, John C., III. (1997). "The Promise and Problems of Meta-Analysis." *New England Journal of Medicine* 337: 559–560.

Cochran, William G. (1954). "The Combination of Estimates from Different Experiments." *Biometrics* 10: 101–129.

Cochrane, Archibald L. (1972). *Effectiveness and Efficiency: Random Reflections on Health Services.* London: Nuffield Provincial Hospitals Trust.

Feinstein, Alvan R. (1995). "Meta-Analysis: Statistical Alchemy for the 21st Century." *Journal of Clinical Epidemiology* 48(1): 71–86. Commentary by Alessandro Liberati, "A Plea for a More Balanced View of Meta-Analysis

and Systematic Overviews of the Effect of Health Care Interventions," is included.

Lazarou, Jason; Bruce H. Pomeranz; and Paul N. Corey. (1998). "Incidence of Adverse Drug Reactions in Hospitalized Patients: A Meta-Analysis of Prospective Studies." *Journal of the American Medical Association* 279(15): 1200–1205.

LeLorier, Jacques; Geneviève Grégoire; Abdeltif Benhaddad; et al. (1997). "Discrepancies between Meta-Analyses and Subsequent Large Randomized, Controlled Trials." *New England Journal of Medicine* 337: 536–542.

Miké, Valerie. (1999). "Outcomes Research and the Quality of Health Care: The Beacon of an Ethics of Evidence." *Evaluation & the Health Professions* 22: 3–32. Commentary by Edmund D. Pellegrino, "The Ethical Use of Evidence in Biomedicine," also appears in this issue, pp. 33–43.

Moher, David. (2001). "QUOROM." [Quality of Reporting of Meta-Analyses] In *Biostatistics in Clinical Trials*, eds. Carol Redmond and Theodore Colton. New York: John Wiley & Sons.

Spitzer, Walter O., ed. (1995). *The Potsdam International Consultation on Meta-Analysis*, special issue *Journal of Clinical Epidemiology* 48(1): 1–172.

Stroup, Donna F., and Stephan B. Thacker. (2000). "Meta-Analysis in Epidemiology." In *Encyclopedia of Epidemiologic Methods*, eds. Mitchell H. Gail and Jacques Benichou. New York: John Wiley & Sons.

Thompson, Simon G. (2001). "Meta-Analysis." In *Biostatistics in Clinical Trials*, eds. Carol Redmond, and Theodore Colton. New York: John Wiley & Sons.

Wenger, Nanette K. (2003). "Menopausal Hormone Therapy and Cardiovascular Protection: State of the Data 2003." *Journal of the American Medical Women's Association* 58: 236–239.

INTERNET RESOURCE

Cochrane Collaboration. Available at www.cochrane.org. Web site of the organization.

MILITARY ETHICS

• • •

Military ethics can mean a wide range of things. It can encompass all aspects of military conduct, from writing performance reviews on subordinates, to relations of military personnel with their civilian leaders, to issues related to war. For the purposes of this entry, however, the discussion will be limited to ethical questions concerning the use of military force for the redress of political disputes. As war becomes increasingly dominated by high technology weaponry (at least in the developed countries), there is an intimate link between developments in science and technology and the questions of appropriate military use of those advances as addressed by military ethics.

Fundamental Issues

Traditionally military ethics has emphasized an approach to just war thinking that has roots in classical and early-Christian sources. In post-Reformation and post-Enlightenment Europe, this ethical and religious tradition found secular and legal codification in the Laws of Armed Conflict (both in international law and in the specific military law of individual nations).

Traditional just war analysis attempts to specify the scope and limits of morally acceptable uses of military force. Two independent sets of judgments are involved. The first, *jus ad bellum* (justice/right *toward* war) considers whether the use of force under a given set of political circumstances is warranted at all. The second, *jus in bello* (justice/right *in* war) frames issues regarding the conduct of military forces in combat.

Jus ad bellum (whether to go to war) questions the extent to which the use of force is justified at all by posing a series of tests. These gauge whether there is a just cause for war, whether there exists a legitimate authority to authorize the use of force, whether there is proportionality in the damage likely to be caused by the use of force measured against the political stakes of the conflict, and whether possibly effective non-military means of resolving the conflict have been exhausted (*last resort*). There is also a *reasonable hope of success* criterion, intended to rule out pointless violence. Because, paraphrasing the great philosopher of war Carl von Clausewitz (1780–1831), war is *politics by another means*, it is important to see whether the desired political result is likely to be attainable. In addition, for a war to be justified, it must be waged for the sake of returning to a better state of peace and conducted with that intention.

It is important to note that although decisions about use of force at this level are clearly military ethics insofar as they are ethical decisions about the use of the military instrument of national power, they are not decisions that involve many military personnel. With the exception of the most senior military advisers to civilian authority, most individuals involved in this level of discussion are the civilian leadership of the nation.

The *jus in bello* (how to conduct war) considers whether care is being taken to be *discriminant* (i.e., to attack directly only military objects and to take precautions against destruction of civilian individuals or objects) and *proportional* (i.e., to expend only the amount of destructive force on a given target that is justified by its believed military importance). Unlike the global assessment of justification and proportionality made at the highest levels of government about whether or not to go to war, these decisions are made at all levels of combat, from the smallest tactical decisions of a rifle squad to the decisions of a theater commander regarding the structure and targets of a strategic bombing campaign.

While both the *jus ad bellum* and the *jus in bello* decisions belong to individual leaders in their official capacities at all levels, the broader society can and does also engage in ethical discourse about those decisions. Especially in a democratic society with abundant technologically mediated public sources of information, citizens as individuals, members of the press, opinion leaders, and so forth all make independent assessments of the ethical quality of the decisions of political and military leaders. Leaders must persuade their citizenry of the justifications for the use of military force in the world, and individual actions of the nation's military (sometimes down to the lowest tactical level) can and do become objects of national scrutiny and ethical assessment. The strength of the connection of the military to the democratic society it serves is decisively influenced by the degree to which the military and the society share a common moral frame of reference and a sufficiently robust common understanding of the realities of military affairs.

An emerging challenge in the area of military ethics and society is that, in large-scale democracies that eschew compulsory military service, fewer members of the society have any direct experience with the military—including pivotal opinion leaders and civilian political leaders. This creates the risk of a diminishing realistic sense of the scope and limits of the capabilities of military power in the society at large and a commensurate risk that the military will be challenged to explain its choices and actions to fellow citizens.

Military Ethics and Technology

Practical military ethics is intimately connected with the military technology available to combatants. Further changes in available technologies have profound ripple effects in the ethical assumptions and accepted ways of behaving of the military—often in ways wholly unanticipated when the technology was introduced and applied. In a phrase commonly attributed to Immanuel Kant, *Ought implies can*, meaning that it is pointless to say someone ought to do something unless the person is possessed of the capability to do it. But in areas of ethics and military technology, at least in some cases, capability calls for use or that *Can implies ought*. As technology makes it possible for military operations to be conducted in novel ways, especially insofar as these come closer to honoring the sprit and letter of the just war criteria, the requirement to do so becomes more stringent. Once acceptable weapons and tactics may, at least for militaries that possess new capabilities, be considered objectionable. With increasing precision in the targeting of air bombardment, it is difficult to imagine militaries possessed of that capability reverting to less precise weapons in any but the most dire of circumstances.

One important theme of the just war tradition is the attempt to make war as humane as possible, even for the combatants. This is manifest in the elaboration of the Geneva Convention rules requiring that combatants who surrender be entitled to *benevolent quarantine* by their captors, including medical care, adequate food and housing, and more. Underlying these rules is the sense that combatant is a temporary status overshadowing the more fundamental common humanity of adversaries. When combatant status is lifted, humanitarian concerns with the suffering and welfare of the individual reassert themselves.

Humanitarian concern, even toward combatants, in the tradition of military ethics is evidenced by periodic attempts to rule out whole classes of weapon technology as inherently inhumane. Such efforts began with medieval Christian church efforts to ban the crossbow as being too accurate and deadly over too long a range. Later the bans on asphyxiating gas weapons, blinding lasers, and hollow-point (so-called *dum-dum*) bullets, and attempts to ban nuclear weapons, all reflected an impulse to identify unethical classes of technology.

A review of these efforts, however, points up their largely ineffectual and erratic character. When each technology first emerged, it presented as a novel and horrific new weapon system. Some bans (most notably that on asphyxiating gas) have held as a matter of customary practice among civilized nations. But it is hard in almost every case to say precisely why certain weapons are uniquely horrific in comparison to other weapons systems developed and deployed later. The ban on gas, for example, may continue in part because of the depth of the historical memory of World War I and the unique horrors gas weapons caused in that conflict, but also because they are not especially effective weapons systems in comparison to alternatives developed later. It is hard to see, from any objective moral perspective,

how being shot with a hollow point bullet (deemed inhumane because of the gratuitous destruction of tissue caused by the tumbling bullet in contrast with the *clean penetration* of a rigid bullet) is less humane than being bombed with a fuel-air explosive that generates tremendous heat and overpressures, and kills by blast and by sucking oxygen out of the environment.

The link between military ethics and technology is not primarily in connections between specific technologies and guiding ethical principles. Specific, technology-by-technology restraints will always be piecemeal, sporadic, and difficult to justify or explain on the basis of a uniform set of moral principles. The connection between military ethics and technology is more subtle and complex. The development of air power is perhaps the clearest example, and worthy of extensive specific review, of the general issues in raised in this regard.

Between the two world wars, a number of air power thinkers developed a theory about the best strategic use of the airplane and bombing. Italian Giulio Douhet and American Billy Mitchell both speculated that long-range bombing would obviate the need for a frontline and trench warfare, both of which were required in World War I. Instead, they argued, the bomber would fly deep into enemy territory and bomb factories, transportation, and other infrastructure essential to the adversary's war effort. They also proposed (without always noting that this was quite another matter) bombing civilians and whole cities directly in the effort to so demoralize the population that the will to continue the war effort would collapse.

The latter proposal ignores, at the most fundamental level, the principle of discrimination that is a cornerstone of the *jus in bello* element of just war thinking. Before World War II, world leaders publicly declared that indiscriminate attacks on cities were completely outside the realm of military ethics and never to be ordered. The U.S. policy of so-called *daylight, precision* bombing was an attempt to maintain the principle of discrimination. Given the technology available at the time and the inherent inaccuracy of bombing from high altitude, it was an effort that had little practical meaning. At the end of the war, all pretense of discrimination was abandoned as the Allied air campaigns culminated in the *conventional* firebombing of Dresden and the atomic bombing of Hiroshima and Nagasaki.

One might have thought that the principle of discrimination in military ethics had effectively been rendered obsolete by this pattern of practice, but it was not. Nuclear weapons, and other weapons of mass destruction of the biological or chemical type would, if used, be impossible to justify under any reasonable interpretation of the just conduct of war. But on the more conventional side of war, the principles reasserted themselves after the end of World War II. The Vietnam era (1964–1975) practice of *free fire zones* in which it was declared that, after notice, all in a given area would be deemed combatants was at least a verbal and legalistic effort to maintain the distinction. More importantly, the introduction late in the Vietnam era of television-guided precision munitions hinted at a whole new connection between technology and military ethics just over the horizon.

In the opening hours of the air campaign of the Persian Gulf War in 1991, the world was introduced to a new manifestation of the link between technology and military ethics. The generation of precision guided munitions (PGMs) that was used held the prospect (only partially fulfilled in that conflict) of *one bomb, one target* accuracy, in which strategic bombing might be conducted even in urban areas with collateral damage to civilians limited to weapons malfunctions and intelligence failures in designating targets incorrectly.

Technologies have only continued to improve. PGMs requiring the risky and difficult laser designation by a pilot during the Persian Gulf War had, by the Kosovo conflict in 1999, been replaced with Global Positioning System-guided weapons that were virtually infallible in finding their targets, without requiring pilot supervision. Targeting mistakes still occurred, of course. But these were largely failures of intelligence and programming rather than of inherently inaccurate or indiscriminate weapons. The Chinese embassy in Belgrade was bombed with great precision, in that the bomb's coordinates were hit precisely; the mistake was in programming those coordinates. Successful conduct of just war has always depended to a large degree on intelligence, of course, because correct identification of legitimate targets rests on intelligence in all but face-to-face encounters between adversaries. However, in combat driven by precision stand-off and robotic munitions of great accuracy, perhaps intelligence will bear the brunt of the moral responsibility for discrimination and proportionality.

Air power is an appropriate focus for a discussion of the connections between military ethics and technology because it is has undergone the most dramatic technological evolution in the post-World War II period. Technological developments for land forces are driven by similar technological and ethical imperatives, however, more in the quest for technologically produced total situational awareness of the battlefield and precisely targeted weapons. Naval forces, too, are increasingly platforms for launch stand-off precision weaponry. The historical review of more than fifty years of the development of air power is instructive in a number of ways, not just for its own sake, but also for what it illustrates regarding the connection between military ethics and technology. Most of that history focused on the ethical test of discrimination. If World War II degenerated into an indiscriminate air war, it was partly out of a misguided strategic idea that bombing civilians would be effective in hastening the termination of conflict and partly from inherent technological limitations of the weapons and platforms available. Subsequent technological development increasingly provided the capability to conduct effective strategic level air bombardment, but to do so in an increasingly discriminate way. So at first glance, here is a clear example of technological development dramatically assisting the abilities of military forces to operate within the boundaries of established principles of military ethics. Further, regarding that development only from the perspective of the ability of the U.S. Air Force and Navy to conduct discriminate strategic air campaigns, technology has provided the capability to meet the requirements of military ethical principles.

The existence of the various technologies of PGMs has, however, generated a number of unanticipated ethical issues as well. Especially stand-off weapons (that is, weapons that can be fired from long distance, placing the operator beyond the range of enemy counterfire such as Air and Sea Launched Cruise Missiles) have already dramatically altered some *jus ad bellum* calculations. The ethical requirement that use of military force be a last resort was always supported by the fact that the decision of a political leader to use force inevitably involved putting the military forces of that nation at risk and almost certainly suffering some casualties. But stand-off weapons hold out the tantalizing prospect of using military force with complete impunity—thereby dramatically lowering the threshold to the use of force. Last resort remains a moral requirement. But without risk to a nation's own forces, the prospect of using missiles *to send a message* might be a political leader's course of action when it would certainly not have been if the possible deaths of aircrews or special forces units had factored into the decision.

The capability that PGMs provide generates ethical issues in another area as well. Because only the United States and a few major high-technology powers possess these capabilities, the entire war convention is challenged when such powers engage in conflict with less technologically advanced states. The Law of Armed Conflict that codifies just war principles is intended to apply equally to and to be observed equally by all combatants. Yet this capability creates a situation in which the United States can scrupulously observe those laws and conduct a highly discriminate air campaign against a lesser adversary that, if it follows those rules, faces only certain defeat. Understandably adversaries equipped only with lower technology weapons come to feel that U.S. forces lack honor in conducting war in this way. To the degree that the respect for the criteria of just war rests on a mutual sense that war can be conducted within those limits and still be a *fair fight*, precision munitions built to honor the principles of discrimination and proportionality may come to undermine respect for those very rules on the part of adversaries.

In practical terms, this asymmetry of capability provides a strong incentive for any adversary to find asymmetrical approaches to offset U.S. capabilities, even if those approaches strain or violate established ethical principles of military conduct. The Iraqis and the Serbs (examples of such lesser powers under attack) have illustrated the consequences of this asymmetry in their use of human shields (their own citizens, captured civilians of the attacking and allied powers, or prisoners of war), deliberate collocation of military and civilian objects (fighter aircraft parked next to mosques, schools, and hospitals), and perhaps dual-use of factories for production of baby formula and chemical weapons (although these cases are less certain).

It is hard to say what exactly follows from these points regarding the status and future of military ethics. It is ironic that weapons developed precisely to return air power to scrupulous respect for the ethical principle of discrimination have the unforeseen and unintended consequence of contributing to undermining the shared respect for those very principles on the part of adversaries. What is clear is the difficulty of predicting nonlinear relationships between developments in military technology and the law and practices of military ethics.

The more general point about the relation of technology and military ethics concerns not a single technology and its implications, but rather the aggregate effect of the overwhelming technological superiority of the United States and, to a much lesser degree, its allies in the whole panoply of military technologies. Taken together, they provide the tools for those militaries to intervene effectively and widely against less technically advanced powers—at least powers whose militaries are conventionally structured. The example of Vietnam and other guerilla wars suggest that some kinds of asymmetry

are relatively immune to high-technology capabilities developed to date, although there too, improved sensor and surveillance technologies offer advantages for land forces as well.

The *jus ad bellum* requirement of just cause has, during the twentieth century come to be restricted to defense against the aggression of others. However, since World War II, a body of human rights law (starting with the Genocide Convention) has begun to sketch out a parallel body of international law that gives less weight to national sovereignty and suggests that the rights of human individuals and groups might provide a basis for legitimate intervention if the state failed to properly protect those rights. Kosovo provided a possible model for the future when the technologically superior powers intervened with relative impunity to protect human rights.

But the existence of the capability also suggests a danger: The superior powers may no longer be constrained by the risks to their own forces and may use their unmatchable technologically-based military power in ways that destabilize rather than stabilize the international system. At its roots, the relatively stable system of mutually respected military ethics developed among the European powers worked, insofar as it did, because powers felt that respecting the rules of military ethics still made it possible to have a fair fight. This asymmetry of capability may make it possible for the technologically superior to operate *in bello* in ways that adhere to the rules of discrimination and proportionality, but within a wider frame *ad bellum* of excessive interventionism.

The Historical Development of Military Ethics

In almost every culture, the warrior class develops some internal sense of appropriate military behavior. While it would be wrong to suggest that the rules are equivalent, the need warriors have to distinguish honorable from dishonorable conduct in war seems nearly if not completely universal.

The specific version of military ethics that evolved into the ostensibly universal principles embodied in the Hague and Geneva Conventions has specific roots. These principles may be traced back to ancient Roman thought and practice, as mediated through history by the European Christian Church and its secularized successors.

Although elements from pre-Christian Roman thought and practice (e.g., Cicero), feed into the origins of just war, the Christian writer Augustine's work is the origin of the unbroken stream of Christian military ethics that leads to the elaborated tradition that exists in the twenty-first century. Augustine wrote during a period when the Roman Empire was collapsing under the weight of barbarian advance and, unlike most of his Christian predecessors, he advocated Christian participation in the military defense of the Empire. While it was far short of Christian religious and ethical ideas, Augustine argued, use of military force to defend the *tranquility of order* provided by the Empire was a legitimate act of Christian love. Military struggle and even death in defense of that order was an act of love for one's neighbors who, if that order were to fall, would endure great suffering.

The Christian soldier is governed by restraints in combat. It is the enemy's misconduct rather than the soldier's wish that brings about the war. The Christian soldier goes to war *mournfully*, accepting the blessing of Jesus as the peacemaker struggling to restore order on behalf of the neighbor. But most importantly, the soldier recognizes the common humanity of the adversary and avoids personal hatred or animus.

Augustine lays the foundation for a tradition that accepts the necessity of coercion and even violent conflict in the name of maintaining order. But it also imposes rules of restraint and caution that are elaborated in subsequent Christian tradition. In the medieval period, for example, Thomas Aquinas and other scholastics developed and elaborated the intellectual framework for military ethics, even as the Code of Chivalry formed the basis of ideal military ethics among the warrior class. During the same period, the idea of a Law of Peoples (*jus gentium*) evolved: a concept that became *customary international law* in later versions of the tradition.

Although the major actors of the Reformation produced their own versions of just war and military ethics in the sixteenth century, the collapse of the unified Christian civilization of Europe and the encounter of Europeans with the inhabitants of the New World spurred the need for a less *religious* and Eurocentric understanding of just war and military ethics. Catholic thinkers such as Francisco Suarez and Franciscus de Vitoria argued that the indigenous peoples of the New World possessed rights. Hugo Grotius, Samuel Pufendorff, and Emmerich de Vattel laid the foundations for a non-religious framework of military ethics and just war, grounded in human reason that would be valid (as Grotius put it) *even if God does not exist*.

The European Enlightenment of the eighteenth century completed the work of secularization. Rationalist thinkers such as Kant argued that ethics generally must be grounded in the nature of human reason alone

and that reason dictated a more rational system than war for the adjudication of international disputes. He envisioned a League of Nations, willing and able to provide world governance on principles better reasoned than the perpetual conflict of interstate rivalry. Such ideas set in motion the hope of a united global community operating in accordance with shared ethical and political principles—an endeavor manifest in the creation of the League of Nations and the United Nations in the twentieth century.

Abraham Lincoln's charge to Francis Lieber to create General Order 100 marked a milestone in the establishment of a state-mandated set of rules for military conduct. Military Codes of Discipline came to replace customary Chivalric Codes as official guidance for governing the conduct of military personnel of the various nations.

At the end of the nineteenth century, under the auspices of the Hague and Geneva Conferences, treaty law governing the conduct of military operations and the treatment of civilians, the rights of neutral powers, prisoners of war, and so on, began to grow. This body of law is the partial codification of the long moral tradition of military ethics, and constitutes customary international law for all states and their militaries.

At the conclusion of World War II, war crimes trials, held in Nuremberg and Tokyo, established the precedent of individual responsibility of commanders and soldiers for war crimes. Although criticized by some as *victor's justice*, they laid the foundation for the idea of individual culpability for war crimes that has evolved into ad hoc war crimes tribunals for Rwanda and the former Yugoslavia. In 1998 the United Nations adopted the Rome Statute calling for the establishment of a permanent standing war crimes court. That treaty received a sufficient number of national ratifications and entered into effect on July 1, 2002; the process of appointing members and establishing procedures was ongoing in the beginning of 2004.

In the early 2000s, the United States was among a small number of states opposed to the creation of the war crimes court due to fears that it would be dominated by political considerations rather than disinterested justice, and by awareness that U.S. forces are more widely deployed (and therefore more likely to be subject to the court's scrutiny) than those of other powers. It is too early to say what the effect of the war crimes court will be. But in intention, it represents the culmination of efforts over many years to give legal shape, form, and enforcement to the fundamental principles of military ethics.

MARTIN L. COOK

SEE ALSO *Airplanes; Baruch Plan; Biological Weapons; Chemical Weapons; Geographic Information Systems; International Relations; Just War; Missile Defense Systems; Limited Nuclear Test Ban Treaty; Weapons of Mass Destruction.*

BIBLIOGRAPHY

Bainton, Roland. (1960). *Christian Attitudes Toward War and Peace: A Historical Survey and Critical Re-Evaluation.* New York: Abingdon Press. The standard historical survey of Christian thought on ethics and war

Best, Geoffrey. (1994). *War and Law Since 1945.* Oxford and New York: Clarendon Press. A superb discussion of the evolution of international law since World War II. Especially helpful in tracing the development of humanitarian and human rights law as a parallel legal stream to the older state sovereignty-Westphalian legal framework.

Best, Geoffrey. (1980). *Humanity in Warfare: The Modern History of the International Law of Armed Conflicts.* London: Weidenfeld and Nicolson.

Christopher, Paul. (1994). *Ethics of War and Peace: An Introduction to Moral and Legal Issues.* Englewood Cliffs, NJ: Prentice-Hall.

Hartle, Anthony. (1989). *Moral Issues in Military Decision Making.* Lawrence: University of Kansas Press. A very useful ethical analysis by a distinguished U.S. army soldier-scholar.

Kelsay, John. (1992). *Islam and War: A Study in Comparative Ethics.* Louisville, KY: Westminster/John Knox Press. The best short discussion of Islamic thought on war in English.

Mendelsohn, Everett; Merritt Roe Smith; and Peter Weingar, eds. (1988). *Science, Technology and the Military.* Dordrecht, The Netherlands, and Boston: Kluwer Academic Publishers. An important collection of papers covering a broad range of topics.

Moorehead, Caroline. (1999). *Dunant's Dream: War, Switzerland, and the History of the Red Cross* New York: Carroll & Graf.

Nardin, Terry, ed. (1996). *Ethics of War and Peace: Religious and Secular Perspectives.* Princeton, NJ: Princeton University Press. A useful attempt to compare Islamic, Jewish, pacifist, feminist and realist ethical traditions to the standard categories of western just war analysis.

Smith, Merritt Roe. (1985). *Military Enterprise and Technological Change: Perspectives on the American Experience.* Cambridge, MA: MIT Press. The results of a discussion among historians of technology that helped initiate contemporary studies of war-technology relationships.

Walzer, Michael. (2000). *Just and Unjust Wars: A Moral Argument with Historical Illustrations,* 3rd edition. New York: Basic Books. The first edition of this influential book was published in 1977.

INTERNET RESOURCE

Mitcham, Carl, and Philip Siekevitz. (1989). "Ethical Issues Associated with Scientific and Technological Research for the Military." *Annals of the New York Academy of Sciences* 577. Available at http://www.annalsnyas.org/content/vol577/issue1/.

MILITARY-INDUSTRIAL COMPLEX

• • •

The *military-industrial complex* is one of a series of ideas that aim to critique the manner in which science, technology, and society have interacted with one another since World War II. The term itself was popularized by U.S. president and World War II general Dwight D. Eisenhower (1890–1969) in a farewell address to the nation on January 17, 1961, in which he warned the American people against "the acquisition of unwarranted influence, whether sought or unsought by [such a] complex" and the corresponding threat it posed to democracy. Although defined as "the conjunction of an immense military establishment and a large arms industry," its influence extends beyond industry and the military (Eisenhower). Often called the *military-industrial-congressional complex*, for instance, it comprises the *iron triangle* of Congress, the Pentagon, and defense industries. Additionally because the military and industry both support and depend upon academic research, another iron triangle has been dubbed the *military-industrial-university complex* (Hughes 2004).

Context and Emergence

The precise origins of the term military-industrial complex are obscure, but the idea is not. During the war, the U.S. government became increasingly dependent on both industrial corporations and scientific research for the production and development of military weapons. Military needs far exceeded those of previous wars. A typical U.S. army division, for example, required 225 times the mechanical horsepower required in World War I (Abrahamson 1983). In response, industry and the scientific enterprise shifted focus to help with the war effort.

Ford Motor Company, for example, manufactured jeeps, general purpose vehicles, and B-24 Liberator aircraft at a rate of one airplane per hour at the peak of production (Grudens 1997). Boeing Aircraft Company designed and built both the B-17 Flying Fortress and the B-29 Superfortress bombers at a rate of up to 362 planes per month. In total, companies produced 303,717 planes during the war—including 18,481 B24s and 12,761 B17s—at a price of $45 billion. According to Henry Stimson, secretary of war under both presidents Franklin D. Roosevelt and Harry S. Truman, "if you are going to try to go to war, or to prepare for war, in a capitalist country, you have got to let business make money out of the process or business won't work" (Higgs 1995, p. 1).

At the same time, the National Defense Research Committee, later the Office of Scientific Research and Development (OSRD), secured vast new resources for scientific research aimed at solving wartime problems. As a result, two new efforts allowed for increased collaboration between large numbers of scientists toward set goals: the centralization and creation of national laboratories, such as Los Alamos and Oak Ridge, and the targeted funding of research projects at universities, such as the Massachusetts Institute of Technology (MIT) Radiation Laboratory and the University of Chicago reactor research.

With the war, funding for large-scale scientific research shifted from industry to government and thus enabled *big science* projects such as the Manhattan Project. The architect of this shift, OSRD chair Vannevar Bush, began a trend to fund and direct scientific research through the military that would last well beyond the end of World War II. New scientific and industrial relationships and institutions begun during the war soon became fixed in U.S. economic and political life with the immediate emergence of the Cold War (1945–1989). It was this entrenchment that Eisenhower sought to highlight as a danger to political life.

Post-Cold War Revival

Throughout the Cold War, increasing military budgets were justified by the Soviet threat. When the Soviet threat disappeared, so too did the justification for large military budgets. Yet neither large military budgets nor the power of the military-industrial complex diminished, they simply reorganized (Hartung). According to Columbia University professor Seymour Melman, the United States has a permanent war economy, having "been at war—somewhere—every year, in Korea, Nicaragua, Vietnam, the Balkans, Afghanistan" since the end of World War II (Melman).

As a result, both scientific and industrial enterprises remain directed toward military ends. The fiscal year 2005 research and development (R&D) budget includes $75 billion for defense R&D and $57.2 billion for nondefense R&D. Defense R&D, therefore, comprises 56.7 percent of the total R&D budget (AAAS 2004). Additionally the fiscal year 2005 defense R&D budget is nearly $20 billion

above what it was at the height of the Cold War, adjusted for inflation but not for growth in the economy.

Defense contractors have gained considerable power and influence because of mergers between previously competing contractors. Because of their size and power, specific contractors—such as Lockheed Martin, Northrup Grumman, and Raytheon—can secure support through sizable congressional contributions. They do so by supporting those candidates with power over their pet programs. Of the forty top recipients of defense contractor campaign donations, thirty-six are on either the congressional Appropriations Committee (the committee with authority over government funds) or Armed Services Committee (the committee with authority over defense programs). As a result, weapons programs, such as the Lockheed Martin F-22 fighter, the most expensive bomber ever built, are not likely to be terminated.

When President George W. Bush was first elected, he and Secretary of Defense Donald Rumsfeld promised a revolution in military affairs in which they would create new, more agile forces. Bush suggested that they might "skip a generation of technology" in certain systems, which would require the elimination of at least one big-ticket system such as the F-22 fighter (Hartung 2001, p. 3). As a testament to the power of the defense industries, this has not happened and in fact "the Pentagon has not shut down a single major weapons production line since the end of the Cold War" (Hartung).

Ethics and Policy Issues

Several scholars have raised concerns about the military-industrial complex throughout the years, including that it is a threat to democracy and to the free market. Lewis Mumford argues that the military-industrial complex threatens democratic processes, because it has become a *megamachine*, a rigid, hierarchical social structure with absolute powers and little outside input (Mumford 1964). In effect, he argues against the authoritarian nature of the military-industrial complex. This echoes Eisenhower's warning that the American people must remain alert and knowledgeable to ensure that the complex "does not endanger our liberties or democratic processes" (Eisenhower).

Seymour Melman argues that the military-industrial complex endangers the free market, because it actually creates a state economy. He contends that appropriations for physical infrastructure, health, and welfare are drying up, and thus "the idea that the U.S. can afford guns and butter without limit is proven false every day" (Melman).

GENEVIEVE MARICLE

SEE ALSO Military Ethics; Science, Technology, and Society Studies.

BIBLIOGRAPHY

Abrahamson, James. (1983). *The American Home Front.* Washington, DC: National Defense University Press.

Hartung, William D. (2001). "Eisenhower's Warning: The Military-Industrial Complex Forty Year Later." *World Policy Journal* 18(1): 1–7.

Hughes, Thomas P. (2004). *Human-Built World.* London: University of Chicago Press.

Mumford, Lewis. (1964). *The Myth of the Machine: The Pentagon of Power.* New York: Harcourt Brace.

INTERNET RESOURCES

American Association for the Advancement of Science (AAAS). "Defense and Homeland Security R&D Hit New Highs in 2005; Growth Slows for Other Agencies." AAAS. Available from http://www.aaas.org/spp/rd/upd1104.htm. Taken from the AAAS R&D Funding Update, November 29, 2004.

Eisenhower, Dwight D. "Farewell Radio and Television Address to the American People." Available from http://wikisource.org/wiki/Military-Industrial_Complex_Speech.

Grudens, Richard. "Henry Ford in WWII." History Net. Available from http://www.thehistorynet.com/wwii/blhenryford/index1.html. Article originally appeared in the January 1997 issue of the journal *World War II.*

Hartung, William D. "Military Industrial Complex Revisited: How Weapons Makers are Shaping US Foreign and Military Policies." Foreign Policy in Focus. Available from http://www.fpif.org/papers/micr/index.html.

Higgs, Robert. (1995). "World War II and the Military-Industrial-Congressional Complex." Independent Institute. Available from http://www.independent.org/newsroom/article.asp?id=141.

Melman, Seymour. "In the Grip of a Permanent War Economy." Bear Left. Available from http://www.bear-left.com/original/2003/0309permanent.html.

MILL, JOHN STUART

• • •

John Stuart Mill (1806–1873) was born in London on May 20. The son of the philosopher James Mill (1773–1836) and the godson of the philosopher Jeremy Bentham (1748–1832). John Stuart Mill was the most influential British philosopher of the nineteenth century, which saw science and technology transform society as significant contributions were made in metaphysics, logic, the philosophy of science, ethics, social and political philosophy, economics, the philosophy of religion, and the philosophy of education. The *System of Logic*

John Stuart Mill, 1806–1873. An English philosopher and economist, Mill was the most influential British thinker of the 19th century. He is known for his writings on logic and scientific methodology and his voluminous essays on social and political life. (*Hulton Archive/Getty Images*)

(1843) and the *Principles of Political Economy* (1848) became canonical textbooks in their fields. Mill died on May 8 in Avignon, France.

Logic

Mill understood his work in technical philosophy as providing a foundation for his social and political philosophy. The purpose of the discussion of the origins of knowledge in the *System of Logic* is to prepare the ground for the social sciences, and the discussion of the social sciences provides the grounds for Mill's moral, political, and economic views.

The first five books of the *Logic* are largely polemical, attacking the philosophical position known as intuitionism, which in the nineteenth century had served as the basis for political conservatism. Intuitionism takes the view that there are innate truths, including moral truths. Innate truths can be known independent of experience, and thus custom and tradition were elevated to the status of timeless truth impervious to empirical refutation. In contrast, Mill wanted to argue that customary practice is often no more than a historical accident or that although it may have been justified in earlier social circumstances, it had outlived its usefulness, and all practice should be subject to revision in light of changing circumstances.

Mill argued that almost every general principle in any domain was the result of an inductive process that began with individual experiences, although Mill conceded a few exceptions. For example, the general principle that nature is uniform seems to be an assumption that people bring to their experience insofar as there are many things people do not understand as examples of uniformity or for which they have no experience, although they continue to subscribe to this belief. There are diseases for which the cause or cure is not known, yet it is presumed despite the failure of past research that the hidden uniformity behind them will be discovered eventually. Mill insisted that these few exceptions had no moral or political implications.

Mill engaged in a protracted controversy with William Whewell (1794–1866), professor of moral philosophy at Cambridge, who had published a *History of the Inductive Sciences from the Earliest to the Present Time* (1837). Whewell coined the term *scientist* in recognition of the idea that traditional "natural philosophy" had become a new form of knowledge. Whewell was a critic of the philosopher Francis Bacon's (1561–1626) conception of the process of induction and wanted to redefine induction as the process by which scientific hypotheses are formulated. He considered this process a creative act rooted in history but not amenable to strict rules. In this he was close to the Kantian view that the most general principles of knowledge were not based on experience but instead were presuppositions. A successful hypothesis starts as a happy guess and evolves over time into a larger structure of thought incorporating both empirical and nonempirical elements. Whewell insisted on the historically evolving nature of scientific hypotheses and laws.

Mill objected on the grounds that Whewell was conflating induction with hypothesis formation and that what mattered was not the original happy guess but the subsequent inductive process by which the guess is confirmed by empirical observation. At this level Mill's dispute with Whewell was merely semantic.

Social Sciences and Technology

Mill contended that there can be a science of human nature and that its basic laws are the psychological laws of association. Moreover, the basic truths about human affairs, including questions of ends, are not part of the content of the psychological laws of association. To explain the basic truths of human action it is necessary to supplement the psychological laws of association with information about the circumstances in which those laws operate.

Human action, unlike physical interaction, cannot be explained in terms of current circumstances. Actions of human beings are not solely the result of their current circumstances but are the joint result of those circumstances and the characters of the individuals; the agencies that determine human character are numerous and diversified. Is it possible to give a systematic account of the circumstances, past as well as present? Mill at one time thought this possible. The science needed to discover and formulate the hypothetical laws of the formation of character he termed *ethology*.

Mill's views on technology are embedded in his historical account of the stages of economic growth. His view owes much to Scottish Enlightenment thinkers such as David Hume (1711–1776), Adam Smith (1723–1790), and Adam Ferguson (1723–1816). Economic and social progress is marked by three stages: savagery, barbarism, and civilization. By civilization Mill meant a modern industrial and commercial society with a liberal culture such as Great Britain. The rise and development of civilization are dependent on "the natural laws of the progress of wealth, upon the diffusion of reading, and the increase of the facilities of human intercourse" ("Civilization," *Collected Works*, Vol. XVIII, p. 127).

The third stage of *civilization*, as described in Mill's essay of that title, is marked economically by industry, politically by limited government and the rule of law, and socially by liberty. Mill saw examples of these combined features in military operations, commerce and manufacturing, and the rise of joint-stock companies. The consequences of the rise of civilization are economic, political, social, and moral. Economically, there has been a vast increase in wealth in which the masses and the middle class have been the primary beneficiaries. Politically, power is shifting from a few individuals to the masses.

Science, Technology, and Politics

Socially, the most important consequence has been the decline of individuality. The future of civilization depends on the masses exercising power in ways that allow the benefits of civilization to continue. Mill did not believe this would happen on its own. The masses must understand and appreciate the moral foundations of liberal culture.

Unlike both classical liberals such as the Philosophic Radicals Jeremy Bentham, James Mill, and orthodox Marxists, Mill was not an economic determinist. The moral world was not a product only of material forces. The functioning of the economy presupposed certain virtues. This explains Mill's economic position in the later *Principles of Political Economy*, the germ of the recommendations in *Representative Government* (1861), and the project that *On Liberty* (1859) would address. The social crisis created by the industrial revolution was class conflict. This crisis was exacerbated in Mill's thinking by the perceived coming of an increasingly democratic society.

Participation in a market economy informed by an individualist moral culture promotes different forms of virtuous behavior. Nevertheless, Mill insisted that there had to be a moral purpose to the technological project. The desire to employ the whole surface of the earth for the production of the greatest possible quantity of food and the materials of manufacture he considered to be founded on a mischievously narrow conception of the requirements of human nature. Among the many things Mill and his father had objected to most vehemently about the new industrial economy was the spoiling of the countryside by the many new and often duplicative railway lines. As hikers, they were sensitive to the destruction of natural beauty and the disappearance of solitude.

Mill also addressed the issue of the stationary state: an economy that no longer grows (a concern for classical economists but not neoclassical economists). Mill did not think that society had arrived at that state, and so more growth was probable. However, he did not consider a stationary state necessarily bad. Wealth is not an end in itself but a means to human fulfillment and individual liberty. Even if there were a stationary state of zero growth, freedom would not necessarily be lost.

Mill was the last major British philosopher to present an integrated view of philosophy and relate the theoretical and normative dimensions of his thought in a direct fashion. Book VI of the *Logic* remains the classic statement of what human science modeled after physical science might be, its limitations and qualifications, and the extent to which it may be useful. As a statement of the aims of and obstacles to the creation of the human sciences, it is unsurpassed.

NICHOLAS CAPALDI

SEE ALSO *Consequentialism; Liberalism; Locke, John; Scientific Ethics.*

BIBLIOGRAPHY

Capaldi, Nicholas. (2004). *John Stuart Mill: A Life.* New York: Cambridge University Press. This is the latest and only biography of Mill in print.

Cohen, Morris R. and Ernest Nagel. (1934). *An Introduction to Logic and Scientific Method.* New York: Harcourt, Brace and Company. The classic restatement of a Millian view.

Schwartz, Pedro. (1972). *The New Political Economy of J. S. Mill.* Durham, NC: Duke University Press. The best discussion of Mill's economics.

Skorupski, John. (1989). *John Stuart Mill.* London: Routledge. A good general discussion of the technical part of Mill's philosophy.

Skorupski, John, ed. (1998). *The Cambridge Companion to Mill.* Cambridge, UK: Cambridge University Press. A useful collection of secondary sources on Mill.

Wilson, Fred. (1990). *Psychological Analysis and the Philosophy of John Stuart Mill.* Toronto: University of Toronto Press. Wilson takes seriously Mill's project in the Logic to explain human beings scientifically.

MINING

• • •

From the moment humans discovered stone tools and salt, they have been extracting and using materials from the Earth. Every American will utilize approximately 2.4 million pounds of mined materials during their lifetime (calculated from Mineral Information Institute statistics). In spite of people's dependence on the products of extractive technologies and their associated sciences, mining is a highly controversial activity surrounded by ethical, political, social, and legal issues. Mining focuses attention on the metaphysical relationship of humans to the Earth, on the impact of their activities on the environment and other species, on issues of equity and sustainability, on human rights and democracy.

Mining is the extraction of metallic or nonmetallic materials from the Earth. The full cycle of mining involves exploration for the material required; mining *sensu stricto,* which is the physical removal of material from the Earth; processing, which is usually required to concentrate or clean the ore; the health, safety, and environmental issues associated with the full cycle of mining activities; and appropriate closure of the site when mining is completed (National Research Council 2002).

Surface mining, where material is separated directly from the surface of the Earth, is the oldest and most common method of mining. Underground mining, where the material is extracted via tunnels dug into the Earth, is used to work deeply buried ores. Mining technology has evolved greatly, but the basic concept of removing rock or minerals from the Earth has remained constant since prehistory. Nonentry mining, by which the valuable components of the rock are extracted without physically removing the surrounding rock, is still at an experimental stage.

The many ethical, social, and political challenges associated with mining can only be addressed within the context of the prevailing philosophical view of the relationship of human beings to the Earth and its resources. From prehistoric time through the sixteenth century, many cultures regarded Earth as animate. Ores grew and matured in the uterus of the Earth; mining was an interference with the natural order and was often accompanied by myths and rituals (Eliade 1962). In the Western world, the organic view of nature was superceded by a mechanical model during the Scientific Revolution: The Earth is inanimate, and its resources should be exploited for the benefit of humans (Merchant 1980). In the late twentieth century, scientists developed holistic syntheses that integrate humans, other living beings, and Earth in an all-encompassing, interdependent Earth system. Some philosophers emphasize the importance of the humanities in understanding the full dimensions of the human–Earth system relationship (Frodeman 2003). These cross-disciplinary concepts are the basis for most modern interpretations of the place and responsibilities of mining.

Polarized positions on the ethics of mining are strongly developed and there have been few true dialogs on the subject. One early-twenty-first century attempt to foster communication is the Mining, Minerals, and Sustainable Development Project, which concluded that economic, social, environmental, and governance issues must be addressed appropriately by all participants in order to meet the conflicting demands of society for the products of mining while still maintaining sustainability (International Institute for Environment and Development, and World Business Council for Sustainable Development 2002). Finding mechanisms whereby all the stakeholders can be involved in negotiating acceptable practices and compensation for mining has proved difficult. Some nongovernmental organizations and companies have promoted formal or informal democratic fora, but they have been difficult to implement in areas lacking good governance or a history of citizen participation.

Mining is inherently inequitable. Earth resources are not distributed evenly, and mines can only be located where there are suitable resources. Many of the social and environmental consequences of mining are concentrated at the mine site even if the consumer or ultimate beneficiary of the mine product, or the wealth it creates, is far away. Resolving these inequities are

Underground mining as depicted in Georgius Agricola's *De Re Metallica* (1556). (© Bettmann/Corbis.)

some of the major ethical and political challenges associated with mining.

A fundamental question concerns ownership and control of the mineral endowment. Does a nation, or a sovereign, or a dictator, own the mineral wealth of a country? Or is it instead the landowner, the owner of the mineral rights, the person or company who discovered the deposit, the artisan miners who may have worked the deposit, or the local community (however defined)? In many cases the owner of a mineral deposit is not competent to mine it. In capitalist societies the high financial risk of mineral exploration and mining is usually borne by corporations that also supply technical expertise, and in return expect a profit from their investment. Almost every country has devised a different formula for regulating mineral ownership and control, for calculating taxes, and for oversight of mining activities and their impact.

The Bingham Canyon copper mine in Tooele, Utah. This mine is the world's largest man-made excavation. Kennecott Utah Copper Corp. produces copper, molybdenum, gold, silver, platinum, and palladium from the century-old mine. (© Bettmann/Corbis.)

A mine may introduce large amounts of capital or people into an area, distorting the economic and social structure. Corruption may become a problem. Wars are fought over the control of resources, and illicit trade particularly in diamonds and columbite-tantalite has funded conflicts, such as those in Angola and Congo, in the twentieth and early-twenty-first centuries. Safeguarding the human rights of workers and local populations is also a concern. Disciplined and transparent governance by governments and companies is necessary to stabilize the impact of mining.

Economic analysis shows that the Earth is unlikely to run out of mineral resources in the twenty-first or twenty-second centuries, which is as far forward as such predictions can be made, but the total cost of mining (including environmental, social, and other external costs) may limit the willingness to produce minerals (Tilton 2003). The role of mining in sustainable development is controversial, and conclusions largely depend on what values or assets one wishes to sustain, and on the scale at which the question is examined. Tilton (2003) argues that mining can contribute to global sustainable development if the products and profits of present-day mining are used to provide other assets of equivalent or greater value to succeeding generations. Analyses that concentrate on preserving the lifestyle, economy, or environment of a particular location are more likely to conclude that mining is a temporary phenomenon which disrupts rather than sustains development.

Technological innovation may lessen the demand for mineral products and lower the environmental impact of mining, but intellectual innovation is also vital to resolve the social and cultural consequences of mining.

MAEVE A. BOLAND

SEE ALSO *Acid Mine Drainage; Development Ethics; Environmental Ethics; Sustainability and Sustainable Development.*

BIBLIOGRAPHY

Eliade, Mircea. (1978 [1962]). *The Forge and the Crucible: The Origins and Structures of Alchemy*, 2nd edition, trans. Stephen Corrin. Chicago: The University of Chicago Press.

Frodeman, Robert. (2003). *Geo-Logic: Breaking Ground Between Philosophy and the Earth Sciences*. Albany: State University of New York Press.

International Institute for Environment and Development, and World Business Council for Sustainable Development. (2002). *Breaking New Ground: Mining, Minerals, and Sustainable Development. The Report of the MMSD Project*. London: Earthscan Publications Ltd.

Merchant, Carolyn. (1980). *The Death of Nature: Women, Ecology, and the Scientific Revolution*. San Francisco: Harper & Row.

National Research Council. (2002). *Evolutionary and Revolutionary Technologies for Mining*. Washington, DC: Committee on Technologies for the Mining Industries, National Academy Press.

Tilton, John E. (2003). *On Borrowed Time? Assessing the Threat of Mineral Depletion*. Washington, DC: Resources for the Future.

INTERNET RESOURCE

International Institute for Environment and Development, and World Business Council for Sustainable Development. "Breaking New Ground: Mining, Minerals, and Sustainable Development. The Report of the MMSD Project." Available from http://www.iied.org/mmsd/finalreport/.

MISCONDUCT IN SCIENCE

• • •

Overview
Biomedical Science Cases
Physical Science Cases
Social Science Cases

OVERVIEW

In the United States the official definition of research misconduct is:

> ... fabrication, falsification, or plagiarism in proposing, performing, or reviewing research, or in reporting research results.... Fabrication is making up of data or results and recording or reporting them. Falsification is manipulating research materials, equipment or processes, or changing or omitting data or results such that the research is not accurately represented in the research record.... Plagiarism is the appropriating of another person's ideas, processes, results, or words without giving appropriate credit. Research misconduct does not include honest error or differences of opinion. A finding of research misconduct requires that: There be a significant departure from accepted practices of the relevant research community; and the misconduct be committed intentionally, or knowingly, or recklessly; and the allegation be proven by a preponderance of the evidence. (Office of Science and Technology Policy 2000, p. 76262)

A somewhat broader definition of scientific misconduct has been put forward by the Wellcome Trust, the largest biomedical charity in the United Kingdom:

> ... [t]he fabrication, falsification, plagiarism or deception in proposing, carrying out or reporting results of research or deliberate, dangerous or negligent deviations from accepted practices in carrying out research. It includes failure to follow established protocols if this failure results in unreasonable risk or harm to humans, other vertebrates or the environment. (Koenig 2001, p. 1411)

Germany (Bostanci 2002) and China (Yimin 2002) have also developed definitions of scientific misconduct that are somewhat broader than the U.S. version.

In all cases, core elements of the definition of misconduct in science (also known as scientific or research misconduct) include fabrication and falsification of research data, and plagiarism (FFP). This reflects both philosophy and history. Researchers depend on the reliability of the published work of others in order to determine how best to design and conduct investigations of research questions. Rather than reproducing all related experiments, investigators expect to be able to build on previous research, not only their own but also that of others. Thus fabrication and falsification undermine the fundamental and central tenets of the scientific enterprise. In addition, researchers expect to be recognized and held accountable for their contribution to a scientific body of knowledge. Plagiarism violates this expectation.

History

Although in retrospect the work of some earlier scientists has been the subject of debate (Broad and Wade

1982), during the seventeeth, eighteenth, and nineteenth centuries the only significant discussion of misconduct among scientists was an isolated work by Charles Babbage (1830), which identified three types of misconduct: *trimming* data to fit expectations; *cooking* data by discarding what did not fit expectation; and the outright forgery or creation of fictitious data. The most famous instance of scientific forgery occurred in the early-twentieth century with the *discovery* of Piltdown Man.

In the 1980s, blatant examples of research misconduct came to light (Broad and Wade 1982, Sprague 1993). As a result congressional committees responsible for oversight of various aspects of science and technology pressured funding agencies to develop policies to address what seemed to be the increasing incidence of scientific misconduct. These agencies, in particular the National Institutes of Health (NIH) and the National Science Foundation (NSF), developed policies designed to explicitly identify and address allegations of scientific misconduct.

In its initial policy, the NIH described misconduct as "serious deviation, such as fabrication, falsification, or plagiarism, from accepted practices in carrying out research or in reporting the results of research" (Public Health Service 1986, p. 2), a definition from which later definitions have derived (Buzzelli 1999). Fabrication, falsification, and plagiarism are clearly provided as examples and the *other serious deviation from accepted practices* (OSD) clause emphasizes the primary role of the scientific community in identifying and setting the ethical standards for its members (Buzzelli 1999). Thus the OSD clause reflects the widespread view that the scientific community has a collective responsibility for establishing and upholding the professional standards of the community (Chubin 1985, Frankel 1993). The OSD clause is a common element of definitions of scientific misconduct found in many policies developed by U.S. funding agencies, universities, and professional societies. Nevertheless, in defining scientific and research misconduct, in the United States, the scientific community has tended to focus on FFP and has opposed the OSD clause (National Academy of Science 1992, Buzzelli 1999).

In 1993 the Commission on Research Integrity (CORI) was formed to advise the U.S. Department of Health and Human Services (DHHS) on ways to improve the Public Health Service response to allegations of misconduct in biomedical and behavioral research. The Commission found that in spite of the community's seeming preference "for a narrow and precise definition centered upon 'fabrication, falsification and plagiarism (FFP)' 'FFP' is neither narrow nor pre-cise" (CORI 1995, p. 8). CORI's report, "Integrity and Misconduct in Research" (1995) clarified the role of intent in research misconduct and reframed the definition in terms of *misappropriation* of words or ideas (specifically including information gained through confidential review of manuscripts or grant applications), *interference* in the research activities of others (i.e., intentionally taking, hiding, or damaging research-related equipment, materials such as reagents, software, writings, or research products), and *misrepresentation* of information so as to deceive, either intentionally or with reckless disregard for the truth (thereby covering both fabrication and falsification). They also identified as other relevant forms of professional misconduct obstruction of investigations of research misconduct and noncompliance with research regulations, and highlighted the need to protect from retaliation those who bring forward good faith allegations of misconduct (commonly known as whistle-blowers). In addition, the Commission emphasized the need for a proactive rather than reactive approach to misconduct in science and recommended that research institutions be required to provide education in research integrity.

Assessment

In the 1980s when concerns about the frequency of scientific misconduct were initially raised, the common response by senior members of the scientific community was that scientific misconduct is rare and in any case science is self-correcting. Given that FFP not only undermines but is inconsistent with the bedrock principles on which scientific research is based, it is not surprising that members of the scientific community would assume that genuine, authentic, and bona fide members of the community would not engage in such practices and that their occurrence would be rare. Indeed the frequency of misconduct continues to be debated. At the same time, it has become clear that the peer review process is largely incapable of detecting fabrication or falsification. What is not in doubt is the serious negative impact of even a single occurrence of misconduct not only for those involved and for those whose work is misdirected by fraudulent research, but also the negative impact on trust both within the scientific community and beyond (Kennedy 2000).

An apparent tension continues with regard to internal (i.e., within the scientific community) versus external governmental control of both the definition of scientific misconduct and of oversight of scientific research. However the tension may be more apparent than real since the scientific community is not homogeneous with regard to its views on research integrity and misconduct. As of 2002, the U.S. government policy regarding scientific misconduct continues to emphasize FFP and reflects vocal opposition by some segments of the scientific community to the OSD clause in spite of the obvious necessary reliance of the clause on the scientific community's own standards and assessment of *accepted practices*. It is nevertheless generally recognized that FFP does not encompass all of the serious deviations from accepted practice that are of concern to the wider scientific community. This is apparent from formal definitions of scientific misconduct like that advanced by the Wellcome Trust, educational programs at research institutions and professional scientific societies, and professional codes of ethics that identify and examine a wide array of other issues that arise in conducting and reporting scientific research, and in training science professionals. These issues include topics considered part of the responsible conduct of research (RCR) such as data management, humane treatment of research subjects whether laboratory animals or human volunteers, conflicts of interest, publication practices, peer review, and mentorship responsibilities. Moreover while the Office of Research Integrity (ORI) is responsible for addressing allegations of scientific misconduct either directly or by overseeing investigations conducted by research institutions, the agency relies on research institutions to conduct inquiries and investigations of allegations of research misconduct brought against their employees and students.

More to the point, the focus of concern both within the scientific community and in governmental agencies (exemplified by the ORI) is evolving (Mitcham 2003). Increasingly the ORI promotes research integrity through education and training in RCR (Pascal 1999). The scientific community, too, places less emphasis on misconduct and is more focused on research integrity and education (Institute of Medicine/ National Research Council 2002). While there is some consensus as to what constitutes the most egregious form of scientific misconduct (i.e., FFP) the concept continues to evolve both within the United States (as a result of the focus on the elements of RCR) and in other countries, for example China and Germany.

STEPHANIE J. BIRD

SEE ALSO *Accountability in Research; Research Integrity; Science, Technology, and Law.*

BIBLIOGRAPHY

Babbage, Charles. (1989 [1830]). "Reflections on the Decline of Science in England and on Some of its Causes." In *The Works of Charles Babbage*, Vol.7, ed. Martin Campbell-Kelly. London: Pickering.

Bostanci, Adam. (2002). "Germany Gets in Step with Scientific Misconduct Rules." *Science* 296: 1778.

Broad, William, and Nicholas Wade. (1982). *Betrayers of the Truth: Fraud and Deceit in the Halls of Science*. New York: Simon and Schuster.

Buzzelli, Donald E. (1999). "Serious Deviation from Accepted Practices." *Science and Engineering Ethics* 5: 275–282. A commentary on "Developing a Federal Policy on Research Misconduct" by Sybil Francis.

Chubin, Daryl E. (1985). "Misconduct in Research: An Issue of Science Policy and Practice." *Minerva* 23(2): 175–202.

Commission on Research Integrity (CORI). (1995). *Integrity and Misconduct in Research*. Washington, DC: U.S. Department of Health and Human Services, Public Health Services.

Frankel, Mark S. (1993). "Professional Societies and Responsible Research Conduct." In *Responsible Science: Ensuring the Integrity of the Research Process*, Vol. 2. Washington, DC: National Academy Press.

Institute of Medicine / National Research Council. (2002). *Integrity in Scientific Research: Creating an Environment That Promotes Responsible Conduct*. Washington, DC: National Academies Press.

Kennedy, Donald. (2000). "Reflections on a Retraction." *Science* 289: 1137.

Koenig, Robert. (2001). "Wellcome Rules Widen the Net." *Science* 293: 1411–1413.

Mitcham, Carl. (2003). "Co-Responsibility for Research Integrity." *Science and Engineering Ethics* 9: 273–290.

National Academy of Sciences (NAS). (1992). *Responsible Science: Ensuring the Integrity of the Research Process*, Vol. I. Washington, DC: National Academies Press.

Office of Science and Technology Policy. (2000). "Federal Policy on Research Misconduct." *Federal Register* 65: 76260–76264.

Pascal, Chris B. (1999). "The History and Future of the Office of Research Integrity: Scientific Misconduct and Beyond." *Science and Engineering Ethics* 5: 183–198.

Public Health Service. (1986). NIH Guide for Grants and Contracts 15(11). Special issue, July 18.

Sprague, Robert L. (1993). "Whistleblowing: A Very Unpleasant Avocation." *Ethics and Behavior* 3: 103–133.

Yimin, Ding. (2002). "Beijing U. Issues First-Ever Rules." *Science* 296: 448.

BIOMEDICAL SCIENCE CASES

Misconduct cases have been more prominent in the biomedical sciences than in the physical and social sciences. This may be because there are more people working in biomedical research than in physical or social science research or because misconduct in biomedical research is more likely to have direct, harmful effects on human beings. Several have been high-profile cases, attracting the attention of the scientific community, independent watchdogs, governmental agencies, and the public at large. The following four cases are some of the best-known instances of alleged misconduct and depict a variety of the ethical issues related to misconduct in biomedical research.

The Sloan-Kettering Affair

In 1974 William Summerlin was at the Sloan-Kettering Institute for Cancer Research, continuing work on a project that he and his supervisor, Robert Good, had begun while the two were at the University of Minnesota. Preliminary data from experiments there had suggested that some tissues, when incubated for several weeks in culture, cease to produce an immune response. If supported, that finding would have dramatic implications for transplantation science, allowing transplants between any two individuals without the risk of rejection.

Summerlin and his coworkers at Sloan-Kettering were having difficulty replicating the results of those initial studies, and as a result, Summerlin had little to show Good in a progress meeting in March 1974. In the elevator on his way to the meeting Summerlin used a marker to draw what appeared to be successful skin grafts on two of the laboratory mice and represented them to Good as successful transplants. Good failed to notice the fraud, but a laboratory assistant caring for the mice discovered the black spots later that day. When he was able to wash the spots away with alcohol, the assistant reported Summerlin. A review committee was established to look into the case.

In the investigations of the affair it was found that Summerlin's data from another transplant experiment conducted in the same period had been falsified. Summerlin had begun a study with two ophthalmologists that was designed to test the same hypothesis: that incubated tissues would not produce an immune response. The protocol required the ophthalmologists to transplant a fresh human cornea onto a rabbit's left eye and then transplant the donor's other cornea into the rabbit's right eye after it had been in tissue culture for several weeks. When Summerlin observed the rabbits, he saw unsuccessful transplants in the rabbits' left eyes and what looked to be successful transplants in their right eyes. He disseminated those remarkable results at several scientific meetings with confidence. In fact, however, the two ophthalmologists had not done the second

transplant on any of the rabbits; therefore, what Summerlin interpreted as successful corneal transplants were actually the rabbits' own corneas. Summerlin later claimed that he was unaware that his coinvestigators had not completed the second half of the protocol.

The institute determined that Summerlin had misrepresented data in both cases. The review committee further concluded that Summerlin had been experiencing emotional problems and placed him on medical leave for a year rather than imposing official sanctions.

Subsequent testing of the only available mouse from the Minnesota laboratory that had undergone a successful skin graft and that had formed the basis for Summerlin's work at Sloan-Kettering revealed that the mouse was a genetic hybrid rather than a purebred mouse, as had been recorded. Because the purebred mouse would have been expected to reject the skin graft but the hybrid mouse would not have, this explains the success of the graft in that case. It is not known whether the hybrid mouse was selected deliberately or accidentally.

The Darsee Case

John Darsee was a prolific and well-liked postdoctoral fellow at the Brigham and Women's Hospital, an affiliate of Harvard Medical School. In May 1981 Darsee's coworkers observed him fabricating data by recording data gathered over several hours so that the data appeared to have been collected over a two-week period. Caught in the act, Darsee apologized and claimed that it had been an isolated incident.

An internal investigation led by Eugene Braunwald, Darsee's supervisor, was conducted, but the incident was not disclosed publicly until several months later, when the investigators uncovered data that Darsee had generated for a multicenter study funded by the National Institutes of Health (NIH). Inexplicable discrepancies between Darsee's data and the results from other participating institutions were found, precipitating an independent investigation and notification of the NIH.

The NIH then launched its own review of Darsee's research. The review committee found problems in five of the papers that Darsee had published and on which Braunwald had been a coauthor and recommended that Darsee be barred from eligibility to receive NIH funding for ten years. The panel condemned Braunwald's supervision of Darsee, stating that his hands-off approach had inhibited the discovery of Darsee's fabrication. In response Braunwald argued that he had followed standard laboratory practices.

Further investigation into work done previously by Darsee at Emory University and Notre Dame uncovered instances of data fabrication and falsification in at least twelve of Darsee's papers that were based on research he had conducted at those institutions.

Harvard Medical School was criticized for its handling of the case and subsequently revised its policies for responding to charges of misconduct. In particular the review committee claimed that the NIH had had a right to know that Darsee, who had continued work on NIH-sponsored research for six months after the incident, had been caught fabricating data.

The Gallo Probe

A well-known controversy involving the isolation of the acquired immune deficiency syndrome (AIDS) virus illustrates a third form of scientific misconduct: plagiarism. In May 1984 a series of four papers appeared in the journal *Science* written by Robert Gallo and his team at the National Cancer Institute, stating that they had identified the virus that causes AIDS and proposing a process for developing a blood test for the virus. Mikulas Popovic, working in Gallo's laboratory, had been able to grow the retrovirus in cells that could survive infection with the virus, a cell line that he called H9. It later was revealed that the H9 cell line had not been developed by Popovic but instead had been cloned from a cell line called HUT78 that had been given to the Gallo laboratory by John Minna's team at the Veterans Administration. Minna's group was not credited in the *Science* papers for that significant contribution.

A second and more high-profile dispute accompanied the Gallo group's accomplishment. In July and again in September 1983 Luc Montagnier's team at France's Pasteur Institute had sent a sample of a viral isolate called LAV to Gallo's laboratory. In spring 1984 Popovic used the H9/HUT78 cell line to grow an AIDS retrovirus, which Montagnier's laboratory had been unable to do because it did not have a cell line that could survive infection with the virus. Gallo was able to produce sufficient quantities of the virus, which he named HTLV-III, to develop a method for testing for the presence of the virus in blood. It was discovered later that HTLV-III and LAV were the same virus, although Gallo had not acknowledged the contribution of the Pasteur Institute. Gallo claimed that the use of LAV was unintentional and must have contaminated his cultures accidentally.

The NIH's Office of Scientific Integrity (OSI) conducted an investigation and found Popovic guilty of four counts of misconduct but held Gallo responsible only for exhibiting a lack of collegiality. In a later investigation by the OSI's successor, the Office of Research Integrity (ORI) at the Department of Health and Human Services (DHHS), Gallo was found guilty of intention to deceive the scientific community about the origin of the materials used to isolate and replicate the AIDS virus. In 1993, however, a federal appeals board cleared Popovic and therefore Gallo of the misconduct charges, citing a lack of evidence that the virus had been stolen.

The Gallo case was significant not only because of the recognition and prestige associated with receiving credit for a discovery of this magnitude. The patent on the blood test for AIDS virus antibodies was lucrative, producing millions of dollars in royalties. It eventually was agreed that those royalties would be split evenly between the United States and France.

The Baltimore Case

Perhaps the most infamous instance of alleged misconduct in the biomedical sciences was the affair that would come to be known as the Baltimore case, even though David Baltimore, for whom the case is named, was not accused of fraud. Baltimore did, however, staunchly defend Thereza Imanishi-Kari against claims that she had fabricated data in a paper on which he was a coauthor. When the accusations were made, Baltimore was a professor of biology at the Massachusetts Institute of Technology (MIT) and the director of the Whitehead Institute. He had been awarded a Nobel Prize in 1975 for his work in virology. Imanishi-Kari was working with Baltimore on a complex project investigating the mechanisms behind the immune response.

Margot O'Toole took a postdoctoral fellowship in Imanishi-Kari's laboratory in 1985, and the two clashed from the beginning. O'Toole was having difficulty getting results consistent with Imanishi-Kari's data and had some problems with the experimental method. When she approached Imanishi-Kari with her concerns, she was dismissed and told that the discrepancies were due to incompetence. While attempting to understand the discrepancies between Imanishi-Kari's results and her own, O'Toole came upon evidence that she believed showed that data in a 1986 *Cell* paper coauthored by David Baltimore had been misrepresented.

O'Toole brought her concerns to senior scientists at both MIT and Tufts University, where Imanishi-Kari had taken a position. Informal investigations were conducted at both institutions. Errors were found in the paper, but the investigators believed that O'Toole's problems with the paper were scientific disagreements and did not demonstrate misconduct. O'Toole and a former graduate student of Imanishi-Kari's continued to push the issue, notifying NIH scientist's Walter Stewart and Ned Feder. In doing so, they sparked parallel investigations by the NIH and by the congressional subcommittee on oversight and investigation with jurisdiction over the NIH that continued for the next six years.

In 1994 a report by the ORI found Imanishi-Kari guilty of numerous counts of fabricating and falsifying data and banned her from receiving NIH funding for a period of ten years. However, two years later the DHHS's Research Integrity Adjudications Panel exonerated Imanishi-Kari of fraud. The panel made note of the many errors in the *Cell* paper as well as the sloppiness of Imanishi-Kari's bookkeeping but stated that solid evidence of intentional misrepresentation was lacking. That was the second ruling by the ORI that had been overturned by an expert panel (the first had been the Gallo ruling), shedding doubt on the office's ability to police scientific misconduct.

The Baltimore case also raised questions about the treatment and protection of whistle-blowers. Throughout the ten years of the ordeal O'Toole was alternately ostracized and praised for her actions and was unable to find work in science. Her experience and the similar experiences of others sparked a movement that resulted in improved protections for whistle-blowers.

Results and Changes

These four cases illustrate a variety of the difficult issues related to scientific misconduct. They raise questions about the high expectations placed on researchers and about authorship requirements, supervision of laboratory work, appropriate attribution of credit, collegiality, transparency of data recording, and treatment of whistle-blowers. These cases also demonstrate that the distinction between honest errors or omissions and intentional fraud is not an obvious one. Significant improvements in the process used to negotiate the murky waters of scientific misconduct have come out of these experiences.

In some cases, a rapid and transparent response to revelations of misconduct may minimize the damage done by those revelations. In 1996, Francis Collins, head of the human genome project at NIH, became aware that data had been falsified by one of his graduate students in five papers that he had coauthored. Collins promptly confronted the student, informed researchers for whom the information would be relevant, retracted

two of the papers and corrected sections of three others. Although this case differs from those above in that the researcher accused of misconduct did not deny the allegations, it may illustrate the advantages of dealing with instances of misconduct quickly and openly.

JANET MALEK

SEE ALSO *Bioethics; Medical Ethics.*

BIBLIOGRAPHY

Anderson, Christopher. (1993). "Popovic Is Cleared on All Charges; Gallo Case in Doubt." *Science* 262(5136): 981–983.

Cohen, Jon, and Eliot Marshall. (1994). "NIH-Pasteur: A Final Rapprochement?" *Science* 265(5170): 313.

Culliton, Barbara J. (1974). "The Sloan-Kettering Affair: A Story without a Hero." *Science* 184(4137): 644–650. First part of a review of the Sloan-Kettering affair.

Culliton, Barbara J. (1974). "The Sloan-Kettering Affair (II): An Uneasy Resolution." *Science* 184(4142): 1154–1157. Second part of a review of the Sloan-Kettering affair.

Culliton, Barbara J. (1983). "Coping with Fraud: The Darsee Case." *Science* 220(4592): 31–35. Review of the Darsee case and its fallout.

Culliton, Barbara J. (1990). "Inside the Gallo Probe." *Science* 248(4962): 1494–1498. First part of a review of the Gallo probe.

Hamilton, David P. (1991). "NIH Finds Fraud in Cell Paper." *Science* 251(5001): 1552–1554.

Kaiser, Jocelyn, and Eliot Marshall. (1996). "Imanishi-Kari Ruling Slams ORI." *Science* 272(5270): 1864–1865.

Kevles, Daniel J. (1998). *The Baltimore Case.* New York: W. W. Norton. Detailed history and analysis of the Baltimore case.

Marshall, Eliot. (1996). "Fraud Strikes Top Genome Lab." *Science* 274(5289): 908.

Palca, Joseph. (1992). "'Verdicts' Are in on the Gallo Probe." *Science* 256(5058): 735–738.

Rubenstein, Ellis. (1990). "The Untold Story of HUT78." *Science* 248(4962): 1499–1507. Second part of a review of the Gallo probe.

PHYSICAL SCIENCE CASES

In the year 2002, two cases of scientific misconduct by physicists received prominent attention. One involved a young scientist at Bell Laboratories named Jan Hendrick Schön, and the other a researcher at Lawrence Berkeley National Laboratory (LBNL) named Victor Ninov. This was a surprising development because nearly all cases that had arisen prior had been in biology, biomedicine, and related fields. The questions arose: Why had the physical sciences previously seemed immune to this kind of misbehavior, and what had suddenly changed?

Qualifications

Before responding to these questions, it is important to consider the scope of misconduct and some charges of historical significance. Misconduct is a narrower concept than ethics in science. There are many ethical issues having to do with conflict of interest, not properly sharing credit, not hyping results or prospects in grant applications, covering up misconduct, reprisals against whistleblowers or malicious allegations of misconduct, violation of due process in handling misconduct cases, treating graduates students fairly, and so on, that are not part of scientific misconduct in the strict sense. During the 1980s and 1990s, after considerable debate, scientific misconduct was carefully defined as fabrication, falsification, or plagiarism (FFP) of results. It is this FFP definition that is most appropriate to bring to bear on considering misconduct in science because, without a well-crafted understanding, many activities can unfairly be called misconduct when they should more properly be called moral weaknesses or improper behavior. This is not to downplay the importance of a host of ethical issues, but simply to be clear in discussion.

Until the two physics cases arose, the fabrication and falsification type of misconduct seemed to be confined to biology and related sciences. A considerable number of such cases surfaced during the 1980s and 1990s. From those cases a pattern emerged of preconditions for such misconduct. First the scientists who commit misconduct are under career pressure. Of course all scientists are almost always under career pressure, but the point is they engage in misconduct for motives more subtle than simple monetary gain. Second scientists do not purposely insert falsehoods into the scientific record, but rather fabricate or falsify data, giving a result they believe to be true without taking the time to do the science properly. In other words, this kind of misconduct is always a violation of the scientific method, never purposely a corruption of the body of scientific knowledge. Such is almost certainly the case because even corrupt scientists believe that science is self-correcting, and a wrong result will eventually be found out. Finally misconduct occurs in fields where reproducibility is not very precise. This last point explains why the physical sciences seemed immune to such behavior while biology did not. If two organisms as identical as they can be made, for example, two transgenic mice, are exposed to the same carcinogen under the same conditions, they are not expected to produce the same tumor, at the same time, in the same place. This is an example of what is known as biological variability. Experiments in biology are not as precisely reproducible as those in physics generally are supposed to be, so a biologist disposed to cheat does not fear that someone else repeating the same experiment will find it out quickly. The two physics cases that arose in 2002 pose a severe test of this pattern.

Ninov Case

Dr. Victor Ninov was a leader of the Berkeley Gas Filled Separator (BGS) group at LBNL. Ninov had joined LBNL in 1997 after a stint at the rival GSI, German acronym for the Laboratory for Heavy Ion Research, in Darmstadt. The BGS is a device designed to sort through the debris of nuclear collisions between a stationary target, and particles that are accelerated in the LBNL 88-inch cyclotron. The Berkeley laboratory has a distinguished history of discovering heavy, radioactive elements by this means. However, although even heavier elements were believed to be possible, it was widely thought that this so called *cold fusion* method of producing new elements had pretty well run its course and entirely new approaches would be needed. This would not have come as good news to Dr. Ninov and the BGS group.

The possibility of a reprieve from this situation arose when a theory published by Robert Smolańczuk predicted a highly enhanced probability of creating superheavy element 118 if projectiles consisting of an isotope of krypton were fired with the right energy into a lead target. The signature of such an event would be a chain of subsequent events, in which the original nucleus shed alpha particles at times and with energies predicted by the theory. This was just the kind of experiment the BGS was designed to do. In May 1999, a paper was submitted for publication, and a few days later a press release was issued by LBNL, both announcing that three instances of decay chains characteristic of element 118 had been observed.

By international agreement, new elements are not official until their discovery has been reproduced. The GSI in Germany and a research group in Japan immediately undertook to reproduce the new result, but both failed. The BGS group did a new series of experiments, and in 2001, produced a fourth signature decay chain. But by now, suspicions had been aroused. A series of investigations ensued determining that the data for all four significant decay chains had been fabricated, and that Ninov was the only person in a position to have done it. The entire BGS group was criticized for not checking the raw data more carefully in what would have been a major scientific discovery, but Ninov alone was found guilty of scientific misconduct. Furthermore the investigations uncovered that in the earlier discovery of element 112 at the GSI in Darmstadt, a discovery that was real and that had been reproduced, data had nevertheless been fabricated, and Ninov had been a member of the group at the time. Ninov was fired by LBNL.

Schön Case

The other physics case involved Jan Hendrick Schön, a young superstar who had recently arrived at Bell Laboratories in Murray Hill, New Jersey, after completing his Ph.D. at the University of Konstanz in Germany. Schön, a postdoctoral member in the research group of a well known and highly respected physicist named Bertram Batlogg, did experiments in which an intense electric field drew electrons to the interface between a semiconducting material and an insulating layer. Such devices are known as MOSFETs (metal-oxide semiconductor field effect transistors) and, using conventional semiconductors such as silicon, they had been the mainstay of the electronics industry for years. The Batlogg group's work involved substituting exotic materials such as organic crystals for the silicon, and using the field effect to alter their properties. Schön's results seemed truly spectacular. In a period of only two years, together with a total of some twenty collaborators, he turned out eighty research papers announcing remarkable breakthroughs that many others had attempted but failed to achieve.

Then questions arose. In some cases, the data just looked too good to be true. In other cases, completely independent curves had identical noise, little glitches in the data that are inevitable in any real experiment, but that should be random, meaning no two experimental curves should be identical to one another. These anomalies and others were reported to the management of Bell Laboratories, which, in May 2002, announced that it had appointed a committee, headed by Malcolm Beasley of Stanford University, to investigate. It also announced that the committee's report would be made public. By contrast, the report of the committee that investigated the Ninov case at LBNL is regarded as a confidential personnel matter, and has not been released to the public.

The Beasley committee, whose report was issued at the end of September 2002, chose to investigate some twenty-four specific allegations, and found that Schön had committed misconduct in at least sixteen of them. They also decided that Schön alone, and none of his collaborators, was responsible. The insulating layer in the MOSFET was the key to the whole affair. The process by which the insulating layer is laid down on the semiconductor is called sputtering. Schön, who started his collaboration with Batlogg when he was still a graduate student, had tried his hand at sputtering an insulating layer on to one of the group's exotic samples in a very modest apparatus at his university, in Konstanz. The insulating layer proved to be much more robust than those that others were able to make. It allowed stronger electric fields to be applied, producing results that no one else could achieve. Because sputtering involves complex processes that are not well understood or controlled, it seemed believable to Schön's collaborators that for unknown reasons, the apparatus in Konstanz could make better insulating layers than could be made anywhere else. Thus it was believable that Schön could get experimental results no one else could produce. People and samples shuttled back and forth between Murray Hill and Konstanz, but all the sputtering was done in the magic machine at Konstanz and Schön alone made nearly all the measurements. The results were, literally, too good to be true. When the Beasley report came out, Schön was immediately fired by Bell Laboratories.

Assessment

The two physics cases of 2002 can be analyzed in light of the pattern, described above, that had emerged from previous cases of scientific misconduct. The three necessary (but certainly not sufficient) factors that seemed to be present whenever misconduct occurred were career pressure, belief in knowing the answer before the experiment was performed, and the expectation that the experiment was not easily and precisely reproducible. All three factors were unmistakably present in the Schön case. The atmosphere at a place like Bell Laboratories puts great pressure on scientists to succeed. The effects that Schön and his collaborators reported were widely believed to be possible, even though no one else had managed to obtain them yet. In fact, in an addendum to the Beasley report, Schön admits that he made some mistakes, but says he still believes all the effects he reported were real. And finally, the field is notorious for its lack of reproducibility. The problem lies not only in the sputtering, but also in the difficulties of preparing good samples of the exotic materials involved. If an experiment fails to reproduce a given result, it does not necessarily show the result was mistaken, it just means the experiment was performed on an imperfect sample Thus a failure to reproduce has no significance at all.

The Ninov case is more subtle, and requires some speculation as to cause. Certainly Ninov and the BGS group were under pressure to produce something new because their measurement technique seemed to have run its course, giving them less leverage to get expensive beam time on the 88-inch cyclotron and perhaps even threatening the continued existence of the group itself. The theory by Smolańczuk gave the group new hope, and quite possibly, Ninov came to believe in it because he needed to. The question of reproducibility appears to pose a contradiction, though, because the field is one in which results must be reproduced before they are official. Ninov seems to have turned the irreproducibility factor upside down. If he believed that element 118 existed, he also must have believed that its discovery could be reproduced, and, when that happened, that he and his group would get credit for the original discovery. This, of course, is exactly what occurred in the discovery of element 112, an experiment he had also been involved in; data had been faked, but the discovery turned out to be real.

These cases demonstrate that the physical sciences never were immune to FFP misconduct, and that nothing has suddenly changed. The necessary factors may line up less often than in some other areas of science, but when they do, misconduct can follow just as in other fields.

DAVID L. GOODSTEIN

BIBLIOGRAPHY

Beasley, Malcolm R., et al. (2002). "Report of the Investigation Committee on the Possibility of Scientific Misconduct in the work of Hendrik Schon and Coauthors." Unpublished.

Broad, William, and Nicholas Wade. (1982). *Betrayers of the Truth.* New York: Simon and Schuster.

Goodstein, David. (1994). "Pariah Science—Whatever Happened to Cold Fusion?" *American Scholar* 63(4): 527–541.

Goodstein, David. (2001). "In the Case of Robert Andrews Millikan." *American Scientist* 89(1): 54–60.

Goodstein, David, and James Woodward. (1996). "Conduct, Misconduct and the Structure of Science." *American Scientist* 84(5): 479–490.

Vogt, R., et al. (2002). "Report, Committee on the Formal Investigation of Alleged Scientific Misconduct by LBNL Staff Scientist Dr. Victor Ninov." Unpublished.

SOCIAL SCIENCE CASES

Issues of scientific misconduct in the forms of fabrication, falsification, or plagiarism (FFP) tend to be most prominently reported in the biomedical area, where fraudulent data may lead to serious consequences for patients receiving treatment. Nevertheless, scientific misconduct in the social sciences may also cause considerable damage—not the least being the undermining of public trust in a scientific endeavor that aims to be of benefit to social decision-making. Among the cases that have been most prominent in this area are those associated with anthropologists and psychologists.

Anthropology Cases

The American anthropologist Margaret Mead (1901–1978) in her famous 1928 study, *Coming of Age in Samoa,* described adolescence in those islands in glowing terms with little cultural competition and easy and frequent sexual activities among teenagers that was not condemned by Samoan society. The only problem with this book, which received high acclaim, was that it was based on a myth, as later documented in detail by Derek Freeman (1983). Reasons for such a vast and almost complete misinterpretation of the facts of the culture, according to Freeman, include the following: Mead could not speak the Samoan language; she lived with an American family while on the islands; she was denied access to the chiefs who determined the laws and customs; she simply overlooked contradictory data to her favorite theories. Freeman's criticisms of Mead have, however, been challenged; for a review of the controversy, see James E. Côte (2000).

Other cases have involved charges that anthropologists have on occasion aided and abetted the mistreatment of indigenous peoples or illegitimately conspired with national governments. Patrick Tierney (2000), for example, charged that during the 1960s the anthropologist Napoleon A. Chagnon was complicit in the fomenting of violence among the Yanomami, a tribal people living in remote areas of Brazil and Venezuela. (He also charged Chagnon's associate, the geneticist James V. Neel, with administering measles vaccine to the Yanomami according to protocols that were not in their best interest.) An American Anthropological Association (AAA) investigation did not sustain the most grievous charges, and in fact argued that Tierney himself, through misrepresentation and sensationalism, failed to practice responsible journalism. Nevertheless, it did admit that the Yanomami are now in such danger as to encourage "anthropologists to reflect deeply upon the ways in which they conduct research" (AAA 2002).

The AAA has also reported on a number of other ethics cases. Among these are the outrage of Franz Boas (1858–1942) at the use of anthropology as a cover for espionage during World War I and debates about the authenticity of the autobiography of the 1992 Nobel Peace Prize winner, the Guatemalan peasant activist Rigoberta Menchú.

Psychology Cases

In psychology, cases are both more numerous and more contentious than in anthropology. One commonly discussed early case in psychology involved the work of John B. Watson (1878–1958), who espoused a strong form of behaviorism. Some people vigorously question the quality of his study, known as Little Albert, that supposedly showed conditioned fear of a stuffed toy rabbit in a baby (Cohen 1979). Whatever the final settlement of the argument regarding Watson's work, there is no doubt that later, starting in the 1980s, such cases would have been judged scientific misconduct by social scientists.

It should be noted, however, that many social scientists were working in biomedical areas. The first such publicized case was that of Stephen E. Breuning, a psychologist studying the effects of psychoactive medications on the behavior of a vulnerable population, the institutionalized mentally retarded, people societies typically strongly protect. Neuroleptic medications, commonly known as tranquilizers, are often given to the mentally retarded to control aggressive and self-injurious behavior. Breuning was conducting studies on these neuroleptic medications, but was collecting little data. Instead he was fabricating data indicating that such medications were harmful to the patients' learning and behavior. Thus, he was strongly suggesting on the basis of fabricated data that removing medications from these vulnerable patients might be helpful to them.

In December 1983 Robert L. Sprague reported Breuning's fraud to the appropriate federal agency that was funding his research, the National Institute of Mental Health. The agency began an investigation that moved with glacierlike speed even in this crucially important health area. Although there were publications in the scientific press about the slowness of the investigation in this important case (Holden 1986), the agency did not issue its first report until April 1987—more than three years after receiving smoking-gun evidence of scientific misconduct (NIMH 1987). Breuning was the first independent scientist with his own federal

research grant to be indicted, tried, and found guilty of fraud in federal court. Considering the seriousness of his offenses, his sentence was light; he served no jail time, but was confined to a halfway house for sixty nights and fined $11,352 (Wilcox 1991).

Another important case in psychology was that of Marion Perlmutter, a psychologist at the University of Michigan who plagiarized the research proposals of Carolyn Phinney, also at Michigan. When confronted with an accusation by Phinney, Perlmutter denied any wrongdoing and the university officials initially supported her (Gordon 1991). When Phinney could not obtain justice through university channels, she was the first victim of scientific misconduct to turn to the courts for relief (Gordon 1993). After a trial in Ann Arbor, Michigan, Phinney was awarded $1 million in damages for theft of intellectual property and research proposals. University officials unwisely followed Perlmutter's request to appeal. The appellate court upheld the trial court and added to the damage award interest because of the years of delay while the appeal process took its course, increasing the award to $1.6 million (Hilts 1997).

Assessment

These are only a few of the more than 300 cases on which data have been collected by Sprague since his discovery and disclosure of the Breuning fraud. Drawing on these and other cases in the social sciences, it is possible to argue three points. First, it is likely that there are more cases of misconduct in the biomedical sciences than in the social sciences (Shamoo and Resnik 2003). One reason for this discrepancy may be that large profits are often involved in research leading to new medications, which is seldom the case in the social sciences. The potential for making large profits seems to bring out the worst in human beings, including scientific researchers.

Second, during the 1990s universities were sluggish in recognizing misconduct problems among their faculty and slow in taking corrective actions. This was as true in the social as in any other sciences.

Third, times have changed, and the situation has improved in the social sciences as elsewhere—though the situation could hardly be termed ideal. There is hope for continued improvement with federally mandated training for graduate students and federal requirements that universities maintain written policies addressing scientific integrity. Furthermore, there has been a sharp increase in the awareness of scientific misconduct among researchers.

Despite this increased awareness, one must be careful to distinguish cases of misconduct in the social sciences from research that is simply controversial. Twin studies, IQ studies, and race studies, for instance, are sometimes mentioned as cases of scientific misconduct. But although research in these areas may have been very controversial, this does not mean that they involved scientific fraud or misconduct. They may have poorly designed or unwise. Still, misconduct and controversy must be distinguished.

ROBERT L. SPRAGUE
CARL MITCHAM

SEE ALSO *Sociological Ethics.*

BIBLIOGRAPHY

American Anthropological Association (AAA). (2002). *Final Report of the AAA El Dorado Task Force.* Arlington, VA: Author. Also available from http://aaanet.org/edtf/index.htm.

Arias, Arturo, ed. (2001). *The Rigoberta Menchú Controversy.* Minneapolis: University of Minnesota Press.

Cohen, David. (1979). *J. B. Watson: The Founder of Behaviourism.* London: Routledge and Kegan Paul.

Côte, James E. (2000). "The Mead–Freeman Controversy in Review." *Journal of Youth and Adolescence* 29(5): 525–538. The introduction to a theme issue on the Mead–Freeman controversy that includes four other articles and some historical documentation.

Freeman, Derek. (1983). *Margaret Mead and Samoa: The Making and Unmaking of an Anthropological Myth.* Cambridge, MA: Harvard University Press.

Gordon, G. (1991). "Academic Fraud Charged at U-M." *Detroit News* 3A, 5A. Newspaper report on the Perlmutter case.

Gordon, G. (1993). "Misconduct Costs U-M $1 Million." *Detroit News.* Another newspaper report on Perlmutter.

Hilts, Philip J. (1997). "University Pays $1.6 Million to Researcher." *New York Times* August 10, p. 10.

Holden, Constance. (1986). "NIMH Review of Fraud Charge Moves Slowly." *Science* 234: 1488–1489. A report on the Breuning case.

"Jury Awards $1.1 Million to Researcher." (1993). *Ypsilanti (Mich.) Press.* On the Perlmutter case.

Mead, Margaret. (1928). *Coming of Age in Samoa: A Psychological Study of Primitive Youth for Western Civilisation.* New York: Morrow.

National Institute of Mental Health (NIMH). Panel to Investigate Allegations of Misconduct. (1987). *Report and Recommendations of the Panel.* Author.

"Psychologist Pleads Guilty to Research Fraud." (1988). *Psychiatric Times,* 21. On the Breuning case.

Shamoo, Adil E., and David B. Resnik. (2003). *Responsible Conduct of Research.* Oxford: Oxford University Press.

Sprague, Robert L. (1993). "Whistleblowing: A Very Unpleasant Avocation." *Ethics and Behavior* 3(1): 103–133.

Tierney, Patrick. (2000). *Darkness in El Dorado: How Scientists and Journalists Devastated the Amazon.* New York: Norton.

Wilcox, B. (1991). "Fraud in Scientific Research: The Prosecutor's Approach." *Accountability in Research: Policies and Quality Assurance* 2: 139–151.

MISSILE DEFENSE SYSTEMS
• • •

Experts have long debated the idea of defending national territory against airborne strategic attack. These debates often conflate feasibility, morality, strategy, and politics, so that each observer must independently weigh such factors even in arguments that seem to be purely technical.

BMD (ballistic missile defense) supporters tend to draw on early strategic theory developed by Wohlstetter (1958) and others from the RAND Corporation (for example Kahn 1970). They generally suggest the following: Nuclear strategy is neither easy nor impossible. It requires repeated analysis and improvement. It should serve national policy, such as deterring enemies (or the nation should change its policy). National leaders have a commitment to preserve and protect the people and the political system as well as they can, which no technical advice can abrogate. Deterrent systems should maximize human control over weapons. Deterrence based on the threat of retaliation against civilians is immoral if targetting enemy weapons is possible. Furthermore, the strongest supporters of BMD tend to have more faith in large-scale technology development and system predictability and performance.

BMD critics, in contrast, generally believe the following. Nuclear weapons are so awful and so difficult to stop that their development should freeze in place and that arms control diplomacy should be relied on to reduce them. Even one or a very few nuclear weapons detonated in war would be as bad as many, so that targeting a few against cities is enough to threaten assured destruction (AD) to any potential attacker, thereby achieving deterrence and stability. No defense is likely to prevent some attackers from getting through to cause unacceptable damage.

In this view, offense dominates defense. Defenses undermine stability and encourage useless competitive arms procurements ("arms races"). Measures to reduce the consequences of nuclear war will only encourage it. These critics accept mutual deterrence based on the threat of retaliation, to the point in the 1970s of considering a policy of [immediate] "Launch-On-Warning" or "Under Attack" (Garwin 1989, pp. 189–198). Therefore, they advocate cooperating with adversaries against a greater and common enemy, the danger of nuclear war, by accepting mutual vulnerability (Carter and Schwartz 1984).

Whatever the value of any of these views on either side, they often have combined technical, strategic, political and ethical beliefs in ways difficult for observers to evaluate. This problem caused one professional society to conduct a formal—and critical—review of the professional standards at work (ORSA 1971).

Historical Development of Arguments

While the debate from the 1960s to the present has focused on ballistic missile threats, strategic defenses may target any air-borne attacker. Strategic defense may use active means such as interceptor weapons and passive means such as hardening (protecting), hiding, and dispersing assets against enemy targeting. After German dirigibles bombarded London during World War I, thinkers from the English writer and futurist H. G. Wells to the U.S. Army Air Corps predicted that future wars would be dominated by air power, which would be "strategic" more than "tactical": It would aim at national will, not forces, by attacking the enemy cities to force the population to demand that its government sue for peace. Defenses could not stop "the bomber always getting through," and only a few bombs would be enough to achieve the strategic goal. Therefore, nations should ignore defenses and rely for security on their own bombers to threaten retaliation. Yet in 1940 the Battle of Britain saw Royal Air Force fighter defenses stop enough German bombers to defeat their strategic purpose, even if many bombs indeed "got through."

The atomic bomb revived the idea that devastating attack was unavoidable, and therefore the only means of stabilizing relations was to threaten retaliation. Some even argued that the *analysis* of the military use of nuclear weapons was immoral and "unthinkable" because it might make nuclear war seem rational.

After World War II, many technologists involved in the development of atomic weapons helped pioneer this debate (Kimball Smith 1965). Radar expert Louis Ridenour, in a collection (Masters and Way 1946) for the nascent Federation of American Scientists immediately after the war, described the great difficulty of countering each aspect of an airborne attack, making defense hopeless. Such arguments have become standard, as in those made by leading assured destruction theorists and BMD critics such as Richard L. Garwin (1989) and others (for example UCS, 1984).

This view reached its peak in 1972, when the United States and the Soviet Union pledged in the Anti-Ballistic Missile Treaty (ABMT) not to defend against each other's missile threat, arguably making further offensive weapon development superfluous. The two powers accompanied the ABMT with a Strategic Arms Limitation Agreement capping offensive forces at some 1,700 U.S. missiles and almost 2,500 Soviet missiles, numbers that diplomats expected to reduce in future negotiation.

These missile levels were more than enough for AD theorists. They saw the only rational use of strategic forces as pure deterrence (threatening cities, if not expecting actually to attack them). They saw military use (targeting forces) as irrational. Cities were good targets because the destruction of enemy cities was easy and of our own, unacceptable. This scenario eliminated both the targeting side's temptation to upgrade its weapons and the targeted side's temptation to make useless, yet still provocative, defenses. A balance of deterrence ensued—"mutual assured destruction" (MAD). Neither side could envision a nuclear war scenario from which it could escape intact. While leaders' acceptance of vulnerability, especially of civilians, might turn ethical traditions on their head, proponents believed they had a better analysis of the dynamics of nuclear peace. With the election of Jimmy Carter to the Presidency in 1976, these views achieved their peak in U.S. policy.

The Soviets, however, frustrated expectations. By 1979, Harold Brown, President Carter's Defense Secretary, told Congress that "Soviet spending has shown no response to U.S. restraint—when we build they build; when we cut they build" (Brown 2003). The Soviets also improved the accuracy of their warheads, which they now mated to their very large boosters. The combination raised the possibility of a disarming first strike—against not U.S. cities but the land-based deterrent forces themselves. Few enough might survive that retaliation would then fall to the submarines and the bombers, in which defense supporters (but not the AD theorists) saw major problems.

BMD Proposals

In 1980 the election of Ronald Reagan to the Presidency signaled a new U.S. skepticism on arms control and assured destruction. President Reagan accepted advice that new technologies based, for example, on directed energy, might create systems that could destroy Soviet missiles, and thereby move zthe basis of deterrence away from mutual threat, toward mutual security. In March 1983, he supported BMD by asking, "Would it not be better to save lives than to avenge them" by countering "the awesome Soviet missile threat … before they reached our own soil or that of our allies"? Was it not "worth every investment necessary to free the world from the threat of nuclear war?" (Reagan 1983).

Despite strong Reagan Administration support, BMD development programs did not receive similar priority from Congress, the military services, or the subsequent presidencies of George H. W. Bush and (especially) Bill Clinton. In December 2001, however, President George W. Bush announced U.S. withdrawal from the ABM treaty and the intention to develop layers of short-, medium-, and long-range interceptors—air-, sea- and space-based—and the systems to manage them (White House 2001).

BMD nevertheless continued to be controversial. U.S. technical experts, pro- and especially anti-BMD, have often demanded that any BMD system reach extremely high levels of effectiveness. Yet often beneath their arguments there lurk basic questions of technology (will it work?) mixed with policy (should it?). These should be made explicit. For example, BMD "effectiveness" makes sense only in terms of some policy goal. A 100-percent-effective shield may be impossible but also strategically excessive. Alternatively, a defense of three independent layers of say 50-percent effectiveness each, defending retaliatory forces, might make any incoming attack prohibitively expensive if not suicidal. It depends on the strategy.

Attitudes outside the United States

The Russian, Chinese, and North Korean governments oppose BMD because it reduces their threat to the United States and its allies. Beginning in the early 1990s Japanese governments, perhaps as worried by their own pacifists as by China and North Korea, engaged in a delicate and muted dance of cooperative BMD development efforts. The problem is that, if Japan lacks both defenses and a tie to a United States that can credibly defend it, it may well face a choice of acquiescence to its neighbors or developing its own retaliatory forces. Either could be a global disaster.

European experts worried that a U.S. defense system might "decouple" the United States from NATO, make nuclear war more thinkable, or remove constraints on conventional war. Yet lacking a Soviet threat to deter in Europe, the United States relies more on conventional forces to support stability, globally. At the same time rogue states and terrorists have pursued their own mass destruction weapons to deter the Unites States from using its forces. Further European objections to U.S. defenses, therefore, seem more related to intra-alliance political jockeying, resentment at the association of BMD with Presidents Reagan and George W. Bush (neither popular in Europe), and a belief that the ABM treaty is worth preserving as a precedent for arms control.

It nevertheless appears that U.S. BMD work will continue, if only to deny future missiles—from China, North Korea, or anywhere else—an unimpeded ride into the United States. Whatever the validity of AD theory that governed U.S. policy during the cold war, the United States is unlikely to continue to pursue that course alone. While seeking peaceful relations with other powers, it is difficult to see how U.S. leaders will not consider protection against the possible worst case, if only to make it less likely. Missile defenses cannot solve all problems. That they nevertheless try to address some significant ones is likely to capture the attention of leaders.

If these trends hold, the role of the scientists and engineers who have challenged BMD will be essential to ensure that missile defense programs achieve technical, programmatic, and strategic soundness. If both the hopes of BMD supporters and the critiques of BMD detractors are more task-focused and less millennial, their debates will be more transparent, professional, and indeed ethical.

THOMAS BLAU

SEE ALSO *Computer Ethics*; *Military Ethics*.

BIBLIOGRAPHY

Brown, Harold. (2003). *Paul C. Warnke Lecture In International Security*. New York: Council on Foreign Relations.

Carter, Ashton B., and David N. Schwartz, eds. (1984). *Ballistic Missile Defense*. Washington, DC: Brookings Institution. See Leon Sloss, "The Strategist's Perspective," 24–48.

Garwin, Richard L. (1989). *Richard Garwin on Arms Control*, ed. Kenneth W. Thompson. Lanham, MD: University Press of America.

Kimball Smith, Alice. (1965). *A Peril and a Hope*. Chicago: University of Chicago Press.

Kahn, Herman. (1963). *Thinking About the Unthinkable*. Croton-on-Hudson: Hudson Institute.

Kahn, Herman. (1970). *Why ABM?* New York: Prentice-Hall.

ORSA [Operations Research Society of America]. (September 1971). *Operations Research* 5. "Guidelines for the Practice of Operations Research."

Ridenour, Louis. (1946). "There Is No Defense." In *One World or None*, ed. Dexter Masters and Katharine Way, 33–38. New York: Whittlesey House, McGraw-Hill.

UCS (Union of Concerned Scientists). (1984). *The Fallacy of Star Wars*, ed. John Tirman. New York: Random House.

White House. (2001). *ABM Treaty Fact Sheet: Statement by the President. Announcement of Withdrawal from the ABM Treaty*. Washington, DC: White House Office of the Press Secretary.

INTERNET RESOURCES

Reagan, Ronald. (1984). Announcement of the Strategic Defense Initiative. Available from http://www.nuclearfiles.org/redocuments/1984/841228-reagan-sdi.html.

Wohlstetter, Albert. (1958) "The Delicate Balance of Terror." P-1472. Santa Monica: The RAND Corporation. Available from www.rand.org.

MODELS AND MODELING

• • •

Models are abstractions of reality, and modeling is the process of creating these abstractions of reality (Wallace 1994). Models take a variety of forms based upon their function, structure, and degree of quantification (Tersine and Grasso 1979). For example the functions of a chart of an organization is to describe and does not provide any predictions or recommendations; a sales forecast predicts the future based upon a particular business strategy; and a procedural manual for a manufacturing process is normative in that it provides advice on how to manage a process. The structure of a model can be symbolic (represented by equations), analog (using graphs to model physical networks), or iconic (physical representations such as scale models). Models are usually thought of as being quantitative, and able to be represented mathematically. However, qualitative models are far more common. For example, mental models play a very important role in the conceptualization of a situation (Crapo, Waisel, Wallace and Willemain 2000) and verbal and textual models are used in the process of communicating mental models. Because reality is near-infinitely complex, all data needs to be processed, which involves a movement from information to knowledge. Models are forms of codified knowledge.

Science can be seen as a model-building enterprise, because it attempts to produce abstractions of reality that help scientists understand the world (Little 1994). Technological advances in computing allow for the development of complex computer-based models in a wide range of fields. These models can be used to describe phenomena observed in the world as well as to provide structure to real or hypothetical experiences described or postulated by individuals or groups. Models play a very important role in formalizing and integrating theoretical principles from science that pertain to the phenomena being studied. For example, the computational models used for weather forecasting integrate scientific principles from a variety of the physical and natural sciences.

As the role of models within society increases, the significance of ethical issues related to the development and use of models also rises. Models are generally designed by experts who may hold privileged positions, yet model users and those affected by models may cover a wide demographic range. Thus, it is ethically imperative that researchers consider the relationships among the modeler(s), the model, the user(s), and those affected by the model.

Models may be developed for a range of purposes, in a variety of domains, including research, education, and applications. This entry begins with a brief overview of the ethics of modeling in each of these domains. The next focus is on ethical issues that span all three domains. Finally, the conclusion provides an assessment of the current status of the ethics of modeling.

Ethics of Modeling in Research

Models play an important role in scientific and engineering research. Scientific researchers seek to better understand the world, and models can serve as a way for them to create these understandings. Engineers try to improve the world by creating new technologies, and modeling allows them to explore their ideas in the abstract before moving on to the concrete. Computer aided design is one example. This technology allows an engineer to create a model design and view the resulting product in a three-dimensional graphical representation. This creative process can be repeated many times with various participants before the physical prototype is produced. In both science and engineering, models serve as tools for understanding the world and the ways in which people can improve that world.

One important ethical issue of modeling in research is the relationship between modeling and the norms of science. John Allison and colleagues (1994) argue that the fundamental ethic of science is an assumption of openness and access to data and methodology that fosters repeatability and verifiability. Yet, they point out that science increasingly relies on proprietary databases that do not allow others to repeat or verify the studies, such as economic analyses that use corporate financial data. They assert that models in this context may pose a danger to society, unless their data and methodology are kept open, as has been overwhelmingly the case over the long history of scholarly scientific research. Thus, it is important to consider not only the ends to which modeling is used in research but also the means through which it is used.

One way to ensure that models for research are used ethically is to develop a code of ethics for modeling within a particular domain of research. Saul I. Gass (1994) explores the codes of ethics for various research fields and organizations. He concludes that a uniform code of ethics should be developed so that researchers within a wide range of specialties can benefit from it.

Ethics of Modeling in Education

Another important use of modeling is for instruction. In education, models can be used to help students better understand a problem. Manipulation of the model—whether it is a formula, a plastic mock-up, or a computer simulation—helps students develop a better understanding of the problem at hand. Similarly, models can be useful in training, potentially allowing trainees to practice techniques and skills in a relatively risk-free environment.

Barbara Y. White and John R. Frederiksen (2000) argue that computer-based models are particularly important in education because they make scientific inquiry potentially accessible to all students. They assert that computer-based models can help students develop the conceptual models necessary for scientific inquiry. These tools allow students to experiment with models in order to better understand naturally occurring relationships captured by theories in physics or other academic subjects. White and Frederiksen further argue that students should be able to use computers not only to learn to apply models but also to create models and understand the principles behind modeling natural systems. Modeling in education can thus include both learning to build models and learning to use models.

Perhaps one of the most ethically intriguing applications of modeling is the use of virtual reality in education and training. Virtual reality models have been used to train and evaluate doctors, pilots, and other professionals. The goal of such models is to provide a safe environment that mirrors the work environment in potentially all ways except the consequences of the actions taken in the simulated environment. One issue requiring further study is the role that consequences play in affecting actions, and consequently, the potential utility of such environments. Another issue is that virtual reality environments may become so realistic that it

becomes difficult or impossible to distinguish between the actual situation and the model of it. In such cases, transparency may be one way to avoid ethical dilemmas. Thorough documentation of the model, delineation of the assumptions the model makes about reality and values, and an explicit representation of the components of the model and how they are linked are all ways to help ensure the transparency of the model.

Ethics of Modeling in Applications

Modeling may also be used in a wide range of applications. Computational models have contributed to developments such as Dupont's discovery and use of ozone-friendly chemicals (Hoffman 1995), structures than can better withstand earthquakes (Booker 1994), and innovations in nanotechnology (Bozman 1993). Computational models are also increasingly being used for public policy-making (Kollman et al. 2003), and as a result they are receiving an increasing degree of attention in the popular press (see for example Ashley 2003). One major application of models is as aids for decision-making. Models used for decision-making may be either primarily descriptive or prescriptive—that is, they may attempt to portray reality as it is or reality as it should be. Neither of these tasks is as simple as it might seem. The design of both descriptive and prescriptive models is influenced by the perspectives of the participants, and thus it requires transparent communication and consensus between the builder(s) of the model and the user(s) of the model (Wallace 1994).

The relationship between the model builder(s) and model user(s) is inherently problematic. John D. C. Little (1994) describes six pitfalls for modelers to avoid:

(1) The user already knows the answer and wants to use the model as a justification for it.

(2) The user wants quick answers and does not give the modeler time to do a thorough study.

(3) The user does not understand the basis for the modeler's results and thus is uncomfortable about using the model.

(4) The user wants a defined, black-and-white outcome from the model.

(5) The user is allowed to put her or his own personal judgments into the model.

(6) The user does not realize that all models are incomplete.

Modelers must find ways to avoid these pitfalls that result from misinformed or misbehaving users.

Deborah G. Johnson and John M. Mulvey (1995) identify three types of relationships between modelers and users. First, they discuss a paternalistic relationship in which the modeler acts as an unquestionable expert with total control of the relationship. Next, they explore a second way of understanding this relationship, the agency model, in which the user has the upper hand in the relationship, and the modeler is merely an implementer of the user's will. They reject both of these views as being unbalanced and failing to ensure that both sides strive to fulfill their roles. They conclude that the fiduciary model is the ideal model for the relationship between the modeler and the user, because under this model, the user and the modeler work together to construct the model and the user's expectations for the model.

Ethical Issues that Connect Modeling in Research, Education, and Applications

In each of these three domains of research, education, and applications, models can be used to either help or replace humans. Models used in research may either assist researchers or take over for them. Educators may either use models or be supplanted by them. Finally, in applications such as decision-making, models may either support human decision makers or automate their roles. Given this stark choice, it is important to consider the ethical implications of both models that help humans and models that replace humans.

Mulvey (1994) argues that models that are used to replace humans, which he refers to as "computerized decision procedures," are ethically problematic because they can easily be misused or abused. Intentional manipulation of a model may be used to serve the will of those that control it, who are often the elites within a society. Thus, models intended to replace humans may be used in antidemocratic and authoritarian ways.

Vincent P. Barabba (1994) points out, however, that models used by humans can also be misused and abused. A model can, for example, be oversold, so that limitations in the accuracy, precision, or scope of the model are underemphasized or completely ignored. In this way, models used by humans may also be used by elites to ensure that their will is achieved.

It is thus important to consider the power dimensions of models and modeling. As discussed above, there are a range of possible relationships between modelers and users, and the best type of relationship appears to be a fiduciary relationship whereby modelers and users each have both responsibilities and expectations as part of the modeling process. It is important that steps are taken to regulate this relationship, to avoid unethical

behavior on either side of the transaction, and to ensure the best outcomes for both modelers and users, as well as for those affected by the model (Leet and Wallace 1994).

Models also present other ethical challenges. Models are designed to make reality more easily understandable, yet these same models may, intentionally or unintentionally, distort reality in important ways. Models may be used to make very value-laden decisions appear "scientific" and "objective." In building and using models, it is thus important to understand their limitations as well as the cultural specificity of the knowledge content and values that are explicitly and implicitly embedded in models (Leet and Wallace 1994).

Assessment of Ethics of Modeling

Richard O. Mason (1994) argues that modelers, as a part of their fiduciary relationship with users, have a professional responsibility for the models they build. To meet this professional responsibility, a modeler must fulfill two covenants: a covenant with reality and a covenant with values. The covenant with reality involves technical and social elements: The faithfulness of a model to reality often depends on highly technical decisions by the modeler, yet it is also a fundamental part of the relationship between the modeler and the user. According to the covenant with values, a modeler must understand and incorporate the user's values into the model in an effective way. These covenants are particularly important because a successful model may become a standard that affects a wide range of users and people affected by the model (see also Carrier and Wallace 1994).

In addition, it is important for the modeling process to be as transparent as possible. Because models always reflect the social and cultural context in which they are created, in both their knowledge content and values, it is most helpful if the model is open and honest about these influences. Models that contain assumptions should make these assumptions clear, rather than masking them as fact. Similarly, the extent to which a model is descriptive or prescriptive should be made immediately obvious to the user. Importantly, allowing the user to see clearly into the model is a way for the modeler to share control and responsibility with the user, allowing the user to make informed decisions based on all relevant data, rather than placing blind faith in a black box.

These three covenants—the covenant with reality, the covenant with values, and the covenant with transparency—can all help modelers and users communicate optimally so that they can mutually benefit from the process of modeling. All three covenants are important, because they make clear what users should be able to expect from designers, allowing designers and users to work as partners. Such cooperation ensures that modeling will be used for ethically responsible uses within the domains of science and technology.

WILLIAM A. WALLACE
KENNETH R. FLEISCHMANN

SEE ALSO *Georgia Basin Futures Project; Operations Research.*

BIBLIOGRAPHY

Allison, John; Abraham Charnes; William W. Cooper; and Toshiyuki Sueyoshi. (1994). "Uses Of Modeling in Science and Society." In *Ethics in Modeling*, ed. William A. Wallace. Tarrytown, NY: Pergamon. These authors address the question of the contribution and interpretation of models affects decisions.

Ashley, Steven. (2003). "Alloy By Design." *Scientific American* 289(1): 24. This brief article documents how the use of computational models led to the invention of a new class of titanium-based alloys, which are now used in medical implants, eyeglasses frames, and spacecraft parts.

Barabba, Vincent P. (1994). "The Role of Models in Managerial Decision Making—Never Say the Model Says." In *Ethics in Modeling*, ed. William A. Wallace. Tarrytown, NY: Pergamon. Barabba propounded Barabba's Law ("Never say the model says!") and discusses ways to limit model builders from making excessive claims for their models.

Booker, Ellis. (1994). "Working Toward a Quake-Proof Design." *Computerworld* 28(5): 86. Describes how computational modeling in an interdisciplinary environment is being applied to the problem of predicting the damage that can result from earthquakes.

Bozman, Jean S. (1993). "Tiny Technology: The Small World of Nanotechnology Opens Possibilities for Molecular Computing." *Computerworld* 27(31): 28. Touches on applications of computational modeling to nanotechnology.

Carrier, Harold D., and William A. Wallace. (1994). "An Epistemological View of Decision Aid Technology with Emphasis on Expert Systems." In *Ethics in Modeling*, ed. William A. Wallace. Tarrytown, NY: Pergamon. Discusses the philosophical foundations of statistics, operations research, and expert systems. Presents a framework to aid decision makers in choosing an appropriate decision technology for solving a particular problem.

Crapo, Andrew; Laurie B. Waisel; William A. Wallace; and Thomas R. Willemain. (2000). "Visualization and the Process of Modeling: A Cognitive-Theoretic View." In *Proceedings KDD-2000: The Sixth ACM SIGKDD International Conference on Knowledge Discovery and Data Mining, Boston, MA*, ed. Raghu Ramakrishnan, Sal Stolfo, Roberto Bayardo, and Ismail Parsa. New York: Association of Computing Machinery. This paper provides a description of the process of modeling and, based upon theories of cognition, shows how visualization can assist in developing computational models for very large, high dimensional datasets.

Gass, Saul I. (1994). "Ethical Concerns and Ethical Answers." In *Ethics in Modeling*, ed. William A. Wallace. Tarrytown, NY: Pergamon. Offers guidelines for professional behavior by modelers.

Hoffman, Thomas (1995). "Making A Difference: Say Goodbye To Ozone-Wrecking Chemicals." *Computerworld* 29 (23): 105. Explains how computational modeling techniques have been used since the 1970s to explore the environmental impacts of ozone-depleting chloroflurocarbons.

Johnson, Deborah G., and John M. Mulvey. (1995). "Accountability and Computer Decision Systems." *Communications of the ACM* 38(12): 58–64. Focuses on the issue of accountability in the design of computer decision systems. Argues that it is important for the designers of computer decision systems to develop and adhere to a set of standards in order to increase public trust in these systems.

Kollman, Ken; John F. Miller; and Scott E. Page, eds. (2003). *Computational Models in Political Economy*. Cambridge, MA: MIT Press. Contains a variety of applications of computational modeling to the fields of political science and economics.

Leet, Edith H., and William A. Wallace. (1994). "Society's Role in the Ethics of Modeling." In *Ethics in Modeling*, ed. William A. Wallace. Tarrytown, NY: Pergamon. Discusses the proposition that values are inherent in any model, and offers suggestions on how to ensure users are cognizant of this fact.

Little, John D. C. (1994). "On Model Building." In *Ethics in Modeling*, ed. William A. Wallace. Tarrytown, NY: Pergamon. Reviews the process of modeling and notes several "pit falls" that can waylay the model builder.

Mason, Richard O. (1994). "Morality and Models." In *Ethics in Modeling*, ed. William A. Wallace. Tarrytown, NY: Pergamon. Presents three covenants that should guide the relationship between the model builder and the user.

Mulvey, John M. (1994). "Models in the Public Sector: Success, Failure, and Ethical Behavior." In *Ethics in Modeling*, ed. William A. Wallace. Tarrytown, NY: Pergamon. Focuses on the interpretation of models and the fact that lack of understanding of the techniques used in modeling hinders a user's ability to ascertain the values inherent in the models.

Tersine, Richard J., and Edward T. Grasso. (1979). "Models: A Structure for Managerial Decision Making." *Industrial Management* 21(2): 6–11. Presents a very comprehensive classification scheme for models.

Wallace, William A., ed. (1994). *Ethics in Modeling*. Tarrytown, NY: Pergamon.

White, Barbara Y., and John R. Frederiksen. (2000). "Technological Tools and Instructional Approaches for Making Scientific Inquiry Accessible to All." In *Innovations in Science and Mathematics Education: Advanced Designs for Technologies of Learning*, ed. Michael J. Jacobson and Robert B. Kozma. Mahwah, NJ: Erlbaum. The authors use the example of physics education to explore how modelng can shape student involvement in science.

MODERNIZATION

•••

Modernization is a slippery term with manifold relations to science and technology. In a narrow sense, it is often synonymous with bringing more advanced science or technology to bear, as in modernizing a construction process or production plant. In a broader sense, social scientists describe modernity as a particular form of culture or society dependent on and supportive of science and technology, with the process of creating such a society defined as modernization. (Related concepts are *urbanization*, the concentration of population into cities, and *secularization*, the recasting of society from a basis in religious beliefs to one based on rationality, science, and technology.)

Insofar as modernization in the broader sense connotes an undermining of traditional values and is presented as a program with its own normative character, it is also of ethical significance, and has been assessed in both positive and negative terms.

Modernization is a somewhat more neutral term for a concept known in the nineteenth century as the "civilizing" process, and during the first half of the twentieth century as "Westernization." The term gained widespread currency in the 1950s, but began attracting substantial criticism during the 1960s.

Positive Assessments

In social, economic, and political theory, modernization is characterized by the achievement of industrialization, high urbanization, secularization, and rationalization. In a 1983 essay submitted for a symposium on Cultural Identity and Modernization in Asian Countries at Japan's Kokugakuin University, Robert M. Bellah analyzed the tension between tradition and modernization, then noted that when these forces successfully collaborate, the results may be remarkable: "A viable tradition should continue to guide individuals and societies in their quest for what is truly good, and modernization should simply supply more effective means for that quest" (Bellah 1983, Internet site). Bellah concludes that although the marriage of tradition and modernization is often over-stated, "the amazingly successful economic modernization" of Japan and the Pacific Rim countries is largely due to "[t]he spirit of the people,

their work ethic, their social discipline, their ability to cooperate ... all ... more or less rooted in one or another aspect of the tradition."

The Cold War vision of modernization as a weapon against the spread of Communism strongly differed from this vision of a consensual and beneficial partnership between tradition and the modern. In an influential 1968 article, Samuel Huntington urged "forced-draft urbanization and modernization which rapidly brings the country in question out of the phase in which a rural revolutionary movement can hope to generate sufficient strength to come to power." Rather than basing modernization on the consent of the governed, Huntington posited that less developed nations could be dragged into modernity—an approach applied in the Strategic Hamlets Program in Vietnam, where populations were forcibly removed at gunpoint to new "modern" surroundings, and their old homes burned.

Thus, proponents of modernization saw the process in two entirely different lights: one as a good that could be forced on subjects regardless of their wishes, and the other as a consensual step, greatly desired by the participants, toward participation in "a world of industrial, competitive nations interacting in a capitalist, free-trade, global framework" (Adas 2003, p. 37).

Critical Assessments

While modernization sounds more neutral than a phrase such as "Westernization," critics complain that it nonetheless carries with it substantial Western baggage. Modernization assumes that the sole criteria of success of a society are gross national product (GNP) and the degree of industrialization. Underlying the theory of modernization is an almost entirely unexamined premise that all other nations should seek to imitate the West, and particularly the United States.

The process of modernization has been described as a cover for the introduction of capitalism without regard for the well-being of local populations. Rather than elevating all nations to equal opportunity participation in free markets, thereby lifting their citizens to higher living standards, critics say that modernization leads perversely to increased impoverishment and greater dependency of former colonies. "Modernization and development have previously been built on considerable exploitation of certain segments of the society and have involved a degree of ruthlessness. Imperialism aided them substantially" (Dube 1988, p. 5). Modernization, of course, also brings with it the glitches experienced by Western capitalist nations, including cycles of recession, inflation, and unemployment.

Other critics question whether it is really an absolute good to eliminate diversity and make people the same everywhere. Ironically, modernization, like Marxism, holds that there is a universal historical process in which "a single modernity" will eventually emerge (Gilman 2003, p. 56).

Modernization has also been said to be based on the premise that science and technology can solve all human problems, rendering unnecessary any specific consideration of ethical implications of their introduction. Yet high technology may lead to high unemployment in third world countries, and therefore modernization theory needs to be modified by the addition of an ethical element, wholly lacking from the work of most writers on the topic. One view is that science and technology should specifically be used to address "social needs ... tempered with distributive justice" (Dube 1988, p. 32).

The countervailing forces to modernization include fundamentalism, anomie, violence, decay of norms, and the dysfunction of social institutions. Aslam Siddiqi offers an interesting critique from a Third World and Islamic perspective in a 1974 work; he says that modernization is an essentially materialistic concept lacking higher ethical value. "Human personality has no sanctity ... Abundance of goods is its greatest achievement, and hedonism is the proper way of life" (p. 13). Siddiqi does not propose the rejection of science and technology, but instead says that it is necessary to "identify the framework for society and to find accommodations between modernization and Islamic requirements" (p. 194).

Conclusion

The term modernization is invested with meanings that are better unpacked and examined individually, to see what assumptions are necessary to support them. While in the early twenty-first century, few people argue that a decentralized, agrarian, low-technology way of life is preferable, there is a consensus that development has moral implications that require close analysis and planning.

JONATHAN WALLACE

SEE ALSO Building Codes; Development Ethics; Enlightenment Social Theory; Green Revolution; Industrial Revolution; Secularization; Sustainability and Sustainable Development; Urbanization.

BIBLIOGRAPHY

Adas, Michael. (2003). "Modernization and the American Revival of the Scientific and Technological Standards of Social Achievement and Human Worth." In Staging Growth: Modernization, Development, and the Global Cold War, ed. David C. Engerman et al. Amherst: University of Massachusetts Press.

Dube, S.C. (1988). Modernization and Development: The Search for Alternative Paradigms. Atlantic Highlands, NJ: Zed Books.

Gilman, Nils. "Modernization Theory, the Highest Stage of American Intellectual History." In Staging Growth: Modernization, Development, and the Global Cold War, ed. David C. Engerman et al. Amherst: University of Massachusetts Press.

Huntington, Samuel (1968). "The Bases of Accommodation." Foreign Affairs 46(4): 642–656.

Siddiqi, Aslam. (1974). Modernization Menaces Muslims. Lahore, Pakistan: Sh. Muhammad Ashraf.

INTERNET RESOURCE

Bellah, Robert M. (1983). "Cultural Identity and Asian Modernization." Available from http://www2.kokugakuin.ac.jp/ijcc/wp/cimac/bellah.html.

MONDRAGÓN COOPERATIVE CORPORATION

• • •

The Mondragón Cooperative Corporation (MCC) is composed of a group of industrial, retail, service, and support cooperatives primarily in the Arrasate-Mondragón valley in the Basque country in Spain. Many scholars have studied Mondragón as a strong example of an industrial cooperative with a longstanding and successful history. From the beginning the MCC has, in its own words, strived for: (1) openness to all; (2) democratic organization; (3) recognition of the importance of work; (4) making capital instrumental and subordinate (people over capital); (5) participatory management; (6) minimal salary differentiation; (7) cooperation with other cooperatives; (8) transformation of society; (9) nondiscrimination in terms of gender, religion, and political affiliation; and (10) education and training for all.

It is a widespread belief among sociologists and economists that an association of producers that tries to develop an alternative to the capitalist model is destined to abandon democratic principles or fail economically. The success of Mondragón challenges this view. Since the first Mondragón cooperative was founded in 1956, the group has grown and continuously increased its profits. In the process it has maintained its cooperative structure almost unchanged. In 2002 the MCC was the seventh largest business group in Spain with a net worth of more than 15 billion euros. In 2003 it

employed more than 66,000 people in 120 firms of three types: financial, industrial, and distribution. The financial group includes the banking activities of Caja Laboral and a social welfare entity, Lagun Aro. The industrial group includes seven divisions: automotive, components, construction, industrial equipment, household appliances, machine tools, and engineering capital goods. The distribution and sales group consists of consumer cooperatives such as Eroski.

History

The project started in 1943 when a newcomer to the area, a young and unorthodox Catholic priest, José María Arizmendiarreta, decided to create a technical school in Mondragón in order to offer new opportunities to young people who had no access to that type of education. Arizmendiarreta never became a member of a cooperative but took part in most of the crucial decisions regarding the MCC project. The technical training school was registered legally in 1948. Eleven of the students in the first class went to the University of Zaragoza to study industrial technical engineering. In 1955 five of them bought a bankrupt firm that had produced heaters and stoves in Vitoria and moved that firm to Mondragón a year later. The firm eventually became Fagor, which was converted to a cooperative in 1958 and in the early twenty-first century is the largest producer of household appliances in Spain.

In 1959 the Caja Laboral Bank was formed with a double aim: to promote savings and to channel funds into other developing cooperatives. In the same year the social welfare entity Lagun Aro was set up to solve the

problem of pensions. Because the government considers them owners, not workers, members of cooperatives cannot be covered by Spain's social security system.

In 1969 the technical school officially became an industrial technical engineering school. The distribution cooperative Eroski was formed in that year. Ikerlan, the first technological research center of the MCC, was started in 1974.

In the late 1970s the organization became more complex, setting up so-called local groups, which bring together sets of cooperatives to do combined activities and optimize results. Beginning in the 1980s, the group increased exports and formed trade missions, and by 2003 it had constructed factories in sixteen other countries. The Caja Laboral has expanded throughout Spain, and Eroski commercial centers and megastores compete successfully with those of multinational firms. In 1990 the group officially became a corporation, and the businesses were organized by sectors rather than geographically.

Throughout its history an important value for Mondragón has been education, both technical and cooperative. In 1997 the University of Mondragón was established, combining all the cooperatives devoted to education: the three industrial technical engineering schools (Mondragón, Txorrieri and Lea-Artibai); Eteo, which is dedicated to business management and administration; and the University College for Teaching.

Another goal of the MCC is to produce its own technological knowledge. In addition to the university the MCC has formed several research centers: Ikerlan, Ideko, Maier Technology Center, Ahotec, Orona EIC, the Business and Organizational Management Research Center (MIK), Modutek, Koniker, and Lortek. In 2002 the Garaia Project developed a research network linking the university, the research centers, and the firms. The objective was to foster the kind of technological knowledge that the cooperatives consider key to their success.

Critical Reactions

Many scholars have tried to explain the extraordinary success of the Mondragón project from different perspectives. Some have seen Arizmendiarreta as a far-sighted leader whose decision-making ability was crucial. Others have pointed to a prior industrial and cooperative tradition in the area. As a result of these and other specific aspects Mondragón often has been presented as a unique experience that would be impossible to reproduce in other places.

A controversial aspect of Mondragón is its supposed relationship with the Basque nationalist movement. The Mondragón area is, along with many others in the Basque country, markedly nationalist, and for this reason it often has been suspected that the MCC has received favorable, protectionist treatment from the regional Basque government, which has always been in the hands of the nationalists. These suspicions have never been substantiated, and it is important to remember that the MCC first developed and achieved economic success during the earlier Spanish dictatorship.

An important problem has resulted from the growth of the cooperatives: Some of them, especially Eroski, require an increasing number of hired employees who are not members of the cooperative. This clearly contravenes the original ideals of the MCC and could be interpreted as leading to a transformation of the cooperatives into firms with a less democratic structure. However, MCC researchers are studying ways to incorporate those workers into the cooperative system.

ANA CUEVAS BADALLO

SEE ALSO *Affluence; Business Ethics; Work.*

BIBLIOGRAPHY

Bradley, Keith, and Alan Gelb. (1983). *Cooperation at Work: The Mondragón Experience.* London: Heinemann Educational Books. Economic analysis.

Kasmir, S. Sharryn. (1996). *The Myth of Mondragón: Cooperatives, Politics, and Working-Class Life in a Basque Town.* Albany: State University of New York Press. Argues that democratic government is not as ideal as claimed.

Whyte, William Foote, and Katheleen King Whyte. (1991). *Making Mondragón: The Growth and Dynamics of the Mondragón Cooperative Complex,* rev. edition. Ithaca, NY: ILR Press. Best source, by a well-known sociologist and his wife.

INTERNET RESOURCE

Mondragón: Corporación Cooperativa. Available at http://www.mcc.es. Translations include English, French, Spanish, and German.

MONEY

• • •

The term *money* derives from the Latin *moneta,* meaning mint or coin, and is most often defined as a medium of exchange and measure of value. Even from its earliest use as a replacement for barter, money was often a tech-nologically produced metal coin and thus associated with developments in the science of metallurgy and metal technology. In the *Nicomachean Ethics* (350 B.C.E.), Aristotle (384–322 B.C.E.) offers a first glimpse of the ethical implications of money as technology when he rejects moneymaking as the proper end of human life on the basis that it has only instrumental value. With the rise of modern scientific economics came efforts to formulate monetary policies for states, and the use and management of money became more closely associated with science, technology, and normative issues. All this is underscored by the German philosopher-sociologist Georg Simmel (1858–1918) who identifies money as the pivotal technological tool that paved the way for the modern technological approach to the world.

Historical Considerations

One of the earliest forms of money was cowrie shells (c. 1200 B.C.E.); based metal (1000 B.C.E. in China) preceded precious metal (700 B.C.E. in the Middle East) coinage. At least as early as Aristotle, whose views have influenced classical and modern discourses on the topic, money was recognized as a medium of exchange and measure of value. Initially simple bartering had sufficed because the goal was subsistence. But even in barter, precise equivalences between bushels of wheat and a cow or a physician's services are difficult to determine, so that questions arose about how to determine a fair exchange or just price. Again in the *Nicomachean Ethics,* Aristotle contends that the just price of a technological product is determined by proportion, with the anchor of proportionality being the status of the producers, as when the shoemaker's product is to the farmer's as the farmer is to the shoemaker. In the *Politics* (350 B.C.E.), he describes how money, usually in the form of precious metals, facilitated exchanges between parties who could not engage in direct transactions. This function of precious metals was further enhanced when they were minted and embossed to attest to their monetary value—generally in excess of the use-value of the metals themselves. With paper or representative money, the disparity between use-value and monetary or exchange value becomes even more pronounced. For Aristotle, the use of money is *contrary to nature* when the exchange is for profit rather than subsistence. The function of money is distorted when it becomes an end-in-itself and the primary measure of wealth.

In the modern period, Adam Smith (1723–1790) continues to distinguish between money and genuine wealth, but goes on to argue that the desire for profit and personal advantage promote private and public good. The profit motive, free competition, and an advanced division of labor that includes the development and use of technology, work together to increase productivity and fuel a "universal opulence which extends itself to the lowest ranks of the people" (Smith 2000, p. 12).

Influenced by Smith, David Ricardo (1772–1823) initially agrees that advances in machine technology benefit all parties—landholders, capitalists, and laborers—but is less sanguine about the alleged advantages for laborers. He eventually concludes that machine technology and labor are in competition and that increased use of the former is often detrimental to the latter. This is by itself insufficient reason to jettison *laissez-faire* principles, for, as Ricardo sees it, government intervention to curtail the use of machine technology to fend off unemployment actually has the opposite effect of driving capital investment offshore and eventually destroying the domestic labor market.

Karl Marx (1818–1883) agrees with Aristotle that legitimate exchange binds human beings together, whereas the profit motive drives them apart. He goes beyond Aristotle, however, when he insists that money is an insurmountable obstacle to genuine human community. In his *Economic and Philosophic Manuscripts of 1844* (1932), Marx argues that money alienates human beings from themselves, from the fruits of their labor, and from each other. In short, money subverts the natural order of things and turns the world upside-down. A return to an authentic mode of human (that is, communal) existence requires the rejection of both private property and money. Only then can one take an optimistic view of the impact of technology on human life. After all, technology has the potential to liberate energy normally expended to obtain the material necessities of life—energy that, once freed, may be redirected toward human cultivation and refinement.

The appearance in 1936 of *The General Theory of Employment, Interest, and Money* by John Maynard Keynes (1883–1946) precipitated a revolution in economics by assigning government a significant role in the economic affairs of free-market states. While the *laissez-faire* approaches of Smith and Ricardo allowed for modest and minimal government involvement in economic matters, Keynes articulated a theory whereby government bears major responsibility for the overall economic health of a nation. According to Keynes, adroit and judicious government intervention in setting fiscal and monetary policies, spending on public works to boost a sluggish economy, and supporting technological innovation would, generally speaking, stabilize the economy,

increase productivity, and foster full employment. The implicit conviction is that eliminating involuntary unemployment and poverty would reduce, if not cure, many of the social ills endemic to failed economic environments.

Keynes's intention was to improve the "technique of modern Capitalism," and he did not challenge the capitalist "dependence upon an intense appeal to the money-making and money-loving instincts of individuals as the main motive force of the economic machine" (Keynes 1963, p. 319). Keynes nonetheless speculates about a day when economic issues will no longer matter. The basic needs of human existence will be met, leisure will be filled with noneconomic activities, and the "love of money as a possession—as distinguished from the love of money as a means to the enjoyments and realities of life—will be recognised for what it is, a somewhat disgusting morbidity, one of those semi-criminal, semi-pathological propensities which one hands over with a shudder to the specialist in mental health" (Keynes 1963, p. 369). With this assessment of the *true value* of money, Keynes, who was arguably the most influential economist of the twentieth century, joined forces with Aristotle and to some extent Marx.

Building on but criticizing Keynes, Milton Friedman (b. 1912) developed a theory of money that argues for measured control of the money supply as a better means than stimulus over the long term. Of course, for both Keynes and Friedman money has become an increasingly abstract phenomenon, far removed from the traditional technologies of coinage and representative money into fiat and credit money that are tied up with new technologies of plastic, computers, and information transfers.

Money and Technology

With the Industrial Revolution, money began to play a central role in the production, exchange, and consumption of all goods and services. During the same period, economic growth became increasingly dependent on and intertwined with technological developments requiring significant capital investment. In other words, money must not lie fallow. The supply of money must be directed at consumption and/or investment. The question is whether money, as a means to an end, is simply a benign technological device requiring no special caution by the user.

Simmel's consideration of money as the *purest form of the tool*, a *pure instrument*, is instructive here. His *Philosophy of Money* (1900) seeks to extrapolate from the "surface level of economic affairs a guideline that leads to the ultimate values and things of importance in all that is human" (Simmel 1978, p. 55). To that end, Simmel pursues two lines of inquiry—the subjective preconditions of economic life and the consequences of using money as the medium of exchange. In this latter inquiry, Simmel formulates his critique of modern technological society.

For Simmel, money enhances human freedom, but this freedom has a price. The overvaluation of money engenders a means-ends reversal whereby money is elevated to the status of an absolute end, while things that are ends-in-themselves are treated merely as means. It is not until money fails to function properly—for example, when money cannot even buy bread—that one remembers which of them has intrinsic value. Simmel also sees a causal connection between money and the modern technoscientific tendency to translate all qualities into quantities so that they can be quantitatively measured and assessed. "The ideal of numerical calculability has been made possible in practical and perhaps even in intellectual life only through the money economy" (Simmel 1978, p. 445). In other words, money is not neutral, and, like all technological artifacts, its use has both positive and negative consequences.

Despite the earlier connections between the exchange value of money and the material substances serving as money, the true nature of money and its socioethical implications cannot be derived from the material in which it is embodied. Just as money was introduced to facilitate bartering, paper money, checks, bank drafts, and credit cards were introduced to facilitate the use of money in commercial transactions. The socioethical implications of money derive from the impact that its use has on people's inner lives and their perceptions of the world.

Like Simmel, who argues that money transforms every quality into a quantity, Jacques Ellul (1912–1994), for instance, maintains in L'Homme et l'argent (Money and power) (1953) that the spiritual power of money transforms every relationship—be it to oneself, to others, or to the world—into one of buying and selling.

Whereas both Aristotle and early modern economists couched their analyses of the use and value of money in ethical and political terms, the view of economics as positive science often appears to treat technical economic issues independently of the broader ethicopolitical dimensions of social life. By embracing the goal of scientific objectivity, economics may obscure how the management of economic systems is never value-neutral. Of course, free market economists such as

Friedman argue forcefully for a positive connection between money and freedom. Money, like all technological artifacts, has important ethical implications. While few in the early twenty-first century would seriously advocate its abolition, one should bear in mind that money surreptitiously shapes self-understanding and valuations of the world.

Naturally, there are people in the field of technology studies who defend the thesis that technology and, *mutatis mutandis*, money are inherently neutral with regard to ethicopolitical values. On this view, technologies are neither good nor bad and are steered in one direction or the other by values that are external to the technologies themselves. And even if one concludes that technologies are value laden, it does not necessarily follow that the relevant values and their consequences are negative.

JOHN E. JALBERT

SEE ALSO *Affluence; Business Ethics; Class; Work.*

BIBLIOGRAPHY

Davies, Glyn. (2002). *A History of Money: From Ancient Times to the Present Day*, 3rd edition. Cardiff: University of Wales Press. A thorough treatment of the history of money with interesting chapters on its early forms and uses. The author emphasizes the need for a broad understanding of money that includes its social and psychological dimensions.

Ellul, Jacques. (1984). *Money & Power*, trans. LaVonne Neff. Downers Grove, IL: InterVarsity Press. English translation of the 1953 French language *L'Homme et l'argent*.

Friedman, Milton. (1992). *Money Mischief: Episodes in Monetary History*. New York: Harcourt Brace Jovanovich.

Friedman, Milton, with Rose D. Friedman. (1962). *Capitalism and Freedom*. Chicago: University of Chicago Press.

Keynes, John Maynard. (1963). *Essays in Persuasion*. New York: W. W. Norton & Company. The general theory is arguably the most significant work in the history of modern economics. Keynes criticizes *laissez-faire* economics and argues for government intervention in the economic affairs of nations. Keynes's economic principles are put to work in the essays on public policy and political economy assembled in essays in persuasion. All are written in nontechnical language and easily accessible.

Keynes, John Maynard. (1997). *The General Theory of Employment, Interest, and Money*. Amherst, NY: Prometheus Books.

Marx, Karl. (1975). "Economic and Philosophic Manuscripts of 1844." In *Collected Works of Marx and Engels*, Vol. 3: 1843–1844. New York: International Publishers.

Marx, Karl. (1996). *Collected Works of Marx and Engels*, Vols. 35–37: *Capital*. New York: International Publishers. *Das Kapital* is required reading, if for no other reason than it is perhaps the most unyielding and influential critique of capitalism to date. In *Kapital*, Marx formulates his labor theory of value and explains profit in terms of the exploitation of laborers.

Ricardo, David. (1996). *The Principles of Political Economy and Taxation*. Amherst, NY: Prometheus Books.

Simmel, Georg. (1978). *The Philosophy of Money*, trans. Tom Bottomore and David Frisby. Boston: Routlege and Kegan Paul. A much neglected work that focuses on the influence of a money economy on social and cultural life. It supplements and provides helpful alternatives to Marx's interpretations of history, value, and alienation. It also directs attention to the positive relation between money and human freedom.

Smith, Adam. (2000). *The Wealth of Nations*. New York: Modern Library. No study of political economy is complete without the inclusion of smith's classic treatise on the subject. Anyone who wishes to understand the development of modern capitalism and the origins of *laissez-faire* economics would do well to begin here. Smith argues that an unregulated economy benefits all of the parties involved in the production, exchange, and consumption of goods.

MONITORING AND SURVEILLANCE

• • •

Monitoring is a general term that refers to the systematic, continual, and active or passive observation of persons, places, things, or processes. By contrast *surveillance* is used to indicate targeted monitoring of activities by police or security officials for specific evidence of crimes or other wrongdoing. Surveillance focuses on individuals, buildings and properties, or vehicles deemed suspicious on the basis of credible information that they are connected in some way to illegal or otherwise inappropriate activity. Surveillance operations carried out by investigators may: (1) be stationary or mobile in nature and require various types of monitoring technologies to enhance the visual or hearing capabilities of officers or operatives doing the surveillance; (2) involve recording of events, locations, days or times, and patterns of behaviors or activities; and (3) include monitoring of telephone or in-person conversations, as well as electronic correspondence such as E-mail or instant messaging notes exchanged between individuals or groups of people. Surveillance is usually carried out in covert ways and with legal authority.

Monitoring typically involves routine recording of activities to warn of trouble or for accounting purposes. Open public spaces such as airports, shopping malls, and other places where large numbers of people gather are monitored to help assure public safety and security. Surveillance is the targeted monitoring of people suspected of committing crimes or other civil wrongdoings. Examples of monitoring tools are smoke detectors and turnstile counters used to determine the number of subway passengers. In contrast, electronic building-access cards have a surveillance element because individuals can be held accountable for improper use of the device. Monitoring systems that are used also as surveillance devices include video cameras in commercial and public spaces. Electronic listening devices that are placed to record conversations of targeted people are surveillance tools. Point-of-sale systems that monitor inventory and customer buying habits may be ethically problematic, but the function of those devices does not have a surveillance aspect as that term is used in this entry.

Spying combines the arts and technologies of monitoring and surveillance along with active intelligence gathering and analysis in order to advance a government or corporate interest. Spying is often commissioned by secretive government agencies in the interest of national security, or by unscrupulous corporations intent on illegally discovering the secrets of competitors. Spying is covert in nature and, if exposed, may have negative legal, political, or financial repercussions for the agencies, corporations, firms, or individuals involved.

The differences between monitoring, surveillance, and spying mostly concern the purposes and sponsors of the activities, and the degree to which they are carried out in relatively covert versus overt ways. The same technologies that are used for monitoring (such as binoculars, night-vision equipment, and listening and recording equipment) can also be used for surveillance and spying. In general, monitoring technologies are used in relatively overt ways in many sectors of society, whereas in surveillance and spying, technologies are used primarily in covert investigations.

Monitoring Technologies in Society

Humans develop their knowledge of monitoring techniques and their skill in using monitoring technologies with age, experience, and training. From childhood, throughout adolescence, and into adulthood, humans combine cognitive skills with sensory perceptions in order to observe, monitor, interact with, and generally function within their environments. In so doing, people learn to decipher patterns, trends, and anomalies and thereby recognize what is ordinary versus unusual regarding places, things, and processes.

Safety and security, as well as efficiency and effectiveness (as in manufacturing processes), are premised on people knowing when things are out of place. For this reason, people are often monitored while driving in traffic, waiting in airports or train and bus stations, working in their places of employment, shopping in malls or detached retail stores, or as they are depositing or withdrawing money from automatic teller machines (ATMs) located at banks or other locations.

Monitoring technologies are combinations of simple and complex tools and techniques that facilitate routine and systematic observation, recording, and analysis of activities or processes in specific locations. Essentially they help people understand what is going on in a given environment. Monitoring technologies encompass a variety of communications, computing, electromechanical, imaging, robotics, and sensing devices and systems. These include but are not limited to closed-circuit television (CCTV) systems, global positioning and tracking devices, and metal or contraband detection devices. Monitoring technologies such as these may also include or integrate various combinations of alarms and warning systems that signal when something unusual occurs, or when a desired state or condition has been met.

Monitoring technologies are used by government agencies; by manufacturing, service, and other businesses; and in fields as disparate as agriculture (for crop and livestock monitoring), astronomy (to track movements of planets, comets, or asteroids), and meteorology (for monitoring and forecasting of weather and climatic patterns). They are used to observe many types of human activities and processes, such as vehicular traffic congestion on public roadways and commercial and military aircraft flight patterns, and to detect malfunctions in manufacturing processes. In medical fields, monitoring technologies are used to check the status of patients on treadmills and to signal problems experienced by those recovering from major surgery (Abrami and Johnson 1990).

Monitoring technologies are employed extensively in security and criminal justice situations (National Institute of Justice 2003). For example, law enforcement officers use all-weather camera systems to observe and record, and also to aid in dispatched responses to, suspicious activities. Intrusion and motion detectors are devices used to detect and signal several conditions. For example, excessive heat or cold indicator devices and warning alarm systems for foreign substances such as smoke, carbon dioxide, and radon are all used to promote safety and security. Virus detection software applications, which are often used in combination with firewalls, help insure computer privacy and security.

Similarly police use cameras mounted in their vehicles to remotely monitor or record interactions between themselves and motorists during traffic stops. Global information system monitoring technologies are used to keep track of the locations of emergency vehicles, or to monitor specific locations and movements of prisoners inside detention facilities or those on supervised release programs. These are just some examples of the various types of monitoring technologies and what they can be used for. In all these situations, monitoring technologies are intended to facilitate detection and warning of unusual and potentially unsafe or threatening behaviors, conditions, or developments.

History of Surveillance

Surveillance is the close observation of a person or group. While technology is not necessary to surveillance, certain technologies greatly facilitate it. Video and computer technologies, for instance, have made surveillance an important feature of modern societies. In many cities—London stands out—the average citizen is captured on video many times each day. Many shops use closed-circuit television to videotape customers and staff and to record transactions. Workplace surveillance is becoming common as well: According to an American Management Association (AMA) survey, "In 2003, more than half of U.S. companies engage in some form of e-mail monitoring of employees and enforce e-mail policies with discipline or other methods.... 22% of companies have terminated an employee for e-mail infractions" (AMA 2003, p. 1).

Following the lead of Michel Foucault (1977), many critics see modern societies as *panopticons*, tending toward Jeremy Bentham's model prison design in which each prisoner is kept under observation by invisible watchers. This metaphor reveals something about the history of surveillance as well as its ethics.

Historically, surveillance has been a labor-intensive undertaking. Bentham's prison was designed to enable a single guard to oversee many prisoners. Short of the severe constraint of a prison environment, this ratio is difficult to attain. For instance, following someone undetected on the street requires a team of several trained agents. Thus widespread covert surveillance of a population would be extremely expensive without technological augmentation. This is also true for reading large volumes of handwritten mail. In both cases, technology has offered possibilities. The automated searching of text, for example, has made it economically practical to read the E-mail of every employee in a firm.

The path of technological development can be expected to influence whether one is exposed to surveillance at a given time. Text is still easier to search than voice or video images. The situation is fluid, however, because technological development is rapid, especially given the widespread security concerns that followed the terrorist attacks of September 11, 2001.

Ethics of Monitoring

Increasingly affordable, interoperable, and compact technologies make possible and help to perpetuate the human desire and willingness to engage in the monitoring of virtually any activity, location, or process. In other words, monitoring technologies make ubiquitous watching possible. George Orwell popularized the fear of omnipresent monitoring and surveillance in his classic novel, *Nineteen Eighty-Four* (1949). Since the book was published, people in developed nations, particularly Americans (who have always been concerned about protecting privacy rights), have become increasingly anxious about the technology-enabled monitoring capabilities of their governments. But, notwithstanding concerns about privacy, widespread and even routine use of monitoring technologies for numerous purposes has become the norm. Indeed given growing worldwide concerns about crime and terrorism, use of sophisticated technologies to support legal surveillance by security and law enforcement officials, and even spying by government intelligence agencies, is often welcomed, if not actually deemed necessary, as a means of enhancing security and safety and reducing fear in both public and private places (SPIE 2002).

While responsible use of monitoring technologies is generally acknowledged as sensible and, therefore, is often encouraged in private property situations, the same is not true for public domains. Law enforcement use of monitoring technologies to observe open spaces is often met with strong criticism from the people who the police or security officials are trying to protect. Resistance to government watchfulness is rooted in the belief that even passive monitoring of public spaces impinges on the privacy and other rights of individuals and groups who are legally present or assembled and are doing nothing wrong.

The controversy and ethical dilemma is twofold. First, will the use of monitoring technologies in public spaces create a social-psychological atmosphere of intimidation versus promoting safety or well-being (Goold 2002)? Second, will increasing legal use of monitoring technologies by authorities lead to collective endorsement of such tactics that, if taken to the extreme, will create conditions resembling a high-tech police state akin to the Big Brother atmosphere conceived by Orwell in *Nineteen Eighty-Four*?

Ethics of Surveillance

Foucault's panopticon metaphor reveals something about the ethics of surveillance:

> The major effect of the panopticon was to induce in the inmate a state of conscious and permanent visibility that assures the automatic functioning of power. So to arrange things that the surveillance is permanent in its effects even if it is discontinuous in its action; ... this architectural apparatus should be a machine for creating and sustaining a power relation independent of the person who exercises it. (Foucault 1997, p. 201)

Thus surveillance creates a new power relationship because those subject to it must always behave as if someone is watching, even if no one is.

While properly focusing on the strategic element in surveillance and pointing out the power differences between watcher and watched, this assessment exaggerates the situation. Though it is true that surveillance need not be continuous to be effective, those being watched have counterstrategies. The simplest is for them not to act as if they are always being watched.

Because surveillance is a dynamic process, unexpected consequences are likely. Consider, for example, radar for monitoring automobile speed, an early form of electronic surveillance. Naively one might think that equipping police with radar would lead all drivers to obey speed limits. But this expectation ignores the strategic element in the situation. Not every road that has a "Speed Controlled by Radar" sign is actually monitored—the police typically follow a mixed strategy and patrol only some of the signed roads. Drivers know this and do not always obey the speed limits. In addition, there are technological countermeasures: Sophisticated radar detectors are cheap and widely used. This situation leads to two kinds of ethical question. First, is this *technological arms race* efficient, once the cost of countermeasures and the failure to control speed completely is taken into account? Second, is surveillance radar fair? Does it catch anyone other than those too poor or naive to participate in the strategic game played out by law enforcement and drivers with antiradar equipment?

Other examples of counterstrategies include obstruction of video surveillance devices and using language ambiguities to confound text-based surveillance. Once the potential of counterstrategies is taken into account, the logic of surveillance goes beyond the panopticon. Most populations are not as confined as prisoners. Most surveillance, to be effective, needs the support of the majority of its subjects (Danielson 2005).

Consider how this plays out in three typical surveillance venues.

STATE AND PUBLIC SURVEILLANCE. Surveillance by agencies of the state is the most familiar model of surveillance. However the Big Brother image is probably out of sync with the practice in many modern technological societies where private surveillance is more prevalent.

State surveillance in democratic states requires public acquiescence. This tends to be forthcoming when events make a security rationale salient, as in states that fear a terrorist attack or experience a great deal of crime. Without this impetus, public outcry has forced liberal states to remove public cameras (e.g., Canada) or subject them to strict regulation (e.g., the United Kingdom).

THE WORKPLACE. Workplace surveillance is distinguished by two features. First, employees are contractually related to employers, so consent, or broad doctrines of implied consent, permit surveillance in the workplace that would be controversial in public places. There are, of course, conflicts over the line between permitted workplace surveillance and protected privacy at work. Surveillance of washrooms and other private spaces has caused controversy, as has intercepting and logging E-mail and personal web browsing.

Second, more computerized jobs expose more workers to surveillance. Computerized surveillance is inexpensive and indiscriminate. New, cheap technologies tend to get overused, beyond their practical and ethical justification. Practically, unwelcome surveillance can undermine employee morale, destroying organizational goals. Ethically, privacy is the value most at risk. For example, widely deployed wireless surveillance cameras effectively broadcast whatever information they pick up, creating an opening for outside interception. This threat is increased by the recent introduction of inexpensive web-based and cell-phone-based cameras.

COMMERCIAL AND INDIVIDUAL SURVEILLANCE. Examples of commercial, individual applications range from the convenience store video camera to the nannycam installed to watch children and caregivers. Because the technology deployed in these contexts is quite primitive, there are additional risks to privacy and other values. In addition, the increased use of surveillance technology in the home challenges traditional lines between public and private spaces (Nissenbaum 1997). People expect to be observed in public and at work—

and adjust their behavior accordingly—but this expectation does not exist for private spaces.

Assessment

The ethics of surveillance is best developed for the workplace. Overall there are three main lessons. First, legitimacy makes a difference by avoiding unwelcome surveillance and lowering the costs of countermeasures. Consent and, as a precondition, education about the technology are obvious ways to increase legitimacy. Second, the ethical risks of surveillance should be conveyed to would-be users, which, hopefully, would limit use to more serious cases. Third, more explicit norms against the incursion of surveillance technology into private spaces may be necessary.

People who object to increased monitoring suggest that quality of life will be unduly, negatively affected by the mere presence of cameras and tracking and recording devices, and that even if people do not have a legal right and expectation of privacy in open spaces, social interactions, unfettered spontaneity, and being able to feel as though one is not being watched are qualities of life that ought not be compromised. Further, if left unchecked, increasing use of monitoring technologies will undermine the freedoms of speech, movement, association, assembly, and religion. Supporters of monitoring usually point out that such devices provide effective deterrence against crimes or other inappropriate conduct, as well as a means to respond to, interdict, and if legally appropriate, apprehend violators. Supporters also point out that the mere presence of cameras and recording devices can make people feel safer, and that persons obeying the law have nothing to hide or fear because police and security officials exist to provide protection and can be held accountable for illegal or inappropriate use of their powers.

Ethical use of monitoring technologies by anyone hinges on circumstances under which people have an expectation of privacy. In general, U.S. courts have ruled that citizens and residents have constitutionally based privacy protection in their homes and other privately owned places. People have considerably less, or no, expectation of privacy, however, as students in private or public schools, in places of employment, or in open spaces or other public places. Proper use of monitoring technologies by private individuals, firms, corporations, or government authorities can improve or lessen quality of life from the standpoint of privacy versus safety and security, and also enhance the quality of manufactured products. Ultimately what constitutes proper use of monitoring technologies is a matter to be resolved on ethical, legal, social, and economic grounds.

SAMUEL C. MCQUADE III
PETER DANIELSON

SEE ALSO *Internet; Privacy; Telephone.*

BIBLIOGRAPHY

Abrami, Patrick F., and Joyce E. Johnson. (1990). *Bringing Computers to the Hospital Bedside: An Emerging Technology.* New York: Springer Publishing.

Bentham, Jeremy. (1969). "Panopticon Papers." In *A Bentham Reader,* ed. Mary Peter Mack. New York: Pegasus.

Constant, Mike, and Philip Ridgeon. (2000). *The Principles and Practice of CCTV,* 2nd edition. Borehamwood, England: Paramount Publishing.

Danielson, Peter. (2005). "Ethics of Workplace Surveillance Games." In *Electronic Monitoring in the Workplace: Controversies and Solutions,* ed. John Weckert. Hershey, PA: Idea Group Publishing.

Foucault, Michel. (1977). *Discipline and Punish: The Birth of the Prison,* trans. Alan Sheridan. New York: Pantheon.

Goold, Benjamin. (2002). "Privacy Rights and Public Spaces: CCTV and the Problem of the 'Unobservable Observer.'" *Criminal Justice Ethics* 21(1): 21–27.

Hunter, Richard. (2002). *World without Secrets: Business, Crime, and Privacy in the Age of Ubiquitous Computing.* New York: Wiley.

Lyon, David, and Elia Zureik, eds. (1996). *Computers, Surveillance, and Privacy.* Minneapolis: University of Minnesota Press.

National Institute of Justice. (2003). "CCTV: Constant Cameras Track Violators." *National Institute of Justice Journal* 249: 16–23.

Nissenbaum, Helen. (1997). "Toward an Approach to Privacy in Public: The Challenges of Information Technology." *Ethics and Behavior* 7(3): 207–219. Develops the public/private distinction.

Orwell, George. (1949). *Nineteen Eighty-Four: A Novel.* London: Secker and Warburg. The classic surveillance dis-utopia. Note that the source of recent surveillance has shifted from governments to include corporations and individuals.

Society of Photo-optical Instrumentation Engineers (SPIE). (2002). *Sensors, and Command, Control, Communications, and Intelligence (C3I) Technologies for Homeland Defense and Law Enforcement.* Bellingham, WA: Author. Conference proceedings of the SPIE, April 1–5, Orlando, Florida.

INTERNET RESOURCE

American Management Association (AMA). (2003). "E-Mail Rules, Policies, and Practices Survey." Available from http://www.amanet.org/research/pdfs/Email_Policies_Practices.pdf. An accessible survey of corporate e-mail surveillance practices in the United States.

MONTREAL PROTOCOL

• • •

The Montreal Protocol on Substances that Deplete the Ozone Layer (MP) of 1987 is an international agreement to protect the stratospheric ozone layer from harmful synthetic chemical compounds. The targets of the MP are synthetic chemical substances that destroy an upper level protective ozone layer of the Earth and whose destructive behavior persists over decades if not centuries, depending on the chemical compound. The MP is considered an exemplary case of science-based policy making, adroit diplomacy, innovative treaty language, and regulatory collaboration, and is the most successfully implemented global environmental treaty in history. It is also the best example to date of global action based on the *precautionary principle*.

The Issue and Efforts Leading to the Montreal Protocol

Ozone is a bluish gas, harmful to breathe, that is composed of three atoms of oxygen. Nearly 90 percent of the planetary ozone is in the stratosphere, an atmospheric region above the troposphere extending from about 10 to 50 kilometers in altitude. In the 1930s, scientists Dorothy Fisk and Charles Abbott discovered how to measure atmospheric ozone, and described the critical role an *ozone layer* plays as a global sunscreen. Stratospheric ozone absorbs a band of ultraviolet radiation (UVb), preventing most of it from reaching the ground where it is particularly harmful to living organisms (causing skin cancer and cataracts, interrupting food chains, and more).

Chlorofluorocarbons (CFCs), now recognized as ozone depleting substances (ODS), were hailed for being safe, friendly, and widely applicable when first invented about the same time that the benefits of the ozone layer were discovered. Besides their original application in refrigeration—where they were both safer and more efficient—CFCs were manufactured for an extremely wide variety of uses: flexible urethane foams (for carpeting, furniture, and automobile seats); rigid polyurethane foams (as insulation for buildings and refrigeration units); blowing agents in non-urethane foams (polyurethane sheet products, foam trays, fast-food wrappers); and refrigerants in automobile air conditioners, and industrial and commercial air conditioners known as chillers. CFCs became an important solvent for the electronics and aerospace industries as a cleaning agent for circuit boards and scientific instruments. Halons, another set of haloge-

FIGURE 1

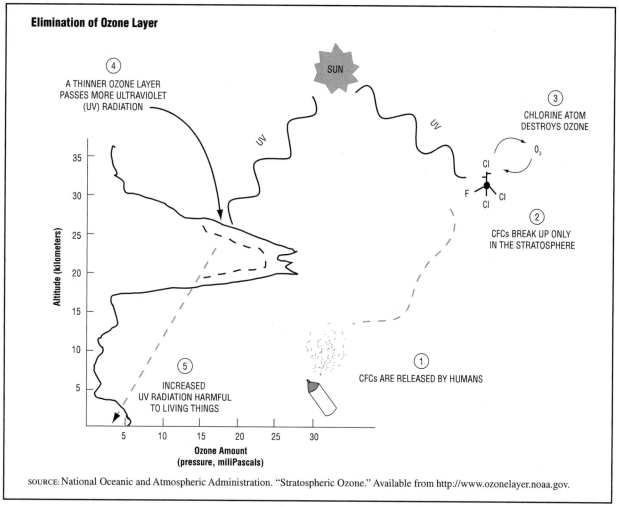

Elimination of Ozone Layer

④ A THINNER OZONE LAYER PASSES MORE ULTRAVIOLET (UV) RADIATION

SUN

③ CHLORINE ATOM DESTROYS OZONE

UV

UV

O_3

Cl

F   Cl

Cl

② CFCs BREAK UP ONLY IN THE STRATOSPHERE

⑤ INCREASED UV RADIATION HARMFUL TO LIVING THINGS

① CFCs ARE RELEASED BY HUMANS

Altitude (kilometers)

35

30

25

20

15

10

5

5   10   15   20   25   30

Ozone Amount (pressure, miliPascals)

SOURCE: National Oceanic and Atmospheric Administration. "Stratospheric Ozone." Available from http://www.ozonelayer.noaa.gov.

nated hydrocarbons, were widely used as flame suppressants in firefighting. Carbon tetrachloride, methylene chloride, and the agricultural chemical methyl bromide used as a soil fumigant and to protect stored agricultural products from pest-related deterioration, are implicated as well.

In the 1970s natural scientists (notably, Richard Stolarski, Ralph Cicerone, Sherwood Rowland, and Mario Molina) questioned whether these chemical compounds were benign in the stratosphere. When CFCs reach the stratosphere, ultra-violet radiation causes them to decompose releasing a chlorine atom that in turn destroys ozone molecules. They concluded that a single chlorine atom released in the stratosphere could eliminate thousands of ozone molecules through a catalytic chain reaction; that this reaction would continue for the life of the chemicals (40–150 years); and that

CFC concentrations in the ozone layer could be expected to reach one to thirty times their current levels with disastrous consequences for the integrity of the ozone layer.

For the next ten years, debates over the science of ozone depletion raged, reflecting different industrial, political, and scientific worldviews and the symbolic resources brought to bear on the issue (Dotto 1978).

Much of the early empirical evidence of a "hole" in the stratospheric ozone layer was discounted by scientists who simply assumed that the extremely low Dobson instrument measurements were due to technical malfunctions. Indeed, the 1982 ozone measurement devices aboard the Nimbus 7 satellite had been programmed to flag low values as erroneous (Gribbin 1988). By the mid-1980s, however, scientists such as Shigeru Chubachi and Susan Solomon provided empirical evidence of stratospheric ozone depletion (Andersen and Sarma 2003). Consensus that an *ozone hole* was swiftly developing left open to debate whether the hole was caused by nature or by invented chemicals. Nevertheless, and importantly, even in the face of continuing uncertainty, the world moved from demands for more research to demands for precautionary regulation in a relatively short period of time.

In 1985, under the auspices of the United Nations Environment Programme (UNEP), Executive Director Mostafa Tolba led Australia, Canada, Finland, Germany, New Zealand, Norway, and the United States to adopt the Vienna Convention (VC). This was the first official version of international understandings and responsibilities regarding the protection of the stratospheric ozone layer. The MP, signed in September 1987, followed (Benedick 1991). By 2003, 184 nations had ratified the MP.

Implementation and Evolution of the Montreal Protocol

Parties to the MP agreed to use national consumption/production figures as a baseline from which to measure targets for phaseout, permitting flexibility so that each nation could determine how best to meet its national phaseout commitment. Article 6 established periodic reviews by scientific and technical experts so that the treaty could be adjusted with the benefit of fast-paced developments in science and technology. With amendments of the MP, the twin principle of *differentiated responsibility/capability* was adopted. Funds, expertise, and technology transfer supported developing countries that were not major contributors to the problem and whose domestic economic priorities were not in line with phaseout. (Article 5 lists 136 such nations in 2003.) The Global Environment Facility took responsibility for helping Countries-With-Economies-in-Transition (high ODS, economically troubled), typically members of the former Soviet Union.

Originally the treaty committed parties to reduce, by 1996, the use of CFCs by 50 percent, using their national 1986 baseline values. Failure to sign the treaty imposed import/export restrictions that encouraged wide participation, especially given that total, worldwide phaseout of the harmful substances meant that non-parties without production capability would not have access to supplies. This also prevented companies seeking to avoid controls on ODS from moving their production facilities to non-parties and exporting back into the countries controlled by the MP (Brack 1996).

By the time the treaty went into force on January 1, 1989, there was already a strong push for amending it, as anticipated. In 1990 the London Amendments provided for a total ban of CFCs by the end of the twentieth century, added other ODS to the list of controlled substances, created the Multilateral Fund (MLFund) to help developing countries phase out, instituted a ten-year grace period for developing country compliance, mandated technology transfer from rich countries, and reclassified hydrofluorocarbons (HCFCs) as *transitional substances*. The Copenhagen Amendments (1992) accelerated the compliance schedule, confirmed the MLFund permanently, and suggested new compounds for the control list, notably HCFCs and methyl bromide. Subsequent adjustments (Montreal 1997, Beijing 1999) replenished the MLFund and tightened control measures.

Administratively the treaty established the MP Secretariat (Nairobi) with K. M. Sarma as the first Executive Secretary and, after 1990, the MLFund Secretariat (Montreal), first headed by Omar El-Arini. The MLFund Executive Committee is composed of equal numbers of developed and developing countries. Four United Nations agencies support the phaseout through activities such as training, information sharing, institutional strengthening, conferences, and consultant services. Each Article 5 country has established a National Ozone Unit; these are strengthened by regional networking activities of the UNEP.

Three autonomous advisory panels—in Science, Environmental Impacts, and Technology and Economic Assessment (TEAP)—report directly to the parties. These volunteer expert review panels are the primary source of the confidence with which the parties have frequently amended the treaty in light of new, credible science and technology.

Over the first decade of the MP implementation, the TEAP, under the collaborative leadership of Stephen O. Andersen and Lambert Kuijpers, rose to preeminence as the worldwide authority on technically credible, economically possible options for speedy phaseout. Other than the Economic Options Committee, the TEAP was organized by industrial sector, and includes divisions such as the Technical Option Committee (TOC) for aerosols, foams, halons, methyl bromide, refrigeration, and solvents. The TEAP found and created new product designs, innovative practices, and industry-wide alterations in production processes that were harmful to the ozone layer.

The TEAP was built on the principle of dynamic collaboration across sciences, technologies, industries, governmental ministries, and citizen groups from around the world. Industries from Canada, Brazil, China, Germany, India, Japan, the Netherlands, Sweden, the United Kingdom, and the United States, among others, contributed more than 50 percent of the approximately 700 TEAP members.

TEAP experts were not required to share the epistemology of precaution. However they were expected to work with disregard of national or industrial interests and toward global solutions with a can-do spirit. They did this by developing strong social bonds of trust and respect (a tight community) and by forging collaborative norms of problem solving, boundary spanning, and information sharing. The effective regulatory community that emerged from the TEAP—largely the result of collaborative leadership as well as linkages to broader constituencies in government, industry, and the academy—became valuable and necessary resources in the creation and transfer of knowledge so essential to MP success (Canan and Reichman 2002).

The one area where phaseout has lagged is addressing the issue of methyl bromide, in which commitment to planetary concerns has not overridden industrial interests and national politics. Nominations for Critical Use Exemptions for methyl bromide have used criteria that differ from the criteria for other ODS. For other ODS, an *essential* use is defined as one that "is necessary for the health, safety or is critical for the functioning of society (encompassing cultural and intellectual aspects)" (Decision IV/25 of the Parties, cited in DeCanio and Norman, 2003). An oft-cited example was the exemption for CFC use for Metered Dose Inhalers (MDIs) having life-and-death criticality. However the MP allows nominations for *critical use exemptions* to the methyl bromide phaseout based on claims that alternatives are not *economically feasible* or that the phaseout would cause *significant market disruption*. As a result, some parties have requested exemptions for a range of methyl bromide applications, including tobacco, pet food, flowers, and golf courses (DeCanio and Norman 2003).

Despite the tremendous progress that has been made accelerating phaseout dates, banning additional chemicals, and identifying, creating, and adopting alternative technologies, the long life of ODS means that restoring the earth's protective stratospheric ozone layer will remain a serious challenge throughout the twenty-first century.

PENELOPE CANAN
NANCY REICHMAN

SEE ALSO *Global Climate Change; International Relations.*

BIBLIOGRAPHY

Andersen, Stephen O., and K. M. Sarma. (2003). *Protecting the Ozone Layer: The United Nations History.* London: Earthscan. An authoritative and exhaustively documented history of the science and diplomacy that led to the Montreal Protocol; contains a detailed account of the contribution of business, industry, and government to develop environmentally sound alternative technologies to restore the ozone layer. Provides a careful record of the role of the media and non-governmental organizations in evolving the global response to the destruction of stratospheric ozone. It is the official UN history of the Montreal Protocol.

Benedick, Richard. (1991). *Ozone Diplomacy: New Directions in Safeguarding the Planet.* Cambridge, MA: Harvard University Press. Provides a detailed account from the perspective of the senior U.S. negotiator of the Montreal Protocol. Features interpretations of the bargaining motivation and stratagems of other countries. Stimulated many others to tell the story from their own perspective.

Brack, Duncan. (1996). *International Trade and the Montreal Protocol.* London: Earthscan. Covers the role of trade as a governing principle for international agreements and how variable national economic positions view trade sanctions as impetus for participation and compliance.

Canan, Penelope, and Nancy Reichman. (2002). *Ozone Connections: Expert Networks in Global Environmental Governance.* Sheffield, UK: Greenleaf. A sociological analysis of the extent and effectiveness of a small number of experts, organized in communities of practice, were instrumental in protecting the ozone layer. Looking systematically at the connection between technology, global environmental policy, and the social connections of experts, the authors focus on the Technology and Economic Assessment Panel of the Montreal Protocol. By combining formal network analysis, biographical interviews and participant observation, they demonstrate that treaty implementation relies on social relations, trust and the collaborative leadership of institutional entrepreneurs.

Dotto, Lydia, and Harold Schiff. (1978). *The Ozone War.* New York: Doubleday and Company. The most comprehensive early account of conflict among scientists, citizens, industry, and political activists. It was published during the decade when UENP and national environmental ministries were created and environmental law was invented, but long before there was much hope of stratospheric ozone protection.

Gribbin, John R. (1988). *The Hole in the Sky: Man's Threat to the Ozone Layer.* New York: Bantam. Documents the triumph of science and diplomacy in securing the Montreal Protocol.

INTERNET RESOURCE

DeCanio, S. J., and C. S. Norman. "Economic Aspects of Nominations for *Critical Use* of Methyl Bromide Under Terms of The Montreal Protocol." Included in UNEP Report of the Technology and Economic Assessment Panel, May 2003 Progress Report. Available at http://www.econ.ucsb.edu/~decanio/papers/MB-UK9r.pdf. A sophisticated analysis of the economics of critical uses of ODS revealing the politics of changing definitions risk.

MORE, THOMAS

• • •

Thomas More (1478–1535) was born in London on February 7 and executed on Tower Hill, London on July 6. He was a lawyer and royal councilor who rose to be Lord Chancellor of England (1529–1532) before falling afoul of Henry VIII over the matter of the king's divorce. Of his voluminous writings, the only one that has anything to say about science and technology is *Utopia* (1516), his vastly influential Latin book about an imaginary island republic somewhere off South America.

To More and those of his fellow humanists who understood the Greek etymology, of the word that he coined for this title *utopia* meant simply *noplace* (*ou + topos*): the word, that is, did not originally have the meaning—an ideal society, or a fictional work about one—acquired in the book's aftermath. Indeed the fundamental interpretive question about the work is whether More intends Utopia as his ideal society. At the least, though, the Utopian commonwealth includes a number of institutions that he clearly regarded as preferable to those of sixteenth-century England and Europe.

The Utopian institutions toward which the book embodies a clearly favorable attitude do not for the most part involve science or technology: England had to wait until 1627, when Bacon's *New Atlantis* appeared, for its prototypical scientific utopia. More finds the principal means to human betterment not in scientific and technological advances but in wiser political, religious, and educational institutions. There are, however, several passages focusing on science and technology, and in all but one the attitude toward these subjects is unreservedly positive.

The account of Utopia is narrated by a fictitious character named Raphael Hythlodaeus, who is supposed to have sailed with Amerigo Vespucci and who now speaks to More and his friend Peter Giles. Just before the account, Hythlodaeus attempts to convince his auditors of the superiority of Utopia to Europe by an historical anecdote. Utopia had had, in about 300 C.E., a previous encounter with Old World visitors, in the form of a company of shipwrecked Romans and Egyptians. The Utopians, Hytholodaeus approvingly observes, profited greatly from this chance event, learning "every single useful art of the Roman empire either directly from their guests or by using the seeds of ideas to discover these arts for themselves ...This readiness to learn is, I think, the really important reason for their being better governed and living more happily than we do, though we are not inferior to them in brains or resources" (1995, p. 107; 2002, p. 39, 40). Later, discoursing again on the Utopians' passion for learning, Hythlodaeus notes that they are "wonderfully quick to seek out those various skills which make life more agreeable" (1995, p. 183; 2002, p. 76). In this instance, having heard in general terms about printing and paper-making from Hythlodaeus and his companions, the Utopians rapidly develop these technologies and use them to reprint the classical Greek and Roman books that Hythlodaeus's group had with them.

Among the ancient books, Hythlodaeus notes, the Utopians were especially pleased to receive works of Hippocrates and Galen, because in Utopia medical science is held in great esteem. In general, the Utopians find science a source not only of practical benefits but of keen intellectual pleasure. Hythlodaeus singles out for special praise their mastery of astronomy, in the pursuit of which "they compute with the greatest exactness the course and position of the sun, the moon and the other stars that are visible in their area of the sky" (1995, p. 157; 2002, p. 65). (For astrology, they have only contempt.) They also regard the exploration of the secrets of nature as a form of worship. God, they suppose, "created this beautiful mechanism of the world to be admired—and by whom, if not by man, who is alone in being able to appreciate so great a thing?" (1995, p. 183; 2002, p. 76).

Another area in which the Utopians are said to be especially inventive is the design of weapons. There is no hint of disapproval in the passage on this subject. (The Utopians avoid war whenever possible, but when it is unavoidable, they excel at it.) Only one passage in More's book intentionally raises the possibility that technological advance may not always be an unmixed blessing. Before reaching Utopia, Hythlodaeus and his companions have occasion to introduce their native South American hosts to the magnetic compass and its navigational benefits. Previously, the natives had "sailed with great timidity, and only in summer." Now, however, they put such trust in the loadstone that "they no longer fear winter at all, and tend to be careless rather than safe." Thus "there is some danger that through their imprudence this device, which they thought would be so advantageous to them, may become the cause of much mischief" (all quotes 1995, p. 49; 2002, p 12). This is as close as More comes to the topic of the ethical

Sir Thomas More, 1478–1535. The life of this English humanist and statesman exemplifies the political and spiritual upheaval of the Reformation. The author of *Utopia*, he was beheaded for opposing the religious policy of Henry VIII. (*The Library of Congress.*)

implications of science and technology—a topic that was, however, to be a major focus of many of the hundreds of utopias (and, latterly, dystopias) that have their prototype in his subtle little book.

GEORGE M. LOGAN

SEE ALSO *Utopia and Dystopia.*

BIBLIOGRAPHY

Baker-Smith, Dominic. (1991). *More's "Utopia."* Unwin Critical Library. London and New York: HarperCollins Academic.

Guy, John. (2000). *Thomas More.* Reputations (series). London: Arnold; New York: Oxford University Press.

Manuel, Frank E., and Fritzie P. Manuel. (1979). *Utopian Thought in the Western World.* Cambridge, MA: Harvard University Press.

More, Thomas. (1995). *Utopia: Latin Text and English Translation,* eds. George M. Logan, Robert M. Adams, and Clarence H. Miller. Cambridge, UK: Cambridge University Press.

More, Thomas. (2002). *Utopia,* rev. edition, eds. George M. Logan, and Robert M. Adams. Cambridge Texts in the History of Political Thought. Cambridge, UK: Cambridge University Press. This edition contains the English translation only (plus annotations and an introduction), omitting the Latin original. While the 1995 edition cited above is the scholarly standard, this 2002 version, a widely-used teaching edition with the same translation, is far handier.

MORRIS, WILLIAM

● ● ●

William Morris (1834–1896) was born in Walthamstow, now part of London, on March 24 and died at Kelmscott House, Hammersmith, London on October 3. During his own lifetime he was best known as a poet, but while his reputation as a poet has continued, his work as a designer with his own firm and as a politically active socialist has been even more enduring. An early love of the Middle Ages helped shape all his activities. He rejected what he saw as the cheap and shoddy ideas and goods of the modern age.

At first Morris thought that social reform was possible through the Anglican ministry. But influenced by the work of social commentator and art critic John Ruskin (1819–1900), especially the fifth chapter of *Stones of Venice* (1851–1853), "On the Nature of Gothic," he turned to art instead. Ruskin convinced him of the need for workers to have a sense of pleasure in their work and surroundings. Morris considered being an architect, then a painter. Moving to London he found no furniture to his liking so he designed his own. He found no house he wished to live in. Turning to his friend Philip Webb, Morris had him design the influential Red House in Bexleyheath outside of London in a simplified red brick Gothic. He formed a design firm to work on the inside of the house and it became a commercial operation.

It was through his work as a designer and a businessman that Morris confronted issues of technology and ethics. He felt that much of the design of the time was ugly and false to nature. Its purpose was not beauty but to advertise the wealth of its purchaser; it was not true to its form; it was not true to Ruskin. Morris believed in talent, not genius, and felt he demonstrated this himself by working in all areas of his firm's production. To modern eyes, many of Morris's designs appear elaborate; in their own time they represented a move toward simplicity. He designed furniture, wallpaper, stained glass, textiles, tapestries, tiles, carpets, and toward the end of his

William Morris, 1834–1896. Morris, one of the most versatile and influential men of his age, was the last of the major English romantics and a leading champion and promoter of revolutionary ideas as poet, critic, artist, designer, manufacturer, and socialist. (*The Library of Congress.*)

life, books for his last enterprise, the Kelmscott Press. His aim, as he wrote in *Arts and Crafts Exhibition Society Catalogue of the First Exhibition*, was "to combine clearness of form and firmness of structure with the mystery that comes of abundance and richness of detail" (p. 27). He wished, in his own words, "to give people pleasure in the things they must perforce *use*, that is the one great office of decoration; to give people pleasure in the things they must perforce *make*, that is the other use of it" (Morris 1882, p. 4).

Morris was aware of being caught in a technological conundrum. He hated what he saw as the low quality of machine products. He is frequently seen as being antimachine. He certainly did not admire the machine but he was perfectly willing to use it as a way of producing his wallpapers and chintzes at lower cost, although his firm's finer work was done by hand. He increasingly came to feel that the reliance on technology was becoming an ethical and political matter and that, to use the modern term, corporate interests would demand cheaper and shoddier production. For instance, he hated the new chemical dyes and insisted on using natural ones.

He became more and more active in politics because he felt that the only way the ordinary person could make and have truly beautiful and useful objects was if socialism were introduced and the economic arrangements of society transformed. He became a convinced Marxist. This did not result in his changing his business methods. Though his workers were well paid, it was not a firm in which he shared the profits. To charges of hypocrisy, he pointed out that his one individual case would not change society and he needed his income to achieve political reform, indeed revolution, for all.

Morris devoted a great deal of his considerable energy to political agitation. The various political groups with which he was associated were the precursors of the British Labour Party, much as he would have disliked it. In his view, society needed to be totally transformed politically if it were to serve the best scientific, technical, and ethical needs of its members. He outlined his utopia in his most famous prose work, *News from Nowhere* (1890). Though he fought for total change, at the same time he had an important influence on contemporary capitalist society. He launched the modern preservation movement through the founding of the Society for the Protection of Ancient Buildings (1877), and he helped create a sensitivity in favor of preserving and protecting the environment. Although in practice he made compromises, he left a legacy of belief in simplicity of form and truth to materials that has had a profound effect on the look, usefulness, and technology of the modern world.

PETER STANSKY

SEE ALSO *Science, Technology, and Literature; Socialism; Utopia and Dystopia.*

BIBLIOGRAPHY

Arts and Crafts Exhibition Society Catalogue of the First Exhibition. (1888). London: London New Gallery.

Briggs, Asa, ed. (1984). *William Morris: News from Nowhere and Selected Writings and Designs.* Harmondsworth: Penguin.

Kelvin, Norman, ed. (1984–1996). *The Collected Letters of William Morris,* 4 vols. Princeton, NJ: Princeton University Press.

MacCarthy, Fiona. (1995). *William Morris: A Life for Our Time.* New York: Knopf.

Morris, May, ed. (1910–1915). *The Collected Works,* 24 vols. London: Longman, Green & Co.; (1973) New York: Oriole Editions.

Morris, William. (1882). *Hopes and Fears for Art.* London: Ellis & White.

Stansky, Peter. (1985). *Redesigning the World: William Morris, the 1889s, and the Arts and Crafts.* Princeton, NJ: Princeton University Press.

Thompson, Edward Palmer. (1976). *William Morris: Romantic to Revolutionary.* New York: Pantheon.

MOVIES

• • •

Motion pictures are one of the most pervasive contemporary technologies, and, since their invention, have been continuously engaged with ethical issues. From the beginning, movies have been accused of corrupting children and adults by communicating godless, overtly sexual, and perverted values. The result has been extensive attempts to control movie content. Even commentators who are against censorship have argued that, independent of any particular content, movies have a morally significant influence. Finally as a new technological medium, films have explored the ethical challenges of new technologies.

Background

In January 1894 inventor Thomas Edison filmed his assistant, Fred Ott, sneezing. Early proponents of the new medium soon began shooting the first fiction films, consisting of only a few scenes. *The Great Train Robbery* (1903) was a milestone, using montage and the point of view of the camera to excite and frighten the audience. By 1907 there were 1 million daily viewers of nickelodeons in the United States. In 1910 the nation had 10,000 movie theaters. Hull House reformer Jane Addams said that "what they [children] saw on the screen was directly and immediately transformed into action." Reverend Wilbur Crafts saw the early cinema as "offering trips to hell for a nickel" (Black 1994, pp. 6, 10). *The Jazz Singer* (1927) popularized the new technology of synchronized sound, allowing actors to speak and sing and writers to create more complex, morally nuanced, and provocative stories.

Later technological developments have not been quite as earthshaking as the addition of sound. Cinemascope, a wide-screen color format introduced in 1953, brought audiences back to the movie theater by creating an experience television could not rival. 3-D films, another 1950s attempt to draw viewers from television, quickly became associated with schlock horror and science fiction efforts and was a mere technological detour rather than a lasting development. The huge-screen IMAX 3-D movies may represent a technological

Charlie Chaplin in a scene from the 1936 film *Modern Times.* The movie explores automation and its repercussions for human beings. (*The Kobal Collection. Reproduced by permission.*)

apex, but the use of digital video instead of film has been more significant in reducing costs of entry for small filmmakers in both the United States and abroad. The digitization of Hollywood films for distribution and projection also reduces costs and makes moviegoing more consistent, eliminating such memorable experiences as the scratchy print and film that breaks during the crucial scene.

Film has long served as a means of advancing scientific understanding, particularly by capturing events that occur too quickly or slowly for the human eye to see (a cheetah running or the growth of a flower), and by archiving scientific information. It has also popularized science to the masses, via such media as IMAX films shown in museums.

Censorship

Attempts to protect citizens by censoring the cinema began at the local level in the United States soon after the nationwide introduction of popular films; states and cities set up their own boards of censorship to determine what could be shown in local theaters. In *Mutual Film Corp.* v. *Ohio Industrial Commission* (1915), the Supreme Court denied First Amendment protection to movies, finding them to be "a business, pure and simple" and therefore not "part of the press" or "organs of public opinion" (*Mutual Film Corp.* v. *Ohio Industrial Commission*, p. 244).

In 1922 production companies launched a preemptive strike against increasingly pervasive state and local censorship by founding the Motion Picture Producers and Distributors Association of America, headed by William Harrison Hays, former postmaster general and chairman of the Republican National Committee. That spring, more than 100 movie censorship bills had been introduced in the legislatures of thirty-seven states. Hays served as a buffer between the producers and public opinion. The studios wanted to police themselves so as to avoid more rigorous censorship from outside. The first *Hays code* prohibited profanity, nudity, drug trafficking, and white slavery, and urged good taste in presenting criminal behavior, sexual relations, and violence.

Compliance with the code was initially voluntary, and Hays frequently threatened public embarrassment as a means of persuading producers to follow his views. Soon enough, the owners of movie theaters would not show films without the seal of approval of the office, making the system mandatory for studios that hoped for national distribution. It was not until the 1950s that films, such as *The Moon Is Blue* (1953) and *The Man with the Golden Arm* (1955), began to be nationally distributed without the seal of the Hays office.

Early in the sound era, Hollywood moved to secure the rights to several popular but controversial novels by respected authors such as Ernest Hemingway, William Faulkner, and Sinclair Lewis, triggering an ethical debate as to whether movies are an art form, mirroring the world like novels, or have a special responsibility to function as "twentieth century morality plays" illustrating "proper behavior to the masses" (Black 1994, p. 39). The Hays office fought to prevent the studios from filming Hemingway's *A Farewell to Arms* (1929), released as a film of the same name in 1932, and Faulkner's *Sanctuary* (1931), filmed as *The Story of Temple Drake* (1933). Failing in these efforts, Hays's people successfully pushed the producers to tone the films down, delete controversial material, and add plot developments or commentary illustrating the negative consequences of *antisocial* behavior.

Until then movie censorship had been primarily a Protestant affair, but in 1930 the Catholic Church proposed its own movie code, which was adopted in large part by the Hays office (Walsh). The possibility of federal censorship of movies was looming. The Catholic-inspired revision of the code, taken literally,

"forbade movies from ever questioning the veracity of contemporary moral and social standards" (Black 1994, p. 41). Producers including Jack Warner and Irving Thalberg rebelled. Movies, they said, are "one vast reflection of every image in the stream of contemporary life." As such, they should be able to present "any book, play or title which had gained wide attention" (Black 1994, p. 41). In 1934 the Hays office was once again reorganized, and code enforcement became much tougher.

The studios were often able to subvert the code by presenting glamorous gangsters and loose women, only to have them pay for their sins by dying at the end of the picture. The *Nation* magazine amusingly referred to this trend as "five reels of transgression, followed by one reel of retribution" (Black, p. 45). The Hayes office intervened in the making of popular gangster films such as *Scarface* (1932), ensuring that the protagonist would die cravenly, not bravely as he did the original script.

Propaganda

During the 1940s, Hollywood and the government entered into partnership for the first time. The Office of War Information asked all filmmakers to consider seven key questions regarding movies made during wartime. The first and most important was, Will this picture help win the war? (Basinger 1998). Hollywood responded enthusiastically with movies calculated to encourage and reinforce patriotic feelings, and engender contempt and hatred for the enemy—in effect, political advertising or propaganda.

An interesting feature of these movies is that they represent the first time Hollywood had both an opportunity and incentive to represent the diversity of American society. Most portrayed a squad or other military group "made up of a mixture of ethnic and geographic types, most commonly including an Italian, a Jew, a cynical complainer from Brooklyn, a sharpshooter from the mountains, a Midwesterner (nicknamed by his state, Iowa or Dakota), and a character who must be initiated in some way (a newcomer without battle experience) and/or who will provide a commentary on the action as it occurs (newspaperman, letter writer, author, or professor)" (Basinger 1998).

The Hays Office nevertheless continued to be an important force in Hollywood into the 1950s, a period during which congressional investigations into communism exerted significant influence over the content of American movies. The Hollywood Ten, directors and screenwriters who went to prison for refusing to name names, and many other writers, directors, and actors saw their careers ruined or, at best, put on hold for many years until the atmosphere changed. Many blacklisted writers continued to work under pen names or through *fronts* (Navasky 1991). Most Hollywood films stayed even more resolutely away from subject matter which could be construed as political; the 1950s was the era of the uncontroversial, extremely traditional, family-centered romance or comedy.

Hollywood, in a second, smaller collaboration with the government, also produced a number of overt propaganda films including *I Married a Communist* (1950), *I Was a Communist for the FBI* (1951), and *My Son John* (1952).

Ratings for Consumer Choice

One important development during the 1950s was the Supreme Court's reversal of the almost forty-year-old decision in *Mutual Film Corp. v. Ohio Industrial Commission*. In the case of *Joseph Burstyn Corp. v. Williams* (1952), the Court granted movies full First Amendment protection. Weakened by new legal protections against state and federal censorship and overwhelmed by the cultural and sexual revolutions of the 1960s, the Hays office was finally discontinued in 1966 and replaced by a new ratings system.

Under the new system, the Motion Picture Association of America (MPAA) assigned an X, R, M or G to every movie. X meant the content of the film was highly sexual or violent; R indicated that the film should be restricted to viewers above a certain age; M advised that the film was appropriate only for mature audiences; and G signified that the film was approved for all audiences. Minors (under 17) could not attend X-rated films, and could only see R-rated ones if accompanied by an adult. In 2005 the revised rating system consists of NC-17 (over 17 years old only); R (under 17 years old only if accompanied by a parent or guardian); PG-13 (may not be appropriate for viewers under 13); PG (parental guidance suggested); and G (general audiences).

From the start, opponents contended that the ratings system was biased: Sexually explicit movies tended to get an X-rating, whereas extremely violent ones frequently received only an R, suggesting a cultural acceptance of violence and disapproval of sex. The ratings system has also been described as a mechanism of political control by the major studios that participate in it. "It's no coincidence that the films given Xs and NC-17s over the years have tended to come from independents, minorities, foreign filmmakers, and women—those out-

side the fold of the seven major studios who are members of the MPAA" (Keough 1999 Internet site).

Influence of the Medium

Since the demise of the Hays office, films have become far more explicit than they were, routinely showing nudity, simulated sex, and increasingly inventive forms of graphic violence (while earning nothing more restrictive than R ratings). F. Miguel Valenti notes that violence and sex sell tickets. An epigraph frequently quoted in film criticism, and usually attributed to Jean-Luc Godard, holds that all that is needed for a movie is a gun and a girl. The debate about whether movies promote violence, immoral or unsafe sexual behavior, or other undesirable acts continues. But some film industry representatives and many consumers deny that movies are a medium of moral expression.

In fact, all films communicate moral ideas, simply by telling stories: ideas about the propriety of certain kinds of social behavior, including sexual and romantic acts, truthfulness and lying, the acceptability of violence; and the mutual rights and responsibilities of various social groups, including wealthy and poor, or police and citizens. Revenge movies, including many Westerns, thrillers, and cop films, show that it is sometimes acceptable to take the law into one's own hands. Many films communicate the idea that those in law enforcement cannot fight crime effectively without disregarding the strictures of the U.S. Constitution.

Thrillers promote a jaundiced or even fearful view of the world. "[T]hey portray a world in which crime, deceit, avarice, intrigue and betrayal are the norm rather than the exception, a film noir world even grimmer than our grimmest perception of daily life" (Dickstein 1981, p. 49). These films may promote "mean world" syndrome, "the feeling instilled in viewers that they live in a dangerous environment" (Valenti 2000, p. 14). However the underlying moral structure renders these films entertaining to audiences. "It is the exposition of moral significance that keeps the audience watching, not the quantity and quality of pyrotechnics on the screen" (Hicks 1995, pp. 106–108).

Peter Bogdanovich, director of *The Last Picture Show* (1971), believes that movies have a profound influence on behavior: "The trouble with portraying any way of life on the screen is that there cannot fail to be an inherent glorification of it, no matter how seamy" (Valenti 2000, Introduction). By contrast, film critic Judith Crist believes that movies have too long a lead time to have much of an influence on American popular

Harrison Ford as Rick Deckard in a scene from the 1982 film *Blade Runner*. The movie explores the definition of humanity in a machine-dominated world. *(The Kobal Collection.)*

culture; because it takes three to five years to make one, "it simply can't be that movies set patterns. They reflect our society" (Thayer 1980, p. 49).

Films aimed at juvenile audiences are widely thought to have a special responsibility to communicate socially acceptable values. Analee R. Ward notes that writers and animators at The Walt Disney Company are aware of their role in forming children's values, but have blind spots. "The role of a female in *The Lion King* is largely that which is associated with love, either romantic or motherly" (Ward 2002, p. 127). She notes possible racism in the portrayal of the hyenas as jive-talking blacks, and homosexual stereotypes in the behavior and mannerisms of the villain, Scar.

Others argue that by giving in to self-censorship, filmmakers often make bland, uninteresting movies. Pediatrician Perri Klass observed that "[I]f children's entertainment is purged of the powerful, we risk homogeneity, predictability and boredom, and we deprive children of any real understanding of the cathartic and emotional potentials of narrative" (Ward 2002, p. 29). Carter Burwell, writing about adult movies, has similarly said that "[I]f people's buttons are pressed in completely

predictable fashion, you're depriving them of the opportunities to have novel and perhaps enlightening experiences" (Valenti 2000, p. 36).

Some critics have noted that movies give a distorted view of historical events. Stephen Fjellman wrote, "What Disney does, perhaps, is kill the *idea* of history by presenting it as entertainment" (Ward 2002, p. 117). Historian Mark Carnes said that films "make the past speak to us with ... complete crystal clarity, so that it speaks to our time. Of course, historians, when they go to the past, don't find that clarity. They find a muted voice in a different language echoing through vast expanse of time" (Public Broadcasting System 1995).

The most provocative criticism, however, is that independent of any particular content, motion pictures have distinctive social and cultural effects that call for ethical assessment. For instance, media analyst Marshall McLuhan's thesis that the *medium is the message* might suggest that because film deals with rich visuals and sounds disembedded from their full physical contexts, it cannot help but make any violence it depicts somewhat attractive. Moreover motion pictures would also seem to have a strong tendency to induce in those who sit in a dark room in front of a large screen the kind of dreamy rootlessness described in Walker Percy's novel *The Moviegoer* (1961). Remarkably, however, there has been little scientific research on the psychological impact of movie watching—certainly nothing like the degree of empirical research devoted to the psychological influence of television.

Movies Examining Science and Technology

From the silent days to the era of huge screens and Dolby sound, films tell stories about new technologies, often in a fantasy or science fiction context. Fritz Lang's *Metropolis* (1927) portrayed a world in which evil rulers used technology to manipulate workers, and in which a woman was impersonated by an evil robot doppelganger. Charlie Chaplin's *Modern Times* (1936) examined the alienation caused by automation. These movies raised the central questions considered by later efforts: What happens when powerful technology evades human control, and what is human as opposed to *other* (Telotte 2001).

The apocalyptic genre (Shapiro 2002) which began in the 1950s with movies such as *Godzilla* (1954), *Them* (1954), and *The Beast From 20,000 Fathoms* (1953) was based on the premise that *there are some things man was not meant to know*. Typically the threat in these movies was a mutant created by radiation from an atomic blast. The *Alien* films (originating in 1979) similarly show humans trying to manipulate forces (the rapacious aliens) that quickly evade their control, with deadly results. The *Terminator* series (originating in 1984) recapitulates a theme, first expressed in movies such as *2001* (1968), *The Demon Seed* (1977), *Colossus: The Forbin Project* (1970), *Westworld* (1973), and *War Games* (1983), in which computers become powerful enough to destroy humankind. More subtle thrillers such as *Minority Report* (2002) portray a future in which humans are punished for overreliance on technology, which never works exactly as planned (the pre-cogs' infallible view of the future can be manipulated).

For McLuhan the popularity of techno-horror and vampire movies reflects more than the simple dominance of science and technology in contemporary culture. Instead they are a collective unconscious articulation of the sense in electronic culture of feeling taken over by technology. "*The Exorcist* [1973] is an account of how it feels to live in the electric age, how it feels to be completely taken over by alien forces and hidden powers" (McLuhan 2004).

But perhaps it is *Blade Runner* (1982) that, though a flawed movie, asks the most interesting question: In a world of machines that can imitate human behavior, even to the point of being indistinguishable from people, how is *human* redefined? Philip K. Dick's *Do Androids Dream of Electric Sheep* (1968), on which the movie was based, answered that the irreducible difference is that humans feel compassion, and machines do not. This powerful idea was drowned out in the movie's pyrotechnics, which transformed it into a more clichéd Hollywood story about eliminating the other.

The fact that so many of these films, which use cutting edge technologies to create their special effects, take an antitechnology stance may be partly due to the requirements of storytelling. A screenplay involves a threat to the protagonist that must be overcome. Though there have always been some films in which a misunderstood hero champions an initially disregarded technology (1930s and 1940s films about inventors and medical innovators; *Lorenzo's Oil* (1992) is a more recent example), audiences prefer stories with more at stake. Technology provides weapons for really frightening villains, or it may actually play the role of the evil adversary. Susan Sontag says that science fiction films are "fundamentally about disaster, which is one of the oldest subjects of art," but which involves an "extreme moral simplification" (Sontag 1986, pp. 213–215).

International Perspectives

The highly influential cinemas of other nations have faced similar ethical challenges. The French New Wave, which introduced a new kind of moral storytelling, set a youth ethic against the morality of an older generation portrayed as hidebound and hypocritical (Marie). New Wave films such as *Breathless* (1960) and *The 400 Blows* (1959) glorified rebels, outsiders, and gangsters. The New Wave continues to resound, almost fifty years later, in the films of contemporary American auteurs including Martin Scorsese and Quentin Tarantino.

Whereas the films of all nations struggle with some degree of government censorship, Soviet cinema developed in an environment in which dissent could mean exile, imprisonment, or even death. Soviet film artists nonetheless evaded censorship by telling stories set in past centuries, sometimes based on the unassailable works of pre-Soviet masters such as Tolstoy and Chekhov, or through movies, such as *Solaris* (1974), based on a novel that is so heavily coded that it escaped the criticism of simpleminded censors. During the upbeat socialism of the Brezhnev era, Soviet films enjoyed a new freedom to portray humans as "inwardly torn by doubt, failing to accomplish anything in life other than the destruction of that which [they] held dear" (Gillespie 2003, p. 18). In the early-twenty-first century, Russian filmmakers, deprived of their former political and social context, are struggling to create a new identity based on shared cultural values and the country's "awesome historical legacy" (Gillespie 2003, p. 122).

Unfamiliar to most Americans and Europeans, India has developed its own powerful cinematic tradition of leisurely told romance and suspense stories interspersed with musical numbers. Colloquially known as Bollywood, the Indian film industry produces 800 films per year, which are shown in 13,000 cinemas and average 11 million viewers daily nationwide. Vijay Mishra notes that Bollywood cinema knits together a widely dispersed Indian diaspora in Western Europe and North America. Expatriate Indians, who through hard work have joined the comfortable middle classes of their adopted countries, inhabit "the desired space of wealth and luxury that gets endorsed, in a displaced form, by Indian cinema itself" (Mishra, p. 236). Bollywood has been "crucial in bringing the 'homeland' into the diaspora ... creating a culture of imaginary solidarity" reaching across India's numerous ethnic groups (Mishra, p. 237).

Conclusion

Movies are simultaneously a reflection of human life and a distraction from it. As such, they are intimately involved with ethics, drawing from and influencing people's views. It is unlikely that any extensive history of the events, mood, or ethics of any modern era will be written without reference to movies of that period.

JONATHAN WALLACE

SEE ALSO *Entertainment; Information Ethics; Popular Culture; Science, Technology, and Literature; Special Effects; Technocomics; Violence.*

BIBLIOGRAPHY

Black, Gregory. (1994). *Hollywood Censored: Morality Codes, Catholics and the Movies.* Cambridge, MA: Cambridge University Press.

Dickstein, Morris. (1981). "The Morality of Thrillers." *American Film* July–August: 49–52, 67–69

Gillespie, David. (2003). *Russian Cinema.* Harlow, England: Longman.

Hicks, Neill D. (1995). "'Fill Your Hand, You Son of a Bitch': The Underlying Morality of Action Adventure Films." *Creative Screenwriting* 2(4): 106–108.

Joseph Burstyn Corp. v. Williams, 393 U.S. 495 (1952).

Keough, Peter. (1999) "An Immodest Proposal: It's Time to Give the Ratings System an X-it." *Boston Phoenix,* August 30. Also available from http://weeklywire.com/ww/08-30-99/boston_movies_1.html.

Marie, Michel. (2002). *The French New Wave: An Artistic School.* Oxford: Blackwell.

McLuhan, Marshall. (2004). "Man and Media." In *Understanding Me: Lectures and Interviews,* eds. Stephanie McLuhan and David Staines. Cambridge, MA: MIT Press.

Mishra, Vijay. (2002). *Bollywood Cinema: Temples of Desire.* New York: Routledge.

Mitchell, Charles P. (2001). *A Guide to Apocalyptic Cinema.* London: Greenwood Press.

Mutual Film Corp. v. Ohio Industrial Commission, 236 U.S. 230 (1915).

Navasky, Victor S. (1991). *Naming Names.* New York: Penguin Books.

Shapiro, Jerome F. (2002). *Atomic Bomb Cinema: The Apocalyptic Imagination on Film.* London: Routledge.

Sontag, Susan (1986). *Against Interpretation.* New York: Doubleday.

Telotte, J. P. (2001). *The Science Fiction Film.* Cambridge, MA: Cambridge University Press.

Thayer, Lee, ed. (1980). *Ethics, Morality and the Media: Reflections on American Culture.* New York: Hastings House.

Valenti, F. Miguel. (2000). *More Than a Movie: Ethics in Entertainment.* Boulder, CO: Westview Press.

Walsh, Frank. (1996). *Sin and Censorship: The Catholic Church and the Motion Picture Industry.* New Haven, CT: Yale University Press.

Ward, Analee R. (2002). *Mouse Morality: The Rhetoric of Disney Animated Film.* Austin: University of Texas Press.

INTERNET RESOURCES

Basinger, Jeanine. (1998). "Translating War: The Combat Film Genre and Saving Private Ryan." American Historical Association. Available from http://www.theaha.org/Perspectives/issues/1998/9810/9810FIL.CFM. Published in *Perspectives Online: The Newsmagazine of the American Historical Association,* October 1998.

Motion Picture Association of America. "Voluntary Movie Rating System." Available from http://www.mpaa.org/movieratings/.

"The Motion Picture Production Code of 1930." Artsreformation.com. Available from http://www.artsreformation.com/a001/hays-code.html.

Public Broadcasting System. (1995). "Interview with David Gergen and Mark Carnes, December 26, 1995." Public Broadcasting System. Available from http://www.pbs.org/newshour/gergen/carnes_12-26.html.

"Red Scare Filmography." The All Powers Project. http://www.lib.washington.edu/exhibits/AllPowers/film.html.

MULTIPLE-USE MANAGEMENT

• • •

Multiple use is a form of natural resource management with ethical dimensions that may have additional implications for other aspects of science and technology by its interdisciplinary nature. In the present case the focus nevertheless remains on natural resource management.

Multiple-use natural resource management is a way of using resources to produce more than one good or service simultaneously. In the U.S. Forest Service this commonly implies managing national forests for such diverse ends as timber production, recreational activities, and environmental protection. Such multiple use easily leads to ethical dilemmas for decision makers. For example, many people living near forests in developing countries make a livelihood out of harvesting timber and non-timber forest products such as honey, nuts, and wild animals on a small scale. Commercial timber operations also have the potential to extract these resources for profit, but only by excluding, at least to some extent, the small-scale harvesters. Decision makers must decide what is the best use of resources: Produce non-timber products to ensure livelihood of communities living near forests? Produce timber to stimulate regional or national economies? Developed nations face similar dilemmas, often compounded by public concern for nearly immeasurable forest benefits, such as recreation, aesthetic beauty, and contribution to global biodiversity.

What Is Multiple Use?

Goods and services produced through a multiple-use management strategy can be complementary, supplementary, or competitive. For example, in Figure 1 the harvest of both timber and non-timber forest products from the same forest are shown to be competitive; the use of standing forest resource to produce timber limits the opportunity to produce non-timber products requiring management decisions. If decision makers decide that timber, for example, is very important and should be harvested at a high level (T1), then by following the curve one can see that non-timber forest products will be harvested at a relatively lower level (NT1). On the other hand, if decision makers think that benefits from non-timber forest products are more important, a management plan might use the NT2 value at the expense of timber interests. A private resource owner could choose a mix of timber and non-timber products that gives the greatest profit. In the context of a public

FIGURE 1

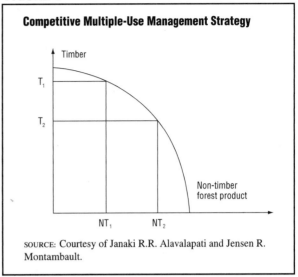

Competitive Multiple-Use Management Strategy

SOURCE: Courtesy of Janaki R.R. Alavalapati and Jensen R. Montambault.

The competing relationship between timber and non-timber forest products in a multiple-use forest management scenario.

resource, however, once single-faceted, often arbitrary, management strategies are abandoned, a variety of involved economic, cultural, political, technological, spatial, and temporal factors raise socioeconomic and ethical dilemmas in multiple-use management.

Socioeconomic, Environmental, and Ethical Issues in Multiple Use

Each forest presents its own medley of site-specific considerations challenging the decision maker to question the fairness of a management plan in terms of how it directly and indirectly affects a variety of stakeholders. Several socioeconomic and environmental justice theories can be applied to exploring the different facets of these issues.

Many ethical problems arise when there is no standard scale for comparing competing issues. For example, it is fairly easy to calculate consistent monetary values for timber. While non-timber forest products are sometimes harvested for a specialized global market, more often they are harvested for household use or local trade in situations in which there exists no market value for these articles. Markets for non-timber forest products, where they do exist, tend toward instability or limited scope. Therefore, taking Figure 1 again as an example, if the management goal is to maximize the monetary gain from a forest, timber would have a distinct advantage over non-timber forest products. In many developing (and some developed) countries, however, non-timber forest products are a major source of income for marginalized or impoverished communities. According to the philosopher John Rawls's theory of social justice (1971), no amount of overall gain is acceptable if it is at the expense of the most disadvantaged. On the other hand, unequal distribution of social goods (rights and liberties, powers and opportunities, income and wealth) is justified if it will help this disadvantaged group. In the case of forest policy, this may mean that a multiple-use strategy is implemented to include both timber and non-timber forest product harvesting at the expense of monetary efficiency because it benefits an otherwise marginalized group.

Basic liberties are not limited to those who are most disadvantaged. Natural resource conflicts frequently arise when the government tries to restrict access by local communities in an area to protect a public good such as biodiversity or the headwaters of a river. This might mean that a local community would lose their livelihood from non-timber forest products or the cultural tradition of family picnics by the river. If communities have a legitimate customary right to use these resources, according to Robert Nozick's theory of social justice (1974), any transfer or exchange is acceptable only if voluntary or without violation of rights. If the communities agree to forego harvesting non-timber forest products or hold their picnics in another area, either out of a sense of altruism or in response to compensation, then it is fair to restrict access to the forest.

The theory of customary rights sometimes conflicts with Aldo Leopold's land ethic philosophy (1949), which argues that all living species and environmental elements, including soils and rivers, for instance, have a basic right to exist at least to some extent in their natural condition. Managers place disproportionate weight on human needs, often ignoring the role these natural functions play in support of the human species. If a community refuses to restrict access to the forest around a river headwater, it might harm the water supply for a much larger human, plant, and animal community downstream. In this case, it becomes difficult to distinguish which is the most disadvantaged group. Followers of an ecocentric philosophy might argue that those species with no voice in the management argument and at great potential risk are actually what Rawls would describe as the most disadvantaged.

Multiple-use natural resource management attempts to address issues of equitability in sharing the benefits supplied by forests, waters, and other resources. The issue of implementing a fair policy, however, is subjective and complex. Economically efficient and ethically acceptable multiple-use management options would be ideal, but very few options pass these criteria simultaneously. In order to ensure a more egalitarian society, it is critical to use these social and ethical principles as binding constraints to maximize efficiency through multiple use of public natural resources.

JANAKI R. R. ALAVALAPATI
JENSEN R. MONTAMBAULT

BIBLIOGRAPHY

Bowes, Michael D., and John V. Krutilla. (1989). *Multiple-Use Management: The Economics of Public Forestlands*. Washington, DC: Resources for the Future. Analyzes Forest management on public lands from an economic perspective to explore the application of multiple-use scenarios.

Leopold, Aldo. (1949). *A Sand County Almanac*. New York: Oxford University Press. This classic text falls into the readable genre of nature writing and also presents seminal arguments for the intrinsic value of wilderness and land ethics.

Nozik, Robert. (1974). *Anarchy, State, and Utopia*. New York: Basic. Focusing on his theory of entitlement, Nozik examines appropriate limits to the function of the state in order to guarantee liberties.

Rawls, John. (1999 [1971]). *A Theory of Justice*. Cambridge, MA: Harvard University Press, Belknap Press. "Liberty" and "difference" principles are discussed in the context of Rawls's moral philosophy.

MUMFORD, LEWIS

• • •

Historian and social philosopher Lewis Mumford (1895–1990) produced a broad critique of modern technology complemented by studies of art, architecture, and urban life. Born in Flushing, New York, on October 19, Mumford studied at the City College of New York (CUNY) but contracted tuberculosis and was forced to leave before earning a degree. In 1919 he became associate editor of the *Dial,* and he later worked as architectural critic for the *New Yorker*. His first book, *The Story of Utopias* (1922), was a literary survey that examined the place of technology in society. This became the main theme in *Technics and Civilization* (1934), which was a founding work in the social history of technology. Although he voiced critical attitudes that sometimes anticipated wider cultural shifts (Hughes and Hughes 1990), Mumford also saw science and technology as positive forces in history. In 1936 he and his wife Sophia settled in rural Amenia, New York, where he died on

January 26 more than half a century later, after a lengthy period of dementia.

Life in Context

In 1915 Mumford discovered the writings of Scottish philosopher Patrick Geddess (1854–1932), from whom he learned to see the built environment and social processes as reciprocal influences. With others he hoped that technology would usher in an era of material abundance, but maintained that such promise would be fulfilled only if technology were subject to social democracy and wise regional planning. Mumford thus fostered a regionalist vision in which the automobile, electricity, and other new technologies would help transform congested cities into balanced and decentralized communities. The Great Depression, however, raised grave doubts, in response to which he argued for new institutions and revitalized values to redirect technology to human ends.

Mumford was an early advocate of World War II, but the loss of his son in the war, the dropping of atomic bombs on Hiroshima and Nagasaki, and the ensuing nuclear arms race left him a fading hope "that a moral transformation may alter the fateful course of technological development" (Hughes and Hughes 1990, p. 6). Many of his later works betrayed a growing pessimism that science and technology were fundamentally irrational and dangerous, which led him to challenge the equation between rationality and modernity. Despite this his stubborn optimism and refusal to lose sight of the human element and submit to technological determinism in the massive waves of sociotechnical change prompted many to consider Mumford one of the last great humanists (Stunkel 2004).

At times, however, Mumford appeared to despair that his cautious utopian vision of an *organic culture* was at odds with an increasingly mechanistic post-World War II society. He rebuked scientists for their alliance with capitalists and the military, but his books in this era received poor reviews. This can be partially explained by his unabashed interdisciplinary holism, which threatened many narrowly specialized academics. As Russell Jacoby noted, he was "a thinker and writer who addresses a literate and general audience about questions and issues undefined or categorized by conventional academic and professional disciplines" (Hughes and Hughes 1990, p. 11). He also remained fiercely independent, declining all employment in institutionalized academia except visiting professorships.

Mumford resolved to react to what he saw as the negative drift of history by analyzing and promoting

Lewis Mumford, 1895–1990. Mumford, an American social philosopher and architectural critic, analyzed civilizations for their capacity to nurture humane environment. He emphasized the importance of environmental planning. (*The Library of Congress.*)

the positive personal and communal forces more in line with his vision. In this work, he influenced U.S. literary studies, architecture, and urban development studies. Unlike John Dewey (1859–1952), Mumford did not emphasize political action as a means of transforming society, but maintained that communities were formed and reformed at the levels of family, church, and workers' associations. His later years were characterized by his ambivalent position that science and technology presented both peril and hope and his determined optimism that the necessary moral and religious transformation could happen and thus alter the course of scientific and technological development. His critique of science and technology continues to influence work in several fields, and his vision for urban renewal and transformation lives on in the Lewis Mumford Center for Comparative Urban and Regional Research, established at the University at Albany, State University of New York (SUNY), in 1988.

Philosophical Anthropology

Mumford is part of the U.S. tradition of this-worldly romanticism that first flowered with Ralph Waldo Emerson (1803–1882) and Walt Whitman (1819–1892). The tradition demonstrates a concern for the preservation of nature and the harmonies of urban life, while insisting that physical matter is not the final explanation of organic activity, especially in its human form. In this sense Mumford represents an even older tradition (stretching back to Aristotle) of a humanities philosophy of technology (Mitcham 1994).

In 1930 Mumford proposed that the machine be considered in terms of both its psychological and practical origins and appraised not just by technical considerations but in ethical and aesthetic terms. This thesis was the germ of *Technics and Civilization*, which sought to integrate the examination of the practical with the good, the true, and the beautiful. The book broke new ground by summarizing technical history for the previous thousand years of European civilization in a way that revealed the reciprocal and many-sided relationships between social values and institutions and the work of inventors, engineers, and industrialists. One popular example is Mumford's treatment of the clock, which is a "piece of power-machinery whose 'product' is seconds and minutes" (Mumford 1934, p. 15). Like Henri Bergson (1859–1941), Mumford saw alienating dangers in the regulating of time by the mechanical clock.

In *Technics and Civilization*, Mumford described the psychological and cultural origins of the machine, explained its material and efficient causes, and outlined a history of machine technics in three overlapping phases: intuitive technics using water and wind (to about 1750); empirical technics of coal and iron (1750–1900); and scientific technics of electricity and metal alloys (1900 to the early-2000s). The last part of the book evaluates social and cultural reactions: "We have seen the machine arise out of the denial of the organic and the living, and we have in turn marked the reaction of the organic and the living on the machine" (Mumford 1934, p. 433). Other civilizations had reached high degrees of technical proficiency and possessed machines, but only the Europeans adapted their entire mode of life to the pace and capacities of *the machine*. Technics (his term for technology) has thus been transformed from mere hardware into a complex sociotechnical system that embodies a way of thinking and being.

Mumford's subsequent writing, insofar as it was an elaboration of *Technics and Civilization*, culminated in the two-volume *Myth of the Machine* (1967, 1970). In it

Mumford argued that humans are not fundamentally to be understood as *Homo faber*, because the human essence is not making but interpreting. The interpretive mind, not the manipulative tool, is the basis of humanity:

> If all the mechanical inventions of the last five thousand years were suddenly wiped away, there would be a catastrophic loss of life; but man would still be human. But if one took away the function of interpretation … man would sink into a more helpless and brutish state than any animal; close to paralysis. (Mumford 1950, p. 8–9)

The elaboration of symbolic culture through language "was incomparably more important to further human development than the chipping of a mountain of hand-axes" (Mumford 1967, p. 8).

Kinds of Technology

On the basis of his philosophical anthropology, Mumford distinguished two basic kinds of technology: polytechnics and monotechnics. The former is the primordial form of making, which is "broadly life-oriented, not work-centered or power-centered" (Mumford 1967, p. 9). Like *appropriate technologies*, polytechnics harmonizes with the many aspirations of human life and functions democratically. Monotechnics is directed toward production, expansion, military superiority, and power.

Although modern technology exemplifies monotechnics, Mumford traced its origins back 5,000 years to the discovery of the *megamachine*, or rigid, hierarchical social organization. Examples include the work crews that built the Pyramids or the Great Wall of China. The center of authority in these ancient megamachines lay in the absolute ruler, whereas in the modern bureaucratically administered megamachine it resides in the system itself. The megamachine and monotechnics produce great material benefit but at the expense of a dehumanizing limitation of human aspirations and the pervasive belief in *the myth of the machine*, or the notion that monotechnics is irresistible and ultimately beneficent. In the 1950s, for example, forecasts predicted that by the year 2000 technology would shorten the workweek to twenty hours. Newly formed institutes of leisure pondered how to spend the resulting free time (Lightman 2003). But in 1990 the average American was actually working 160 hours longer than twenty years earlier (Schor 1991). For Mumford this phenomenon illustrates the enthrallment to the myth of the machine.

But the megamachine can be resisted, especially because it is not ultimately beneficial. Mumford attempted to demythologize monotechnics and to make a plea against losing sight of humanity, its purposes, and its dreams. He called for a reevaluation of the machine in order to master it and put it to work in the service of life. Technology should be promoted when it enhances human meaning and the *personal* aspect of existence, but not when it restricts life in the service of power.

Mumford explored as well the positive technologies of art and urban life, and his *The City in History* (1961) won a national book award. The second volume in his four-volume *renewal of life* series (1954) championed a technology modeled on patterns of human biology and a *biotechnic economy*. In *Art and Technics* (1952) Mumford contrasted art as a symbolic communication of inner life with technology as a power-manipulation of external objects. He did not seek a simpleminded rejection of technology but wanted to complement the Promethean myth of human beings as tool-using animals with the story of Orpheus. The animal became human "not because he made fire [a] servant, but because he found it possible, by means of his symbols, to express fellowship and love, to enrich [a] present life with vivid memories of the past and formative impulses toward the future, to expand and intensify those moments of life that had value and significance" (Mumford 1952, p. 35).

CARL MITCHAM
ADAM BRIGGLE

SEE ALSO *Science, Technology, and Society Studies*.

BIBLIOGRAPHY

Hughes, Thomas P., and Agatha C. Hughes, eds. (1990). *Lewis Mumford: Public Intellectual*. New York: Oxford University Press. Sixteen papers by an assortment of scholars, plus a synoptic introduction by the editors. Reveals the breadth of Mumford's interdisciplinary social criticism.

Lightman, Alan. (2003). "The World is Too Much with Me." In *Living with the Genie: Essays on Technology and the Quest for Human Mastery*, ed. Alan Lightman, Daniel Sarewitz, and Christina Desser. Washington, DC: Island Press. Explores the intangible human losses from the fast-paced nature of modern life.

Miller, Donald L. (1989). *Lewis Mumford: A Life*. New York: Weidenfeld and Nicolson.

Mitcham, Carl. (1994). *Thinking through Technology: The Path between Engineering and Philosophy*. Chicago: University of Chicago Press. Some material from the analysis of Mumford, on pages 40 through 44, has been adapted for this article.

Mumford, Lewis. (1922). *The Story of Utopias*. New York: Boni and Liveright.

Mumford, Lewis. (1934). *Technics and Civilization*. New York: Harcourt, Brace. A 1963 reprint includes a new introduction by the author plus corrigenda.

Mumford, Lewis. (1950). *Man As Interpreter*. New York: Harcourt, Brace.

Mumford, Lewis. (1952). *Art and Technics*. New York: Columbia University Press.

Mumford, Lewis. (1954). "Technics and the Future of Western Civilization." In *In the Name of Sanity*. New York: Harcourt, Brace. This is a lecture presented at the 100th anniversary meeting of the American Association for the Advancement of Science (AAAS) in 1948.

Mumford, Lewis. (1961). *The City in History: Its Origins, Its Transformations, and Its Prospects*. New York: Harcourt, Brace & World.

Mumford, Lewis. (1967, 1970). *The Myth of the Machine*, 2 vols. New York: Harcourt Brace & World.

Mumford, Lewis. (1973) *Interpretations and Forecasts: 1922–1972: Studies in Literature, History, Biography, Technology, and Contemporary Society*. New York: Harcourt Brace Jovanovich. Forty-two selections from his work, by the author.

Mumford, Lewis. (1982). *Sketches from Life: The Autobiography of Lewis Mumford*. New York: Dial Press. The first volume of an unfinished autobiography. See also Mumford's more schematic *My Works and Days: A Personal Chronicle* (New York: Harcourt Brace Jovanovich, 1979).

Schor, Juliet B. (1991). *The Overworked American: The Unexpected Decline of Leisure*. New York: Basic Books. Chronicles the increased workloads and stress faced by U.S. citizens since World War II.

Stunkel, Kenneth R. (2004). *Understanding Lewis Mumford: A Guide for the Perplexed*. Lewiston, NY: Edwin Mellen Press. Organized by concept such as *architecture* and *megamachine*.

MURDOCH, IRIS

• • •

Philosopher and novelist (Jean) Iris Murdoch (1919–1999) was born in Dublin, Ireland on July 15 and educated at St. Anne's College, Oxford, where she also taught from 1948 to 1963. She won the 1978 Booker Prize for her novel *The Sea, The Sea*, which provocatively opens with the protagonist's project of "learning to be good, after a life of egoism, art and power." Murdoch is especially renowned for reviving the classical humanistic philosophy of Plato. She makes Plato's philosophy of ideal truth, beauty, and goodness timely and accessible to general readers, articulating a view of human life as love's labor in journeying from illusion to truth. This vision is especially challenging in a world dominated by scientific reason and technological

Iris Murdoch, 1919–1999. The works of this novelist and philosopher portray characters whose warped and often dreamlike perceptions of reality create suffering among those whose lives they attempt to dominate. (*The Library of Congress.*)

pursuits of material goods. Murdoch died on February 8 in Oxford, England.

Murdoch's uniqueness as a twentieth-century novelist-philosopher is found in *Acastos* (1987), her two Platonic dialogues on love and religion. Like Plato, Murdoch writes philosophically about aesthetics and moral values, arguing that close connections between facts and values in the creative arts and the sciences are necessary to enable humans to live better and more wisely. For Murdoch, the critical difference between creativity in the arts versus the sciences is that the arts, especially literature, represent humanity in the world of relationships, reflected through the creative mind in play with the unlimited, unconscious self. In Murdoch's writings, individuals aim to refine human desires and longings for unreachable goodness through their interpersonal relations of love, and are not satisfied with the more abstract beauty and goodness prominent in the sciences. In thus reinventing literary art and ethics, Murdoch explores the quest of the passionate self for a goodness beyond any individualistic center of self. This indefinable, sublime good that humans seek can become destructive when desires and relationships are based

more upon obsessive loves and fantasies about oneself and others, than upon moral and spiritual goodness and love. Unlike basically selfish, egotistical humans, goodness represents a necessary, ideal *otherness* that transcends the human ego.

For self and society to move toward the good is to be rescued from vices of deception and self-deception in the search for beauty, truth, and the virtues of self-knowledge, humility, and compassion. Beauty is the one good to which humans are attracted as if by instinct, and is what galvanizes the creative pursuits of new technologies as well as arts. Yet without developing a purer sense of self, and humility based on knowledge of oneself and others, humans fail in their creativity to find or experience the very things they yearn for, love and happiness, acceptance and understanding.

Murdoch draws inspiration not only from Plato, but also from related philosophers such as Immanuel Kant (1724–1804). For Plato, the ideal forms are distinct from the physical universe, and the form of the Good is even "beyond being" (*The Republic*, Book VI, 509b). For Kant, the dualism lies in the contrast between the rational free will and the determinism of the natural world known by sense experience and laws of causality. Murdoch drew further influence from central twentieth-century philosophers such as the existentialist Jean-Paul Sartre, on whom she wrote the first book in English, and the philosopher of language Ludwig Wittgenstein, with whom she shared a mistrust of written words and language as unable to express full wisdom. With Sigmund Freud she also shared the view that the source and impetus toward knowledge and achievement is sexual.

Murdoch's achievements as both novelist and Platonist argue the importance of living well, ethically, and wisely. By breaking away from barriers to female philosophers and novelists in her own time and place, Murdoch reinvigorated the Idea of the Good for an era dominated more by laws and rules than by the creative works of arts and sciences, to reveal and embody material progress toward ideal truth, beauty, and goodness.

MARY LENZI

SEE ALSO *Consequentialism; Deontology; Virtue Ethics.*

BIBLIOGRAPHY

Bayley, John. (1999). *Elegy for Iris.* New York: St. Martin's Press. Also see the film *Iris* (2001), based on her husband John Bayley's memoir of Murdoch's struggle with and death from Alzheimer's disease.

MacIntyre, Alasdair. "Which World Do You See?" *New York Times Review of Books*, January 13, 1993. A review of *Metaphysics As A Guide to Morals*, by Iris Murdoch.

Murdoch, Iris. (1983). *The Philosopher's Pupil.* New York: Viking Penguin Press, Inc.

Murdoch, Iris. (1987). *Acastos: Two Platonic Dialogues.* New York: Viking Penguin Press.

Murdoch, Iris. (1992). *Metaphysics As A Guide to Morals: Philosophical Reflections.* London and New York: Allen Lane Publisher.

Murdoch, Iris. (1998). *Existentialists and Mystics: Writings on Philosophy and Literature*, ed. Peter Conradi. New York: Penguin Books.

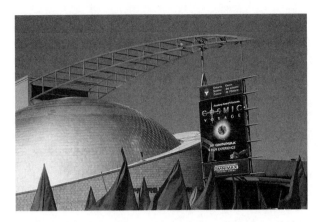

The Ontario Science Centre in Toronto. Since opening its doors in 1969, the center's 600-plus exhibits have fascinated more than 37 million visitors. (© *Dave G. Houser/Corbis.*)

MUSEUMS OF SCIENCE AND TECHNOLOGY

• • •

Science and technology museums have the power to inspire and educate millions of visitors each year. As mediators between expert scientists and the general public, museums have the responsibility to provide informed and balanced exhibits. Ethics are embedded in museum decisions, from determining what objects to collect to what exhibits to mount and what to say about them. This discussion examines the long history of science and technology museums and raises some of the ethical questions museums face, particularly how an educational mission is defined by the competing tensions of representation, political influence, funding, and entertainment.

From Cabinets of Curiosities to Science and Technology Centers

As showcases for scientific discoveries, technological marvels, and natural wonders, museums became popular across Europe during the sixteenth and seventeenth centuries. These palaces of the muses began as private collections for acquiring physical knowledge and became displays of individual wealth and power. As explorers brought back new curiosities from around the world, these collections were a systematic attempt to organize the explosion of new knowledge. A complete cabinet of curiosity would have one of everything in the world, organized and displayed in a continuum from the ordinary to the exotic, sometimes even including the imaginary.

Natural history dominated scientific representation in museums for several centuries. From the mid-eighteenth century, collections of ornithology, entomology, paleontology, and geology formed the basis for large public museums. These museums were organized by Linnaean classification with hierarchal representations of human progress. When curators began including technology exhibits in museums in the late nineteenth century, the exhibits were also organized as a reflection of human progress. A typical framework included *synoptic series* that traced the evolution of a particular technology—for example, a series on sailing from rafts to steamships.

In addition to permanent museum facilities, the public had opportunities to see the latest in science and technology at temporary shows and traveling exhibits. The "great exhibition of the works of all industry of all nations" opened in 1851 at the Crystal Palace in London and ushered in an age of world's fairs. Cities sponsored these year long celebrations to showcase top standards in industry and national pride in technical achievement. In the early twentieth century, several companies turned their exhibits into traveling shows that toured cities after the fairs closed, allowing even more people to see their wares. Many factories even offered tours of their facilities, giving visitors an inside look at working in different industries.

In 1969, the year a human being first walked on the moon, an innovation in science and technology museums occurred: the launch of the first hands-on science and technology centers. San Francisco's Exploratorium and Toronto's Ontario Science Centre forged a new path for exhibiting science. Frank Oppenheimer, a Ph.D. physicist who worked on the Manhattan Project (headed by his brother J. Robert Oppenheimer), founded the exploratorium to supplement science curricula. He wanted to combine invention and play in order to encourage students to look at science from a new perspective. Science and technology were no longer tied to national or history museums, and curators began interpreting objects using new exhibiting techniques in a variety of non-traditional museum settings. In the 1980s industrial archaeology gained momentum, displaying technology in the physical spaces of abandoned factories.

As the notion of what constituted a museum expanded, traditional methods of exhibiting objects also changed. Throughout the twentieth century, museums began showing science and technology within social and cultural contexts. Natural history exhibits began placing animals in realistic groups representing predator-prey relationships and biodiversity within the environment. Technology ceased to be represented as a forward march of progress, and the complicated relationships among science, daily life, and the environment began to be explored. These changes in exhibit practices set the stage for the ethical questions for museums of science and technology.

Ethical Questions of Museum Exhibitions

Museum practitioners are well aware of the ethical dilemmas posed by every acquisition or exhibition. The museum studies literature often raises extended blocks of questions, such as Sharon Macdonald's introduction to *The Politics of Display*, a collection of essays addressing ethics in science and technology museums:

> Who decides what should be displayed? How are notions of "science" and "objectivity" mobilized to justify particular representations? Who gets to speak in the name of "science," "the public" or "the nation"? What are the processes, interest groups and negotiations involved in constructing an exhibition? What is ironed out or silenced? And how does the content and style of an exhibition inform public understanding?

The museum community has not reached a reasonable consensus on any of these questions.

The literature in the field has traditionally addressed these questions through case studies, but the analysis of individual museums or exhibitions does not often lead directly to changes in collection and exhibition practices. The difficulty in assessing effective exhibitions and implementing guidelines for future directions is that the ethical dilemmas museums face are a tangled knot of competing interests. Frequently, each new exhibit struggles with the same fundamental questions, hoping to maintain a balance among the diverse tensions of exhibit design.

At the core of the debate is the fundamental question: What is the purpose of museums? For many museums this can be generally answered under the aegis of education. Most museums exist to collect and share information, but how this mission is interpreted highlights the ethical dilemmas museums face: What should be collected? How should the objects be displayed? Who is the intended audience? What should they learn?

Although many science and technology centers have similar exhibits demonstrating scientific principles, are these fair representations of scientific practice? Science is a coordinated practice of trial and error: state a hypothesis, create trials, collect data, analyze the results, draw conclusions, and repeat as necessary. However, museums often display science as a finished product. Where are the experiments? Where are the failures? Even the popular hands-on interactive exhibits do not reflect the dynamic nature of science because they fail to show the evolution of scientific thought and practice.

Interactive science centers frequently push the boundaries of an educational environment. Techniquest in Cardiff, Wales, is billed as the largest hands-on science center in Great Britain, but the cacophony of children running in every direction raises the question: Is any active learning taking place? Advocates for science centers argue that stimulation of multiple senses encourages learning. They also argue that interacting with science in a fun and entertaining manner encourages students to continue studying science at more advanced levels. As funding for school trips to science centers grows, teachers must ask at what point does the balance shift from education to entertainment, and museums must make their positions clear.

In developing countries where non-scientific world views persist and significant portions of the population remain illiterate, do science museums have different education responsibilities? Armalendu Bose, retired director of the National Council of Science Museums in India, sees museums as having "the responsibility of educating the masses—literate, semiliterate, or even illiterate—about the social benefits of science and the need to imbibe a value and [to] practice a way of life imbued with scientific outlook." This brings an explicit value judgment to bear on exhibit design, raising a host of new questions: Where should museums position themselves along the spectrum of education to avocation? Do museums have the responsibility to explain the effects of policy decisions on scientific research? Should they be forums for debate? Can they be advocates for policy change? These questions in turn become questions of representation and interpretation.

Museums make choices at each stage in designing an exhibit. From what objects to include to what descriptions to write, curators craft a specific experience for the museum visitor. Until the end of the twentieth century, the voice of interpretation was the anonymous museum authority, but in the mid-1990s two exhibitions by the Smithsonian Institution brought the question of museum authority to center stage. The highly controversial exhibits *Science in American Life* and *The Crossroads: The End of World War II, The Atomic Bomb and the Origins of the Cold War* garnered international attention and sparked what would become known as the "history wars." Science in American life, which was funded in part by the American chemical society, explores the interaction between science and society. Criticism of the exhibit came from scientists who felt that it trivialized scientific achievements while emphasizing negative outcomes of scientific research. The debate over the crossroads exhibit centered on the Enola Gay, the airplane that dropped the atomic bomb on Hiroshima. Should a museum attempt to ask critical questions of wartime actions, as original plans for the exhibit did with a section describing the aftermath of the bombing? Or should museums allow interested parties, such as veterans groups or members of congress, to write a heroic narrative of the events? The debate, amplified by the media, eventually led to the cancellation of the exhibit. The battle over the exhibits sparked debate over who controls the information presented to the public. Is it the museum? Is it the donor? Is it the person or company featured in the exhibit? Is it the media? Is it a political party? Is it a scientific expert? Who speaks for science in history museums? How do you represent a heterogeneous group of scientists? These questions forced the museum community to reflect on the purpose of museums and their ethical responsibilities to the variety of audiences they serve.

As a reaction to the controversies, many museums have shied away from politically sensitive exhibits. This limits the amount of contemporary scientific research that is exhibited to visitors and makes museums artifacts of science history. One suggestion for mounting exhibits without offering potentially controversial interpretations is to let the objects speak for themselves. Unfortunately, this presents a dilemma leading back to the educational mission of museums. Lacking any explanations, museum visitors may not understand the exhibit's content unless they are already informed on a particular subject matter. Another approach is to allow all interested parties a platform for explaining their views, but this can make an exhibit cumbersome and likewise confuse the visitors.

Tied to questions of representation and interpretation are questions regarding museums' responsibilities to their donors. Museums operate on a precarious business model; proceeds from visitors rarely cover operating expenses. Museums rely on grants, donations, and government funds to maintain and expand their collections, and these monies rarely come with no strings attached. Should donors have any input into the content of an exhibit? Historically, this has not been an ethical dilemma. In the 1910s the Smithsonian's curator of mineral technology built the collection by soliciting corporate donations and relinquishing control of exhibit labels to company copyeditors, making it explicitly clear that the company's name "would be conspicuously present." But as critics began noticing the increased advertising in museums during the 1990s and suggested that corporations had undue influence on exhibit development, museum directors began reforming exhibit policies. Curators in the early 2000s attempt to make clear breaks between funding and content, acknowledging financial contributions but attempting to limit influence on exhibit design.

Possibilities for the Future

It is unlikely that any of the questions raised here will be resolved decisively. Rather, museums will continue to attempt to balance the competing internal tensions inherent in exhibit design. As institutions of learning, museums need to evolve to reflect changes in current scientific practices while being mindful of their histories. In tackling current ethical questions and uncovering fresh ones, here are a few suggestions for possible directions for future exhibits at science and technology museums.

- Museums should reflect current scientific practice. Boston's Museum of Science has started the Current Science and Technology Center to highlight leading edge research and science in the news. Following this model, museums could become educational centers for sharing scientific research with the public, and museums could position themselves as forums for debate.

- Museums should tackle complex scientific problems. If museums are intended to be institutions for life-long learning, they should not be built exclusively for children. Exhibits should aim for a range of intellectual audiences, ranging from the uninformed novice to the educated non-expert.

- Museums should highlight the multifaceted and interdisciplinary nature of modern science. Science is no longer neatly divided into disciplines, and museums should not be either. An example would be exhibits showing the interactions among biologists, engineers, and doctors in the development of new medical devices. Exhibits could also explore the relationships between science and other disciplines, such as the law or business. Both of these intersections would be shown in an exhibit on technology and the patent system.

- Museums should take advantage of new technologies to share their collections with a wider audience. Visitors used to have to travel to museums to see wonders, but the Internet has brought these wonders into the home, office, and classroom. The Science Museum of London has started an ambitious program to catalogue its collection online. If other museums follow suit, the diffusion of knowledge could reach tremendous numbers of people.

ALLISON C. MARSH

SEE ALSO *Activist Science Education; Education; Interdisciplinarity; Science, Technology, and Society Studies.*

BIBLIOGRAPHY

"America's Museums." (1999). *Daedalus: Journal of the American Academy of Arts and Sciences* 128(3): entire issue; contains 15 essays.

Conn, Steven. (1998). *Museums and American Intellectual Life, 1876–1926.* Chicago: University of Chicago Press.

Findlen, Paula. (1994). *Possessing Nature: Museums, Collecting, and Scientific Culture in Early Modern Italy.* Berkeley: University of California Press.

Henderson, Amy, and Adrienne L. Kaeppler, eds. (1997). *Exhibiting Dilemmas: Issues of Representation at the Smithsonian.* Washington, DC: Smithsonian Institution Press.

Lilenthal, Edward T., and Tom Engelhardt. (1996). *History Wars: The Enola Gay and Other Battles for the American Past.* New York: Henry Holt.

Macdonald, Sharon, ed. (1998). *The Politics of Display: Museums, Science, Culture.* London: Routledge.

Society for the History of Technology. (1965). *Technology and Culture* VI(1).

Spaulding, Julian. (2002). *The Poetic Museum: Reviving Historic Collections.* Munich: Prestel.

MUSIC

• • •

The history of Western music involves a transition from music understood as reflecting the harmony of the cosmos to the industrial production of desegregated sounds. From classical antiquity until the sixteenth century, music was a way to cultivate the senses for the good of a specific ethos. Politicians and physicians still talked about music when they searched for the right mixture of powers in politics or the right mixture of bodily humors in medicine. But with the demise of cosmological harmony manifested in Pythagorean proportionality, music became the disembedded art of sound production. Since the nineteenth century, modern music has been influenced substantially by scientific progress and its technological fallout. The technogenic production of sound reflects the disappearance of the traditional deep ethical relevance of music.

Music and Ethics

From ancient times to the sixteenth century, philosophers, musicians, physicians, and politicians understood music as an art intimately associated with ethics. In Greece as in other cultures, music and dance were significant threads in the fabric of every day life. Hymns were sung to praise and address the gods of its ethos. Outside religious rituals, music accompanied weddings, funerals, harvests and wars: Most social occasions not only had their own time but also were marked by their own musical instruments and modes. Authors such as Plato (*Republic*), Aristotle (*Politics*), Boethius (*De musica*), and Isidore of Seville (*Etymologiae* [Etymologies]) considered the influence of specific modes of music, rhythms, and musical instruments on body and soul (Anderson 1966, West 1992). It was believed that music not only mirrored the cosmos but also influenced the constitution of both individuals and society (Lippman 1992). Plato in *The Republic* describes different modes and rhythms with regard to their ethical effects (books 2, 3, and 7) and stresses their importance for education. Along with arithmetic, geometry, and astronomy, *musiké* was used as a sensible route to the appreciation of appropriate correspondences: "musical training is a more potent instrument than any other, because rhythm and harmony find their way into the inward places of the soul. ... he who has received this true education of the inner being will most shrewdly perceive omissions or faults in art and nature" (Book III, paragraph 401).

The demonstration of the harmonic order of the cosmos with the help of a monochord, a rectangular sound box with a single stretched string, was the cornerstone of Greek ethical education. Teacher introduced pupils to the proportions of the musical consonances. According to one legend, it was Pythagoras of Samos who discovered the connection between the first four numbers and the musical consonances (octave 2:1, fifth 3:2, and fourth 4:3) and thus became the founder of music. For Pythagoras and his successors, musical consonances mirrored the harmonic order of the world. The first four numbers, the so-called *tetraktys*, were considered harmonic because they symbolized the four seasons, the four directions, and the four humors. This doctrine was handed down to the Middle Ages through Boethius's *De musica*, which distinguishes between *musica instrumentalis* (music that can be heard), *musica mundana* (music of the heavens), and *musica humana* (harmonic mixture of the bodily humors).

The Disembedding of Music

In the sixteenth and seventeenth centuries, the notion of music as the reflection of a given harmony started to fall apart. Doubts about the authority of the Pythagorean legend and new technologies such as musical printing questioned the millennia-old assumption of the embeddedness of music in a cosmological order and its ethical relevance. In a diverse range of treatises such as Gioseffo Zarlino's *Le istitutioni harmoniche* (1558; The harmonic foundations) and Johannes Kepler's *Harmonices mundi* (1619; The harmony of the world). *musica instrumentalis* is still considered an echo of the music of the spheres and the body. But at the same time, authors appear who complain that music has lost its power to form an ethos: Antonio de Ferraiis' s *De educatione* (1505) and Richard Pace's *De fructu qui ex doctrina percipitur* (1517; The benefit of a liberal education), for example, on the education of princes, follow the tradition of Plato and Aristotle by emphasizing the ethical value of a musical education. But they lament the loss of the sense for harmony, a sense that, from the pre-Socratics to their contemporaries, was fundamental to recognition of the good.

For music that was played in the Middle Ages, the technique of musical notation was understood as a memory aid for its performance. This changed during the sixteenth century when the German composer Nikolaus Listenius (dates unknown) claimed that a composition should be an *opus perfectum et absolutum*, an independent piece of art. He rejected the traditional notion of composition as the expression of God's creation (Kaden 1992). Notation did not serve as a blueprint for musical performance, but for the production of an autonomous, timeless piece of art made of composed tones. The Swiss scholar Henricus Glareanus (1488–1563) explicitly declared notated musical tones the foundation of music. The technological invention of musical printing in the late fifteenth century fostered this new understanding of composition as the production of a piece of art and made possible its conservation and reproduction. The musical artifact, namely, musical tones aesthetically arranged according to the tastes of the time, became the quintessence of music. Music now was understood as an art that fosters the individuality of its creators.

At the same time, philosophers and mathematicians, such as Giovanni Benedetti, Galileo Galilei, Marin Mersenne, and Isaac Beeckman, made the musical tone and its acoustic foundations an object of empirical research. These figures were the first to examine the validity of the canonical tradition of Pythagoras, rather than seeking to demonstrate its truth. Their experiments refuted the doctrine that the *tetraktys* was the harmonic foundation of music. In his *Discorsi e dimostrazioni matematiche, intorno a due nuove scienze* (1638; Dialogues concerning two new sciences), Galileo proved that the traditional assumption that the same ratios produce musical consonances when they "expressed relative weights of hammers, weights attached to strings, or the volume enclosed in bells or glasses" was wrong (Palisca 1961, pp. 128–129). What before had been considered a universal law reflecting a universal harmony was suddenly demystified as an empirical fact true only "for strings with the same thickness, length, and quality, and stretched to the same tension" (pp. 128–129). In his *Harmonie universelle* (1636; Universal harmony), Mersenne developed a mathematical formula for calculating the relation between the frequency of oscillation and the pitch of a string. By replacing the length of a string segment (e.g., 2:1 for the octave) with the frequency of oscillation (1:2), he anticipated the shift from cosmology to science (Cohen 1984).

Music as an Object of Scientific Research

The invention of measuring devices in the eighteenth century transformed musical qualities to calculable quantities. The tuning fork, developed by the trumpeter and lutenist John Shore in 1711 and Étienne Loulie's *chronomètre* (1696) gave a technological impetus to the quantification of pitch and tempo. Loulie's apparatus was almost 2 meters high, and although considerably improved by the French mathematician Joseph Sauveur at the beginning of the eighteenth century, was used only by music theorists and scientists. But in 1816 Johann Nepomuk Maelzel began manufacturing his version of the metronome (invented circa 1812 by Dietrik Nikolaus Winkel). With Maelzel's successful commercialization of the metronome, which was soon adopted by Beethoven (who retroactively marked metronome beats in his compositions, though these are sometimes questioned) and other composers, timekeeping became common in musical practice. The Italian tempo indication (for example, adagio, allegro, or presto), common since the seventeenth century, had determined the characteristics of a piece. The metronome fixed those characteristics to defined units per minute, replacing the description of qualities with quantifiable measurements of speed.

At the beginning of the eighteenth century, Sauveur founded the science of acoustics, a discipline designed to explore sound the same way optics analyzed light. Unlike his predecessors Mersenne, Kepler, or Galileo, who still searched for the harmonic principles of music, Sauveur did not distinguish between music and noise; he treated both as kinds of physical sound. This new scientific perspective on music created the foundations for musical acoustics, which, within one and a half centuries, would transform musical theory. In his *Génération harmonique* (1737; Harmonic generation), the French composer Jean-Philippe Rameau became the first to use Sauveur's research to support his own musical theory by referring to its acoustical foundations (Palisca 1961). Jean Jacques Rousseau (1712-1778), the French philosopher, introduces "acoustics" into the terminology of music with his dictionary of music. Whereas instrument makers used discoveries in the field of acoustics to improve musical instruments such as the piano and the violin; musicians, composers and musical scholars mostly neglected the importance of acoustics for their own work.

During the nineteenth century, music became the object of systematic scientific research in the laboratories of physicists and physiologists. In order to exchange and compare results within the scientific community, they had to develop standardized parameters. The acoustical examination of the tone required a universal point of reference. In 1834, following a suggestion of the German acoustician Johann Heinrich Scheibler, a convention of physicists in Stuttgart adopted Scheibler's standard pitch of A above middle C = 440 hertz. Fifty years later an international committee agreed on a standard pitch with global validity. A professionally defined and bureaucratically prescribed standard did away with the diversity of pitches that had been characteristic of each place and its ethos. The millennia-old art of attuning oneself to the appropriate and good of a certain place was replaced by submitting to experts' guidelines.

The German physiologist Hermann von Helmholtz (1821–1894) was the towering figure in acoustical research on music in the second half of the nineteenth century. In his study *On the Sensations of Tone* (1863), he reformulated the Pythagorean interpretation as a scientific problem and presented his new physiological,

psychological, and physical foundations of musical theory. Helmholtz was an advocate of "objectivity," a new scientific paradigm of his time that was based on the use of scientific instruments. By developing scientific instruments that made not only the analysis but also the technical synthesis of sounds of different musical instruments possible, he revolutionized the understanding of music. Since Helmholtz, the axioms and technological fallout of the acoustical laboratory frame the understanding and meaning of musical instruments, hearing, consonance, and tone.

Music as the Production of Sound

At the beginning of the twentieth century, Helmholtz's laboratory notion of music as sound production became an everyday assumption. Without his acoustical research, the inventions of the phonograph by Thomas Edison in 1877 and the telephone by Alexander Graham Bell in 1875 would have been unthinkable (Peters 2004). The phonograph was commercially exploited by organizing concerts where real musicians had to compete with the machine. The audience was supposed to recognize that the machine was able to mimic musicians (Thompson 1995). In the early telephone days—the late nineteenth and early twentieth centuries—the new technology of analyzing and synthesizing sounds was primarily used to transmit concerts, operas, and variety shows to marketplaces, bars, hotels, or the parlor. Radio, which debuted in 1920, replaced the telephone as a device for broadcasting music.

At the same time new technologies made music an industrial product, the sound of industrial machines such as airplanes and trains entered theaters and concert halls. Arthur Honegger's *Pacific 231* (1923), a musical dedication to the then strongest American Locomotive Kurt Weill's *Der Lindberghflug* (1929; The Lindbergh flight), or Frederick Converse's *Flivver Ten Million* (1926), praising the 10 millionth Ford car, document how music reflected the industrial age and its technological innovations (Braun 2002). The Italian futurists even used the noise of steam engines and other machines together with conventional musical instruments in order to create industrial soundscapes. Electronic instruments gave birth to innumerable new sounds. The aetherophone, or theremin (1921) by Leon Theremin, the Sphräophon (1926) of Jörg Mager, and Maurice Martenot's Ondes Martenot (1928) produced artificial sounds that were enthusiastically welcomed by concert and movie audiences. Machines for synthezising sounds were introduced in 1929 and became commercially viable with the synthesizer invented by Robert Moog in 1964.

The invention of the triode vacuum tube by the American inventor Lee de Forest in 1906 and of the transistor in 1947 opened up the possibility of amplifying and modifying sounds. It was the avant-garde of popular musicians who, in the 1950s and 1960s, were fascinated by the new technological potential and started to use amplifiers, microphones, and loudspeakers. With the help of electrified musical instruments such as the electric guitar, music groups invented and produced their own characteristic sounds, that is, their individual "trademark sound," which facilitated commercialization in popular as well as in classical music. Since then, sound engineers behind the scene have become the ones who produce the sounds adapted to the taste of different consumer groups. Technicians operating recording machines, filters, and mixers determine the musical output on records and in concert halls. Musicians and composers used machines such as the tape recorder, the vocoder, the synthesizer, or the sampler to design new sounds or to imitate the sound of musical instruments. Computer programming, tape recording, the "playing" of turntables or musical instruments were equally used as means for sound production.

The technological imperative of contemporary music was discussed controversially among composers, philosophers, and musicologists after World War II. In "Music and Technique" (1959), Theodor W. Adorno expressed disapproval of contemporary composers who incorporated technology into their works. He called their search for a new kind of music based on the electronic generation of sound a banality that would raise engineers to composers and lower composers to technicians. According to him, music without notation and interpretation would be nothing but a technogenic production and reproduction of something audible. In contrast to Adorno, apologists of electronic music such as Karlheinz Stockhausen (b. 1928), John Cage (1912–1992), and Pierre Schaeffer (1910–1995) praised its new forms of expression that overcame the outdated limits of traditional music. They sought a new kind of music that would provide the technological society with its appropriate musical expression.

In the 1980s the computer ushered in the era of boundless possibilities of sound production. In the early twenty-first century new sounds are generated, conventional ones are simulated, and all types of sounds are mixed arbitrarily regardless of their historical and cultural meanings (Théberge 1997). With little fanfare, sound designers and artists use noises and artificial sounds as well as plainchants venerating the Madonna or pop songs by the American singer Madonna as a resource for their artistic productions. Be it songs of African shamans in the supermarket or classical symphonies in a parking lot—disembedded sounds have become the background music of a technogenic society.

MATTHIAS RIEGER

SEE ALSO *Entertainment; Popular Culture; Science, Technology, and Society Studies.*

BIBLIOGRAPHY

Adorno, Theodor W. (1999). "Music and Technique." In *Sound Figures*, trans. Rodney Livingstone. Stanford, CA: Stanford University Press. Translation of "Music und Technik," originally published, 1959. A collection of articles originally published in the 50s and 60s on the social aspects of musical experience. In the last essay, music and technology, Adorno discusses the impact of the electronic production and reproduction of music on its aesthetics.

Anderson, Warren D. (1966). *Ethos and Education in Greek Music.* Cambridge, MA: Harvard University Press. Surveys the contribution of musicians and music-making in Greece from archaic to the graeco-roman periods. Covers topics such as musicians' dress, affinities with shamans and gods, musical instruments, notation and scales. Other cultural musical practices are cited as suggestive parallels.

Braun, Hans Joachim, ed. (2002). *Music and Technology in the Twentieth Century.* Baltimore and London: John Hopkins University Press. A collection of articles by musicologists and historians of technology. Covers the area of electronic music and instruments, the history of sound recording, aesthetics, education and the echo of technological inventions such as the railroad and the airplane in twentieth century music.

Cohen, H. Floris. (1984). *Quantifying Music: The Science of Music at the First Stage of the Scientific Revolution, 1580–1650.* Dordrecht, Netherlands: D. Reidel. Main study on the transition from the scientific analysis of music in terms of numbers to an approach on an essential physical basis, covering the acoustical research of Galilei, Beeckman, Descartes and Kepler.

Helmholtz, Hermann von. (1954 [1863]). *On the Sensations of Tone as a Physiological Basis for the Theory of Music,* trans. Alexander J. Ellis. New York: Dover Facsimile edition of the translation by Ellis (1875), which offers a comprehensive introduction into musical acoustics as well as into nineteenth century musical aesthetics and music history.

Kaden, Christian. (1992). "Abschied von der Harmonie der Welt" [Farewell to the harmony of the world]. In *Gesellschaft und Musik: Wege zur Musiksoziologie* [Society and music: approaches to the sociology of music], ed. Wolfgang Lipp. Berlin: Duncker und Humblot. Discusses the shift from a cosmological understanding of music to the modern notion of music as an independent art that developed between the fifteenth and the seventeenth centuries.

Lippman, Edward A. (1992). *A History of Western Musical Aesthetics*. Lincoln: University of Nebraska Press. Provides a brief introduction into the role of music concerning ethics in antiquity and the middle ages. Focuses on musical aesthetics from the sixteenth century (music as fine art) to the twentieth century.

Palisca, Claude V. (1961). "Scientific Empiricism in Musical Thought." In *Seventeenth Century Science and the Arts*, ed. Hedley H. Rhys. Princeton, NJ: Princeton University Press. Deals with the transformation of music from an ethical art to aesthetics. Referring to the main scientific discoveries in musical acoustics in the sixteenth and seventeenth centuries, Claude describes the influence of scientific thinking on musical theory taking the example of the theories of overtones and the sympathetic vibration of strings.

Peters, John Durham. (2004). "Helmholtz, Edison, and Sound History." In *Memory Byte: History, Technology, and Digital Culture*, ed. Lauren Rabonovitz and Abraham Geil. Durham: Duke University Press. Examines the influence of Helmholtz's acoustical work on Edison and Bell's inventions.

Plato. *The Republic*. Books 1–5, trans. Paul Shorey. Cambridge, MA: Harvard University Press 1930.

Sauveur, Joseph. (1984). *Collected Writings on Musical Acoustics*, ed. Rudolf Rasch. Utrecht, Netherlands: Diapason Press. Collection of writings by Joseph Sauveur on different aspects of musical acoustics such as harmonics, beats, tuning and standard pitch.

Sterne, Jonathan (2003). *The Audible Past: Cultural Origins of Sound Reproduction.*Durham, NC: Duke University Press. Explores the cultural origins of sound production and reproduction in the nineteenth century; covers the social and historical preconditions for the invention of technical devices such as the telephone, microphone, phonograph, and radio.

Théberge, Paul. (1997). *Any Sound You Can Imagine: Making Music, Consuming Technology*. Hanover, NH: Wesleyan University Press. Offers a sustained analysis of social conditions of technological innovations in the design of music technologies, focusing on digital technologies in the production of popular music.

Thompson, Emily. (1995). "Machines, Music, and the Quest for Fidelity: Marketing the Edison Phonograph in America, 1877–1925." *Musical Quarterly* 79(1): 131–171. Describes the marketing of the Edison phonograph by competitions between the phonograph and live music.

West, Martin L. (1992). *Ancient Greek Music*. Oxford: Clarendon Press. Introduces different aspects of ancient Greek music such as instruments, rhythms, scales, notation, and the role of music in religious and everyday life.

N

NAGASAKI

SEE *Hiroshima and Nagasaki*.

NANOETHICS

• • •

Nanoscience, nanoengineering, and nanotechnology involve the study, design, and manipulation of natural phenomena, artificial products, and technological processes at the nanometer level. Because a nanometer is one-billionth of a meter (10^{-9} meter), this effectively means research, design, and operations at the atomic and molecular levels. Nanoethics aims to promote critical ethical reflection in this relatively new field. It complements other efforts to explore the moral dimensions of the scientific and technological transformations in human action such as nuclear ethics (dealing with very large scale power generation and its challenges), biomedical ethics (focusing on the bioscientific and bio-technological aspects of medicine), and computer ethics (emphasizing the technological redefinition and processing of information).

Background and Prospects

Early inspiration and vision for the pursuit of nanoscience and nanotechnology is widely credited to physicist Richard P. Feynman's (1918–1988) talk "There's Plenty of Room at the Bottom" at the 1959 annual meeting of the American Physical Society. He concluded that speech with a financial challenge, offering $1,000 to the "first guy who can take the information on the page of a book and put it on an area 1/25,000 smaller in linear scale in such a manner that it can be read by an electron telescope" (http://www.its.

caltech.edu/~feynman/plenty.html). In 1982 Gerd Binnig and Heinrich Rohrer invented the scanning tunneling microscope (STM), which made Feynman's challenge technically feasible and essentially marked the technological beginning of nanoscience and nanotechnology research. International Business Machines (IBM) patented the invention, and demonstrated the microscope's incredible power by writing the initials *IBM* with thirty-five individual xenon atoms.

Thirty years after Feynman's talk, President Bill Clinton, at a 2000 appearance at Feynman's home institution, the California Institute of Technology, announced the U.S. National Nanotechnology Initiative. Other initiatives were subsequently launched in many other countries indicating significant political and economic motivations to promote this new area of scientific knowledge and to accelerate nanoscale technical understanding and control of the physical world. Together with private funding from corporations and venture capital investors, support for nanoscience and nanotechnology initiatives is anything but small.

K. Eric Drexler's *Engines of Creation* (1990) provided the one of the first dramatic visualizations of possible nanotechnology futures general overview of nanotechnology. Subsequent developments led to the production of rapidly produced nano-scaled devices, such as nanoscale storage and nanotube transistors; molecular transistors and switches; atomic force microscopes; focused ion and electron beam microscopes; novel materials; nanowires and nanostructure-enabled devices; non-volatile RAM, nano-optics, nanoparticle solubilization, and nano-encapsulation for drug delivery. Products already on the market by the early 2000s included sunscreens, fabrics, sports equipment, house paint, and medical devices.

A report by the National Science and Technology Council claims that "the emerging fields of nanoscience and nanoengineering are leading to unprecedented understanding and control over the fundamental building blocks of all physical things. This is likely to change the way almost everything from vaccines to computers to automobile tires to objects not yet imagined is designed and made" (http://www.wtec.org/loyola/nano/IWGN. Public.Brochure/IWGN.Nanotechnology.Brochure.pdf). Endorsements of the U.S. National Nanotechnology Initiative refer to the possibilities of miniaturized drug delivery systems and diagnostic techniques, positive environmental impacts through drastic reductions in energy use and the rebuilding of the stratosphere, extending and repairing deficits in the human senses, and security systems smaller than a piece of dust. One nanotechnology visionary, whose ideas are controversial, Drexler envisions that molecular assemblers could make possible low cost solar power; cures for cancer and the common cold; cleanup of the environment; inexpensive pocket supercomputers; accessible space flight; and limitless acquisition and exchange of information through hypertext.

Concerns and Criticisms

Some dismiss these claims as *hype,* not grounded in scientific reality. Nobel Laureate in chemistry (1996) Richard Smalley disagrees with Drexler about the ability to create self-replicating, self-assembling devices. Harvard University chemist George Whitesides concurs, arguing that there exists no concept of how to design a self-sustaining, self-replicating system of machines. There is a great deal of speculation and debate over future applications, and no one knows if the machines created will be able to do the things hoped for, such as to remove molecules from their environments, cause them to reproduce themselves in new environments, and use them to create devices such as molecular robots for engineering purposes.

Extreme reactions, such as those expressed in Michael Crichton's novel *Prey* (2002), where swarms of nanobots aggressively and intelligently seek to eat human flesh, reflect fear that scientists will not have complete control over the products of nanotechnology. These opinions call for moral reflection about the inevitability of nanotechnology development, the risks and harms imbedded in precise, atomic manipulation by humans, and potential inability to undo harmful technological advances. Aside from the more dramatic concerns expressed in science fiction (such as nanobots) are questions pertaining to (a) equity and access; (b) envir-

onmental safety; (c) irreversible and mysterious changes to food, water and air; (d) privacy and security; and (e) the philosophical considerations of introducing mechanical systems into biological organisms.

One cause of concern that ensued early in the emergence of nanotechnology developments was over the idea of grey goo; the possibility that nanoscaled robots (nanobots) originally designed for specific manufacturing processes might make copies of themselves, atom by atom, replicate endlessly and consume large areas of matter, even the world. Although the debate over grey goo has lessened over time, the idea still occasionally surfaces in public debates and science fiction.

The Canadian based Action Group on Erosion, Technology and Concentration (ETC), a nanotechnology watchdog organization, is concerned that nanotechnology development is moving too quickly, without any real oversight regarding environmental safety, public heath, and other societal concerns. The ETC identifies three phases of nanotechnology development. The first (which is already well underway) involves bulk production of nano-scale particles for use in products such as sprays, powders, coatings, and fabrics. In these applications, nanoparticles contribute to lighter, cleaner, stronger, more durable surfaces and systems. In the second phase, scientists seek to manipulate and assemble nanoscale particles into supra-molecular constructions for practical uses. The third phase would be mass production, possibly self-replicating nanoscale robots, to manufacture any material, on any scale. Ultimately, according to the ETC, nanomaterials will be used to affect biochemical and cellular processes, such as for engineering joints, performing cellular functions, or combining biological with non-biological materials for self-assembly or repair.

Ethical Issues and Analysis

The rapid development of nanoscience and nanotechnology is not simply a technological initiative, but has social aspects as well. While fueled by scientific ingenuity, it is also motivated by political pressures, competition for new international markets, venture capital ambitions, and competing conceptualizations of the public good. There is a sense of urgency that because of potential dangers (such as freely migrating carbon nanotubes penetrating plant, animal, and human cells, or uncontrollable self-assemblers) science must learn how to respond effectively and proactively to avert any consequential and irreversible social and environmental harms.

In this vein, some have called for implementation of a precautionary principal and a moratorium on further nanotechnology pursuits. Bill Joy reflected upon the potential dangers of genetics, nanotechnology, and robotics, and stated that "These possibilities are all thus either undesirable or unachievable or both. The only realistic alternative I see is relinquishment: to limit development of the technologies that are too dangerous by limiting our pursuit of certain kinds of knowledge" (Joy 2001, p. 11).

Joy's writing unleashed vigorous debate, and was strongly criticized by nanotechnology proponents such as Christine Peterson of the Foresight Institute. In the interest of providing safe opportunities for the development and commercialization of molecular manufacturing, the Foresight Institute has written a set of self-regulation guidelines for the development of nanotechnology, and argues that, if adopted by research scientists and the industries involved, those guidelines should suffice in addressing ethical concerns over the development of nanotechnology. Others defend the continued pursuit of nanoscience, nanoengineering, and nanotechnology on moral grounds, contending that they are relatively benign enterprises, representing a good and natural evolution in scientific inquiry, and further, that any restraint on development of nanotechnology will inhibit the improvement of humankind. Many important questions remain unanswered regarding the prevention of potential environmental accidents and abuses, or threats to human health and safety that may result from the release of nano-scaled devices into the atmosphere, waterways, the food chain, and medicine.

The use of nanotechnology to design improved surveillance systems raises the issue of the privacy rights of individuals. The potential of nanotechnology to produce powerful and precise new weapons calls into question the purposes of advanced and redefined forms of military combat and intervention. Miniaturization and hybridization of commonly used electronic devices tests the assumption that faster and cheaper is equal to better, and demands examination of how market imperatives could supercede other social goods and respected human values.

Scientists have a moral responsibility to be conscientious in their research because nano-scaled science and engineering fundamentally entail risk taking with novel, unpredictable, relatively untested new materials and devices in the realm of public and environmental safety. Of course, as with any new technology, responsibility for the ethical development of nanotechnology also lies with those who make public policy and society in general. The more philosophical questions will be answered not by scientists, but in the public domain: How does society identify what is *the good* or *the harm?* What new materials and processes should society be exposed to? What values can be sacrificed in the attempt to achieve precise human control and manipulation of matter?

Policy Responses

Matters to be resolved include how government is to be held accountable for funding stipulations that influence actual nanoscience research, timeline and reporting of results, the ethics of basic research questions that grantees study, and the technologies they are asked to develop. Provisions for access to education and technical training in this new field is also a matter of public policy. Who will pay for and provide the specialized retraining needed for teachers, or for the equipment, facilities and supplies needed for the schools? How will society assure democratic inclusion and full public access in this fast moving new initiative?

In the United States, the NSF has taken a leadership role in consideration of social and ethical issues in the development of nanotechnology. The NSF sponsors major conferences and panels for the purpose of considering the societal and ethical issues involved in nanotechnology. It allocates funding for individual researchers, and has established major centers of research. The European Commission regularly releases sponsored reports on issues related to nanotechnology health and safety. The European Parliament has held public hearings on nanotechnology, and sponsored various other public forums for widespread discussion of the emerging concerns. Yet because nanoscience and nanotechnology are still in an early stage of development, there is a significant lack of international consensus over distinctions of fact and fiction in their potential, and few clearly agreed upon articulated nodes of ethical concern.

There are multiple questions to be considered and new policies to be debated regarding who will receive the benefits of nanotechnology developments, and at what cost and to whom. Ownership, power, and control issues regarding devices and processes that are fundamentally invisible to the human eye present interesting ethical challenges both legally and socially. Some political rhetoric uses the language of competition, describing the international climate of nanoscience initiatives as a race. The very notion of a race raises the questions of why science is in such a hurry and to what end. The issues of who will *win* this race, and how world powers will implement and control the applications of nanotechnology have not as yet been effectively examined. Public policy must also respond to the potential for private individuals to gain access to the raw materials of nanotechnology, such as carbon nanotubes, or eventually, assemblers. Who, then, will oversee or control the use individuals make of those materials, such as for the building of experimental devices or weapons of mass destruction? To protect society from possible harm, external controls may have to be put in place to regulate and govern the types of nanotechnology that corporations can develop. Moral responsibility dictates that corporations adhere to rigorous self regulation, abide by widely adopted rules, principles and codes, such as those proposed by the Foresight Institute, and/or become involved in public policy, citizen review groups, and the like.

Public policy must also address the management of nano-related toxicity, release and control of nano-scaled, self-replicating artifacts, subtleties of nano-scaled surveillance mechanisms, inequities in access to power, and other unpredictable nano-related implications for society.

Conclusion

Through the tools now available, extensions of human hands and eyes (such as the atomic force and atomic probe microscopes) allow scientists to observe and manipulate atoms directly, move them, rearrange them, and reconfigure them. The resultant potential, to create atomically built hybrids of synthetic, mechanical, and biological components and turn them into novel devices, suggests that society is embarking on an incredibly powerful, tremendously exciting, but possibly dangerous undertaking. The development of nanotechnology could mean fundamental and beneficial changes to our relationships with the physical world, as human beings gain greater power to manipulate their bodies and environment. Where might such awesome abilities lead? What will happen when nanoscience and nanotechnology advance enough to achieve the results aimed for by scientists? What society does with this new knowledge may determine the changing substance of the physical, social, cultural, economic, moral, and perhaps even spiritual lives of humankind. Are people fully cognizant of and fully prepared to accept and adapt to those changes? Are science and the public in general proceeding with conscientious commitment? The ethical challenges are as daunting as the technical ones.

ROSALYN W. BERNE

SEE ALSO *Bioengineering Ethics; Biotech Ethics; Environmental Ethics; National Science Foundation; Posthumanism; Science Fiction.*

BIBLIOGRAPHY

Crichton, Michael. (2002). *Prey*. New York: Harper Collins.

Crandall, B. C., ed. (1999). *Nanotechnology: Molecular Speculations on Global Abundance*. Cambridge, MA: MIT Press.

Drexler, K. Eric; Chris Peterson; and Gayle Pergamit. (1990). *Engines of Creation: The Coming Era of Nanotechnology*. New York: Anchor Books.

Drexler, K. Eric; Chris Peterson; and Gayle Pergamit. (1991). *Unbounding the Future: The Nanotechnology Revolution*. New York: William Monroe and Co.

Editors at Scientific American, eds. (2002). *Understanding Nanotechnology*. New York: Warner Books, Inc.

Gross, Michael. (1999). *Travels to the Nanoworld: Miniature Machinery in Nature and Technology*. New York: Plenum Trade.

Kaku, Michio. (1997). *Visions: How Science will Revolutionize the 21st Century*. New York: Anchor Books.

Mnyusiwalla, Anisa; Abdallah Daar; and Peter A. Singer. (2003). "Mind the Gap: Science and Ethics in Nanotechnology." *Nanotechnology* 14(3): R9–R13.

Nelson, Max, and Calvin Shipbaugh. (1995). *The Potential of Nanotechnology for Molecular Manufacturing*. Santa Monica, CA: Rand.

Ratner, Mark, and Daniel Ratner. (2003). *Nanotechnology: A Gentle Introduction to the Next Big Idea*. Upper Saddle River, NJ: Pearson Education, Inc.

Regis, Edward. (1995). *Nano: The Emerging Science of Nanotechnology*. Boston: Little, Brown and Company.

Roco, Mihail C., and William S. Bainbridge. (2001). *Societal Implications of Nanoscience and Nanotechnology*. Dordrecht, The Netherlands: Kluwer Academic Publishers.

Smalley, Richard E. (2001). "Of Chemistry, Love and Nanobots." *Scientific American* 285, no. 3 (September): 76–77.

Stephenson, Neil. (1995). *The Diamond Age*. New York: Bantam Books.

Whitesides, George. (2001). "The Once and Future Nanomachine." *Scientific American* 285, no. 3 (September): 78–83.

INTERNET RESOURCES

ETC Group. Home page at http://www.etcgroup.org/.

Feynman, Richard P. (1959). "There's Plenty of Room at the Bottom." Available at http://www.its.caltech.edu/∼feynman/plenty.html.

Foresight Institute. (2000). "Foresight Guidelines on Nanotechnology." Available at http://www.foresight.org/guidelines/current.html.

Joy, Bill. (2000). "Why The Future Doesn't Need Us." *Wired Magazine*, issue 8.04. Available at http://www.wired.com/wired/archive/8.04/joy.html

NANOTECHNOLOGY ETHICS

SEE *Nanoethics*.

NATIONAL ACADEMIES

• • •

The U. S. National Academies are a consortium of four organizations. They are composed of the National Academy of Sciences, the National Academy of Engineering, the Institute of Medicine, and the National Research Council.

History and Structure

The National Academy of Sciences (NAS) was founded in 1863 to provide scientific and technical advice to the government. It is a membership organization of leading scientists, and new members are selected by the current membership. The membership decides how many total members to admit, and the number as of 2004 was about 1,800. In 1916 NAS realized that it could not meet the demand for advice from its members alone and therefore organized the National Research Council (NRC) to make it possible to enlist the larger scientific and technical community in its mission of providing expert advice to government. The National Academy of Engineering (NAE) was formed in 1964, and the Institute of Medicine (IOM) in 1970. Like NAS, NAE and IOM are membership organizations of the most respected engineers and medical professionals respectively. NAE has about 1,900 members, and IOM has approximately 1,200. The three organizations jointly manage NRC, which is the operating arm of the Academies.

The National Academies are not government organizations. The federal government chartered NAS, but the Academies are private organizations. The Academies do, however, receive federal funds to conduct studies at the request of Congress or federal agencies. State governments, foundations, and private companies also support studies, but industry can provide no more than 50 percent of the cost of a study. NAS, NAE, and IOM can each conduct studies independently, but NAS with support from NAE conducts most of its studies through the NRC. IOM is not a formal part of the structure of NRC, but its program must be approved by the NRC Governing Board and its reports must meet the requirements of the NRC Report Review Committee. The NRC issues about 250 reports per year, and at any given moment has roughly 6,000 volunteers serving on 600 study committees. In 2003 the National Academies had a staff of 1,200 and a budget of about $225 million.

The National Academies have a long and distinguished history of involvement in a wide range of activities related to science, technology, and ethics. The typical process for any Academies activity begins with a request from the federal government to conduct a study and issue a report on a specific topic. The Academies then select a committee of experts from relevant disciplines to perform the study. Once the committee members are named, there is a period of public comment to make certain that there is no bias or conflict of interest within the committee. Then the committee is formally appointed. The committees usually include some NAS, NAE, or IOM members, but most committee members are not members of these institutions. The expertise needed for these studies include law, ethics, and other nonscientific disciplines, and the individuals come from think tanks, advocacy groups, and industry as well as the universities. All committee members are volunteers; they are assisted by Academies staff.

The committees usually work for about eighteen months to produce a consensus report. All reports are subjected to rigorous review by the Report Review Committee, which appoints reviewers who are independent of the institution, have had no role in preparation of the report, and are unknown to the committee. Once the study committee has satisfied the reviewers that the report is fair and accurate, the report is published by the National Academies Press and is available for public purchase.

A list of sample NRC studies follows:

Science and Human Rights (1988)

The Responsible Conduct of Research in the Health Sciences (1989)

Shaping the Future: Biology and Human Values (1989)

Extending Life, Enhancing Life: A National Research Agenda on Aging (1991)

The Social Impact of AIDS in the United States (1993)

Women and Health Research: Legal and Ethical Issues of Including Women in Clinical Studies (1994)

Society's Choices: Social and Ethical Decision Making in Biomedicine (1995)

Biotechnology: Scientific, Engineering, and Ethical Challenges for the 21st Century (1996)

Xenotransplantation: Science, Ethics, and Public Policy (1996)

Non-Heart-Beating Organ Transplantation: Medical and Ethical Issues in Procurement (1997)

Cells and Surveys: Should Biological Measures Be Included in Social Science Research? (2001)

Integrity in Scientific Research: Creating an Environment That Promotes Responsible Conduct (2002)

Research Ethics in Complex Humanitarian Emergencies: Proceedings of a Workshop (2002)

Responsible Research: A Systems Approach to Protecting Research Participants (2002)

Scientific and Medical Aspects of Human Reproductive Cloning (2002)

The Experiences and Challenges of Science and Ethics: Proceedings of an American-Iranian Workshop (2003)

Guidelines for the Care and Use of Mammals in Neuroscience and Behavioral Research (2003)

Unequal Treatment: Confronting Racial and Ethnic Disparities in Health Care (2003)

Ethics Related Activities

In addition to producing studies at the request of others, the Academies sometimes use their endowment funds to prepare studies and organize activities at their own initiative. One such project began in the 1980s when there were a number of prominent cases of scientific fraud. NAS decided that it had a responsibility to make certain that all scientists understood the rules and responsibilities of scientific research. In 1989 NAS published *On Being a Scientist: Responsible Conduct in Research*, which provides a detailed discussion of the norms governing the proper behavior of scientists. More than 200,000 copies were distributed, and an expanded version was published in 1995. NAS distributed 70,000 copies of the new edition free to graduate students.

The Academies also operate the Joseph Henry Press, which publishes books by independent authors on a variety of scientific subjects. One title is *The Common Thread: A Story of Science, Politics, Ethics, and the Human Genome* (2002) by Georgina Ferry and 2003 Nobel laureate John Sulston. IOM also publishes books by independent authors, such as *Science and Babies: Private Decisions, Public Dilemmas* (1990) by Suzanne Wymelenberg.

Finally NAS publishes the scholarly journal *Proceedings of the National Academy of Sciences* (1914–present) and co-publishes with the University of Texas at Dallas the quarterly policy magazine *Issues in Science and Technology* (1984–present). *Proceedings* includes scientific research articles, but some of these touch on ethical as well as scientific concerns. An example is Paul R. Ehrlich's "Intervening in Evolution: Ethics and Actions" (2001). *Issues* is an independent magazine that provides a forum where individuals can express their views on a wide range of subjects. It regularly publishes articles and book reviews that address ethical and social concerns.

Although NAS conducts most of its activities through the NRC, it maintains direct control of the Committee on Human Rights, which was formed in 1976 to protect human rights, particularly of scientists, throughout the world. NAE and IOM became cosponsors in 1994. The committee uses the prestige of the institutions to defend scientists, engineers, and health professionals who are unjustly detained or imprisoned for behavior that is protected by the Universal Declaration of Human Rights. The committee investigates suspected violations, appeals directly to governments when appropriate, offers moral support to prisoners and their families, and works to make the public aware of the need to protect human rights. The committee serves as the secretariat for the International Human Rights Network of Academies and Scholarly Societies, which includes organizations from fifty countries. The committee has had numerous successes in obtaining the release of people being unfairly detained.

Because of their reputation and renown, the Academies are able to attract leaders from government, academia, and industry to events that provide a forum for discussion of controversial issues. The NAS Building in Washington, DC, is the site of numerous workshops, conferences, and symposia at which experts and decision makers debate the critical ethical issues related to science and technology. Examples include a series of workshops on regulatory issues in animal care and use as well as several meetings about human reproductive cloning and the treatment of human subjects in research.

The National Academies have enormous influence in all aspects of science and technology because of their long history of providing guidance to government, the rigorous review process through which all reports must pass, and the widely recognized expertise of committee members. NRC reports are regularly featured in the popular press, and committee chairs are often invited to testify before Congress or to brief administration officials. The full text of all reports is available for free on the Academies Internet site, which makes the site a valuable source of information for scholars, journalists, and government officials.

KEVIN FINNERAN

SEE ALSO *National Institutes of Health; National Science Foundation; Royal Society.*

BIBLIOGRAPHY

Cochrane, Rexmond C. *The National Academy of Sciences: The First Hundred Years, 1863–1963.* Washington, DC: National Academy Press, 1978.

Committee on the Preparation of the Semi-Centennial Volume. *A History of the First Half-Century of the National Academy of Sciences: 1863–1913.* Washington, DC: National Academy Press, 1913.

INTERNET RESOURCE

The National Academies: Advisers to the Nation on Science, Engineering, and Medicine. Available from www.nationalacademies.org.

NATIONAL AERONAUTICS AND SPACE ADMINISTRATION

• • •

The National Aeronautics and Space Administration (NASA) is the principal civilian space agency in the United States, and the leading space science agency in the world. Its scientific and technological activities pose a variety of ethical issues, from setting program priorities to environmental impacts and risk–safety tradeoffs. NASA decisions, however, rarely turn on explicitly ethical considerations (see, for example CAIB 2003, PCSSCA 1986). Common influences on NASA decisions include interest-group lobbying, Congressional politics, and intra-agency competition for resources.

NASA's Mission and Other Space Activities

Legislation created NASA in 1958, building on existing civilian aviation research activities of the National Advisory Committee for Aeronautics (NACA). The core of NASA's mission is space exploration, divisible into human exploration and space science. Human exploration includes, for example, the space shuttle and the International Space Station (ISS) in Earth orbit and the Apollo missions to the Moon. Space science includes astronomy and robotic planetary exploration missions; the Hubble Space Telescope (HST) is the most visible example of the former, while the Mars rover missions of 2004 exemplify the latter. Exploration and science overlap: Astronauts installed instruments on the Moon, and scientific experiments are conducted on the ISS and shuttle. Other NASA programs include earth science (satellites that look down at the earth) and practical applications such as communication satellites. In 2004 President George W. Bush called for human planetary exploration.

Other U.S. agencies with space activities include the National Oceanic and Atmospheric Administration (NOAA) and the Department of Defense. NOAA operates satellites to gather data in support of its missions (weather forecasting, for example). The Defense Department and intelligence agencies support their missions with satellites for surveillance, communication, and navigation. Private commercial activities, some virtually independent of NASA, include launch services and satellites for communications and Earth observation.

As an independent agency, NASA reports directly to the U.S. president. Although managed from a Washington, DC, headquarters its operations are decentralized in two ways: First, the great majority of NASA employees work at eight field centers such as the Johnson Space Center near Houston, Texas. Second, private-sector contractors do most of NASA's work, and most of its scientific research is conducted through grants to universities. In 2002 the NASA budget was around $15 billion, supporting 18,000 civil service employees and a contractor workforce several times as large.

NASA's involvement with science and technology is extensive: Virtually all its missions embody advanced technology (although some long-lived missions use yesterday's state-of-the-art technology). It developed the Saturn launch vehicle for Apollo, and the shuttle as a general-purpose, reusable launch vehicle. It created the HST, perhaps the most productive scientific instrument ever, and its series of missions to other planets were the basis for the new field of planetary science.

Ethical Issues

Broadly speaking, many justify space exploration primarily in terms of human adventure and scientific knowledge. A strong version of this position is that humans have an innate need to explore and learn about the world around them. In this view, humans leaving Earth is a straightforward extension of the species' past spread across Earth. Further in this vein, certain images from space, such as Earth seen from *Apollo 11* and the violent galaxies captured by HST, show how fragile and lonely this beautiful planet is, inspiring efforts to preserve it. A somewhat more modest justification holds that, regardless of human history, today humans want to go into space essentially because they can.

Against this background, and to some extent because of it, NASA activities raise a diverse set of ethical issues. These run from whether space exploration can or need be justified in terms of human history, anthropology, and psychology, to the dangers of planetary cross-contamination, risks to astronauts, and honesty in justifying and describing particular programs.

The possibility of life on other planets has animated reflection across much of human history. Search for evidence of life is an important aspect of many planetary missions. But if missions that land on Mars carry with them microbes from Earth, the Earth microbes may confuse the results. Future generations may be misled. Humankind may have "polluted" another planet. (Against this possibility NASA sterilizes spacecraft before launch.)

Further, many scientists want to bring back to Earth a Mars sample for study more complete than can be done remotely on Mars. If life exists there, a returned sample or dust on the returning spacecraft might contain organisms threatening to life on Earth. The threat is remote because NASA will take steps to isolate any returned spacecraft and sample, but given human ignorance it still raises the issue of whether NASA programs might cross-contaminate planetary life-forms. NASA recognizes the issue and therefore ended the Galileo mission in 2001 by crashing it into the atmosphere of Jupiter, which was intended to extinguish all Earth-life aboard it. If humans "colonize" Mars, however, cross-contamination is probably inevitable.

Another form of contamination is the debris missions leave in orbit. A collision with even a small object can disable a spacecraft. Thus early missions leave risks for following ones. Debris in low Earth orbit will slowly reenter because of residual atmospheric drag, but debris in higher orbits remains for centuries. The vastness of space dilutes the risks, but they remain real. Recognizing this, NASA and the world's other space agencies are working to minimize debris from future missions.

Risks to Life

The loss of life in space transportation accidents dramatically raises questions of risks. For example, what purposes justify risking astronaut lives in space missions? In the *Challenger* and *Columbia* space shuttle disasters risk became loss.

In the past NASA dismissed the risks of shuttle flight, claiming at one time that the accident rate would be one shuttle lost in 100,000 flights. Empirically it is roughly 2 in 100. Reliability of 98 percent is good for a launch vehicle—perhaps the best possible, and perhaps acceptable for professional astronauts on valuable missions. What about amateurs: a "teacher in space," members of Congress, scientists? Do the experiments done on the ISS justify the risk to astronauts tending them? Is returning the HST to the Smithsonian Institution at the end of its life worth the risk of a shuttle mission to retrieve it? Do seven astronauts have to be sent up for this mission? Perhaps the science done by the HST justifies the risk of the missions flown to keep it operating, but a mission to retrieve it for the Smithsonian seems questionable.

The death of seven astronauts in each of two shuttle accidents makes clear that one way to reduce the potential loss is to reduce the number of crew on each mission. The first accident involved a "teacher in space" who was to inspire young students. In order to decide if the risk she took was appropriate, one would have to ask hard-to-answer questions such as whether inspiration was likely, and whether students most needing inspiration would be positively affected. Another dimension is whether an amateur could give adequately informed consent to the risk.

Risk issues become entangled: The HST will eventually reenter Earth's atmosphere. Being massive, it will not burn up; large pieces are expected to reach the ground, presenting an involuntary risk to people on Earth. Guiding the HST down to a remote ocean area would greatly reduce that risk, but it has no capability for a guided reentry because NASA originally planned for shuttle retrieval. A mission to install a reentry package could also service HST to lengthen its scientifically productive life. Several incommensurate considerations are thus involved: The risk to professional astronauts, the risk to bystanders on Earth, and the value of HST science. Balancing these risks calls for ethical discussion. One proposed solution involves the use of robots to service HST.

Promoting and Justifying Programs

A different ethical problem arises in the description and justification of programs. NASA began as a geopolitical response to the Soviet Union's launch of *Sputnik 1*, to demonstrate that U.S. technical capability was superior to that of the USSR. The program, however, was promoted as space exploration—as the realization of humanity's drive to explore and gain knowledge. In reality space exploration was the means for the end of demonstrating U.S. prowess. From the beginning there has been a mix of motives, of ends and means. The ISS is variously justified and described as space exploration and as a science laboratory in space. But these are both problematic: As the station goes around and around Earth, the incremental exploration on additional orbits becomes vanishingly small, while the risk to astronauts remains the same. Second, there are questions as to whether the science on the ISS is worth what it costs. That is, if the justification is scientific, one must ask whether the same funds could support better science, for example in space astronomy (SSB 2003).

Similarly, NASA's justification of a program to develop a nuclear power reactor in space is questionable. The public justifications are that nuclear power would enable new activities, including scientific missions. Nuclear power is probably necessary for missions outside the solar system, and perhaps for extended human exploration missions within the solar system. Nevertheless, to justify the nuclear program a scientific mission to study Jupiter's moons, which had been endorsed by the scientific community and which could be done without nuclear power, has been adopted as the nuclear program's first mission, to give the technology development a clear target. The adopted mission had to be redesigned to require nuclear power; a scientific mission became a nuclear mission. That is, from the time of adoption forward the criteria for making decisions about the mission became nuclear first, science second. Scientific questions no longer drive the mission; rather the driver is developing and demonstrating nuclear power in space—science is a stalking horse. It would be more honest to call this a nuclear program using a science mission to demonstrate possibilities.

Of course a program to put a nuclear reactor into space faces all the ethical problems of nuclear programs on Earth, if in a different form. First are the hazards in the development program and the hazards of launching fissile material. Further, when its fuel is exhausted the reactor will become both another bit of nuclear waste and another bit of space debris. Where and how will it be "disposed of"? Typically, such questions are considered technical, not ethical.

RADFORD BYERLY, JR.

SEE ALSO *Apollo Program; Space Exploration; Space Shuttle Challenger and Columbia Accidents; Space Telescopes.*

BIBLIOGRAPHY

Bilstein, Roger E. (1989). *Orders of Magnitude: A History of the NACA and NASA, 1915–1990.* Washington, DC: National Aeronautics and Space Administration. NASA history as seen from the inside.

McDougall, Walter A. (1985). *The Heavens and the Earth: A Political History of the Space Age.* New York: Basic. The best general history of space activity.

National Research Council. Space Studies Board (SSB). (2003). *Factors Affecting the Utilization of the International Space Station for Research in the Biological and Physical Sciences.* Washington, DC: National Academies Press. Scientists describe the scientific usefulness of the space station.

U.S. Columbia Accident Investigation Board (CAIB). (2003). *Report of the Columbia Accident Investigation Board,* Vol. 1. Arlington, VA: Author. The official, comprehensive report.

U.S. Presidential Commission on the Space Shuttle Challenger Accident (PCSSCA). (1986). *Report of the Presidential Commission on the Space Shuttle Challenger Accident.* 5 vols. Washington, DC: Author. Comprehensive report from the executive branch.

INTERNET RESOURCES

National Research Council. Space Studies Board (SSB). "On Scientific Assessment of Options for the Disposal of the *Galileo* Spacecraft." Available from http://www7.nationalacademies.org/ssb/galileoltr.html. Letter, June 28, 2000, from SSB Chair Claude Canizares and Committee on Planetary and Lunar Exploration Chair John Wood to Dr. John Rummel, NASA planetary protection officer. Describes efforts to prevent planetary contamination.

National Research Council. Space Studies Board (SSB). "Factors Affecting the Utilization of the International Space Station for Research in the Biological and Physical Sciences." Available from books.nap.edu/openbook/NI000492/html/.

U.S. Columbia Accident Investigation Board (CAIB). "The CAIB Report." Available from http://www.caib.us/news/report/.

U.S. Presidential Commission on the Space Shuttle Challenger Accident (PCSSCA). "Report of the Presidential Commission on the Space Shuttle Challenger Accident." Available from http://history.nasa.gov/rogersrep/51lcover.htm.

NATIONAL GEOLOGICAL SURVEYS

• • •

National geological surveys provide scientific knowledge about a nation's lands, natural resources, and natural hazards within particular political, social, and legal contexts. At any given time, the work done by a geological survey reflects the public good as governmentally defined. Regardless of specific activities, however, geological survey scientists have special responsibilities as public scientists to maintain high standards of scientific inquiry and to remain credible irrespective of shifting priorities and pressures. Historical review of the U.S. Geological Survey (USGS) illustrates how one major national geological survey has sought to address priorities of the public it serves and to contribute to the common good.

Historical Review

During the nineteenth century, many nations recognized the importance of understanding the nature and distribution of their natural resources and thus established national geological surveys. The British Geological Survey (BGS, established 1835) and the Geological Survey of Canada (GSC, established 1842) were the earliest of these organizations that have operated continuously since their founding. Initially, the BGS, the GSC, and subsequent sister geological surveys in other countries focused on supporting the mineral needs of industrialization. Because countries equated their security and standing in the world with economic viability, the ability to locate raw materials for industrial development became the first major justification for beginning or continuing national geological surveys.

In the United States, mapping and science explorations, reconnaissances, and surveys sponsored by the federal government began in 1804 and continued thereafter under the aegis of the War Department, the Treasury Department, and/or the Department of the Interior (established in 1849), which was responsible for the stewardship and management of federal lands and their resources. In 1879, Congress and the President discontinued three competing mapping and science surveys of the public domain (Rabbitt 1979); their activities in biology passed principally to the Commissioner of Agriculture. In place of these surveys, Congress and the President established the USGS as a bureau of practical geology within the Department of the Interior to respond to pressing national needs for minerals for construction and currency.

The USGS was made responsible for "the classification of the public lands and examination of the geological structure, mineral resources, and products of the national domain" (U.S. Statutes at Large, v. 20, p. 394, March 3, 1879), but its operations were confined to the 1.2 billion acres in the public domain lands, most of which was acquired during westward expansion of the nation and lay west of the 100th meridian. The General Land Office, established in 1812 and transferred to Interior at the Department's founding in 1849, continued its land-parceling (cadastral) surveys and classifications—including mining, grazing, timbering, and agriculture—as the basis for disposition and title as a source of revenue and public good. To conduct the *scientific* classification of the public domain, USGS Director Clarence King planned a series of land maps to provide information for agriculturists, miners, engineers, timbermen, and political economists (Rabbitt 1980). In 1882, Congress implicitly extended USGS responsibilities to include the entire country, not just public lands, when it authorized preparation of an improved geologic map of the United States and by necessity a national geographic base map (Rabbitt 1980, Nelson 1999).

By the last decade of the nineteenth century, the increasingly recognized consequences of the rapid exploitation of lands and their resources spawned the first significant conservation movement in the United States. The USGS responded to these concerns between 1888 and 1902 by gaining statutory approval to study surface and ground water (which led to a national stream-gauging network), to map forest reserves, and to conduct reclamation investigations.

Studies by the USGS in support of the exploitation of natural resources continued well into the twentieth century, work spurred by concerns for economic growth, public needs, and national defense. The mineral industry had supplanted agriculture as the U.S. principal business activity in 1859. Raw materials needed during the Civil War, postbellum national development, and the emergence of the United States as a world power between 1898 and 1918 justified the view that resource studies were critical to the economic well-being and security of the nation (Rabbitt 1980, 1986; Cloud 1980). Beginning in 1938 and 1939, the USGS increased its critical- and strategic-minerals program for national defense. During World War II, the USGS increased its minerals and water-resource investigations and its mapping for military purposes; the agency also founded a Military Geology Unit for terrain-intelligence studies at home and in combat theaters. These activities, along with energy programs and the study of uranium and other radioactive materials also begun during World War II, continued and expanded during the subsequent Cold War based on much the same rationale: providing the nation with a better understanding of these resources as aids to exploration and development for economic and military security (Rabbitt 1989).

After World War II, it was generally believed that good science automatically created societal benefits (Sarewitz 1996), and USGS scientists pursued research goals within broad programmatic guidelines to generate new science to apply. At the same time, the USGS responded directly to societal needs as they arose by adding new missions. By the mid-1960s, for example, USGS personnel studied the effects of underground nuclear explosions, mapped the Moon, helped to train astronauts for the manned space program, and established long-term cooperative projects with government agencies in Brazil, Pakistan, Saudi Arabia, and other countries (Rabbitt 1989).

The environmental movement of the 1970s also influenced the direction and scope of USGS activities. Land-use choices no longer were viewed from a wholly exploitative standpoint. The USGS response to environmental issues included a greater emphasis on water quality (including the development of a toxics-hydrology program and the implementation of a National Water-Quality Assessment), investigation of the environmental effects of resource extraction such as acid mine drainage, and studies of climate change, including global assessment of changes in glaciers and the monitoring of permafrost. USGS studies of uranium in the 1970s focused on deposit models and assessment of resources, but the research emphasis later shifted to addressing the appropriate disposal of low- and high-level radioactive wastes at sites such as Yucca Mountain in Nevada. The USGS had provided the nation and the world with classic work in ore-deposit modeling (thereby advancing exploration, development, and science), but society's concerns shifted to the consequences of extraction and the USGS responded by modifying the emphasis of its mineral-resource activities.

Toward the end of the twentieth century, several national and global trends combined to influence USGS priorities and change its role and that of earth scientists. The rapid development of information technology fueled societal expectations for more information. At the same time, population growth in the United States affected regions previously sparsely settled, and ever-larger segments of society were exposed to the dangers of coastal storms, earthquakes, floods, landslides, volca-

nic eruptions, and wildfires. It became clear to the USGS that its studies of the impact and causes of these events would have to be linked more closely to emergency response needs and yield more rapid results. To have the most significant influence on decisions of public safety, the information needed to be available in a timely manner and thus required a response capability of twenty-four hours per day, seven days per week. The availability of real-time data expanded the public and municipality demand for innovative products. By using rainfall amounts and stream-gauging hydrographs, the USGS has predicted the severity and duration of flood events for emergency response efforts. Emergency managers and industry began to use USGS products that showed them the intensity of ground shaking within minutes of an earthquake, enabling them to make quick-response decisions. The engineering community began to use these same products to assess the behavior of structures during earthquakes and to develop more precise building codes.

Remote sensing and satellite operations such as Landsat and their archives became major activities within the USGS. The development of the Internet and the digital revolution enabled the USGS to respond to public demand for a diversity of real-time data, geospatial products, and scientific interpretations through use of the World Wide Web. In the early 2000s, the USGS implemented *The National Map*, an effort to make up-to-date digital topographic maps available to the public via the Internet.

In 1996, the National Biological Service, founded within the Department of the Interior three years earlier, became part of the USGS. This broadened the mandate of the USGS beyond the geographic, geologic, and hydrologic sciences. The USGS became a natural science organization, unique among the national geological surveys of the world because of the breadth of capabilities within the agency. The USGS began to focus on a more integrated approach to its scientific work to address the complex issues facing society.

Global Cooperation

National geological surveys are increasingly aware of the global nature of their efforts. This awareness is manifested through their increasingly global activities and through organizational partnerships and alliances. In the 1990s, the International Consortium of Geological Surveys (ICOGS) was formed to address the public perception that the missions of the national surveys were completed and that their services were no longer needed in the twenty-first century. ICOGS has worked to increase awareness of the importance of the earth sciences for the public and for policymakers. The International Union of Geological Sciences (IUGS) and the United Nations Educational, Scientific, and Cultural Organization (UNESCO), as well as numerous professional societies, have also addressed the awareness issue through major education campaigns. In addition, individual national surveys have formed a number of strategic alliances that improve their quality and effectiveness. One example is the partnership among the USGS, the GSC, and the Consejo de Recursos Minerales (CRM) of Mexico that has resulted in continental-scale efforts and products of mutual interest, such as geophysical maps, standards, geochemical surveys, and the geologic map of North America. Other groups such as the Coordinating Committee for Geoscience Programmes in East and Southeast Asia (CCOP) and the Circum-Pacific Council also reflect an emphasis on addressing earth science issues through a collaborative process of multiple national surveys, academia, and the private sector.

The program activities of most national geological surveys are also adopting a more global view. The BGS, the Australian Geological Survey Organisation (AGSO), the French survey (BRGM), the South African Council for Geoscience, and the USGS all have active programs providing earth-science support to the developing world. In addition to these dominantly cost-sharing activities, there is an increase in global assessments and information gathering. For example, the USGS operates a global seismographic network that provides high-quality information on seismic events to researchers and the public. Because resources such as minerals, oil, and gas are such vital commodities and have profound economic implications, the USGS conducts global assessments of these resources. In addition, the USGS reports on the demand for more than 100 mineral commodities, both domestically and internationally for approximately 180 nations. The USGS also receives and processes data from the Landsat satellites and provides images of the earth available to all biweekly. National surveys are also playing an expanded role in diplomacy. The USGS has cooperated with the Geological Survey of Ireland and the BGS on a possible mineral assessment in the border area of Ireland and Northern Ireland, has collaborated with nations in the Middle East relative to the region's seismic hazards, and has worked in Cyprus relative to hydrologic and seismic-hazard issues.

Future Directions

Population pressures challenge Earth's capacity to sustain a viable human society without deleterious effects. The common good has, over time, been redefined to include other values in addition to economic growth, and the public arena is fraught with competing and often conflicting values. Appropriate choices by decision makers and society require scientific insights about complex natural systems and the probable consequences of any proposed decision. Society demands pertinent and reliable scientific information in forms useful for decision-making. Science alone, however, is not the determining factor in most decisions—social, economic, and aesthetic values enter in as well. The tradeoffs inherent in societal choices, and the variable confidence in which knowledge is held at any given juncture, also need to be communicated. In the early twenty-first century, the USGS began a focused effort to improve and expand the use of its scientific results to inform the public and support decision making at all levels of society by exploring the problem of incorporating science into value-laden societal decisions. Ultimately, society will decide which tradeoffs are acceptable based on the its values, but the USGS can provide the critical scientific understandings that can help inform the nation about these choices.

As scientists strive to define their research goals by focusing on the decision context of the information needed, many recognize that it will be difficult to sustain their impartiality and integrity. Before the twenty-first century, many research scientists maintained a significant distance between their research and the decisions that might be based on their results. The challenge will be to bridge the gap between scientists and decision makers without compromising impartiality. The law that established the USGS in 1879 required that "the Director and members of the Geological Survey shall have no personal or private interests in the lands or mineral wealth of the region under survey, and shall execute no surveys or examinations for private parties or corporations" (U.S. Statutes at Large, v. 20, p. 394, March 3, 1879). These ethical requirements remain important ones, as society looks to the USGS for honest, impartial, and useful analyses of difficult choices ahead. All societies need the insight of public earth scientists and their engagement in issues of great societal importance.

Throughout their history, national geological surveys, including the USGS, have reflected the priorities and values of the nations they serve. Although the issues that determine the scope of their missions change over time, three principal activities are conducted: (1) long-term monitoring of the earth and its processes; (2) assessment and applied studies; and (3) basic research and understanding of physical, chemical, and biological processes. In the future, the national geological surveys will face societal challenges that increasingly involve the complex interactions of humankind and the natural world. Among the most important challenges will be the mitigation of natural hazards; an increased demand for water, mineral, and energy resources; the consequences of human activities with respect to earth's ecosystems; and the implications of climate variability. As people expand their definition of quality of life to include human and ecosystem health, decision makers will need insights based on the most reliable knowledge to make informed choices.

P. PATRICK LEAHY
CHRISTINE E. TURNER

SEE ALSO Expertise; Geological Information Systems; Modernization; Science Policy.

BIBLIOGRAPHY

Cloud, Preston. (1980). "The Improbable Bureaucracy: The United States Geological Survey, 1879–1979." *Proceedings of the American Philosophical Society* 124(3): 155–167. A brief overview of the history of the U.S. Geological Survey and the scientific leaders who contributed significantly to the development of the agency.

Nelson, Clifford M. (1999). "Toward a Reliable Geologic Map of the United States, 1803–1893." In *Surveying the Record: North American Scientific Exploration to 1930*, ed. Edward C. Carter II. Philadelphia: Memoirs of the American Philosophical Society 231: 51–74. A history of the national portrayal of U.S. geology from four geognostic maps published between 1803 and 1832 to ten geologic maps issued between 1834 and 1893, with a brief review of subsequent maps through 1974.

Rabbitt, Mary C. (1979). *Minerals, Lands, and Geology for the Common Defence and General Welfare*, Vol. 1. Washington, DC: U.S. Government Printing Office. A detailed history of public lands, federal science and mapping policy, and the development of mineral resources in the United States to the founding of the U.S. Geological Survey in 1879.

Rabbitt, Mary C. (1980). *Minerals, Lands, and Geology for the Common Defence and General Welfare*, Vol. 2. Washington, DC: U.S. Government Printing Office. A detailed history of geology in relation to the development of public-land, federal-science and mapping policies, and the development of mineral resources in the United States during the first twenty-five years of the U.S. Geological Survey.

Rabbitt, Mary C. (1986). *Minerals, Lands, and Geology for the Common Defence and General Welfare*, Vol. 3. Washington, DC: U.S. Government Printing Office. A detailed history of the period 1904–1939 in the U.S. Geological Survey and the relation of geology to the development of public-land, federal-science and mapping policies, and the development of mineral resources in the United States.

Sarewitz, Daniel. (1996). *Frontiers of Illusion: Science, Technology, and the Politics of Progress*. Philadelphia: Temple University Press. A perspective on modern science policy in terms of what types of science should be pursued, who makes such choices, and how science can be evaluated in the context of broader social and political goals.

INTERNET RESOURCE

Rabbitt, Mary C. (1989). "The United States Geological Survey, 1879–1989." *U.S. Geological Survey Circular* 1050. Available from www.usgs.gov. A brief history of the relation of geology during the first 110 years of the U.S. Geological Survey to the development of public-land, federal-science and mapping policies, and the development of mineral resources in the United States.

NATIONAL INSTITUTES OF HEALTH

• • •

The National Institutes of Health (NIH) is the biomedical research agency of the U.S. federal government. Located in Bethesda, Maryland, a suburb of Washington, DC, the NIH funds intramural (federal employee) scientists and extramural (outside the federal government) researchers across the country. Eighty percent of the NIH budget goes to grants to outside universities and laboratories. Research conducted at the NIH or with NIH funds leads to a more complete understanding of human health and to developing preventions, cures, and therapies for disease. At the same time, the agency has been forced to address and respond to ethical questions regarding research subjects, research topics, and scientific conflicts of interest.

From the Marine Hospital Service to the Hygiene Laboratory

In 1798 President John Adams signed legislation that started the United States on a long road of funding health-related activities by creating a Marine Hospital Service (MHS), an agency that would fund hospitals in ports to care for military personnel when they fell sick at sea. However the U.S. government did not support health-related research for much of the nineteenth century.

Widespread acceptance of the germ theory in the 1870s led to an increase in the number of American scientists doing research on disease. The basic idea that one specific germ caused one specific disease led to an explosion of new studies on microbes, immunity, and vaccines. Scientists began to trace diseases back to a particular *vector* such as water, milk, insects, or healthy human carriers. The government began to organize and enforce quarantines to curb epidemics—and such measures worked. Scientists soon identified the bacteria that caused diphtheria, tuberculosis, typhoid fever, anthrax, and malaria.

In the 1880s, the government decided to expand the role of the MHS to include bacteriological research. A laboratory was set up at the MHS facilities in Staten Island, New York, and Joseph Kinyoun was appointed to run it, since he was one of the few MHS officers who had studied the new science of bacteriology. Kinyoun called his facility a *laboratory of hygiene* and it soon became known as the Hygienic Laboratory. Kinyoun's first paper described his methods for making a positive diagnosis of cholera using his microscope and bacteriological methods.

In 1891 the laboratory moved to a more prestigious government location: near the U.S. Capitol Building in Washington, DC. Kinyoun and his associates began to manufacture vaccines and antitoxins (known collectively as biologics) for diphtheria, rabies, and smallpox. After a tragedy caused by contaminated diphtheria antitoxin in St. Louis, Congress passed the 1902 Biologics Control Act, putting the Hygienic Laboratory in control of regulating biologics for the entire country.

In 1902 Congress expanded the Marine Hospital Service to the broader Public Health and Marine Hospital Service and reorganized the laboratory. The four new divisions were Pathology and Bacteriology, Zoology, Chemistry, and Pharmacology. Ph.D.s were hired alongside the M.D.s, introducing new scientific techniques and expertise. In 1912 the name Marine Hospital Service was dropped entirely. Activities of the newly conceived laboratory included exploring noncontagious diseases and conducting studies of dairies, pollution, and water filtration systems to identify causes for disease. Scientists showed that their research could assist public health officials in preventing epidemics and keeping the public safe.

Epidemiology became an important function of the Hygienic Laboratory in the early twentieth century. Scientists would be dispatched to a location where there was an outbreak of a disease such as yellow fever or typhoid. They would investigate the cause of the disease by finding the vector that passed it along or identifying problems of diet or pollution. Joseph Goldberger, for example, traveled across the South for many years doing studies on populations with outbreaks of pellagra. By closely observing the people who had the disease, experimenting with different diets, and exhaustively searching for possible causes among the populations of small towns, institutions, orphanages, and prisons, he correctly identified the disease as a dietary deficiency, rather than a contagious disease. Goldberger's discovery eventually led to the elimination of pellagra as a dangerous disease that once plagued an entire region.

The staff of the Hygienic Laboratory grew. Rocky Mountain spotted fever studies spawned a new outpost in Montana to better study the insect vector in that region. While the Pathology and Bacteriology division researched diseases such as typhoid fever, the Zoology division studied a new species of hookworm known to cause disease and the Chemistry division studied the role of stomach acid and the chemistry of blood. The Pharmacology division studied toxicity of alcohols and the effect of certain drugs on blood pressure.

NIH: 1930s to 1950s

In 1930 Congress approved more funding for a new building and expanded the role of the Hygienic Laboratory—renamed the National Institute of Health by Senator Joseph Ransdell—to fund scientists in new fields such as the study of chemicals used in warfare. As the Great Depression further eroded private support of scientific research, scientists increasingly looked for help from the federal government. The Ransdell Act of 1930 ushered in a new era of expanded government support for scientific research.

Cancer became the first disease to generate enough public panic to build support for legislation funding scientific research. In 1937 every senator in Congress cosponsored a bill creating the National Cancer Institute (NCI), which would (in 1944) become a subset of the NIH. The 1937 bill was important for another reason: It authorized the NCI to award grants and fellowships to outside scientists conducting research. This granting (extramural) program would become a fundamental part of the NIH's work.

With more responsibilities came the demand for more space. In 1935 the government accepted a gift of land from suburban estate owners Luke and Helen Wilson. The original plan was to use the land for the growing number of animals used in research, but by 1941 all the scientists had packed up their laboratories and joined the animals in a newly built complex of six brick buildings.

During World War II NIH scientists studied many problems among workers in the war-related industries in the United States. For example, they studied the levels of toxicity incurred by people working in industries such as synthetic rubber, ships, tanks, munitions, and airplanes. Disease research focused on malaria, yellow fever, and typhus, all of which proved devastating to the troops abroad. NIH scientists also studied the oxygen needs of pilots at certain altitudes.

In 1944 Surgeon General Thomas Parran and NIH Director Rolla Dyer helped pass a new law, the Public Health Service (PHS) Act, which revitalized the NIH. The act authorized more granting authority to the NIH and also allowed for spending on clinical research. Additionally the NIH was required to prepare public materials to inform the general public about its research and how that research affected people's health.

The next decades were a period of rapid growth for the NIH. Based on the success of the National Cancer Institute, many of the new institutes focused on the study and cure of certain diseases. During World War II many recruits were deemed unfit for service due to poor mental health or poor dental health, leading to the creation of new institutes to ease the effect of these problems in the population. These included the National Heart Institute, the National Institute on Mental Health, and the National Institute of Dental Research.

In 1953 the promise of clinical research was realized in the NIH Clinical Center, then the largest research hospital in the world with 540 beds. The special design of the Clinical Center ensured that the scientists and physicians kept in close contact while studying the effects of certain drugs or therapies on patients. Doctors referred many of the patients to the Clinical Center from all over the country. Other patients were *normal volunteers*, whose participation in studies produced baseline information about how healthy people reacted to proposed therapies. This data could then be compared with that from ill patients.

The clinical center was opened in the shadow of Nazi medical experiments. Its initial ethics rules, guided by the Nuremburg code, mandated informed consent from the human subjects of research and instituted an internal review process. In 1974, after the abuses of the Tuskegee syphilis study were made public, congressional action required the creation of Institutional Review Boards (IRBs) to oversee human research projects and an office of protection from research risks at NIH to oversee the IRBs.

In the decades after World War II, the proliferation of institutes—most of them still linked to a particular disease or body part—brought hundreds and then thousands of scientists, laboratory technicians, and support staff to the Bethesda campus, which also grew to accommodate more buildings.

NIH: 1960s to 2000s

Virus research was one major area of study at the NIH in the 1950s and 1960s. Developing new ways to grow and identify viruses, scientists identified dozens of new virus strains, leading to better and more effective ideas for curbing outbreaks.

One major line of research that has carried through in dozens of NIH laboratories in the second half of the twentieth century is genetics. In the late 1950s and early 1960s, scientists working with Marshall Nirenberg deciphered the genetic code. Building on this basic research, researchers in the 1960s and early 1970s learned how to cut and recombine DNA. In the 1980s the Human Genome Project was launched with the goal of charting the human genome, a goal that was reached in 2003 by scientists at the National Human Genome Research Institute. NIH scientists have also been leaders in experimental clinical research such as gene therapy.

In the 1970s, fields such as genetics research advanced rapidly because of new molecular biology techniques and instrumentation. The scientific competition that resulted also led to misconduct by some scientists. Beginning in the early 1980s, NIH led investigations into this issue. It also sponsored studies on how to ensure research integrity and programs to incorporate ethics training in the graduate education of scientists.

Research on chronic disease has been a mainstay of NIH research, from the earliest days of cancer and heart research to diseases such as diabetes, arthritis, and drug and alcohol addiction. A long-term NIH-funded study in Framingham, Massachusetts, provided evidence about heart health that has led to major education campaigns about the importance of exercise, low-fat eating, and smoking cessation. Dental research led to the mass fluoridation of water as it was shown to reduce the number of dental caries in children. In the 1980s infectious disease again took center stage when NIH researchers began studying AIDS.

In the 1990s the NIH continued to expand. New institutes funded studies of the aging population and the effects of nursing on patient care. Thousands of NIH scientists conduct research on the main campus in Bethesda, and at outposts such as the National Institute of Environmental Health Research in North Carolina and the Rocky Mountain Laboratory in Montana. In 1998 Congress voted to double the NIH budget over five years, and this money funded scientists all over the country and even around the world.

The NIH is proud to claim five Nobel Prize winners who did their major work on the Bethesda campus. In addition, dozens of members of the National Academy of Sciences have worked at the NIH. Over 100 scientists based at other institutions won the Nobel Prize based on research conducted with NIH funding in fields as varied as chemistry, physiology, medicine, and economics.

Several centers founded in the 1980s and 1990s complement the basic science at the NIH. For example, the Center on Research in Women's Health tracks the inclusion of women in clinical trials of drugs and disease therapies, and the Office of Technology Transfer encourages partnerships between scientists and industry. The National Center on Minority Health and Health Disparities monitors the NIH and works to eliminate health disparities. The National Center for Research Resources helps link scientists with the resources they need to make their projects work. These and other components of the NIH help ensure that the mandate to inform and protect the public from disease is carried out.

Ethics and Politics

In 1977, the NIH added a bioethicist to its staff. In 1995, bioethics was expanded to an entire program that supports training of new bioethicists and conducts research that seeks to inform public policy in health research. Though many argue that disease knows no politics, the NIH has had to deal with many issues that have divided bioethicists and the agency's supporters along political lines. Certain choices about which diseases to study and which patients to admit (such as AIDS patients in the 1980s) aroused controversy. Stem cell research worried many Americans and Congress members in the late 1990s and at the turn of the twenty-first century. New allegations in 2003 about NIH scientists accepting funding from pharmaceutical companies led to congressional calls for stricter rules about consulting and stock ownership. The regulations, similar to those imposed on scientists at regulartory agencies, sharply curtailed participation by NIH scientists on boards, committees, and participation in professional associations as well as their ownership of health-related stocks. In 2005, the new rules led to intense debate among NIH staff and scientists, who feared that they would result in driving top people away from federal service. Though for the most part scientists can work quietly in their laboratories without worrying about politics, institute directors must testify before Congress about how they are spending taxpayers' money. In 2005 with 27 institutes and centers and an annual budget of $27.9 billion, the NIH would be barely recognizable to Joseph Kinyoun, its first director.

SARAH A. LEAVITT

SEE ALSO *Complementary and Alternative Medicine; Health and Disease; National Academies; Research Integrity.*

BIBLIOGRAPHY

Harden, Victoria A. (1986). *Inventing the NIH: Federal Biomedical Research Policy, 1887–1937*. Baltimore: Johns Hopkins University Press. This volume describes the research conducted at the Hygeinic Laboratory and the expansion of its research program in the early-twentieth century before the NIH moved to Bethesda.

Kraut, Alan M. (2003). *Goldberger's War: The Life and Work of a Public Health Crusader*. New York: Hill and Wang. This engaging work explores the epidemiological work of Dr. Joseph Goldberger, a PHS scientist who solved the mystery of the cause of pellagra in the early twentieth century. The book outlines the policies and research goals of the early Hygeinic Laboratory scientists in Washington, DC.

Park, Buhm Soon. (2003). "The Development of the Intramural Research Program at the National Institutes of Health After World War II." *Perspectives in Biology and Medicine* 46(3): 383–402. This article outlines the research plans of the NIH as it grew astronomically in the decades after World War II. New goals are examined, the work of particular scientists is explored, and the overall mission of the NIH in this period is explained.

Hannaway, Caroline; Victoria Harden; and John Parascandola, eds. (1995). *AIDS and the Public Debate: Historical and Contemporary Perspectives*. Amsterdam and Washington, DC: IOS Press. This edited volume includes an essay by former Surgeon General C. Everett Koop in which he describes his positions on AIDS-related sexual education topics in the 1980s. The volume is an important addition to the story of how U.S. scientific and medical communities responded to the AIDS epidemic and how the NIH was influential in the early years of research and clinical studies.

INTERNET RESOURCE

National Institutes of Health. Available from www.history.nih.gov. This web site links visitors to a brief illustrated history of the NIH, along with more specialized web exhibits on certain research initiatives, such as genetics research, particular individual scientists such as Martin Rodbell and Marshall Nirenberg, and particular scientific achievements at the NIH. The site also provides bibliographies, copies of unpublished and out-of-print sources on NIH history, and much more.

NATIONALISM

• • •

Nationalism is a dominating political concept while being at the same time theoretically and practically problematic. In relation to science and technology, it is common to talk about national styles—French science and engineering are more rationalist, English science and engineering more empirical—and to see science and technology as having different national impacts. Certainly the scientific community in United States is able to marshal a greater percentage of gross domestic product (GDP) for research investment than similar communities in any other developed country, and U.S. culture is the most high-tech saturated in the world. Nationalism both energizes scientific and technological communities and has served as a justification for behavior that has been argued to violate scientific standards of conduct with regard, for instance, to research involving human subjects and to the sharing of knowledge. The scientific community has on occasion also seen itself as opposed to nationalism (and able to replace it with the "republic of science"), while nation-states have suspected scientists of disloyalties and seen them as a threat to national security. The following analysis of nationalism is thus designed to provide a basis for further exploration of such issues.

Nationalism as Theoretical Enigma

Nationalism is among the most problematic concepts in the social sciences. At the core of this enigma is the frequently observed discrepancy between the emotive and politically mobilizing *power* of nationalism and its flimsy or minimal *content* when analyzed as a political ideology. Primarily because nationalism is a political "ism," it is readily classified alongside other political/ideological isms, such as conservatism, liberalism, and socialism. But in contrast with adherents of these other ideologies, nationalists do not seem to be required to take on many, or indeed any, substantive intellectual commitments (Anderson 1983). One popular definition of nationalism, for example, holds that it is the doctrine "that the boundaries of the state and of the nation should always be congruent" (Gellner 1983, p. 1), but apart from this being minimalist in the extreme, there are still many enthusiastic nationalists in states that *long ago* realized such a doctrine But this does not appear to have diminished the nationalist enthusiasms and commitments that still emerge in such states from time to time.

This discrepancy between the appeal and content of nationalism has led some theorists to argue that it is not a political ideology at all, but a more emotive phenomenon, closer to religion than to politics. Others,

while not endorsing this view, have suggested that its intellectual vacuity is precisely the secret of its mobilizing power. For while other political ideologies, precisely because of the substantive commitments they entail, will necessarily *divide* populations, nationalism *unifies*, through its broad emotivism, people who would otherwise differ—whether by socioeconomic status and interests, ethnicity, gender, or philosophical and value commitments.

Nationalism and Identity

Such an observation, while true and important, still does not explain the broad, unifying emotive appeal of nationalism. The response most favored by scholars of nationalism is that nationalist politics are a form of *identity politics* and that the emotive power of nationalism comes from its being one of the most common ways, at least in the modern world, in which people identify themselves and others. Because a threat to personal identity is one of the most profound, it is therefore not surprising that nationalist conflicts often call up deep, even hysterical, emotions in those involved.

The sociology and psychology of identity, while assisting in explaining the emotive power of nationalism, generates its own difficulty. It is a commonplace of the literature on identity that any human being always possesses *multiple* identities, of which a national identity is but one. Thus a person may be an American, a Christian, a woman, a feminist, a wife, a Democratic voter, a computer programmer, a keen basketball fan, and an enthusiastic gardener; which identity she emphasizes at any moment will depend on the context and situation. In certain circumstances (say, when she is traveling abroad, or when America is at war) her American identity may be uppermost. But it remains to be explained why people will kill and die for a national identity more readily than a gender, religious, or domestic political identity, or any other numerous possible identities. It is in addressing this question above all that theorists of nationalism divide into two broad groups or camps.

Modernist Theories of Nationalism

The great debate in the literature on nationalism concerns both the historical antiquity and the fundamental roots or sources of its appeal and power. Modernist theorists argue that national identities are of relatively recent origin and political construction. In essence, they maintain that since the late eighteenth century states or state elites have politically constructed national identities among the mass of their populations. For modernist theorists the two seminal historical events in the construction of nationalism as a political ideology and identity were the American (anticolonial) Revolution of 1776 and the French (antimonarchical) Revolution of 1789. These two events were seminal because between them they dethroned the predominant principle of political legitimacy of the premodern period (the divine right or divine status of hereditary rulers) and installed the modern democratic principle (rule in the name of, and with the assent of "the People") in its stead. In both these revolutions the boundaries of the legitimacy-bestowing "People" were taken to be coterminous with the boundaries of "the Nation." That is, "the People" who made up the new democratic citizenry were generally defined as those living within a certain geographical space, speaking a certain language, and sharing a common culture (Hobsbawm 1992, Gellner 1983).

Nevertheless, while the above was the core of the mobilizing and legitimizing ideology of both these modern revolutions, in practice the actual people—the population—occupying the geographical territories involved (the thirteen American colonies, the former kingdom of France) were often *not* possessed of the characteristics with which they were predicated in the noble rhetoric of "the People." Thus many were illiterate, quite a number did not support their respective revolutions at all, and in France many did not even speak French, at least not the Parisian variety in which the revolution was conducted. Thus in the postrevolutionary situation state elites set about *turning* populations into "The People" through the imposition of mass education systems using a single language or (in the French case) a particular regional version of a language. Such education systems not only turned "peasants into Frenchmen" (Weber 1976) or American colonists into Americans, they also specifically introduced the newly educated masses to the national symbols of identity and loyalty (flags, anthems, and constitutional principles) and inculcated them with a nationalist version of history in which they could find a sense of pride in their new identity.

These state-led "nation-building" practices became even more vital when, in the later years of the nineteenth century, France began its own industrial revolution and millions of non-English-speaking European and other immigrants flooded into the industrializing United States. The need of industrializing countries for a skilled and literate labor force, and the emergence of other economic and social institutions to shape that force (large industrial towns and cities, modern mass communications and infrastructure), gave further impetus to this state-directed process of "nationalizing" the masses. On this account the creation of national identities and

identifications is simply part of the political and economic modernization of states and their populations, a process that introduces this new political identity (that of being a free and equal "citizen" of a "nation-state") as it also introduces a range of other new economic, occupational, and social identities (Gellner 1983).

Whatever its historical merits, this "modernist" theory of nationalism still leaves certain crucial questions unanswered. First, while it can and does account for the creation of "mass" nationalism, it has to assume the preexistence of a state elite nationalism that it does not itself explain. This is especially a problem in the case of English nationalism, which, as an elite or upper-class phenomenon, predates both the American and French Revolutions by a hundred years (Colley 1992, Newman 1997), and is therefore radically anomalous in the modernist account (Smith 1998). Second, precisely insofar as modernist theory emphasizes that national political identities are only *one* of the identity changes brought about in human populations by modernization, it still leaves unexplained the singular emotive power of national political identities specifically, relative to the many other modern identities ("worker," "employer," "liberal," "socialist," "feminist," "Yankees fan") created in the course of modernization.

One "modernist" attempt to deal with the latter question is found in Benedict Anderson's seminal book *Imagined Communities* (1983). Anderson suggests that it is significant that nationalism both borrows a great part of its emotional power from religious feelings and symbolism, and that it originated in a place and time when conventional religious belief was coming under widespread and systematic challenge. Anderson emphasizes the role of religious or quasi-religious symbols in national identification (cenotaphs, tombs of unknown soldiers, hymnlike national anthems). He also suggests that nationalism provides a sort of secularized version of immortality to replace explicitly religious notions coming under challenge. Thus, though any individual citizen lives and dies, "the Nation" itself lives on, and, in making the "ultimate sacrifice" for his or her nation on the field of battle, the individual citizen ensures the continuity/immortality of the national collective.

Although suggestive, such an interpretation is hardly conclusive. First, the late-eighteenth-century origin remains assumed. Second it is not clear that nationalism *did* replace or supplant conventional religiosity. It is true that conventional religious belief came under widespread challenge in the eighteenth-century Enlightenment, in certain restricted circles. But it nonetheless remained powerful and important as a mass phenomenon, and many who have killed and died for their nation have *also* seen themselves as killing and dying for God. That is, nationalist sentiment seems far more often to combine with, or even coattail upon, conventional religious conviction as to supplant it.

Antimodernism and "Ethnicity"

The essential view shared by all antimodernist theorists is that, in some way or another, modern nationalism is a "politically transformed" version of a much older, even primordial, phenomenon in human life, ethnicity. This latter concept is not without its problems, but in its original meaning at least, ethnicity is a biological or putatively biological concept. An "ethnic group" is a group of people claiming descent from a common ancestor, with such groups varying considerably in size and having many and various names in different languages. In English terms such as *tribe, clan,* and *family,* and indeed *nation* itself, are all terms with such an original biological or "kinship" meaning. Historically, people who have claimed a common biological descent have also shared a common language and have often had important customs and beliefs (religious, magical, sexual, etc.) in common.

In essence then, antimodernist theorists of nationalism claim that the creation of nationalism is best conceived as the modern "political transformation" of much older ethnic identities. Nationalism turns group identities that people possess but are not conscious or aware of (ethnic identities) into conscious, self-aware political identities (national identities). On this account there have for centuries been people who were ethnically English living in the geographical space known as England, but they became consciously, politically English only sometime in (most likely) the seventeenth century. Likewise, there have for millennia been people who were ethnically Chinese in the area of the world known as China, but they became consciously, politically Chinese only sometime in most likely the late nineteenth or early twentieth century (Smith 1986).

The clear advantage of this concept (which can embrace language, religion, and culture as well as biology) is that it explains why it was relatively easy for state elites to create mass nationalist loyalties (they were "only" making conscious what in some sense or other had long existed) and why (conversely) it may be difficult to create fervent, self-identifying nations across boundaries of biology, language, or culture. It also explains where "elite nationalists" come from. Elite nationalists are just the first people *within* an old ethnic group to make their ethnic identity a conscious political

identity. This is a feat made easier by certain aspects of elite privilege (for example, greater leisure time and education and greater capacity to travel—and thus to see other peoples and cultures *and* to see them as "other" than "their own").

But antimodernist theories of nationalism are not without their problems. First, the existence and flourishing of such immigrant-based nation-states as the United States, Canada, Australia, New Zealand, and (indeed) Brazil or Argentina demonstrates that, while it may be more difficult to create national identities and solidarities across ethnic/cultural boundaries, it is by no means impossible to do so, given enough time and the appropriate political will. While most modern nation-states may indeed be dominated by a politically transformed ethnic core nation, not *all* are. This in turn implies the following: *All* modern nation-states that are now ethnically plural as the result of global population movements will not necessarily be imperiled by ethnic divisions between old host ethnic core groups and new arrivals. This may happen, as the result of political failures of one sort or another, but the relatively successful creation of the above-mentioned multiethnic, immigrant-based "nation-states" of the nineteenth century suggests that there is nothing inevitable about it.

Second, as the more sophisticated antimodernist theorists readily admit, ethnic identities are not themselves in any way fixed, static, or "unchanging" (Smith 1986). Human beings can, and have, changed even their biological group characteristics (physiogamy, skin tones, etc.) very considerably over long historical periods through interbreeding. Moreover, although ethnic groups usually *claim* to be biological entities, virtually none of them are, or are exclusively. That is, virtually all social anthropologists and historians who have studied large or largish human kin-based groupings (past and present) have emphasized that they operate through what is called "fictive" as well as real (biological) kinship. Slaves, war captives, or simple peaceful adoptees may be incorporated into a kinship group by the use of kinship terminology (by being treated as "uncles," "brothers," "cousins," etc.), and then, over time, this original adoption is forgotten and the people in question both claim to be and are accepted as "real" kin. (They or their descendents may even become so through interbreeding.)

Third, the above observations imply that *no* human ethnic groups existing today are in fact ethnic groups in the narrow biological sense (that is, actual biological descendents from a common ancestor). All of them are really linguistic and (to an extent) cultural groupings and therefore more or less open to any adoptees who are accepted into them. Therefore, although there *is* a linguistic and cultural "ethnic nation" of English people in England, they long ago ceased to be a biological descent group. (They are in fact a mélange of many such groups including Celts, Angles, Saxons, Normans, Danes, and others.) Moreover, although these composing groups "happen" to share broadly "Caucasian" physiogamies and skin tones (so that ethnically English people are "white" people), there is no reason why, in the future, this shared biological fact may not change markedly as the result of widespread interbreeding with non-Caucasians. What is true of the English ethnic nation is equally true of the Chinese, French, German, or any other ethnic nation.

Ethnic versus Civic Nationalism

These are not merely abstract historical considerations. They have vital contemporary implications. Because in a world of massive global population movements, the central political issue now facing all states is that of the relationship between civic national identities and ethnic national identities. While an ethnically Chinese person who settles in (say) Australia can readily, even "instantly," become a *civic* Australian citizen by going through a "naturalization" ceremony, being issued an Australian passport, and so on, this person will become *ethnically* Australian only by learning the English language well, adopting Australian cultural mores, and so on. All historical and contemporary evidence suggests that converting civic national identities into ethnic national identities (if that is what the people in question wish to do) will be a *much* slower process than formal civic incorporation—a process possibly requiring many generations to occur. But such evidence also suggests that there is nothing impossible about it, if the right "open" political and cultural conditions exist. Moreover, if the right conditions exist civic nationality can be turned into cultural or ethnic nationality quite quickly, as for example in the United States.

Nationalism and Globalization

This review of the major modern theories of nationalism has done little to dispel its enigmatic quality. All its theorists and theories are able to do *some* justice to this extraordinarily slippery phenomenon, while none do it *total* justice The reason may be relatively simple. It may be that *any* theorization of human identity in general (and not just of national identity) must come to terms with an important but frequently overlooked paradox. This is that (1) all human identities are parasitic upon

notions of "difference" or "otherness"—for example, "male" identity on "female" identity, "white" identity on "colored" or "black" identity, and "liberal" identity on "conservative"—which are structurally "fixed," or apparently fixed, over relatively short historical periods. This semantic parasitism of identity on otherness has the possibility of conflict built into it, if the right (or wrong) political and historical conditions arise during those periods. (2) Despite this, all human identities are also, to a greater or lesser degree, plastic or changeable in the long run. Thus, human identity differences that at one historical moment can seem both immutable and inherently conflictual can (and indeed have) come at another time either to cease to exist altogether or to be regarded as perfectly and peacefully compatible, even mutually enriching. In this perspective then, ethnicity theories are strong in telling why national identities are slow to change but weak in explaining how and why they *have* changed over long periods and will no doubt continue to change. Conversely, modernist theories are good at laying out one important means and mechanism of change (manipulation by political elites) but weak in explaining why *some* identities seem much easier for such elites to manipulate than others.

If this is the case, then the central question facing all theories of nationalism concerns what political and institutional conditions tend to fix or "reify" the currently existing global pattern of ethnic/cultural differences and what political and institutional conditions tend to encourage the change or mutability of that pattern. When the matter is put this way, its implications for both bodies of theory are clear. At a certain period in modern history (roughly from the eighteenth century onward), a given global pattern of human ethnic grouping was (as both the modernists *and* the antimodernists assert) made conscious (through political mobilization by state-related elites), then further fixed and reinforced by such measures as the laying down of spatially exact and controlled state borders, the issuing of passports and citizenship papers, and the creation of a single "national" education system in a single "national" language.

In a word, some ethnic nations were "statized"—turned into so-called nation-states—whereas others were enforcedly incorporated into these state-dominant ethnicities or simply subordinated, as "second-class citizens," within these states. In many of the latter cases such subordinated groups also had their own demands for statehood denied or suppressed. Seen from a contemporary perspective this historical statization of *some* ethnicities was an enormously powerful force in politically fixing a particular historical ethnic pattern and making it seem both "natural"—the only possible pattern—and difficult, if not impossible, to change.

In a world that is globalizing rapidly—not only economically and technologically but also, to a degree, culturally—this political "fixing" of identity by statization may now be the central problem facing humans. There is no doubt that statizing (some) nations has made it more difficult (more difficult, that is, than would other more open and flexible political arrangements) for all human beings—whether members of dominant or "statized" ethnicities or not—to deal effectively with the unique problems posed by globalization *and* more difficult for all of them to take full and proper advantage of the economic and other opportunities it affords. This is because the principal socioeconomic and cultural differences and disparities that globalization creates are *not* ethnic or national differences at all, but differences that deeply cut across both ethnicity and nationalism. And there is strong reason to believe that those human beings who recognize this first, and act accordingly, will therefore be (and indeed already are) those who benefit most from globalization, whereas those who remain mired in ethnicity and national identity (and through the early twenty-first century this is the vast majority of humankind) will also be those least well equipped either to take advantage of globalization's opportunities or to solve its problems (Kitching 2001).

Conclusion: Nationalism, Science, and Technology

As indicated at the beginning, nationalism, science, and technology exists in some tension with each other. As an enigmatic form of identity that is dependent on otherness and historically plastic, nationalism has also been able to oppose and be opposed by various forms of science and technology. Obvious examples of opposition have involved the Nazi German rejection of "Jewish science," the Communist criticism of "bourgeois science," and Islamist efforts to simultaneously reject and transform infidel science and technology. The failures of such efforts in the past may nevertheless suggest some of the limits of nationalism as a transforming process.

Historically, nationalism was also associated with the Treaty of Westphalia (1648) that granted to nations sovereignty, that is, ultimate powers of life and death, within certain geographic borders. To some degree the positive character of these boundary conditions reflected the limits of early modern technology (especially forms of transportation and communication) and depended on them (the state did not have at its disposal mid-twentieth century means of propaganda nor a virtually unlimited ability to kill large numbers of any ethnically diverse population). Late twentieth century criticisms of nationalism in the name of internationalism in many instances reflect changes in technologies and the new forms of communication and power they place in the hands of some political elites that would statize certain pre-national identifies. The international opposition to statization by the Chinese in Tibet, the Serbs in Kosovo, or the Sunnis in Iraq all reflect a willingness to subject nationalist plasticity joined to technological power to transnationalist criticisms. In such cases science and technology themselves may likewise be seen as paradoxical promoters and delimiters of nationalism.

GAVIN KITCHING

SEE ALSO *Democracy; Science Policy.*

BIBLIOGRAPHY

Anderson, Benedict. (1983). *Imagined Communities: Reflections on the Origin and Spread of Nationalism.* London: Verso. One of the most original and imaginative treatments of the issue. Especially good on the relationship between Western and non-Western forms of nationalism.

Colley, Linda. (1992). *Britons: Forging the Nation, 1707–1837.* New Haven, CT: Yale University Press. A "modernist" treatment of British nationalism. Very good on the political construction and manipulation of ethnicity in the British context.

Gellner, Ernest. (1983). *Nations and Nationalism.* Oxford: Blackwell. One of the clearest and most accessible texts on the subject. Very orthodoxly modernist.

Hobsbawm, E. J. (1992). *Nations and Nationalism since 1780: Programme, Myth, Reality*, 2nd edition. Cambridge, UK: Cambridge University Press. An extremely skeptical treatment of nationalism by a great historian. Nationalism is here treated explicitly as an ideological phenomenon and an extremely duplicitous one.

Kitching, Gavin. (2001). *Seeking Social Justice through Globalization: Escaping a Nationalist Perspective.* University Park: Pennsylvania State University Press. Review of the various nationalist responses to globalization.

Newman, Gerald. (1997). *The Rise of English Nationalism: A Cultural History, 1740–1830.* New York: St. Martin's Press. Useful contrasting text to Colley. Much more insistent on the ethnic roots of nationalism.

Smith, Anthony D. (1986). *The Ethnic Origins of Nations.* Oxford: Blackwell. Important pioneering treatment of this very important topic.

Smith, Anthony D. (1998). *Nationalism and Modernism: A Critical Survey of Recent Theories of Nations and National-*

ism. London: Routledge. An extremely cogent and well-balanced review of this central debate.

Weber, Eugen. (1976). *Peasants into Frenchmen: The Modernization of Rural France, 1870–1914.* Stanford, CA: Stanford University Press. Fascinating text for the exploration of the differences and interrelationships of elite and mass nationalism.

NATIONAL PARKS

● ● ●

A national park, as distinct from a landscaped urban park, is a place set aside to preserve a natural geology or ecology deemed to possess significant inherent value. The concept of a national park thus constitutes a practical effort to place a specific ethical limit on technological development, sometimes for scientific as well as public benefit.

Historical Origin

Shortly after northwest Wyoming was annexed as part of the Louisiana Purchase in 1803, mountaineers and trappers began returning from their adventures in the American West with stories of a strange and mysterious place where steaming water bubbled from the ground and geysers shot like clockwork into the sky.

Rumors swirled for decades, until, in 1870, several expeditions were organized to explore the area around the Yellowstone River. The first expedition was so awed by the hissing, cauldron-like landscape that upon return members began a campaign for the creation of the world's first national park.

In response the federal government funded a second, scientific expedition, which was led by Dr. F. W. Hayden, then head of the U.S. Geological Survey. The group also included photographer William Henry Jackson, whose photographs (often developed *on location* in the hot springs) would prove the existence of a national treasure to skeptical Easterners and convince the country that Yellowstone needed to be set aside for the ages. Another participant, Lieutenant Gustavas C. Doane testified before Congress about what he had seen:

> [This land] is without parallel; as a field for scientific research, it promises great results; in the branches of geology, mineralogy, botany, zoology, and ornithology, it is probably the greatest laboratory that nature furnishes on the surface of the globe. (Everhart 1972, p. 6)

On March 1, 1872, President Ulysses S. Grant signed the Yellowstone Act. With the creation of Yellowstone National Park, 2 million acres were established "as a public park or pleasuring ground for the benefit and enjoyment of the people." The act went on to stipulate that regulations would be put in place to provide for the preservation "from injury or spoliation, of all timber, mineral deposits, natural curiosities, or wonders ..."

Many in Congress voted for the creation of Yellowstone National Park because they did not want to see it destroyed by the type of crass commercialization and over-building that had occurred in New York's Niagara Falls. Preservation for its own sake was not a foundational idea. Indeed ideas such as Manifest Destiny and abundance were hallmarks of the frontier sensibility. Few thought the bounty of America had limits. Fewer still thought the government had any business interfering with their right to exploit the scenic wonders and natural resources at the frontier.

Though a great park was created, with visitors came despoliation. By 1886 the cavalry had to be called in to protect the park from vandalism, logging, and hunting. By the time a bill passed, creating a National Park Service within the Department of Interior to administer public lands, there were thirty-one national parks and monuments in the United States and a growing awareness that some type of protection was critical to the survival the nation's wild and scenic places. With the passage of the National Park Service Act of 1916, the debate between conservationists and preservationists over just how to protect the parks was settled: Conservationists arguing for the *wise use* of the natural resources in the national parks lost out to the preservationists who argued that wilderness areas should remain untouched and unexploited.

One of the authors of the National Park Service Act was landscape architect Frederick Law Olmstead, Jr.—of New York City's Central Park fame—who supported the notion of preserving places that would provide a contrast to and respite from the pace of the modern world. He envisioned parks where ordinary citizens could rest mind, body, and soul. From spiritual uplift to scientific research, recreation, and education, national parks were seen as a way to enhance the lives of the general public. The spread of the national park idea—that large tracts of wilderness should be protected for all time—could arguably be called one of the great contributions from the United States to world civilization.

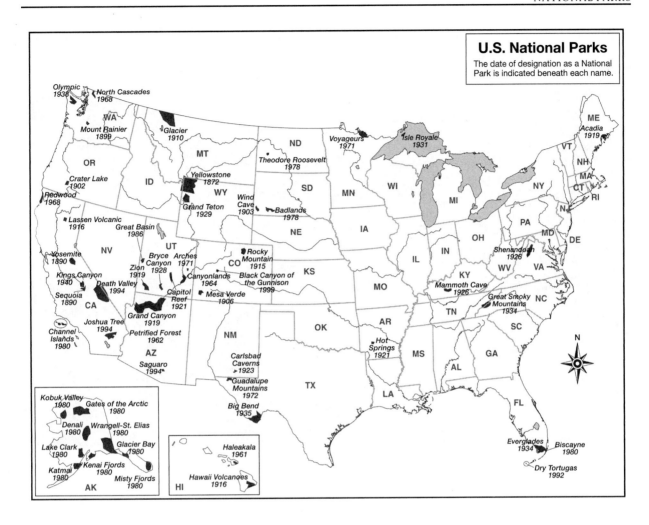

U.S. National Parks
The date of designation as a National Park is indicated beneath each name.

Outside the United States

By the outbreak of World War I, Canada, Australia, New Zealand, Mexico, and Sweden had adapted the American concept of national parks to their own lands and needs. (In many of these countries, the primary motive for establishing national parks was the protection of native peoples rather than the flora, fauna, and natural wonders of the area.) In 1914 Switzerland created a national park, but dedicated it to scientific research rather than recreation.

In the inter war period, news of the massive slaughter of African wildlife led to the 1933 London Conference for the Protection of African Fauna and Flora. The conference helped inspire the creation of large national parks in eastern and central Africa to protect game populations and preserve areas for scientific study, but its ideals and goals had much wider influence and were used as a blueprint to help establish national parks worldwide.

As the national parks idea took root, an awareness developed of the need for some type of world organization that could promote nature conservation. In 1948 at a conference sponsored by the United Nations, the International Union of Conservation of Nature and Natural Resources (IUCN) was founded.

In the early twenty first century the IUCN, in coordination with the United Nations Environment Programme, is a self-described *green web* in which 140 countries, more than 750 non-governmental organizations, and 10,000 internationally renowned scientists generate environmental conventions, global standards, and scientific knowledge. It has become the voice, and often the instrument, for worldwide action to protect the biodiversity of species, ecosystems, and landscapes. The IUCN also monitors and maintains a database of National Parks and Protected Areas.

A *protected area* was defined at the Fourth World Congress on National Parks and Protected Areas (Caracas,

Bison grazing near hot springs in Yellowstone National Park. Yellowstone is the first and oldest national park in the world and covers 3,470 square miles. The park is famous for its various geysers, hot springs, and other geothermal features and is home to grizzly bears, wolves, and free-ranging herds of bison and elk. It is the core of the Greater Yellowstone Ecosystem, one of the largest intact temperate zone ecosystems remaining on the planet. (© Michael S. Lewis/Corbis.)

Venezuela, 1992) as "land and/or sea especially dedicated to the protection and maintenance of biological diversity, and of natural and associated cultural resources, and managed through legal or other effective means" (World Conservation Monitoring Centre Internet site). Though in the early 2000s there are more than 100,000 protected areas worldwide, not all of them are national parks—defined by the IUCN as a "natural area of land and/or sea, designated to (a) protect the ecological integrity of one or more ecosystems for present and future generations, (b) exclude exploitation or occupation inimical to the purposes of designation of the area and (c) provide a foundation for spiritual, scientific, education, recreational and visitor opportunities, all of which must be environmentally and culturally compatible" (UNEP Protected Areas Programme: "Definition of a Protected Area" Internet site).

In the IUCN category of national parks, there are more than 3,300 worldwide. Other protected areas categories include Strict Nature Reserves/Wilderness Areas:

protected areas managed mainly for science or wilderness protection; Natural Monuments: protected areas managed mainly for conservation of specific natural features; Habitat/Species Management Areas: protected areas managed mainly for conservation through management intervention; Protected Landscape/Seascape: protected areas managed mainly for landscape/seascape conservation and recreation; and Managed Resource Protection Areas: protected areas managed mainly for the sustainable use of natural ecosystems. Though fewer in number than the other protected areas, national parks account for 30 percent of the global network of protected areas, due to the fact that they are often much larger in size.

Ethical Defense of National Park Concept

As the global population passes 6 billion, pressure increases for human occupation of national parks as well as the exploitation of their natural resources. But the arguments for protection remain strong: It is important to preserve the genetic resources and diversity of species found in the world national parks in order to preserve the strains from which our modern and increasingly vulnerable food crops derive. These areas also serve as a repository of edible and medicinal plants and for vital watersheds that provide water to urban and agricultural regions. And they protect cultural, archeological, and natural monuments.

Visits to national parks can also revivify a sensitivity to nature and perhaps even strengthen an environmental ethic that is so essential for human survival and the continuation of all species. Finally the national parks can, as former U.S. National Park Service Director George Hartzog, Jr., so eloquently put it, help us "better understand, or perceive, our place in the universe" (Everhart 1972, foreword by George Hartzog.)

MARILYN BERLIN SNELL

SEE ALSO *Conservation and Preservation; Deforestation and Desertification; Environmental Ethics; Environmental Regulatory Agencies; Nature.*

BIBLIOGRAPHY

Everhart, William C. (1972). *The National Park Service.* New York: Praeger.

Green, Michael J. B., and James Paine. (1997). *State of the Worlds Protected Areas at the End of the Twentieth Century.* (Paper presented at IUCN World Commission on Protected Areas Symposium on "Protected Areas in the 21st Century: From Islands to Networks.") Cambridge, UK: World Conservation Monitoring Centre.

Quigg, Philip W. (1978). *Protecting Natural Areas: An Introduction to the Creation of National Parks and Reserves.* New York: National Audubon Society.

Whitman, Sylvia. (1994). *This Land is Your Land: The American Conservation Movement.* Minneapolis, MN: Lerner.

INTERNET RESOURCES

Pitcaithley, Dwight T. (2001). "Philosophical Underpinnings of the National Park Idea." National Park Service. Available from http://www.cr.nps.gov/history//hisnps/NPSThinking/underpinnings.htm.

World Conservation Monitoring Centre. Protected Areas Programme. Available from http://www.wcmc.org.uk/protected_areas/data/cnppa.html.

World Conservation Monitoring Centre. Protected Areas Programme. "Definition of a Protected Area." Available from http://www.wcmc.org.uk/protected_areas/data/sample/iucn_cat.htm.

NATIONAL SCIENCE FOUNDATION

• • •

The U.S. National Science Foundation (NSF) is a federal independent (non-cabinet) agency, established by the National Science Foundation Act of 1950, as amended, and related legislation, passed by the U.S. Congress and signed into law by the president. Its funds come through appropriations in the federal budget each year. Its budget in fiscal year 2003 was approximately $5.5 billion. These funds go mostly in the form of grants to the nation's colleges and universities for research and educational projects in all the sciences and engineering.

In the fall of 1975, NSF began a program to support research projects and related activities on ethics and science, technology, and society. The program continues in the early twenty-first century and, with continuing attentiveness, it could continue for many more years. This entry highlights some of the adventures in its survival and identifies past and continuing challenges.

Initial Stages (1972–1976)

In the early 1970s, NSF program officers began discussing ideas for research activities that would examine ethical issues associated with new developments in science and technology. Biologists in particular recognized that people would raise questions about the social implications of their research and findings, and that such questions were thus worthy of study. Because NSF supported research and educational projects, support for these activities seemed appropriate. Not all NSF staff agreed that these issues merited NSF consideration or support; some were concerned that such questions did not lend themselves to scientific study; others were concerned that such a program might be too inclined to accentuate the negative. Correspondingly, there was also disagreement about how best to organize such an effort. Should it be a separate program with its own funding authority, or should decisions about ethics projects be left to the other research programs?

In association with the National Endowment for the Humanities (NEH), the NSF organized an advisory committee to consider what should be done. After several years of deliberation and attempts to have existing programs solicit and review proposals, the advisory committee recommended that NSF establish a separate program. Using processes for review similar to other NSF programs, the new program could cooperate with the NEH in considering proposals; some could be funded individually by each agency, and some could be funded jointly.

Middle Years (1977–1985)

What was originally called the Ethical and Human Values in Science and Technology (EHVIST) program made its first awards as a program in fiscal year 1976. NSF and NEH cooperated in support of projects through 1980; in that year, NEH decided to focus on questions about science, engineering, and technology of interest in basic humanities research, and the cooperation ended. While planning for EHVIST was underway, the effort was housed in the office of the NSF director. When the grants program began, the foundation decided to place the new program in the Directorate for Science and Engineering Education, which was hospitable to the idea. An intellectual rationale was that the new program would support research to examine ethical issues in all of the sciences and engineering; thus, it would not be appropriate to house the program in one of the research directorates. At this time, NSF had three research directorates: Physical and Mathematical Sciences and Engineering; Astronomical, Atmospheric, Earth, and Ocean Sciences; and Biological, Behavioral, and Social Sciences.

Shortly after, Congress authorized and appropriated funds for another program at NSF, one to provide scientific assistance to citizens' groups (Hollander 1984). The foundation decided to house both programs in an organizational unit with the Public Understanding of Science (PUS) program (by the early 1990s known as the Informal Science Education Program) and programs involving state and local governments. As other bureaucratic reorganizations developed, the head of this unit made the case that these programs, minus PUS, which was clearly educational, would be better placed in the Directorate for Scientific, Technological and International Affairs (STIA). STIA housed international programs and statistical studies of science and technology and could also include these other special activities. Thus, EHVIST moved to STIA in 1980. It was a fortuitous move, because the administration of President Ronald Reagan zeroed out the budget for the Education Directorate for fiscal year 1981. Had ethics funds still been part of that directorate's budget, the program might have easily vanished and been very difficult to resurrect. As it was, given general budget difficulties, the administrators of the NSF might have decided not to continue the program. Although the administrators did indeed cut the program budget, they listened to numerous voices in the scientific and other scholarly communities and kept the program alive.

At the time the program began, social and intellectual movements in the United States and abroad were focusing on issues of science, technology, and society (Dickson 1984). These movements recognized the need to examine ethical and value dimensions in that interaction. The concerns are international, not just national, although they take distinctive shapes in different parts of the world. Interest among scientists and engineers in these kinds of problems might be said to have developed prominence with the nuclear bomb and the founding of the Pugwash conferences on science and world affairs in 1957. World War II also posed challenges to biologists and physicians, with the atrocities of Nazi scientists in the name of eugenics, and biologists and physicians began to recognize that new developments in genetics would pose increasing ethical questions. Environmental hazards and climate change issues, problems of scientific misconduct, new monitoring technologies—the list of concerns continues to grow. In addition science and technology create ethical opportunities that are worthy of study. They range from uses of forensics in criminal justice to uses of computer technology for disabled populations. The distinctive and increasingly powerful roles of engineering, science, and technology in modern life assure that subject matter for careful research will not be lacking anytime soon.

In these middle years, the program shortened its name slightly, from Ethical and Human Value Implications of Science and Technology (EHVIST) To Ethics and Values in Science and Technology (EVIST). It distributed more than 150 awards, ranging from a high of twenty-three in the year the program began, to a low of eight in 1983, as a result of the diminution of the program budget that year. (In the early twenty-first century, the program averages twenty-five to thirty awards per year; average award amount is $80,000 for an award of twelve to eighteen months.)

The awards covered a wide range of topics. One major area was environmental and hazards issues, which when coupled with agricultural ethics issues, formed a grouping representing about 20 percent of the total awards. These kinds of projects focus on the ethical and value dimensions of interactions of science, technology, and society, and these interactions continue to be a prominent area for program support. For awards made after 1985, this same grouping represents about 25 percent of the total. Areas that began to emerge—such as the use of animals in research, university–industry–government relations, and publication ethics—primarily examined issues in the conduct or practice of science and engineering. Awards in 1978–1979 to investigators Judith P. Swazey, Karen Seashore Lewis, and Melissa S. Anderson, for example, resulted in some of the first and most complete reports on perceptions of misconduct among science and engineering faculty and graduate students. This trend toward awards for such studies has continued and grown as societal concerns about professional accountability have increased. In the early twenty-first century, the program invests considerable resources in both research and educational projects in areas of research ethics. For example, support to the Association for Practical and Professional Ethics in 1999 through 2003 provided training in research ethics for graduate students and postdoctoral fellows in science and engineering. While the number of awards from 1989 to 2001 approximately doubled from the earlier period of 1976 to 1987, the number of awards in research and publication ethics increased eight times.

Principal investigators on ethics projects come from humanities, social science, and natural science and engineering fields. Over the period 1976 to 1987, the split among the three groups was almost even. By the early twenty-first century, a greater proportion of awards was going to social and behavioral scientists. During the period 1976 to 1987 the ratio of male to female investigators was about three to one and improving. This trend continues to be an area of program strength: From 1990 through 2002 the ratio of principal investigators was two males to one female. Support for minority and handicapped investigators in the earlier period was low. It is increasing, especially for Hispanic investigators, and there have been a few notable efforts; for instance, the award to a deaf historian of science, Harry G. Lang, resulted in the 2000 book, *A Phone of Our Own: The Deaf Insurrection against Ma Bell*, which won a number of awards. The program made a grant in 1998 to the American Philosophical Association to sponsor panels at its meetings, as well as to award small grants for investigators to research the implications for diversity of developments in science, engineering, and technology. A book containing some of the presentations and research results has been published (Figueroa and Harding 2003). Despite such examples, much room for improvement in this area remains. Finally, as with other NSF programs, investigators at universities granting postbachelor degrees received and continue to receive the majority of grants.

Years of Trial (1986–1992)

The EVIST program's support for new projects reached a low in 1983 because of the Reagan administration budget cutbacks. It was struggling back up when another blow hit. This time, the attack came from within. NSF Director Eric Bloch was looking for funds to support

more large-scale projects such as engineering research centers, and he concluded that the million or so dollars per year that went to ethics should be shifted to help in this effort. Thus the NSF budget request to Congress for fiscal year 1986 contained no funds for ethics.

When news of this plan filtered out, the program's supporters, particularly grantees, members of the panel that reviewed proposals to EVIST, and officers and staff at the American Association for the Advancement of Science and other professional societies, protested to members of the Congressional committees that oversaw the NSF budget. They were able to make a persuasive enough case that the legislators insisted that NSF maintain support for ethics projects. The foundation heeded this advice, while deciding to manage its support for the activity in a new way—as a foundation-wide responsibility (Hollander 1987).

The Directorate for Biological, Behavioral, and Social Sciences agreed to assume primary responsibility for the program, but funds to support projects would have to come from all the foundation's directorates. Fortunately, directorates at NSF had been multiplying during this time. New ones included the Directorate for Engineering and the Directorate for Computer and Information Science and Engineering. They agreed to participate, effectively increasing the program's budget by 33 percent. The number of new projects being supported rose to twenty-one by 1987.

The rationale for supporting ethics across the foundation may have had both intellectual and control components. On the former, it was supposed to assure the involvement of the scientists and engineers who managed various NSF research programs. On the latter, it could provide oversight of the treatment of science and engineering in the senses of accuracy and circumspection, as these fields perceived those attributes. These goals were, to some extent, met. Any potential for good, however, was more than outweighed by the management problems. For instance, what could be done when more good proposals came in, say, in biology rather than fields in other directorates? Fortunately, the directorate housing the program provided a greater amount of funds so that adjustments could be made. And what could be done when a program officer simply did not want to be bothered, or did not think that such activities should be supported through NSF? Luckily, sympathetic division directors with a few loose dollars could often be found. But the management headaches—for the program manager—were numerous.

Other organizational changes affected the program at this time. It joined forces with NSF's History and

Philosophy of Science (HPS) program; both program directors argued that the two programs, with similar interests in science, technology, and society relations, should be housed in one unit. Management agreed. Separately, the social and behavioral sciences argued for a directorate of their own. This happened in 1992, and the two programs—with HPS now called Science and Technology Studies (STS) and EVIST now called Ethics and Values Studies (EVS)—moved to the new Directorate for Social, Behavioral, and Economic Sciences (SBE).

The increasing complexity of managing a program across all the directorates at NSF became ever more apparent and, finally, upper management agreed that funds for the effort should once again be consolidated. All the directorate heads signed a memorandum to that effect, and the NSF budget to Congress for fiscal year 1994 included funding for an independent ethics program. One unusual component remains: Because the Directorate for Education and Human Resources (EHR) has its own line in the foundation's appropriation, its support for ethics education projects remains separate. It has continued its support, and EHR, Engineering, and SBE are taking the lead in developing a foundation-wide program for 2005, on ethics education for scientists and engineers, especially science and engineering students.

In the midst and irrespective of this bureaucratic turmoil, the academic interest in the ethical and value dimensions of science and technology continued to develop. Growing numbers of journals, programs at colleges and universities, and professional associations indicated increased institutionalization of the field. New ethics centers and courses in issues of ethics for the professions continued to appear at the nation's colleges and universities. A 1990 article by Nicholas H. Steneck and Rachelle D. Hollander reviewed the EVS program. One major research areas highlighted in the report was engineering ethics. From 1999 to 2003 several of the nation's engineering colleges established chairs in engineering ethics. The Association for Practical and Professional Ethics (APPE) was founded in 1991. Besides its individual members, who represent many different disciplines and fields, APPE has more than 100 institutional members. In 1995 the journal *Science and Engineering Ethics* was established, with coeditors in the United States and the United Kingdom. An even more recent example, *Ethics and Information Technology* was founded in 1999; its editors are from the United States, the United Kingdom, and Europe. The affiliation of the two NSF programs—EVS and STS—reflects broader synergies as departments of science and technology studies are becoming more numerous at colleges and universities in the United States and elsewhere.

Years of Consolidation and Challenge (1993–)

Basically, the characterization of research topics and methods in the Hollander and Steneck article from 1990 remain appropriate for the program. Many projects fall into more than one category. Research methods remain diverse. Approaches involve individual investigations as well as collaborative research and workshops. Research includes analytical or conceptual philosophical analysis, case study or issue-oriented research, empirical research in the social and behavioral sciences, and science and technology assessment. A research approach of increasing importance is that of science and technology studies. The program supports numerous educational activities and has helped other NSF programs include ethics education in their activities.

Within NSF, EVS began a successful effort to incorporate ethics activities into the NSF Research Experiences for Undergraduates (REU) Sites projects in the early 1990s. All the research directorates support these summer programs, which bring small groups of undergraduate science and engineering majors to campuses, where they participate with faculty in research projects. The sites projects encourage promising undergraduates to continue their science or engineering education, expose them to interesting research, and promote diversity among undergraduates and in the science and engineering professions. The ethics component began with a successful pilot effort in chemistry in 1992. By the next year, the other NSF directorates had signed on and the next REU program announcement indicated that projects were eligible for small amounts of funding specifically for ethics education as part of their summer programs. Each year since the beginning of the new century, more than twenty-five projects receive ethics funds. The field with the most REU projects with an ethics component funded through EVS is biology, but all of the directorates participate, and the Engineering directorate funds many of these projects on its own.

In 1997 NSF began a foundation-wide program called Integrative Graduate Education and Research Training (IGERT). This program supports interdisciplinary graduate education projects around a research theme. These large awards, for amounts in excess of $2.5 million, extend over five years. EVS succeeded in incorporating ethics activities into IGERT. The program requires that these projects include ethics in their curricula; the announcement for IGERT states that "The graduate experience should ... equip students to understand and integrate scientific, technical, business, social, and ethical issues to confront the challenging problems of the future" and that IGERT projects must include specifically "integrated instruction in ethics and the responsible conduct of research" (NSF, "IGERT Program" Internet site). EVS is undertaking a small-scale initial evaluation of these efforts.

In 1995 NSF management asked EVS to merge with a small NSF program called Research on Science and Technology (RST). RST supports projects that examine the role of public investments in science, engineering, and technology. After consultation with its panel and the broader communities of EVS and RST investigators, the program agreed. With neither group wishing to lose its name, both were placed under the more general rubric, Societal Dimensions of Engineering, Science, and Technology (SDEST). In the late 1990s and into the twenty-first century, the SDEST/EVS-RST budget stabilized at about $2.5 million per year, augmented by another $500,000 in assistance from other programs for ethics projects. Given NSF emphasis on foundation-wide priorities, and general constraints in the federal discretionary budget, the program is unlikely to see much direct budget expansion. One way to overcome this problem is to infuse ethics research and educational activities into other interdisciplinary research areas now getting NSF attention, such as information technology research and nanotechnology. While this is not easy, it is possible, and seems to be increasing.

Discussion was underway in Fall 2004 among the Science and Technology Studies program and the SDEST/EVS-RST program to consolidate their activities under the rubric Science and Society. The newly inclusive program would have four components:

- Ethics and Values in Science, Engineering, and Technology

- History and Philosophy of Science, Engineering, and Technology

- Social Studies of Science, Engineering, and Technology

- Studies of Policy, Science, Engineering, and Technology

The change in names is intended to assist applicants in determining where to apply. It may encourage further development of connections with the sciences and engineering programs, which are increasingly aware of the need to address social shaping of science and technology, and its implications. This increased recognition can be seen more broadly in federal funding for research on ethics and the human genome and the call for similar funding for ethics and nanotechnology.

EVS research faces problems similar to those facing other interdisciplinary or transdisciplinary areas in which NSF wants to encourage research: fostering interdisciplinary communication, defining researchable issues, and finding outlets where results will be recognized as valuable. Identifying the need for recognition captures an aspect of the difficulty. EVS research has distinctive frameworks, and investigators cite the prior literature. It is difficult, however, both to train new EVS researchers and to make the results visible for new and established researchers in the disciplines and fields that EVS researchers study.

Progress is being made. The wide variety of educational activities is making EVS results more accessible in the research communities to which they are relevant. All fields of science and engineering recognize the relevance of issues of ethics as they related to the practice of science and engineering. Most recognize the relevance of issues of ethics in connection with interactions among science, engineering, and society. This is a significant change from the situation in the early 1970s, when the thought of an ethics program at NSF was barely a gleam in one or two people's eyes.

RACHELLE D. HOLLANDER

SEE ALSO *Engineering Ethics; Nanoethics; National Academies; Science Policy.*

BIBLIOGRAPHY

Dickson, David. (1984). "Science and Society: Public Participation vs. Democratic Control." In his *The New Politics of Science.* New York: Pantheon. Provides an interesting view of the beginnings of the EHVIST program.

Figueroa, Robert, and Sandra Harding. (2003). *Science and Other Cultures: Diversity in the Philosophy of Science and Technology.* New York: Routledge. Collection of essays about the influences of science and technology on and from cultures, persons, and traditions.

Hollander, Rachelle D. (1984). "Institutionalizing Public Service Science." In *Citizen Participation in Science Policy,* ed. James C. Petersen. Amherst: University of Massachusetts Press. Includes a history of the rise and fall of the NSF program to provide scientific assistance to citizens' groups.

Hollander, Rachelle D. (1987). "In a New Mode: Ethics and Values Studies at the NSF." *Science, Technology, and Human Values* 12(2): 59–61. Describes the decision to manage EVS as a "distributed" program.

Hollander, Rachelle D., and Nicholas H. Steneck. (1990). "Science- and Engineering-Related Ethics and Values Studies: Characteristics of an Emerging Field of Research." *Science, Technology, and Human Values* 15(1): 84–104. Comprehensive review of the awards made in the programs first decade.

INTERNET RESOURCES

National Science Foundation (NSF). "Integrative Graduate Education and Research Traineeship (IGERT) Program: Program Solicitation." Available from http://www.nsf.gov/pubs/2000/nsf0078/nsf0078.htm. Information on program goals and application procedures.

National Science Foundation (NSF). "Research Experiences for Undergraduates (REU): Program Announcement." Available from http://www.nsf.gov/pubsys/ods/getpub.cfm?nsf02136. Information about program goals and application procedures.

National Science Foundation (NSF). "Societal Dimensions of Engineering, Science, and Technology (SDEST) Program." Available from http://www.nsf.gov/sbe/ses/sdest/. Includes examples of program awards via link to award abstracts.

NATIONAL SCIENCE FOUNDATION: SECOND MERIT REVIEW CRITERION

• • •

In the early twenty-first century, science finds itself caught in a dilemma that is arguably of its own making: Its very success in terms of understanding and controlling nature means that it has given birth to powers that transcend the traditional boundaries between science and society. Rather than being viewed as essentially neutral in terms of values, society increasingly views scientific knowledge as leading to various types of winners and losers. The review criteria for National Science Foundation proposals offer an instructive case study of this increasingly prominent dynamic.

Background

Established in 1950, the National Science Foundation (NSF) is the only federal agency dedicated to the support of education and basic research across all scientific and engineering disciplines, except for the biomedical sciences (which are handled by the National Institutes of Health). Although no authoritative definition exits, it is generally agreed that basic scientific research is oriented chiefly toward the discovery and creation of new knowledge, without regard for its eventual employment.

In 1993 Congress passed the Government Performance Results Act (GPRA). The purpose of GPRA was to increase the focus of federal agencies on improving and measuring "results," which would in turn provide congressional decision makers with the data they require to assess the effectiveness and efficiency of federally funded programs. In effect, GPRA sent the message that federal funding is contingent on attaining and demonstrating results. Partly in response to such demands for demonstrable results, in 1995 the NSF adopted a new strategic plan: *NSF in a Changing World* (NSF 95-24). NSF's new strategic plan included among the long-term goals of the foundation the promotion of the discovery of new "knowledge in service to society."

In 1996 the National Science Board (NSB) established the NSB-NSF Task Force on Merit Review to examine and evaluate NSF's generic merit review criteria, which had been in effect since 1981, in light of the new strategic plan. In its "Discussion Report" (NSB/MR-96-15) the task force recommended replacing previous review criteria with two simple questions: (1) What is the intellectual merit of the proposed activity? (2) What are the broader impacts of the proposed activity? The simplification was proposed to help connect NSF investments to societal value while preserving an ability to select proposals on the basis of scientific excellence. Such criteria were more clearly related to the goals and strategies of *NSF in a Changing World*. NSF published the recommendations of the task force on the web, through press releases, and through direct contact with universities and professional associations, and received around 300 responses from the scientific and engineering community.

In light of these responses, in 1997 the task force published its "Final Recommendations" (NSB/MR-97-05). The responses raised several concerns about the new criteria, including what the task force termed the issue of "weighting" the criteria: Criterion 1 was perceived by respondents as more important than 2, or criterion 2 was perceived as irrelevant, ambiguous, or poorly worded. Moreover, respondents expressed concern that for much of basic research it is impossible to make meaningful statements about the potential usefulness of the research. Ultimately, however, the task force recommended that the new criteria be adopted. Later in 1997, NSF issued Important Notice No. 121, which announced NSB approval of the new merit review criteria, effective October 1.

The NAPA Report

In 1998, and again in 1999, Congress directed NSF to contract with the National Academy of Public Administration (NAPA) to review the effects of the changes in NSF's merit review criteria. NAPA is an independent, nonpartisan organization chartered by Congress to help federal, state, and local governments improve their effectiveness, efficiency, and accountability. In 2000 NSF commissioned the NAPA study.

The NAPA study reviewed relevant legislation, reports by external review committees, interviews with NSF personnel, and interviews with members of the scientific and engineering community. In addition, the NAPA study analyzed sample projects funded under both the old and the new criteria, as well as the intentions of those reviewing proposals using the new criteria. Published in February 2001, the NAPA report provides a history of the development of NSF's new merit review criteria, compares the 1997 criteria to the 1981 criteria, and details many of the challenges faced by the merit review process during the period from 1997 to 2000. The NAPA report offers several recommendations to help NSF improve the merit review process, among which is a recommendation to address the "philosophical issues" raised by the new criteria, in particular criterion 2.

The latter recommendation was based in part on its observation of the diverse interpretations of and reactions to the new merit review criteria among members of the scientific and engineering community. Although the NAPA report fails to delineate explicitly what it considered to be the philosophical issues, it nevertheless provides an excellent source from which those issues can be gleaned. Such issues include:

- whether criterion 2 is inconsistent with criterion 1
- whether criterion 1 is more important than criterion 2
- whether criterion 2 is in need of conceptual clarification
- whether interpretations of criterion 2 are discipline-dependent
- whether reactions to criterion 2 rely on one's conception of scientific inquiry.

These issues are, of course, interrelated: A physicist committed to a strict division between basic and applied scientific research might interpret the criteria as inconsistent, whereas a geologist whose research in plate tectonics might one day lead to predictive capabilities might not; said geologist might nonetheless view criterion 1 as significantly more important than criterion 2.

Moreover, consideration of such issues also raises philosophical issues in the realm of science policy. Is NSF moving away from its emphasis on basic research? If so, is NSF offering a new conception of scientific inquiry? If so, what is this new conception? Is this new conception coherent? If not, should NSF change its

merit review criteria? Should criterion 2 be abandoned? If so, must NSF's strategic plan be reconceptualized? What impact would such a reconceptualization have on NSF's compliance with GPRA? Should NSF still receive federal funding? If so, how much and for what?

In attempting to incorporate intellectual merit and broader societal impacts more fully, NSF's 1997 merit review criteria raise a host of philosophical issues. Demands for federal agencies to show results in order to receive funding show no signs of vanishing. It remains to be seen how such issues will be addressed.

J. BRITT HOLBROOK

SEE ALSO *National Science Foundation*.

BIBLIOGRAPHY

INTERNET RESOURCE

National Academy of Public Administration (NAPA). "A Study of the National Science Foundation's Criteria for Project Selection." A report by a panel of the National Academy of Public Administration for the National Science Foundation, issued February 2001. Available from http://www.napawash.org/publications.html.

NATIONAL SOCIETY FOR PROFESSIONAL ENGINEERS

SEE *Professional Engineering Organizations*.

NATURAL DISASTERS

SEE *Building Destruction and Collapse*.

NATURAL LAW

• • •

Central to natural law theories of morality is the idea that there are guiding principles for human conduct higher than those of personal self-interest, particular social custom, or positive governmental statute. Such a higher law is characteristically thought to be objectively true, accessible to reason, and universally obligatory. This law is *natural* in the sense that the goods it defines are logically related to the rational nature of human beings. Though many advocates are theists, typically from the Catholic tradition, who ground the content of natural law in divine will, the tradition includes non-theistic theorists as well.

The norms of natural law must be distinguished from *laws of nature*, which are purely descriptive propositions identifying causal relations between material entities, events, or phenomena. Yet because of its appeal to nature, the conceptualization and understanding of which has been deeply affected by modern natural science, and subject to major technological transformation, natural law has been both challenged by and sometimes taken as a challenge to science and technology.

Types of Natural Law Theories

There are two kinds of natural law theories: natural law theories of legality and natural law theories of morality. Natural law theories of legality argue there are necessary moral constraints on the content of law. Natural law theories of morality are concerned with the character, grounds, and principles of morality. Although many who subscribe to natural law ethics also subscribe to natural law jurisprudence, the two theories are logically independent. Someone who accepts the theory of law may not accept the theory of morality, and a natural law moral theorist could consistently hold that, unlike morality, law is essentially conventional in character.

Although ethicists disagree about how best to characterize natural law theories of morality, nearly every natural law ethicist accepts the following four theses: (a) moral principles are either objectively true or objectively false; (b) the truth value of a moral principle is determined, in part, by whether it accurately reflects the facts of human nature or can, in some sense, be derived from the facts of human nature; (c) at least one moral principle is objectively true; and (d) the principles of morality can be discerned by reason. Many, but not all (e.g., Moore 1996), natural law ethicists are theists who relate the content of natural law to God as the creator of human nature. All natural law theories of morality thus include meta-ethical claims (theses a, b, c), normatively ethical claims (thesis b), and epistemic claims (thesis d).

Substantive natural law theorists are generally concerned with identifying the natural goods and principles that should guide rational human behavior. At its highest level of abstraction, natural law simply requires persons to pursue what is good and avoid what is bad. But a full understanding of obligations requires identifying what is good and bad in relation to human nature. Such goods are typically argued to include the following: spirituality, life, health, inner peace, knowledge, friendship, the *marital good*, aesthetic experience, play, pleasure,

intellectual creativity, and justice. Further, because human beings may respond in problematic ways to what is good, natural law ethicists also often distinguish between defective and non-defective responses; many theorists, for example, identify homosexual relations as a defective response to the marital good. Taken together, a catalogue of natural goods and a comprehensive account of what distinguishes defective from authentic responses to such goods will fully define the content of the natural law: Human beings are obligated to pursue such goods in non-defective ways. Natural law ethicists commonly believe that such pursuit will culminate in the development of virtuous character traits. As it relates to science and technology, natural law theory would evaluate science and technology according to whether they respond in an authentic way towards the basic natural goods.

Though early natural law moral theorists understood laws of nature and laws of morality as being related, modern theorists distinguish the two. Laws of nature are both descriptive and empirical in character, stating mechanistically causal regularities between various material entities or events. In contrast laws of morality are normative in character and seek to guide the behavior of persons who can freely choose to violate such laws. While natural law theorists are likely to accept that laws of nature and laws of morality ultimately both reflect the true nature of things, natural law theories are properly concerned only with explicating the norms, laws, principles, and rules that should constrain human behavior.

Some critics have argued that natural law theory cannot consistently posit a normative teleology for humans without positing a normative teleology for all other entities. On this line of reasoning, natural law theorists cannot without contradiction (a) derive both the laws of nature and the laws of morality from the natural law but (b) hold that the laws of nature are descriptive while the laws of morality are normative. If humans are subject to a normative teleology, then all entities must be.

The natural law theorist can respond in the following way. Whether or not any particular entity is subject to a normative teleology of some kind is determined by the kinds of property it instantiates. Human beings are governed by a normative teleology that posits moral standards they are obliged to satisfy because humans are moral agents in virtue of having the properties of rationality and free will. Other entities lack these properties and hence are not subject to such standards that prescribe behavior; it makes little sense to think that, in the literal normative sense, a quark ought to behave in this or that particular way.

Other living things are, of course, fairly characterized as having interests. For example, cows are sentient and hence have an interest in being free from suffering. These interests are not implausibly characterized as "goods" towards which the behavior of non-rational living beings is typically oriented. However, it is clear that goods of this kind do not define standards that *prescribe* behaviors for those other living things. Although humans, qua rational moral agents, might be obligated by a law that requires a respect of the interests of other living beings, those living beings could not be *obligated* to do anything.

By means of such reasoning, the natural law theorist attempts to reconcile the differences between rational agents, non-rational living beings, and other material beings while the claim that the movements and behaviors of all existing entities are defined and governed by the natural law. Moreover, such arguments allow natural law theory to highlight the importance of both scientific and ethical inquiry: Scientific inquiry allows humans to determine the interests of other living things, while ethical inquiry allows humans to determine the extent to which they are obligated to respect and promote those interests.

Historical Overview

Although Aristotle (384–322 B.C.E.) is frequently cited as the first natural law theorist because of his view that human behavior should be directed toward the natural function of living well or flourishing, the Stoics subscribed to a greater number of the distinguishing tenets of natural law theory. According to the Stoics, the cosmos alone is complete and hence ordered and good As rational creatures, human beings are obligated to partake of this good by deploying reason to grasp the order and goodness of the universe. Those who succeed in doing so and in living their lives in ways that cohere with these qualities of the universe cosmos will achieve happiness and fulfill their function of living well. Notable Stoics include Zeno (336–264 B.C.E.), Cleanthes (331–232 B.C.E.), Chrysippus (280–206 B.C.E.), Panetius (185–110 B.C.E.), Posidonius (135–51 B.C.E.), Epictetus (55–135 C.E.), and Marcus Aurelius (121–180 C.E.).

The most influential of Stoics was Marcus Tullius Cicero (106–43 B.C.E.), whose definition of law deeply influenced subsequent natural law thinkers. In Cicero's words, "Law is the highest reason, implanted in Nature, which commands what ought to be done and forbids the opposite" (*De Legibus*, I. 18). Implicit in this definition

are most of the core tenets of natural law: Law is defined by nature, has highest authority, is accessible to reason, and directs rational beings toward what ought to be done (what is good). Like Aristotle, Cicero believed that human beings have a function, built into human nature, and that achieving this function produces true happiness and virtue. Unlike Aristotle, Cicero explicitly attributes natural law to a divine influence in human affairs.

Historically the most influential of all natural law moral theorists is undoubtedly Thomas Aquinas (1225–1274). Like many twelfth- and thirteenth-century philosophers, Thomas worked to bridge the core elements of Christian theology and Aristotelian philosophy.

Thomas saw the universe as the created material embodiment of God's perfect rationality and distinguished four types of law: eternal, natural, divine, human. Determined by divine will, the eternal law consists of the set of timeless, objective truths that govern the movement of all things in the universe, including non-human things, and includes what science calls the laws of nature. Eternal law is thus similar to what science calls the laws of nature. Natural law is a subset of eternal law that applies to the behavior of human beings. Divine law consists of the subset of eternal law pertaining to the ultimate fate of human beings following divine judgment, and is found in revelation. Human law consists of those norms that have a human source and are consistent with natural law.

Because the first precept of natural law requires "that good is to be done and pursued, and evil is to be avoided" (*Summa Theologica* I–II, Q.94, a.2), Thomas must give an account of the relevant goods. Accordingly he distinguishes three kinds of good: (a) those goods that humans share with all other entities, such as the inclination to preserve their being in accordance with their nature (b) more specific goods that humans share with other animals, such as the desire to mate; and (c) goods that are valued because of the human capacity for rationality, such as a desire to live in society and to pursue knowledge. These latter goods, on Thomas's view, include moral goods, such as honesty, integrity, and more. The natural law, then, consists in principles that direct human beings toward the pursuit of those goods that are distinctly human and hence define standards of human virtue.

The distinctly modern period in natural law history began with Hugo Grotius (1583–1645) and his famous argument that, contra Thomas, the content of natural law does not depend on God's existence. A Christian, Grotius nonetheless took the position that natural law reflects goods that are valuable independent of God's will. As Plato might express the point, it is not the case that natural law is good because God chooses it; rather, God chooses natural law because it is good. Because it is the value of these goods that explain God's choosing them (and not the other way around), God could not have changed the content of the natural law

Grotius rejected the view that the binding force of natural law depends on God's existence or on the threat of a divine sanction. Because the content of natural law is grounded in timeless principles of reason rather than divine volition and because human beings have a rational nature, natural law binds humans because its content is rational and not because it is backed by a divine sanction. Grotius subsequently developed a social contract theory of state legitimacy that was grounded in his views about natural law. Though subsequent social contract theorists were influenced by Grotius, some rejected his views about the foundations of natural law. John Locke (1632–1704), for instance, grounded his social contract theory in the idea that natural law governs life in the state of nature, but argued that its content is grounded in divine will.

Contemporary Natural Law Theory

Natural law theorizing is currently enjoying a revival due primarily to the work of various Catholic thinkers, including Germaine Grisez, John Finnis, and Robert George. Finnis develops a comprehensive theory of natural law that begins with an analysis of the concept of law. Finnis conceives of natural law as explicating the basic principles of what he calls *practical reasonableness*. He grounds an identification of these basic principles, which express fundamental human goods, partly on empirical observations of what is universally valued. For example, he notes that all human societies show a concern for the protection of human life, restrict sexual activity, display a concern for truth, know friendship, have some conception of property, and value recreation (Finnis 1980). These goods are protected by principles.

Natural law theory should not, however, be equated with Catholicism. First, many other religious traditions incorporate ideas that figure prominently in natural law theory. C.S. Lewis, for example, has pointed to various elements in the Dao that are suggestive of natural-law commitments. Some Buddhists see a natural teleology in all existing beings and sometimes describe "dharma" as being like the natural law, which is discovered by means of introspective meditation. Second, while many of the most influential contemporary natural law theorists are catholic, not all are. For example,

Leo Strauss (1937–1973) is famous for his disdain for modern philosophical and political theorizing, as well as for his views that (a) life should be led in accordance with the natural order of humanity's being and (b) theorizing of all kinds should be subordinate to theology.

Much late-twentieth century work in natural law theory applies the principles of natural law to issues of sexuality, such as abortion, contraception, and homosexuality. The intrinsic value of sexuality (the marital good) consists in its capacity to create "a two-in-one-flesh communion of persons" that constitutes two persons as "becoming ... one organism" (George 1999, p. 168). Because the unitive capacity of sexual activity is grounded in its reproductive function, sexual intercourse is legitimate only if performed by a man and a woman in a lawful marriage without contraceptives. As is readily evident, natural law theorizing on sexual morality tends to reflect the substantive Catholic doctrines to which its chief proponents subscribe.

Natural Law Assessments of Technology: General Considerations

It is sometimes thought that natural law theories imply that any technology is presumptively problematic. On this line of reasoning, natural law theories equate *good* with *natural* and *bad* with *unnatural*. Because, by definition, human technologies are artifactual and hence not natural (that is, unnatural), it follows that any human technology and its intended uses should be presumed morally problematic until an adequate moral justification for it can be given.

This reasoning misrepresents the natural law theory account of the good. While natural law theory holds that the good is defined by human nature, this does not imply—or even suggest—that artifacts are necessarily unnatural in any relevant sense. There is nothing in any plausible account of human nature that would justify believing that the development and use of artifacts is, as a matter of principle, contrary to human nature. This would imply, absurdly, that the use of food utensils is contrary to human nature.

Indeed, if anything, most mainstream natural law theories would suggest that the intended uses of technology should be presumed good until shown to be morally problematic. The moral evaluation of any particular technology will require a nuanced analysis of two issues: (a) whether the intended use of a technology promotes a fundamental moral good; and (b) whether the intended use of a technology responds in a non-defective way to some fundamental moral good. Just as natural law jurisprudence subjects positive law to assessment by a higher law, so natural law moral theory of technology would assess technology by a higher law. But just as natural law ethics evaluates positive law according to whether its content conforms to a higher law, so natural law ethics evaluates technology according to whether particular uses conform to a higher law. And just as natural law ethics begins with the rebuttable presumption that positive law is legitimate, so too it begins with the rebuttable presumption that technology is legitimate.

But most, if not all, technological advances satisfy. Serious technological research is generally focused on developing technologies designed for uses that further important human interests such as life, health, play, and other goods. In free economies, the market incentives are simply insufficient to support technological research that is not connected with basic human goods. It is true, of course, that any particular technology may respond defectively to one of the basic goods. Arguably, violent video games are a defective response to the basic human good of recreation. But in a market economy, private resources will typically be directed at producing technologies that respond in some direct (and marketable) way to the basic human goods. Accordingly in the absence of some obvious problem with a particular technology (or intended use), it may reasonably be characterized as presumptively good.

This, of course, is not to deny either that technologies can be misused or that the intended uses or functions of some technologies are themselves morally problematic. It is clear, for example, that any weapons technology can be used for wrongful purposes. Indeed one may plausibly argue that the very function of any weapons technology is morally problematic; while possession of a weapons technology may be used to deter violence, its characteristic function is to inflict injury on other living beings—a function that is presumptively problematic. Nuclear weapons and other weapons of mass destruction are especially problematic in this regard.

The point is that, as an empirical matter, most (as opposed to all) technologies are intended to be used—and are characteristically used—in ways that promote some important human interest. Thus a complete natural-law evaluation of any particular technology will usually turn on whether it satisfies (b) above (i.e., responds in a non-defective way to the relevant goods). If it responds defectively to the good, then it must be rejected as morally problematic. As the Pontifical Academy for Life explains, "[i]t is never licit to do evil intentionally in order to achieve ends that are good in themselves" (Pontifical Academy for Life, Art. 9).

In any interesting case, however, this issue will be far more difficult than the issue of whether a particular technology promotes some basic good. Consider the difficulties in giving a natural law analysis of intellectual property and digital file-sharing technologies. On the one hand, copyright protection promotes a variety of interests that are plausibly characterized as basic moral goods. Copyright protection promotes intellectual innovation and knowledge by providing a material incentive to create content. Further, by protecting inventors' material interests in their creations, copyright protection promotes physical health and well being; after all, property interests are valuable as a means to these more important ends. On the other hand, copyright protection restricts the free flow of useful information—which can be consumed by all persons at once without reducing its supply. As is readily evident, the issue of whether this feature of information warrants characterizing copyright protection as a defective response to the basic moral goods that it intends to promote is exceptionally difficult.

It is worth noting that such epistemic difficulties lead some proponents to believe that while natural law theory may guide behavior in most instances, it is indeterminate with respect to some moral issues. Natural law theory is not, on this view, intended to provide some sort of determinate decision procedure for resolving ethical issues. Rather it provides a catalogue of general considerations that point the way toward the good life.

Biotechnology

Although one would expect natural law theorists to devote considerable energy to assessing new technologies, they tend to focus on issues of sexual and reproductive morality. Because many natural law theorists belong to the Catholic Church, which has made propagation of its views on such matters a high priority, it is not surprising that so much energy is devoted to these issues. But given the importance of the various moral issues arising in connection with many new technologies, it is regrettable that natural law literature on these emerging technologies is so comparatively thin.

Most natural law research on technology has focused on biotechnology. As a general matter, natural law theorists are unanimous in affirming the need for biotechnological research to promote the vital natural goods of human health and human knowledge, but emphasize the need to focus on technologies that produce those goods in non-defective ways. Only research that responds nondefectively to the goods of knowledge and health is encouraged as morally legitimate under the natural law.

One important issue in determining whether a particular biotechnological inquiry or application responds nondefectively to some good is whether it respects the integrity of the human person. The use of human embryos in research or in a technology designed to treat a disease is condemned as failing to recognize the integrity of such lives. According to the Pontifical Academy for Life, "The attitude some adopt concerning the legitimacy of sacrificing the (physical and genetic) integrity of human beings at the embryonic stage in order to destroy them ... to benefit other human individuals is ... totally unacceptable" (Pontifical Academy for Life, Art. 9). Such research and applications are problematic because they treat intrinsically valuable human beings as mere receptacles of instrumental value, namely, as objects to be used to benefit other human beings.

Natural law theorists also converge in condemning technologies that assist a terminally ill person in committing suicide on the ground that such technologies fail to respect the moral integrity of the person. Although suicide itself should not be punished, the use of these technologies to assist a suicide should. As David Novak puts the point, "because suicide itself is prohibited, those assisting in a suicide, not being its victim, are to be punished on the grounds that *there is no agency for sin*" (Forte 1998, p. 20). Though a patient might consent to physician-assisted suicide, such consent is not morally effective because one cannot waive the integrity of one's person.

Natural law theorists criticize efforts to develop technologies that can be used to clone human beings or to select for various genetic characteristics in one's offspring for somewhat different reasons. Such technologies may be defective responses to natural goods because they fail to respect the integrity of human persons, but they are also defective for other reasons. For example, one theorist worries that "cloning and asexual reproduction may contribute to the erosion of our sense of the gift of procreation, of our role as parents, ... and of our understanding of sexual intercourse and love" (deBlois 1994, p. 213). While understanding the truth about the human genome, technologies that lend themselves to such applications are unacceptable: "Cloning with a view to the reproduction of human beings is a practice contrary to human dignity and should not be allowed" (Holy See).

Natural Law, Technology, and the Environment

Impact on the natural environment is another relevant issue in assessing a technology under natural law theory. Many technologies obviously affect the environment in deleterious ways that are potentially significant from an ethical point of view. The contribution of any particular technology to pollution, species extinction, and depletion of natural resources is important in evaluating the acceptability of that technology under natural law theories, at the very least, because all these effects may negatively impact the pursuit of basic human goods that are at least as important as the interests the technology seeks to advance.

Central to a natural-law evaluation of the environmental impacts of technology, however, is the issue of whether the theory posits a direct obligation on humanity's part to respect and promote the interests of other non-human natural beings that is grounded in the idea that such beings are deserving of respect for their own sakes. A natural law theory that posits a direct obligation to this effect assigns some measure of moral standing to non-human beings whose interests must then be taken into moral consideration. A natural law theory that does not posit such a direct obligation assigns no measure of moral standing to non-human beings. On this latter view, the only obligations to respect and promote nature are owed to other human beings and are grounded in nature's value in promoting human flourishing.

Natural law theories differ in their evaluation of a technology's effects on the environment depending on whether they assign moral standing to other beings. An anthropocentric theory that assigns moral standing to only human beings is, other things being equal, less likely to reject a technology on the strength of its environmental impacts than either an animocentric theory that assigns standing to sentient non-human animals or a biocentric theory that assigns standing to all living beings. The smaller the moral community, the fewer beings whose interests or goods count in evaluating any particular behavior. Still, it is important to note that more expansive versions of natural law theory have sufficient resources to ground a very strong ethical commitment to the environment.

KENNETH EINAR HIMMA

SEE ALSO *Aristotle and Aristotelianism; Biotech Ethics; Christian Perspectives; Human Nature; Libertarianism; Science, Technology, and Law; Thomas Aquinas.*

BIBLIOGRAPHY

deBlois, Jean; Patrick Norris; and Kevin O'Rourke. (1994). *A Primer for Health Care Ethics: Essays for a Pluralistic Society*. Washington, DC: Georgetown University Press.

Finnis, John. (1980). *Natural Law and Natural Rights*. Oxford: Oxford University Press.

Forte, David, ed. (1998). *Natural Law and Contemporary Public Policy*. Washington, DC: Georgetown University Press.

George, Robert. (1999). *In Defense of Natural Law*. Oxford: Clarendon Press.

Grisez, Germaine. (1983). *The Way of the Lord Jesus*, Vol. 1: *Christian Moral Principles*. Chicago: Franciscan Herald Press.

Grotius, Hugo. (1964). *The Rights of War and Peace*, trans. F. W. Kelley, et. al. New York: Oceana.

Haakonssen, Knud. (1996). *Natural Law and Moral Philosophy: From Grotius to the Scottish Enlightenment*. Cambridge, England: Cambridge University Press.

Moore, Michael. (1996). "Good without God." In *Natural Law, Liberalism, and Morality*, ed. Robert George. Oxford: Oxford University Press.

INTERNET RESOURCES

Pontifical Academy for Life. IX General Assembly. "Concluding Communique on *The Ethics of Biomedical Research for a Christian Vision*" Vatican official site. Available from http://www.vatican.va/roman_curia/pontifical_academies/acdlife/documents/rc_pont-acd_life_doc_20030226_ix-gen-assembly-final_en.htm.

Holy See. "Observations on the Universal Declaration on the Human Genome and Human Rights." Vatican official site. Available from http://www.vatican.va/roman_curia/pontifical_academies/acdlife/documents/rc_pa_acdlife_doc_08111998_genoma_en.html.

NATURE

• • •

Thinking about science, technology, and ethics easily raises questions about nature. Science considers whether and how nature can be understood. Technology considers whether and how humans can control nature. Ethics considers whether and how science and technology can be guided by standards of right and wrong that might be rooted in nature. One of the most common objections to science and technology is to argue that they go against nature, just as one of the strongest defenses is to present them as eminently natural.

Nature and Reason

The English word *nature* is derived from the Latin word *natura*, which is related to the verb *nasci* (to be born) and the noun *natus* (birth). The Latin *natura* corresponds to the Greek *phusis*, of which the root is *phu* (growing, becoming, being). Nature is the original birth or coming into being of something. More generally, nature is concerned with the "first things," the origins of things.

The idea of nature seems to have been discovered or invented first by ancient Greek philosophers and scientists. Aristotle (384–322 B.C.E.) identified the "first philosophers" as "humans who spoke about nature" in looking for the "principles" or "beginnings" of all things (*Metaphysics* 983b5–19). These Greek philosophers thought of *phusis* as the beginning or coming to be of something. But more often *phusis* meant the sort or kind or description of something—the essential character of a thing or a class of things. The nature of something could be what it is at birth or what it grows into at maturity, what it is at its beginning or at its end. "Nature is an end," Aristotle explained, "because whatever anything is like when its growth is completed, that we call the nature of each thing" (*Politics* 1252b33–35). These Greek philosophers began by asking about the nature of each thing, what each thing is like. And then they asked what everything was like. Thus, the Greek philosopher Parmenides (c. 515 B.C.E.) could write a book with the title *On Nature*, which considered the "nature" of everything.

When nature becomes everything, it is impossible to define. But generally nature is a term of distinction, and so its meaning may be clarified by asking what is its opposite. In ancient Greece, "nature" (*phusis*) was most commonly set in opposition to "custom" (*nomos*) or "art" (*techne*). Custom and art are human products. By contrast, nature is what arises on its own without human interference. Nature is what is not customary or artificial.

Philosophy or science arose in ancient Greece when a few thinkers noticed that customary practices and beliefs varied across human societies. This led them to doubt the authority of human customs and to look for what was universally true by nature as opposed to what was believed to be true by human custom. Whatever arises by human custom or artfulness is changeable, but what arises by nature, it was argued, is unchangeable and thus more real than the perishable products of human activity.

The ultimate justification for customary practices and beliefs is the claim that they are divine, that they originated from the commands of gods or god-like ancestors. But when Greek philosophers and scientists explained the "first things" as natural rather than customary or artificial, this suggested that even the gods might be artificial or human-made, as being products of storytelling. The natural was opposed to the divine or the supernatural. Consequently, as indicated by the Athenian trial and execution of Socrates (469–399 B.C.E.), who was charged with impiety, the philosophic discovery of nature implied a questioning of the gods.

Revelation and Nature

The religious believer could respond by denying the idea of nature as the autonomous order of the world and affirming that whatever exists is what it is only through the creative activity of the gods or God. The Hebrew scriptures contain no word that corresponds to *nature*. In the Greek scriptures, the word *phusis* does not occur except in the letters of Paul, who was influenced by Greek philosophy.

Yet the medieval scholastic tradition of Biblical theology adopted the Greek idea of nature insofar as God was understood to be the creator of nature. Indeed, this assumption allowed Thomas Aquinas (1225–1274), for instance, to interpret the order found in the cosmos (which he termed *lex aeterna* or eternal law, because absent revelation the world was seen as eternal) and in human nature (which he termed *lex naturalis* or natural law) as both rational and normative in character. The natural law of what it is to be human was manifest in three levels of natural inclination or desire: for physical life, for family and children, and for political and rational experience.

In the late medieval period, as creation itself increasingly came to be conceived in technological terms, this nevertheless led to nature being thought of as God's artifice. As a divine construction, nature could stand on its own and was governed by its own "secondary laws." Although God ultimately remained the transcendent "first cause" of all things, the divine necessarily began to be pushed to the margins of scientific investigations.

The founders of early modern science such as Galileo Galilei (1564–1642), Francis Bacon (1561–1626), Robert Boyle (1627–1691), and Isaac Newton (1642–1727) adopted this medieval teaching in defending the science of nature as the study of "secondary causes," while increasingly delimiting the higher authority of Biblical theology as the study of God as "first cause." Nature was the book of God's works, and the Bible was the book of God's words. The book of nature was written in the language of mathematics, which was more pure and more progressive than theological disputes concerning historical revelations. To understand nature, scientists were thus encouraged to discover those mathematical principles of nature that constituted the "laws of nature."

The mathematical and observational methods of modern science have succeeded in uncovering the laws of nature in a sense much more expansive and less normative than for Thomas Aquinas. Does this advance in the scientific understanding justify the control of nature? Does the possession of power convey the legitimacy of its use? Bacon, René Descartes (1596–1650), and other early modern proponents of science certainly projected that their new science would conquer nature for human benefit. Beginning in ancient Greece, philosophers and scientists had striven for a theoretical comprehension of nature. Modern scientists under the banner of Bacon and Descartes strove for power over nature. The point was not just to understand nature but to change it, so that modern science from its beginnings exhibited an inherently technological orientation.

Organism versus Machine

The contrast between traditional and modern concepts of nature may also be presented as a contrast between visions of nature as an organism and as a machine. For the Greek philosophers and medieval theologians, nature was primarily manifest as a something that is born and grows. Even for premodern materialists such as Lucretius (c. 99–c. 55 B.C.E.), nature seems to be a super organism with a consequent sacred or awe-inspiring character. Although he seeks to remove all religious superstition from the world and present nature as devoid of gods, his poem *De rerum natura* opens with praise of sky and earth as the father and mother of all living things. In the presence of such a reality— indeed, as part of such a reality—humans are called upon to accept and to live in harmony with it. And for Plotinus (204–270 C.E.), throughout "the air, the earth and sea, there are advents of terrestrial, aquatic, and aerial gods [so that] the world is throughout filled with deity; and on this account is according to the whole of itself the image of the intelligible" (Proclus, Platonic Theology, 7.2).

For modern philosophers such as Descartes, however, nature was primarily manifest by inanimate entities such as rocks that can nevertheless interact as carriers of energy to create complex structures. For Descartes, even living things are complex machines— plants, animals, and human bodies (including the human brain and nervous system) are all machines.

Such a view of nature as machine undercuts the traditional distinction between nature and artifice. The science of nature as machine yields a technology by which nature as technology can be further molded by human beings to serve human purposes. When Bacon declared that "nature to be commanded must be obeyed," he transforms the premodern basic end in itself of obedience to nature into a mere means (*Novum organum* I, 3). Although he argues that all humans can do "is to put together or put asunder natural bodies" with "the rest [being] done by nature working within" (*Novum organum* I, 4), for him nature as a mechanical process has already ceased to exhibit much in the way of intrinsic value. From the eighteenth century romantic poets to contemporary deep ecologists, humans have worried that the science and technology of nature as machine brings about first in theory and then in practice, in Bill McKibben's phrase, "the end of nature": a wholly artificial world controlled by human will with no room left for natural spontaneity or wildness.

In response to this Romantic notion of nature and technology in conflict, some people have defended technology as itself natural. All organisms alter their environments in adaptive ways, and many animals build artificial structures: Beavers construct dams, bees fabricate hives, and leaf-cutter ants cultivate fungus gardens and herd aphids. Charles Darwin (1809–1882) contended that tool-making was common in the animal world, and human technology differed in degree not in kind. Some biologists argue that human technology expresses "niche construction," which is a trait found generally in the living world, because organisms do not just adapt to fixed environments, they also change environments to construct their own niches. There is no fixed "balance of nature," because nature is constantly in flux from the ever-changing forces of both physical and organic causes. For example, the present concentration of oxygen in the atmosphere has arisen from the production of oxygen by photosynthetic organisms. As a consequence, many organisms have evolved a capacity for aerobic respiration and other traits as adaptations to this atmospheric increase in oxygen levels over the course of geological time. Without such a change in the atmosphere brought about by ancient photosynthetic organisms, human beings could never have evolved.

The Problematic Appeal to Nature

Despite the modern replacement of nature as divine with nature as machine, and outside the more extreme Romantic attempts to re-valorize nature, it is nevertheless the case that the appeal to nature exerts a popular influence. On the one side, one of the most common criticisms of genetically engineered foods or bioengineered human-machine hybrids is that they are in some sense unnatural. On the other, one of the most common general forms of praise for science and technology is that they are natural and thus improperly delimited. The so-called naturalistic fallacy is found across the spectrum of discussions about relations between science, technology, and ethics.

Among those who have criticized this appeal to nature as a ground of moral judgment, it is common to distinguish two senses of nature. When scientists speak of the laws of nature, they mean nature as the collective whole of everything that exists or could exist, including humans. When non-scientists speak of nature they more common refer to whatever is spontaneous or not the result of human contrivance.

Insofar as nature covers the entire order of things, argued John Stuart Mill (1806–1873) in a classic modern criticism of the appeal to nature, the moral injunction to "follow nature" makes no sense; humans have no choice in the matter. Everything people do must conform to nature in this abstract, all-encompassing sense. On the other hand, if nature is the spontaneous order of things free from human influence, then "following nature" would be irrational and immoral. It would be irrational, because any human action would alter the course of nature and would thus be unnatural. And it would be immoral, because natural phenomena often have evil effects. Mill declares in his essay "Nature": "Either it is right that we should kill because nature kills; torture because nature tortures; ruin and devastate because nature does the like; or we ought not to consider at all what nature does, but what it is good to do." Morality requires that we go against the impulses of nature.

So morality is not natural, Mill concludes. Rather, it is nature artificially perfected by human cultivation and artifice to satisfy the moral concerns of human beings. Those who argue for a natural moral law mistakenly assume that what *is* can be the rule and standard for what *ought* to be. Natural science can reveal the natural facts of existence, but morality must tell humans about the moral values of human life.

This distinction between is and ought, or between facts and values, supports the common distinction between nature and culture. Morality is assumed then to arise not from nature but from culture, because moral norms of right and wrong, good and bad, are products of human cultural artifice. Through science, people can understand nature. And through technology, people can control nature. But to judge the moral ends of scientific understanding and technological control, one must go beyond nature and enter the realm of culture, which is an artificial world of human social contrivance set apart from the natural world. As Remi Brague (2003) has shown, Mill's essay on nature manifests the shift from the premodern idea that nature is a model for human action to the modern idea that nature needs to be corrected, not imitated.

The proponent of natural moral law might respond by saying that although cosmic nature might be indifferent to moral distinctions, human nature is not. If one can identify some human desires and inclinations as natural and not merely conventional, one can say that the naturally good human life is one that satisfies those natural desires and inclinations. Variable moral customs of culture can then be judged as good or bad, depending on whether or not they conform to those natural desires and inclinations. So, for example, if human beings have natural desires for life, for parental care, and for social bonding, then one can judge those beliefs and practices that satisfy these desires as naturally good.

Even Mill accepts this in his utilitarian morality, when he claims that the ultimate good for human beings is the attainment of happiness, which is the satisfaction of their natural desires. For example, humans' moral duties to others arise from their natural sentiments as social animals who care for their fellow creatures (Mill 1991). Of course, as Mill insists, people's moral virtues do not spring spontaneously from their human nature, because they need to be cultivated through individual habituation and social customs. But still, as Aristotle said, the cultivation of such virtues is made possible by our natural desires and inclinations (*Nicomachean Ethics*, 1103a14–26).

And so reflections about science, technology, and ethics lead to complex questions about the meaning of nature. To ponder such questions is part of human nature.

LARRY ARNHART

SEE ALSO *Bioengineering Ethics; Christian Perspectives; Descartes, René; Earth; Environmental Ethics; Environmentalism; Nature versus Nurture; Rousseau, Jean-Jacques; Sierra Club; Thoreau, Henry David.*

BIBLIOGRAPHY

Brague, Remi. (2003). *The Wisdom of the World: The Human Experience of the Universe in Western Thought.* Chicago: University of Chicago Press.

Klein, Jacob. (1985). "On the Nature of Nature." In *Lectures and Essays,* ed. Robert B. Williamson and Elliott Zuckerman. Annapolis, MD: St. John's College Press.

Lewis, C. S. (1967). "Nature." In *Studies in Words,* 2nd edition. Cambridge, UK: Cambridge University Press.

McKibben, Bill. (1989). *The End of Nature.* New York: Random House.

Mill, John Stuart. (1969). "Nature." In *Collected Works of John Stuart Mill,* Vol. 10., ed. John M. Robson. Toronto: University of Toronto Press.

Mill, John Stuart. (1991). *On Liberty and Other Essays,* ed. John Gray. Oxford, UK: Oxford University Press.

Odling-Smee, F. John; Kevin N. Laland; and Marcus W. Feldman. (2003). *Niche Construction: The Neglected Process in Evolution.* Princeton, NJ: Princeton University Press.

Strauss, Leo. (1953). *Natural Right and History.* Chicago: University of Chicago Press.

Torrance, John, ed. (1992). *The Concept of Nature.* Cambridge, UK: Clarendon Press.

Wigner, Eugene P. (1984). "The Unreasonable Effectiveness of Mathematics in the Natural Sciences." In *Mathematics: People, Problems, Results,* ed. Douglas M. Campbell and John C. Higgins. Belmont, CA: Wadsworth.

NATURE VERSUS NURTURE

• • •

This familiar expression indicates a division between those who offer biological explanations for some human behaviors and those who insist on environmental explanations. The root of the problem is a basic uncertainty about the causes of human physical and psychological traits. Some traits are obviously inherited in a biological sense, such as having a four-chambered heart or the ability to learn to talk. Such characteristics are said to belong to humans by *nature,* from a root word meaning *birth.* Other traits are not inherited, but are a result of environmental influences. A person can inherit a parent's hair color, but not his or her tattoo; and a person must learn the French language in order to speak it. Acquired traits are said to be due to *nurture,* which in this context indicates any influence other than biological inheritance.

Distinguishing in Specific Cases

In analyzing physical characteristics, it can be difficult to tease nature and nurture apart. Why is Steve eight inches taller than Ric? Perhaps this difference is only natural because Steve's parents are taller than Ric's parents. But the difference in stature could quite literally be due to nurture: Perhaps Ric was starved as an infant. Then again, it may be the result of both: Steve picks up five inches from his Mom and Dad, and another three at the dinner table. The problem is much more difficult when analyzing behavior because the range of possibilities is greater. Natural influences on behavior might be quite strong, so that culture plays only a marginal role. At the other extreme, it may be that human beings are born with almost no instincts or innate ideas. Perhaps the only significant influence on any person's behavior is the behavior of other persons, living and dead.

This uncertainty regarding the relative weight of nature and nurture quickly becomes a controversy when discussing behaviors of greater significance. For example, suppose boys are more aggressive at play and girls more caring. One explanation is that society creates this gender difference by giving toy guns to boys and baby dolls to girls. It is possible, however, to argue the opposite: Because girls are already inclined toward motherhood, they receive the dolls they want; and because boys are more aggressive from birth, they select toys that look like weapons. This sort of question tends to divide scholars into hostile camps. Naturalist Edward O. Wilson has dubbed those who offer the second explanation *hereditarians*. Those who insist on the former, he calls *nurturists*.

Hereditarians vs. Nurturists

The opposing beliefs of hereditarians and nurturists colors almost all contemporary discussions of human behavior and its causes. A minor cause of the debate is an old turf war between the social sciences and the humanities, on one side, and the physical sciences, on the other. Some sociologists and English professors see explaining what people think, say, and do, in part by reference to genes, proteins, and neurons, as an invasion of their territory by physicists and biologists. This invasion is especially unnerving because fields such as psychology or history can never hope to match the precision, clarity, and predictive power of the hard sciences. Nurturism is an attempt to carve out a space in which the soft sciences do not have to compete with physics.

The major cause of the conflict between the soft and hard sciences arises from the political, ethical, and aesthetic implications of the hereditarian's view. Any influence that biology is allowed over human behavior seems to come at the expense of moral responsibility. Many facts about existing societies strike people as unjust: sexual inequality, crime and war, economic inequality, among others. To the extent that these social ills are due to nature, society cannot blame anyone for them. Nurturists prefer arguing that the hereditarian view always justifies the status quo. It certainly seems to undermine the indignation that might drive any fundamental change.

The hereditarian view also seems to place limits on the range of possible reforms. If male aggression and desire for status are natural, then every society will suffer from some measure of crime and inequality. If human beings have an instinct to divide themselves into mutually hostile groups, as do chimpanzees, then no society will be free from ethnic, racial, or religious conflict. If women naturally desire to care for their own children, then there is little hope of transforming childrearing into an altogether collective activity, as many utopian communities have attempted to do. The hereditarian view does not deny the possibility of reform, but it does suggest that the best societies will be only marginally better than the ones that have always existed. Nurturists tend to be offended by this idea.

Proposed Resolutions

If an intellectual impasse goes unresolved long enough, some will inevitably grow tired of it and look for a way out. The oldest peace plan is a form of dualism involving the construction of a demilitarized zone between the study of nature and the study of culture. Natural scientists would be allowed to study all natural processes, including human evolution; but should resist any temptation to explain such things as human social and political behavior, history, art or literature by reference to nature. The study of culture should be regarded as an autonomous and independent field of inquiry.

Another attempt to resolve the issue involves a holistic approach to nature and nurture. Much of the anxiety over natural explanations of behavior relies on an overly simplistic view of genetic causation. In that view causation works one way: Genes create proteins that in turn create organisms. A person's nature is fixed from the beginning, and there is very little that can be done about it. The holistic approach is based on a more complex view. Many genes spend their time switching other genes on and off, often in response to external information. A person's genetic code may be fixed, but genetic *nature* is not: It molds itself in response to the environment. Moreover many genes cannot function without information from the environment. Human beings are born with a capacity to learn language, but they must be exposed to a language during certain critical periods in development in order to learn it. Here culture is as much a part of nature as are genes.

Both dualism and holism present themselves as compromises, but are in fact attempts to win by default. Only those who believe that biology has almost no influence over individual personalities will take dualism seriously. Likewise although holism presents a very flexible version of human nature, it nevertheless makes hereditarian assumptions about the influence of biology on behavior. The argument between nurturists and hereditarians does seem likely to wind down for the simple reason that hereditarians are winning. There is little doubt remaining that genes do influence significant behaviors, and that in many cases—twin studies for example—biological inheritance is a much better predictor of an individual's life course than social environment.

Ethical and Political Significance

The moral and political significance of the difference in opinion between nurturists and hereditarians is more difficult to decipher. If the expression of genes really does change in response to the environment, culture may be as difficult to change as nature. Almost every child will easily master a first language, but few people learn a second language well enough to pass for a native. Perhaps this is because one's first language shapes the mind in more or less permanent ways. Evidence suggests that the infant mind is primed to learn language, and much the same thing may be true of morality and other aspects of political culture. Similarly acknowledging that people are naturally disposed to certain behaviors probably makes them more, rather than less, responsible. An individual who recognizes a personal propensity to alcoholism or spousal abuse, is better able to take responsibility for the condition.

The hereditarian view may be liberating in a much more profound way. For example the debate over admitting women into the military has usually turned on whether one believes that sexual differences are mostly due either to socialization or to nature. However the opposite should also be true. Males not only make up most of the soldiers in every society, they also commit almost all the violent crimes. If women serve in large numbers in the military, society must ask what effect this will have on their behavior after their military service is concluded. There is no great need to worry if psychological dimorphism is natural because no change in social environment will make women as dangerous as men. But if these behavior patterns are socially constructed, introducing women into the military might have disastrous consequences. If women learn to behave like men, not only on the battlefield but back home, the crime rate in a society could easily double. Contrary to popular belief, the hereditarian view may be friendlier to social reform than the nurturists view.

The tension between nature and nurture is at least as old as Plato's *Timaeus*. According to premodern nat-

ural philosophy, nature was largely fixed, and was superior in dignity and authority to any product of technology; only nurture was in large measure subject to human control. In this view the role of such sciences as agriculture, medicine, or politics was to tend nature as one had tended the god, in order to promote human flourishing.

The early moderns rejected this approach, and chose to view nature as a "rich storehouse" of materials, as English philosopher Francis Bacon (1561–1626) said, to be manipulated "for the relief of man's estate." The distinction between nature and nurture was relatively unimportant: Given the right technologies, either can be brought under the yoke of human will. Human beings thus acquire an unprecedented sense of responsibility for their own destiny.

Some of that early modern confidence remains in the early 2000s; however, it has been tempered by other considerations. For example the human genome project promised to provide a powerful new tool for the diagnosis and treatment of disease; however, about 5 percent of its budget was devoted to exploring the ethical and social consequences of this project. This was in part political: The public neither fully understands nor trusts innovative technologies. But it also recognizes the limits of engineering as a metaphor for technology. Much of nature as well as human behavior remains stubbornly resistant to technoscientific ambitions. This may be because human life rests on a vast array of interactions between biology and culture, an array that is too complex ever to be mastered. Perhaps an approach to nature and nurture that combines modern science and technology, with at least a dose of ancient piety, is necessary.

KENNETH C. BLANCHARD, JR.

SEE ALSO *Aggression; Ethology; Homosexuality Debate; Nature; Sociobiology.*

BIBLIOGRAPHY

Bacon, Francis. (2001). *The Advancement of Learning.* New York: Modern Library.

Fukuyama, Francis. (2002). *Our Posthuman Future: Consequences of the Biotechnology Revolution.* New York: Picador.

Lewontin, Richard. (2001). *The Triple Helix: Gene, Organism, and Environment.* Cambridge, MA: Harvard University Press.

Pinker, Steven. (2002). *The Blank Slate: The Modern Denial of Human Nature.* New York: Viking.

Ridley, Matt. (2003). *Nature Via Nurture: Genes, Experience, and What Makes Us Human.* New York: Harper Collins.

Rose, Hilary, and Steven Rose, eds. (2000). *Alas, Poor Darwin: Arguments against Evolutionary Psychology*. New York: Harmony.

Wilson, Edward. O. (1988). *On Human Nature*. Cambridge, MA: Harvard University Press.

NAZI MEDICINE

• • •

Medical research and practice under Germany's National Socialist regime (1933–1945) has come to serve as an archetype for the immoral uses to which science and technology can be applied. In many instances appeals to science were used to justify evil actions, and independent reflection failed to criticize unethical research protocols and medical interventions. Without diminishing the horrors that resulted, it is nevertheless important to place such actions in context in order not to so distance them that they offer no lessons from which others might learn.

Social Context

Genetics and related eugenic claims were at the heart of Nazi racial ideology that ultimately led to genocide in Europe during World War II. Although the study and application of eugenics did not begin with National Socialism, it was in Nazi Germany that eugenics became a central component of state policy. The same academic and research institutions that were so critical in the development of modern medicine, medical science, and medical education were also directly complicit in the most massive program of human destruction in history.

Henry Friedlander (1995) nevertheless cautions that the murderous application of eugenic and racial principles by German physicians must be understood in terms of the motivations of other professions. German physicians were professionals who, like all professionals, sought financial security, career advancement, and professional recognition. Motives certainly varied, but these physicians were all German nationalists who generally subscribed to the racist and eugenic components of National Socialism. While providing a rationalization for their actions, ideology was nevertheless probably not the primary motivation for most physicians.

Studies by Michael H. Kater (1989) reveal that German physicians tended to be more closely associated than other professionals with Nazi Party organizations such as the National Socialist Physicians' League, the SA (Sturmabteilung, the military arm of the Nazi party

founded in 1921, but disarmed and neutralized by Hitler in 1934) and the SS (Schutzstaffel, initially recruited from the SA in 1923 as Hitler's personal bodyguard, and the embodiment of Nazi racial ideology that developed into a vast police, military, and economic empire). About a third of all physicians were members of the National Socialist Physicians' League, and by 1939, almost 45 percent of physicians in Germany were members of the Nazi Party, figures substantially higher than those of other professions (such as lawyers [25%], teachers [24%], and musicians [22%]). Moreover, 7 percent of all physicians in Germany were members of the SS. Their professional needs seemed to find relatively more satisfaction within the context of the growing power of the SS and its extraordinary role in matters of life and death.

Under National Socialism, German medical science soon identified individual Germans considered by the state to be inferior and expendable. Acknowledging the influence and experience of American eugenicists and compulsory sterilization laws in the United States, the Nazis on July 14, 1933, enacted the Law for the Prevention of Progeny of Sufferers from Hereditary Diseases. Hundreds of thousands of Germans and, later, Austrians were sterilized without their consent after being medically diagnosed with conditions deemed hereditary and undesirable. These conditions included "feeblemindedness," schizophrenia, and manic-depressive disorder, among others.

T4 Policies

The genocidal policies of the Nazi regime commenced shortly after the outbreak of war in September 1939, with the decision to exterminate the handicapped in Germany. Friedlander identifies the first victims as disabled children and adults who were in institutions. Under the euphemism of *euthanasia*, the killers described their task as "destruction of life unworthy of life." Hitler's Chancellery, with the support of the health division of the Ministry of the Interior, directed the killings. It established various front organizations, headquartered in Berlin at Tiergartenstrasse No. 4, and known as T4. Physicians and psychiatrists, hospital directors and bureaucrats, directed the T4 killings and served as medical experts in the selection of victims they never saw. In addition to starving some patients to death, these physicians murdered patients with overdoses of Luminal (a sedative) and Veronal (sleeping tablets), and also morphine-scopolamine.

In the spring of 1941, the T4 killings were expanded to include concentration camp prisoners. This new task was designated Special Treatment 14f13. In late 1941 and 1942, T4 methods and technology were transferred to the east where the SS established extermination centers at Chelmno, Auschwitz, Treblinka, Belzec, Sobibor, and Majdanek, modeled on the T4 centers, for the extermination of Europe's Jews and Gypsies. There, physicians supervised the registration of the arriving victims, administered the gas, pronounced the victims as dead, and participated in looting the corpses. Besides extracting gold teeth for the Reich treasury, physicians performed countless autopsies on the bodies of their victims in order to provide younger physicians with training and academic credit, as well as to recover organs, especially brains, for scientific study at medical institutes.

SS physicians tolerated unhygienic conditions, inadequate food, and inhuman working conditions in the camps. Moreover, they were complicit in inhuman corporal punishment when they certified that prisoners were healthy enough to undergo beatings. SS physicians also participated in the murder of prisoners in most camps, using lethal injections and other medications to kill their victims.

At Auschwitz-Birkenau, with its assembly line methods of killing, medical officers selected those destined for the gas chambers. In addition, most SS physicians at Auschwitz participated in cruel and unethical medical experiments on human beings. Many were young and inexperienced physicians who wanted to learn, and who did these experiments in order to obtain degrees or to secure some publications. SS physicians performed the function of both concentration camp medical officer, a position that had existed since the early 1930s, and extermination center physician, a position that materialized early on in the war as part of the T4 operation.

In the end, the T4 physicians and SS physicians at Auschwitz volunteered for their positions. They could have refused to participate but did not. There is general agreement among scholars that they became murderers because they were consumed with ambition while remaining, at the same time, more or less loyal to the racist ideology of Hitler's regime.

Historical Consequences

American military courts conducted a series of twelve trials at Nuremberg between December 1946 and April 1949, which included the trial of a group of twenty-three Nazi physicians and members of the German medical establishment for T4 ("euthanasia") killings and medical experiments. These trials generated an in-depth search for ethical rules to be observed before initiating experimental therapy with human beings. Beginning with the creation of the Nuremberg Code in 1947, which condemned medical abuses in experimentation on human beings, a body of ethical guidelines has accumulated over the years.

The Nazi medical establishment also produced some *good* science, according to Robert N. Proctor (1999), within the larger eugenic and racial context of Nazi medicine and its agenda of systematic murder of the handicapped, Jews, and Gypsies. Under National Socialism, German epidemiology was probably the most advanced in the world. Before World War II, for example, German medical science established the relationship between tobacco use and lung cancer. This reflected the regime's goal of improving the overall public health of the German people, of which its *racial hygiene* policies constituted a significant part. As Proctor concludes, the campaign against tobacco provides a compelling insight into the complex nature of the racially based public health initiatives of Nazi Germany, responsible as they were for both better nutrition and forced sterilizations, for both genocide and campaigns against smoking.

William E. Seidelman (2000) has written that the legacy of Nazi medicine included an *amnesia* that conditioned the postwar German and Austrian medical establishments until the late twentieth century particularly with regard to the continued use of the fruits of Nazi medical practice. The links between Nazi ideology, the cruel and exploitative medical experiments that German physicians conducted on the victims of that ideology, and the sterilization, euthanasia, and extermination policies conducted by physicians under Nazi authority, raise questions that have immediate relevance to contemporary controversies over the nature and course of research in human genetics and biotechnology.

FRANCIS R. NICOSIA

SEE ALSO *Eugenics; Euthanasia; Holocaust; Human Subjects Research; Race; Research Ethics.*

BIBLIOGRAPHY

Burleigh, Michael. (1994). *Death and Deliverance: "Euthanasia" in Germany, c. 1900–1945.* Cambridge, UK: Cambridge University Press. Links the Nazi mass murder of the Jews, Gypsies, and the handicapped by revealing the redeployment of "euthanasia" personnel to the mobile killing units and the extermination camps after 1941.

Friedlander, Henry. (1995). *The Origins of Nazi Genocide: From Euthanasia to the Final Solution.* Chapel Hill: University of North Carolina Press. Traces the rise of racist and eugenic ideas in Germany and elsewhere in the early twentieth century, their contribution to the Nazi "euthanasia" program beginning in 1939, and to the policy of extermination of Jews and Gypsies beginning in 1941.

Kater, Michael H. (1989). *Doctors under Hitler.* Chapel Hill: University of North Carolina Press. Focuses on the institutional framework of the medical profession in Nazi Germany and examines how German physicians participated in crimes against Jews and other victims.

Lifton, Robert Jay. (1986). *The Nazi Doctors: Medical Killing and the Psychology of Genocide.* New York: Basic. Demonstrates how many German physicians were transformed from healers to murderers, and the overall role they played in Nazi genocide.

Nicosia, Francis R., and Jonathan Huener, eds. (2002). *Medicine and Medical Ethics in Nazi Germany: Origins, Practices, Legacies.* New York: Berghahn Books. Six distinguished scholars consider the various ways in which German physicians and the German medical establishment were complicit in crimes against humanity, and the disturbing legacy they left in post-Holocaust Germany and Austria.

Proctor, Robert N. (1999). *The Nazi War on Cancer.* Princeton, NJ: Princeton University Press. Focuses on the discovery of the link between tobacco smoking and lung cancer in Nazi Germany, and demonstrates how the positive health activism of the Nazis derived from the same roots as their medical crimes against humanity.

Seidelman, William E. (2000). "The Legacy of Academic Medicine and Human Exploitation in the Third Reich." *Perspectives in Biology and Medicine* 43(3): 325–334. Examines the continued use in post-World War II Germany and Austria of some of the results of Nazi medical experiments during World War II.

NEGATIVE EUGENICS

SEE *Eugenics.*

NEOLIBERALISM

• • •

The term *neoliberalism* is used to characterize the dominant economic policies pursued in the United Kingdom, the United States, and some developing countries such as Chile since the late 1970s or early 1980s. It is noteworthy that during this same period governmental policies toward the support of science and technology were undergoing important critical assessments. On the one hand, the scientific community proclaimed its autonomy but, on the other, sought increased governmental support for its research. In the United States, however, the Federal Technology Transfer Act of 1986, requiring national laboratories to promote technology transfer and to promote partnerships, was part of the deregulation and privatization of government activities. During this same period, the disclosure of instances of misconduct in scientific research raised questions about the ability of an autonomous scientific community to govern itself.

Genesis

Neoliberal policies, first identified with the Conservative government of Prime Minister Margaret Thatcher (1979–1990) in the United Kingdom and the Republican administration of President Ronald Reagan (1981–1989) in the United States, represented a sharp break with the so-called Keynesian consensus that had dominated both domestic and international economic policy-making from the end of World War II to the late 1970s. (The consensus was called *Keynesian* because it was based on the theories of the British economist John Maynard Keynes [1883–1946] and his followers.) At the heart of that consensus had been the view that, unless continually "guided" and "pump-primed" by governments, free market or capitalist economies were unable to provide either full employment or a stable pattern of economic growth. Generated as a reaction to the Great Depression of the 1930s, and reinforced by a successful experience of strong state management of economies in the war years, Keynesian theories and policies appeared unable to cope with the so-called stagflation that marked the early 1980s—the combination of high unemployment with high inflation that hit virtually all industrial economies at the end of the postwar "long boom" in the world economy.

Keynesian policies had come under criticism from a minority of economists even before the stagflation period. Such policies were seen as having encouraged strong structural inflexibilities and rigidities in market economies, rendering them both less technologically and commercially innovative than they would otherwise have been, and making them particularly vulnerable to problems of inflation as productivity increases failed to keep pace with increases in wages and other costs. Such critiques had not been very politically effective previously, but became more so when the chronic inability of all industrial economies to absorb the 1970s oil price increases—and the double-digit inflation and sharply reduced profit rates that arose as a result in most of them—seemed to confirm the very "rigidity" and "inflexibility" of which opponents of Keynesianism had warned (Armstrong, Glyn, and Harrison 1984).

Characteristics

Neoliberalism involves a crucial reversal of the fundamental policy premise of Keynesianism. For Keynesians the fundamental problem of free market or capitalist societies was the possibility and actuality of "market failure" and the need for state intervention to prevent or correct such failures; for neoliberals the fundamental problem is that state interventions in markets fail far more frequently than they succeed, or even when they do succeed in their particular policy goals (such as full employment) have unanticipated consequences in other areas of market functioning—consequences that ultimately undermine their supposed successes. Neoliberals therefore return to the fundamental premise of Adam Smith's *Wealth of Nations* (1776): that economic policy should, in general, err on the side of *laissez-faire*, of "letting alone," of allowing "market forces" to function unimpeded by state action—unless there is some very strong reason *not* to do so. Their fundamental policy premise therefore is that in capitalist or market economies "state failure" is a much greater problem and danger than "market failure." According to John Williamson (2002), the only "strong reasons" that neoliberal economists will usually countenance as justifications for state action are the enforcement of legal contracts (requiring a judicial system and a police force) and the requirements of state external defense (requiring a state-funded military apparatus).

Because of its reversal of Keynesian policy premises and of the "burden of proof" for state intervention, neoliberalism undoubtedly received an enormous political impulse from both the collapse of communism in the USSR and Eastern Europe in the early 1990s and the failure—or perceived failure—of the state-led economic development (often referred to as import substitution industrialization) that dominated many parts of the Third World from the 1960s through the 1980s. Both phenomena could be seen as classic examples of "state failure"—of the failure of state-dominated economic policies to generate economic innovation and development and to raise mass living standards—relative to the performance of more "free market" economies (Stiglitz 2002). Neoliberal economists and policymakers are particularly given to seeing the success of economic development efforts in certain parts of the former Third World—in China and East Asia most notably—as examples of "free market" success. This neoliberal view of the so-called Asian Tiger economies, or newly industrializing economies, however, has been strongly contested by opponents of neoliberalism, as described further below.

Originality

Questions arise about the originality of neoliberalism—and in particular about its relationship to classical nineteenth-century economic liberalism. Some analysts have denied that neoliberalism, as an economic doctrine, is in any way original, and have seen it simply as a return to the fundamental *laissez-faire* policy premises of both the classical and neoclassical economists of the nineteenth century. Others have denied this and sought to justify the prefix *neo* in a variety of ways: neoliberals are much more concerned with market exchanges, and in particular with legally guaranteed ("contractual") monetary exchanges of goods and services, than were their nineteenth-century predecessors who (so the argument goes) were much more concerned with "real," "material" production processes and with monetary exchanges only as a part or aspect of these real processes (Treanor Internet article). Neoliberals are "neo" precisely because they are in general more politically conservative, especially on social issues, than their nineteenth-century predecessors, who were politically as well as economically liberal (Shah Internet article). They are *neo*liberals because they are generally more nationalistic than classical nineteenth-century liberals. It has even been argued, that, in practice, neoliberals actually support disguised modern forms of "mercantilist" economic policy (the kind of nationalistic economic policy expressly attacked by Adam Smith). They do so because, so it is alleged, they use "free market" and (especially) "free trade" ideas to justify and reinforce the economic power and domination of the rich nations of the world—especially the United States (Shah Internet article).

None of these justifications of the *neo* prefix seem especially convincing for two reasons. First, all these characterizations come from neoliberalism's opponents. In fact, with very rare exceptions (DeLong Internet article), economists and politicians who are referred to by their opponents as neoliberals do not use this term themselves. Generally speaking, people who are tagged as neoliberals refer to themselves simply as "conventional economists" or "believers in free markets" or even "economic pragmatists." Second, all the above justifications are empirically doubtful, in the following ways:

1. If modern neoliberals can be attacked as disguised economic nationalists or even as apologists for economic imperialism, then so can classical nineteenth-century liberals (and especially British liberals) (Kitching 2001).

2. Although some neoliberals are undoubtedly very nationalistic (Thatcher comes immediately to mind), others seem just as "globalist" or "internationalist" in their outlook as any nineteenth-century liberal, and have indeed not infrequently been attacked for justifying or defending "free trade" policies that lead to job losses in the United States, Europe, or elsewhere.

3. While it probably is true that modern economic theory in general is even more "abstract/mathematical" and "monetary-exchange" oriented than its nineteenth-century predecessors, this is probably much more a reflection of the changing structure of capitalist markets in the contemporary period than a mark of any major theoretical or ideological shift.

4. While some neoliberals may be politically or socially conservative (Thatcher again comes to mind, along with her economic "guru" Friedrich Hayek and the American economist Milton Friedman), a number of others are almost anarchistic in their support for "free individual choice" in social issues. Others, still tagged "neoliberal" by their opponents, are in fact advocates of a rather wider range of state interventions (often on social or equity grounds) than the majority of market-oriented economists. Joseph Stiglitz (2002, 2003), former chief economist of the World Bank, frequently espouses such "modified Keynesian" views now, as does the neoliberal (or former neoliberal?) trade economist Paul Krugman.

On balance then it seems most accurate to ignore the *neo* prefix or to see it as simply a synonym for *new* or, perhaps better yet, for *revived*. Neoliberalism is in fact simply a revived form of nineteenth-century "free market" economic liberalism adapted in specific ways to the changed economic context of the late twentieth and early twenty-first centuries, but not theoretically or ideologically new in its fundamentals. Insofar as part of the changed economic context involves the increased importance of science and technology, the proper relation between science, technology, and economic liberalism is one *neo*liberalism issue.

Merits and Demerits

Neoliberalism's merits include:

1. Its acute, and to a large degree empirically accurate, analysis of the severe shortcomings of state economic policymaking both in the former communist countries and in many parts of the Third World. In particular, neoliberals have revealed the very peculiar cultural assumptions about the values and actions of state power holders that were built into Keynesian economics and into the Keynesian-influenced "development economics" of the 1950s to 1970s. Working in a European context Keynes and his followers felt able to ignore classically Smithian questions about the corruptibility of state power holders. But there are many parts of the world where such questions *cannot* be ignored, or are ignored only at the peril of total policy failure. Neoliberalism seems most justified when arguing, in line with Adam Smith, that free markets should be preferred to state economic policymaking in many contexts *not* because the former are perfect, or even near perfect in their results, but because they are *less* radically *imperfect* (in social and political, as well as economic, terms) than the only alternative can offer (DeLong Internet article).

2. Its insistence that mass standards of living can rise substantially *only* in countries and societies that have a dynamic involvement in world trade. Neither attempted economic autarchy nor attempts at minimization of involvement in the world trade system can or will lead to anything other than economic stagnation and impoverishment. Moreover, this is true even when the pattern of world trade is "biased" or "distorted" in various ways in the interests of strong or dominant nations and economic interest groups (Mandle 2003).

Neoliberalism's principle weaknesses are:

1. A chronic inability to grasp that human activities and interactivities that in one intellectual framework may be termed "economic" can equally well (and equally accurately) in another intellectual framework be conceived as "social" and/or "political." This is a weakness built not into neoliberalism specifically but into economics as such, as an intellectual discipline. The most common confusion in which it results is the supposition that because there are processes in the real world that are "simply" or "purely" economic (and not "social" or "political"), governments and states can then also make and implement policies that are "purely" economic (and not "social" or "political"). But this is a delusion. *All* economic processes are simultaneously social and/or political, and *all* economic policies have social or political dimensions or aspects. Significantly it is those economists who, for one reason or another, transcend their training enough to grasp this, and grasp it firmly, who usually move to become "modified" or "critical" neoliberals (Stiglitz, Krugman, and J. Bradford DeLong, for example, all fall into this category).

2. A tendency for neoliberal economists in particular to *ignore* the less than optimal political context in which current capitalist markets operate in the *developed* as well as the underdeveloped world. These include: protection or subsidization of special-interest groups for domestic electoral reasons (Stiglitz 2003); global economic regulatory bodies whose functioning is hamstrung by the insistence of powerful states that such interests be protected (Stiglitz 2002); the political "muscle" of large international firms and the way this effects their competitive behavior; and above all the socially polarizing and politically destabilizing effects of market-produced inequalities. Neoliberals most frequently justify their ignoring of such issues by claiming that these are "social" or "political" issues (and not "economic" ones) and therefore beyond their compass. Such weaknesses lead to allegations that neoliberalism is simply a justifying ideology of "capitalist imperialism" and in particular of the rich capitalistic elites of the Western world (Martínez and García Internet article). Though such allegations are an oversimplification, they are perfectly understandable given the obtuse or "head-in-the-sand" behavior described above. In addition,

3. If one accepts the "anti-neoliberal" account of the success of the Asian Tigers, viz. that these economies developed through carefully and cleverly *state-guided* forms of industrialization and trade policy (Wade 1990, Amsden 1989), then it follows that the powerful "minimalist" argument for the market over the state, though it may hold in many cases, does not hold in all. This opens up the possibility that the difference between countries that successfully develop economically and those that do not is *not* a simple difference between those that are market oriented and those that are state oriented in their economic philosophies and policies. Rather it is simply a difference between those that make appropriate and effective state economic policies and those that do not.

Finally, the degree to which successful economic development can be explained solely as a free market phenomenon, questions arise about the productive importance of science and technology. It would be interesting to known whether different levels of public and private investments in science and technology among countries with similar liberal economic policies can be associated with different rates of economic growth.

GAVIN KITCHING

SEE ALSO *Communitarianism; Liberalism.*

BIBLIOGRAPHY

Amsden, Alice H. (1989). *Asia's Next Giant: South Korea and Late Industrialization.* New York: Oxford University Press. Excellent empirical and theoretical account of South Korea's industrial "miracle," emphasizing both the domestic and international political background to this success.

Armstrong, Philip; Andrew Glyn; and John Harrison. (1984). *Capitalism since World War II.* London: Fontana. Compendious account of post-war capitalism utilizing a mass of statistical material and an extremely original theoretical framework.

Kitching, Gavin. (2001). *Seeking Social Justice through Globalization: Escaping a Nationalist Perspective.* University Park: Pennsylvania State University Press. An account of globalization sympathetic to the economics of the neoliberal view but highly critical of its social and political perspectives.

Krugman, P. (1997). *Pop Internationalism.* Cambridge, MA: MIT Press. A very good defense of neo-liberal free trade doctrines written to be readily comprehensible to the non-economist.

Mandle, Jay R. (2003). *Globalization and the Poor.* New York, Cambridge, UK: Cambridge University Press. Takes a rather similar theoretical view of globalization to Kitching but focuses much more tightly and empirically on the issue of poverty.

Stiglitz, Joseph E. (2002). *Globalization and Its Discontents.* New York: Norton. Excellent critical account of the workings of the World Bank and the International Monetary Fund in the current global economy, by a former "insider."

Stiglitz, Joseph E. (2003). *The Roaring Nineties: A New History of the World's Most Prosperous Decade.* New York: Norton. Stiglitz's most forthright condemnation of the application of neoliberalism to U.S. domestic policy-making.

Wade, Robert. (1990). *Governing the Market: Economic Theory and the Role of Government in East Asian Industrialization.* Princeton, NJ: Princeton University Press. Probably the most well-known text arguing that the success of the Asian tigers is not a vindication of free market economic doctrines.

INTERNET RESOURCES

DeLong, J. Bradford. "'Globalization' and 'Neoliberalism.'" Available from http://www.j-bradford-delong.net/Econ_Articles/Reviews/alexkafka.html.

Martínez, Elizabeth, and Arnoldo García. "What Is Neoliberalism?" CorpWatch. Available from http://corpwatch.radicaldesigns.org/article.php?id=376.

Shah, Anup. "A Primer on Neoliberalism." Available from http://globalissues.org/TradeRelated/FreeTrade/Neoliberalism.asp.

Treanor, Paul. "Neoliberalism: Origins, Theory, Definition." Available from http://web.inter.nl.net/users/Paul.Treanor/neoliberalism.html.

Williamson, John. "The Washington Consensus as Policy Prescription for Development." World Bank, 2002.

Available from http://www.worldbank.org/etools/bspan/ PresentationView.asp?PID=1003&EID=328. All the above are relatively brief, readily accessible but moderately sophisticated Internet discussions of neoliberalism. Only DeLong's piece, however, is totally free from any kind of axe grinding.

NETWORKS

• • •

Networks are particular types of human relations or technological creations, sometimes compared to systems and webs, that establish unique exchanges between human beings and spaces. Since the 1700s, and especially since the invention of the Internet, networks have been subject to scientific analysis. Insofar as they define or influence human behavior they may be subject to ethical assessment.

Network Types and Influences

In mathematics a network is commonly defined as a directed graph with vertices (or nodes) and weighted edges (also called arcs or links). As such networks come in different structural types: bus, ring, star (hub and spoke), mesh (web), and more (see Figure 1). Networks can be further distinguished in terms of numbers of vertices and edges. Each structure has its own intrinsic properties, which can be enhanced or modified by giving different weights or strengths to the various links, as when (for instance) one link in a star network is weighted more heavily than another.

Throughout history networks have provided the foundation and infrastructure for humans to conduct wide-ranging economic and social activities. Well-known physical networks in which nodes correspond to locations in space and links to appropriate connections with associated flows include transportation and communication networks. Transportation networks have evolved over the centuries through advances in science and technology and come in a myriad of forms: road, rail, air, or waterway, with a variety of associated modes of travel. They traverse physical distances to facilitate business transactions, military conquest, and visits among colleagues, clients, friends, and family, as well as enabling people to explore new areas and to expand horizons. Communication networks, in turn, allow exchanges of information not only within communities but also across regions and national boundaries by means of postal services, telephones, radio, television,

computers, satellites, and microwaves that carry written messages, video, and/or electronic data. Energy networks, as another example, provide the necessary fuel to support many transportation and communication network transactions.

In addition, more abstract networks such as financial networks, a variety of logistical networks (e.g., supply chains), as well as knowledge and social networks (based on transportation and communication networks) play new and not yet completely understood roles in societies and economies. The reliability, efficiency, and accessibility of such networks enhance production and distribution, facilitate the exchange of information and knowledge, and add to the diversity and richness of goods and services. At the same time, the structure of such networks and the connectivity provided by them may yield insights and advantages for particular individuals and organizations.

Organizations today, be they local, regional, national, or global in scope and as diverse as businesses, educational institutions, or governments, are highly dependent on networks, which are becoming increasingly interrelated. Indeed, individuals may now be able to conduct financial transactions electronically and to shop globally from their places of employment and have the products delivered to the desired destinations. They may also, in certain circumstances, be able to work from home or other chosen locations depending on the management of the underlying networks, their utilization and availability, and the auxiliary ethical character of network designs, accessibility, and usage.

Fascinatingly, the structure of social relationships may also be represented as a graph/network, and the study of social relationships has given rise to the multidisciplinary topic of social network analysis. In such a context, important measures include the number of connections for an individual (represented by nodes in the network), the strength of these connections, the centrality of various individuals, and the existence of cliques and subgroups. Moreover, one can calculate the degrees of separation. Clearly, the existence and structure of social networks also affects the usage of physical networks, notably transportation and various communication networks. The latter networks, in turn, play pivotal roles in the evolution of social networks.

The Science of Networks

The topic of networks and network management dates to ancient times with classical examples including the publicly provided Roman road network and the time-of-day

FIGURE 1

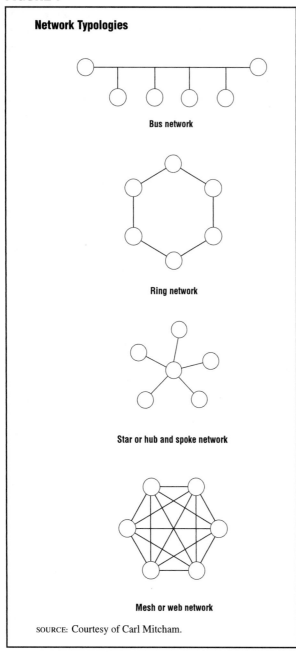

Network Typologies

Bus network

Ring network

Star or hub and spoke network

Mesh or web network

SOURCE: Courtesy of Carl Mitcham.

Interestingly, not long thereafter, François Quesnay (1694–1774), in his *Tableau économique* (1758), conceptualized the circular flow of an economy as a network. Gaspard Monge (1746–1818), who had worked under Napoleon Bonaparte in providing the infrastructure support for his army, published what is probably the first paper on the transportation network model in 1781. Much later, and following the first book on graph theory by Dénes König in 1936, works by the economists Leonid V. Kantorovich (1939), Frank L. Hitchcock (1941), and Tjalling C. Koopmans (1947) considered the network flow problem associated with the classical transportation problem. Thus the study of network flows, primarily in a transportation context, preceded the development of even optimization theory and such elegant algorithmic techniques as the simplex method (see Dantzig 1948).

Indeed, the emergence and evolution of a plethora of physical networks over space and time, coupled with realizations of the importance of abstract networks, and the effects of human decision-making on networks through their utilization and management, has given rise to the development of rich and powerful theories that are rigorous, scientific, and network-based. The novelty of networks lies in that they are pervasive and fundamental and provide the fabric for the connectivity of societies and economies. At the same time, methodologically, network theory has developed into a powerful and dynamic medium for abstracting complex network-based problems. Many contemporary networks (including the Internet) are characterized by a large-scale structure, complexity of interconnections and interrelationships, congestion, and distinct behavior of the users. One illustrative phenomena is the Braess paradox (1968), in which the addition of a new road in a transportation network—or a link in a communications network such as the Internet (see Korilis, Lazar, and Orda 1999)—makes all users of the network worse off. Methodologies for the formulation and analysis of network systems are thus of wide practical significance (see Ahuja, Magnanti, and Orlin 1993; Nagurney 1999; Nagurney and Dong 2002).

Today it is possible, through advances in scientific models, theories, and computational tools, to predict optimal routes on networks from different origins to destinations both from a system-optimized perspective, in which there is a central controller of the network flows, and from a user-optimized one, in which users of the network select their optimal routes in what may be viewed as a selfish manner (see Beckmann, McGuire, and Winsten 1956; Dafermos and Sparrow 1969; Nagurney 1999; Nagurney and Dong 2002). In addition, it is possible to optimize financial portfolios from a network

chariot policy, whereby chariots were banned from the ancient city of Rome during particular times of the day (Nagurney 2000). The topic of networks as a subject of scientific inquiry originates in a 1736 paper by the Swiss mathematician Leonhard Euler (1707–1783), which is considered the earliest paper on graph theory, where a graph in this context is meant as an abstract or mathematical representation of a system by its depiction in terms of vertices (nodes) and edges (or arcs) connecting various pairs of vertices.

perspective (Nagurney and Siokos 1997; Nagurney 2003), to predict the profit-maximizing production and shipment patterns between tiers of network decision-makers (Nagurney and Dong 2002), and to even determine information flows in an organization (Wu et al.).

More recently, social networks have been integrated with economic networks, in the form of supply chains, through the theory of supernetworks (see Walkobinger and Nagurney 2004) in order to capture relationship levels as flows in addition to product shipments. Such complex networks not only synthesize and integrate the structure of the underlying social and economic relationships but also capture human behavior and decision-making and the associated impacts. Moreover, the dynamics of the interactions between the various decision makers as well as how their relationships evolve over time (and how they compete and/or cooperate) can be modeled, along with the optimal product flows and prices.

There are nevertheless many questions of ethical significance concerning networks, their operation and management, and their accessibility and usage.

Accessibility and Ethics

In regard to accessibility, consider transportation and communication networks. Accessibility concerns the design of the network itself. The number of nodes and the number of links connecting the nodes determine the network topology, whereas the quality of the links affects the ultimate accessibility and usage. For example, well-built roads will support travel and trade, whereas an impoverished transportation network infrastructure can seriously impede development and growth. At the same time, the availability of alternative modes of transportation may enhance employment because workers can reach their (possible) places of work. Similarly, those who cannot drive or cannot afford car ownership may be able to use cost-appropriate transportation modes (if such are available).

The interrelationships between networks in this context also have ethical implications. For example, it is now well-established that transportation and especially vehicular transportation on congested urban networks not only results in a loss of productivity but has serious consequences for the environment because of pollution emissions (Nagurney 2000). Moreover, these emissions are not necessarily local but are often *transported* over political boundaries. Hence, the choices made by an individual in terms of route/mode selection can negatively affect distant populations. Although there may be economic approaches to ameliorating some of these negative effects through, for example, tolls or pollution charges, there may also be incentives put in place that appeal to humans' individual sense of ethics.

In terms of communication networks, notably the Internet, the accessibility issue has received a great deal of attention especially from a variety of government organizations. Indeed, terms such as the *digital divide* have become part of the popular lexicon. In certain fields, particularly science, the essentialness of accessibility to the Internet for research, information, and knowledge dissemination is well known (Alberts; Newman 2001). Less emphasized and as important is to increase the connectivity in less-developed and developing nations, which not only may have poor communication infrastructures but may suffer from substandard energy networks, as well.

Not only do scientists benefit from accessibility to communication networks such as the Internet, but educational systems throughout the globe can only be enriched through reliable and efficient Internet connections.

Usage and Ethics

Increased access to interconnected networks also raises ethical issues. For example, given that information on individuals can be retrieved in seconds by anyone with appropriate computer connections, there are serious questions concerning privacy of the information and the right of individuals to check the correctness of the data and information concerning themselves. Moreover, the regulation of the content of what is circulating on the Internet, given its huge and immediate reach, is a subject of both ethical and legal importance. In addition, such computer-based crimes as hacking and computer piracy are examples of illegal and unethical usage of communication networks. Such activities can have serious financial as well as personal consequences (see, e.g., UNESCO).

The Internet, by helping to span the globe and enhancing people's right to communicate, has given freedom to many voices. It has played a major role in social and economic transformations and has helped in the internationalization of trade, especially through electronic commerce and the globalization of nations' economies. In addition, the Internet has allowed new social networks to evolve, oftentimes between individuals and among groups who have never even met face-to-face. Freedom, however, must come with responsibility, a sense of ethics, and solid judgment of the consequences of one's actions on others. Never has the subject of networks and ethics been more timely and relevant.

ANNA NAGURNEY

SEE ALSO *Communication Ethics; Computer Ethics; Digital Divide; Information Society; Internet; Radio; Roads and Highways; Telephone.*

BIBLIOGRAPHY

Ahuja, Ravindra K.; Thomas L. Magnanti; and James B. Orlin. (1993). *Network Flows: Theory, Algorithms, and Applications.* Englewood Cliffs, NJ: Prentice Hall.

Barabási, Albert-Laszló. (2002). *Linked: The New Science of Networks.* Cambridge, MA: Perseus.

Beckmann, Martin J.; C. B. McGuire; and Christopher B. Winsten. (1956). *Studies in the Economics of Transportation.* New Haven, CT: Yale University Press.

Braess, Dietrich. (1968). "Über ein Paradoxon der Verkehrsplanung" [On a paradox of traffic planning]. *Unternehmenforschung* 12: 258–268.

Dafermos, Stella C., and Frederick T. Sparrow. (1969). "The Traffic Assignment Problem for a General Network." *Journal of Research of the National Bureau of Standards* 73B(2): 91–118.

Dantzig, George B. (1948). "Programming in a Linear Structure." Washington, DC: Comptroller, U.S. Air Force.

Euler, Leonhard. (1736). "Solutio problematis ad geometriam situs pertinentis" [The solution of a problem relating to the geometry of position]. *Commetarii Academiae Scientarum Imperialis Petropolitanae* 8: 128–140.

Hitchcock, Frank L. (1941). "The Distribution of a Product from Several Sources to Numerous Localities." *Journal of Mathematics and Physics* 20: 224–230.

Kantorovich, Leonid V. (1939). *Mathematicheskiye metody organizatsiyi i planirovanya proizvodstva.* Leningrad: Publication House of the Leningrad State University. Translated in 1960 as "Mathematical Methods in the Organization and Planning of Production." *Management Science* 6: 366–422.

König, Dénes. (1936). *Theorie der Endlichen und Unendlichen Graphen* [Theory of limited and unlimited graphs]. Leipzig, Germany: Teubner.

Koopmans, Tjalling C. (1947). "Optimum Utilization of the Transportation System." In *Proceedings of the International Statistical Conferences.* vol. 5: 136–145. Reprinted in *Econometrica* 17 (supplement) (1949): 136–145.

Korilis, Yannis A.; Aurel A. Lazar; and Ariel Orda. (1999). "Avoiding the Braess Paradox in Non-cooperative Networks." *Journal of Applied Probability* 36(1): 211–222.

Monge, Gaspard. (1781). "Mémoire sur la théorie des déblais et des remblais" [Memorandum on the theory of excavation and embankment construction]. In *Histoire de l'Académie Royale des Sciences,* avec les Memoires de Mathematique et de Physique, pour la Meme Annee. Paris: Tires des Registres de Cette Academie.

Nagurney, Anna. (1999). *Network Economics: A Variational Inequality Approach,* 2nd edition. Boston: Kluwer Academic.

Nagurney, Anna. (2000). *Sustainable Transportation Networks.* Cheltenham, UK: Edward Elgar.

Nagurney, Anna, ed. (2003). *Innovations in Financial and Economic Networks.* Cheltenham, UK: Edward Elgar.

Nagurney, Anna, and June Dong. (2002). *Supernetworks: Decision-Making for the Information Age.* Cheltenham, UK: Edward Elgar.

Nagurney, Anna, and Stavros Siokos. (1997). *Financial Networks: Statics and Dynamics.* Heidelberg, Germany: Springer-Verlag.

Newman, M. E. J. (2001). "The Structure of Scientific Network Collaborations." *Proceedings of the National Academy of Sciences of the United States of America* 98: 404–409.

Quesnay, François. (1895 [1758]). *Tableau économique.* British Economic Society. Reprint, with an introduction by H. Higgs.

Van Alstyne, M., and E. Brynjolfsson. (1996). "Could the Internet Balkanize Science?" *Science* 274(5292): 1479–1480.

Walkobinger, Tina, and Anna Nagurney. (2004). "Dynamic Supernetworks for the Integration of Social Networks and Supply Chains with Electronic Commerce: Modeling and Analysis of Buyer-Seller Relationships with Computations." *Netnomics* Amherst: Virtual Center for Supernetworks, Isenberg School of Management, University of Massachusetts at Amherst.

INTERNET RESOURCES

Alberts, Bruce. "Science and Human Needs." President's address, 137th annual meeting of the National Academy of Sciences, Washington DC, May 1, 2000. Available from http://www.nationalacademies.org/president/alberts.html.

United Nations Educational, Scientific and Cultural Organization (UNESCO). "UNESCO and an Information Society for All." Available from http://www.unesco.org/webworld/telematics/gis.htm. Published in 1996.

Wu, Fang; Bernardo A. Huberman; Lada A. Adamic; and Joshua R. Tyler. "Information Flow in Social Groups." Information Dynamics Laboratory, Hewlett-Packard Labs, Palo Alto, CA. Available from http://www.hpl.hp.com/shl/papers/flow/. Published in 2003.

NEUMANN, JOHN VON

SEE *von Neumann, John.*

NEUROETHICS

• • •

Neuroethics is the area of bioethics that focuses on issues unique or especially relevant to neuroscience. It is a relatively new term that has been used in a variety of more restricted ways referring to: (1) ethical issues associated with neurology (the subfield of medicine focused on disease and injury of the nervous system) (Pontius 1993); (2) ethical issues associated with the technological advances of neuroscience (Farah and Wolpe 2004); and (3) the neurological basis of ethical thought and behavior (Caplan 1983, Roskies 2002). While attention has primarily focused on the potential applications of technological development, all of these topics appropriately fall under the purview of neuroethics.

Neuroscience is that field of the biological sciences that examines the structure and function of the nervous system. It includes all stages of development from initial differentiation of cells that will become part of the nervous system in the developing organism, through senility and brain death. Topics of investigation range from the submicroscopic level, that is, ions and molecules that are involved in nerve cell function and the genes that are uniquely expressed in the brain, to mental activity and behavior. It includes, but is not limited to, the fields of neurochemistry, neurophysiology, neuropharmacology, neuroanatomy, neuroendocrinology, psychoneuroimmunology, neurology, psychiatry, psychology, and cognitive science.

Neuroscience, directly or indirectly, examines the underpinnings of thought, feeling, and behavior. Neuroethics is concerned with ethical, legal, social and or public policy implications of neuroscience research findings, as well as with the character of the research itself. The neurosciences are rapidly evolving and advances in science and technology have made possible ever more detailed examination of the nervous system and its activity, and of behavior and mental processes. As a result, what were once merely hypothetical situations and potential ethical issues and concerns are increasingly more real and immediate.

History

The term neuroethics seems to have first been coined in 1993 (Pontius 1993), though widespread usage of the term followed a seminal conference in 2002 (Marcus 2002). However the concept has a long history: The tension between notions of free will and determinism and the seeming duality of the mind and body have been of substantial interest to ancient as well as modern philosophers and increasingly among neuroscientists themselves. In the 1950s and before, concerns associated with prefrontal lobotomy and brainwashing as techniques for altering or influencing brain function received increasing attention (Valenstein 1986). In the 1960s some proposed psychosurgery as a method of social control, which created considerable controversy (Chorover 1979, Valenstein 1980). Beginning in 1983 the Society for Neuroscience, the primary professional society of neuroscientists in the United States, initiated annual social issues roundtables aimed at examining the ethical, legal, and social implications of neuroscience research. These symposia examine a wide array of topics including research into possible sex differences in the brain and the application of that research, therapeutic and nontherapeutic use of cognitive enhancers, neurotoxicity of food additives, brain death, the use of fetal tissue to treat neurological diseases, and the role of neuroscience research into drug addiction in the development of health and public policy. In 1983, the Office of Technology Assessment (OTA, a former congressional agency whose mission was to provide legislators with information about scientific findings relevant to the development of public policy) commissioned a report on the societal impacts of neuroscience (OTA 1984). Thus while the term neuroethics is relatively new, the field that it names is not. Rather it is a long-standing area of interest given new life with a new name and new tools.

Features of the Nervous System

Four characteristics of the nervous system with important implications for neuroethics are (1) its complexity, (2) its plasticity, (3) the dynamic, interactive quality of its elements, and (4) the remarkable variation in structure and function from one individual to the next. Although the brain is widely thought of as an organ of the body analogous to the heart, kidney, or liver, the brain and associated elements of the nervous system are more complex than the rest of the body. More genes are uniquely expressed in the brain (Hahn, Van Ness, and Chaudhari 1982) and more different types of cells are found in the brain than in the rest of the body. In addition, cells are interconnected, sending and receiving electrical and biochemical communications from nearby cells as well as cells in distant parts of the brain and the body. As a result, cell circuits extend the complexity of the brain.

The nervous system is remarkably adaptive. The interconnectivity of the cells of all components of the nervous system including the brain and sense organs, (and indeed connections with the endocrine, immune, and other physiological systems) lead to dynamic, interactive communication that makes it possible for brain cells to be sensitive to, and responsive to, changes both internally within the organism, and in its external environment. The interactive communication between cells also results in short-term and sometimes long-term changes in the cells themselves that, for example, may make the individual organism more, or less, responsive to a particular external stimulus.

Technological advances reveal increasingly detailed information about molecular and cellular mechanisms of perception, emotion, cognitive function and behavior. At the same time, the complexity and adaptive nature of the nervous system result in a certain fluidity of information about the brain. Theories of brain structure and function continue to evolve and however much is known, much remains to be discovered.

Ethical Issues

The concerns that are encompassed in the domain of neuroethics are associated uniquely or especially with the practice or conduct of neuroscience research or with the application of neuroscience findings.

CONDUCTING NEUROSCIENCE RESEARCH. All areas of research share some ethical issues associated with the nature of research itself. Integrity of the research process affecting reliability of results, appropriate allocation of credit, and management of potentially conflicting interests are among the many issues that are common to all areas of research to one degree or another, and do not fall exclusively into the purview of neuroethics. However even topics that are common to many fields, such as the humane treatment of research subjects and controlling for bias in research design, have special relevance to research in the neurosciences.

As an example, one of the ethical principles fundamental to research involving humans is respect for persons and its corollaries of autonomy and informed consent or decision making. Among the implications of these principles are that individuals must voluntarily choose to participate in research (i.e., they cannot be coerced, deceived, or manipulated into participating), and that they can discontinue their involvement at any time during the research. One broad area of neuroscience research explores the causes and mechanisms that underlie dementia, including Alzheimer's disease, with a primary long-term goal of developing treatments and a cure. Participation or involvement of individuals with early symptoms can be invaluable to various lines of research into any disease. However the capacity of ill individuals, even those who are healthcare professionals, to make a fully informed decision to participate in research is debatable. Moreover unlike most ill individuals, for example those with heart disease, patients with dementia may have a diminished capacity to fully comprehend the ramifications of consent to research participation depending upon the extent of their disease. As an example, agreement to provide a monthly blood sample may seem less onerous when an individual can comprehend an altruistic goal of developing a cure for Alzheimer's disease. As the disease progresses the individuals understanding of the research may become little more than the awareness of a painful needle. While the clinical research community has developed proxy or surrogate consent as a strategy that allows family members or other legal guardians to give consent for the patient, the notion of research participation as a fully informed choice becomes questionable and problematic.

Neuroscience research with laboratory animals also poses special concerns. Required for both the ethical and scientific justification of the use of laboratory animals in research is that the work has the potential to provide valuable insights into biological structure and/or function that lay the foundation for the understanding, and ultimately treatment, prevention, and/or cure of disease. The companion expectation is that research with animals can be carried out with minimal or no pain, suffering, or distress to the animals. Some areas of neuroscience research challenge these two concepts. For example, when research focuses on mental conditions like schizophrenia or elements of cognition like intentionality, investigators must make assumptions about the similarity between the brain activity of laboratory animals and humans. The reliability of those assumptions and their implications for the understanding of human brain function and disease can be questioned. Moreover when the focus of research is pain or stress, then pain and/or stress are unavoidable elements of the research itself. Indeed, paradoxically, the more like humans a research animal is, the more informative is the research yet, one could argue, the less reasonable the justification for conducting the research in animals because it is unethical to investigate the phenomenon in humans. Institutional Animal Care and Use Committees (IACUCs), in particular, and, to a lesser degree, the peer review process consider the ethical issues associated with the use of animals in research. However the special problems posed by neuroscience research may not always be explicitly or fully considered.

Controlling for bias in research design, while always an important aspect of research ethics, is of particular relevance and concern in neuroscience research because of the extent, nature, and implications of findings in this field. Assumptions that underlie research questions may not be adequately investigated themselves. Yet they are likely to reflect conscious or unconscious bias that arises from long-standing socially determined beliefs. For example, it is widely assumed that some differences in male and female behavior reflect anatomical and physiological differences in the brains of males and females. While this may be true, it is not clear whether biological differences relevant to behavior result from the presence of different sex-related genes or molecules, or from differences in the myriad external factors that shape interactions with others from birth, or a combination of both. Whatever the basis of sex differences in behavior, the extent to which they are linked to biology and perceived as predetermined and immutable can have far-reaching ramifications for education, employment, healthcare, and other areas of social and public policy.

APPLICATION OF RESEARCH FINDINGS. The ethical issues associated with the use of research findings are linked to the particular application: Who uses the information, how is used (e.g., to monitor brain activity, to manipulate behavior, etc.), and for what purpose (e.g., therapy, enhancement, etc.). In addition, whether the information is about the general population or about a particular individual, the accuracy and reliability of the information is always an important consideration, as is accurate presentation of its limits because it directly affects the capacity of individuals to make informed decisions.

Individuals may seek information for self-knowledge, therapy, or self-enhancement. If the information is general and benign, with noninvasive applications (e.g., mnemonic techniques for remembering names) the accuracy and reliability of the research findings are less critical than if the information may expose an individual to risk (e.g., research that suggests a particular dietary supplement is an effective sleep aid although it has the potential for inducing heart arrhythmias). When research findings provide information specific to a particular individual, the accuracy and reliability of the information is critically important depending on the nature of the information and the purpose for which it is being gathered. Thus the reliability of predictions of a test for a debilitating hereditary neurological or mental illness is key. If test findings are perceived to be consistent indicators (markers) for the disease (i.e., individuals with a positive test result inevitably get the disease), then the actual reliability and limits of the test (and the research upon which the test is based) are critically important so that individuals being tested can make adequately informed medical and personal decisions. At the other end of the continuum, if the test is an indication of a predisposition for a mental illness

(a much more common occurrence), then additional ethical concerns arise.

In particular, given the dynamic and interactive nature of the human mind, knowledge of the identification of a biological element that is neither necessary nor sufficient for a mental illness but rather indicates a predisposition for that condition can become a contributing factor in its own right, and a self-fulfilling prophecy. Thus ethical concerns regarding information about predispositions to disease are related not only to the accuracy and reliability of the test, but also to the nature of the nervous system and the independent power of the information itself. In addition, given the continuing social stigma associated with mental illness, provision of test results to third parties, whether health insurance providers, employers, family members, or others, may also contribute to stress and the development, expression, and manifestation of disease. As a result, information about mental function poses risks as well as benefits because it is provided in a personal and social context with which it interacts. Technological advances can improve the accuracy of the information but may not have much impact on the contexts in which it is provided.

When neuroscience research yields scientific information and technological developments that make possible access to the brain activity of others, additional ethical concerns arise. Fundamental to this is the actual and perceived correlation between brain activity and mental activity. The possibility of monitoring the mental activity of others raises concerns about privacy and notions of individual integrity. In general, respect for the individual includes the right to privacy and exceptions are only allowed when the health, safety, and welfare of that individual, or others, is threatened. The extent of the invasion of privacy (and attendant harm to the principle of respect for persons and potential harm to that individual) is balanced against the seriousness and certainty of the harm or threat to be averted. An obvious setting in which such privacy might be invaded is in the criminal justice system. It is well-established that eyewitness testimony is unreliable. The potential for conflicting interests among experts as well as the concerns of a hostile or threatened witness can also call into question the reliability of courtroom testimony. Thus, if and when it is possible and in the putative interest of justice, authorities might seek to access directly the memories of a witness or an accused to determine what actually happened. Similarly they might seek access to the mental activity of a perpetrator in order to determine the individual's intentions.

Increasingly, advances in technology also make possible direct intervention in brain function in an even more nuanced and refined way. In the mid- to late-twentieth century, brainwashing, electroconvulsive shock therapy (ECT), and psychosurgery were used to alter brain function and behavior. These procedures are relatively crude and invasive. Current psychosurgery methods, referred to as stereotaxic surgeries, use heat or radiation to destroy very specific tissue identified using brain imaging techniques. Compared to earlier forms of psychosurgery (also known as functional neurosurgery for psychiatric disorders), such as prefrontal lobotomy, stereotaxic surgeries are relatively less invasive, success rates are high, and complications are minimal. Nevertheless the procedures are irreversible, and surgeries (and electroconvulsive shock therapy) are employed in therapy only as a last resort for treating serious mental illness that has not responded to other forms of treatment.

With increased understanding of brain chemistry, physiology, and pharmacology has come the development of pharmacological agents targeted to particular biochemical pathways because research indicates that the neurotransmitter systems associated with these pathways are associated with particular mental activity. These pharmacological agents are primarily designed to be prescribed to treat an individual's self-report of dysfunction. Issues of benefit versus risk, patient expectations and informed decision making, and allocation of resources are ethical issues that arise with any therapy. However because brain dysfunction and mental illness are often at the extreme ends of normal brain function, some therapeutic agents may be able to enhance normal function. For example, some treatments for Alzheimer's disease or other forms of dementia may be able to enhance normal cognitive function. The use of pharmaceuticals for nontherapeutic enhancement rather than therapy not only changes the benefit versus risk analysis, and alters discussions of the fair allocation of scarce resources, but also raises questions regarding who is being enhanced, by whom, and for what purpose.

Computer Brain Interfaces. In the early twenty-first century research is exploring the possibility of electrochemical implants that can serve as a brain computer interface (BCI). These, too, are initially designed to be therapeutic (e.g., to overcome physical limitations or visual deficits). However there is a distinct and important difference between the BCI that makes the brain of a quadriplegic a *transmitter* that can manipulate the external environment (e.g., move a cursor on a computer screen) and an implant that makes the brain a

receiver either for information about the outside world or for altering brain function (e.g., to treat obsessive-compulsive disorder).

While manipulation and control of others are always ethically problematic because they violate the basic bioethical principle of respect for individuals and their autonomy, two primary considerations are (a) the degree of invasiveness and (b) the extent to which the individual being controlled is aware of, and consents to, the control (Dworkin 1976). The degree of invasiveness is a fluid notion since education and subliminal suggestion while not physically invasive like pharmaceuticals and BCIs can permeate one's thinking with long-term, widespread effect (e.g., educational programs that include evolutionary theory and/or creationism or that exclude reference to or acknowledgment of the Jewish holocaust and/or Chinese *comfort women*). Moreover the conscious intent of manipulation or control may well be in the eye of the beholder. Thus education while not physically invasive is potentially manipulative, subliminal suggestion is not physically invasive but is designed to be manipulative, and psychoactive agents and BCIs are invasive but can be perceived as manipulative or not. Scientific and technological advances that reflect new or refined understanding of brain structure and function have the potential for making possible more specifically targeted monitoring and manipulation of individual or group perceptions and function, but the ethical concerns are akin to those raised regarding con artists, rabble rousers, propaganda, and deceptive advertising.

Issues of Self-Knowledge. More complicated are the ethical issues associated with the scientific and technological advances in neuroscience that make possible increased nontherapeutic self-knowledge, modification, and enhancement. While insights into one's own motivation, self-understanding, personal growth, and development are generally lauded, artificial means for obtaining such insights, for example, through psychoactive *recreational* drugs, is often frowned on primarily because of the potential risks associated with psychoactive drugs and their uncertain benefits. Yet it is possible that techniques in brain imaging may reveal individual traits or thought patterns similar to (or different from) those revealed by less scientifically or technologically dependent approaches (e.g., psychotherapy, meditation, or prayer). Psychotherapeutic agents that modify brain chemistry to treat mental conditions (e.g., anxiety, depression, or schizophrenia) are prescribed, and taken, to modify brain function, mental activity, and behavior. Individuals taking these agents may feel *more like themselves* or conversely *not themselves*. This not only prompts the philosophically interesting question of how one defines and recognizes *the self*, but also raises ethical concerns regarding the extent to which peer and/or societal pressures may lead an individual to modify his or her mental processes, behavior, or other elements of the self in order to conform to the expectations of others or to internalized social norms. In addition, artificial enhancement of performance, whether mental or physical, is highly controversial, and the potential development of cognitive and/or emotional enhancers to gain personal advantage raises issues of respect for persons (i.e., the self and others) and informed-decision making, risk versus benefit, the fair allocation of resources, and fairness in competition.

Neurobiology of Ethics

The other side of the conceptual coin of neuroethics is the neurobiological underpinnings of ethical thought and practice (Caplan 1983, Roskies 2002). The cognitive and emotional elements that contribute to ethical reasoning and behavior are relatively unexamined. Nevertheless ongoing and future neuroscience research is likely to contribute to an intellectual understanding of moral development, the processes of moral reasoning and decision making, and the mechanisms by which ethical decisions are expressed in behavior. How society understands notions of free will and moral agency will be influenced by the findings of neuroscience research. Of necessity this understanding will reflect recognition of the limits of human capabilities: "it simply makes no sense to talk about ethical ideals that are beyond the reach of human conduct, motivation and behavior" (Caplan 1983, p. 106).

However a potential pitfall, as with research in neuroscience in general, is the way that conscious and unconscious assumptions may introduce an inappropriate bias into research design, analysis, or reporting. For example, it is widely assumed that moral reasoning is a rational rather than emotional process. This assumes a potentially false dichotomy in brain processing. Thus the ethical issues that are likely to be raised by future investigations of the neurobiological basis of ethics will be complex and dynamic like the nervous system itself.

Controversies

As suggested, a critical element in identifying and examining some ethical issues associated with neuroscience hinges on the relationship between brain activity and mental activity. While the consensus of the neuroscience community is that, at least in humans, brain and mind are two sides of the same coin, there is considerable controversy and disagreement regarding the degree to which mental activity can be correlated with, identified as, and reduced to brain activity. An early notion was that each individual memory was embodied in a single cell so that, for example, every individual has a specific cell dedicated to his or her grandmother (hence the name *grandmother cell theory*). That particular concept of memory has been discredited. Moreover, the view that patterns of brain activity detectable with imaging technologies or by monitoring electrical changes can be identified with specific cognitive functions is not universally accepted. The reliability of this correlation is central to the ethical concerns associated with the scientific and technical developments in neuroscience.

There is much more to be learned about the structure and function of the nervous system. It is clear that the ethical issues inherent in the practice, applications, and implications of this area of research will continue to become apparent.

STEPHANIE J. BIRD

SEE ALSO *Bioethics; Consciousness; Emotion; Medical Ethics; Research Ethics.*

BIBLIOGRAPHY

Caplan, Arthur L. (1983). "Out with the 'Old' and In with the 'New'.—The Evolution and Refinement of Sociobiological Theory." In *Ethical Questions in Brain and Behavior: Problems and Opportunities*, ed. Donald W. Pfaff. New York: Springer-Verlag.

Chorover, Stephan L. (1979). *From Genesis To Genocide*. Cambridge, MA: MIT Press.

Dworkin, Gerald. (1976). "Autonomy and Behavior Control." Hastings Center Report 6, no. 2: 23–28.

Farah, Martha J., and Paul Root Wolpe. (2004). "Monitoring and Manipulating Brain Function: New Neuroscience Technologies and Their Ethical Implications." Hastings Center Report 34, no. 3: 35–45.

Hahn, William E.; Jeffrey Van Ness; and Nirupa Chaudhari. (1982). "Overview of the Molecular Genetics of Mouse Brain." In *Molecular Genetic Neuroscience*, eds. Francis O. Schmitt, Stephanie J. Bird, and Floyd E. Bloom. New York: Raven Press.

Marcus, Stephen J., ed. (2002). *Neuroethics: Mapping the Field: Conference Proceedings, May 13–14, 2002, San Francisco, California*. New York: Dana Press.

Office of Technology Assessment. (1984). *Impacts of Neuroscience*. Washington, DC: Congress of the United States.

Pontius, Anneliese A. (1993). "Neuroethics vs. Neurophysiologically and Neuropsychologically Uninformed

Influences in Child Rearing and Education." *Psychological Reports* vol. 72, pp. 451–458.

Roskies, Adina. (2002). "Neuroethics for the New Millenium." *Neuron* 35 (July 3): 21–23.

Valenstein, Elliot S. (1986). *Great and Desperate Cures: The Rise and Decline of Psychosurgery and Other Radical Treatments for Mental Illness*. New York: Basic Books.

Valenstein, Elliot S., ed. (1980). *The Psychosurgery Debate*. San Francisco: Wh Freeman.

NEUROSCIENCE ETHICS

SEE *Neuroethics*.

NEURALITY IN SCIENCE AND TECHNOLOGY

• • •

The fundamental relationship among science, technology, and ethics is often claimed to be one of neutrality. After all, science and technology can be put to good or bad uses by good or bad people; they are thus value-neutral. It is sometimes implied paradoxically that this neutrality constitutes the special value of science and technology. In contrast, critics have argued that assertions of neutrality are attempts to escape responsibilities for the specific consequences of various scientific and technological projects. How can weaponized anthrax spores designed to kill people be described as value-neutral? This entry attempts to reference some of these claims and counterclaims and provide an analysis for their assessment.

Preliminary Distinctions

It is important to note that neutrality may be modified not just by moral or ethical but also by political, aesthetic, religious, epistemological, ontological, or any number of other qualifiers. Most discussions deal with issues of what are called axiological neutrality, that is, some form of value. The following discussion of value neutrality thus aims to cover questions of not just of moral or ethical but also political, aesthetic, religious, and related senses of neutrality, though not epistemological, ontological, and other forms of neutrality.

With regard to value neutrality a distinction should be made between the antecedent values that motivate the realization of science and technology, and the value that science and technology have once they are realized. Claims about neutrality and antecedent values focus on the value judgments that moti-

vate scientific and technological activity: Science or technology is neutral with respect to a set of values if its processes and products are not informed by those values. Claims about the value of science and technology once realized focus on the consequences of scientific and technological activity and the value of those consequences. In this context those who make claims about neutralism assert that scientific and technological activities merely create possibilities but do not cause any specific possibilities to be realized. To actualize any of those possibilities, other events beyond science (the investigation of phenomena) and technology (the creation of specific objects, or "artifacts") are needed, and those other events are not conditioned, required, or determined by science or technology. On this view, the value neutrality of science and technology is a product of their causal neutrality, of their not being sufficient in themselves to bring about either good or bad consequences.

Neutrality of Science and Technology With Respect to Antecedent Values

A simple interpretation of the claim that science is neutral is that science is value-free. Science is the impartial search for truth without regard for the interests of those affected. If scientists are allowed to work without external hindrance, they will provide objective answers to questions such as whether tanning booths cause cancer and whether humans have evolved from nonhumans.

This position is informed by a fundamental presupposition: The world is independent of how humans might want it to be. The natural order is not determined by human interests. If people want to get as close as possible to understanding how things really are, they must leave their values—expressions of what they want—out of that effort.

This view overlooks the fact that although the natural order may not be influenced by human interests, science is. What people are interested in is an expression of their values, and one of the things they want is to understand how things are. Science, like every other human activity, is driven and influenced by human values. The idea of neutrality with respect to values must be modified to account for this argument.

The standard modification is to divide the antecedent values motivating science into two categories. On the one hand there are the external or contextual values that direct scientific work. These values include the political, economic, and cultural interests that scientists bring to their practice. On the other hand there are the internal or constitutive values that direct science. These are the scientific values of scientists. Patrick Grim (1982) identifies the most fundamental internal values as *truth* and *demonstration*. Scientists want to find out which claims are true and which are false, and they insist on some kind of demonstration as the means of sorting true from false claims.

The idea of the neutrality of science with respect to values can be reformulated as follows: Although some set of external values is always present and may play a role in determining which problems a scientist will work on, once scientists begin their work, those external values should play no role in guiding procedure or determining findings. Instead, internal values should take over and guide the application of methods, the determination of results, and the reporting of both.

Critics have challenged this view, arguing that contextual values are present even in the application of method and the determination of findings (Longino 1990). However, the idea of scientific neutrality cannot be eliminated as an ideal, for it is *what people want from science*: People do not want contextual values to determine scientific results. Suppose the question is whether exposure to ultraviolet rays in tanning booths increases the risk of contracting skin cancer. For many people the reports of findings generated by tanning booth manufacturers would not be sufficient to answer the question even if the internal norms of truth and demonstration were values strongly held by the manufacturers' scientists. The context of that research raises suspicions. People would want independent verification by scientists with different contextual values, preferably values that are neutral with regard to the investigation at hand. Thus, people recognize the distorting power of contextual values and try to minimize that distortion; that is, people seek to get science as close to the ideal as possible.

In regard to the idea that technology is neutral with respect to values, it again becomes clear that this notion cannot be maintained in the form of a strict absence of values. Technology, like science, is a human endeavor that necessarily is guided by values: conceptions of what is good or desirable for humans to be or do.

One approach to maintaining a form of freedom from values in technology parallels the case of science. An external-internal distinction can be made, with all the political, ethical, social, and other values on the external side and the values of *effectiveness* and *efficiency* seen as the internal, constitutive values of technology.

Just as truth, the fundamental constitutive value of science, is independent of human aim, so too is effectiveness. Effectiveness is the degree to which an action achieves its end. Given an end, a technological means to that end is either effective or not effective, and that effectiveness is independent of people's values (what people want). To this extent the independence of technology from external values parallels that of science.

Efficiency, however, is problematic. As Alex Michelos (1972) points out, efficiency is not an unanalyzable basic value but a relationship between other values, specifically a ratio between what people value as benefits and what they value (negatively) as costs. Judgments of the efficiency of an action depend on what is counted as its benefits and costs, and the decision about what to count as benefits and costs is external to technology. Consider, for example, the different assessments of efficiency that can be obtained for a technology such as a poultry-eviscerating line if in one assessment the physical and psychological costs borne by those working on the line are excluded whereas in another assessment those costs are included. Efficiency is a value derived from external, non-technological values. As one description would have it, efficiency is a socially constructed value.

Neutrality of Science and Technology with Respect to Consequences

The second form of value neutrality is founded on two claims: (1) there is always more than one possible use for the products of science or technology, and (2) the activities or products of science and technology do not determine if or how those products (knowledge or artifacts) will be used.

The claim that there are multiple uses for every piece of knowledge or artifact seems correct in the case of basic science: Because the knowledge that the basic sciences provide is general knowledge of the most fundamental composition, structure, and events of the natural world, it seems that there are always several possible applications. For example, knowledge of elements and their atomic structure can be applied in metalworking, firefighting, criminology, cooking, and so on. A more specific piece of knowledge, such as knowledge of geologic fault lines, can be used to predict earthquakes and set insurance rates for homeowners.

The applied sciences, however, seek to focus basic science on materials of and processes for possible use; thus, applications are already "in mind." In some cases the range of applications is wide, such as with knowledge about the electrical properties of ceramics. In other cases the range of uses is more narrow: Knowledge about the microstructure of oil-bearing shale seems to have only one application.

However, a neutralist might contend that there could be other applications of a piece of specific knowledge that have not yet occurred to anyone. Rather than known applications in the sense of current, technologically feasible applications, a neutralist might contend that the range of applications is the set of logically and materially possible applications, including those not yet conceived. On this view the range of applications for any piece of knowledge is unknown, although in principle there would still be a finite range of uses for every piece of scientific knowledge.

With regard to technology, the claim that artifacts can serve ranges of uses needs clarification. If one focuses on an artifact's use in the sense of what that artifact does—its *function*—it is clear that many artifacts have more or less specific functions built into them. A canoe transports people and goods over water; that is what it does, and it does nothing else. Although a canoe may be turned upside down on land to provide shelter, that is not the purpose for which it was designed, and a canoe is ill suited to that purpose. Similarly, the function of a wool topcoat is to shield one's body from the cold; it is not well suited to serve as a blanket or a painting dropcloth. To this extent the neutralist case regarding multiplicity of purposes is overstated.

A second sense of use is the *purpose* served by artifacts in performing their functions. This sense of the word points to *why* humans make artifacts do what they do. Purposes generally come in hierarchies: People do A in order to get B, want B in order to get C, and so on. If this is the meaning of the neutralist claim that artifacts can serve multiple purposes, that claim is true but trivial. However, the neutralist claim here is that artifacts are flexible with respect to their *immediate* purpose: A carpenter's hammer can perform its functions of driving and pulling nails in serving the purpose of hanging a picture or constructing gallows; a bicycle can perform its function of moving people over land, for the purpose of making deliveries or getting exercise. The history and sociology of technology tend to highlight this phenomenon. Alexander Graham Bell thought that the telephone would be used for business communication only, never imagining its use for personal communication. The sociologist Michel de Certeau (1984) has noted numerous creatively adept technologies.

Assessing this version of neutralism, it must be granted that people use canoes and hammers and bicycles to serve multiple purposes. The same thing is true of machine tools and electrical power grids. Yet there are many artifacts that can serve only one purpose in performing their functions. A bomber flies off and drops bombs in order to damage people and things. That is the only immediate purpose a bomber serves. A bulletproof vest shields one's body from a bullet (its function) so that one may survive a shooting (its purpose). Washing machines and raincoats are other examples of single-purpose artifacts. If this argument is correct, the claim that artifacts can serve multiple purposes is false as a universal proposition: The question of the neutrality of artifacts with respect to the range of purposes they serve must be decided on a case-by-case basis.

The second neutralist claim regarding consequences—that science and technology do not determine *that* their products be used or to *what* use those products will be put—is most plausible in the case of pure science. The activity of pure science is removed from the context of practical use in terms of both the content of the activity and the intent of the practitioners. Indeed, there may not be currently possible uses.

Technology has a different relationship to practical context. Although it is correct to say that humans can decide not to use an artifact they have created, the whole point of technological activity is use. Human needs are insufficiently met by the unmediated interaction of people with nature: People *must* make and use artifacts in order to live. Although people are free to choose not to use a particular artifact, they are never free to choose to use no artifacts.

A focus on artifact use reveals one way in which technology is not always value-neutral. Artifacts determine *how* they are used: All artifacts, from saws to computers, impose methods of operation on would-be users, and people who effectively use artifacts for *any* purpose—good, evil, or neutral— use them in accordance with their operational functions. One cannot cut a board effectively by holding on to the blade of a saw. Artifacts determine what behaviors must be brought to bear by humans in order to operate them.

At least in some cases the exercise of those behaviors is directly beneficial or detrimental to the agent independently of the purposes served, objects made, or payment gained. Using a computer for any purpose causes eyestrain. In such cases the artifact used is a causal condition of positive or negative value regardless of human intentions regarding its use or its instrumental consequences.

This argument may apply to scientific activity as well. To the extent that such activity produces satisfying or dissatisfying experiences, science may have value independently of the values that constitute it or its instrumental value.

An argument raised against neutralism is that in choosing to use a certain technological object or system one is simultaneously, if unconsciously, making a commitment to a certain form of social organization. Lewis Mumford (1964) and Langdon Winner (1986) have argued, for instance, that nuclear power plants typically require a hierarchical social organization with authoritarian relationships of command and control. Such forms of organization are certainly not politically neutral. Empirical research on the deployment of specific artifacts in specific organizations (Liker, Haddad, and Karlin 1999) raises serious questions about the generalizability of this argument. The evidence suggests that although artifacts determine task characteristics such as skill variety, the nature of organizational governance and control over technological activity is a matter of human choice.

RUSSELL J. WOODRUFF

SEE ALSO Biological Weapons; Critical Social Theory; Efficiency; Existentialism; Values and Valuing.

BIBLIOGRAPHY

Certeau, Michel de. (1984). The Practice of Everyday Life, trans. Steven Rendall. Berkeley: University of California Press.

Grim, Patrick. (1982). "Scientific and Other Values." In Philosophy of Science and the Occult, ed. Patrick Grim. Albany: State University of New York Press. Argues that all science is shaped by values, and distinguishes its inessential, background values from its essential values of truth and demonstration.

Liker, Jeffrey K.; Carol Haddad; and Jennifer Karlin. (1999). "Perspectives on Technology and Work Organization." Annual Review of Sociology 25: 575–596. Reviews the literature on the relation between technology and the nature of work, and concludes that technology's impact is contingent on a large number of non-technical factors such as the reason for introducing the technology, management philosophy, labor-management contracts, and the degree of agreement about technology and work organization.

Longino, Helen. (1990). Science as Social Knowledge: Values and Objectivity in Scientific Inquiry. Princeton, NJ: Princeton University Press. Challenges the claim that the scientific practice is shaped by constitutive or cognitive values while being routinely insulated from contextual values.

Michalos, Alex. (1972). "Efficiency and Morality." Journal of Value Inquiry 6: 137–143. Argues that efficiency is not an unanalyzable technical value, but is rather a ratio of benefits to costs that is always constructed by reference to norms outside technology.

Mumford, Lewis. (1964). "Authoritarian and Democratic Technics." Technology and Culture 5(1): 1–8. Makes the case that two types of technologies have existed side by side since the late Neolithic era: democratic technologies—small-scale, employing concrete, tacit knowledge, controlled by the individuals directly engaged, and powered by individual humans or animals; and authoritarian technologies—large-scale, employing abstract symbolic knowledge, controlled by disengaged authorities, and powered by mass armies or machines.

Rosenbrock, Howard. (1999). "Engineers and the Work That People Do." In The Experience of Work, ed. Craig Littler. New York: Palgrave Macmillan. Describes how assumptions guiding the construction of technical apparatus tend to implicitly devalue humans and result in inefficient use of the humans interacting with the apparatus.

Rudner, Richard. (1953). "The Scientist Qua Scientist Makes Value Judgments." Philosophy of Science 20(1): 1–6. Argues that ethical judgments are indispensable elements of scientific practice: Because no scientific hypothesis is ever completely verified, the decision to accept a scientific hypothesis as sufficiently warranted always depends on a judgment about how ethically significant a possible mistake would be.

Winner, Langdon. (1986). "Do Artifacts Have Politics?" In The Whale and the Reactor. Chicago: University of Chicago Press. Argues that technological objects can be political in two ways: some are designed to control or channel the decisions and actions of people; others are elements of socio-technical systems that either require or are strongly compatible with specific forms of social organization, typically authoritarian, hierarchical social organization.

Woodruff, Russell. (1997). "Artifacts, Neutrality, and the Ambiguity of 'Use.'" Research in Philosophy and Technology 16: 119–127. Distinguishes "use" as function, purpose and method, and uses these distinctions to discuss the neutrality of technological objects.

NEW ATLANTIS

SEE Atlantis, Old and New.

NEWTON, ISAAC

• • •

A central figure in the foundation of modern physics, mathematics, optics, and the scientific method, Sir Isaac Newton (1642–1727) was born in the Lincolnshire hamlet of Woolsthorpe on December 25. Newton matriculated at Trinity College, Cambridge in 1661, receiving there his B.A. (1665) and M.A. degrees (1668). He became a Fellow of the College in 1667, and in 1669, at the age of twenty-six, was appointed Lucasian Professor of Mathematics. Election to the Royal Society followed in 1672. In 1696 Newton relocated to London, where he became Warden and then Master of the Royal Mint.

Sir Isaac Newton, 1642–1727. An English scientist and mathematician, Newton made major contributions in mathematics and theoretical and experimental physics and achieved a remarkable synthesis of the work of his predecessors on the laws of motion, especially the law of universal gravitation. (© Bettmann/Corbis.)

He was elected President of the Royal Society in 1703 and knighted in 1705. He died on March 20 in London.

Newton's greatest discoveries and innovations came during his Cambridge years. In the mid-1660s he developed the calculus. His 1672 paper on colors confirmed the heterogeneous nature of light. In the early 1670s Newton constructed the first practical reflecting telescope. In the following decade, the mathematical physics of the *Principia mathematica* (1687) yielded spectacular results: the laws of motion, the inverse-square law of universal gravitation, elegant mathematics to underpin astronomy and physics, and the unification of terrestrial and celestial mechanics. In the three editions of this work, he also developed principles of an inductive method that still serve science in the early twenty-first century. The *Principia* is the grandest achievement of seventeenth-century mechanical philosophy and one of the most revolutionary books in the history of science. Newton's *Opticks* (1704) codified earlier research and placed optics on a firm footing; later editions helped establish an experimental agenda for the subsequent decades. As President of the Royal

Society, Newton reinvigorated the organization's experimental program. His first curator of experiments, Francis Hauksbee, Sr., developed an electro-static machine that helped foster the study of electricity in the eighteenth-century. His second curator, John Theophilus Desaguliers, exemplified the Baconian ideal of producing useful knowledge through liaising with proto-industrialists, developing mine ventilation machines, and during his employment as a waterworks engineer on the Thames.

Enlightenment Image and Correction

Despite his popular association with a deterministic and purely mechanical cosmos, Newton's image as a rationalist proponent of a clockwork universe is a wishful construction of Enlightenment apologists who re-crafted him in their own mold. Newton's natural philosophical ethos conforms more closely to Renaissance ideals. He was committed to the goal of recovering the *prisca sapientia* (ancient wisdom), believing that the ancients possessed superior forms of knowledge that could and should be recovered. Newton's public and private writings show that he rejected the idea of a mechanized universe, holding instead to a providentialist view in which God periodically intervenes to keep Nature on course. Newton's supporter Samuel Clarke (1675–1729) eloquently defended these ideas in his famous correspondence of 1715 to 1716 with the German philosopher Gottfried Leibniz (1646–1716). Newton also worked to reintroduce spirit into natural philosophy. Further his surviving papers reveal that he was not only a practicing alchemist, but that he devoted more time and energy to the study of theology and prophecy than to natural philosophy.

These commitments did not remain in a separate intellectual sphere, but played a role in shaping Newton's metaphysics and his natural philosophical style. An example of this is his adherence to a form of epistemological dualism in which knowledge is divided into two categories. Lower, relative forms of knowledge are accessible to the vulgar, while higher, absolute forms of knowledge can only be penetrated by the adept—a distinction seen in the thought of the Pythagoreans, Plato, Maimonides, in the alchemical tradition, and Newton believed, in the Bible. Accordingly Newton emulated the coded literary style he believed was used by the Hebrew prophets and the Pythagoreans in order that only the *wise* would understand his meaning (Daniel 12:10). This helps explain why so many had so much difficulty understanding his *Principia*. Newton once explained that "to avoid being baited by little Smatterers in Mathematicks ... he designedly made his Principia abstruse; but yet so as to be understood by able Mathematicians" (Newton in Snobelen 2001, p. 205).

The distinction between the relative and the absolute plays a role in Newtonian physics as well. In the "Scholium to the Definitions" at the beginning of the Principia, Newton distinguishes relative space and time from absolute space and time. Absolute space is rigid and immovable, while "absolute, true and mathematical time" flows evenly and uniformly; both exist "without reference to anything external" (Principia, p. 408). In contrast, the space and time of sensation and measurement are relative or relational. Thus he writes in the "Scholium": "Accordingly, those who there interpret these words [time, space, place, motion] as referring to the quantities being measured do violence to the Scriptures. And they no less corrupt mathematics and philosophy who confuse true quantities with their relations and common measures" (Principia, p. 414). By alluding to biblical hermeneutics, Newton hints at a link between theology and science. For Newton, absolute space and time are predicates of God's omnipresence and eternal duration, an idea he developed from biblical theology, Stoicism, Philo, and Rabbinical thought. As a reflection of this, Newton suggested in private that God's omnipresence might be the cause of gravity, something that would help explain the universal nature of the phenomenon.

Newtonian Method

In the "Rules of Reasoning" laid out in the Principia, Newton advocates an inductive approach to the study of Nature. This approach is also commended in the "General Scholium," in which he expresses a disdain for discussions about substance and states that his natural philosophy does not extend beyond a description of the phenomena. Newton was satisfied with his ability to describe the phenomenon of universal gravitation mathematically; as for the ultimate cause of gravity, he famously declares: "I feign no hypotheses" (hypotheses non fingo). Both the inductive method and the derogation of frivolous hypotheses are outlined in Query 31 of the Opticks: "As in Mathematicks, so in Natural Philosophy, the Investigation of difficult Things by the Method of Analysis, ought ever to precede the Method of Composition. This Analysis consists in making Experiments and Observations, and in drawing general Conclusions from them by Induction, and admitting of no Objections against the Conclusions, but such as are taken from Experiments, or other certain Truths. For Hypotheses are not to be regarded in experimental Phi-

losophy" (Opticks, p. 404). Natural philosophical reasoning should be a posteriori rather than a priori.

But Newton does not reject the use of hypotheses outright; instead, he eschews dreaming up vain and unwarranted hypotheses, especially those that lead to system building. This approach is a pointed attack against the French philosopher René Descartes (1596–1650). For Newton, as for his most passionate disciples, there are also moral corollaries to scientific method. When Roger Cotes, Cambridge's Plumian Professor of Astronomy, wrote the preface to the second edition of the Principia, he contrasted the Newtonian inductive method with the speculative-hypothetical approach: "Those who take the foundation of their speculations from hypotheses, even if they then proceed most rigorously according to mechanical laws, are merely putting together a romance, elegant perhaps and charming, but nevertheless a romance."

Similarly, Colin Maclaurin, the Scottish Newtonian and professor of mathematics at Edinburgh, compares Newton's inductivism with "that pride and ambition, which has led philosophers to think it beneath them, to offer anything less to the world than a complete and finished system of nature; and, in order to obtain this at once, to take the liberty of inventing certain principles and hypotheses, from which they pretend to explain all her mysteries" (Maclaurin, Account of Newton's Discoveries, p. 7). Maclaurin likens this method to beginning "at the summit of the scale, and then, by clear ideas, pretend[ing] to descend though all its steps with great pomp and facility, so as in one view to explain all things" (p. 18). Instead Newton's experimental method, which begins with analysis before progressing to mathematical synthesis, is the better approach to truth in natural philosophy, even though "the beginnings are less lofty" because "the scheme improves as we arise from particular observations, to more general and most just views" (p. 18).

Right science must be preceded by and coupled with right method. Natural philosophical arrogance and presumption leads to error, corruption, and systems constructed out of thin air. Newton's followers championed the inductive method that prioritized gathering empirical evidence as a humble technique in contradistinction to what they saw as the intellectual hubris.

Newton was convinced that similar methods would also lead to a recovery of true, biblical doctrine and the teachings of the primitive Christians. Rather than shape Scripture to fit a priori theories, Newton believed God's truth should be drawn directly from a close reading of the Bible. This project led him to reject several central orthodox teachings as doctrinal corruptions, including the Trinity and the immortality of the soul. Newton disdained the fourth-century hypothetical and ontological discussions of the substance of God that distorted the unipersonal God of the Bible into the Trinity—a doctrine that he saw as little better than polytheism. By the standards of his day, such conclusions made him a heretic and brought the need for caution and circumspection. Nevertheless, Newton covertly attacked the Trinity in his "General Scholium." That this attack appeared with an overt challenge to Cartesian planetary vortex theory shows that Newton believed that corruption in natural philosophy was linked to corruption in religion. The inductive approach extended to his prophetic interpretation, and there are striking parallels between his "Rules of Reasoning" and a series of prophetic rules he developed earlier in the 1670s.

Newton applied an inductive approach to his natural theology as well, writing in one manuscript "God is known from his works" (Newton in McGuire 1996, p. 119) Newton was convinced that an inductive program in natural philosophy would lead to God. Near the end of Query 28 in the *Opticks* Newton argues that "the main Business of natural Philosophy is to argue from Phænomena without feigning Hypotheses, and to deduce Causes from Effects, till we come to the very first Cause, which is certainly not mechanical" (*Opticks*, p. 369). Likewise, at the end of his discussion of God in the General Scholium, Newton asserts that "to treat of God from phenomena is certainly a part of natural philosophy" (*Principia*, p. 943).

Assessment

The recovery of this pre-enlightenment understanding of Newton disrupts common contemporary notions of Newton as an advocate of completely mechanical and deterministic universe. Newton may not have anticipated the degree to which the ethical and religious corollaries would be separated from his natural philosophy after his death by Enlightenment thinkers and later by positivists. Yet it is clear that he attempted to found a science that is thoroughly infused with a religious understanding of nature and that emphasizes the need for moral virtue on the part of its practitioners. While most in science in the early twenty-first century accept the Enlightenment reading of Newton's legacy, Newton himself would have seen the development of the study of nature after his death as another corruption to be deplored.

Although Newton recognized disciplinary distinctions, ultimately for him there were no impermeable barriers between philosophy, physics, and faith. Because Newton was committed to the topos of the Two Books, namely, that God had *written* both the Book of Nature and the Book of Scripture, he believed that truth ultimately comes from the same divine source and thus is one. Consequently Newton highlights moral and religious corollaries to the study of Nature in the conclusion of his *Opticks*: "And if natural Philosophy in all its Parts, by pursuing this Method, shall at length be perfected, the Bounds of Moral Philosophy will be also enlarged. For so far as we can know by natural Philosophy what is the first Cause, what Power he has over us, and what Benefits we receive from him, so far our Duty towards him, as well as that towards one another, will appear to us by the Light of Nature" (*Optics*, p. 405). For Newton, advances in natural philosophy were completely bound up with moral and religious concerns. These, in turn, related to right method: a humble empiricism. Whether in science or religion, Newton believed that the inductive method led to purity and truth.

The recovery of this pre-Enlightenment understanding of Newton poses at least two challenges. The first is whether Newton himself appreciated the extent to which his science could in succeeding generations be cut free from religious and ethical perspectives. The failure to recognize the degree to which his work could so easily be reinterpreted by his Enlightenment followers may raise some doubts about the sagacity of Newton's own self-understanding. The second is whether the severing of the ties that Newton experienced is justified, that is, whether it in truth represents a purification or a corruption of modern natural science. Although the general consensus is, of course, that it represents a purification, and that Newton was in fact mistaken about the connections he experienced between science and religion, a full appreciation of Newton himself might be a stimulus to question such a position.

STEPHEN D. SNOBELEN

SEE ALSO *Descartes, René; Royal Society.*

BIBLIOGRAPHY

Alexander, H. G. (1956). *The Leibniz-Clarke Correspondence.* Manchester, UK: Manchester University Press.

Cohen, I. Bernard, and George E. Smith, eds. (2002). *The Cambridge Companion to Newton.* Cambridge, UK: Cambridge University Press. A valuable collection of essays on Newton's mathematics, physics, philosophy and religion.

Cohen, I. Bernard, and Richard S. Westfall, eds. (1995). *Newton: Texts, Backgrounds, Commentaries.* New York: Norton.

Dobbs, Betty Jo Tetter. (1991). *The Janus Faces of Genius: The Role of Alchemy in Newton's Thought*. Cambridge, UK: Cambridge University Press.

Fauvel, John; Raymond Flood; Michael Shortland; and Robin Wilson, eds. (1988). *Let Newton Be! A New Perspective on His Life and Works*. Oxford: Oxford University Press.

Force, James E., and Richard H. Popkin, eds. (1999). *Newton and Religion: Context, Nature, and Influence*. Dordrecht, The Netherlands: Kluwer Academic.

Leshem, Ayval. (2003). *Newton on Mathematics and Spiritual Purity*. Dordrecht, The Netherlands: Kluwer Academic.

McGuire, J. E. (1978). "Newton on Place, Time, and God: An Unpublished Source." *The British Journal for the History of Science* 11: 114–129.

McGuire, J. E. (1996). *Tradition and Innovation: Newton's Metaphysics of Nature*. Dordrecht, The Netherlands: Kluwer.

Maclaurin, Colin. (1968). *An Account of Sir Isaac Newton's Philosophical Discoveries*. New York and London: Johnson Reprint Corporation. Originally published in 1748.

Newton, Isaac. (1999). *The Principia: Mathematical Principles of Natural Philosophy*, trans. I. Bernard Cohen, and Anne Whitman, with the assistance of Julia Budenz. Berkeley: University of California Press. The standard English translation of the *Principia*.

Newton, Isaac. (1952). *Opticks: or a Treatise of the Reflections, Refractions, Inflections & Colours of Light*. New York: Dover Publications. The standard English edition of the *Opticks*.

Snobelen, Stephen D. (2001). "'God of Gods, and Lord of Lords': The Theology of Isaac Newton's General Scholium to the *Principia*." *Osiris* 16: 169–208.

Stewart, Larry. (1992). *The Rise of Public Science: Rhetoric, Technology, and Natural Philosophy in Newtonian Britain, 1660–1750*. Cambridge, UK: Cambridge University Press.

Westfall, Richard H. (1980). *Never at Rest: A Biography of Isaac Newton*. Cambridge, UK: Cambridge University Press. The leading and most comprehensive biography of Newton

NEW ZEALAND PERSPECTIVES

SEE *Australian and New Zealand Perspectives.*

NIETZSCHE, FRIEDRICH W.

• • •

Friedrich W. Nietzsche (1844–1900) was born in Röcken, Prussia, on October 15. He attended the prestigious boarding school at Pforta, where he was educated in the classics, literature, poetry, and the arts. He went on to study classical philology, first at the University of Bonn, and later at the University of Leipzig. His scholarly promise was so great that he was appointed in 1869 as professor extraordinarius of classical philology at the University of Basel (Switzerland). Following a brief and debilitating tour of duty in the Franco-Prussian War, he returned to Basel and produced his first major work, *The Birth of Tragedy out of the Spirit of Music* (1872). The book was so poorly received that its publication effectively signaled the end of his academic career. He finally resigned from the university in 1879 and lived on a small pension awarded him by the Swiss government.

Nietzsche's most influential work was *Thus Spake Zarathustra*, published in four parts between 1883 and 1891. In this ambitious work, he depicted the fictitious Zarathustra as a charismatic teacher whose appearance heralds the redemption of the modern world. Zarathustra is best known for his controversial teaching of the Übermensch (or "overman"), whom he proposes as "the meaning of the earth." Were his auditors to embrace this untimely teaching, Zarathustra insists, they would be prepared finally to emerge from the shadow of the dead God and take their rightful place as the legislators of the future. In doing so, they would shed the burden imposed on them by the resentful, ascetic morality that they have inherited from its twin sources, Christianity and Platonism. Zarathustra's teaching of the Übermensch thus conveys the promise of a life predicated on a love of the body and an aspiration to noble values.

Nietzsche intensifies his attack on conventional morality in his next two books, *Beyond Good and Evil* (1886) and *On the Genealogy of Morals* (1887). In both works he rehearses his influential distinction between *master* (or *noble*) *morality* and *slave morality*. Whereas the master morality takes its shape and direction from an originating act of self-affirmation, by means of which the master deems "good" everything about and pertaining to him, the slave morality originates in the slave's designation of his tormentors as "evil." Only as an afterthought, and in contrast to his "evil" oppressors, does the slave deem himself "good." According to Nietzsche, the master morality celebrates passion, commitment, struggle, and immediacy, whereas the slave morality honors the virtues of suffering, deprivation, passivity, and psychological cunning.

In both books, Nietzsche advances the controversial thesis that contemporary European (or Christian) morality is in fact descended from a slave morality. Although freed from the material conditions of slavery, modern people have become habituated to serve as their own slave masters. Burdened by guilt and wearied by relentless self-surveillance, moderns impose upon themselves the defining values of slavery. Nietzsche further conjectures that protracted adherence to a descendant version of the slave morality may have crippled moderns beyond repair, such that a renaissance of nobility may no longer be possible.

In *On the Genealogy of Morals*, Nietzsche extends his critique of conventional morality to include the scholarly practice of science (*Wissenschaft*). Here he investigates the role of science in the reign of the ascetic ideal, hoping to expose contemporary practitioners of science as unwittingly honoring the values of declining life—even as they increasingly turn their research to matters related to health, evolution, leisure, and longevity. The problem with the contemporary practice of science, he explains, lies in its failure thus far to determine the actual *value* of truth; the scientific enterprise thus remains stubbornly unscientific with respect to itself. He consequently asserts that the otherwise unimpeachable "will to truth" masks a more basic expression of *faith* in truth. It is in this sense that science serves the ascetic ideal, for it proceeds under the uninterrogated assumption that possession of the truth will redeem humankind, which implies that humankind stands in *need* of redemption. Although science continues to sponsor exciting discoveries, its dependence on the ascetic ideal implicates all such discoveries in the ongoing assault on our beleaguered affects. This assault in turn hastens the advent of the "will to nothingness," which Nietzsche identifies as the will never to will again.

Nietzsche said little about emerging technologies, despite availing himself of railways, typewriters, experimental drugs, postal systems, and other innovations of the late nineteenth century. He was deeply suspicious, however, of the rise of technology in general, which he regarded as symptomatic of advancing cultural decay. He was particularly critical of the technologies marshaled in support of European imperial expansion. He regarded the aspiration to empire as an organized distraction from the crisis of European culture. In his view, the pursuit of imperial possessions would not solve the problem of European decadence but simply export it across the globe.

Nietzsche's productive philosophical career ended in 1888. At the beginning of the next year he suffered a nervous breakdown. After a brief stay in a Jena sanitarium, he was placed in the care of his mother, who relocated him to her home in Naumburg. He lived there in a state of catatonic silence, which was broken only by occasional piano improvisations and infrequent bursts of babble. Following the death of his mother in 1897, he was relocated to Weimar by his younger sister, Elisabeth Förster-Nietzsche, the widow of a prominent

anti-Semite and Aryan supremacist. Elisabeth succeeded not only in fashioning her now-famous brother into a kind of cult figure, but also in forging a connection between his philosophy and the rising tide of reactionary politics in Germany. Following his death in Weimar on August 25, his sister continued her appropriation of his philosophical teachings, eventually steering them into convergence with the ideology that soon would inform National Socialism. That Nietzsche would have repudiated any such alliance did not deter Elisabeth from presenting her brother's ideas as providing the philosophical inspiration for Hitler's *Reich*.

DANIEL CONWAY

SEE ALSO *Alienation; Existentialism.*

BIBLIOGRAPHY

Aschheim, Steven E. (2001). *In Times of Crisis: Essays on European Culture, Germans, and Jews.* Madison: University of Wisconsin Press.

Kaufmann, Walter. (1968). *Nietzsche: Philosopher, Psychologist, Antichrist.* New York: Vintage Books.

Moore, Gregory. (2002). *Nietzsche, Biology, and Metaphor.* Cambridge, UK: Cambridge University Press.

Nietzsche, Frederich. (1966). *Beyond Good and Evil: Prelude to a Philosophy of the Future*, trans. Walter Kaufmann. New York: Vintage Books.

Nietzsche, Friedrich. (1999). *Thus Spake Zarathustra*, trans. Thomas Common. Mineola, NY: Dover.

Nietzsche, Friedrich. (2000). *Basic Writings of Nietzsche*, trans. and ed. Walter Kaufmann. New York: Modern Library Classics.

NIGHTINGALE, FLORENCE

• • •

The founder of modern secular nursing, a social activist, and a pioneer in the use of social statistics, Florence Nightingale (1820–1910) was born on April 12 in Florence, Italy, the child of a wealthy, prominent English family. Given a classical education by her father, the serious, devout young woman was drawn to caring for the sick, but nursing was then a form of menial labor that was considered inappropriate for members of her social class. Nightingale persisted; for years she visited and gathered information on hospitals in England and abroad, sought training in Germany, and in 1853 became superintendent of a nursing home in London, where she undertook reforms to improve patient care.

After the start of the Crimean War (1854–1856) the public reacted with outrage to newspaper reports of the horrid conditions endured by British soldiers wounded in battle, and Nightingale was appointed to bring nursing care to the military. Arriving at the hospital in Scutari, Turkey, with a team of thirty-eight nurses, including fourteen Anglican and ten Roman Catholic sisters, she found overcrowding, filth, infestation, and disease. Far more soldiers were dying of cholera and typhus than were dying of their wounds. Against the objections of the hospital staff, Nightingale took firm administrative measures, set up sanitary kitchen and laundry facilities, and procured supplies with private funds. The death rate fell from 42.7 percent to 2.2 percent in six months. An international heroine at age thirty-six, Nightingale was immortalized by Henry Wadsworth Longfellow, as "a lady with a lamp" making her nightly rounds on the hospital wards, in his 1857 poem "Santa Filomena."

Nightingale used her Crimean experience to lobby for the reform of medical care in the army, publishing an 800-page book, *Notes on Matters Affecting the Health, Efficiency, and Hospital Administration of the British Army* (1857). She included documentation that the death rate of army recruits in peacetime was nearly twice that of the comparable civilian population. Queen Victoria, to whom Nightingale had been presented as a debutante, supported her aims, as did friends in influential positions. Despite resistance within the bureaucracy, reforms followed. A Royal Commission for the Health of the Army was set up in 1857, and a similar commission was established for the army in India in 1859. Nightingale wrote *Notes on Nursing* (1859) and *Notes on Hospitals* (1859) and founded the Nightingale School of Nursing at St. Thomas's Hospital in London (1860). Nurse training programs based on her system were established during her lifetime in twenty countries, including a thousand in the United States alone.

Florence Nightingale was called the "Passionate Statistician" because her spirited campaigns for reform were anchored in carefully compiled data to convince those in power of the validity of her cause. Fascinated by mathematics since childhood, she found guidance in the social physics of Adolphe Quetelet (1796–1874), a Belgian astronomer and pioneer of sociology who developed the notion of the *average man* to show that observed regularities in the traits and behavior of groups could be characterized by the laws of probability. She devised graphic techniques to convey her politically explosive findings and was aided in her analyses by William Farr (1807–1883), a physician and the founder of British vital statistics. She urged the introduction of

Florence Nightingale, 1820–1910. The English nurse was the founder of modern nursing and made outstanding contributions to knowledge of public health.

statistics into higher education and with the help of the scientist Francis Galton (1822–1911) sought to establish a university chair in statistics.

After 1857 Nightingale lived as an invalid and rarely left her home. According to a comprehensive biography (Dossey 1999), her disability was consistent with chronic brucellosis, an infection contracted in the Crimea, but equally significant was the central role of religion in her life. Much is revealed in Nightingale's journals and the thousands of letters she wrote. Since the age of seventeen Nightingale felt that she had been called by God for a special mission. Well versed in the tradition of Western mysticism, she was inspired by strong women such as Saint Catherine of Siena (1347–1380), Saint Catherine of Genoa (1447–1510) and Saint Teresa of Avila (1515–1582), whose intense spiritual lives found expression in service to humanity. In her daily life, coping with illness and engaged in widespread reform activities through her writing and personal contacts, she accommodated the contemplative's need for solitude, guided to the end by her inner vision. She died in London on August 13, 1910.

It was Florence Nightingale's mission to lessen human suffering through better healthcare and the prevention of disease. Her novel approach was the use of statistical evidence to show the way: quality data on which to base policies to serve the common good, with a call for the education of administrators as well as the public to help them understand. The study of statistics was for her a moral duty.

VALERIE MIKÉ

SEE ALSO Bioethics; Medical Ethics.

BIBLIOGRAPHY

Cohen, I. Bernard. (1984). "Florence Nightingale." *Scientific American* 250 (3): 128–137. Article assessing Nightingale's contributions contains samples of her statistical graphics methodology

Dossey, Barbara Montgomery. (1999). *Florence Nightingale: Mystic, Visionary, Healer.* Springhouse, PA: Springhouse Corporation. A large-size, richly illustrated volume, a thoroughly researched biography that includes analysis of the spiritual motivation of Nightingale's life and work.

Kopf, E. W. (1916). "Florence Nightingale as Statistician." *Journal of the American Statistical Association* 15: 388–404. Reprinted in *Studies in the History of Statistics and Probability*, vol. 2, ed. M. G. Kendall and R. L. Plackett (1977). New York: Macmillan.

Nightingale, Florence. (1969). *Notes on Nursing: What It Is and What It Is Not.* New York: Dover. A manual of nursing care, revolutionary in its day.

Nightingale, Florence. (2002–2005). *Collected Works of Florence Nightingale*, 8 vols., ed. Lynn McDonald. Toronto: Wilfrid Laurier University Press. The first 8 in a projected series of 16 volumes.

Walker, Helen M. (1929). *Studies in the History of Statistical Method, with Special Reference to Certain Educational Problems.* Baltimore: Williams & Wilkins. Includes discussion of Nightingale's efforts to promote statistics education.

NONGOVERNMENTAL ORGANIZATIONS

• • •

Nongovernmental organizations (NGOs), as independent of both governments and corporations, are the major components of an international or global civil society. The term first came into official use in the Charter of the United Nations (1945), Chapter 10, Article 71, in order to acknowledge a consultative role for non-state actors in the Economic and Social Council. Since then the term has broadened to include, in the World Bank definition, "private organizations that pursue activities to relieve suffering, promote the interests of the poor, protect the environment, provide basic social services, to undertake community development" (Operational Directive 14.70). In common usage, NGOs are simply non-profit organizations that, even as they have become increasingly professionalized, remain dependent on donations, voluntarism, and appeals to ethical ideals.

Although it is difficult to provide exact numbers, in 2000 there were certainly more than 25,000 NGOs operating worldwide. The rapid development of NGOs since the 1970s has been stimulated in part by scientific and technological developments, especially in communication, while NGOs also play increasingly significant roles in promoting the ethical uses of science and technology.

Classifications of NGOs

NGOs can be divided into different overlapping categories according to both form and content. Formally it is useful to distinguish between operational NGOs that seek to realize various projects, and advocacy NGOs that seek to raise consciousness about some particular cause. The International Red Cross/Red Crescent is an example of an operational NGO; Amnesty International an example of an advocacy NGO. Of course, many NGOs include both operational and advocacy activities, for example the American Association for the Advancement of Science (AAAS), which promotes both professional development within the technical community and seeks to educate the general public about the importance of science.

NGOs may also be classified in terms of their interests. From the perspective of interests, NGOs may focus on humanitarian relief such as the Médicins Sans Frontières (Doctors without Borders) or humanitarian development such as Habitat for Humanity; emphasize human rights or environmental issues; exhibit religious or secular bases; and promote professional, trade, or social developments. NGOs are also sometimes distinguished as primarily community-based, national, or international organizations.

Environmental NGOs

One type of NGO that is especially relevant to science, technology, and ethics issues is environmental NGOs, which will be considered here in more detail in order to illustrate relevance, strengths, and weaknesses. Environmental NGOs have formed in direct response to the impact of an increasingly technological world and the increased exploitation of the world's natural resources. Again, although many groups fall under this broad category, environmental NGOS are not uniform in mission, priorities, strategies, or activities. NGOs range from small, grassroots organizations to large nonprofit corporations with boards of directors and professional staffs. Many specialize in particular areas of advocacy or activity and tend to focus their work either geographically or topically. Some are located primarily in North America and work mainly on local or national issues. Others are headquartered in the North, but focus their attention on issues primarily involving developing countries. Still other NGOs have a global focus with affiliated groups active in many different countries.

Local environmental groups are concerned with specific issues such as protection of a local water supply or a site-specific contamination problem. Some of the larger organizations tend to focus on broad areas of national or global concern, such as the Wilderness Society, the National Audubon Society, and The Nature Conservancy, which are concerned with wildlife and habitat protection. Other national groups emphasize the public health threats associated with pollution. Many organizations focus more comprehensively on environmental quality, linking concern for public land and wildlife with pollution and public health issues.

Environmental NGOs attempt to bring about change in a variety of ways. Some engage in public protest marches and demonstrations, civil disobedience, and other participatory public actions and media events to draw attention to specific concerns. Some groups prepare and distribute educational materials and sponsor public educational events. Some environmental NGOs are actively involved in lobbying efforts to ensure appropriate policy solutions to environmental problems. These groups may also act as watchdogs, to ensure that those subject to environmental regulations comply with requirements. Some NGOs pursue environmental remedies through legal action. Other groups work directly on issues such as protecting biodiversity by purchasing land to protect endangered habitats for plants and wildlife. Most NGOs employ a variety of strategies to accomplish their objectives.

Brief History of Environmental NGOs

The conservation movement, in the mid- to late-1800s, gave rise to the first notable environmental NGOs in the United States, many of which remain active in the twenty-first century. This era is often referred to as the "first wave of environmentalism." Influenced by the growth of scientific knowledge that revealed the conse-

quences of more than two centuries of unchecked human exploitation of the environment, Americans began to understand the costs of losing vast expanses of land and resources. Conservationists challenged the notion that America's resources were inexhaustible.

Several influential writers and activists during this period inspired the forming of the first environmental NGOs in the United States. For example, in 1886 George Bird Grinnell (1849–1938) proposed a society for the protection of the nation's birds; this idea gave rise to the Audubon Societies. The Boone and Crockett Club, founded in 1887 by Theodore Roosevelt (1858–1919) and other well-heeled sportsmen, brought attention to the wasteful slaughter of big game animals.

Early conservationists tended to take an anthropocentric or human-centered view of the environment. The underlying philosophy was the efficient use and conservation of resources for human benefit. By the late 1880s, a second strand of thinking emerged. In 1892 John Muir (1838–1914), a Scottish-born immigrant and advocate for the preservation of nature, founded the Sierra Club. While Muir did not dispute the conservationist notions of resource management, he believed that certain natural areas should be treated as sacred realms and protected from all resource exploitation. Muir advocated the preservation of nature for its own sake, and for the preservation of vast areas of land through public ownership.

During the first half of the twentieth century, hunting and fishing organizations, primarily elite organizations of affluent white men, were the most active and influential NGOs. In 1922, a group of Midwestern sportsmen formed the Izaak Walton League of America to advocate for the protection of wildlife habitat. The National Wildlife Federation was formed in 1936 as a clearinghouse for conservation issues.

In 1935, naturalist Aldo Leopold (1886–1948) founded the Wilderness Society based upon a "land ethic" in which humans are viewed as part of nature rather than conquerors of nature. Like Muir, Leopold believed that nature has value in its own right.

The second wave of environmentalism did not emerge in the United States until in the 1960s. For almost 100 years, environmental NGOs were concerned primarily with preserving wilderness or conserving natural resources. The second-wave environmental movement grew out of many concerns. The industrial growth of the United States following World War II produced prosperity, population growth, and pollution. Increased public attention on the problems of pollution, population, consumption, and waste enlarged the environmental agenda

TABLE 1

A Representative List of Environmental NGOs and Founding Dates

Environmental Organization	Date Founded
Audubon Society —became the *New York Audubon Society,* the precursor organization to the *National Audubon Society.*	1886
Boone and Crockett Club—"promotes the management of big game and associated wildlife in North America and maintain all aspects of sportsmanship in big game hunting."	1887
Sierra Club—"encourages the exploration, enjoyment and protection of the wild places of the earth and practices and promotes the responsible use of the earth's ecosystems and resources; seeks to educate and enlist humanity to protect and restore the quality of the natural and human environment, uses all lawful means to carry these objectives."	1892
American Scenic and Historic Preservation Society—no longer in existence.	1895
National Audubon Society—"to conserve and restore natural ecosystems, focusing on birds, other wildlife, and their habitats for the benefit of humanity and the earth's biological diversity."	1905
National Parks and Conservation Association—"to protect and enhance national parks for present and future generations."	1919
Izaak Walton League—"to conserve, maintain, protect and restore the soil, forest, water and other natural resources of the United States and other lands; to promote means and opportunities for the education of the public with respect to such resources and their enjoyment and wholesome utilization."	1922
The Wilderness Society—"deliver to future generations an unspoiled legacy of wild places, with all the precious values they hold."	1935
National Wildlife Federation—"educating and empowering people from all walks of life to protect wildlife and habitat for future generations."	1936
Ducks Unlimited—"conserves, restores, and manages wetlands and associated habitats for North America's waterfowl."	1937
Defenders of Wildlife—"the protection of all native wild animals and plants in their natural communities; programs focus on the accelerating rate of extinction of species and the associated loss of biological diversity, and habitat alteration and destruction."	1947
The Nature Conservancy—"preserve the plants, animals and natural communities that represent the diversity of life on Earth by protecting the lands and waters they need to survive."	1951
World Wildlife Fund (now known as *WWF*) "to stop the degradation of the planet, natural environment and to build a future in which humans live in harmony with nature, by conserving the world's biological diversity and ensuring that the use of renewable natural resources is sustainable."	1961
Environmental Defense Fund—"links science, economics and law to create innovative, equitable and cost-effective solutions to society's most urgent environmental problems."	1967
Friends of the Earth—"international network of grassroots groups in 70 countries. Defends the environment and champions a healthy and just world."	1969
National Resources Defense Council—"safeguard the Earth, its people, its plants and animals and the natural systems on which all life depends; to restore the integrity of the elements that sustain life — air, land and water — and to defend endangered natural places; to establish sustain ability and good stewardship of the Earth as central ethical imperatives of human society."	1970
Clean Water Action—"national citizens' organization working for clean, safe and affordable water, prevention of health-threatening pollution, creation of environmentally-safe jobs and businesses, and empowerment of people to make democracy work."	1971
Greenpeace—"an independent, campaigning organization that uses non-violent, creative confrontation to expose global environmental problems, and force solutions for a green and peaceful future. Greenpeace's goal is to ensure the ability of the Earth to nurture life in all its diversity."	1971
Zero Population Growth (now known as *Population Connection*)—"educates young people and advocates progressive action to stabilize world population at a level that can be sustained by Earth's resources."	1972
Cousteau Society—"to educate people to understand, to love and to protect the water systems of the planet, marine and fresh water, for the well-being of future generations."	1973
Worldwatch Institute—"through accessible, and fact-based analysis of critical global issues, informs people around the world about the complex interactions between people, nature, and economies; focuses on the underlying causes of and practical solutions to the world's problems, in order to inspire people to demand new policies, investment patterns and lifestyle choices."	1975
Earth First!—loosely affiliated with the tenets of deep ecology, "seeks to encourage a more harmonious relationship between nature and humans."	1980
People for the Ethical Treatment of Animals—"dedicated to establishing and protecting the rights of all animals; operates under the simple principle that animals are not ours to eat, wear, experiment on, or use for entertainment."	1980
Citizens Clearinghouse for Hazardous Waste (now known as the *Center for Health, Environmental and justice*)—"provides technical information and training for local citizens to hold industry and government accountable and to work towards a healthy, environmentally sustainable future."	1981
Earth Island Institute—"develops and supports projects that counteract threats to the biological and cultural diversity that sustain the environment. Through education and activism, these projects promote the conservation, preservation, and restoration of the Earth."	1982
Conservation Fund—"forges partnerships to protect America's legacy of land and water resources. Through land acquisition, sustainable programs, and leadership training, the Fund and its partners demonstrate effective conservation solutions emphasizing the integration of economic and environmental goals."	1985
Rainforest Action Network—"campaigns for the forests, their inhabitants and the natural systems that sustain life by the global marketplace through grassroots organizing, education and non-violent direct action."	1985
Rainforest Alliance—"to protect ecosystems and the people and wildlife that depend on them by transforming land-use practices, business practices and consumer behavior.	1986
Conservation International—"to conserve the earth's natural living heritage, global biodiversity, and to demonstrate that human societies can live harmoniously with nature."	1987

SOURCE: Courtesy of M. Ann Howard.

and gave new impetus to the work of environmental NGOs. During this second wave, national organizations such as the Sierra Club and the Wilderness Society used the new public concern for environmental issues to educate the public and expand membership. In addition, an average of eighteen new NGOs were forming each year during the period 1960 to 1980.

National NGOs were effective lobbying organizations, compelling political action in a variety of areas such as wilderness protection, pollution control, and management of hazardous chemicals. The United States Congress responded to the new public concern through a complex array of statutes. New environmental laws such as the National Environmental Policy Act (1970), the Clean Air Act (1970), and the Clean Water Act (1972) widened public access to the courts, allowing legal challenges to federal agency actions. A new category of environmental NGOs appeared during this period. Although some of these groups were offshoots of the older, more traditional organizations, these new organizations, such as the Environmental Defense Fund (1967) and the Natural Resources Defense Council (1970), used the courts to bring attention to serious environmental problems. Many of the new federal environmental laws gave environmental NGOs and their issues standing in the courts, leading to a whole new field of law and environmental advocacy.

Third wave environmentalism emerged in the 1980s and was characterized by the "mainstreaming" of environment issues. The largest national NGOs grew significantly in the early 1980s, in large measure due to growing public pessimism about the state of environment, in spite of the legislative initiatives of the 1960s and 1970s. For example, the Wilderness Society grew by more than 140 percent between 1980 and 1983, and the Sierra Club increased its membership by 90 percent during the same period. Toward the end of the 1980s, most of the larger NGOs experienced additional growth in membership as the public grew more concerned about global environmental problems such as ozone depletion and global climate change.

By the mid 1980s, the national environmental NGOs were shifting their strategies from legal challenges and anti-business lobbying to a more collaborative problem-solving stance working directly with corporate interests. During this time, many of the larger national NGOs began working with government and industry to fashion "market-based" solutions to environmental problems.

Not all NGOs embraced cooperative strategies. More radical environmental activists encouraged "direct action" and more controversial activities. For example, Greenpeace, founded in 1971, was one of the most visible environmental groups in the early 1990s because of its highly publicized protests against polluting companies. Critics often described the actions of some of these groups as "ecoterrorism." Earth First!, a splinter group of the Wilderness Society, practiced tree-spiking, driving nails into trees with the intent of damaging chain saws in opposition to cutting down trees in major forest areas.

Grassroots environmentalism was a significant force during the 1980s and 1990s, and remains so in the twenty-first century. In contrast to the larger national NGOs that tend to be very centralized and led by mostly white, well-educated, middle-class professionals, grassroots organizations are comprised of people who cut across racial, class, and educational lines. Inspired by the efforts of Lois Gibbs at Love Canal in the 1970s, the grassroots movement began as a populist movement against toxic waste. Although most of the grassroots organizations operate independently of the mainstream organizations, a number of national networks, such as the Citizens' Clearinghouse for Hazardous Wastes, provide organizational skills and technical assistance to local groups.

Part of the growth of grassroots environmentalism included the emergence of environmental justice groups. These groups have coupled environmental issues with other social issues associated with poverty, racism, and classism. These organizations are concerned with distributive justice and remedying past injustices (based on race and class) and focus on a variety of issues including waste disposal, worker health and safety, housing, pesticides, and facility siting. Some of the larger NGOs have taken up environmental justice causes; however, most local groups, wary of the larger NGOs, tend to work outside the mainstream organizations.

The International Environmental Movement

The international environmental NGOs emerged in the 1990s, almost a century following the appearance of the first wave of American environmentalism. During the 1970s and 1980s, the global implications of environmental issues became more evident. A growing body of scientific knowledge brought to life the damage caused by worldwide exploitation of natural resources by the relatively few industrialized nations. Most of the serious problems of global air and water pollution were directly attributable to the activities of the developed countries. The watershed was the 1992 United Nations Conference on Environment and Development—called the

"Earth Summit"—held in Rio de Janeiro. While official country representatives met under the auspices of the UN conference, more than 30,000 individuals representing several thousand environmental groups, many from the developing world, held a global forum to draw attention to issues impacting people and the environment around the world. The Earth Summit had a catalytic effect on NGO growth and network building throughout the world. NGOs in developing nations perform somewhat different roles than the NGOs of developed countries. They may fill a void due to ineffective or nonexistent government programs or they may supplement the work of government agencies.

Analysis

As the history of NGOs suggests, these organizations can be instrumental in organizing public pressure on environmental issues at the local, national, and international levels. NGOs have played an important role in bringing new issues to the public agenda and have sponsored innovative solutions to key environmental issues. The NGO presence heightens public scrutiny of government decision making on critical environmental issues. Historically, NGOs had a different stake in power politics and were able effectively to serve as a counterpoint to other political or economic interests. However, as NGOs have become more mainstream and engaged in working relationships with government and industry, many have observed the changing nature of the NGOs.

Decision-making structures within environmental NGOs vary widely. At the heart of grassroots organizations is a strong commitment to citizen participation. The process within these organizations is often very participatory and direct stakeholders decide upon agendas and strategies. In contrast, mainstream environmental NGOs are often criticized for their undemocratic practices. In many, central staff or the board of trustees has the final say on issues and strategies, often without the advice or consent of members or regional chapters. Some have grown so large that more democratic decision making is not feasible.

The national NGOs must deal with the tensions caused by the conflicts associated with preserving the organization and preserving the environment. Many of the nationals have been criticized for excessive deference to industry in effort to reach collaborative solutions. They also are criticized for abandoning grassroots interests in favor of organizational protectionism.

Most national NGOs rely on member contributions to fund their activities. Some groups hire consultants to determine what issues would elicit the highest donations.

Fundraising activities and newsletters often are primarily designed to maximize contributions rather than to inform membership. Some groups have been criticized for exaggerating or overexploiting potentially harmful problems such as asbestos or pesticides, in order to enlarge memberships or increase member contributions.

Most of the larger NGOs must also raise funds from outside sources. Most do not have memberships large enough to be financially autonomous, especially to support professional administrators, lawyers, and scientific experts. NGOs raise funds from foundations, governments, other NGOs, and private corporations. Often funding interests are represented on governing boards. This may lead to questions of cooptation. Critics argue that organizational priorities may be more influenced by the interest of the funders rather than environmental quality. Even large foundations have directed the priorities of mainstream NGOs, favoring cautious reform and noncontroversial strategies such as public education. Some large foundations tend to shut out organizations that take more radical positions such as zero-cut policies in public forests or zero discharge of contaminants.

Some critics note that the largest industrial polluters have become the largest donors to the bigger environmental NGOs. Because of this, some suggest that while national NGOs may be better positioned to influence national policy, grassroots organizations will have a greater impact on industry practices and corporate interests in the future because they are willing to openly confront industry's management of pollution and hazardous waste, the siting of hazardous waste facilities, and private sector exploitation of resources.

M. ANN HOWARD

SEE ALSO *American Association for the Advancement of Science; Bioethics Centers; Professional Engineering Organizations; Sierra Club.*

BIBLIOGRAPHY

Dowie, Mark. (1995). *Losing Ground: American Environmentalism at the Close of the Twentieth Century.* Cambridge, MA: MIT Press.

Lafferty, William M., and James Meadowcraft, eds. (1996). *Democracy and the Environment: Problems and Prospects.* Cheltenham, UK: Edward Elgar.

Redclift, Michael, and Graham Woodgate, eds. (1997). *The International Handbook of Environmental Sociology.* Cheltenham, UK: Edward Elgar.

Shabecoff, Philip. (1993). *A Fierce Green Fire: The American Environmental Movement.* New York: Hill and Wang.

Taylor, Dorceta E. (2000). "The Rise of the Environmental Justice Paradigm." *American Behavioral Scientist* 43(4): 508–580.

Thiele, Leslie Paul. (1999). *Environmentalism for a New Millennium: The Challenge of Coevolution*. New York: Oxford University Press.

NORMAL ACCIDENTS

• • •

The concept of *normal accidents* was formulated by sociologist Charles Perrow in *Normal Accidents: Living with High Risk Technologies* (1984), but is related to a number of other analyses of accidents in complex, technological societies. Perrow used the concept to describe a type of accident that inevitably results from the design of complex mechanical, electronic, or social systems. The theory has had extended influence on subsequent analyses of accidents and errors related especially to advanced technologies.

Perrow's Normal Accidents

The unexpected and interactive failure of two or more components is not sufficient to cause a normal accident when there is enough time to solve the problem before it becomes critical. Instead normal accidents in Perrow's sense occur only in systems that, in addition to being *complexly interactive*, are also *tightly coupled*. One example would be two components whose failures start a fire while silencing the fire alarm. Intervention by system operators in the early minutes or hours of such an incident often makes things worse, as when manual fire alarm activation might open doors that allow the fire to spread.

Perrow believes that normal accidents are an inevitable consequence of human reliance on complex and tightly coupled systems. By confronting the causes of normal accidents, the designers, users, and potential victims—in fact, society as a whole—can make appropriate practical and ethical decisions about the systems involved. Once one understands why normal accidents occur, and also the fact that they are almost inevitable in complex systems, Perrow suggests that "we are in a better position to argue that certain technologies should be abandoned, and others, which we cannot abandon because we have built much of our society around them, should be modified" (Perrow 1984, p. 4).

In *Normal Accidents*, Perrow provides several examples to flesh out his argument. One case involves the loss of the two square mile Lake Peigneur in Louisiana. The lake was in simultaneous use by shipping companies (via a canal connected to the Gulf of Mexico), fishermen, tourists (the Rip van Winkle Live Oak Gardens was on its banks), and oil companies (Texaco was drilling for oil in a part of the lake only three to six feet deep). Under the lake was a salt mine operated by the Diamond Crystal Company. Texaco's oil rig penetrated the mine and vanished from sight, after which all of Lake Peigneur drained into the mine, creating a whirlpool that pulled in several barges, a tug, and sixty-five acres of the Rip van Winkle Gardens. The canal to the Gulf reversed course, creating a 150-foot waterfall as the lake drained away. An underground natural gas well ruptured and bubbles floated to the surface, caught fire and burned. In just seven hours, Lake Peigneur was gone—without, however, taking a single life.

The accident was caused by the fact that the lake, oil rig, and mine were complexly interactive and tightly coupled. Subsystem operators understood none of the relationships and did not communicate adequately with one another. The Peigneur Lake incident illustrates another of Perrow's points, about the social allocation of responsibility. Instead of analyzing the system as a whole with an eye to reducing complexity or ameliorating the tight coupling, each of the players held the others responsible, Texaco accusing Diamond Crystal and vice versa. In analyzing the near-meltdown at the Three Mile Island nuclear plant in 1979, Perrow noted that the equipment vendor and the system operators blamed each other. Systems of adversarial litigation can in such cases militate against the solving of system problems.

Another phenomenon analyzed by Perrow is that of *non-collision course collisions*, in which ships on parallel, opposite courses suddenly turn and hit one another at the last moment. Perrow tells the story of the Coast Guard cutter *Cuyahoga*, operating at night. Although lookouts correctly interpreted the three lights visible on the *Santa Cruz II* to mean the ship was headed toward them on a parallel course, they did not inform their captain, because they knew he was aware of the other ship. What they did not realize was that the myopic captain had noted only two lights on the *Santa Cruz II*, interpreting these to mean that it was a smaller fishing vessel, sailing ahead of the *Cuyahoga* and in the same direction. As the *Cuyahoga* came closer to the freighter, the captain turned to port, to pass outside the other ship. In reality, since the *Santa Cruz II* was headed toward him, he turned out of a parallel course, which would have passed the *Santa Cruz II* without incident, right into its track, causing a collision with the loss of eleven lives.

Ruins on Lake Peigneur. The generally accepted cause of the disaster is that a miscalculated oil probe punctured the roof of a salt shaft, creating a drain for the lake. The lake then proceeded to drain into the hole, as the mine was evacuated. The giant whirlpool created sucked in the drilling platform, eleven barges, many trees, buildings, and some of the surrounding terrain. (© Philip Gould/Corbis)

Perrow argued that in the relatively brief moments available, operators who must function rapidly in real time construct a simplified view of the environment based on available, often incomplete, information. Once this has been accomplished, all contradictory information is excluded. A related problem is the extremely authoritarian command structure used at sea; in which first mates are much less comfortable questioning their captains than copilots are in the air. Such non-collision course collisions are common and, according to Perrow, constituted a majority of the cases he studied in which ships hit other ships.

Perrow noted the differences in social factors between air and sea travel and transport that promote the much larger percentage of accidents at sea. The differing factors include levels of government regulation, pressure to meet schedules, communication between captains and crew, and social status of air versus sea travelers. He concluded that much of the technology developed to make aviation safer, such as traffic control systems, is not used at sea, though it could be.

Perrow also analyzed cases in which safety devices encourage people to engage in more risky behaviors. For example, the installation of new braking systems in trucks, decreasing the possibility of failure on mountain roads, has not resulted in a decline in the number of accidents. Truck drivers who believe they have safer brakes will drive faster because it can save time and money. Similarly in some industries such as marine transport, insurance may make owners complacent, as the real cost of upgrading ships to prevent loss may exceed that of replacing them. Studies of these and related phenomena of automobile accidents (and even business and financial management) have resulted in development of the concept of risk homeostasis, in which increases in safety tend to be complemented by changes in behavior that once again increase risk to a certain acceptable level (Wilde 2001, Degeorge, Moselle, and Zeckhauser 2004).

Perrow's analysis is largely confirmed by high profile systems accidents that have occurred since the book was published. The loss of the *Challenger* (complex interaction, tight coupling between the fragile o-rings and the explosion potential of the fuel tanks) and the *Columbia* (complexity plus coupling between the disposable tanks, off which ice or insulation might fall, and the fragile tiles on the wings which could be damaged by them) space shuttles are two cases in point. In both instances the communications failures highlighted by Perrow are visible (the engineers on the *Challenger* knew that o-rings fail at freezing temperatures, but could not get their managers to postpone the launch; those on the *Columbia* launch wanted to get military spy satellite photos of the wing tiles but could not get their supervisors to agree). A blackout in the eastern and central United States and part of Canada in August 2003 is another example: The highly interdependent utilities failed to function as part of one system, the malfunctioning problem-detection software failed to warn of the overload in one provider, resulting in cascading failures of turbines, and the providers and utilities failed to warn others of known problems.

Competing Analyses

Since Perrow's work a number of studies have both criticized and extended his arguments. Among the most influential are Scott Sagan's *The Limits of Safety* (1993) and Dietrich Dörner's *The Logic of Failure* (1996). Sagan examines two competing theories on safety, normal accident and high reliability, for their ability to explain historical experiences in the control of nuclear weapons. In opposition to normal accident theory, high reliability theory posits that systems can be made safe by employing redundancy measures, decentralizing authority so that those nearest a problem can make quick decisions, and rigorously disciplining operators. It is an optimistic belief that well-managed and designed organizations can be perfectly safe.

Sagan shows that nations such as the United States and Russia use high reliability theory to manage their nuclear weapons. He then provides several examples of accidents and near-accidents that challenge the central assumption of this theory, namely, that nuclear systems can be made safe. Sagan argues that the normal accident theory better explains nuclear weapons systems, which are so complex and tightly coupled that accidents, although rare, are inevitable. He points to such limitations on high reliability theory as conflicting goals and priorities, constraints on learning, limitations on leaders' ability to control the human and technical components of the system, and pressure to turn memories of failures into successes. Sagan concludes that more out-

side reviews and information sharing, changes in organizational cultures (including less faith in redundancy), complete nuclear disarmament, and decoupling interactions are all alternatives to increase the safety of nuclear weapons systems. None of these alternatives, however, is very likely to occur.

Dörner claims that our main shortcomings when faced with complex problems are a tendency to oversimplify and a failure to conceive of a problem within its system of interacting factors. Failure does not necessarily result from incompetence. For example, the operators of the Chernobyl nuclear reactor were experts, and in fact ignored safety standards precisely because they felt that they *knew what they were doing*.

Dörner identifies four habits of mind that account for the difficulty in solving complex problems: (a) slowness of thinking; (b) a desire to feel confident and competent; (c) an inability to absorb and retain large amounts of information; and (d) a tendency to focus on immediately pressing problems and to ignore the problems that solutions are likely to create. Dörner's work highlights the area of normal accident theory dealing with cognitive and psychological factors (i.e., human error) in accidents.

In line with Dörner's analysis, Keith Hendy's Systematic Error and Risk Analysis (SERA) software tool investigates, classifies, and tracks human error in accidents. It employs a five-step process that guides investigators through a series of questions and decision ladders in order to determine where errors occurred (Defence Research and Development Canada 2004).

Perrow's initial work has thus sparked continuing analyses of complex technological systems and the causes of their failures, so that debates about the risks and benefits of technology are regularly influenced by normal accident theory. The results of such debates are nevertheless mixed. In fact, Perrow maintained that some systems, like nuclear power, should be abandoned, while others, like marine transport, require significant modification, but can be made reasonably safe.

Perrow's book, though presented as a narrow study of the functioning of technological systems, is also a study of the psychology of human error, which could be fatal even in low-tech systems, and is much more dangerous today given the speed, size, and clout of modern technology. Perrow's work deserves continuing recognition because he was arguably the first to introduce the concept that accidents, rather than being a lightning bolt from the blue, are inherent in the nature of complex systems, and that human provisions to avoid the consequences may actually engender more danger.

JONATHAN WALLACE
ADAM BRIGGLE

SEE ALSO Unintended Consequences.

BIBLIOGRAPHY

Degeorge, François; Boaz Moselle; and Richard J. Zeckhauser. (2004). "The Ecology of Risk Taking." Journal of Risk and Uncertainty 28(3): 195–215.

Dörner, Dietrich. (1996). The Logic of Failure: Recognizing and Avoiding Error in Complex Situations. New York: Perseus Books Group.

Perrow, Charles. (1984). Normal Accidents: Living with High Risk Technologies. New York: Basic Books. Reprinted, Princeton, NJ: Princeton University Press, 1999. The leading analysis of the primary role of human error in high technology accidents.

Sagan, Scott. (1993). The Limits of Safety: Organizations, Accidents, and Nuclear Weapons. Princeton, NJ: Princeton University Press.

Wilde, Gerald J. S. (2001). Target Risk 2: A New Psychology of Safety and Health. Toronto: PDE Publications. Revised and expanded edition of a book first published in 1994.

INTERNET RESOURCE

Defence Research and Development Canada. "Using Human Logic and Technical Power to Explain Accidents." Available from http://www.drdc-rddc.dnd.ca/newsevents/spotlight/0402_e.asp.

NUCLEAR ETHICS

• • •

Industrial Perspectives
Weapons Perspectives

INDUSTRIAL PERSPECTIVES

There are powerful undercurrents in motion that seek to change the way people work with and think about the nuclear industry. The nuclear energy industry is capable of transforming terrestrial life for better or worse. Never has an industry possessed such awesome forces, and never has there been a greater need for an ethics to guide the way an industry develops. To this end it is useful to review the history of the industry and the highly diverse influences that have produced it. In particular there are two main influences. One is associated with military policy and focuses on geostrategic decisions related to nuclear war, whether offensive or defensive.

The second is located within the civilian area, includes both nuclear medicine and nuclear power generation, and touches on issues of safety, environmental pollution, and economics. The focus here will be on the civilian aspects of the industry.

The Discovery of Radioactivity

At the end of the nineteenth century scientists were examining the properties of cathode ray tubes. These consisted of an enclosed glass vessel that had two electrodes set into the glass at opposite ends of the chamber. When almost all the air in the chamber had been removed and one of the electrodes was heated while the other electrode was given a positive charge (the anode), it was noticed that rays were emitted from the hot (cathode) electrode. In 1895 in Würzburg, Germany, Wilhelm Conrad Röntgen (1845–1923) noted that a plate coated with barium platinocyanide held in front of a functioning cathode-ray tube fluoresced and emitted light. What was more, when he placed a light-opaque material between the plate and the tube, the fluorescence did not cease. Clearly the rays derived from the tube, which he called "X rays," by passing through an opaque material, had done something that visible light rays did not do.

The next year in Paris, Antoine-Henri Becquerel (1852–1908) noted that certain minerals fluoresced when they were exposed to ultraviolet light and that they were capable of fogging an adjacent photographic plate, even when that plate was covered by a double layer of light-opaque paper. One such mineral was uranyl potassium sulfate crystal. He later showed that the effect was largely due to the metal component, uranium. While most of the interest at the time focused on the X rays, Marie Curie (1867–1934) and Pierre Curie (1859–1906) showed that other elements were capable of making penetrating radiations and in the process discovered the elements radium and polonium.

Becquerel, however, made one further vital discovery. After putting a sample vial containing the Curies' radium into his vest pocket, he noted some time later that his skin in the region covered by the pocket became burned. He thus discovered the biological effects of radiation, a phenomenon that was soon put to medical use for a wide variety of ailments, although most such treatments led to a worsening of the condition being treated. (Both the Curies and Becquerel received Nobel Prizes for their discoveries; Marie Curie became the first person to receive two such prizes for her discoveries in the chemistry of radioactive elements.)

Types of Radiation

From these beginnings it became clear that the radiation could be divided into several clear types. X rays and later c-rays (gamma rays) were shown to behave like light rays, being part of the electromagnetic spectrum, whereas β-rays (beta rays) were shown to be streams of negatively charged electrons and α-rays (alpha rays) were helium atoms without the electrons (that is, helium atomic nuclei consisting of two protons and two neutrons). Each of these radiations can be made to generate point sources of light for each energetic emission. From this it has been observed that each gram of radium emits some 3.7×10^{10} emissions per second—or 1 curie of radioactivity, a baseline parameter. By comparison, all humans are exposed to both cosmic rays from the sun and to radioactivity from rocks and gases of the earth to a level that varies between 20 detectable emissions per second to about 200 in special areas of such countries as India, Iran, and Brazil. In term of other units of measure, such radiations give normal background levels of 3–600 millisieverts (mSv or mGray) per year.

To acquire a concept of the properties of such radiations it is useful to note that

- In terms of emissions, exposure to X rays, β-rays, or c-rays is less damaging than the equivalent amount of α-rays by a factor of about 20.

- Exposure to 10 sieverts (Sv) in one day is normally lethal to one human.

- Exposure to 10 Sv over one year would have a chronic effect on one human, such as cancer.

- Workers or sailors involved in the nuclear industry or the nuclear-powered navy are allowed to be exposed to 2.2 mSv/day.

- The 541 atmospheric tests of nuclear weapons set off between 1945 and 1980, which exploded the equivalent of 440 megatons of TNT, have increased the normal background radiation by 0.04 mSv/year.

- The additional radiation from all the world's nuclear power stations amounts to 0.002 mSv/year.

- A medical or dental X ray delivers, in seconds, 0.4 to 10 mSv.

- A modern CAT scan exposes a person to some 10 mSv.

- To achieve a biological effect the amount of radiation that has to be delivered has to exceed a certain "threshold" level.

Radioactivity in the Laboratory and Medicine

The civilian nuclear industry has, as a by-product, made available many radioactive materials that find uses in the laboratory or medical diagnostic facilities. Elements such as tritium (H^3 or hydrogen with one proton and two neutrons), carbon-14, sulfur-35, and phosphorous-32 are all β-ray or -particle emitters, while iodine-131 is a c-ray emitter. People who work with chemical compounds containing such isotopes need not be unduly worried about the effects of radioactivity on their persons, because β-rays travel only a few millimeters and do not penetrate the walls of glass tubes or containers. By contrast, 3 million c-rays and 250,000 β-particles emanating from natural sources pass though an individual human every minute.

These radioactive isotopes have enabled scientists to map out the route taken during the chemical transformation of food materials to cellular components and wastes and have unlocked the mysteries that surrounded the process of photosynthesis on which advanced life depends. In the medical area, the use of X rays for diagnosis is widespread, and the use of radioactive iodine in immunoassays for the detection of micrograms of materials per milliliter of sample is a powerful tool in measuring hormone and other metabolites of interest in medical and veterinary applications. A more recent use of radioactive isotopes has been in assay systems that enable determining the sequence of the bases in molecules of nucleic acids. Such assays have been used to acquire knowledge of the full sequence of the human genome and identify particular genes that cause inherited defects.

Most ethical debate on the use of genetic engineering techniques for the correction of defects in single gene disorders (typically, cystic fibrosis or immune disorders caused by a faulty enzyme, amino deaminase) has taken the view that such efforts are worthy and should be encouraged. It is also held, however, that only the phenotype should be affected and efforts to correct the defect in gametes should not be allowed. When it comes to the use of genetic engineering to effect enhancements of individuals (eye, hair and skin color, intelligence, musical and athletic abilities, etc.), ethical arguments are adduced to prevent such efforts, although the use of the growth hormone gene may be applied to correct a pathological condition, dwarfism, but not to produce basketball players.

Cancer treatments based on radiation (X rays, c-rays, and β-rays) are many and varied. Whole-body radiation of 10 Sv (10,000 times the annual background exposure) will cause the cessation of the development of bone marrow. Cancer treatment is based on the need to kill cells whose replication control mechanism has become ineffective. There is, however, the risk of killing other (collateral) cells and also of causing a cancer as a result of damaging nucleic acid molecules (genes) in neighboring tissues. Therefore the basis of successful therapies is to engineer treatments to maximize the therapeutic effects while minimizing the chances of coincident damage.

From Nuclear Energy to Electrical Power via the Atomic Bomb

The route from radiation to the atomic bomb came via the demonstration of the fission of atomic nuclei in 1938 by the German chemists Otto Hahn and Fritz Strassmann, which was followed by the separate investigations of Niels Bohr and Enrico Fermi on the fission of uranium atom nuclei. Experiments of all four led to the understandings of the crucial position of the uranium-235 isotope as opposed to the more abundant version of that element, uranium-238. The separation of these isotopes occupied the scientific and engineering acumen of many in both the United Kingdom and the United States.

In 1941 the work done in the United Kingdom influenced Vannevar Bush in the United States to authorize the construction of a subcritical experimental nuclear reactor or "pile." President Franklin D. Roosevelt backed the program in October of that year. In April of the next year, Fermi relocated to the University of Chicago, where he built a larger and more active reactor in the Stagg Field squash courts; calculations regarding the amount of material that would be needed to make a bomb were set in motion. Using the mental and physical understandings and skills of tens of thousands of scientists and engineers who were given an unlimited budget the outlines of the nature of an atomic bomb emerged. In January 1945 after much empirical experimentation and theoretical calculation, the scientists concluded that some 10 kilograms of plutonium or 40 kilograms of uranium-235 would be the minimal amounts of material necessary to set off an atomic explosion. The first such explosion took place on July 16, 1945, at the Alamogordo bombing range in New Mexico, while the second was over the city of Hiroshima, Japan, twenty-one days later. In 1952 the first deuterium-(H^2-) based fusion bomb (in which protons fused together to make the nucleus of an atom of higher atomic weight [lithium] than the original atoms [hydrogen]) was exploded at Enewetok Atoll in the Pacific Ocean, releasing power 100 times greater than that of the fission bombs—the equivalent of some 10 million tons of TNT.

Now that the genie had left the bottle, the way was open for both the peaceful and military use of nuclear energy by any country that could afford the time, expertise, and money. The first use of a nuclear reactor for the production of electrical energy occurred onboard a submarine, namely the USS *Nautilus*, completed in January 1954. As of 2005 there were over 150 ships (mainly submarines, aircraft carriers, and icebreakers) powered by more than 220 small nuclear reactors.

Land-based nuclear reactors that were designed to generate usable power in the form of electricity had the dual function of also making plutonium as a result of the nuclear reactions that occur when the fissile uranium generates heat. The uranium provided the electricity for national power grids, while the plutonium was added to the material that could be used for the production of bombs. The first such station to have this dual function was built in the United Kingdom at Calder Hall; it went commercial in October 1956. Since then, some 440 commercial nuclear power reactors and 284 research reactors have been built. They operate in 56 countries and supply some 16 percent of the world's total electricity base load. In Lithuania and France over 70 percent of the electricity supply is derived from nuclear reactors.

Assessment

Despite the large number of facilities that contain a nuclear reactor, the number of casualties that have resulted are relatively few. From the late 1950s to the early 2000s casualties directly associated with nuclear reactors numbered less than fifty. This is many fewer than the fatalities caused by other methods of generating electrical energy during the same period. There have been six serious events in which radioactivity has spilled over into the environment, the most damaging being that of the Chernobyl explosion in 1986 near the city of Kiev in Ukraine (then part of the USSR). Thirty-one people died and 1,800 children had to be provided with antidotes to thyroid cancer. Almost a million people were evacuated, and 10,000 square kilometers of land were designated as unfit for use. There was no evidence of other radiation-induced illnesses in the local population, which began moving back into the vacated area in the late 1990s.

There have been many studies examining the relationship between the incidence of leukemia and cancer and the locality of a power-generating nuclear reactor. Thorough examination of such data leads to the conclusion that although from time to time some radioactive material may have leaked from such establishments there has not been a noticeable and definitive increase in cases of cancer in the vicinity of such power stations.

Nevertheless, because nuclear reactors are associated with bombs, the fear of this technology has been disproportionate to its actual lethality. Paul Slovic's book on the perception of risk (2000) provides data that shows that while nuclear energy is *perceived* as generating the greatest risk, the *actual* risk is less than one chance in a million that a person who lives within five miles of a nuclear reactor for fifty years will die by an accident related to the reactor (a risk equivalent to that provided by smoking 1.4 cigarettes). Additionally, much has been made of the costs and dangers of decommissioning nuclear power reactors and of handling radioactive materials from this operation as well as the waste materials from the processing of spent fuel rods. The technology of radioactive waste storage has progressed, yet it is necessary to annually remove from circulation relatively small quantities (several tons) of highly radioactive material that retains its radioactivity for tens of thousands of years or longer. Were such material buried, as is suggested, there remains a danger that the containers may rupture, allowing seepage of radioactive material into the local groundwater. Nevertheless, sites for the indefinite storage of such materials held in a glass matrix within metal containers may be found in deep abandoned mines located in geologically stable areas.

The real terrors of the nuclear industry are in the area of bombs, a complex issue in and of itself. On the one hand, the end of the cold war (1945–1989) led to an overall decrease in the total number of nuclear weapons and agreements concerning the disposition of the remainder. On the other hand, China, India, Pakistan, and other countries have developed their own nuclear weapon capabilities. The expansion of trade will at least in some instances promote nonbelligerent conditions. And regardless of the connection between the nuclear power industry and nuclear weapons, one day oil and gas supplies will run out, and energy will still be needed.

At that time both worldwide population and its average rate of energy consumption are likely to have increased considerably. Although the energy of winds, rivers, tides, waves, and solar photons are likely to be increasingly captured and converted to distributed electrical power, it is unlikely that such supplies will satisfy human needs. The nuclear power option will increase in importance as conventional sources of energy are used up. It could be prudent to create the conditions for such an eventuality while the opportunity still exists to experiment without the pressures of urgent needs.

If in fact humanity turns to the nuclear power option, the issue of safety will need to be addressed. Modern societies have developed extensive systems of rules and regulations to protect the health and safety of those working with dangerous procedures, chemicals, or physical conditions. It may be expected that a parallel suite of regulations already in use in the nuclear industry will be extended and refined for a future, enlarged nuclear industry.

A related issue herein is that of global warming (or climate change). It is widely believed that the anthropogenic (human) production of carbon dioxide is, at the least, partly responsible for the increase in temperatures that has been observed around the planet. Many believe that this has been caused by human combustion of fossil fuels (coal, methane gas, and oil) for generating electricity and powering vehicles. An approach to militate against further increases in carbon dioxide proposed by James Lovelock, the initiator of the Gaia hypothesis, and others, is to use more nuclear reactors for the production of electricity. This electricity in turn could be used to generate hydrogen from the electrolysis of water to provide fuel for vehicles fitted with hydrogen-based fuel cells that generate electricity for onboard motors. This approach does not add to the carbon dioxide in the atmosphere and is safe, clean, and cost effective; it is possible to obtain 2.5 million times more energy from a gram of uranium than from the same amount of coal. A nuclear power program could be used in conjunction with other environmentally friendly approaches to energy generation, including wind, wave, biomass, and solar power.

Conclusion

The history of the development of the nuclear industry provides a paradigm of the emergence of a powerful technology from the observation of natural phenomena at the level of the individual scientist. At each stage the emerging new knowledge coupled with the development of techniques and equipment brought humanity to a more reliable understanding of the way nature worked and how humans operated. When the survival of the nation state was threatened as never before (after the devastating attack on Pearl Harbor, Hawaii, on December 7, 1941) America poured unlimited resources into the building of the atomic bomb. Could the scientists and engineers have decided not to develop atomic weapons at that time on the basis that the expression of the capability to develop such weapons could jeopardize the future survival of humanity? The question remains how humanity would respond to a similar challenge if it occurred again. In the end, humans have acquired awesome capabilities. It is perhaps thanks to the ethical strictures that humans have also built up over the ages that, for the most part, the use of the new and powerful technology has been restrained to beneficial ends. Such ethics are predicated on the bending of all human efforts to achieve the enhancement of the survival of humans on this planet, and they are perhaps encompassed in the following ethical statement by Hans Jonas: "Act so that the effects of your action are compatible with the permanence of genuine human life" (Jonas 1984, p. 11).

It might also be noted that there have been prominent scientists (Albert Einstein and Robert Oppenheimer in particular) who, having surveyed the results of their decisions in the heat of wartime, later recanted their enthusiasm for the project on which they worked so hard. Such retroactive evaluations may serve as a teaching device, but they do not help solve the problems that humans face in the early twenty-first century.

Energy released from nuclear reactions has the potential of providing almost unlimited amounts of virtually clean power into the indefinite future. It may also power spaceships, enable humans to colonize other planets of the solar system, and resolve medical pathologies. If it ever becomes feasible to progress to the harnessing of fusion power as demonstrated in the hydrogen bomb, then issues of power generation would no longer distract humanity from efforts to enhance the personal and social lives of all human beings. Yet, as with all the tools developed by humankind over the last 2.5 million years, it must be recognized that nuclear energy may be used to cause harm as well as provide benefits. Humanity's efforts, therefore, have to be directed at developing and practicing those ethics and morals that prevent harmful uses while enabling and encouraging beneficial deployments. The future of the human species depends upon the success of this endeavor.

RAYMOND E. SPIER

SEE ALSO *Chernobyl; Nuclear Waste; Three-Mile Island.*

BIBLIOGRAPHY

Jonas, Hans. (1984). *The Imperative of Responsibility: In Search of an Ethics for the Technological Age*, trans. Hans Jonas and David Herr. Chicago: University of Chicago Press.

Slovic, Paul. (2000). *The Perception of Risk*. London: Earthscan Publications.

Spier, Raymond E. (2001). *Ethics, Tools, and the Engineer*. Boca Raton, FL: CRC Press.

Spier, Raymond E. (2002). "Ethical Issues Engendered by Engineering with Atomic Nuclei." In *Science and Technology Ethics*, ed. Raymond E. Spier. London: Routledge.

INTERNET RESOURCES

Lovelock, James. "Nuclear Power Is the Only Green Solution." Available from http://www.perfect.co.uk/2004/05/james-lovelock-nuclear-power-is-the-only-green-solution. Lovelock's comment on the need for the activation of the nuclear power option.

Uranium Information Centre. "Nuclear Power in the World Today." Available from http://www.uic.com.au/nip07.htm. Provides information on commercial nuclear reactors that provide electricity to national grids.

Washington and Lee University. "Alsos Digital Library for Nuclear Issues." Available from http://alsos.wlu.edu. Provides a wide range of annotated references for the study of nuclear issues.

WEAPONS PERSPECTIVES

Ethical and political reflection on nuclear power was initially stimulated by the dangers of nuclear weapons. Even as the possibility of the atomic bomb began to be imagined in the 1930s, physicists became worried about its social, political, and ethical implications. By the time the first bomb was exploded in 1945, and even more as the nuclear arms race took hold in the 1950s, scientists, engineers, military professionals, politicians, and the attentive public became increasingly concerned about nuclear research and development, testing, and deterrence policy. As much as any other science and technology during the twentieth century, nuclear weapons have challenged ethical reflection. Although such weapons present major benefits—otherwise they would not have been invented, produced, and used—they also have built-in disadvantages that are not always easy to assess. As Albert Einstein remarked in 1946, the problem created by nuclear weapons is "not one of physics but of ethics."

Communities of Reflection

Nuclear weapons and ethics have been discussed in three overlapping communities of reflection. As the community of discovery and inventive origins for both nuclear science and weapons technology, scientists and engineers have played a major role in promoting ethical criticism. As the community that pioneered the use of nuclear weapons, the military has analyzed from its own perspective many ethical and political aspects of nuclear weapons. Finally, as the primary source of funding and ultimate beneficiary (and victim) of nuclear weapons, citizens and their democratic leaders have sought to place nuclear weapons in the broadest ethical context. Nuclear ethics and weapons issues may thus conveniently be considered in relation to the interacting discourses opened up by these three communities.

THE SCIENTIFIC-ENGINEERING COMMUNITY. In the 1930s scientists in Great Britain and the United States promoted nuclear weapons research because of the threat that Nazi Germany might develop such weapons. In 1945, when it became clear that Germany had not come close to developing the atomic bomb, some scientists at Los Alamos National Laboratory, where the bomb was being designed and fabricated, argued that such work was no longer justified. The majority view, however, was that work should go forward in order to demonstrate to the world the possibilities of such weapons, to complete a challenging technoscientific project, and perhaps in order to contribute to the continuing war effort against Japan.

After the bombing of Hiroshima and Nagasaki a group of scientists and engineers involved with atomic bomb development took the initiative to promote public education about the awesome power of nuclear weapons and lobbied for their international control. This ethical work led to three institutional initiatives—the Federation of Atomic (later American) Scientists (founded 1945), the *Bulletin of the Atomic Scientists* (first published in 1945), and the International Pugwash movement (founded 1957)—each of which became critical of the subsequent nuclear arms race, especially in the form of atmospheric testing and later proliferation.

Generally speaking, scientists and engineers felt a strong moral responsibility to educate politicians and the public about both the benefits and dangers of nuclear weapons. Yet a divide developed within the technical community between those who maintained the benefits outweighed the dangers and those who argued the dangers outweighed benefits. In the early 1950s this came to a head in a dispute between J. Robert Oppenheimer, who opposed hydrogen bomb development, and Edward Teller, who supported it. For Oppenheimer, the atomic bomb was sufficiently powerful for any conceivable military purpose, whereas for Teller the threat that the Soviet Union might develop a hydrogen bomb was sufficient to justify its pursuit. Among scientists one of the basic disagreements was and has continued to be over when enough is enough, and what precisely scientific responsibility entails.

THE MILITARY COMMUNITY. Among those involved with the military both as professional soldiers and policy analysts, questions arose primarily in relation to strategic policies. In the military there was never any sense that German defeat should undermine the justification of nuclear weapons work. From an early date the military saw nuclear weapons as a means of exercising military power and set about formulating appropri-

ate strategies to take advantage of its unique features. The major result was development of the concept of nuclear deterrence—a strategy that nevertheless gave rise to a number of important and well-explored ethical quandaries.

One quandary concerned whether nuclear weapons should be directed toward military or civilian targets. Although traditional just war theory argued against "countervalue" targeting of civilians, to limit nuclear weapons targeting to "counterforce" assets might, especially during a crisis, actually encourage an enemy toward a preemptive nuclear strike in order to try to avoid the loss of its nuclear capabilities. Counterforce targeting also tends to encourage a nuclear arms race for increasingly accurate weapons. The policy question then becomes: What is the most ethical way to target nuclear weapons?

Another quandary considers in what sense it is ethically permissible to threaten what it would not be ethically permissible to do. There is little disagreement that it would be ethically wrong to use nuclear weapons against a large civilian population in an enemy country, especially because the results would affect large numbers of people in other, neutral countries, and be likely to rebound even on the attacking country. But what if the best way to avoid the actual use of nuclear weapons is to threaten their use on civilian populations? What, then, is the most ethically defensible policy, especially in relation to a totalitarian country or a regime ruled by someone whose behavior may not be rational?

Finally, insofar as there are prima facie justifications for defending oneself against attack from nuclear weapons, to threaten a country with nuclear retaliation seems legitimate. But insofar as there are prima facie prohibitions against threatening innocent people, and given that nuclear weapons cannot but harm innocent people, to threaten the use of nuclear weapons seems equally illegitimate. Prima facie or deontological arguments thus both support and oppose the development and use of nuclear weapons.

THE POLITICAL COMMUNITY. The political community is divided into two groups: the established political community and the oppositional political community. Each form of the political community has sought to overcome the quandaries elaborated within the military community.

From the beginning the established political community, in alliance with the military community, sought ways to use nuclear weapons to pursue political ends (especially in relation to the nuclear standoff with the

Soviet Union). For the United States especially, nuclear weapons allowed the country to counter a Soviet superiority in ground troops in Europe in a way that was politically tolerable (that is, without maintaining a large standing military and at a relatively low annual financial burden). The solution to the ethical quandaries was to promote technological fixes in the form of civil defense and/or the development of some kind of defensive missile system.

By contrast, the oppositional or alternative political community, in alliance with a vocal segment of the scientific community, argued for a political fix to the quandaries of nuclear deterrence. One such political fix comprised proposals for the internationalization of nuclear weapons control. An even more radical proposal argued for unilateral nuclear disarmament. In the middle, the alternative political community actually succeeded in 1963 in getting the major nuclear powers to sign the Limited Nuclear Test Ban Treaty halting nuclear weapons testing in the atmosphere. Later voluntary and reciprocal moratoriums were developed among some powers with regard to underground nuclear testing. But such U.S.–USSR agreements have had only marginal influences on many other countries. And oppositional efforts to limit nuclear proliferation have been problematic at best.

Further Issues

Disagreements among the three communities of reflection have carried over into a number of closely related issues. Among such issues are questions of the moral probity of civilian defense and defensive missile systems, the effectiveness of such systems (especially missile defense systems that rely on complex, automated responses to information that can itself be quite problematic), the problem of how to respond to worries in civilian populations affected by nuclear weapons industry sites, and the difficulties of nuclear waste disposal.

Three ethical issues that have received only marginal discussion may also deserve notice. First, there is a somewhat suppressed debate regarding whether many of the fears about nuclear weapons have been well founded. After all, since 1945 nuclear weapons have not been used except as features of deterrence strategies. Are worries about the dangers of nuclear weapons misplaced? Or have the expressions of fear had the salutary effect of helping to keep mistakes from being made? Second, some have suggested that the shift in nuclear testing to an increasing reliance on computer simulations may deprive nuclear scientists and engineers, not to mention soldiers and politicians, of a direct experience

of the destructive powers of nuclear weapons that itself has also had a salutary effect on their handling and use. Third, with the advent of the possible use of nuclear devices by nonstate actors and terrorists, new questions arise about the responsibilities of those who have developed and are continuing to develop nuclear weapons.

Finally, it might be suggested that despite initial appearances, many of the issues with regard to nuclear weapons only present in especially dramatic form questions that relate to modern science and technology in general. Science and technology in general place in human hands enormous power for transforming the world, many of which entail quandaries similar to those associated with nuclear weapons. The pollution of the natural environment and the burning of fossil fuels, which seem necessary to pursue benefits for present generations, may have negative impacts on future generations in ways that mirror the deterrence targeting of enemy populations (which benefit the targeting populations at the potential expense of the targeted populations). Thus it can be argued that ethical reflection on nuclear weapons should not be isolated from ethical reflection on other technologies, or that the results of ethical reflection in regard to both nuclear and nonnuclear technologies should be compared and contrasted for the benefit of science, technology, and ethics as a whole.

CARL MITCHAM

SEE ALSO *Baruch Plan; Just War; Limited Nuclear Test Ban Treaty; Military Ethics; Nuclear Nonproliferation Treaty; Nuclear Waste.*

BIBLIOGRAPHY

Cohen, Avner, and Steven P. Lee, eds. (1986). *Nuclear Weapons and the Future of Humanity: The Fundamental Questions.* Totowa, NJ: Rowman and Allanheld. The single best collection of articles on this topic.

Einstein, Albert. (1946). "The Real Problem Is in the Hearts of Men." *New York Times Magazine,* 23 June, 7 and 42–44. An interview with Michael Amrine.

Hashmi, Sohail H., and Steven P. Lee, eds. (2004). *Ethics and Weapons of Mass Destruction: Religious and Secular Perspectives.* Cambridge, UK: Cambridge University Press.

Hollenbach, David. (1983). *Nuclear Ethics: A Christian Moral Argument.* New York: Paulist Press.

Jaspers, Karl. (1961 [1958]). *The Future of Mankind,* trans. E. B. Ashton. Chicago: University of Chicago Press. One of the first philosophical reflections on the need for "an essentially new way of thinking" if human beings are to survive the invention of nuclear weapons.

Kavka, Gregory S. (1987). *Moral Paradoxes of Nuclear Deterrence*. Cambridge, UK: Cambridge University Press. A tightly argued exploration of the major quandaries.

Lee, Steven P. (1993). *Morality, Prudence, and Nuclear Weapons*. Cambridge, UK: Cambridge University Press.

Nye, Joseph S., Jr. (1986). *Nuclear Ethics*. New York: Free Press. An important analysis of deterrence policy by a U.S. policy analyst.

Shrader-Frechette, Kristin S. (1983). *Nuclear Power and Public Policy: The Social and Ethical Problems of Fission Technology*, 2nd edition. Dordrecht, Netherlands: Reidel Publishing. The first and still influential philosophical assessment of nuclear power, which also makes some relations to the nuclear weapons–nuclear power connection.

Shrader-Frechette, Kristin S. (1993). *Burying Uncertainty: Risk and the Case against Geological Disposal of Nuclear Waste*. Berkeley and Los Angeles: University of California Press. Although this book focuses on Yucca Mountain, Nevada, the arguments developed here are of quite general significance.

NUCLEAR NON-PROLIFERATION TREATY

• • •

The nuclear Non-Proliferation Treaty (NPT) is the only legally binding multilateral agreement that commits signatory states to an active pursuit of disarmament. It is a major example of an attempt to govern the development and use of technology, in this case, one of the most powerful technologies ever developed.

Historical Development

Early post-World War II efforts to contain the proliferation of nuclear weapons were unsuccessful. The United States (1945) was followed rapidly by the Union of Soviet Socialist Republics ([Soviet Union] now Russia, 1949), United Kingdom (1952), France (1960), and the People's Republic of China (1964) as nuclear weapons states (NWS), quickly dissipating assumptions that nuclear technology was difficult to both acquire and master. In fact the increasing construction of nuclear reactors introduced a sense of urgency for a multilateral treaty that would halt and eventually reverse the proliferation of nuclear energy and weapons technology. The NPT was therefore designed to strike a balance between the NWS, the five states who manufactured and or exploded a nuclear weapon prior to January 1, 1967, and non-nuclear weapon states (NNWS), in ways that would diminish and eventually eradicate the use of nuclear weapons.

Throughout the 1950s, there were a series of initiatives by both NWS and NNWS to check the proliferation of nuclear technology. Although there were fundamental disagreements between the United States and the Soviet Union on the specifics of these initiatives, these efforts nevertheless set the precedent for a multilateral treaty that would include non-dissemination and non-acquisition principles as its fundamentals.

These NPT negotiations took place in three distinct phases. Phase one consisted of bilateral talks in the late 1950s and early 1960s between the United States and the Soviet Union. Although both countries favored nonproliferation, there were serious divergences on how to implement it. The United States, along with Canada, France, and the United Kingdom, submitted a package to the United Nations in August 1957 that included a non-transfer commitment. The Soviets objected on the grounds that it still allowed for the deployment by a nuclear power of its weapons under the justification of self-defense, and wanted to add a clause prohibiting the stationing of nuclear weapons in foreign countries.

The main sticking point continued to be the U.S. proposal for a North Atlantic Treaty Organization-based (NATO) Multilateral Nuclear Force (MNF), which the Soviets argued constituted proliferation. Although the U.S. draft treaty sought to clarify collective defense arrangements by maintaining that the United States would hold a veto on deployment of U.S. weapons, the Soviets would not agree to such a provision. However both countries ultimately agreed on the premise that nuclear nonproliferation was of the utmost importance. The United States conceded on the collective defense MNF and in the end of 1966 the Soviet and U.S. chairmen of the Eighteen-Nation Disarmament Committee (ENDC) reached a tenable agreement on the basic premises of the proposed NPT.

Phase two of deliberations occurred between the United States and its NATO allies. The NNWS members of NATO expressed significant concern over the planning of nuclear defense within the confines of their region without their full consent. The United States sought to clarify how a non-proliferation treaty would support collective defense obligations. The U.S. interpretation of the draft treaty stated that while nuclear weapons and the framework of the treaty covered explosive devices, delivery systems were not included. Therefore the treaty did not prohibit planning of nuclear defense between the NATO allies, nor deployment of U.S. controlled and operated nuclear weapons on territory of non-nuclear NATO members. The Soviets did not object to this interpretation, as the United States would maintain full control over their nuclear weapons throughout Europe, specifically where nuclear weapons were deployed.

Phase three of the negotiation took place throughout the 1960s in the United Nations and occurred simultaneously with the bilateral U.S.–U.S.S.R. talks as well as the NATO negotiations. It began when the United Nations General Assembly adopted the Irish resolution in 1961 calling for all states to enter into a nonproliferation treaty that would outlaw the transfer and acquisition of nuclear weapons. Following the Irish resolution was UN resolution 2028 in 1965, which codified five principles necessary for a non-proliferation treaty: Both NWS and NNWS states would be prohibited from proliferation of any kind; NWS and NNWS would share the responsibilities of the treaty; the goal of the treaty would be nuclear disarmament but also general and complete disarmament; there would be practical policies in place to ensure the effectiveness of said treaty; and the establishment of nuclear weapon free zones should not be hindered by the treaty. Resolution 2028 provided the fundamental framework for the final version of the NPT and it was from this document that the United States and Soviet Union began to develop an actual codified multilateral treaty to end proliferation of nuclear weapons.

Finally on August 24, 1967, the United States and the Soviet Union submitted identical but separate drafts of the treaty to the ENDC. After many revisions, the treaty was approved by the UN General Assembly and opened for signature on June 12, 1968, to the depositary governments of the United States, the United Kingdom, and the Soviet Union. The treaty went into effect on March 5, 1970. France and China eventually signed on as did 183 NNWS.

NPT Commitments: Successes and Failures

The NPT commits signatory NWS to not transfer their nuclear weapons to NNWS, or assist them in acquiring nuclear weapons. NNWS signatories agree to renounce nuclear weapons, and to remain open to inspections of their nuclear materials and activities by the International Atomic Energy Agency (IAEA). The NPT further commits states to hold conferences every five years in Geneva, Switzerland, to review the implementation and effectiveness of the treaty. In 1995, twenty-five years after the formal commencement of the treaty, the review conference voted to extend the agreement indefinitely, as opposed to holding five year reviews.

The NPT is important in that it is the legal basis for the nonproliferation and disarmament regime and the only universal arms control treaty. Countries throughout the world have been able to develop nuclear power for peaceful purposes without threatening neighbors or enemies. The NPT has had such an impact that, more than thirty-five years later, there are only eight countries that possess nuclear weapons (United States, Russia, France, England, China, India, Pakistan, and North Korea), a far cry from the hundred that was once predicted. Several countries, including South Africa, Argentina, and Brazil, have even been convinced to give up nuclear capabilities based on the strength of the regime.

While all the successes of the NPT may never be known, there are also some negatives to the regime. Critics contend that the larger share of the responsibility falls on NNWS, and that they face a military disadvantage because they are required to submit their programs to IAEA inspections while NWS are not. Non-aligned NNWS—that is, countries who are not part of a military alliance with the NWS states—sought security assurances that the NWS would not use weapons against them, but this was never explicitly confirmed in the final draft of the NPT.

Others claim that IAEA safeguards are oftentimes ineffectual, as was the case with Libya, which denied, and North Korea, which continues to deny access for IAEA inspection. India, which first tested a peaceful nuclear device in May 1974; Pakistan which tested a nuclear weapon in May 1998 following a test by India; and Israel are not party to the NPT, but all have nuclear weapons. For them to join, they would have to dismantle their nuclear weapons, as South Africa did in 1991. The world continues to encourage these countries to renounce their nuclear program and join the NPT. However each of the three nations is known to have nuclear weapons, as is North Korea, proving that despite the strides made by the NPT, proliferation is still possible and a valid threat to international security. Nevertheless those states party to the NPT continue to endeavor to strengthen the effectiveness of the NPT, and remain committed to securing nuclear free zones, and checking the proliferation of nuclear weapons.

The NPT, in both its successes and failures, exemplifies efforts to develop mechanisms of international governance for technologies of international significance. In this respect it may be compared to the Montreal Protocol for the reduction of the emissions of chlorofluorocarbons (CFCs) or the Kyoto Protocol for the reduction of green house gas emissions. Comparisons might also be made with the Law of the Sea Treaty for international sharing in the exploitation of seabed mineral resources and treaties to demilitarize space. The need for multinational governance of science and technology is clearly an important issue about which greater sophistication will only be developed by trial and error learning.

JESSICA L. COX
MARGARET COSENTINO

SEE ALSO *Baruch Plan; Just War; Limited Nuclear Test Ban Treaty; Military Ethics; Nuclear Ethics.*

BIBLIOGRAPHY

Bailey, Emily; Richard Guthrie; Daryl Howlett; and John Simpson. (2000). *Programme for Promoting Nuclear Non-Proliferation Briefing Book Volume I: The Evolution of the Nuclear Non-Proliferation Regime,* 6th edition. Southampton, UK: Mountbatten Centre for International Studies.

Sokolski, Henry. (2001). *Best of Intentions: America's Campaign against Strategic Weapons Proliferation.* Westport, CT: Praeger Publications.

INTERNET RESOURCES

Monterey Institute of International Studies. "Center for Nonproliferation Studies Non-Proliferation of Nuclear Weapons Treaty Tutorial." Available from http://cnsdl.miis.edu/npt.

United Nations. "Treaty on the Non-Proliferation of Nuclear Weapons." Available from http://www.un.org/Depts/dda/WMD/treaty/.

United States Department of State. "Treaty on the Non-Proliferation of Nuclear Weapons." Available from http://www.state.gov/www/global/arms/treaties/npt1.html.

NUCLEAR REGULATORY COMMISSION

• • •

The U.S. Nuclear Regulatory Commission (NRC) is an independent agency of the federal government with a mission to protect public health and safety from the hazards of the civilian use of nuclear energy technology. The NRC oversees nuclear power reactors, radioactive waste disposal, and medical, industrial, and academic uses of nuclear materials. The NRC was created through the Energy Reorganization Act of 1974, which divided the Atomic Energy Commission (AEC, created in 1947) into the NRC and the Energy Research and Development Administration (ERDA). The ERDA was subsequently subsumed into the Department of Energy (DOE).

Prior to the reorganization, the AEC both regulated and promoted nuclear energy technology as well as managed the nuclear weapons complex. Having both regulatory and promotional functions for nuclear technology within the AEC led many in Congress and in the general public to charge the AEC with a conflict of interest. Recognizing that low public confidence in the AEC to regulate objectively would slow and perhaps paralyze the growth of nuclear technology, Congress created the NRC solely to regulate the nation's civilian nuclear energy activities. The DOE meanwhile had assumed the former AEC nuclear technology promotional functions and its management of the U.S. nuclear weapons complex.

Five commissioners, with one as chair, lead the NRC. All are presidential appointees who require Senate confirmation to staggered five-year terms. In addition to the staff officers who perform the daily work of the NRC, there is the Atomic Safety and Licensing Board Panel (ASLBP), which is the adjudicatory arm that conducts public hearings primarily on licensing and enforcement actions.

Regulatory Practice

When a reactor license holder decides to decommission a reactor and terminate the license, any member of the potentially impacted public may request an adjudicatory hearing. Depending on the regulatory action, a formal or informal hearing may be held. Formal hearings are trial-like proceedings with discovery of evidence, sworn testimony, and cross-examination. The determination of whether a hearing will be informal or formal is codified in NRC rules, Title 10 of the Code of Federal Regulations, Part 2, but there is allowance for some discretion. Although ASLBP members are employees of the NRC, there are rules under the Administrative Procedures Act that enable panel members' judicial independence. There are some stakeholders, however, who believe that the ASLBP cannot truly preside over an unbiased hearing because it is part of the NRC. ASLBP decisions ultimately may be appealed to the U.S. Supreme Court. There also exist three external advisory committees to the NRC, one on reactor safeguards, one on nuclear waste, and one on the medical uses of isotopes. These committees are made up of nuclear professionals from industry, academia, and government.

The NRC also regulates the production and use of source, special, and by-product nuclear materials. Source materials are the elements of thorium and uranium not enriched in the isotope uranium-235. Source material may be converted into special nuclear material, which is capable of undergoing the fission reaction (splitting of the atom). Special nuclear materials are uranium isotopes 233 and 235 and plutonium. By-product materials are made in the process of producing or using special nuclear material. By-product materials are used in medical, industrial, or research applications, such as carbon-14 for radioactive dating. By-product materials are also wastes from reactor operations, such as spent nuclear fuel, and from mining operations, which are called mill tailings. Collectively, source, special, and by-product materials are referred to as AEA (Atomic Energy Act) materials.

To produce or use AEA materials requires an NRC license. In general licenses may be considered either reactor or material licenses. For instance, a nuclear power plant owner holds at least two licenses, a reactor license for the operation of the power plant and a material license for possession of the fuel. The NRC regulates all aspects of a license, from initial licensing through termination. It is primarily license fees that fund the NRC. The NRC also has a fee-based certification and quality assurance program. In lieu of issuing a license, the NRC will certify some products, such as spent-fuel shipping casks. NRC certification then enables the potential user of these products to begin using them as long as the product meets the certification standards. Certification enables the NRC to expedite the regulatory process because standardization of design assures regulatory compliance without the burden of determining whether an individual case meets its regulatory criteria.

NRC regulations are promulgated through a formal rule-making process. Petitions for rule-makings may come from any of the NRC stakeholders, such as industry, nongovernmental environmental organizations, or individual citizens. The proposed rule is published in the *Federal Register*, and the public is invited to comment. In promulgating its final rule, also published in the *Federal Register*, the NRC explains how it had considered the public's comments. In general, the NRC does not hold hearings about proposed rules. Additionally, the NRC has an electronic rule-making forum, RuleForum, where the public may assess information and documents related to a rule, such as the comments of other stakeholders. An electronic reading room that contains all the NRC's public documents is also available. All NRC public documents are physically located in the reading room at headquarters in Rockville, Maryland. The NRC also performs research to support its regulations. Other activities include international cooperation regarding safety and security.

Regulatory Philosophy

Beginning in the mid-1990s, the NRC started to adopt a general regulatory philosophy across all its activities that is more risk-informed as well as performance-based.

To implement this philosophy requires the NRC to ask three questions of its regulatory activities, the so-called risk triplet: What may go wrong? What are the consequences? How likely are these consequences to occur? Since its inception, the NRC had focused primarily on the consequences of what may go wrong and had prescribed "defense-in-depth" measures to effectively manage consequences; such measures include redundancy in emergency systems and engineering margins of safety. Asking how likely or probable a technology failure is requires carefully examining the relationships among the constituent elements and considering how each element contributes to the performance of the whole. This process enables the NRC to identify critical areas that may need more attention to safety. It may also find that a marginal decrease in resources in some areas is warranted because the decrease has no measurable effect on safety. This is one way risk information contributes to regulatory decision-making.

Enabling the public to better understand the reasons why the NRC believes a particular course of action poses no undue risk to the public health and safety is a major, continuing challenge. Indeed recognizing that the public's confidence in its regulatory integrity is critically dependent on the transparency of the decision-making process, the NRC continues to explore opportunities for open communication.

DARRYL L. FARBER

SEE ALSO *Nuclear Ethics; Three-Mile Island.*

BIBLIOGRAPHY

Balogh, Brian. (1991). *Chain Reaction: Expert Debate and Public Participation in American Commercial Nuclear Power, 1945–1975.* Cambridge, UK: Cambridge University Press.

Mazuzan, George T., and J. Samuel Walker. (1985). *Controlling the Atom: The Beginnings of Nuclear Regulation, 1946–1962.* Berkeley and Los Angeles: University of California Press.

Walker, J. Samuel. (1992). *Containing the Atom: Nuclear Regulation in a Changing Environment, 1963–1971.* Berkeley and Los Angeles: University of California Press.

Walker, J. Samuel. (2000). *Permissible Dose: A History of Radiation Protection in the Twentieth Century.* Berkeley and Los Angeles: University of California Press.

INTERNET RESOURCE

"U.S. Nuclear Regulatory Commission." Available from http://www.nrc.gov.

NUCLEAR WASTE

• • •

Disposal of nuclear waste has been a contentious issue both in the United States and elsewhere in the world. Difficult questions are involved, including: (1) where should one put the waste? (2) How long must such waste be stored before it does not pose a hazard to society? (3) What confidence can be placed in estimates of long-term confinement, and how great are the uncertainties? Because of the differing views on these topics and their complexity, their treatment here will necessarily be limited.

The focus here will be high-level radioactive waste produced at nuclear power plants. Excluded is any discussion of defense-related radioactive waste, or low-level radioactive waste generated from nuclear power, medical applications, industrial applications, and research. The basic issues for these other types of waste are related to and can be informed by the present analysis. There are, as well, books and lengthy articles providing more comprehensive treatments, which are included in the references.

Nuclear Waste Itself

In the United States there are two types of radioactive waste produced at nuclear reactors: low-level waste (LLW) and high-level waste (HLW). While low-level nuclear waste represents most of the waste volume, high-level waste represents most of the radioactivity. For this reason HLW presents the major problem.

High-level waste in the United States (and also Sweden and Finland) comprises the used nuclear fuel elements, called spent fuel. In France, Great Britain, and Japan, where fuel is reprocessed to remove unused uranium fuel and plutonium (which represents 95 percent of the material in spent fuel), HLW primarily includes fission products and long-lived radioactive materials called actinides. (Russia and China are developing reprocessing capability, and Germany, the Netherlands, Switzerland, and Belgium reprocess their spent fuel elsewhere.) These are incorporated into radiation-resistant glass to produce blocks that can be placed into a temporary storage facility or a permanent underground facility. In the United States there is no reprocessing, so the HLW is in the form of solid fuel elements that contain all the products mentioned above.

High-Level Waste Disposal Facilities

In the early twenty-first century, all spent fuel in the United States was stored on the sites of the nuclear power reactors because no long-term storage facilities were available. This included sixty-four reactor sites in thirty-one states. When spent fuel is initially removed from the reactor it generates considerable heat from radioactive decay so that initial storage is in pools of water. After the spent fuel has been stored for a minimum of five years it can be moved to specially designed steel and concrete aboveground casks, approved by the Nuclear Regulatory Commission (NRC), that rely on air cooling to remove the heat. No accidents with spent fuel elements have occurred in which radiation has been released to the public. The HLW generated at a reactor in forty or more years of operation can be stored on-site, indicating that the volume of HLW generated at each reactor is quite low. Indeed, as Kristin S. Shrader-Frechette (1993) has argued at length, aboveground monitored retrieval storage may be a defensible option.

Political Processes for High-Level Radioactive Waste Disposal

The U.S. Department of Energy (DoE) has the ultimate responsibility for permanent disposal of high-level waste in the United States. Based on a strong consensus of international expert opinion, the best place for permanent storage of HLW is in a geologic repository deep underground in an environment that is both geologically stable and exceptionally dry. The Nuclear Waste Policy Act (NWPA) of 1982 chartered the DoE with the responsibility to develop a permanent geological repository for HLW. The NWPA also charged the Environmental Protection Agency (EPA) with the responsibility for developing environmental standards and the NRC with responsibility for evaluating whether the repository design submitted by the DoE meets these standards.

Initially three potential sites were identified for detailed study as possible repositories. The law was amended in 1987, however, to focus on a single site at Yucca Mountain in Nevada. Through these amendments, Congress also established an independent advisory group of experts, the U.S. Nuclear Waste Technical Review Board (NWTRB), to evaluate the technical and scientific validity of the DoE's efforts to develop a repository. The NWTRB issues annual reports to Congress and the secretary of energy with their evaluations.

Under the DoE plan, solid nuclear waste would be placed in extremely durable containers—called waste packages—that would be put into deep underground tunnels in dry, stable, volcanic rock. The safety concern is that, over time, enough water would come in contact with

FIGURE 1

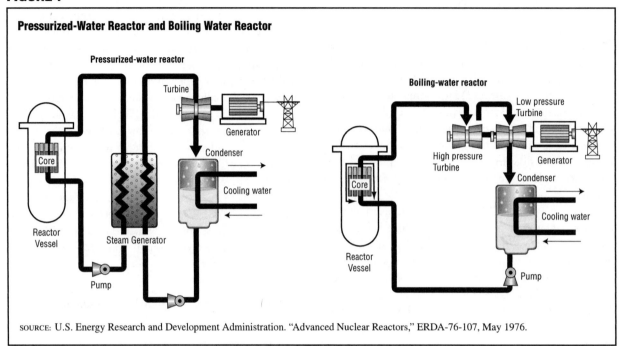

Pressurized-Water Reactor and Boiling Water Reactor

SOURCE: U.S. Energy Research and Development Administration. "Advanced Nuclear Reactors," ERDA-76-107, May 1976.

the waste to cause the release of radioactive elements and the transport of these materials to the water table. The proposed Yucca Mountain repository is about 1,000 feet below the land surface and 1,000 feet above the water table.

In 2002, after fifteen years of study, the DoE issued reports concluding that the Yucca Mountain site was suitable for a geologic repository for HLW. The DoE that year submitted to the president a recommendation for approval to proceed with the development of the Yucca Mountain repository. The NWTRB did not make a judgment regarding this recommendation because acceptability involves public policy issues that are beyond the board's mandate. The board did note that no scientific or technical factor had been identified that would eliminate Yucca Mountain as a permanent repository site, but also that there were gaps in data and basic understanding that result in important uncertainties in performance estimates. In essence, although sophisticated models have been used to predict whether the waste can be safely stored to meet EPA and NRC criteria, there remain uncertainties in the accuracy of the models and in the predictions. How much certainty is required to make a decision? And are the criteria for leak rates or confinement times the appropriate ones to use? These critical issues are difficult for experts to evaluate and for the public to understand. In the end, a political decision on acceptability is required.

Notwithstanding the concerns indicated above, President George W. Bush approved the recommendation and sent it to Congress, which then voted to approve it as well. The DoE's goal is to begin storing spent fuel beginning in 2010. A minimum of fifty years has been specified for studies of the repository performance before it can be closed. The DoE then has to apply to the NRC for a license to close the repository. During this time the repository will be monitored to enhance the understanding of the processes taking place in the repository, to determine if the behavior is in agreement with predictions of the models, and to correct any problems that are identified.

Yucca Mountain

The approvals to proceed with the repository at Yucca Mountain were highly controversial. The citizens and government of Nevada have strongly opposed the repository, regardless of whether the site is suitable. They contend that the benefits of nuclear power are primarily obtained elsewhere in the nation, but Nevada is expected to accept the risks for any kind of problem or accident related to handling or disposing of spent fuel at the repository. Because this is a national issue it is probably inevitable that there would be a conflict between federal and state interests. In December 2001 Nevada filed suit in federal court against the decision to proceed based on several technical and legal issues. In July 2004

Drums of radioactive waste lying in a trench at Hanford Nuclear Reservation. (© Roger Ressmeyer/Corbis.)

the U.S. Court of Appeals ruled that the U.S. Environmental Protection Agency (EPA) illegally set its radiation release standards for groundwater for the proposed high-level radioactive waste dump. Two months later, the Nevada attorney general initiated a new lawsuit, claiming the DOE lacked the authority to make many of the decisions required to continue the project.

Some opponents of Yucca Mountain repository argued that outstanding scientific questions remained that should be answered before one could be reasonably confident that the safety criteria can be met. They called for further research and a resolution of some of the technical uncertainties. Antinuclear groups, such as Greenpeace, expressed opposition to any solution to the waste problem, including Yucca Mountain. This strikes at the core issue of acceptable risk and how the United States, as a society, is to deal with wastes. Not doing anything is simply a different kind of solution and may not be the best one for society. Furthermore, the waste is a by-product of a technology that was introduced for the benefit of society, in this case to produce electricity

without the environmental problems of fossil fuels. Ultimately to gain the benefit it is necessary to address and solve the waste problem. But can one find a solution in which all the stakeholders are satisfied? Certainly Nevada and its citizens were not satisfied. Whether the Yucca Mountain decision achieves fairness and acceptability will continue to be debated by groups with differing opinions.

Another issue that affects public acceptability is whether spent fuel can be shipped safely to the site or whether such shipments pose an unacceptable hazard. What about accidents or terrorist attacks? The transport of spent fuel would occur on railway cars or in trucks in specially designed casks. These casks, designed to meet requirements of both the NRC and the U.S. Department of Transportation, are tested to demonstrate they can withstand crashes, fire, water immersion, and puncture. A truck carrying such a cask was crashed at 80 miles per hour into a concrete barrier. Although the cask was damaged it did not leak. Moreover, shipments of spent fuel are not new. From the early 1960s to the early 2000s, about 3,000 such

Entrance to a tunnel into Yucca Mountain in Nevada. The mountain is the site of the proposed Yucca Mountain Repository, a U.S. Department of Energy terminal storage facility for spent nuclear fuel and other radioactive waste. (© Dan Lamont/Corbis.)

ore within a couple thousand years, whereas spent fuel requires considerably longer because of the plutonium. Furthermore, the plutonium that is recovered through reprocessing is incorporated into fuel, thus reducing the total inventory of plutonium. But reprocessing also carries risks of proliferation, because reprocessed plutonium might be diverted or stolen to produce nuclear weapons. Initially the United States was committed to reprocessing, but in the late 1970s President Jimmy Carter decided not to proceed with reprocessing in the hope that other nations would follow the U.S. example. This would have limited the opportunity to clandestinely obtain plutonium that was produced in nuclear power reactors. Carter's effort proved unsuccessful because neither the Europeans nor the Japanese showed any interest in following suit. Independent of the security argument over reprocessing there is no economic incentive in the United States to revive reprocessing unless the price of uranium fuel rises significantly.

High-Level Waste in Other Countries

High-level waste disposal is required for every country that has nuclear power. Active research programs for deep geologic storage are under way in many countries, including Sweden, Finland, Germany, France, Switzerland, Great Britain, Russia, and China. Only Finland has committed politically to a specific disposal site. Other nations are carrying out research at one or more sites and have yet to complete the selection process. In December 2003 the European Union decided to evaluate the possibility of regional repositories, primarily to assist smaller countries. Based on the experience in the United States and in many of the above nations, it may be a difficult and contentious process before a final decision is reached.

Assessment

While critical issues have been decided in creating the Yucca Mountain repository there are many outstanding issues still to be resolved. Scientific studies that support critical engineering design decisions are still needed. The issuance of an NRC license, which will include extensive public hearings and most likely legal challenges, is also ahead. Furthermore, numerous construction activities must be completed. With expected appeals, it will be a daunting task for the Yucca Mountain repository to be ready to receive spent fuel by 2010.

Finally, creating a permanent repository will be a very expensive undertaking. As of September 2002, the

shipments covered more than 2.7 million kilometers (1.7 million miles) of U.S. roads and railways without any radioactive material being released as a result of an accident. Regarding terrorist acts, there are factors that make such shipments undesirable targets. The casks are massive and weigh many tons; and the trucks and trains that carry them are guarded and tracked via satellite communication. Even a shoulder-mounted rocket would be unlikely to crack the cask, and if it did little radioactivity would be released to the environment because the fuel is solid. The implication by anti–Yucca Mountain groups that the transported fuel represents a serious hazard is not supported by experience or analyses.

The plutonium that is in the spent fuel presents a different type of issue. Reprocessing reduces the volume of waste by about 75 percent and slashes the amount of time that the waste needs to be stored; reprocessed HLW will return to the radioactivity levels of mined

fund to pay for design, development, and ultimate storage of spent fuel has accumulated $23 billion and grows by $1 billion per year because of the Congressionally mandated 0.1 cent per kilowatt-hour charge on nuclear-generated power.

Nevertheless, the full cost to society of safely disposing of nuclear waste must factor in the damages avoided or benefits because of the noncarbon emissions from nuclear power. In other words, depending upon how the damages to the environment from fossil fuel plants are valued, the cost of disposing of nuclear waste may be a real bargain.

Future generations will judge whether the nation acted responsibly and appropriately in its decision regarding the disposal of spent fuel at an HLW repository at Yucca Mountain. If the decision is reversed, long-term monitoring would be needed to assure that this repository has solved a problem and not created a new one.

EDWARD H. KLEVANS
DARRYL L. FARBER

SEE ALSO *Nuclear Ethics; Nuclear Regulatory Commission; Waste.*

BIBLIOGRAPHY

Carter, Luther J. (1987). *Nuclear Imperatives and Public Trust: Dealing with Radioactive Waste.* Washington, DC: Resources for the Future.

Murray, Raymond L. (2003). *Understanding Radioactive Waste,* 5th edition, ed. Kristin L. Manke. Columbus, OH: Battelle Press.

Organisation for Economic Co-operation and Development. Nuclear Energy Agency. (1995). *The Environmental and Ethical Basis of Geological Disposal of Long-Lived Radioactive Wastes.* Paris: Author. Also available from http://www.nea.fr/html/rwm/reports/1995/geodisp.html.

Shrader-Frechette, Kristin S. (1993). *Burying Uncertainty: Risk and the Case against Geological Disposal of Nuclear Waste.* Berkeley and Los Angeles: University of California Press.

INTERNET RESOURCES

"Nuclear Science and Technology." American Nuclear Society. Available from http://www.aboutnuclear.org/. Contains information on waste types and disposal methods.

"Office of Civilian Radioactive Waste Management." U.S. Department of Energy. Available from http://www.ocrwm.doe.gov/. Includes information on the Yucca Mountain Project, including reports relating to the recommendation submitted to President George W. Bush.

"Radioactive Waste." U.S. Nuclear Regulatory Commission. Available from http://www.nrc.gov/waste.html. Contains explanations of waste, its regulation and transportation.

"Radioactive Wastes." World Nuclear Association. Available from http://www.world-nuclear.org/info/inf60.htm.

"RadWaste.org." Herne Data Systems. Available from http://radwaste.org/. General website with a great deal of information on radioactive waste.

U.S. Nuclear Waste Technical Review Board. "Report to the Secretary of Energy and the Congress." Available from http://www.nwtrb.gov/reports/reports.html. Report from March 2003 summarizing the board's activities during 2002.

NUTRITION AND SCIENCE

• • •

Although awareness of the relationship between food and health has a long history, the science of nutrition developed out of discoveries in modern chemistry and medicine. The professionalization of nutrition science resulted in its directly influencing food production, preparation, and consumption. Increased influence also meant increased responsibility to the public and the food industry as well as to governmental and international agencies. Such issues as the safety of food and its just distribution led to both controversies and codes of ethics. Nutrition scientists have not always recognized and analyzed the relationship between their work and moral values, but with the growth of the world's population and the increase in knowledge of what constitutes a healthful diet, the ethical debates surrounding nutrition science are likely to multiply and deepen.

Historical Developments

As with their animal ancestors, members of *Homo sapiens,* in order to survive, learned about edible fruits, vegetables, and animals from experience, and to increase the quality of their lives in difficult conditions, humans domesticated crops and animals for food. Though the process was slow, they also learned how to improve plants and animals by selection and hybridization. Before the science of nutrition developed, humans had discovered how to exploit such microorganisms as yeast and bacteria to manufacture such new foods and beverages as cheese and beer. By the use of salt and desiccation, they could also preserve foods to sustain life during times of shortages.

In the nineteenth century, as chemistry became a sophisticated discipline, knowledge of complex organic compounds allowed researchers to pinpoint foodstuffs essential for good health and to reveal insalubrious or fraudulent foods. Technological changes associated with food production during this period were much more dramatic than in the previous millions of years of *Homo sapiens'* evolution. During the nineteenth century problems associated with rapid industrialization, such as polluted water, adulterated food, and inadequate sanitation in overcrowded slums, led to the public health movement in England, Germany, France, and the United States. Stimulated by ethical concerns, public health officials alerted citizens to the dangers of foods that were nutritionally inadequate, sometimes even dangerous. For example, milk, traditionally viewed as a nutritious food for most children, could be the carrier of disease, until Louis Pasteur introduced a procedure, pasteurization, whereby heating milk killed infectious microorganisms. Advocates of pasteurization, however, were often opposed by members of the food industry who wished to avoid additional costs. In the 1880s in the United States, the agricultural chemist Stephen M. Babcock attacked milk adulteration by discovering an efficient test to determine milk's fat content, which did more, according to a Wisconsin governor, to promote ethical behavior among dairymen than reading the Bible.

The most dramatic developments in the science and technology of nutrition occurred in the twentieth century when new knowledge, techniques, laws, governmental agencies, and public policies contributed to extending life expectancies in many industrialized nations in Europe and America by over twenty-five years. But at the beginning of the century nutrition science remained in its infancy as, for example, many physiologists believed that what kind of food people ate mattered little, as long as diets supplied enough energy (calories) and sufficient materials (proteins) for the body's growth and maintenance. Only slowly did scientists discover the importance of trace nutrients for ideal health.

During the first third of the twentieth century researchers found that such diseases as rickets, beriberi, and scurvy had a specific dietary origin. Lack of small amounts of vitamins (water-soluble or fat-soluble organic substances essential for good health) caused these diseases. Concurrently American and German scientists, studying the roles of amino acids in nutrition, found several of these compounds essential to good health. Furthermore, work done largely in the United States indicated that, besides vitamins and amino acids, the healthy survival of experimental animals and humans required in their diets such inorganic elements

as sodium, potassium, calcium, magnesium, iron, zinc, and phosphorus. This accumulated knowledge was so important that, by the time of World War II, the National Research Council published a set of Recommended Dietary Allowances (RDAs) for foods, vitamins, and minerals. Though concerns growing out of wartime food rationing prompted this list, RDAs proved so helpful that they continued to be periodically issued, with modifications based on up-to-date nutritional research.

Nutritional Professionals and Ethics

The communication and multiplication of discoveries by nutrition scientists was facilitated by the formation of professional organizations and journals. For example, the American Institute of Nutrition began publishing its *Journal of Nutrition* in 1928, and in 1939 several important nutrition scientists in the United Kingdom formed the Nutrition Society, whose official publication was the *British Journal of Nutrition*. In the decades after World War II, agricultural chemists discovered high-yield crops that enabled farmers to produce enough food to nourish every person on Earth (the "green revolution"). Nevertheless, hundreds of millions of people remained malnourished, presenting concerned authorities with profound ethical problems, because the United Nations (UN) as well as various religious organizations maintained that every human had an inalienable right to be free from hunger and deficiency diseases.

At the UN, the Standing Committee on Nutrition established an Intergovernmental Working Group to develop guidelines for the implementation of the right to adequate food, as recommended by the 2002 World Food Summit. As international and national agencies and various professional societies became sensitive to the ethical implications of food and nutrition, so did trade associations involved with the production of foods and dietary supplements. For example, the Council for Responsible Nutrition, a trade organization, developed a code of ethics "dedicated to enhancing the health of the American public through improved nutrition, including the appropriate use of dietary supplements." This organization, founded in 1973, played an important role in several laws passed by the U.S. Congress in the last quarter of the twentieth century regulating nutritional substances.

Despite these ethical codes, laws, and world conferences, the numbers of the malnourished, according to reports issued by the UN's Food and Agriculture Organization, continued to increase in the decades after World War II. Some believed that the problem could be solved only in terms of development, that is, providing poor countries with the scientific and technical know-how to grow the food they needed. Others criticized this approach, because technology, while it could help to increase crop yields and improve food distribution, could prevent neither natural disasters nor political turmoil. Still others believed that malnutrition, which occurred not only in developing but also in developed countries, is a complex problem involving science and technology as well as economics, politics, and culture. These people held that what was needed was a multifaceted program that, while introducing new foods and technologies, also paid attention to economic growth, health education, and regional ecologies and cultures.

Safety and Equity

Unlike early nutrition scientists, who were able to link various diseases to specific dietary causes, modern researchers have discovered that such diseases as cancer and arteriosclerosis have multiple causes. According to some researchers, foods high in saturated fats will increase blood cholesterol levels, and many nutrition scientists agree that elevated levels of low-density lipoproteins (LDLs), which carry most of the cholesterol, increase the risk of coronary artery disease. Based on this evidence some criticized McDonald's and other fast-food restaurants for selling unsafe foods. Indeed, some critics went so far as to attack the majority of American food-production companies for reducing the consumption of fresh fruits and vegetables and increasing high-fat, high-sugar artificial foods, thus being partially to blame for such health problems as obesity, diabetes, and heart disease.

Cultural groups and their associated ideologies often influence dietary practices. Some traditional foods, such as green tea in Asian countries, have proven to be beneficial, but other practices, such as Latin American mothers' withholding milk and eggs from sick children, are harmful. Several cultural groups have practiced vegetarianism for a variety of social, religious, economic, or nutritional reasons. Some prominent nutritionists have attacked vegetarianism, insisting that meat is needed to avoid deficiencies of such essential substances as vitamin B_{12}. Others have pointed out, however, that there are hundreds of millions of Hindus, most of whom do not consume any animal products throughout their lives, and few of them exhibit B_{12} deficiencies and they generally have reduced incidences of heart disease, colon cancer, and diabetes.

A principal aim of the ethics of nutrition is to improve the food habits of people, and an important component of this good work is to understand a country's culture. Equity requires that every human being in every culture has the right to be properly nourished. Consequently developed countries, with their surpluses of food, have a duty to the undernourished in developing countries. Even in developed countries citizens have the right to be provided with good food, but in the United States, for example, many consumers have either wasted their money or harmed their health by various food and diet fads. Many nutrition scientists consider it unethical for "medical quacks" to be making large amounts of money in this way from gullible Americans.

Nutrition Controversies

While many believe that science and technology should be an important part of the solution of such problems as malnutrition, others see science and technology as part of the problem. For example, scientists invented various herbicides to aid farmers in food production, but some of these herbicides were used in the Vietnam War to deprive people of food. This was certainly not the first war in which participants used starvation as a weapon, as the siege of Paris during the Franco-Prussian War (1870–1871) and the siege of Leningrad by the Germans during World War II make clear.

Controversies also exist about what constitutes a balanced diet and whether or not dietary supplements should be used. For example, medical researchers and nutrition scientists seem to have reached agreement that Americans should reduce fats in their diets, an assertion repeatedly confirmed by the U.S. Department of Agriculture's dietary guidelines and in its widely disseminated Food Guide Pyramid, in which fats and sweets occupy a tiny area at the pyramid's top, indicating that fats, oils, and sweets should be consumed only "sparingly." But recent critics of the "dogma of the deadliness of dietary fats" have pointed out that the data are ambiguous on the benefits of low-fat diets. Despite the proliferation of reduced-fat food products, obesity and diabetes have actually increased. Furthermore, epidemiological studies of countries such as France, where animal fat consumption has risen, have shown that heart-disease death rates have declined.

Mainly in Western countries, recent controversies have centered on anorexia nervosa, a self-imposed starvation disorder, and bulimia, a binge–purge eating disorder. Scientists are divided over the roles played by society and the media as well as by a person's genetic makeup, psychological state, and physiology in fostering such conditions. Other controversies over vitamins, herbs, and fiber in the diet have revealed the complex interrelationships existing among professional nutritionists, members of the natural-foods movement, food producers, and various scientists outside the nutritional field. Ethical issues are inextricably bound into these controversies because of various conflicts of interest. For example, the work of some nutrition scientists has been supported by food producers, but advocates of megavitamin therapy for health problems ranging from the common cold to cancer have accepted contributions from companies manufacturing these vitamins. Some who express concern over the unregulated sale of herbs and nutritional supplements want the government to control their use the way they do prescription drugs, but those who consider these substances as foods see such actions as infringing their freedom of choice.

Advances in medical technologies have also raised concerns about the nutrition of the elderly and dying. Many religious ethicists distinguish between ordinary and extraordinary means of treatment, claiming that a moral obligation exists to use ordinary means (food and water) to maintain life but no strict obligation exists to use extraordinary means (respirators). Others hold that no obligation exists to continue feeding a patient when only biological, not mental, life remains; still others argue that this assessment exhibits an impoverished view of human personhood. Ethical issues raised by feeding the world's poor, sick, and dying are certainly controversial and complex. Scientific knowledge and new technologies can help solve some of these problems, but they may exacerbate others. Further complexities will confront humankind in the future, because nutrition is an evolving science. As research generates new knowledge and technologies, ethicists, as they have in the past, will have to take into account this expanded understanding in making their moral judgments.

ROBERT J. PARADOWSKI

SEE ALSO Agricultural Ethics; Food Science and Technology.

BIBLIOGRAPHY

Brogdon, Jennie, and Wallace C. Olsen, eds. (1995). *The Contemporary and Historical Literature of Food Science and Human Nutrition*. Ithaca, NY: Cornell University Press. A bibliographical survey of primary journals, core monographs, and historical literature in nutrition science with chapters covering 1850 to 1950, 1950 through the early 1990s, and important recent developments in food science.

Brown, Lester Russell, and Erik P. Eckholm. (1974). *By Bread Alone*. New York: Praeger. Brown, who began his presidency of the Worldwatch Institute in the year this book was published, is concerned with the ethics of global food production and distribution and their relationship to population growth and resource depletion.

Lacey, Richard W. (1994). *Hard to Swallow: A Brief History of Food*. Cambridge, UK: Cambridge University Press. Lacey wrote this popular account of the production, processing, and healthfulness of foods to stimulate readers to think about the nature of what they eat.

Mather, Robin. (1995). *A Garden of Unearthly Delights: Bioengineering and the Future of Food*. New York: Penguin. A science journalist analyzes the controversial new field of food bioengineering and the movements against it.

Maurer, Donna, and Jeffrey Sobal, eds. (1995). *Eating Agendas: Food and Nutrition as Social Problems*. New York: Aldine de Gruyter. A sociological analysis of food safety, biotechnology, vegetarianism, and other issues by experts from around the world.

Mayer, Jean. (1972). *Human Nutrition: Its Physiological, Medical, and Social Aspects*. Springfield, IL: Charles C. Thomas. Surveys the field of human nutrition in its scientific as well as social and political contexts.

McCollum, Elmer Verner. (1957). *A History of Nutrition: The Sequence of Ideas in Nutrition Investigations*. Boston: Houghton Mifflin. McCollum, who has been called "America's most eminent nutrition scientist," presents a knowledgeable survey of the principal advances.

Sanjur, Diva. (1982). *Social and Cultural Perspectives in Nutrition*. Englewood Cliffs, NJ: Prentice-Hall. A sociocultural analysis of food, with an emphasis on conceptual frameworks and methodological options.

Shue, Henry. (1996). *Basic Rights: Subsistence, Affluence, and U.S. Foreign Policy*, 2nd edition. Princeton, NJ: Princeton University Press. This new edition of a well-received book updates the author's thesis that justice requires developed nations to share their knowledge and wealth with the chronically malnourished nations.

OFFICE OF RESEARCH INTEGRITY

• • •

The United States Office of Research Integrity (ORI) has broad responsibilities for monitoring investigations of misconduct and promoting integrity in research supported by the Public Health Service (PHS). It is administratively located in the Office of Public Health and Science (OPHS) within the Office of the Secretary of Health and Human Services (OS, HHS) and reports to the Secretary of HHS through the Assistant Secretary for Health (ASH). The scope of its responsibilities extends to about four thousand research institutions worldwide. Although separate from the major PHS research funding agencies, such as the National Institutes of Health (NIH), the Centers for Disease Control (CDC), and the Food and Drug Administration (FDA), ORI works with these agencies as well as other government agencies to promote responsible conduct in federally supported research.

Origin and Development

The origins of ORI extend back to the early 1980s when Congress began formal investigations into a number of widely reported cases of misconduct in research. The federal agencies that supported the research and the research community initially assured Congress that misconduct in research was rare and appropriately handled through professional self-regulation. However, after more cases emerged, some involving high profile researchers, Congress intervened and passed the Health Research Extension Act of 1985 requiring PHS to establish a formal definition of and provisions for investigating misconduct in PHS-funded research.

FIGURE 1

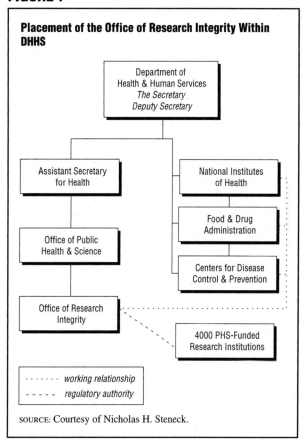

Placement of the Office of Research Integrity Within DHHS

SOURCE: Courtesy of Nicholas H. Steneck.

In response to the Congressional call for action, PHS published an Interim Policy on Research Misconduct in 1986, followed in March 1989 by the announcement that two offices would be established to investigate and adjudicate research misconduct cases: the

Office of Scientific Integrity (OSI) in the Office of the Director of NIH and the Office of Scientific Integrity Review (OSIR) in the Office of ASH (OASH). Five months later in August 1989, in a so-called Final Rule for research misconduct, PHS outlined the responsibilities of the new offices as well as those of research institutions accepting PHS funds for research. The two offices were combined in May 1992 to form the ORI, located in OASH.

During its early years, ORI focused the majority of its efforts on research misconduct, including the investigation of individual cases, the development of an assurance program for institutional misconduct policies, and the organization of programs designed to help research institutions develop expertise for handling their own misconduct cases. In the early twenty-first century, spurred in part by a reorganization plan published in May 2000, more attention has been given to understanding the factors that influence research integrity and ways to foster responsible conduct in research. These efforts are promoted through both funding and professional support for conferences, educational programs, and research projects.

Relations to Science, Technology, and Ethics

The ORI role in the discussion of the relationships between science, technology, and ethics is concerned with actual researcher practices and whether these practices conform to the standards and/or ideals for responsible conduct in research. Accordingly its efforts generally *do not* encompass the consideration of broader ethical questions, such as the appropriateness of particular research topics or the ethical dilemmas posed by human- or animal-subject research. ORI is also concerned principally with biomedical and behavioral research. Its work, however, relies on methods and advice from the social sciences, natural sciences, humanities, and relevant professions.

ORI has played a prominent, if at times controversial, role in stimulating the national debate about the importance of integrity in research and the adoption of policies to promote responsible conduct in research. During the 1990s, ORI and its counterpart agency in the National Science Foundation (NSF), the NSF Office of the Inspector General, assumed the lead in defining research misconduct and establishing procedures for its investigation. The three inappropriate behaviors that were identified by PHS and NSF as antithetical to responsible conduct in research—fabrication, falsification, and plagiarism (FFP)—quickly became community standards and were adopted by many research institutions as the basis of their misconduct policies. The common federal definition of research misconduct, formulated in December 2000, begins with FFP. During the prolonged discussion of the definition of research misconduct in the 1990s other options were suggested, but none received wide acceptance by the research community.

ORI has also played an important role in encouraging the research community to think of integrity in research as more than simply avoiding misconduct. Others have contributed to this effort. In 1992 a National Academy of Sciences (NAS) Report, *Responsible Science*, argued that along with misconduct researchers must be concerned with other *questionable research practices*, such as the failure to maintain adequate records, improper or undeserved authorship, or the inappropriate use of statistics. Through its conference programs and research on research integrity grants, ORI continues to encourage serious discussion of and research on the many factors that foster and detract from integrity in research.

Finally ORI is deeply involved in efforts to foster education on the responsible conduct of research (RCR). National recognition of the importance of RCR can be traced to the 1989 Institute of Medicine Report, *The Responsible Conduct of Research in the Health Sciences*. Within a year, NIH made RCR education a requirement for all Training Grant (T-32) applications and in 2000 ORI proposed, but later suspended, a requirement that would have made RCR education mandatory for key personnel on all PHS-funded research. Whether or not ORI ever issues a final RCR policy/requirement, it is committed to and is providing resources for developing and assessing ways to improve integrity in research through education.

The pressures and public concerns that led to the formation of ORI are unlikely to disappear in the near future. While the number of cases remains small in comparison to the size of the research community, research misconduct remains a problem that continues to undermine public confidence. Moreover, as the financial, political, and social stakes of research outcomes grow in importance, the significance of questionable research practices takes on new meaning. Improper or undisclosed conflicts of interests have been discovered in the deaths of subjects enrolled in clinical trials and the biased reporting of research results. Research data are sometime improperly hoarded, taken, or used. Authorship standards vary widely and are frequently abused. As long as the *human side* of research remains an important factor in shaping both practice and outcomes, ORI, its companion agencies elsewhere in government, and institutional research offices should continue to play an important role in protecting the public's investment in research.

NICHOLAS H. STENECK

SEE ALSO *Misconduct in Science; Research Integrity.*

BIBLIOGRAPHY

Bivens, Lyle W. (1994). "ORI and Misconduct Investigations." *Science* 263: 593.

Commission on Research Integrity, U.S. Department of Health and Human Services. (1995). *Integrity and Misconduct in Research: Report of the Commission on Research Integrity.* Washington, DC: Author. This *Report* began the internal review process that led to the reorganization plan published in May 2000.

Hallum, Jules V., and Suzanne W. Hadley. (1990). "OSI: Why, What, and How." *ASM News* 56(12): 647–651.

Institute of Medicine, and Committee on the Responsible Conduct of Research. (1989). *The Responsible Conduct of Research in the Health Sciences.* Washington, DC: National Academy of Sciences. The first major report to urge research institutions to institute training in the responsible conduct of research.

National Academies of Science. (1992). *Responsible Science: Ensuring the Integrity of the Research Process.* Report of the Committee on Science Engineering and Public Policy, Panel on Scientific Responsibility and the Conduct of Research. Washington, DC: National Academy Press. A careful study of the issues raised by misconduct in research that influenced later discussions even though its proposed definition of research misconduct was not adopted.

Pascal, Chris B. (1999). "The History and Future of the Office of Research Integrity: Scientific Misconduct and Beyond." *Science and Engineering Ethics* 5(2): 183–198.

Pascal, Chris B. (2000). "Scientific Misconduct and Research Integrity for the Bench Scientist." *Proceedings of the Society for Experimental Biology in Medicine.* 224(4): 220–230.

Price, Alan R. (1994a). "The 1993 ORI/AAAS Conference on Plagiarism and Theft of Ideas." *Journal of Information Ethics* 3(2): 54–63.

Price, Alan R. (1994b). "Definitions and Boundaries of Research Misconduct—Perspectives from a Federal-Government Viewpoint." *Journal of Higher Education* 65: 286–297.

Steneck, Nicholas H. (1994). "Research Universities and Scientific Misconduct—History, Policies, and the Future." *Journal of Higher Education* 65: 310–330.

Steneck, Nicholas H. (1999). "Confronting Misconduct in Science in the 1980s and 1990s: What Has and Has Not Been Accomplished?" *Science and Engineering Ethics* 5(2): 1–16.

Steneck, Nicholas H., and Mary D. Scheetz. (2002). *Investigating Research Integrity: Proceedings of the First ORI Research Conference on Research Integrity.* Washington, DC: Office of Research Integrity.

OFFICE OF TECHNOLOGY ASSESSMENT

• • •

The U.S. Congress established the Office of Technology Assessment (OTA) in 1972, late in the administration of President Richard Nixon (1969–1974). The brainchild of Representative Emilio Q. Daddario, a Connecticut Democrat (1959–1971), the OTA would become, along with the Library of Congress (established 1800) and the Congressional Budget Office (established 1974 one of three federal agencies providing advice directly to Congress rather than to the executive branch of government. Envisioned as an "early warning" mechanism that would alert lawmakers to the unwanted side effects of developing technologies, it also aimed to provide Congress with expertise somewhat analogous to that provided by presidential science advisors since the administration of Franklin Delano Roosevelt (1933–1945).

Historical Context

The OTA emerged in an era when science and technology were, on the one-hand, undergoing rapid expansion thanks in large measure to government sponsorship of research and development. On the other hand, science and technology during the 1960s and early 1970s had also come under increasing scrutiny and criticism in such works as Rachel Carson's *Silent Spring* (1962), Jacques Ellul's *The Technological Society* (1964), Ralph Nader's *Unsafe at Any Speed* (1965), Theodore Roczak's *The Making of a Counter Culture* (1967), and Charles Reich's *The Greening of America* (1970). Issues ranging from the unwanted side effects of pesticides to unsafe automobiles and the escalating arms race between the United States and the Soviet Union all contributed to an increasing awareness of, and concern about, the direction of modern technological society.

Against this backdrop, Representative Daddario, as chair of the Science, Research, and Development Subcommittee of the U.S. House Committee on Science and Astronautics, began exploring possibilities for equipping lawmakers with a mechanism through which the unwanted side effects of the burgeoning technological revolution could be foreseen and, thereby, forestalled. In essence, he envisioned arming the federal government with "a method of analysis that systematically appraises the nature, significance, status, and merit of a technological program" (Daddario 1967, p. 8). To perform this task, he recommended establishment of a technology assessment board, which, with its apt acronym, "TAB," would remain alert to the potential dangers and benefits of new technologies.

Concern over unwanted side effects of technological development, however, composed only one-half of the OTA mandate. In the face of an expanding federal budget for science and technology, members of Congress increasingly expressed concern that the legislative branch of government was being outstripped by the executive branch, thus making it difficult for Congress to fulfill its duties in the appropriations process and in the oversight of executive agencies. Specifically, members of Congress began demanding that the legislature have its own source of technical advice, independent from the executive branch.

These two distinct functions—namely, an early warning mechanism and independent scientific and technical advice—came together as Daddario's subcommittee completed the OTA legislation. The marriage, however, was an uneasy one. At the outset, Congress ensured that its membership would retain tight control over both the overall direction of the office and its specific tasks. A bipartisan Technology Assessment Board (TAB) governed the OTA. In addition to the nonvoting director, it consisted of six senators and six representatives, with an equal number of Democrats and Republicans. Assisting the TAB was the Technology Assessment Advisory Committee (TAAC). Comprising scientific and technical experts appointed by the TAB, the TAAC was charged with making recommendations to the TAB on the operations of the office and on specific assessments—but only on TAB request.

Success and Failure

Under Daddario, who served as the first director from 1973 to 1977, the OTA managed to navigate the tensions of its dual mission of providing independent advice and assessing the negative impacts of technology. Under Daddario, the office earned a reputation for providing timely and high-quality, if rather low-profile, studies in response to committee requests. But in contrast to the initial vision of the OTA as a bold early-warning apparatus, the office, in those early years, failed to fulfill its role as an assertive policy-influencing mechanism. As such, in the eyes of some critics, the office proved a stark disappointment.

The second director, Russell Peterson (a former Republican governor of Delaware), had grander visions. Rather than dodging controversy and serving as mere adjunct of congressional committees, Peterson sought more autonomy for the OTA. In concert with the original early-warning idea, he envisioned the office as a leading force in defining federal technology policy. To that end, he had the OTA promulgate its own list of priority areas in need of attention. These ranged from "Applications of Technology in Space" to "Impacts of Technology on Productivity, Inflation, and Employment" to various environmental issues. Peterson's initiatives, while truer to the original technology-*assessment* idea, failed to reckon with the other *raison d'être*: the desire for experts beholden to the legislature, independent of the executive branch. Not surprisingly, members of the TAB bristled at his attempt at autonomy, and Peterson's tenure lasted barely a year.

In contrast to Peterson and his idea of defining a broad agenda to influence national policy, the third director, John Gibbons, a former research director at oak ridge national labs, moved the office back into a more reserved role as obedient respondent to congressional committee requests and reliable information source for Congress. Under Gibbons, the office consciously avoided making policy recommendations in its reports. Small by federal government standards, the office had about 200 employees and an annual budget of approximately $20 million. The OTA stabilized and survived for the next fifteen years under Gibbons and its final director, Roger Herdman, who took over when Gibbons left in 1993 to become science adviser to President Clinton. But, in forsaking a role as a policy advocate "assessing" alternatives, its leaders sowed the seeds of its eventual demise.

In early 1995, fresh off victory in the 1994 elections, fiscally conservative members of Congress sought to reduce the federal budget. Precisely because the OTA had defined itself as an objective information agency rather than a more autonomous and assertive policy advocate, it became hard to defend the office against charges that its functions could be subsumed into the legislature's much larger source for independent information, the Congressional Research Service (CRS) of the Library of Congress. Not persuaded that the OTA offered something that set it apart from the more traditional research capabilities of the CRS, Congress eliminated funding for the OTA in 1995.

In its twenty-three year history, the OTA produced some very solid and reputable studies in response to congressional committee requests. These included approximately 750 reports on topics ranging from energy to transportation to health. In the broader scheme of things, however, the OTA is perhaps more noteworthy insofar as it sheds light on an interesting attempt by U.S. lawmakers to equip government with an ability to foresee technological development and how, because of the executive–legislative tensions existing in the U.S. federal government, that initiative became configured and constrained by broader political and institutional dynamics.

GREGORY CLIFFORD KUNKLE

SEE ALSO *Constructive Technology Assessment; Discourse Ethics; Technology Assessment in Germany and Other European Countries.*

BIBLIOGRAPHY

Bimber, Bruce. (1996). *The Politics of Expertise in Congress: The Rise and Fall of the Office of Technology Assessment.* Albany: State University of New York Press. Professor Bimber uses the OTA as a case study to explore the politicization of expert advice in congressional agencies, concluding that the pluralistic nature of Congress results in more objective advice than that which prevails in executive agencies.

Daddario, Emilio Q. (1967). *Technology Assessment.* Statement prepared for the Subcommittee on Science, Research, and Development of the House Committee on Science and Astronautics. 90th Cong., 1st sess. Committee Print.

Kunkle, Gregory C. (1995). "New Challenge or the Past Revisited: The Office of Technology Assessment in Historical Context." *Technology in Society* 17(2): 175–196. The author examines the early history of the OTA to shed light on the then-current debate about abolishing the OTA, arguing that proponents of the OTA had to redefine the mission of the agency as something more than mere independent information in order for the OTA to survive.

U.S. Office of Technology Assessment. "The Ota Legacy, 1972–1995." CD-ROM, Sn 052-003-01457-2. Pittsburgh, PA: U.S. Government Printing Office. This CD contains a collection of all of the official OTA reports prepared for Congress. The reports are also available at http://www.wws.princeton.edu/%7eota/ns20/pubs_f.html.

OIL

• • •

The word *oil* is derived from the Greek *elaia* by way of the Latin *oleum*, both of which mean olive. Olive oil and other clear oils derived from plants have been closely associated with civilization and health for thousands of years. In *Genesis* a dove brought an olive leaf to Noah as a sign that the biblical flood was over and humans could reinhabit the earth. The Greeks considered the olive tree to be a symbol of victory and purification. Oils have been part of diverse traditions of medical practice in many parts of the world. They have been used to treat wounds and for general care of the body. In addition, oils have been integral to the preparation of foods and fine cuisine. Three types of oil are recognized in the early twenty-first century: vegetable oil, animal oil, and rock oil. Olive oil is a vegetable oil, whale oil is an animal oil, and petroleum is rock oil.

From Oil for Health to Oil for Energy

From the perspective of modern science and technology, oil is liquid petroleum. Petroleum is composed primarily of hydrocarbon molecules with some inorganic impurities. It can exist in the solid, liquid, or gas phase. The phase depends on composition, temperature, and pressure. The average molecular weight of hydrocarbons in oil is usually greater than the average molecular weight of hydrocarbons in gas at the same temperature and pressure. Natural gas is predominantly methane.

People have used petroleum for thousands of years. As early as 3000 to 2000 B.C.E., Middle Eastern civilizations such as those in Egypt and Mesopotamia used oil to construct buildings, waterproof boats and other structures, and mummify bodies. During that period, small amounts of oil were collected from surface seepages. Arabs used oil to create incendiary weapons as early as 600 C.E. By the 1700s, oil produced from shale oil was being used in Europe to light streets in Modena, Italy, and to make paraffin wax candles in Scotland (Shepherd and Shepherd 2003).

American George Bissell has been called the person most responsible for creating the modern oil industry (Yergin 1992). Bissell realized in 1854 that rock oil—as oil was called in the nineteenth century to differentiate it from vegetable oil and animal fat—could be used in lighting and cooking. Bissel formed the Pennsylvania Rock Oil Company of Connecticut in the mid-1850s and named James M. Townsend president.

Bissell and Townsend believed that rock oil could be produced from below the surface of the Earth in the same way that water was produced using water wells. Townsend commissioned Edwin L. Drake to drill a well in Oil Creek, near Titusville, Pennsylvania, where many oil seepages had been observed. The project began in 1857 and struck oil on August 27, 1859.

The value of oil increased dramatically as a result of the success of Drake's well. The abundant supply of rock oil served as a substitute for whale oil, which was growing scarce and expensive, and reduced the need to hunt whales for fuel. Within fifteen months of Drake's strike, Pennsylvania was producing 450,000 barrels per year from seventy-five wells. By 1862, 3 million barrels of oil were being produced and the price of oil dropped to ten cents per barrel (Kraushaar and Ristinen 1993).

The invention of the electric light bulb caused a drop in the demand for kerosene in 1882 and a corresponding drop in the demand for rock oil. The drop did not last long, however, because the rapidly expanding automobile industry needed oil for fuel and lubrication.

By 1900 Standard Oil, a company founded by John D. Rockefeller in 1870, held a virtual monopoly over oil production in the United States. Congress passed the Sherman Antitrust Act to reintroduce competition in the oil industry. By 1909 the United States was producing 500,000 barrels of oil per day, which was more oil than the combined production of all other countries. The United States produced more than half of the world oil supply in the first half of the twentieth century.

The Politics and Ethics of Oil

Discoveries of large deposits of oil in Central America, South America, and the Middle East in the early 1900s eventually led to increased production outside of the United States. Production in the continental United States peaked in 1970 and has since been declining. Oil demand has continued to grow, however, in both the United States and the rest of the world. Since 1948 the United States has imported more oil than it exports. In the early-twenty-first century, the United States imports about half of its oil (Deffeyes 2001).

Petroleum has been an internationally traded commodity since the end of the nineteenth century. International and multinational petroleum companies have appeared as a result of the global distribution of oil and its importance to societies around the world. These companies are based in a home country, but must operate within the regulatory framework of each host country. Relationships between oil producing companies and host countries vary widely. Most host countries issue licenses or leases to production companies.

Until 1973 oil prices were influenced by market demand and the supply of oil that was provided in large part by a group of oil companies called the *Seven Sisters*. In 1960 Saudi Arabia led the formation of the Organization of Petroleum Exporting Countries (OPEC). OPEC became a major player in the oil business in 1973 when it raised the price of oil exported by its members. This rise in price became known as the first oil crisis as prices for consumers in many countries increased significantly.

In the early-twenty-first century, nations around the world are concerned about the global dependence on finite resources and the environmental impact of fossil fuel combustion. For example, how should the supply of oil be distributed? Should developed nations encourage less developed nations to seek self-sufficiency? Or should all nations seek an equitable distribution of energy to prevent social turmoil? As another example, measurements of ambient air temperature have shown a rise in the average temperature of the Earth's atmosphere. The rising temperature is called global warming and is attributed in large part to the emission of fossil fuel combustion byproducts into the atmosphere. The need to address these concerns is motivating an international effort to implement a sustainable development policy as the world undergoes a transition from an energy mix dominated by fossil fuels to a broader energy mix that depends on a range of energy sources.

JOHN R. FANCHI

SEE ALSO *Energy; Environmental Ethics; Global Climate Change.*

BIBLIOGRAPHY

Deffeyes, Kenneth S. (2001). *Hubbert's Peak: The Impending World Oil Shortage.* Princeton, NJ: Princeton University Press. Describes a particular methodology for forecasting the supply of oil and predicts an impending oil shortage. Designed for a general readership.

Fanchi, John R. (2004). *Energy: Technology and Directions for the Future.* Boston: Elsevier. Reviews the science and technology of energy sources, the environmental impact of energy sources, and energy forecasts. Assumes the reader has a technical background.

Kraushaar, Jack J., and Robert A. Ristinen. (1993). *Energy and Problems of a Technical Society,* 2nd edition. New York: Wiley. Explains the technological problems of society to a non-technical audience.

Shepherd, William, and D. W. Shepherd. (2003). *Energy Studies,* 2nd edition. London: Imperial College Press; River Edge, NJ: World Scientific. Reviews several energy sources primarily for a general readership.

Yergin, Daniel. (1992). *The Prize.* New York: Simon and Schuster. Provides a historical introduction to the modern oil industry.

OPEN SOCIETY

• • •

The term *open* has a special salience in such phrases as "open markets," "open records," "open government," and "open-ended" discussion or project. In such contexts it denotes both freedom and transparency, two fundamental values of a democratic society. Indeed, the term *open society* has itself become almost synomous with democracy, and is sometimes used to name the ideal of both the scientific and the non-scientific social orders.

Although Henri Bergson (1859–1941) first employed the term *open society* in *The Two Sources of Morality and Religion* (1935) and Eric Voegelin (1901–1985) made Bergson's interpretation a key concept in his philosophy of history, it was *The Open Society and Its Enemies* (1945) by Karl R. Popper (1902–1994) that gave the phrase wide currency. The concept of the open society has since sparked numerous scholarly debates as well as practical applications. Although based on core values such as equality in social relations, freedom of inquiry and speech, and transparancy in decision making and knowledge production, the precise meaning of an open society has never been settled. Furthermore, globalization and the increasing threat of terrorism are reshaping conventional understandings of closed and open societies.

Bergson and Popper

From the earliest articulations of the concept by Bergson and Popper, there have been important differences in the ways in which the open society has been interpreted and used. Bergson's concept was more a vertical openness to the ground of being or the transcendent. Popper's openness was primarily within the framework of secular liberalism; it was a horizontal openness to the experimental trial and error method. As one commentator remarks, Bergson's openess was centered on his "theocentric humanism," whereas Popper's was based on his "anthropocentric humanism" (Germino 1974, p. 14).

For Bergson, the primitive closed society attached strict obligations to custom and operated under the rules of "Authority, Hierarchy, and Immobility." It was warlike, dominated by a religious dogma, and controlled by an elite. Bergson envisioned the open society as an ideal yet to be wholly realized. Although the spread of Western values in the process of globalization may approximate his vision, it is important to note that Bergson's open society went beyond material and political conditions. Central to his conception was a spiritual openness to the rhythm of the cosmos and the interrelatedness of life. One way to sum up Bergson's account of closed and open societies is to see the former as emphasizing impersonal *orders* as the source of morality, whereas the latter emphasizes the source of morality found in "*appeals* made to the conscience of each of us by *persons* who

represent the best there is in humanity" (1935, p. 84). The closed society is bound by static laws and conventions, whereas the open society is best represented by heroes and mystic saints who break with the strictures of their group in a dynamic fashion. Thus, the two sources of morality are *dogma* (which can include science and its static, mechanistic ideal) and *inspired intuition* (and its ideal of dynamic, free creativity).

Unlike Bergson's work, Popper's critique of closed societies came with the benefit of hindsight by which to characterize and judge the brutal totalitarianism of the Nazi regime. Although initially lenient and even approving with regard to the Soviet Union, Popper eventually categorized Stalinism as a closed society. For Popper, a closed society is marked by the rigidity of its customs and their irrational acceptance by the masses. An open society, by contrast, is one in which citizens face personal choices and moral responsibilities (both absent in closed societies). Open societies are marked by personal interaction, whereease closed societies present only abstract, impersonal, and anonymous human relations. Open societies replace saturating social conventions with personal freedom, rationality, and critical thought.

Finally, it should be noted that for Popper, the concept of the open society flowed naturally from his philosophy of science. Both rely on fallibalism: Scientific progress is made by subjecting theories to critical scrutiny, and progress in an open society can be sustained only if individuals are free to critically evaluate governmental decisions and engage in "piecemeal social engineering." Disputes in scientific communities and open societies should be resolved by critical discussion rather than force.

Despite their differences, both Bergson and Popper agreed that there was a general historical trend toward democracy and openness. However, both explicitly denied any inherent momentum or logic to history, insisting rather on its open-endedness based on the historical engine of human choice. Both also warned that a relapse to the condition of closed societies is always possible, because the natural will to power can never be completely erased by the virtuous conventions of open societies. In fact, their very openness and tolerance ensure that these societies will remain vulnerable to such a relapse. A rational (Popper) or enlightened (Bergson) citizenry can always be duped by a strong-willed leader or clan.

Open Society Debated

Popper did not associate his concept of open society with any particular political or economic philosophy.

His refusal to define the concept in this manner has fueled critical and theoretical debates. Dante Germino and Klaus von Beyme collected a wide-ranging series of essays on *The Open Society in Theory and Practice* (1974) that touches on its implications for work, education, politics, religion, and other fields of human experience. The book exposes the plurality of viewpoints and contested meanings of the open society. Many papers raise doubts about the ability of modern industrial or post-industrial society, with its emphasis on technological rationality, to foster openness. Some in this camp call for radical departures from prevailing assumptions about humans and nature. Others argue that it is precisely and only within the modern, secular world of western liberalism that values of openness can prevail. This debate signals the durability of the original fissure underlying the Bergsonian and the Popperian uses of the term. The former critics call for a new consciousness focused on deep experiences, which have been marginalized by the scientific and secular world-view. The latter insist that reason and (properly demarcated) science are essential for the flourishing of open societies.

In *Popper's Open Society After Fifty Years: The Continuing Relevance of Karl Popper* (1999), Ian Jarvie and Sandra Pralong collect fifteen essays that introduce Popper (including an interview with Popper on his ninety-second birthday), critique the central ideas of *The Open Society*, and apply those ideas to later social, political, and philosophical concerns. Some contributors argue that Popper's arguments have lasting value but need restating away from the particular instances of Plato (427–347 B.C.E.), Georg Wilhelm Friedrich Hegel (1770–1831), and Karl Marx (1818–1883) toward more general critiques of authority, community, and bureaucracy. Others criticize Popper for practicing the very historicism he attacked. Still other essays take up the relation between Popper's philosophy of science and his thoughts on the open society. The work concludes with several reflections on the implications of Popper's work, especially for Eastern European countries.

In *The Governance of Science: Ideology and the Future of the Open Society* (2000), Steve Fuller argues that the increasing scale of the scientific enterprise has eroded the ideal of science as an open society. He connects this claim with three political theories of science, and argues that "[t]he open society is possible only in a republican regime, where, unlike liberal or communitarian regimes, a clear distinction is drawn between staking an idea and staking a life. This distinction underwrites the fundamental principle of the open society: the right to be wrong" (p. 5). Fuller also traces the opposing pulls of liberalism (capitalism) and communitarianism (multiculturalism) in the governance of science by the university. He concludes with a look toward the future of the social contract with science, which he argues is best reformed by continuing the process of decoupling state power from the authorization of knowledge claims. In this, Fuller echoes one of Popper's central concerns, namely, that scientific claims and the direction of scientific research always remain open to public debate.

Related Concepts

Popper's open society was based on a critique of two practices in the philosophy of history. First, he criticized historicism, or the belief that history develops according to certain intrinsic principles toward a determinate end. Second, he challenged holism, or the belief that societies are greater than the sum of their members. Popper argued instead that history is open-ended and driven by individual choices.

Popper's analysis was anticipated by previous examinations of the social order within science (see the work of Robert Merton) and echoed by other post-World War II concerns for scientific freedom (see the work of Michael Polanyi). More generally, while never explicitly referencing the open society, holism, or historicism, Hannah Arendt develops a critique of totalitarianism and an analysis of the human condition (1958) that can be interpreted as supportive of Popper's basic argument against the "making" of history, although she would question any sanguine interpretation of individual autonomy.

A much more radical promotion of open society principles is found in the work of Popper's student, Paul Feyerabend, and his arguments for "epistemological anarchism." For Feyerabend, Popper is too limited in the application of his openness ideal, and in *Science in a Free Society* (1978) argues that the movement that once led to the separation of church and state should now bring about a separating of science and state. Science should be disestablished as the rational norm in advanced technological societies; society should not just be free for science but freed from science, that is, open to more than science.

In *The Closing of the American Mind* (1987), Allan David Bloom distinguishes between two types of openness in modern Western societies. First, there is the openness of reason that refuses to equate the good with one's own way of life, but takes the further step of using reason to inquire into nature in order to discover truth, beauty, and goodness: "Nature should be the standard by which we judge or own lives and the lives of peoples" (p. 38). Second, however, is the openness of indifference.

This openness denies reason's ability to find a standard for right living in nature or models of right conduct in history. Instead, it slips into moral nihilism and cultural relativism.

Bloom thus suggests that the open society at once presents the chance to discover an a-cultural, transhistorical, natural truth and the possibility that such a search will compel its members into another type of closed society, closed within the culture of relativism. People must escape their contingent cultural conventions to be fully human, but such an escape leads to a closed indifference if they cannot use reason to discover stable and more universal standards of conduct. His argument also hints at Stanley Rosen's (1989) distinction between the ancients and the moderns. In a sense, the ancients represent closed societies that offer security and order at the risk of tyranny. The moderns represent open societies that offer freedom and choice at the price of nihilism and licentiousness. Building off of this latter possibility, Bloom maintains that "Openness used to be the virtue that permitted us to seek the good by using reason. It now means accepting everything and denying reason's power. The unrestrained and thoughtless pursuit of openness [equals] closedness" (pp. 38–39).

The notion that openness reveals mere contingency and meaninglessness is challenged by Richard Rorty (1989). Rorty would accuse Bloom of the metaphysical assumption that reason must provide "an order beyond time and change which both determines the point of human existence and establishes a heirarchy of responsibilities" (Rorty 1989, p. xv). Rorty's utopia is one of "liberal ironists"—liberal in that they aspire to personal excellence and social justice, ironical because they recognize such goods are not guaranteed by a stable ontological order. For Rorty openness is retained by means of nominalist cultural narratives that construct compassion rather than by the seeking of moral formulas for action based on theory.

Open Society Applied

In the construction of such narratives, perhaps the Open Society Institute (OSI) is the largest concrete application of Popper's notion. The philanthropic activist George Soros founded OSI in 1993 as a way to synthesize initiatives that began in Central and Eastern Europe as early as 1984 to encourage the transition to democracy. Since then, the Soros network has expanded to include initiatives throughout the world, including the United States, to promote open societies through legal, governmental, and economic reform. It also supports education, media, public health, and human rights initiatives. The OSI seeks to diminish and prevent the negative consequences of globalization. In this sense, it recognizes the threats posed to open socieities by global capitalism in addition to those posed by more traditional forms of authoritarian rule.

Other concrete (if perhaps unconscious) manifestions of Popper's notion are found in the open source and free software movements, and in the promotion of open access in scientific publishing. The claim that the source code for programs should be open to all users, thus enabling them to identify weaknesses in the code and correct them—as is the case with the software that makes possible the World Wide Web on the Internet (a program that Tim Berners-Lee, its designer, explicitly declined to patent)—exemplifies Popperian principles. The argument that basic software utilities should be freely available rather than controlled by a quasi-monoply such as Microsoft is a natural extension of these principles. Finally, the promotion of open access scientific publication—that is, publication that allows all users a free, worldwide right of access to read, copy, and distribute the results of scientific research—constitutes a further effort to institutionalizes practices in harmony with open society ideals.

Globalization and Terrorism

The globalizing reach of modern science, technology, and production forces as well as Western values and political associations can be interpreted as the intrusion of the open society on "traditional" or more "closed" cultures. Ethics is not as easily globalized as science and technology. Although a simplification, something similar is true with regard to the economic globalization of markets versus the political globalization of democracy. Modernizing forces do not produce any uniform transition from closed to open societies, which is a mixed blessing for all involved. Diverse movements from wars of independence to environmental and human rights activism have tried to respond to the dislocations that can result from this selective globalization. But perhaps the most serious backlash against modernization and globalization, and the one that best illustrates the contemporary relationship between closed and open societies, is terrorism.

Although an ancient tactic, terrorism (especially those attacks carried out by extremists who justify their actions by appeal to Islamic ideologies) has taken on heightened global importance since the attacks against the United States on September 11, 2001. The potential to utilize the machines and weaponry of modern technoscience has increased the threat posed by terrorists to the citizens of open societies. Just as important, however, is the vulnerability to terrorist attacks created by the very ideals of an open society. Personal and civil liberties, tolerance, and multiculturalism all inhibit the leadership of open societies in their efforts to thwart terrorist plots. Terrorists are also able to capitalize on the freedom of information presented by the Internet. Thus, relatively loose networks of people bounded by a set of beliefs can organize and commit complex, integrated attacks due in large measure to modern telecommunication technologies. This form of "closed society" retains the dogmatic, hierarchical, and ideological characteristics criticized by Bergson and Popper, even though it now lacks the geographical and political organizing structures and avails itself of "open" streams of information.

The controversy over the Patriot Act signed by President George W. Bush in 2001 "to deter and punish terrorist acts in the United States and around the world" demonstrates the tension that terrorism presents between closed and open societies. It is an open question whether an effective war against terrorism requires the curtailment of certain civil liberties in order to more effectively control and monitor suspects. If so, however, at a certain point, such tactics may jeopardize the very ideals of the open society that they aim to defend.

Shortly after the September 11 attacks, Atef Ebeid, the Egyptian Prime Minister, criticized human rights groups for defending the human rights of potential terrorists. "You can give them all the human rights they deserve until they kill you," he said. "After these horrible crimes committed in New York and Virginia, maybe Western countries should begin to think of Egypt's own fight against terror as their new model" (Remnick 2004, pp. 75–76). In the war against terror, the leadership of Egypt maintains that all pretenses to an open society must be discarded, thus suggesting that democratic states run the danger of winning one war by losing another.

ADAM BRIGGLE
CARL MITCHAM

SEE ALSO *Governance of Science; Political Economy; Science Policy.*

BIBLIOGRAPHY

Arendt, Hannah. (1958). *The Human Condition.* Chicago: University of Chicago Press.

Bergson, Henri. (1935). *The Two Sources of Morality and Religion.* Garden City, NY: Doubleday. The original French edition appeared in 1932.

Bloom, Allan. (1987). *The Closing of the American Mind.* New York: Simon and Schuster.

Feyerabend, Paul. (1978). *Science in a Free Society.* London: NLB.

Fuller, Steve. (2000). *The Governance of Science: Ideology and the Future of the Open Society.* Philadelphia: Open University Press.

Germino, Dante, and Klaus von Beyme, eds. (1974). *The Open Society in Theory and Practice.* The Hague: Martinus Nijhoff.

Jarvie, Ian, and Sandra Pralong, eds. (1999). *Popper's Open Society after Fifty Years: The Continuing Relevance of Karl Popper.* London: Routledge.

Merton, Robert. (1938). "Science and the Social Order." *Philosophy of Science* 5: 321–337.

Polanyi, Michael. (1946). *Science, Faith, and Society.* London: Oxford University Press.

Popper, Karl. (1945). *The Open Society and Its Enemies.* 2 vols. London: Routledge.

Popper, Karl. (1961). *The Poverty of Historicism*, 2nd edition. London: Routledge. Revised book version of original articles of 1944.

Remnick, David. (2004). "Letter from Cairo: Going Nowhere." *New Yorker* July 12 and 19: 74–83.

Rorty, Richard. (1989). *Contingency, Irony, and Solidarity.* Cambridge, UK: Cambridge University Press.

Rosen, Stanley. (1989). *The Ancients and the Moderns.* New Haven, CT: Yale University Press.

OPERATIONS RESEARCH

• • •

Operations Research (OR) is defined, according to the International Federation of Operational Research Societies, as a scientific approach to the solution of problems in the management of complex systems. Unlike the natural sciences, OR is a science of the artificial in that its object is not natural reality but rather human-made reality, the reality of complex human-machine systems. OR involves not just theoretical study but also practical application. Its purpose is not only to understand the world as it is, but also to develop guidelines about how to change it in order to achieve aims or to solve certain problems. Ethical considerations are thus crucial to almost all aspects of OR research and practice.

Origins

OR as a specific scientific discipline dates back to the years immediately preceding World War II. First in the United Kingdom and later in the United States, interdisciplinary groups were constituted with the objective of improving military operations through a scientific approach. A typical example is the British Anti-Aircraft Command Research Group, better known as Blackett's Circus, which consisted of three physiologists, four physicists, two mathematicians, one army officer, and one surveyor.

Experience with OR in the military context during the war was the basis for new applications in industry afterward. The development of complex, large, and decentralized industrial organizations together with the introduction of computers and the mechanization of many functions required novel scientific approaches to decision making and management. This need led to the establishment, not only in industry but also in academia, of what formally became known as *operational research* in the United Kingdom, and *operations research* or *management science* in the United States (these last two terms are often used synonymously).

The first national OR scientific society was founded in 1948 in the United Kingdom. The U.S. societies, Operations Research Society of America (ORSA) and the Institute of Management Science (TIMS), which later merged as the Institute for Operational Research and the Management Sciences (INFORMS), followed a few years later. In 1959 the International Federations of Operational Research Societies was established.

Optimization plays a major role in OR methodologies: Problems are formulated by means of a set of constraints (equalities or inequalities) and an objective function. The maximization or minimization of the objective function subject to the constraints provides the problem's solution.

Codes versus Principles

Ethics in any applied science develop along two complementary lines. First scientific or professional codes of ethics can be created. These are typically sets of rules, sometimes well-defined, sometimes generic. Useful as they are, ethics codes are external directives not evolved from any individual's ethical beliefs and may lead to double standards. Some evidence suggests that people apply ethical standards at work that are often different and significantly lower than those they follow in their private lives. Although no major national OR society has a formal ethics code, the codes of related scientific disciplines may be applied to OR.

A second way to develop a particular ethics is through an individual approach based upon principles and values instead of rules that govern behavior. According to philosopher Hans Jonas (1903–1993), the following principle can be the basis of an ethical discourse: People have a responsibility toward others, be it humankind (past, present, and future generations) or nature. Another principle of responsibility complements this general rule: Knowledge in all forms must be shared and made available to everyone; cooperation rather than competition should be at the basis of research activity. The latter is called the sharing and cooperation principle (Gallo 2003). These principles are basic to confronting two issues that are crucial to the survival of society: increasing societal inequalities and sustainability.

Models and Methods

If the above are accepted as appropriate principles of responsibility, they can be applied to OR, and, in particular, to model building, which is the fundamental OR activity. The first issue in this regard is determining whether ethics has anything to say about model construction. In his excellent book on ethics and models, William A. Wallace (1994) reports a consensus in the OR research community to the effect that "one of the ethical responsibilities [of modelers] is that the goal of any model building process is objectivity with clear assumptions, reproducible results, and no advocacy" (Wallace 1994, p. 6), and on the "need for model builders to be honest, to represent reality as faithfully as possible in their models, to use accurate data, to represent the results of the models as clearly as possible, and to make clear to the model user what the model can do and what its limitations are" (Wallace 1994, p. 8).

But might responsibility also arise at an earlier stage, when choosing the methodology to create the model? In other words, are methodologies (and hence models) *value neutral*? This is a controversial issue. It can be argued that behind the role of optimization in OR and the parallel development of optimality as a fundamental principle in the analysis of economic activities and in decision making related to such activities, there are assumptions with ethical implications. Among these is whether self-interest is the only motivation for individual economic choices; whether maximization of the utility function is the best formal way to model individual behavior; and whether, by applying the proper rate of substitution, anything can be traded for anything else, with the consequence that everything can be assigned a monetary value.

These considerations have led some, including J. Pierre Brans (2002), to advocate the use of multicriteria approaches in order to balance objective, subjective, and ethical concerns in model building and problem solving. Such approaches do not reduce, by weighting, different, often noncommensurable, criteria (including those derived from ethical considerations) to one single criterion. Instead each criterion maintains its individuality, leading to a solution that is acceptable to or appropriate for the parties, rather than one that is objectively optimal.

Another issue in the application of principles of responsibility is that optimization-based models are often solution oriented: The final goal of the model is the solution, for instance, the recommendation of action to be made to the client. Some argue that the process is more important than the solution: creating a learning process in which all parties involved acquire a better understanding of the problem and of the system in which the problem arises, with its structure and its dynamics, and have a say in the final decision. These concerns, which call for a broader sense of responsibility not only with respect to the client but to all stakeholders, have led to divisions in the OR community. The development of alternative approaches such as systems thinking and soft operational research are some results.

Clients and Society

Another important question concerns the kind of clients served. As pointed out by Jonathan Rosenhead (1994), OR practitioners "have worked almost exclusively for one type of client: the management of large, hierarchically structured work organizations in which employees are constrained to pursue interests external to their own" (Rosenhead 1994, p. 195). Yet these are not the only possible clients. Other types of organizations exist, operating by consensus rather than chain of command, and representing various interests in society (health, education, housing, employment, environment). Such organizations usually have limited resources though the problems they face are no less challenging for the OR profession.

This fact has ethical relevance. Because the use of models constitutes a source of power, the OR profession runs the risk of aiding the powerful and neglecting the weak, thus contributing to the imbalance of power in society. A positive but rather isolated example of OR assistance outside the sphere of big business is community operational research in the United Kingdom. This initiative has allowed many OR researchers and practitioners to work with community groups, such as associations, cooperatives and trades unions.

Another way OR may contribute to power imbalances at the international level is in the strict enforcement of patents and intellectual property rights. Wide dissemination of methodologies and software, in accordance with the sharing and cooperation principle

mentioned above, might reduce the technology divide between rich and poor countries.

GIORGIO GALLO

SEE ALSO *Management; Models and Modelling.*

BIBLIOGRAPHY

Brans, J. Pierre. (2002). "Ethics and Decisions." *European Journal of Operational Research* 136: 340–352.

Gallo, Giorgio. (2003). "Operations Research and Ethics: Responsibility, Sharing and Cooperation." *European Journal of Operational Research* 153: 468–476.

Rosenhead, Jonathan. (1994). "One-Sided Practice—Can We Do Better?" In *Ethics in Modeling*, ed. William A. Wallace. Tarrytown, NY: Pergamon.

Wallace, William A., ed. (1994). *Ethics in Modeling.* Tarrytown, NY: Pergamon.

OPPENHEIMER, FRANK

• • •

Frank Oppenheimer (1912–1985) was born in New York on August 14, the younger brother of J. Robert Oppenheimer. Like his brother he became a physicist, but with a focus on experimental work rather than theory. As a physicist he contributed to the development of the atomic bomb, and then in 1969 became a leader in science education by founding the interactive San Francisco Exploratorium. He died of lung cancer in Sausalito, California, on February 3.

After earning a B.S. in physics from Johns Hopkins University in 1933 he studied for a time in Europe before going to the California Institute of Technology where he earned his PhD in 1939. In 1941 he began work at Oak Ridge, Tennessee, on separating Uranium–235, the fissile isotope, from the more common Uranium–238, then subsequently became special assistant to his older brother at the Los Alamos National Laboratory where the atomic bomb was being designed and constructed. Like many other scientists he was upset by the use made of the atomic bomb at the end of World War II, and became involved in efforts to educate the public about the new dangers of nuclear weapons.

Immediately after the war he held teaching appointments first at the University of California, Berkeley, then at the University of Minnesota. When the U.S. Congress House on Un–American Activities Committee exposed the fact that he and his wife had for a time during the 1930s been members of the Communist Party, he was forced to leave university teaching. For the next decade he became a cattle rancher in southern Colorado. Then in 1957 he took a job teaching high school science in Pagosa Springs, Colorado, where he became an enthusiastic and creative educator, moving shortly thereafter to the University of Colorado in Boulder. There he created the "Library of Experiments" to pioneer the kinds of interactive techniques that eventually became the hallmark of the Exploratorium.

The idea for the Exploratorium gestated during a 1965 Guggenheim fellowship in which Oppenheimer studied science museums in Europe, and became convinced of their need as a form of public science education. Although invited to work at the Smithsonian Institution in Washington, DC, he chose to start from scratch in San Francisco, where he proposed to create a new kind of science museum in the abandoned Palace of Fine Arts near the San Francisco marina. He served as its director until his death.

CARL MITCHAM

SEE ALSO Atomic Bomb; Education; Museums of Science and Technology; Oppenheimer, J. Robert.

BIBLIOGRAPHY

Oppenheimer, Frank. (1972). "The Exploratorium: A Playful Museum Combines Perception and Art in Science Education," *American Journal of Physics*, vol. 40, no. 7 (July), pp. 978–984.

INTERNET RESOURCES

Oppenheimer, Frank. (1984). "Interview with Frank Oppenheimer," Oral History Project, California Institute of Technology Archives, Pasadena, California. Available from http://oralhistories.library.caltech.edu/69/.

Frank Oppenheimer web site at the San Francisco Exploratorium. Available from http://www.exploratorium.edu/frank/index.html.

OPPENHEIMER, J. ROBERT

• • •

J(ulius) Robert Oppenheimer (1904–1967) was born in New York City on April 22 of a privileged, assimilated German-Jewish family. Known widely as the "father of the atomic bomb," Oppenheimer also thought that physicists had special responsibilities as a result of their contributions to this development. He argued for

J. Robert Oppenheimer, 1904–1967. The American physicist made fundamental contributions to theoretical physics and was director of the atomic energy research project at Los Alamos, New Mexico. *(National Archives and Records Administration.)*

international control of nuclear weapons and against the U.S. development of the hydrogen bomb. He died of throat cancer in Princeton, New Jersey, on February 18.

Education and Career

Oppenheimer received a liberal and wide-ranging education in New York City, at Harvard University, and at several leading scientific centers in Europe, receiving his Ph.D. under Max Born in 1927. His most creative scientific work was performed in the period 1927–1942, first at Göttingen, Germany, with Born, and then at the California Institute of Technology and, primarily, at the University of California Berkeley. His first major contribution was the *Born-Oppenheimer approximation*, a seminal recipe for dealing with molecular interactions. He subsequently published important papers on nuclear and particle physics. He also studied astrophysical phenomena, involving general relativity, neutron stars, and gravitational collapse.

At Berkeley Oppenheimer became arguably the most important and certainly the most charismatic American-born physics theorist. His close association with Ernest O. Lawrence helped spread his fame as a theoretical physicist capable of understanding and working with the most advanced high energy experiments. In 1942 he became scientific director of the Los Alamos center of the Manhattan Project, where the atomic weapons of World War II were designed, built, and finally delivered for use over Japan in August 1945. Resigning from Los Alamos after the war, he became director of the Institute for Advanced Studies at Princeton, where he once again demonstrated his talents as an organizer and scientific leader.

Politics and Ethics

As a result of his spectacular accomplishment with the atomic bomb, Oppenheimer was elevated to a position of extraordinary prestige and power in both the scientific and the political worlds. He became an international celebrity and governmental adviser, raising questions of conscience for the scientific community and arguing for United Nations (UN) control of nuclear weapons. In 1947, at the Massachusetts Institute of Technology, he gave a talk in which he made the comment that as a result of their development of the atomic bomb physicists had *known sin* and thus had a responsibility to help educate other scientists, politicians, and the public about the devastating power of these new weapons.

Early in his Berkeley years Oppenheimer became involved in political activities. He supported many organizations and interest groups that could be identified as leftist. Such activities and associations later caused Oppenheimer difficulty during the period of intense anti-communist sentiments that gripped the United States in the early days of the Cold War, and an Atomic Energy Commission (AEC) hearing resulted in the removal of his secret security clearance in 1954.

The denial of Oppenheimer's clearance was based on several factors. One was his unswerving opposition to the efforts of the U.S. government to develop a hydrogen bomb. Another was his past associations with left wing and pro-Soviet groups, and also the fact that at one time in 1943 he did not reveal a discussion with Haakon Chevalier, a friend and French professor at Berkeley, about the possibility of personal contacts between American and Soviet scientists outside official channels. The reason for not reporting this incident may have been his unwillingness to betray a friend, whom he felt was innocent of venal motive. As for his opposition to the hydrogen bomb, in retrospect Oppenheimer appears to have been punished for a dissenting view on a controversial topic, a state of affairs that is part of the normal democratic decision making process. In any case President John F Kennedy ordered what amounted to his rehabilitation in 1963 by awarding him the Enrico Fermi prize, the highest honor granted by the AEC.

Oppenheimer was an aesthete; a consummate scholar of languages, ancient cultures, and literature; as well as an accomplished physicist. He had refined tastes, supported by his inherited wealth. He was a self-proclaimed lover of the *common man*, exemplified in his espousal of liberal and leftist causes. Yet he worked on military weapons and projects. He did not oppose *research* on the hydrogen bomb, only on its development as a deliverable weapon. In telling testimony before the U.S. Congress, he once commented that such development was so *sweet technically* that it could not but be tried. Although known for acerbic remarks at scientific presentations, he was admired, even loved, by students and junior colleagues. Although loyal to friends, in the Chevalier case he caused irreparable damage to a career when he did belatedly describe their conversation. While his scientific productivity was outstanding, he missed producing any single contribution that would have placed him in the first ranks. In sum he was a scientist, teacher, scientific administrator, and public figure, whose flaws prevented him from achieving the highest level in the intellectual pantheon, and yet who raised important ethical issues for the scientific community and public.

BENJAMIN BEDERSON

SEE ALSO *Atomic Bomb; Oppenheimer, Frank; Science Policy.*

BIBLIOGRAPHY

Herken, Gregg. (2002). *Brotherhood of the Bomb.* New York: Henry Holt and Company.

Schweber, Silvan S. (2000). *In the Shadow of the Bomb: Bethe, Oppenheimer, and the Moral Responsibility of the Scientist.* Princeton, NJ: Princeton University Press.

INTERNET RESOURCE

Bethe, Hans A. "Biographical Memoirs: J. Robert Oppenheimer." National Academy of Sciences. Available from http://nap.edu/html/biomems/joppenheimer.html.

ORGANIC FOODS

• • •

At the most basic level, organic food is grown or raised without the use of synthetic chemicals. In the production of vegetables and fruits, no synthetic pesticides or fertilizers may be used, and no hormones or antibiotics

Fresh tomatoes with a "No Pesticides" sign. Organic vegetables are grown without the chemical herbicides and pesticides used in conventional agriculture. *(Nancy R. Cohen/Getty Images.)*

may be used in the rearing of livestock or poultry. The concept of organic food, however, remains fuzzy. Beyond restricting the use of synthetic chemicals, other issues sometimes incorporated into the idea of organic food include: no sewage-sludge fertilizers, no food irradiation, no genetically modified organisms, humane conditions for livestock and poultry, sustainable land use practices, and just treatment of workers in the food production process.

Until the twentieth century, all human food was organic. At the dawn of World War II, the few pesticides in use were derived from plants (for example, nicotine, rotenone, pyrethrum) or minerals (for example, arsenic and sulfur compounds). Paul Müller's 1939 discovery of the insecticidal properties of DDT, in conjunction with military needs to control infectious disease, propelled the chemical industry to full-scale production, which continued after the war as DDT and other pesticides were put to agricultural use. Pesticides and chemical fertilizers, along with new crop hybrids, farm machinery and irrigation techniques, enabled industrial agriculture, which aims to increase agricultural yield while decreasing the costs of production in order to maximize both

food production and profits. Following World War II, the United States exported industrial agriculture across the globe for humanitarian and economic purposes. The green revolution began in the 1940s according to Norman Norlaug, Nobel laureate and widely recognized father of the green revolution. New hybrid crops were only one part of the green revolution, new agricultural techniques, including the heavy use of synthetic fertilizers, pesticides, irrigation techniques, and new farm equipment played a significant role in both the green revolution and the viability of the new plant hybrids.

Meanwhile, Lady Eve Balfour of England investigated, practiced, and promoted organic farming starting in 1938. She published *The Living Soil* in 1943, which led to the 1946 formation of the Soil Association, still the United Kingdom's leading organic foods organization. In the United States J. I. Rodale popularized organic gardening through the soil and health foundation, founded in 1947. He created several publications including *Organic Farming and Gardening* (est. 1942) and *Prevention Magazine* (est. 1950). His son, Robert, expanded this work by establishing the Rodale Institute and Rodale Press to promote the healthy land/healthy human connection. It was not until the environmental movement began in the 1960s, however, that organic foods flourished. In 1962, Rachel Carson's *Silent Spring* called attention to the public health and environmental consequences of industrial agriculture and unchecked pesticide use. The resulting concern over public health and the environment created a demand for organic food throughout the industrialized world.

In 1990 the United States Congress passed the U.S. Organic Foods Production Act, mandating the U.S. department of agriculture "(1) to establish national standards governing the marketing of … organically produced products; (2) to assure consumers that organically produced products meet a consistent standard; and (3) to facilitate interstate commerce in fresh and processed food that is organically produced." But the debate over the development of organic standards between the initial 1994 recommendations and the final rules implemented in October 2002, and the global debate more generally, exposed significant ethical and scientific disagreements.

Organizations such as the Soil Association (est. 1946), Organic Trade Association (est. 1985), and the Organic Consumers Association (est. 1998) claim that organic foods promote a healthy, safe, and sustainable system of food production. But critics such as the Hudson Institute Center for Global Food issues and the American Council on Science and Health point out that no scientific evidence exists that organic foods are significantly more nutritious, safer, or tastier than conventionally grown foods. These critics have suggested that government promotion of organic foods undermines confidence in conventionally grown foods to the detriment of the poorest members of society and perpetuates a kind of fraud whereby organic food producers charge extra for products with no significant benefit.

Arguments for industrial agriculture rest on efficiency and the elimination of hunger, while arguments for organic food emphasize environmental and sometimes social sustainability. Some people accuse advocates of organic agriculture of elitism in prioritizing the environment over the needs of the poor. At the same time, organic advocates accuse industrial agriculture of prioritizing profits over environmentally and socially sustainable agriculture. Issues over how to define organic standards, how to enforce standards in an international food market, the appropriate burden of proof for the organic foods industry, and the relative importance of feeding the poor versus creating a sustainable system of food production pervade the organic debate.

Underlying this debate is the critical issue of global population growth. The Green Revolution succeeded in the sense that it prevented the starvation catastrophe predicted by Thomas Robert Malthus (1766–1834) and Paul R. Ehrlich (b. 1932). But with rapid increases in agricultural yield diminishing, one must explicitly consider the roles of organic and conventional food production in a world with a still burgeoning population.

JASON M. VOGEL

SEE ALSO *Agricultural Ethics; Food Science and Technology; Genetically Modified Foods; Nutrition and Science; Vegetarianism.*

BIBLIOGRAPHY

Balfour, Eva B. (1943). *The Living Soil and the Haughley Experiment.* London: Faber & Faber.

Lipson, Elaine Marie. (2001). *The Organic Foods Sourcebook.* New York: McGraw-Hill.

Trewavas, Anthony. (2001). "Urban Myths of Organic Farming." *Nature* 410: 409–410.

U.S. Congress. *Organic Foods Production Act of 1990.* Public law 101–624.

INTERNET RESOURCE

Organic Consumers Association. Available at http://www.organicconsumers.org/

ORGANIZATION FOR ECONOMIC CO-OPERATION AND DEVELOPMENT

• • •

The Organization for Economic Co-operation and Development (OECD) was born in 1961 as the successor to the Organization for European Economic Co-operation (OEEC), which itself was created after World War II to administer the United States' Marshall Plan funding of European recovery. OECD is related in its structure, antecedents, and goals to other post–World War II international agencies such as the International Monetary Fund and the World Bank.

Brief History

In December 1959 the presidents of the United States and France, the West German chancellor, and the British prime minister, meeting in Paris, issued a communiqué calling for the industrialized countries to cooperate to help the less-developed world and to "pursu[e] trade policies directed to the sound use of economic resources and the maintenance of harmonious international relations, thus contributing to growth and stability in the world economy and to a general improvement in the standard of living" (OECD 1961, p. 11).

The OEEC member countries agreed on several principles. First, Europe's economic recovery was complete, and the OEEC was no longer needed. Second, it had become more evident in the postwar years that "the policies of any individual country had a direct and unavoidable influence for good or bad on economic conditions in every other country" (OECD 1961, p. 9). Third, the member nations, acting together, could use the new organization as a forum for giving "a higher priority than in the past to the problems of helping the less-developed countries of the world." The convention establishing the OECD was signed on December 14, 1960, and it went into effect on September 30, 1961.

Late in 1961, the twenty member nations met for the first time in Paris. They set a joint target of 50 percent growth in gross national product for the period 1960 to 1970, in support of the thesis that the industrialized countries could support the developing world only by sustaining their own growth at the same time. And they reaffirmed their commitment to interdependence: "[I]ndependent pursuit by each country of its legitimate objectives could not only aggravate existing disequilibria in the world economy but might also prevent the attainment of its objectives" (OECD 1961, p. 21).

The U.S. Senate, which under the Constitution was called upon to "advise and consent" to the convention creating the OECD, reacted with some anxiety. To some isolationist senators and their constituents, the OECD seemed like a Trojan horse containing the elements of a new, international executive organization that would usurp Congress's legislative powers. In the February 1961 hearings of the Senate Foreign Relations Committee, chaired by Senator William Fulbright, a Democrat from Arkansas, the senators were gradually lulled by witnesses from the executive branch reassuring them that the OECD would neither supplant the United Nations nor infringe upon the powers of Congress. "[T]he impression might be left," Senator Fulbright said, "that the OECD does not do anything." No, countered the State Department witness, OECD is an "important instrumentality . . . providing for the first time for an opportunity for full and free discussion . . . [in an] atmosphere of complete candor."

Structure

The OECD Council, the organization's supreme governing body, includes representatives of all the members. The Council meets occasionally at a higher "ministerial" level, and more regularly holds gatherings of the permanent representatives. The Council acts through the issuing of decisions, agreements, recommendations, and resolutions. Because there must be complete unanimity for the issuance of any of these, the negative vote of any member is sufficient to veto any OECD action.

The Executive Committee, consisting of representatives of ten members, meets every week. Other entities appointed by OECD are committees on economic policy, technical cooperation, and trade. OECD's broadranging interests include nuclear power, immigration, capital flows, science, technology, tourism, fisheries, and education.

In January 1960 the OEEC created the Development Assistance Group. Under the newly formed OECD, this entity was renamed the Development Assistance Committee (DAC) in October 1961, and it was given a key role in OECD's efforts to aid the Third World. The original DAC members were Belgium, Canada, France, Germany, Italy, the Netherlands, Norway, Portugal, the United Kingdom, the United States, the European Economic Community organization, and Japan. DAC has never disbursed funds of its own, but acts "as a centre for the exchange of information and experience in this field" (OECD 1961, p. 22). Its members are the source of 90 percent of the total flow of private and public capital to developing nations. The DAC nations recognized that aid would be wasted unless the recipient countries were able to increase their own exports as a result. DAC and OECD therefore took on the subsidiary mission of providing "expanding markets for the products of the less developed countries and to remedy the instability of their export earnings" (OECD 1961, p. 23).

Assessment

The DAC has been highly criticized for its failures. DAC and OECD treat aid as an absolute good, without ever confronting the variety of existing definitions and implementations, let alone the political underpinnings. Most aid relationships are based on "historical circumstances or some particular interest. . . . [T]he work of DAC must to a large extent be an exploration of the margins within which a joint or common policy exists or can be created" (*International Organizations*, p. 235).

OECD and DAC both were explicitly created lacking any legislative or executive power; their effectiveness is limited to "mutual exhortation" of the member countries. "One might well get the impression," says *International Organizations* dryly, "that much of its work must have been in vain" (p. 236). The overall volume of capital flow from the industrial nations to the developing ones has almost stagnated since DAC's creation.

DAC's main strength has been in the gathering and reporting of information. Its Annual Aid Review is a comprehensive collection of data and also serves as an "exercise in shame tactics, exposing behavior of those countries which give least or do so with the most demands and conditions." However, all the data is provided by the states surveyed; there is little independent collection or assessment of the data. According to critics, it is "difficult to identify individual improvements of aid policies clearly attributable to" the Annual Review (*International Organizations*, p. 237).

Very occasionally, specific solutions to problems are proposed at DAC meetings, but most such proposals have come to nothing, and "for the most part hopes of actual coordination have been dashed, and even the best-prepared meetings have remained exchanges of uncertain usefulness" (*International Organizations*, p. 238).

DAC's official view is that aid is the bounty of rich countries to Third World nations, and the self-interestedness of most aid is "never alluded to in its publications" (*International Organizations*, p. 239). DAC members, used to working behind the scenes without more comment or criticism than the organization's lack of authority warranted, were undoubtedly surprised to be the target of developing nations' anger at the 1964 conference of the United Nations Conference on Trade and Development (UNCTAD). Third World resentment of DAC's high-handedness led the organization to set new, possibly retaliatory, standards under which aid would be tied to the performance of the recipient.

DAC's contributions are difficult to evaluate. Its members never wanted it to be an executive agency—it is a forum only. DAC's already minimal clout has diminished as world aid policies, effected through treaties and other forums, have stabilized. "As a 'rich man's club,' it attracts suspicions of a power which it does not possess. ... Theories of the conspiratorial neo-colonialist character of Western aid are certainly not confirmed in [OECD] deliberations" (*International Organizations*, p. 245).

The Cold War (1945–1989), and the West's desire to counteract Soviet influence, was a major motivation for aid to developing countries from the 1950s on. The dissolution of the Soviet Union in 1989 may have made OECD's work even less significant. During the cold war, there was a struggle between "realism" and "liberalism" within DAC: Is the purpose of aid to counter the spread of Communism, or is it the developed world's humane

obligation to help? DAC has never officially decided to concentrate on either the neediest countries or those that do the best job promoting democracy.

After 1989, OECD was active as a consultant to former Soviet satellites liberalizing their economic systems. An OECD delegation sent to advise Poland announced that "radical changes in attitude" among Polish workers and enterprises would be necessary for Poland's ambitious program to succeed (Greenhouse 1990).

In more recent years, OECD has again found itself on the receiving end of public anger, this time as a "fellow traveler" of globalizing forces. In October 1997 the staid and reclusive organization was astonished to be confronted by a coalition of antiglobalization activists and nongovernmental organizations, which asked it to suspend negotiations on a proposed Multilateral Agreement on Investment. OECD complied, placing a hold on talks, and France then withdrew, ending the effort entirely because of OECD's unanimity requirement.

OECD's Directorate for Science, Technology and Industry (STI) has studied and reported on ethical issues in the use of technology. In a 2001 policy brief titled "Sustainable Development: Critical Issues," the organization asked, "How can we meet today's needs without diminishing the capacity of future generations to meet their own?" It concluded that government must protect the environment and the resources available to future generations by "internaliz[ing] the costs" of bad behavior. For example, taxes on polluters, or a pollution permit trading system, align the market with the goals of sustainable development by causing the polluters to pay the actual costs of their activities, rather than making the public do so. The directorate regards this approach as more effective than a regulation-based one. STI has also done substantial work on biotechnology, including patent issues with a strong ethical component.

Another OECD crusade has been against the bribery of public officials, particularly by companies wishing to obtain international trade contracts. The organization proposed an Anti-Bribery Convention, which by December 2003 had been ratified by thirty-five nations (OECD, "Steps Taken").

OECD has also focused on the issue of decreasing the military expenditures of developing countries, in order to free resources for redeployment to sustainable development and other areas of concern. In 1997 DAC commissioned a series of case studies of military expenditures, noting that the majority of the funds borrowed by certain developing countries were for military purposes (OECD, "Final Report").

Another of DAC's significant concerns has been high population growth, which DAC links to the "vicious circle of underdevelopment [which causes] poverty, malnutrition, illiteracy and environmental degradation" (OECD, "DAC").

Conclusion

In its more than forty years of existence, OECD has kept a low profile consistent with its lack of executive power. Most references to it in research databases concern the organization's own publications or lead to the phrase "OECD countries," which is commonly used as shorthand for industrialized, aid-giving nations. OECD is entirely uncritical when it comes to the motives and modalities of international aid, avoiding the ethical questions raised in worldwide debates on modernization and globalization. The general impression given in the literature is of a publicly funded think tank busily producing valuable statistics and research reports, but with minimal impact on real-world policymaking.

JONATHAN WALLACE

BIBLIOGRAPHY

Greenhouse, Steven. (1990). "World Economic Aide Sees Trouble for Warsaw Policy." *New York Times* January 19.

Ohlin, Goran. (1968). "The Organization for Economic Cooperation and Development," in *International Organization* 22, no. 1 (Winter 1968) Cambrige: MIT Press.

Organisation for Economic Co-operation and Development (OECD). (1961). *Organization for Economic Cooperation and Development*. Paris: OECD Publications.

INTERNET RESOURCES

Organisation for Economic Co-operation and Development (OECD). "DAC: Development Co-operation in the 1990s." Available from http://www.oecd.org/LongAbstract/0,2546,en_2649_201185_1896179_1_1_1_1,00.html.

Organisation for Economic Co-operation and Development (OECD). "Final Report and Follow-Up to the 1997 Ottawa Symposium." Available from http://www.oecd.org/dataoecd/16/48/1886718.pdf.

Organisation for Economic Co-operation and Development (OECD). "Steps Taken and Planned Future Actions by Participating Countries to Ratify and Implement the Convention on Combating Bribery of Foreign Public Officials in International Business Transactions." Available from http://www.oecd.org/dataoecd/50/33/1827022.pdf.

Organisation for Economic Co-operation and Development (OECD). "Sustainable Development: Critical Issues." Available from http://www.oecd.org/document/55/0,2340,en_2649_34499_1890487_119696_1_1_1,00.html.

ORGAN TRANSPLANTS

• • •

From the first successful kidney transplant in 1954, organ transplantation has advanced radically to become one of the greatest technological achievements in medicine. As of the early twenty-first century, doctors have successfully transplanted six different organs: the liver, kidney, pancreas, heart, lung, and intestine, as well as several different types of tissue. Simultaneous transplantation of multiple organs is possible as well. The possibility of organ transplant offers hope to thousands of patients suffering from organ failure who may have no other option. However, as the technique improves, the number of people waiting for an organ increases rapidly. More people die on the waiting list each year as the organ shortage escalates. Based on OPTN data as of November 26, 2004, there are 86,876 people on the United Network of Organ Sharing (UNOS) waiting list in need of a transplant and approximately 7,983 individuals died in 2003 while waiting for an organ.

Process and Costs

Cadaveric organ transplant is currently the most popular form of transplantation. However, living donation from both related and non-related donors is widely accepted for kidneys and increasingly more common for liver patients. In the United States, UNOS functions as a centralized system for the allocation of available organs. When an organ becomes available, UNOS is contacted by a local Transplant Coordinator and determines which candidate is the most suitable for the organ, based on clinical factors such as tissue matching, blood type, length of time on the waiting list, immune status, and geographical location. For heart, liver, and intestine transplants, the medical necessity of the potential recipient is also considered (United Network of Organ Sharing Internet site).

As organ transplant becomes a more routine procedure for those suffering from organ failure, it is important to recognize that there continues to be risks involved in this type of surgery. Transplant success has historically hinged on whether or not the recipient's immune system would attack the foreign organ, jeopardizing the effectiveness of the transplant. To limit this, antigen matching between the donor and recipient is a primary concern of UNOS. In the early 1980s, cyclosporine became the first of many drugs to effectively suppress the human immune system to prevent organ rejection. Although not perfect, immunosuppression has become critical to further advancements in transplantation. The intensity

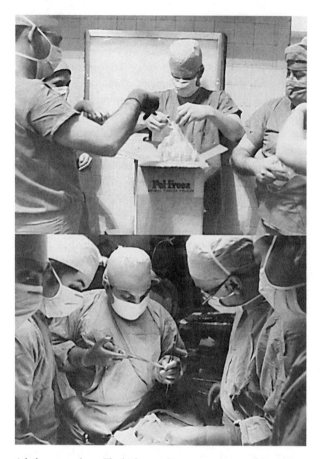

A kidney transplant. The high cost of immunosuppressant therapy for kidney recipients has become a subject of Congressional debate. (© UPI/Corbis-Bettmann.)

of the immunosuppressant treatment can leave recipients susceptible to potentially life-threatening infections.

Immunosuppressant drugs are a lifetime commitment for organ recipients; unfortunately, they are expensive. Kidney recipients spend an average of $10,000 to $14,000 on such medications each year. Congress has struggled with how to pay for this expensive therapy since its conception. Numerous policies have been passed since 1972 to aid in the cost of kidney transplantation as well as immunosuppressant medications for recipients of kidney, liver, and heart transplantations who qualify for Medicare at the time of transplantation and extends for limited time post-transplant. Despite much effort, many transplant recipients still struggle with the increased cost of post-transplant medication critical for their survival. Noncompliance rates due to inadequate finance for organ recipients has been difficult to determine, but may be a common cause of graft failure (Kasiske, Cohen, Lucey, and Neylan 2000). Ethical debates have arisen on this issue. Some believe giving an organ to a patient for whom it is financially

impossible to continue treatment is wasting an organ that could save another life. Others argue it is unethical to deny the life-saving procedure to those of lower socioeconomic class.

Allocation Issues

The allocation of organs has been the source of extensive ethical and political concern. Organs are considered a precious and limited resource because few are available for transplantation, and because of the altruistic nature of the gift of an organ. Many question whether there ought to be standard psychosocial criteria added to the evaluation process to prevent various types of discrimination. Providing prisoners with a transplantable organ has prompted a significant public debate. This was highlighted by the controversy surrounding a prisoner in California who received a heart transplant in January 2001. The debate is centered on the question of who should be given the power to determine whether one individual is more worthy of an organ transplant; beginning this type of preferential treatment is what many ethicists consider a "slippery-slope."

Xenotransplantation

Xenotransplantation is one potential method of attacking the organ shortage. The prefix *xeno-* means "foreign"; a xenotransplant refers to the process of transplanting a cell or organ from a foreign species. After consideration of factors such as availability, anatomy, and familiarity with the animal, pigs have emerged as the most promising donor option. Genetic engineering offered opportunity to modify the donor animal to more closely resemble the human recipient; coupled with improvements of immunosuppressant therapy, the chance of organ rejection could potentially be significantly decreased (Sachs, Sykes, Robson, and Cooper 2001). At the beginning of the twenty-first century, there had been little success in xenotransplantation, and much debate on the ethics and policy involved with the field. One primary concern with the development of xenotransplantation is the potential for an epidemic caused by previously unknown animal diseases being transferred to humans. Some believe this risk is too dangerous and that xenotransplantation should not be tested.

Another concern that arises with xenotransplantation, a discussion also relevant for certain allotransplant policy, is the commodification of the human body. Organ donation in the United States is considered an altruistic gift. However, policy proposals for financial incentives and some international policies for the buying and selling of organs puts a monetary value on organs. Organs for xenotransplant will be controlled by commercial companies; a recipient will have to purchase an organ. Because these organs will be genetically modified to resemble human organs, commercialization of the organs may have implications for socioeconomic equality. It would also create a rhetoric of human body parts as a purchasable commodity, a concept with which many ethicists have been skeptical (Bach, Ivinson, and Weeramantry 2001).

The benefit of organ transplantation for those suffering from organ failure is virtually undisputed. Unfortunately due to the complexity of the procedure, availability of organs, and the many other variables that factor into an organ transplant, there is still enormous debate surrounding transplantation.

SHELDON ZINK

SEE ALSO *Bioethics; Medical Ethics.*

BIBLIOGRAPHY

Bach, Fritz A.; Adrian J. Ivinson; and Christopher Weeramantry. (2001). "Ethical and legal issues in technology: Xenotransplantation." *American Journal of Law & Medicine* 27(2–3): 282–300. Discusses the risks and benefits of technology in general with a focus on xenotransplantation. The authors examine some of the legal, ethical and human rights considerations that this innovation raises.

Fishman, Jay A., and Robert H. Rubin. (1998). "Medical Progress: Infection in Organ-Transplant Recipients." *The New England Journal of Medicine* 338(24): 1741– 1751. Examines the risk of infection for transplant recipients. Provides detailed information regarding the most common infections faced by patients and the points at which various infections arise. Also addresses ways to potentially reduce the risk of infection, through pre-transplant screening of both the donor and the recipients, and treatment of infection with antimicrobial therapy post-transplant.

Kasiske, Bertman L.; David Cohen; Michael R. Lucey; and John F. Neylan. (2000). "Payment for Immunosuppression After Organ Transplant." *Journal of the American Medical Association* 283(18): 2445– 2450. Deals with the problem of payment for immunosuppressant therapy for organ recipients. The government has struggled with this issue and does cover some immunosuppressant medication for eligible donors; the authors discuss the need for expanding this coverage.

Sachs, David H.; Megan Sykes; Simon C. Robson; and David K. C. Cooper. (2001). "Xenotransplantation." *Advancements in Technology* 79: 129– 205. Provides a detailed overview of the field of xenotransplanation. Includes a history and explanation of the science and mechanism of transplanting an organ from an animal into a human, and discusses the possibilities and implications of using xenotransplantation in the future.

INTERNET RESOURCE

United Network of Organ Sharing. "Who We Are." Available from www.unos.org. A section of the UNOS Internet site that provides a detailed description of the history and membership of the organization.

ORTEGA Y GASSET, JOSÉ

• • •

José Ortega y Gasset (1883–1955) was born in Madrid on May 8 and became the most influential Spanish philosopher of the twentieth century, with a reputation and influence that extended from Spain to Latin America and beyond. Ortega was the first professional philosopher to make technology an explicit theme for critical reflection. He died in Madrid on October 18.

Ortega in His Circumstances

Ortega earned a doctorate at the University of Madrid in 1904, after which he did postdoctoral work in Germany. His course of study included not only philosophy but also comparative literature, law, biology, and psychology. Having been influenced by the Generation of 98 (1898, the year in which Spain lost the last of its colonies to the United States and a period in which Miguel de Unamuno [1864–1936], Pío Baroja [1872–1956], and other writers responded with new visions of the nation), Ortega became a leading figure of the Generation of 27 (1927, the year of the emergence of a literary and artistic avant garde that included Federico García Lorca [1898–1936] and Pablo Picasso [1881–1973]).

Outside the academic world Ortega worked as a journalist, publisher, and politician and served as a member of parliament between 1931 and 1933, during the Second Spanish Republic. After the Spanish Civil War (1936–1939) he went into exile, initially in Argentina, but in 1945 he settled in Portugal and then returned to Spain in 1948 to found the Institute de Humanidades, where he lectured until his death.

The basic theme of Ortega's philosophy was announced in *Medicaciones del Quijote* [Meditations on Quixote] (1914), in which he argued for understanding human beings in relation to their circumstances. "Yo soy yo y mi circunstancias" [I am myself and my circumstances] was the formative statement with which he placed *razón vital* (living reason), a kind of existentialist vitalism, at the center of philosophical reflection. It was in an attempt to understand living reason at work in his own circumstances that Ortega, over the course of his

José Ortega y Gasset, 1883–1955. The Spanish philosopher and essayist is best known for his analyses of history and modern culture, especially his penetrating examination of the uniquely modern phenomenon "mass man." (*NYWTS/The Library of Congress.*)

philosophical career, analyzed the historical condition of Spain (*España invertebrada* [1921]), the character of modern art (*La deshumanización del arte* [1925]), the transformation of politics (*La rebelión de las masas* [1930]), the dynamics of history (*Historia como sistema* [1936]), and the post–World War II destiny of Europe (*Meditación de Europa* [1949]).

Ethics and Technology

Ortega's philosophy is a critique of the rationalism that has been dominant since the eighteenth century. As an affirmation of life that nevertheless acknowledges the essential character of reason in human beings, his philosophy is fundamentally ethical in its orientation. The primordial reality is life, in which individuals find themselves as castaways struggling not to drown. This is the basic human activity: not contemplation or science but rather "staying alive," with one of the instruments in the struggle being technology.

It is this perspective that Ortega brought to bear on technology in a number of works but especially in a

1933 university course that appeared in book form under the title *Meditación de la técnica* [Meditation on technics] (1939). More partial contributions to this analysis can be found in works as diverse as *The Revolt of the Masses*; *En torno a Galileo* [Around Galileo] (1933), translated as *Man and Crisis*; *La idea principio en Leibniz* [The idea of principle in Leibniz] (published posthumously in 1958); *Una interpretación de la historia universal* [An interpretation of universal history] (published posthumously in 1959); and lectures such as "Goethe sin Weimar" [Goethe minus Weimar] (1949) and "El mito del hombre allende la técnica" [The myth of humans outside technics] (1951).

Meditación de la técnica begins with a prophetic pronouncement about the future of philosophy and technology: "One of the themes that in the coming years is going to be debated with the most determination is the sense, advantages, dangers, and limits of technics" (*Obras completas* 1946–1983, Vol. V, p. 319). According to Ortega, technology does not so much help humans adapt to and be able to live in the natural world that surrounds them as it is an instrument that permits them to adapt nature to the satisfaction of their needs. Those needs include not only those of the primary type (food, shelter, etc.) but also those, which produce well-being, not just life but a vision of the good life. For example, the bow is an invention created both to hunt and to play music.

Whereas an animal can live only in a manner that is dependent on nature, humans are capable of distancing themselves from nature, becoming introspective, and, from the point of this self-absorbtion, performing the act of inventing. Technological innovation creates a "supernature" that becomes a mediator between humans and nature. In the historical development of this technology Ortega distinguishes three stages: accidental technology, crafted technology, and the technology of the technician.

In the first stage technology appears in limited and rudimentary forms; human beings view technological innovation as the result of chance, not of their capacity for invention. In the second stage craft techniques have a greater presence and complexity, although invention and production are not clearly distinguished. More important, humans do not realize their capacity for invention because the technical advances they produce are considered not innovations but variations within a craft tradition.

In the third stage humans finally recognize that technology is the fruit of their ability to invent. They dissociate the moment of invention, which belongs to the inventor or engineer, from the act of application, which belongs to the worker. In this stage humans begin to create not only instruments or tools but also machines that replace human work: the set of "invention factories" (as the inventor Thomas Edison [1847–1931] called his laboratory) and systems for research and development leading to new and imaginative technologies.

It is in this third stage, Ortega argues, that humans now find themselves and in which they discover a horizon of unlimited possibilities. Before the modern period most people were limited by the circumstances in which they both inherited a vision of how to live and adopted the apparently unchanging technical means to realize it. In the contemporary world, however, with the emergent ease of external technical invention, human attention is distracted by ever more superficial activity. In Ortega's words, in the modern world "before having some particular technics one has technics itself" (*Obras Completas* 1946–1983, p. 369).

However, at this point human beings must face two temptations. On the one hand, they tend to lose interest in the science on which technology depends because it seems so readily available that producing it does not seem to be required any longer. On the other hand, they specialize, thus abandoning any comprehensive view of reality that might provide a basis for orienting or focusing technological developments. Able to become anything they want, they cease to want to become anything at all.

Ortega presents a defense of technology as an element that makes human life human. However, he points out that the capacity, in principle unlimited, that technology now offers to humans may tempt them to believe that they live from technology and not with it, that they are merely forms of technological life, not creatures that use technology to live. Insofar as human beings allow themselves to give in to that temptation, human life eventually will become meaningless and living reason will wither and die.

Implications

More than other seminal philosophers of technology in the European tradition, such as Martin Heidegger (1889–1976), Herbert Marcuse (1898–1979), and Jacques Ellul (1912–1994), Ortega appreciated the positive aspect of technology, its intimate engagement with what it means to be human. At the same time, more than some people today who enthusiastically celebrate the achievements of technology, he recognized the dangers of what might be called "technology only technology." Whether and to what extent Ortega's thought can be brought to bear in specific discussions about science, technology, and ethics remains to be seen.

VINCENTE BELLVER CAPELLA
TRANSLATED BY JAMES A. LYNCH

SEE ALSO *Conservatism; Existentialism; Spanish-Language Perspectives.*

BIBLIOGRAPHY

Gray, Rockwell. (1989). *The Imperative of Modernity: An Intellectual Biography of José Ortega y Gasset.* Berkeley: University of California Press.

Ortega y Gasset, José. (1946–1983). *Obras completas,* 12 volumes. Madrid: Alianza. Volume V contains *Meditación de la técnica* in *Ensimismamiento y alteración* (1939). An incomplete translation is available as "Thoughts on Technology" in *Philosophy and Technology: Readings in the Philosophical Problems of Technology,* ed. Carl Mitcham and Robert Mackey. New York: Free Press, 1972, pp. 290–313. Though Ortega's reflections on technology can be found in many of his writings, *Meditación de la técnica* gives a comprehensive vision of his thoughts on the matter.

OZONE

SEE *Montreal Protocol.*

P

PARETO, VILFREDO
• • •

The unique contributions of Vilfredo Pareto (1848–1923) to mathematical economics as well as sociopolitical theory were predicated on a remarkable background and education. The son of Raffaele Pareto (a minor Italian noble, civil engineer, political refugee, repatriated professor, and then government minister) and a Frenchwoman, Marie Métenier, Fritz Wilfried Pareto (renamed in 1882 Vilfredo Frederico Damaso Pareto, ultimately the Marchese of Parigi), was born in Paris on July 15. The household was bilingual, but after his father's political safety was assured, the family removed to his native Genoa (1855–1859); spent several years in Casale Monferrato, Piedmont, so his father could improve his professional position as a government administrator of mines and industry; then went to Turin; and finally settled in Florence. In 1889 Pareto married Alessandrina "Dina" Bakounine (Bakunin; not from the anarchist's family), who left him in 1901. He lived with Jeanne Régis from 1906 and married her in 1923 (after relocating in order to divorce Bakounine), and adopted her daughter, Marguerita Antoinette Régis. He died on August 23 in Céligny, Switzerland, where he had lived since 1900, spurning honors bestowed on him *in absentia* by the new Italian fascist government.

Pareto was rigorously educated in mathematics and the natural sciences, as well as the classics, partly in the school where his father taught—he imitated his father by pursuing mathematics, physics, and engineering. He precociously finished his doctorate in 1870 with a thesis on the then-new applications of differential equations to the question of elasticity and equilibrium in solid bodies, a work he always valued.

Vilfredo Pareto, 1848–1923. The Italian sociologist, political theorist, and economist is chiefly known for his influential theory of ruling elites and for his equally influential theory that political behavior is essentially irrational. (*The Library of Congress.*)

His subsequent management positions (with the Rome Railway and then the Italian iron industry, 1870–1889) compelled him to travel throughout Europe to learn practical business matters, and eventually to loathe the seedy deal making that accompanied the job. His anti-government lectures to laborers were shut down, and repelled by the plutocratic government; he ran for a Florentine seat in the legislature. He wrote 167 political articles for newspapers and magazines between 1889 and 1893, arguing that the Italian aristocracy had ruined the national economy through protectionism, cronyism, and graft.

Barred from a professorship in Italy, he accepted Léon Walras's (1834–1910) vacated chair in political-economics at Lausanne, Switzerland in 1893. He retired at fifty with a substantial inheritance from his uncle and perfected his quest for a quantifiable social science inspired by his reading of Auguste Comte (1798–1857) (another prodigy who had studied mathematics and engineering and coined the term *sociology*). It was his extraordinary proficiency in applied mathematics that facilitated Pareto's cardinal contributions to early econometrics, to equilibrium and systems theory in sociology, and, by redefining cyclical patterns to rulership, to political science.

Pareto is a neglected genius of the modern period. Living coterminously with Max Weber (1864–1920), Émile Durkheim (1858–1917), Georg Simmel (1858–1918), and Sigmund Freud (1856–1939), he shared none of their posthumous fame—except for a brief period in the 1930s when he was lionized, especially among Harvard intellectuals. This is probably more a quirk of history than a sound judgment about the quality of his ideas and research. In his autobiography, Mussolini claimed to have attended Pareto's lectures on political economy at Lausanne (with many other students), and a link was forged in the popular mind between fascism and Pareto's theory of *the circulation of elites*. The connection is artificial because Pareto detested any form of authoritarian rule, including fascism. Yet his ideas have suffered as a consequence of this unsavory historical connection.

The arguments of Pareto's *Course of Political Economy* (1896), which features *the Pareto optimality or ophelimity principle* are nevertheless referenced in every economics textbook. Moreover his *Socialist Systems* (1902), the *Manual of Political Economy* (1904), and his million-word *Mind and Society* (1916) evidence a level of speedy productivity and creativity that has few rivals. These works have not been seriously reconsidered, except in Italy and France, during the entire post-World War II period.

Like other gifted scientists and technicians who since the Enlightenment have turned their analytic tools toward social analysis, Pareto realized that economics alone, even if elegantly quantitative in design, could not explain the great bulk of human behavior because people do not generally behave to *maximize their utilities*. Even though he claimed to rely on the *logico-experimental method* in all his socioeconomic analyses, he thoroughly understood its limitations. Pareto's complex typological analysis of the role of nonrational, nonlogical, or irrational behaviors (*resides, derivations,* and *sentiments,* as he called them) in individuals and social groups has not been equaled in scope and depth. Yet the pessimistic conclusions he drew from his dogged historical and cultural research repels most readers today who are understandably, given recent history, more interested in ameliorative than in denunciative social theory.

What makes Pareto so difficult to embrace is his clear-eyed insistence on examining history and contemporary events through the scientist's lens, free of any idealized notions of what *ought to be* or *might have been*. Intensely idealistic when young, he soured on the *illusions of the epoch* (e.g., nationalism, Marxism, socialism, anarchism, imperialism, among others), viewing all of them as delusionary systems enabling social actors to feign rational behavior while hiding their real motives behind baroque structures of excuses and ideological justifications (derivations). Pareto never read Freud, but his work could be viewed as adding a macroanalytic dimension to the microanalysis common to psychoanalysis. Similarly when economists now speak about the *irrational exuberance* of stock markets, they are unknowingly speaking in Pareto's terms, and could well put to use his analysis of the socioeconomic environment. The same goes with regard to many discussions of science and technology policy that propose benefits from cancer research or space exploration that lack sound justifications.

ALAN SICA

SEE ALSO *Comte, Auguste; Efficiency; Engineering Ethics; Italian Perspectives; Management.*

BIBLIOGRAPHY

Pareto, Vilfredo. (1935). *Trattato di sociologia generale* [Mind and Society]. 4 volumes, trans. Andrew Bongiorno and Arthur Livingston, with the advice and active cooperation of James Harvey Rogers. New York: Harcourt, Brace, and Co.

PARSONS, TALCOTT

• • •

The leading anglophone social theorist between about 1940 and 1965, Talcott Parsons (1902–1979), who was born in Colorado Springs, Colorado, on December 13, was a tireless synthesizer of ideas from classical social and economic theory, functionalist anthropology, psychoanalysis (in which he was trained), and psychology. Though he did not create pathbreaking scientific concepts or procedures, nor contribute formally to ethical reasoning, he did succeed in grafting a robust affection for scientific method (as his generation understood and venerated it) onto the massive edifice of classical social theory in a way that no one else had managed.

Parsons was the youngest child of an early feminist mother (who could trace her ancestry to Jonathan Edwards [1703–1758], the American "divine") and a Congregational minister who became president of Marietta College. Parsons first studied biology at Amherst College, then shifted to political economy of the German-historical type. After a year at the London School of Economics (1924–1925), he moved to the University of Heidelberg, receiving his doctorate there with a dissertation on "'Capitalism' in Recent German Literature: Sombart and Weber." After teaching one year at Amherst he became an economics instructor at Harvard University, where he became a full professor in 1944 and where he remained until his retirement in 1974. He married Helen Bancroft Walker on April 30, 1927, and with her produced three children, Anne (an anthropologist of Italian culture), Charles (an economist), and Susan. Diabetic since the age of fifty-six, he died at seventy-six while on a trip to Heidelberg, on May 8, 1979, while celebrating his formative academic experience in that town fifty-three years earlier.

In 1946 Parsons helped form a new department, Social Relations, which brought together anthropology, political science, social psychology, and sociology. His keen attention to the claims of progressive, liberalizing science, coupled with an ever-present desire to understand the ethical meaning of social action (individually and collectively) were provoked by his parentage and upbringing, plus the special context of Harvard between 1927 and 1974, where he worked closely with a galaxy of gifted students and colleagues. His fascination with the proper role for "the professions," and how groupings of professionals could serve as a bulwark against the deadening routine of bureaucracy, on the one hand, and the self-serving market scramble of the capitalist on the other, was a theme adopted straight from Émile Durkheim's 1892 book, *Division of Social Labor*. It dove-

Talcott Parsons, 1902–1979. The American sociologist analyzed the socialization process to show the relationship between personality and social structure. His work led to the development of a pioneering social theory. (*AP/Wide World Photos.*)

tailed perfectly with the strict Protestant morality, left-leaning in its politics, that he had absorbed while a boy. Parsons was also president of the American Sociological Association in 1949.

At Harvard, Parsons educated four self-aware generations of enterprising sociologists who carried his structural-functionalist scheme around the country and the world, particularly during the 1950s and 1960s (with a small renaissance in the early 1980s). His leadership of the theory wing of American sociology began to wane with C. Wright Mills's (1916–1962) famous attack on "grand theory" in *The Sociological Imagination* (1959) and was ended by Alvin Gouldner's (1920–1980) rhetorical masterpiece, *The Coming Crisis of Western Sociology* (1970).

Of Parsons' fourteen books, his first one, *The Structure of Social Action* (1937), remains of paramount interest. In this large study of Max Weber (1864–1920), Durkheim, Vilfredo Pareto (1848–1923), and the English economist Alfred Marshall (1842–1924), Parsons claimed to have discovered a "convergence" of ideas among four geniuses that culminated in Parsons's own ideas about the nature of normatively ordered social action. He was especially interested in how societies deal with the "Hobbesian problem of order," which is understandable given the history of the twentieth century to that point. But he was equally dedicated to updating the perennial question first systematically presented by Durkheim in 1892: What is the proper balance between the rights of individuals to express their uniqueness and the needs of the larger society to constrain these egocentric rights through normative controls? Fascinated with normative "consensus" and the avoidance of costly societal conflict, Parsons created his own sociological glossary, including such terms and concepts as voluntarism, pattern variables, the AGIL scheme of action (1963), and universalistic versus particularistic norms, as well as a large assortment of two-by-two tables that illustrated the personality/social structure dialectic in terms that seemed to validate his way of seeing the world.

Parsons's statements about science and technology now seem banal because he uncritically echoed the great enthusiasm for Big Science that so much infected the post–World War II period. A comment from his 1971 book, *The System of Modern Societies*, is typical:

> Applied science did not begin to have a serious impact upon technology until the late nineteenth century. But technology has now become highly dependent upon research "payoffs," involving ever-wider ranges of the natural sciences, from nuclear physics to genetics, and also the social or "behavioral" sciences, perhaps most obviously economics and some branches of psychology. The social sciences share with the natural sciences the benefits of some striking innovations in the technology of research. (p. 96)

His most important work in this regard is a little-known empirical study he conducted with many collaborators between 1946 and 1948, "Social Science: A Basic National Resource." Here he argued that the new National Science Foundation ought to support the social sciences (contrary to the desires of President Franklin Roosevelt), because of its "scientifically based" contribution to the war effort. He wrote op-ed pieces for the *New York Times* making the same point, and led the fight for equal funding for social science because of its basic importance to national security, as well as its pivotal role in the general acquisition of knowledge.

Parsons was rediscovered briefly in the 1980s by a new generation of theorists, both in the United States and in Europe, but the "neofunctionalism" that briefly carried his banner has since become moribund. His future importance will probably turn around his first book, and he will be remembered as a great systematizer in an era that no longer cared for the presentation of knowledge in such "grand" synthetic gestures.

ALAN SICA

SEE ALSO *Durkheim, Émile*.

BIBLIOGRAPHY

Gouldner, Alvin W. (1970). *The Coming Crisis of Western Sociology*. New York: Basic Books.

Mills, C. Wright. (1959). *The Sociological Imagination*. New York: Oxford University Press.

Parsons, Talcott. (1937). *The Structure of Social Action: A Study in Social Theory with Special Reference to a Group of Recent European Writers*. New York: McGraw-Hill.

Parsons, Talcott. (1986 [1948]). "Social Science: A Basic National Resource." In *The Nationalization of the Social Sciences*, ed. Samuel Z. Klausner and Victor M. Lidz. Philadelphia: University of Pennsylvania Press.

Parsons, Talcott. (1971). *The System of Modern Societies*. Englewood Cliffs, NJ: Prentice-Hall.

PARTICIPATION

• • •

Participation can mean different things depending on context. In the context of science, technology, and ethics, the concept of participation points toward questions of how technologies might be developed to promote political interaction among democratic citizens, and issues of how technoscientific expertise may itself be related to democratic decision making. The present analysis focuses on the second issue by examining three philosophical perspectives on participation in preface to making a sociological argument—an argument that will (a) refine how the problem of participation ought to be conceptualized and (b) consider all the normative aspirations of philosophy to work in conjunction with empirical studies for the purpose of offering citizens and scientists alike greater reflexive purchase on their collective decision making.

Preliminaries

Before turning to the three perspectives on participation, it is useful to note a few things in general about the problem of participation. *Expertise* is a term that is not only associated with knowledge, skill, and authority, but also with hierarchy—elitism, paternalism, and power. On the one hand, hierarchy is an essential component of representative democracy. If a government encouraged people to vote about everything without regard to expert knowledge, it would promote self-destructive mob rule rather than democracy. On the other hand, the existence of hierarchy can threaten the possibility of democracy. Democracy is practically synonymous with *equality*: Democratic citizens are equal in the legal, rights-based context of public involvement and participation in their own governing.

Because an essential condition of a democratic society is arguably the right of its citizens to participate in the public processes that directly affect them, and because justice demands that vulnerable groups who might be adversely impacted by such decisions be represented in the decision-making process, it is difficult to determine the proper relation between expertise and democracy. Stakeholders who represent different values find it difficult to agree on exactly who should participate in establishing the policies influencing what scientific and technological research is pursued, how it should be conducted, and the methods for disseminating results. Even when consensus exists over who has the right to participate, debate continues over the extent of the participation that different groups are justified in expecting. Sometimes a demand that marginalized voices be heard is associated with expectations that lay perspectives be given preferential treatment. The problem of participation is therefore one that ultimately concerns the politics of inclusion and exclusion. It is not only about how science and technology are mobilized, but also about how they are practiced and who benefits.

Three Circumspect Views

The gateway to a deeper appreciation of the problematics of democratic participation in science and technology is through three existing perspectives on relations between scientists and non-scientists: (a) Some believe that if enough patience is exerted, experts and laypeople can simply resolve any disagreements through dialogue and further experimentation, (b) others argue that laypeople simply cannot provide better answers for technical questions than experts can, and (c) still others maintain that technoscientific research ought to be self-governing because only experts are competent enough to decide how technoscientific inquiry ought to proceed. Each of these positions will be considered in turn, and objections noted. (Given the almost religious authority granted to science in the early-twenty-first

century, it is common to refer to nonscientists as the laity. The terminology is adopted here simply for convenience and with no intent to accept its subordinating implications.)

POLANYI'S REGULATIVE IDEAL. Michael Polanyi (1891–1976) best articulates the first view: If enough patience is exerted, experts and laypeople can solve their disagreements through dialogue and further experimentation. Polanyi calls attention to the fact that the popular authority of science is challenged in many circles, and he thus raises the question: How can conscientious citizens in a free society competently decide rival interpretations of nature? Recognizing that participants who espouse fundamentally different views cannot resolve their differences if they frame their discussion as if it were taking place within one organized branch of knowledge, Polanyi appeals to the democratic process of free discussion and respect for civil liberties, thus placing particular emphasis on *fairness*. He defines fairness as trying to state one's case objectively, and *tolerance* as the capacity to discover whatever sound points an interlocutor espouses. Polanyi insists that striving for fairness and tolerance can further the end of resolving controversies only if people who make strong epistemic commitments endorse these virtues.

In order for a community to effectively promote free discussion, its members must not only be committed to believing that there is such a thing as objective truth but must feel obligated to pursue it, and indeed they must believe themselves capable of acquiring it. Polanyi's solution to the problem of participation thus rests upon a regulative ideal, one that a community must autonomously choose to pursue. His solution also rests on the conviction that participation in the common enterprise of science by scientists, which is to say, the devotion of all scientists qua scientists to scientific ideals, is itself a model for political democracy because the basic ideals that guide the cognitive ambitions of science are democratic ideals. For example, the ideal of the equality of all observers—genuine scientific research must be replicable by anyone who has the appropriate scientific training and the appropriate technical apparatuses—and the ideal of publication and open dissemination—the results of scientific investigation belong to humanity—are ideals that accord with a democratic vision of the politics of science.

MESTHENE'S NEW DEMOCRATIC ETHOS. Emmanuel Mesthene, who at one time directed the Harvard University Program on Technology and Society (1964–1972), puts forth the second view: Laypeople simply cannot provide better technical answers to technical questions than experts. Mesthene points out that experts are becoming integrated into all phases of the process of government. The information that needs to be gathered and analyzed in order to make so many of our modern choices depends on the successful coordination of the experts who have mastered technological devices and scientific knowledge. Mesthene claims that as a result of the pervasive practice of deferring issues that were once the subject of public debate to experts in particular fields—experts who function almost independently of the democratic political process—traditional democratic aspirations are eroding. Under the assumption that the expert-lay divide will only continue to grow as the gap enlarges between the technoscientific experts who actually guide policy and the citizens who are in principle charged with establishing it, Mesthene contends that people need to revise their understanding of what democracy is and accept a new democratic ethos adequate to the demands and structure of the modern technoscientific society.

What might this new democratic ethos look like? On the one hand, Mesthene insists that the experts who gather the information that society needs to shape its policies should be ultimately accountable to the electorate; the freedom of the general populus to express its opinions and preferences must somehow be preserved. Presumably what Mesthene has in mind is a process that would allow elected representatives to act as proxies for public opinion by helping to establish federal research funding priorities. On the other hand, he famously declares that "no amount of *participation*—in the popular sense of the term—can substitute for the expertise and decision-making technologies that modern government must use" (Mesthene, p. 81).

POLANYI'S AUTONOMOUS RESEARCH. Polanyi articulates the third view: Technoscientific research ought to be self-governed because only experts are competent enough to decide how such inquiry ought to proceed. He succinctly expresses this point when he writes, "The choice of subjects and the actual conduct of research is entirely the responsibility of the individual scientist, [and] the recognition of claims to discoveries is under the jurisdiction of scientific opinion expressed by scientists as a body" (Polanyi 1951, p. 53). This view found its way into the public sphere when presidential science advisor Vannevar Bush challenged Senator Harley Kilgore's populist position by arguing for a *social contract* with science, one that protects the right of scientists to autonomously pursue basic research.

Objections

Pragmatist philosopher John McDermott (1969) objects to Mesthene's view that laypeople cannot do better than experts in providing technical answers to technical questions. Although Mesthene's view of the expert-lay divide may appear to rest on a realistic understanding of just how deep the differences between them in knowledge, skill, and ability are, McDermott contends that Mesthene's view also reinforces the technocratic position that the greatest resource of a society is its experts and not the general populus. In McDermott's classic critique of Mesthene, he points out that what Mesthene must presuppose idealistically, in order to have as much confidence in the technical elites as he does, is that the people who pursue technoscientific careers are *altruistic bureaucrats*: that they lack a generalized drive for power; that they gain advantage and reward only to the extent that they bring technical knowledge to bear on technical problems; and that they remain shielded from the bias of ideology because their commitment to solving technical problems rules out subjective forces of influence.

While Polanyi characterizes scientists in a free society as tolerant, and Mesthene envisions them to have no general interests antagonistic to those of their problem-beset clients, Paul Feyerabend (1975, 1978), by contrast, resolutely declares that modern scientific experts have become *ideologues*: The more time and energy they devote to advancing a position, the more difficult it becomes for them to be open-minded to points of view that challenge their core beliefs. Noting how students in the natural sciences are instructed in the technical dimensions of a scientific field but only minimally exposed to the historical arguments against the theories that make the contemporary conventional wisdom seem true or useful, Feyerabend insists that scientists have become overconfident about how to conduct research properly and how to set the boundaries for generating accurate conclusions; as a result, they are prone to uncritically dismissing alternative research methods and conclusions. In order to break the hold of expert ideology, Feyerabend argues that nonexperts ought to be institutionally empowered to judge expert viewpoints and agendas. The view of participation that Feyerabend puts forth is that duly elected committees of laypeople should regulate all scientific research that can affect the public sphere. Because the exalted authority of science is incompatible with any legitimate democracy, experts ought to be regarded first and foremost as public servants. Were this result achieved, Feyerabend argues, laypeople would realize that they have more to contribute to the pursuit of knowledge than experts who distort the value of their own achievements.

Although Feyerabend's reputation in philosophical circles remains mixed, his principal message concerning the need to better investigate the inflated authority of expertise, notably the ways in which technical decision making can be value-laden, has been enormously influential. But Feyerabend's commitment to a certain vision of democracy drives him to dichotomize the world into experts (elites) and laypeople (commoners). He is thus criticized for being insensitive to the possibility that laypeople are a disparate lot, with a variety of background skills, who do not share a common aptitude for regulating—and, as he sometimes suggests, for criticizing—expert advice. Ultimately, by classifying laypeople reductively, Feyerabend unwittingly licenses the possibility for the opposite trend to occur, for scientific elites to reduce laypeople to a mass of ignorance. Hence it was possible for Paul Gross and Norman Levitt to respond: "Scientific decisions cannot be submitted to a plebiscite; the idea is absurd. Applied to science education, for example, letting people vote on what should be taught would give us countless schools in which *creation science* would replace evolutionary biology" (Leavitt and Gross, p. B2). Even Philip Kitcher (2001) felt justified in referring to the possibility of laypeople making direct decisions on matters of science policy as *vulgar democracy* leading to the *tyranny of the ignorant*.

The consequence of using extreme positive or negative terms to caricature all laypeople and all scientific elites, and all the options by which they might participate politically, cannot escape the astute observer. The insistence that laypeople should have absolute and sole regulatory authority over technoscientific practice, or that they should have no right whatever to intervene in important technoscientific decisions, obscures the plausible ways of legitimately increasing citizen involvement in technoscience.

Interdisciplinary Research

Moving beyond the reductive expert-lay dichotomy requires that theorists focus on more subtle categories, as for example Don Ihde (1996) does in his discussion of *well-informed amateurs*. Ihde suggests that such an amateur would have the critical advantage of being neither a complete insider nor complete outsider to the domain of technoscience under dispute. Ihde's analysis further suggests that in order for philosophers to better address the problem of participation they need to become more empirically oriented. They need to have a more concrete understanding of how different constituencies interact with scientists, engineers, and policy makers. What Ihde can be interpreted as advocating, then, is that in order to better pursue normative projects, it is

necessary for philosophers to do empirical fieldwork or to have more felicitous interdisciplinary exchanges with anthropologists, historians, and sociologists.

If philosophers were more empirically focused, what would they study? Perhaps what philosophers should do is carefully study different instances of technoscientific negotiation, noting, for example, what has *enabled* or *prevented* successful encounters. Robert Crease, a philosopher of science and technology who is also the official historian for Brookhaven National Laboratory, provides an exemplary instance of this type of empirically oriented philosophical research. In "Fallout: Issues in the Study, Treatment, and Reparations of Exposed Marshall Islanders," he examines a failed account of expert-lay negotiation. Crease's essay concerns the actions of U.S. doctors, politicians, and activists, all of whom sought to aid Marshallese inhabitants accidentally exposed to fallout in the wake of a nuclear weapons test in 1954. He thus investigates a classic case of Western intervention in non-Western culture during a period in which a politically volatile climate was conducive to traditional technoscientific experts losing their authority. Crease argues that it would be a mistake to pigeonhole this kind of story into traditional social movement narratives involving victimization or oppression, the civil rights struggle, the struggle against cultural imperialism, or the Tuskegee syphilis experiments, among others. He demonstrates that the only way to explain the distrust that the Western scientists experienced is to concretely examine the context in which specific forms of participation were prohibited.

An example of successful expert-lay interaction suggests the kinds of cases that deserve further study. Theorists such as Steven Epstein, who have written on the AIDS pandemic, noted how people with HIV and AIDS were capable of developing credibility with scientists researching the issue despite being initially marginalized. This expert-lay alliance was hard to forge. It required that activists: (a) learn about the culture of medical science, including not only its dominant assumptions, but also how to speak its language; (b) successfully present themselves as representing a potential clinical-trial subject population (that is, people with HIV or AIDS); (c) provide compelling epistemological, moral, and political arguments; and finally, (d) form strategic alliances with scientists by taking advantage of preexisting personal, political, and epistemological tensions. Ultimately this alliance depended on the creation and maintenance of an interdependent and overlapping discourse. It depended on what Crease calls *impedance matching* between networks of science groups and networks of stakeholders.

As philosophers working in conjunction with the interested and affected constituencies come to inquire empirically into which provisions and circumstances have successfully promoted both better participation and more socially successful technoscience, they will be better placed to address the normative question: Which provisions should be instituted, and under what circumstances, to allow laypeople to have greater legitimate participation in technoscientific affairs? Extrapolating from existing research, it seems likely that successful solutions to the problem of participation will be ones in which theorists refrain from positing an ideal intermediary to serve as an arbitrator between experts and the putatively lay public. Participation therefore remains an important philosophical topic because it is a classic example of how philosophy (in principle at least) can assist in the practice of public affairs.

EVAN M. SELINGER

SEE ALSO *Democracy; Expertise; Georgia Basin Futures Project; Polanyi, Michael; Stakeholders.*

BIBLIOGRAPHY

Barber, Benjamin. (1984). *Strong Democracy: Participatory Politics for a New Age.* Berkeley: University of California Press. Critical assessment of the limits of liberal theory with respect to the possibility of genuine democratic practice.

Borgmann, Albert. (1984). *Technology and the Character of Contemporary Life.* Chicago: University of Chicago Press. Seminal analysis of how modern technological practices tend to undermine participation in skillful endeavors and communal activities.

Crease, Robert. (2003). "Fallout: Issues in the Study, Treatment, and Reparations of Exposed Marshall Islanders." In *Exploring Diversity in the Philosophy of Science and Technology,* eds. Robert Figueroa and Sandra Harding. New York: Routledge. One of the first philosophical analyses of problems associated with the authority of technical expertise in a multi-cultural context that is firmly rooted in empirical history.

Epstein, Steven. (2000). "Democracy, Expertise, and AIDS Treatment Activism." In *Science, Technology, and Democracy,* ed. Daniel Lee Kleinman. Albany: State University of New York Press. One of the definitive works on the sociology of aids activism.

Feenberg, Andrew. (1995). *Alternative Modernity: The Technical Turn in Philosophy and Social Theory.* Berkeley: University of California Press. Definitive demonstration of the value of Frankfurt-school critical theory with respect to discussions of public participation in technical decision-making.

Feyerabend, Paul. (1975). *Against Method.* London: Verso. One of the first monography length critiques of the logical posivist's treatment of scientific inquiry.

Feyerabend, Paul. (1978). *Science in a Free Society.* London: Verso. Highly influential work in which the authority of technical experts is condemned on ideological grounds.

Goldman, Steven. (1992). "No Innovation without Representation: Technological Action in a Democratic Society." In *New Worlds, New Technologies, New Issues,* ed. Stephen H. Cutcliffe, Steven L. Goldman, Manuel Medina, and José Sanmartín. Bethlehem, PA: Lehigh University Press. A succinct account of the fundamentally political character of technical innovation.

Hickman, Larry. (1992). *John Dewey's Pragamatic Technology.* Bloomington: Indiana University Press. The authoritative secondary source on John Dewey's philosophy of technology.

Ihde, Don. (1996). "Why Not Science Critics?" *International Studies in Philosophy* 1: 45–54.

Ihde, Don. (1998). *Expanding Hermeneutics: Visualism in Science.* Evanston: Northwestern University Press. A provocative philosophical attempt to get beyond the epistemic impasse of the expert-lay divide. Includes the 1996 article above.

Kitcher, Phillip. (2001). *Science, Truth, and Democracy.* New York: Oxford University Press. A normative theory of the ideal relation between public participation and scientific inquiry.

Kleinman, Daniel Lee, ed. (2000). *Science, Technology, and Democracy.* Albany: State University of New York Press. One of the few edited volumes on the topic.

Levitt, Norman, and Paul Gross. (1994). "The Perils of Democratizing Science." *The Chronicle of Higher Education* October(5): B1–B2. An important contribution to the contemporary "science wars" from the perspective of natural scientists.

McDermott, John. (2003). "Technology: The Opiate of the Intellectuals, with the Author's 2000 Retrospective." In *Philosophy of Technology: The Technological Condition,* ed. Robert Scharff and Val Dusek. Malden, MA: Blackwell Publishing. An influential critique of technocracy.

Mesthene, Emmanuel. (1970). *Technological Change: Its Impact on Man and Society.* Cambridge, MA: Harvard University Press. An influential defense of technocracy.

Mitcham, Carl. (1997). "Justifying Public Participation in Technical Decision Making." *IEEE Technology and Society Magazine* 16(1): 40–46. Perhaps the clearest presentation of normative arguments for the principle of public involvement in technical decision-making available.

Polanyi, Michael. (1951). *The Logic of Liberty: Reflections and Rejoinders.* Chicago: University of Chicago Press. One of the earliest treatments of the problem of participation in the philosophy of science.

Polanyi, Michael. (1962). "The Republic of Science." *Minerva* 1: 54–73.

Sclove, Richard. (1995). *Democracy and Technology.* New York: Guilford Press. An influential plea for strong democracy in which communities are empowered to participate in decisions concerning the adoption of technologies.

PARTICIPATORY DESIGN

• • •

Participatory design (PD) is an approach to engineering technological systems that seeks to improve them by including future users in the design process. It is motivated primarily by an interest in empowering users, but also by a concern to build systems better suited to user needs. Traditionally, PD has focused on the design of information systems, though the same approach has been applied to other technologies. In order to respect the social contexts in which users work, PD practitioners explicitly consider the practical demands workers must meet in order to do their jobs, as well as the political relationships that exist between workers, their management, and technology designers. As a design subdiscipline, PD directly addresses both technological and ethical issues in the design of systems. Because of this, some people have argued that PD can be used as a model for the "democratization of technology."

History

Participatory design has its roots in northern Europe with the combination of two research programs studying the empowerment of workers with respect to technology. It is generally seen as developing from the Scandinavian "collective resources" research program that focused on union empowerment in contract bargaining situations through the education of union officials and members about various production technologies (Bjerknes, Ehn, and Kyng 1987). The other program, "socio-technical systems design," was pursued primarily by British researchers at the Tavistock Institute and focused on the design of technologies to empower individual workers by enabling and supporting autonomous workgroups (Mumford 1987). Both research programs had in fact grown out of the Norwegian Industrial Democracy Project begun in the 1960s, though the British contribution to PD is often overlooked (Emery and Thorsrud 1976).

The second generation of the Scandinavian approach was marked by the Swedish-Danish UTOPIA project in 1981, the first recognized development project. Conceived in response to the discouraging results of the earlier trade union projects, which had found that existing technologies limited the possibilities of workers to influence workplace organization, UTOPIA targeted technology development as a prospective site for user involvement and influence. In cooperation with the Nordic Graphic Workers Union, the UTOPIA (both an acronym and an ideal) project studied a group of news-paper typographers working without computer support in order to develop a state-of-the-art graphics software product for these skilled graphics workers. The objective was to create a commercial product that the unions could then demand as an alternative to the deskilling technologies available in the market. In doing this, their goals came into alignment with socio-technical systems research. By 1985, the British and Scandinavian traditions had rejoined under a common banner of democratizing technological systems design. The consequence was a new focus on the participation of workers in technological design discussions, and this was to be the essential feature of the PD tradition from that point on (Greenbaum and Kyng 1991).

Politics of Participation

PD has come to be defined by its attempts to involve users in the design of information technologies, and research in the field has examined the various challenges that these attempts have faced. Depending on the different features of the various workplaces that have been engaged, problems of communication, workplace politics, and design politics have received the most attention. The differences in work contexts range across unionized and non-unionized workplaces, democratic and non-democratic countries, small and large organizations, public and private institutions, commercial and non-profit organizations, volunteer and paid workers, and various configurations of labor and management. The design projects also differ in the extent to which they try to use existing or off-the-shelf technologies, as opposed to custom tailored systems. Finally, the roles and responsibilities of design engineers and workers in the process of systems design can vary widely, thus influencing the politics of design.

The principle method used by PD to involve users in design is to have them participate in meetings with design engineers. It is this simple idea that makes this approach "participatory." Participation in this sense is usually taken to mean participation in discussions about a technology, as opposed to actual participation in the construction of a system as engineers or builders. While this might seem simple, it turns out that there are various sorts of problems that arise in these meetings, due mainly to problems of communication between people of differing knowledge and perspectives.

Simply allowing users to sit in on design meetings is insufficient to achieve participation because the politics of both the workplace and the design process can intervene. Sometimes managers are considered to be part of the user group, even though only the workers below them will ever deal directly with the technology in question. The politics of the workplace can then impinge on the process to the extent that managers may resist the participation of low-level workers, or intimidate them in the meetings, or act to discount their authority, skill, and knowledge. Even when managers are not present, the users themselves may not be fully aware of how best to articulate their knowledge of the workplace or what they need and desire from the new technology, or they may underestimate the value of their own skills and knowledge. The politics of the design process often gives engineers, with their expert knowledge, much greater authority in making design decisions. As such, it can be difficult for users to express themselves and not simply defer to the authority of these expert designers. All of these political forces tend to silence the voice of users in the design meetings, and a serious effort must be made to counteract these tendencies.

Design engineers can also find it difficult to communicate effectively with users. Engineers tend to express themselves in technical language, and usually discuss design ideas in terms of nuts-and-bolts internal operations, rather than how a technology relates directly to a user. As such, it can be a daunting task for an engineer to describe design alternatives in a way that users are able understand and respond to them with informed opinions. As a result of these problems, a great deal of energy is expended in PD to create visualizations and mock-ups of proposed systems so that they can be evaluated by users. It is also common to send designers to the workplace to observe users, or even train them to do the work of the users of a proposed system.

Gender poses an additional set of problems to effective participation in design. In many work contexts, the positions traditionally occupied by women are often viewed as being of lower value by management and unions. This undervaluing of women's work easily overflows into inequalities of participation in design activities, especially when combined with social prejudices that view technological design as a masculine pursuit. Moreover, unless gender issues in the design process are recognized and dealt with, there exists a strong possibility of reproducing these gender politics through the technology (Green, Owen, and Pain 1993). Even though PD shares many of its organizational ideals and goals with feminist philosophies and organizations, researchers have found special challenges to utilizing PD in feminist organizations. Ellen Balka (1997) reports that common features of feminist organizations such as decentralized organizational structures, high dependence on volunteer and transient workers, lack of ade-quate funding and resources, and lack of technological training among organization members pose particular problems for implementing PD in these organizations.

Ultimately, PD does not consist of a set of strict rules or methods for how to go about designing systems. Instead, PD prescribes an attitude of including users, encouraging their thoughtful participation, and being sensitive to the political and ethical challenges facing designers. Specifically, it encourages designs that empower users, respects and encourages their skills and job satisfaction, and protects their individual autonomy as much as possible given their jobs and work environment. It also provides case studies and techniques that have worked to varying degrees in various specific design projects as a resource to draw upon in future design projects. Several conferences and journals have brought together the results of many such projects (Bloomberg and Kensing 1998). For more on the politics of representing work, see Liam Bannon's 1995 article, "The Politics of Design."

Democratizing Technology

Some authors, such as Langdon Winner (1995), have proposed that PD stands as an example of a new kind of technological citizenship. Under the current forms of citizenship, there is very little room for individual voices to shape the design of the technologies that permeate society. Private companies driven primarily by commercial interests produce most of these technologies. PD does not offer universal participation, or democratic control over all technologies, but it is argued to be a step in the right direction by allowing some non-commercial values to influence some technologies.

It is crucial to note that arguments such as Winner's hold out a procedural notion of justice as the political ideal. It is the very participation of people in design that is democratic, just as the right of all citizens to vote makes a government democratic. Thus, democratizing technological systems raises many of the same problems facing democratic governmental systems. Just as the people in a democracy are free to elect a tyrant and the majority might use the system to exploit and repress minority groups. It is not clear that universal participation actually leads to a society or technology that is free or empowering. What PD can do is bring designers, users, and the technology itself into a process through which the technology can develop in useful ways.

A more detailed history of PD, its connections to broader social movements such as the quality of working life movement and Total Quality Management, and a consideration of the ethical and political issues it raises can be found in Peter M. Asaro's 2000 article, "Transforming Society by Transforming Technology."

PETER M. ASARO

SEE ALSO *Design Ethics.*

BIBLIOGRAPHY

Agre, Philip. (1995). "From High Tech to Human Tech: Empowerment, Measurement, and Social Studies of Computing." *Computer Supported Cooperative Work* 3(2): 162–195.

Asaro, Peter M. (2000). "Transforming Society by Transforming Technology: The Science and Politics of Participatory Design." *Accounting, Management and Information Technologies*, Special Issue on Critical Studies of Information Practice, 10: 257–290.

Balka, Ellen. (1997). "Participatory Design in Women's Organizations: The Social World of Organizational Structure and Gendered Nature of Expertise." *Gender, Work and Organizations* 4(2): 99–115.

Bannon, Liam. (1995). "The Politics of Design: Representing Work." *Communications of the ACM* 38(9): 66–68.

Bjerknes, Gro; Pelle Ehn; and Morten Kyng, eds. (1987). *Computers and Democracy: A Scandinavian Challenge.* Brookfield, VT: Avebury.

Bloomberg, J., and F. Kensing. (1998). "Special Issue on Participatory Design." *Computer Supported Cooperative Work: The Journal of Collaborative Computing* 7: 3–4.

Ehn, Pelle, and Morten Kyng. (1987). "The Collective Resource Approach to Systems Design." In *Computers and Democracy*, ed. Gro Bjerknes, Pelle Ehn, and Morten Kyng. Brookfield, VT: Avebury.

Emery, Fred, and Einar Thorsrud. (1976). *Democracy at Work: The Report of the Norwegian Industrial Democracy Program.* Leiden, Norway: Martinus Nijhoff.

Feenberg, Andrew. (1991). *Critical Theory of Technology.* New York: Oxford University Press.

Green, Eileen; Jenny Owen; and Den Pain; eds. (1993). *Gendered by Design? Information Technology and Office Systems.* London: Taylor & Francis.

Greenbaum, Joan, and Morten Kyng, eds. (1991). *Design at Work: Cooperative Design of Computer Systems.* Hillsdale, NJ: Erlbaum.

Holtzblatt, Karen, and Hugh R. Beyer, eds. (1995). "Special Issue on Requirements Gathering: The Human Factor." *Communications of the ACM* 38(5).

Mumford, Enid. (1987). "Sociotechnical Systems Design: Evolving Theory and Practice." In *Computers and Democracy*, ed. Gro Bjerknes, Pelle Ehn, and Morten Kyng. Brookfield, VT: Avebury.

Mumford, Enid, and Don Henshall. (1979). *A Participative Approach to Computer Systems Design.* New York: Wiley.

Suchman, Lucy. (1995). "Making Work Visible." *Communications of the ACM* 38(9): 56–64.

Schuler, Douglas, and Aki Namioka (1993). *Participatory Design: Principles and Practices.* Hillsdale, NJ: Erlbaum.

Winner, Langdon. (1995). "Citizen Virtues in a Technological Order." In *Technology and the Politics of Knowledge,* ed. Andrew Feenberg and Alistair Hannay. Bloomington: Indiana University Press.

PASCAL, BLAISE

• • •

Mathematician, physicist, inventor, philosopher, religious thinker, and writer, Blaise Pascal (1623–1662) was born in Clermont-Ferrand, France, on June 19, 1623, the second of three children of Étienne Pascal, a government official and man of wide learning. His mother died in 1626, and in 1631 the family moved to Paris. His exceptional talents evident early on, Pascal was educated entirely by his father, who in 1635 introduced him to Marin Mersenne's newly founded *académie*, where the latest problems in mathematics, science, and philosophy were being discussed. At sixteen he wrote an original work on conic sections. At nineteen he invented a calculating machine, the *Pascaline*, that was awarded an early form of patent; a series of further machines were built and a few have survived. (Some letters were discovered in 1935 and 1956, written in 1623–1624 by the German scientist Wilhelm Schickard, which contained a description and sketch of a mechanical calculator he had developed, and the news that his model was destroyed in a fire.) There is now a programming language called *Pascal.*

Technology, Experiment, Theory

Hearing about Evangelista Torricelli's experiment with the barometer (a glass tube of mercury inverted in a bowl of mercury), Pascal undertook in 1646 to carry out variations of the experiment and then explained the results, showing that atmospheric pressure decreases (the mercury level drops) with increasing altitude. He discovered the basic principle of hydrostatics, *Pascal's Law:* In a fluid at rest in a closed container, a pressure change in one part is transmitted without loss to every portion of the fluid and the walls of the container. (The SI unit of pressure is the *pascal.*) He invented the syringe and the hydraulic press.

These developments had revolutionary impact on scientific thought, as they refuted the Aristotelian doctrine that there is no vacuum. Pascal asserted that in studying nature careful experiment and logical thinking must take precedence over respect for authority (*Preface*

Blaise Pascal, 1623–1662. The French scientist and philosopher was a precocious and influential mathematical writer, a master of the French language, and a great religious philosopher. (*The Library of Congress.*)

to the *Treatise on the Vacuum* [1651]). He gave a detailed exposition of scientific method, with the following thesis: A hypothesis is false if contradicted by a single experimental result, and only possible or probable if all observations are consistent with it (*New Experiments Concerning the Vacuum* [1647]).

A 1654 correspondence of Pascal with the mathematician Pierre de Fermat (1601–1665) concerning a gambling problem marked the birth of probability theory, the study of patterns of chance events and the formulation of laws governing random variation. Pascal solved the problem by means of the arithmetic triangle, a numeric structure that now bears his name, and in the process introduced the binomial distribution for equal chances and the method of mathematical induction (reasoning by recurrences) applied to expectations. In his studies of the cycloid, the curve traced by a point on a circle that rolls along a straight line, he anticipated the calculus. His 1658 treatise *On Geometrical Demonstration* shows that he was also far ahead of his time in recognizing the importance of the axiomatic method in mathematics.

Religion and Decision

At the forefront of science in technology, experiment, and theory, Pascal was drawn to religion in 1646, when his family came in contact with Jansenism, an austere Catholic movement with its center at Port-Royal, near Paris. On November 23, 1654, he had a profound religious experience that became the dominant force in his life; a parchment record of it, called *Pascal's Mémorial*, was found sewn into his coat after his death. He never formally joined the Jansenist community, but Port-Royal was henceforth his spiritual home. His *Provincial Letters* (1656–1657) were written in defense of a Jansenist theologian accused of heresy, and as such are mainly of historical interest. Their enduring popularity rests with the brilliance of Pascal's style, which set the tone for the development of modern French prose.

But Pascal is best known for his *Pensées*, a collection of nearly 1,000 fragments of writing for a projected defense of Christianity. With an incisive portrayal of the human condition in all its glory and misery, Pascal explores the limits of reason and the hope offered by faith in revelation. Especially famous is the fragment known as "Pascal's Wager," intriguing, but often misunderstood.

Pascal introduces mathematical concepts to address a theological issue. The question of God's existence is to be answered as if by the toss of a coin at the end of life. By analogy with a game of chance, Pascal presents an existential dilemma that calls for a decision, and in his approach foreshadows modern existentialist thought. In an imagined dialogue with a worldly skeptic, he employs what are key elements of decision theory, a product of the twentieth century concerning courses of action in the face of uncertainty.

Pascal proposes betting on God's existence and acting accordingly. If God exists, the gain will be eternal bliss in the hereafter—infinite gain for a finite risk. But he goes further, on the theory that practice yields insight. He submits that even if God does not exist, the rewards of a life of virtue will lead to the realization that nothing has been risked. At the end of the "Wager," Pascal explains that the arguments he used were inspired by his own faith, his passionate desire to show others the way.

In frail health from childhood, Pascal was, in his final years, too ill for sustained intellectual effort. He gave his belongings to the poor. In the last year of his life, he designed and inaugurated the first public transportation service, leaving the proceeds to charity. He died in Paris on August 19, 1662.

Pascal is a major figure in the history of ideas because of the range and intensity of his interests and his thought-provoking response to the uncertainties revealed in the expanding world of the seventeenth century. He accepted the skeptic's view that it is impossible to prove first principles. But he stressed the role of intuition, and across the spectrum of human experience pleaded for the full use of reason as the ethical norm: "Man is only a reed, the weakest in nature, but he is a thinking reed.... All our dignity consists in thought.... Let us then strive to think well; that is the basic principle of morality" (Pascal 1995, p. 66).

VALERIE MIKÉ

SEE ALSO *French Perspectives; Scientific Revolution; Virtue Ethics.*

BIBLIOGRAPHY

Davidson, Hugh M. (1983). *Blaise Pascal.* Boston: Twayne Publishers. Short general biography in Twayne's World Authors Series.

Guardini, Romano. (1966). *Pascal for Our Time,* trans. Brian Thompson. New York: Herder and Herder. Analysis of Pascal's life and work from a broad cultural and psychological perspective by a leading philosopher of religion and literature interpreter of the twentieth century.

Mesnard, Jean. (1952). *Pascal, His Life and Works,* trans. G. S. Fraser. New York: Philosophical Library, Inc. Classic biography by a Pascal scholar and editor.

Miké, Valerie. (2000). "Seeking the Truth in a World of Chance." *Technology in Society* 22: 353–360. Discusses the relevance of Pascal to problems of contemporary culture.

Pascal, Blaise. (1952). *The Provincial Letters; Pensées; Scientific Treatises.* In *Great Books of the Western World,* Vol. 33. Chicago: Encyclopedia Britannica. Contains six major scientific treatises, including those cited in the article, and the extant correspondence with Fermat on the theory of probability.

Pascal, Blaise. (1995). *Pensées* [Thoughts], trans. A. J. Krailsheimer. London: Penguin Books. Fine modern English translation, with introduction by the translator.

Pascal, Blaise. (2000 [1998]). *Oeuvres Complètes* [Complete Works], 2 vols., ed. Michel Le Guern. Paris: Gallimard. New edition of Pascal's complete works.

PAULING, LINUS

• • •

Linus Carl Pauling (1901–1994) was born in Portland, Oregon, on February 28, and his two Nobel Prizes symbolize his contributions to science and ethics: His Nobel Prize in Chemistry (1954) was awarded for his research

Linus Pauling, 1901–1994. The American chemist was twice the recipient of a Nobel Prize. He revealed the nature of the chemical bond, helped to create the field of molecular biology, founded the science of ortho-molecular medicine, and was an activist for peace. (*The Library of Congress.*)

on the chemical bond and the structures of complex molecules, and his Nobel Peace Prize (1962 but awarded in 1963) was given for his campaign to halt the atmospheric testing of nuclear weapons. Pauling's early life was spent in Oregon, where he received a bachelor's degree from Oregon Agricultural College and met Ava Helen Miller, his future wife, who would have an important influence on his ethical development. Pauling's education continued at the California Institute of Technology, from which he received a doctorate in 1925.

In the first two decades of his career Pauling made significant contributions to structural chemistry that included determining the structures of many molecules by using the techniques of X-ray and electron diffraction. He also developed a theory of the chemical bond based on the new field of quantum mechanics. In the 1930s he became interested in hemoglobin and antibody molecules. Pauling was conventionally patriotic during World War II, and for his military contributions, such as an oxygen meter widely used in submarines and airplanes, he was given the Presidential Medal for Meritin 1948.

Because of the development of nuclear weapons during the war Pauling, like many other scientists, became sensitive to the ethical consequences of scientific discoveries. At the urging of his wife, he included attacks on war and pleas for peace in his public speeches. After winning the Nobel Prize for Chemistry he began to use his increased prestige to convince people that nuclear testing was immoral because it caused birth defects and cancer. In the late 1950s Pauling became increasingly involved in the debate over nuclear fallout, especially through the Scientists' Bomb-Test Appeal, which he wrote and helped circulate. That appeal, along with his lawsuits and other activities, helped bring about the partial test-ban treaty of 1963. When the treaty went into effect, Pauling received the news that he had won the Nobel Peace Prize.

In the final decades of his life Pauling founded the new field of orthomolecular medicine to investigate the connection between good health and the proper proportion of various molecules, especially vitamins, in the body. That advocacy had an ethical component because Pauling felt that it was immoral for researchers and government agencies to keep that knowledge from the public, whose suffering could be minimized and whose health could be maximized by the correct intake of different vitamins. Like his stand against nuclear testing, Pauling's campaign for megavitamin therapy was controversial; many nutritionists believed that a balanced diet without vitamin supplementation was sufficient for good health. Ironically, both Ava Helen and Linus Pauling died—she in 1981 and he thirteen years later on August 19—of cancer despite their hope that their high vitamin intake would help them avoid that disease. Pauling died at his ranch on the Big Sur coast of California.

Orthomolecular medicine has enthusiastic advocates and opponents, but Pauling's contributions to science are incontrovertible. His discoveries in structural chemistry, molecular biology, and molecular medicine have been called the most illuminating body of scientific work of the twentieth century. His crusade for the nuclear test ban has resulted in smaller amounts of radioactive materials in the environment, with a consequent improvement in the health of many people. Finally, his example as an activist scientist inspired many others to use their scientific knowledge for the betterment of humanity.

ROBERT J. PARADOWSKI

SEE ALSO *Weapons of Mass Destruction.*

BIBLIOGRAPHY

Divine, Robert A. (1978). *Blowing on the Wind: The Nuclear Test Ban Debate, 1954–1960.* New York: Oxford University Press. Though concerned with the arguments of such participants as Linus Pauling and Edward Teller, Divine emphasizes the scientific and ethical complexities of this turbulent public debate.

Hager, Thomas. (1995). *Force of Nature: The Life of Linus Pauling.* New York: Simon & Schuster. The most detailed of the several Pauling biographies.

Pauling, Linus. (1983). *No More War!* (25th Anniversary Edition). New York: Dodd, Mead. The book that best reflects Pauling's views on war and peace in general and on the test ban debate in particular.

PEER REVIEW

• • •

Peer review is a term of art covering a set of practices that collect and apply the judgement of expert reviewers (identified as "expert," not just "knowledgeable"—so the designation is a political justification as well as a substantive one) to decisions about which manuscripts to publish, which proposals to fund, and which programs to sustain or trim. Peer review and its variants are preferred in science not only because they bring appropriate expertise to bear on decisions, but also because they assert the professional autonomy of scientists. The review of original ideas grounded in acceptable evidence certifies the accuracy, validity, and heuristic value of results. Peer reviewers are collegial critics who contribute uniquely to this competitive negotiation process by allocating scarce resources—money, time, space—and the career capital they help to generate. Outcomes based on peer review thus concentrate or disperse available resources over a pool of eligible competitors, advancing collective knowledge and practice, on the one hand, and individual careers, on the other.

Although peer review is a highly valued process, it nevertheless lacks careful or rigid definition. What constitutes a "peer" may be disputed, the factors to be considered by reviewers may vary, and the weights accorded their judgments are likely to be unequal. Moreover, there is probably an inverse relationship between knowledge and conflict of interest: the smaller a circle of peers the more sound and nuanced their knowledge of an area, but the more likely that these peers are friends or maintain potentially compromising relations with those being reviewed. How those relations are restrained to preserve balanced judgment is a challenge to peer review procedures.

For example, there are no hard-and-fast rules about how a peer reviewed journal, versus a non–peer reviewed journal, will decide what to publish. The former, however, are valued for the presumed standards of rigor and fairness that often carry scientific and academic prestige. Likewise, some peer review processes for grants are blind to reviewer and reviewed (proposer) alike, others only to the reviewed. Reviewers may vary widely in number and characteristics (demographic, intellectual, national, or organizational context) and may shade reviews in unanticipated ways. Finally, the collective judgment represented by peer review is sometimes deemed unassailable, often just advisory. Corporations and government agencies also employ peer review—internally or with external reviewers mixed in—to assess the quality of the science destined for reports, decisions, or policy recommendations. Peer review in scientific and technological contexts has been most subject to analysis, but also diminished in level of detail by preserving the anonymity and confidentiality promised by editors and agency stewards.

Origins and Purposes

Peer review of scientific manuscripts dates back to the *Philosophical Transactions of the Royal Society* of the mid-seventeenth century. The origins of peer review for grants are more recent and murky. The National Advisory Cancer Council, established in 1937, was authorized to review applications for funding and to "certify approval" to the surgeon general. The Office of Naval Research developed an informal variant of peer review, which may have been brought to the National Science Foundation (NSF) when Alan T. Waterman became its first director. Peer review is not mentioned in the NSF founding legislation, but the agency is known as its foremost practitioner (England 1983). The more widespread development and application of peer review processes has occurred episodically since the 1960s.

Understanding peer review requires reflection on both its purposes and values. Peer review circulates research ideas in their formative stages to key gatekeepers in a field. Sometimes this signals others to avoid duplication of effort. Other times it calls attention to a problem that is promising, attracting other researchers and setting off a race for priority (for example, work on cancer genes). Thus, by the time new research is finally published, aspects of its findings and methods may be generally familiar to many in the field, speeding its acceptance and utilization while drawing constructive criticism.

Peer review may also bring values beyond scientific or technical quality to research funding decisions. These values may be overriding or subtle, and they relate ways in which peer review is grounded in a democratic context. History attests to the political contamination of science and other forms of malpractice, such as Nazi attempts to control science in Germany or the manipulation of genetics in the Soviet Union by Trofim Lysenko (1898–1976) (Chubin 1985). Indeed after the cold war, many postsocialist countries sought to replicate peer review practices used in the West. In most cases, a system of government distribution of research funds was sought that favored quality of ideas over professional stature alone. In contrast, during the same period, the United States fine-tuned its peer review practices to achieve other goals. For example, NSF program officers try to balance their portfolios by taking account of geographic distribution, age, gender, or ethnicity of investigators; research participation of four-year colleges or historically black colleges and universities; or the hotness of a topic or method. At the National Institutes of Health (NIH), advisory councils are empowered to recommend some proposals for funding because they address urgent national needs (U.S. GAO 1999).

With the Government Performance and Results Act of 1993 requiring U.S. research agencies to show that their investments yield societal benefits, some wonder if scientific experts are the best qualified reviewers to render such judgments (NAPA 2001). At NSF, reviewers now must address two merit review criteria: scientific merit and broader social impact (two other criteria were dropped because they were routinely ignored or deemed too difficult to measure). The latter encompasses educational benefits ranging from precollege outreach to increased participation of students from underrepresented groups and enlarged undergraduate research experiences to ways to enhance public understanding of the scientific content of workaday processes and outcomes.

A relatively recent innovation allows more direct citizen participation in scientific and technical allocation decisions. The Dutch Technology Foundation, for example, has augmented traditional peer review with lay review by citizens. In the United States, activist and support groups for various diseases have applied similar pressure, especially at NIH (which uses a quantitative scoring system that leaves little room for study section or institute director discretion). Other federal agencies, such as the Office of Naval Research, the Defense Applied Research Projects Agency, and parts of the Department of Energy and the Department of

Agriculture, limit their use of external peer reviewers to the identification of more risky but potentially highly rewarding areas of research and development. In the end, who participates in the process redefines *peer* and alters the purpose of the review.

Peer review allows scientists to make recommendations in a privileged zone, apart from the general public. It creates the expectation that the principles of fair and ethical behavior embedded in professional culture will be observed. This may seem inconsistent with the principle of public participation, but should be understood as reflecting the role of peer review as a boundary process that demarcates the limits of authority based on credentials or power. When participation crosses borders, participants carry the distinctive characteristics of their professional region (Gieryn 1999, Guston 2000). A good review system thus preserves professional autonomy while permitting lay participation. This balances deference to expert evaluation against sensitivity to societal needs and extrascientific values (concerning research applications, risks and benefits to whom, and long- versus short-term consequences) (Atkinson and Blanpied 1985).

Ethical Dimensions

Precisely because peer review is a highly valued process that spans the boundaries of several social worlds—science and policy, research and practice, academe and bureaucracy, public and private—its purposes and meaning may be understood differently across communities and at different times in the history of a single community. Focusing primarily on peer review as a process for managing scientific publication and grant funding, what follows is a brief review of some of the value and ethics-related dimensions that often manifest themselves as competing understandings and aspirations. (For elaboration, see Chubin and Hackett 1990.)

OPENNESS AND SECRECY. Peer review is in principle open to the community of qualified scientists as proposers or reviewers. The process of peer review, as procedures, criteria, rating scales, and such, is knowable, transparent (or at least translucent), and held to account for its workings and outcomes. But the criteria are themselves seldom discussed.

Peer review is also secret. Confidentiality is sacrosanct, and anonymity is assured throughout much of the process. Meetings are typically closed, with proposals, reviews, and panel discussions deemed privileged information. To outsiders, who participates and how they are chosen can seem mysterious, and the identities of the reviewers—who represent the intellectual community-at-large—are generally not disclosed.

EFFECTIVENESS AND EFFICIENCY. Peer review is asked to be effective—to recommend projects that would advance knowledge and confer social benefit. But it is also asked to be efficient, to operate at low cost (e.g., for travel and reviewer compensation) and minimize the burden imposed on proposal writers and reviewers alike.

How realistic are these expectations? A thorough review might take half a day, but reviewers are usually not paid for their services. Of course, the reviewer is partly compensated by learning what constitutes a fundable proposal and gaining access to unpublished ideas and data.

Nonetheless, a low success rate—10 to 20 percent in many agencies these days—reduces the expected return (to proposers and agencies) for the investment of effort. Hence the invention of a two-stage proposal process with the first a preliminary proposal that can be screened into or out of the more competitive second stage.

SENSITIVITY AND SELECTIVITY. The peer review system is asked to be highly sensitive and highly selective of research projects at the same time. A sensitive review system would detect the merit in every worthwhile proposal, whereas a selective system would filter out all projects of dubious quality or significance.

But scientific research can be risky, and given the difficulties in communicating original ideas clearly and persuasively, it is possible that the phenomenon of interest may itself be in question (e.g., the Higgs process, the top quark, prions). A system acutely sensitive to scientific merit would probably support some projects that do not work out. One so selective that only projects beyond skepticism are chosen for funding would surely ignore some good ideas along with the rest. And inevitably, some researchers write better than others. Still others construct better proposals than conduct the research once funded. What is the review rewarding?

INNOVATION AND TRADITION. Peer review couples what Thomas Kuhn (1977) terms an "essential tension" between originality and tradition in science with what Robert Merton (1973 [1942]) defines as the norm of "organized skepticism." Promising new ideas are tested against the cumulative store of shared knowledge and established theory. Peer review challenges whether new ideas are truly novel and worth pursuing, and purports to distinguish between sound innovation and reckless speculation.

Reviewers defend tradition against claims of originality when they reject novel ideas as impractical, unworkable, or implausibly inconsistent with the established body of knowledge. Sharp disagreements among reviewers about the merits of an idea may indicate a promising but risky new research path. Consensus, in contrast, might indicate an insufficiency of important problems left to solve, the grip of a school of thought, an overbearing conservatism, or just plain risk-aversion.

An innovative review system would reward novelty and risk taking, whereas a traditional system would sustain the research trajectory established in the body of accepted knowledge by restraining bold excursions. Peer review is expected to identify, encourage, and support frontier work but to screen out fads and premature ideas (Stent 1972).

MERIT AND FAIRNESS. Peer review is expected to be meritocratic, judging proposals and manuscripts in accordance with the stated criteria. NIH instructs proposal reviewers to evaluate all the science, only the science, and nothing but the science. The rendered judgment is to extract the science from speculation, rhetoric, common sense, practical benefit, and whatever else the proposer orchestrated in the document.

Peer review is reputed to apply standards of fairness to ideas apart from consideration of a scientist's reputation, personal characteristics, or geographic or academic position; the economic potential of the proposed work; or its relevance to pressing national needs. Nevertheless, advantages accumulate over the course of a career, making it increasingly difficult to judge what one does apart from who one is (or has accomplished). In this way, the Matthew Effect prevails: In recognition and influence the rich get richer, the poor poorer (Merton 1973 [1968]).

It may thus be wrongheaded to assume that the best science simultaneously serves one's career, one's discipline, and the welfare of the nation. Just as the principle of equitable distribution might indicate that decisions at the margin should favor investigators who currently have inadequate funds, similar arguments could be advanced for criteria such as growing research capacity, increasing educational or economic investments, or making politically savvy allocations. Such decisions deviate from strictly meritocratic principles, yet are entertained by participants much of the time, leading to charges of earmarking, log-rolling, cronyism, and elitism (U.S. Congress 1991, Chubin 1990).

RELIABILITY AND VALIDITY. As an assessment tool, peer review must be both reliable and valid, that is, have little random error and measure what it is supposed to measure. To be reliable, ratings should show high levels of agreement between raters and consistency from one group of raters to another. To be valid, a measure must take account of the scientific merit of a proposal in all its complexity without becoming distorted by other properties of the proposal. But merit is both abstract and multifaceted. A valid evaluation of a proposal, therefore, is said to derive from the combined assessments of several diverse experts. How their reviews are weighed depends on the steward—the program manager or journal editor—and the mission that he or she serves.

Evaluating a proposal or manuscript from several divergent perspectives, not surprisingly, may yield low inter-rater agreement; different experts reach different judgments about quality as seen through their particular set of cognitive lenses (Cicchetti 1991, Harnad 1982). In this sense, peer review builds sound inferences upon a broad foundation. Given the limited number of reviews that can be elicited for any one proposal and the range of reviewer backgrounds necessary to cover the intellectual content of the proposal, divergent recommendations can result. Stewards and editors act on those recommendations when they decide whether or not to fund or publish (or to defer a decision until a revision addresses criticisms).

Conclusions

Clearly, peer review does many things and serves many values, but it cannot simultaneously deliver on all things equally well. Which purposes and which values are most important for which sorts of science? Who is to decide?

Similarly, involving the best researchers in the review process probably leads to better and more legitimate reviews—those that will be accepted by the community. But such experts are also the most likely proposal writers. Because it is unwise to allow people to review proposals for a competition in which they are also contestants, strategies for handling such conflicts of interest must be accepted by the community, or the legitimacy of the process will erode.

Because peer review sometimes can straddle disciplines, it may also cross the boundaries of knowledge production and professional practice, of research and policy. At one extreme, it will be highly particularistic by restricting the competitors to those with certain characteristics (through what is known as set-asides by age, gender, discipline, prior accomplishment, or location at an institution with a track record or facility to conduct the research). At the other extreme, peer review will be highly universalistic, resembling a lottery with the criteria of choice seemingly random and unrelated to properties of the chosen projects. In practice, review processes fall between these polar extremes, which competitors usually find to be fair and the outcome justified enough so they try again even after an unsuccessful submission.

Developing a review process that has widespread legitimacy entails building responsibilities, relationships, and trust. Together, these qualities add research findings to a body of knowledge, introduce conjectures into theories, and socialize researchers into a community that has moral as well as intellectual authority. In the end, peer review is expected to demand rigor and integrity, while stimulating new knowledge that ultimately makes a difference in people's lives. To do so, it must be responsive to emerging needs and possibilities. Ultimately, the flexibility of human judgment and the quality of collective imagination will determine which values and purposes are served by peer review.

DARYL E. CHUBIN
EDWARD J. HACKETT

SEE ALSO *Accountability in Research; Expertise.*

BIBLIOGRAPHY

Atkinson, Richard C., and William A. Blanpied. (1985). "Peer Review and the Public Interest." *Issues in Science and Technology* 1(4): 101–114.

Chubin, Daryl E. (1985). "Open Science and Closed Science." *Science, Technology, and Human Values* 10 (Spring): 73–81.

Chubin, Daryl E. (1990). "Scientific Malpractice and the Contemporary Politics of Knowledge." In *Theories of Science in Society*, ed. Susan E. Cozzens and Thomas F. Gieryn. Bloomington: Indiana University Press.

Chubin, Daryl E., and Edward J. Hackett. (1990). *Peerless Science: Peer Review and U.S. Science Policy.* Albany: State University of New York Press.

Cicchetti, Dominic V. (1991). "The Reliability of Peer Review for Manuscript and Grant Submissions: A Cross-Disciplinary Investigation." *Behavioral and Brain Sciences* 14(1): 119–186.

England, J. Merton. (1983). *A Patron for Pure Science: The National Science Foundation's Formative Years, 1945–57.* Washington, DC: National Science Foundation.

Gieryn, Thomas F. (1999). *Cultural Boundaries of Science: Credibility on the Line.* Chicago: University of Chicago Press.

Guston, David H. (2000). *Between Politics and Science: Assuring the Productivity and Integrity of Research.* London: Cambridge University Press.

Harnad, Stevan, ed. (1982). *Peer Commentary on Peer Review: A Case Study in Scientific Quality Control.* Cambridge, UK: Cambridge University Press.

Kuhn, Thomas S. (1977). *The Essential Tension.* Chicago: University of Chicago Press.

Merton, Robert K. (1973 [1942]). "The Norms of Science." Reprinted in *The Sociology of Science*, ed. Norman W. Storer. Chicago: University of Chicago Press.

Merton, Robert K. (1973 [1968]). "The Matthew Effect in Science." Reprinted in *The Sociology of Science*, ed. Norman W. Storer. Chicago: University of Chicago Press.

National Academy of Public Administration (NAPA). (2001). *A Study of the National Science Foundation's Criteria for Project Selection.* Washington, DC: Author.

Stent, Gunther S. (1972). "Prematurity and Uniqueness in Scientific Discovery." *Scientific American* 227(6): 84–93.

U.S. Congress. Office of Technology Assessment. (1991). *Federally Funded Research: Decisions for a Decade.* Washington, DC: U.S. Government Printing Office.

U.S. General Accounting Office (GAO). (1999). *Federal Research: Peer Review Practices at Federal Science Agencies Vary.* Washington, DC: Author.

PEIRCE, CHARLES SANDERS

• • •

Charles Sanders Peirce (1839–1914), pronounced "purse," was born in Cambridge, Massachusetts on September 10, and died in Milford, Pennsylvania on April 19. In the year of his birth, the first electric clock was built, ozone was discovered, and the growth of cells was charted, while the year of his death saw Robert H. Goddard (1882–1945) inaugurate his rocket experiments and J. H. Jeans (1877–1946) publish a paper on "Radiation and Quantum Theory." Peirce graduated from Harvard College in 1859, the year English naturalist Charles Darwin's (1809–1882) *On the Origin of Species* appeared. Peirce's life was thus framed by significant scientific and technological developments; its fruits included a multifaceted contribution to early twenty-first century philosophical understanding of scientific investigation and other human achievements. Trained as an experimental scientist, Peirce worked in this capacity for both the Harvard College Observatory and the U.S. Coast and Geodesic Survey. His contribution, however, was far more that of a philosopher than a scientist.

Philosopher of Semiotics and of Science

Peirce is best known in philosophy as the founder of pragmatism and, outside that discipline, as the theorist who, at roughly the same time as the Swiss linguist Ferdinand de Saussure (1857–1913), envisioned a comprehensive study of signs. But Peirce did far more than envision the possibility of such an investigation: He systematically elaborated, yet left ultimately unfinished, a theory of signs designed to provide indispensable resources for a normative account of objective inquiry and, beyond this, for a systematic analysis of the myriad forms of meaning—not just those observable in the practices of experimental or objective investigators. Saussure coined the word *semiologie* to designate this study, whereas Peirce used the term *semeiotics* (now more commonly spelled *semiotics*).

But the scope of Peirce's concerns is inadequately conveyed by calling attention to his role in the founding of pragmatism and semiotics. He tended to identify himself as a logician, but he vastly expanded the scope of logic. Moreover, he devoted considerable energy to defending an evolutionary cosmology informed by the monumental achievements of such classical metaphysicians as Plato, Aristotle, and Friedrich Schelling as well as by what he judged to be the most important implications of the greatest scientific discoveries of his own day.

While Peirce devoted a great deal of his intellectual energy to an understanding of science, he tended to ignore questions specifically concerning technology. This might seem ironic, given his pragmatic commitments. He tended, however, to draw a sharp distinction between theory and practice. He believed in a strict division of intellectual labor and that the very best work required a steadfast concern with a more or less delimited object of investigation. However, he conceived theory itself to be a historically evolved and evolving practice (or, more accurately, a family of such practices). Indeed, Peirce was keenly interested in preserving the integrity of theoretical practices, defining them ultimately in terms of the objective of simply discovering truths not yet known. At the heart of his pragmatism, then, one finds not only a refusal to subordinate theoretical practices to other forms of practices but also an insistence that theory itself is a unique form of human practice.

Peirce's account of science is distinguished by a number of factors, but most importantly by the role he accords abduction in the conduct of inquirers and the attention he pays to the history of science as a resource for understanding science. He identified abduction as one of the three modes of inference (deduction and induction being the other two). Abduction is that mode

Charles Sanders Peirce, 1839–1914. Peirce, one of America's most important philosophers, made important contributions in both philosophy and science. His work in logic helped establish the philosophical school of thought known as pragmatism.

by which hypotheses are formulated or initiated. In classifying it as a form of inference, Peirce was refusing to leave the formulation of hypotheses as a mysterious, psychological process. The work of scientists involves the complex interplay of all three modes of inference, but abduction is clearly central to this work. Long before Thomas Kuhn's *The Structure of Scientific Revolutions* (1962), Peirce was acutely aware of how an adequate conception of science must be based upon a detailed acquaintance with the actual development of diverse experimental practices. Such acquaintance reveals the intimate relationship between theoretical discoveries and technological innovations. Thus, whereas Peirce did not make technology in general a focal object of his theoretical concern, he did devote attention to how technology operates *within* science.

The Normative Sciences

Somewhat late in his life Peirce came to an appreciation of the importance of what he called the normative sciences (logic, ethics, and esthetics) and, within this cluster of sciences and his broader classification of human practices, an appreciation of the pivotal role of ethics as both a cultural inheritance and a normative science. He came to see logic as a species of ethics. Whereas ethics offers a normative account of self-controlled conduct, logic provides a normative account of a species of such conduct, namely, self-controlled thought or inquiry. Just as logic in this sense depends upon a more general theory of self-controlled action, so ethics depends upon a critical theory of the intrinsically admirable or worthwhile ends of action. Peirce proposed *esthetics* as the name for this theory of the ends of action. A critical determination of the ends one espouses is at least as important as a critical assessment of the variable means available for the realization of a given objective.

Peirce's historically informed understanding of experimental inquiry is, arguably, one of the most complete, nuanced, and adequate accounts of science yet articulated. The centrality he accords to abduction distinguishes his account of science from most others and, in addition, more intimately connects his theoretical understanding of scientific investigation to the actual practices of scientific investigators than do rival accounts. Though he did not specifically concern himself with technology, his philosophy of science and theory of signs provide resources for illuminating numerous aspects of the diverse phenomena studied by philosophers of technology and others interested in such phenomena. His classification of the theoretical sciences is, in fact, embedded in a more comprehensive classification of human practices; this classification offers important suggestions for how to understand the relationships between the theoretical and technological undertakings of humankind.

Finally, even though he did not explore ethics or esthetics as deeply as he studied logic, his general conception of the normative sciences and his specific treatments of ethics and esthetics are sites yet to be mined by contemporary inquirers, especially ones interested in the interconnections among science, technology, and ethics.

VINCENT COLAPIETRO

SEE ALSO *Pragmatism; Semiotics.*

BIBLIOGRAPHY

Anderson, Douglas R. (1995). *Strands of System: The Philosophy of Charles Peirce.* West Lafayette, IN: Purdue University Press.

Eisele, Carolyn, ed. (1985). *Historical Perspectives on Peirce's Logic of Science: A History of Science.* New York: Mouton.

Kuhn, Thomas. (1962). *The Structure of Scientific Revolutions.* Chicago: University of Chicago Press.

Peirce, Charles S. (1957). *Essays in the Philosophy of Science.* New York: Liberal Arts Press.

Peirce, Charles S. (1998). *Chance, Love, and Logic; Philosophical Essays,* ed. Morris R. Cohen. Lincoln: University of Nebraska Press.

Peirce, Charles S. (1998). *Charles S. Peirce: The Essential Writings,* ed. Edward C. Moore.

Peirce, Charles S. (1998). *His Glassy Essence: An Autobiography of Charles Sanders Pierce,* ed. Kenneth Laine Ketner. Nashville: Vanderbilt University Press.

PERSISTENT VEGETATIVE STATE

• • •

Persistent vegetative state (PVS) was identified by that name in 1972 by the neurologists Bryan Jennett and Fred Plum (Jennett and Plum 1972). Both the name and the state have been a source of controversy since that time.

General Description

PVS results from the total lack of function of the cerebral cortex, the large outer part of the human brain. The size of the cortex in different species of vertebrates correlates with their respective levels of intelligence, with primates having the largest cortex among all genera and humans having the largest among all primates. Cortical activity is necessary for all types of cognitive states, from sight and hearing to speech and thought. The most common causes of loss of cortical function are traumatic injuries and anoxic-ischemic injuries. Traumatic injuries include those seen in car or motorcycle accidents, and anoxic-ischemic injuries include those seen in strokes, drowning accidents, and cardiac arrest, in which there is a loss of oxygen (anoxia) or blood flow (ischemia) to the brain. Either cause can lead to the same outcome, but because that outcome occurs by different routes, there are some distinctions in the diagnostic criteria.

Whether the origin of a brain injury is traumatic or anoxic-ischemic, the initial result of a severe injury is a coma. Patients in a coma look as if they were asleep, although they never open their eyes or have sleep-wake cycles. In fact, they are not in a sleeplike state but are deeply unconscious, as is evidenced by the fact that they cannot be awakened by even the most painful stimuli and do not exhibit reflex responses to such stimuli. However, comas are usually a temporary stage of response to injury. Generally a patient is in a coma for no more than two weeks. After that time coma patients progress to one of three alternatives: They regain consciousness, die (most commonly as a result of swelling of the brain that causes herniation of the brain stem and loss of brain stem function), or enter a vegetative state.

Some patients improve after emerging into a vegetative state. They subsequently may regain a normal level of consciousness or improve slightly and enter a minimally conscious state. However, the longer they remain in a vegetative state, the less likely it is that they will ever improve. Thus, a PVS is defined as having been in a vegetative state for a length of time that makes further improvement highly unlikely. If the cause is anoxic-ischemic, in which case there is a fairly uniform causal pattern of neural death, one needs to wait three months to make the diagnosis. If the cause is trauma, which has greater variability of intermediate causes of neural death, one needs to wait a year to achieve the same degree of certainty and thus make a diagnosis of PVS. The exact location of the blow, the degree of force, and even factors such as the condition of the brain and the skull at impact can be variables in the degree of brain damage.

The Concept of Vegetative

Why is the term *vegetative* used? In the classic terminology dating back to Aristotle humans are defined as uniquely rational, with emotional (or irrational) traits being shared with animals. Purely physiological functions such as digestion are called vegetative; they are neither rational nor irrational, and they have nothing to do with social interaction at any level. It is only these physiological functions that are preserved in patients in whom the brain stem is the only surviving part of the brain.

Therefore, in contrast to cases of death diagnosed by brain criteria, the vegetative state is characterized by the presence of all brain stem functions (autonomic nervous system regulation of body temperature, pulse, blood pressure, breathing, reflexes, and sleep-wake cycles) without any of the cortical functions. Thus, most or all brain stem reflexes typically are intact in PVS patients: cold calorics (cold water in the ear canal causes lateral eye movement toward that ear), papillary (response to light), corneal (light tough of the eyeball causes a blink), threat (a quickly approaching object causes blinking), gag, and painful stimuli (usually a sternal rub or pressure on the fingernail beds causes withdrawal). For all these reasons the verbal slip of calling a PVS patient brain-dead is a mistake that threatens family members' trust in doctors and other health-care professionals.

Although the definition of PVS is made clinically, that is, empirically, it is possible to use neuroimaging techniques such as computed tomography (CT) scans, functional computed tomography (FCT) scans, and positron emission tomography (PET) scans to build confidence in the prognosis at an earlier time. In cases such as an observed loss of oxygen for thirty minutes or when there is a loss of cortex replaced by cerebrospinal fluid that is documented on a CT scan, experienced neurologists may feel confident in making the diagnosis of PVS in less than the three or twelve months recommended in the American Neurological Association Task Force report (American Neurological Association 1993). For some families that do not want to wait, this can be very helpful. However, others may feel rushed and may become skeptical if they discover that the neurologist is making a diagnosis sooner than is recommended in the consensus statement.

Causes of PVS

The largest numbers of cases of PVS are caused by anoxic-ischemic injuries, and this diagnosis has increased in frequency. This is the case because it takes only four or five minutes without oxygen for a patient to begin to have permanent brain damage in the cortex, which requires very large amounts of oxygen. However, the inner parts of the brain, the brain stem and midbrain, require less oxygen and can return to function after much longer periods of oxygen deprivation. (One might picture the cortex as a softball wrapped around the golf ball–sized brain stem.) Thus, anything that causes the loss of some or all oxygen to the brain for more than five minutes may lead to PVS. The most common cause of that loss occurs when a patient "codes," that is, when the patient's heartbeat or breathing stops.

Why is this cause the source of a growing number of cases of PVS? In the United States and many other countries after the invention of cardiopulmonary resuscitation, it became routine for all patients to be "full code" unless they specifically requested otherwise. When a patient is discovered unconscious as a result of acute loss of cardiac or pulmonary function, a "code" is begun, starting with clearing the airway and beginning chest compressions and ending with cardioversion/defibrillation and endotracheal intubation and mechanical ventilation. The code ends either when a heartbeat is restored or when the physician who is running the code decides to "call" it (that is, to call an end to the code), which will be the time when death is declared.

A code typically is run for thirty to forty-five minutes. However, it is up to the physician, using clinical judgment, to determine how long to wait before calling, or ending, a code. In light of the nature of the brain, if a pulse does not return after fifteen to thirty minutes, there is the risk of permanent brain damage, including global loss of cortical function. The length of time a code is run cannot be determined precisely to avoid all cases of PVS because there is usually some oxygen going to the brain during the code as a result of the chest compressions applied by the physician. However, because of the nature of cardiopulmonary resuscitation (CPR) as an acute and heroic effort to save a life that is being lost, it is antithetical to try to "call" codes more conservatively to minimize the number of cases of PVS at the cost of not maximizing the number of lives saved.

In contrast, the number of cases of PVS resulting from trauma has decreased as a result of the greater use of seat belts and air bags in cars and the wearing of helmets by bicyclists and motorcyclists. There is no registry of patients in PVS, and so the number cannot be known with any degree of certainty. The most common guess is that there are 10,000 people in the United States in a PVS, although the number could be half or twice that.

Ethical Issues

The ethical issues raised by PVS are as complex as the neurology is. For example, three of the most publicized and controversial cases in medical ethics involved young women who were in a PVS: Karen Ann Quinlan in New Jersey in the 1970s, Nancy Cruzan in Missouri in the 1980s, and Terri Shiavo in Florida in the early 2000s. In each case the patient's family wanted to make the decision to stop life-sustaining treatment once their loved one's grim prognosis became evident.

At least two factors make decisions regarding PVS patients very difficult. First, observing these patients is an unnerving experience: Although awake during the day, they have some movements of the arms, back, neck, and head, including grimaces and smiles, and make sounds such as moans and grunts. This makes it almost inevitable that the family will have doubts about the diagnosis and about whether the patient may show improvement eventually. Second, although these patients require extraordinary around-the-clock nursing care to avoid bedsores and infections, they need relatively little medical intervention except a feeding tube to provide artificial nutrition and hydration. If this care is provided and the occasional infection is treated with antibiotics, PVS patients can have a normal life span. Thus, some have been kept alive for three or four decades. These two factors make it very difficult for families to stop the life-sustaining treatment for patients in a PVS even when they are confident that the patient would not want to live in such a condition.

When these issues first were addressed by the bioethics community in the 1980s, many people argued that feeding tubes and artificial nutrition and hydration should be considered a necessary component of humane treatment and be required to demonstrate respect for human dignity, comparable to being kept clothed and given some privacy. This view has become less common but still is held by some theologically oriented bioethicists in the Roman Catholic and Orthodox Jewish traditions. Support for the position that artificially provided nutrition and hydration constitutes necessary medical treatment was called into question as the nature of PVS became understood and, simultaneously, the hospice movement began to promote the idea of death with dignity. Although it still is not universally accepted, there is a broad consensus among clinicians, lawyers in the field of health law, and ecumenical and secular bioethicists that artificial nutrition and hydration should be consented to or refused on the basis of an evaluation of its benefits and burdens to patients on a case-by-case basis.

This is the ultimate controversy regarding PVS: determining how a patient would want to live. Perhaps the best philosophical clarification of the issue came when James Rachels (1986) summed up the sentiment that family members of PVS patients had expressed by saying that the life of PVS patients was over years before they died. Rachels distinguished between life in a biological sense and life in a biographical sense; put more colloquially, PVS patients no longer "have a life" even though they are still alive. Thus, the use of a living will or an advanced medical directive may be the only way to determine how a patient would want to be treated if found in a persistent vegetative state.

In light of the controversy surrounding PVS, it is clear that some medical conditions are not as easy to manage as others. Although the definition of PVS is relatively straightforward, the ethical issues are not. PVS continues to be an area of much debate, both ethically and legally, and the issues surrounding it are not easy to resolve. Because of this PVS will continue to be researched and discussed to help ease the discomfort involved in making decisions about patients in a persistent vegetative state.

JEFFREY P. SPIKE

SEE ALSO *Bioethics; Brain Death; Medical Ethics.*

BIBLIOGRAPHY

American Neurological Association Committee on Ethical Affairs. (1993). "Persistent Vegetative State." *Annals of Neurology* 33(4): 386–390. A succinct summary of both the clinical and ethical issues raised by PVS.

Brody, Baruch. (1992). "Special Ethical Issues in the Management of PVS Patients." *Law, Medicine, and Health Care* 20(1–2): 104–115. Discusses a wide range of ethical issues, including futility and the just allocation of resources.

Childs, Nancy L., and Walt N. Mercer. (1996). "Brief Report: Late Improvement in Consciousness after Post-Traumatic Vegetative State." *New England Journal of Medicine* 334(1): 24–25. There have been a few reports of improvement of patients in PVS such as this case.

Jennette, Bryan, and Fred Plum. (1972). "The Persistent Vegetative State after Brain Damage: A Syndrome in Search of a Name." *Lancet* 1: 734–737.

Lynn, Joanne, ed. (1989). *By No Extraordinary Means: The Choice to Forgo Life-Sustaining Food and Water*. Bloomington: Indiana University Press. A collection representing both sides of the debate on whether medical provision of nutrition and hydration through invasive procedures should be considered optional life-sustaining medical treatment or ethically required humane treatment. Much has been written since, but this may have appeared just at the point in the history of the debate when the tipping point was reached and a new consensus was developed.

Multi-Society Task Force on PVS. (1994). "Medical Aspects of the Persistent Vegetative State." *New England Journal of Medicine*, Part I: 330(21): 1499–1508; Part II: 330(22): 1572–1579. This report by the American Association of Neurology (AAN) and the American Academy of Neurology (ANA) defines the current standard of practice.

Rachels, James. *The End of Life*. Oxford: Oxford University Press, 1986. Chapter three, "Death and Evil," includes his famous distinction of biological versus biographical life, and other interesting observations on the conceptual and ethical problems that result when a life lacks the proper elements of a narrative.

PESTICIDES

SEE *Carson, Rachel; DDT.*

PETS

• • •

In contrast to wild animals living in zoological settings that have scientific value as representatives of particular species, or with domesticated work animals and those kept for their value as commodity producers, a pet is any domesticated animal or wild animal living in a domestic setting that is cared for, enjoyed, and valued for a unique set of characteristics that differentiates it from other members of the same species. Mitochondrial canine DNA evidence suggests that humans have kept dogs, the first animals to emerge as pets, for tens of thousands of years. Many pets are valued solely for their role as companions, treated by those who care for them with affection as though they were friends or family members. Some pets are also working animals: They hunt, herd, perform search and rescue operations, control traffic, protect homes from pests and strangers, or otherwise serve to extend human capacities, in versions of what Aristotle called *living tools* (Nicomachean Ethics VIII, 8). These animals, though, can be considered pets if and only if they are also valued for their companionship to humans.

Ethics of Pets

While numerous theories have been developed that focus on human beings and the environment as objects of ethical concern, no ethical theory deals specifically with how companion animals ought to be valued and viewed. The animal rights perspective popularized by philosopher Peter Singer, as well as theories of bioethics, may serve as starting-points for thinking about the ethics of keeping animals in zoos or consuming their meat; they are much less serviceable for thinking about human-pet relationships. Ethical relations toward pets as human companions can better be understood from within such moral perspectives as the Humean doctrine of moral sentiment, the obligation to *do no harm*, religious-based ethics, or the ethics of care. To the extent that scientific discoveries and technological innovations increase the mutual quality of life between pets and their human caregivers, they may be described as ethically positive; to the extent they lead to treating pets with decreased respect, kindness, and concern for the quality of their lives, as ethically negative.

Traditionally, ethical behavior toward pets has been synonymous with treating them humanely, that is, with care in providing them with healthful living conditions, compassion stemming from their status as dependent creatures, and respect for their dignity and well-being. Advances in science and technology have served this end in a number of ways. For instance, the growth of information technology has had a positive impact in extending and bettering the lives of pets. The use of radio collars and microchips with identifying information implanted under the skin now helps to reunite lost pets with their human companions; in the future, it is possible to imagine using devices with Global Positioning System (GPS) capabilities or nanoscale sensors to track the whereabouts of pets. Passports for pets, in effect in European Union countries from July 2004, include microchip identifiers that serve as a mechanism to allow cats, dogs, and ferrets with valid anti-rabies vaccination certificates to travel with their human companions without raising concerns about risks to public health.

The growth of the Internet has also led to improved living conditions for pets. Internet sites such as petfinders.com help facilitate the placement of abused and homeless pets, while other animal welfare organizations use the Internet to increase attention to the plight of former working animals (such as race horses and greyhounds) whose adoption as pets could prevent them from being destroyed, and of specific breeds of pets in need of a rescuing hand. By publicizing the needs of pets as well as opportunities for their adoption, the resources of the Internet have contributed to bettering the treatment of animals in shelters by leaving fewer unwanted pets to be euthanized, as well as to increasing global awareness of animal welfare and environmental ethics issues related to pets. These issues include international trade in exotic birds and other species, as well as the ongoing trade in and consumption of dog meat in some Asian countries despite government regulations making such practices illegal.

Raising Ethical Standards through Science and Technology

At the same time, advances in science and technology have been instrumental in encouraging a higher ethical standard for the treatment of pets than the minimal ethical standard of humane treatment or regard for animal safety and welfare. Because of the value that pets have in the eyes of their human companions, scientific or technological developments that help to extend the lives of individual pets or make these lives better are generally perceived as having a positive ethical dimension. Research in veterinary medicine and the application of human-related medical research to the veterinary sphere have led to the development of many measures to improve pet health and well being. These measures encompass a wide spectrum, including new types of immunizations, radiation treatments for pets with thyroid conditions, cataract operations to restore the sight of blind pets, MRI imaging technology and laser surgery for pets, and medications to prevent heartworm and other common but life-threatening parasites. Another development related to these innovations can be seen in the availability of therapeutic pet foods designed for animals with special health needs.

Such developments are not, however, always seen as morally benign. For example, the expense involved with obtaining many innovative health-related measures for pets has raised the question of whether it is moral to take such costly measures to extend the life of a single pet rather than applying the same resources to reduce human need and suffering.

Just as the birth of the cloned sheep Dolly in 1997 generated interest in cloning other kinds of livestock, it also sparked research into how cloning techniques might be applied for the purposes of cloning cats and dogs. In 2002 the first cloned kitten, aptly named "cc," was born at Texas Agricultural and Mechanical University to an adult cat acting as a surrogate mother implanted with cloned embryos formed by fusing denucleated feline egg cells with DNA from the nuclei of cumulus cells belonging to the original cat. Some researchers involved in this project have also been engaged in a similar but thus far (in 2004) unsuccessful endeavor, named the Missyplicity project, intended to clone Missy, a mixed-breed dog.

These efforts, and the potential for such cloning technologies to be transferred to commercial ventures, have prompted considerable moral controversy. Those who argue against cloning pets on ethical grounds have claimed that it is immoral to clone pets when there are so many homeless pets in shelters available for adoption, that the desire to clone a pet is based on the misguided idea that cloning could give it immortality, and that a cloned pet could be seen as less valuable than the original pet. At the same time, those who claim there is nothing morally wrong with pet cloning research stress that it could equally lead to more loving relationships between humans and their pets, as well as have important collateral benefits, such as the creation of better seeing-eye and search-and-rescue dogs. For those who see animals as technological devices, cloning, once perfected, could be perceived as merely a more effective production method.

Other attempts to apply new reproductive technologies to pets, such as genetic engineering, also give rise to ethical concerns. The development of zebra fish engineered with a sea coral gene so that they appear fluorescent under ultraviolet light raises the issue of whether, given that such fish are primarily appealing for their entertainment and novelty, it is consistent with respect for pets to commercially breed and market them. Additionally such fish might have potentially negative impacts on ecosystems should they (as have genetically engineered salmon) find their way into natural waterways.

Ethical issues further surround some conventional practices of breeding pets for certain characteristics such as small nostrils or prominent eyes that make them attractive representatives of particular breeds of animals at the expense of their health and welfare. The European Convention for the Protection of Pet Animals, first open for signature in 1987 and subsequently signed and ratified by a number of members of the Council of Europe, restricts the breeding of pets in such a way that would pass along defects, such as extremely small size, hairlessness, and other hereditary characteristics, that put them at risk for physical and mental diseases. A potentially suggestive avenue for the development of transgenic animals is one that could lead to pets whose blend of traditional aesthetic appeal with mitigated risk for disease might serve to allay ethical concerns regarding their existence.

While scientific and technological innovations have by and large been instrumental in enhancing the coexistence of humans and their pets they do not always lead to mutual flourishing, particularly in advanced industrialized countries. As in these settings the role of pets as companions to humans has grown, so has human interest in having pets whose welfare is otherwise not in question conform to human expectations and living patterns at the expense of their animal "otherness." In some locations, protective public policies have been introduced to prevent pets from being declawed, devoiced, or otherwise medically altered to accommodate the largely urban lifestyles of their owners. This interest can also lead humans to respond emotionally to their companion animals in ways that overemphasize their role as companions. For example, dogs historically bred to be involved in physical work alongside humans can suffer when, in a society dominated by developments in science and technology, their need to work goes unrecognized and unrewarded. In advanced technological society, insuring that human-pet relationships are to the mutual benefit of both partners can be seen as an ongoing ethical challenge.

Technology itself, in the form of robotic dogs and cats, or "cyberpets" such as the tamagotchi handheld video games that simulate feeding, training, playing, and other aspects of pet ownership, may serve as a means for meeting this challenge and for meeting at least some of the human needs now met by animal pets. If so, these technological pets might be a positive development. Many animal pets are mistreated by their owners, and this would not be a problem with mechanical pets under the assumption, which is probably safe at least for the near future, that they are not sentient. Furthermore, mechanical pets would presumably not breed, hence eliminating some of the vast number of killings of unwanted dogs, cats, and other companion animals.

DIANE P. MICHELFELDER
WILLIAM H. WILCOX

SEE ALSO Agricultural Ethics; Animal Rights; Animal Welfare.

BIBLIOGRAPHY

Beck, Alan M., and Aaron Katcher. (1996). *Between Pets and People: The Importance of Animal Companionship.* West Lafayette, IN: Purdue University Press. The authors' research concentrates on the significant role pets can play as therapeutic agents in bettering human physical and mental health.

Haraway, Donna. (2003). *The Companion Species Manifesto: Dogs, People, and Significant Otherness.* Chicago: Prickley Paradigm Press. Haraway looks at companion species who, much like the cyborgs in her earlier *Cyborg Manifesto* (1985) bring together technoscience and nature in novel ways, and suggest how people might better learn to live with significant otherness.

Journal of Applied Animal Welfare Science 5(3), 2002. This issue contains reflections, both pro and con, on the ethics of cloning pets. Contributiors include philosopher Hilary Bok ("Cloning Companion Animals is Wrong"), Mark Green ("New Dog, Old Tricks"), and Lou Hawthorne ("A Project to Clone Companion Animals"). Hawthorne serves as the lead scientist on the Missyplicity project.

Midgley, Mary. (1983). *Animals and Why They Matter.* Athens, GA: University of Georgia Press.

Rowan, Anthony, ed. (1988). *Animals and People Shaping the World.* Hanover, NH: University Press of New England. Includes articles on "Pet-Keeping In Non-Western Societies" and "The Emergence Of Modern Pet-Keeping."

Serpell, James. (1986). *In the Company of Animals: A Study of Human-Animal Relationships.* Oxford: Basil Blackwell. In this work, Serpell, director of the Center for the Interaction of Animals and Society at the University of Pennsylvania, takes a comprehensive look at the relationships between humans and their pets.

Tuan, Yu-Fu. (1984). *Dominance and Affection: The Making of Pets.* New Haven, CT: Yale University Press.

Varner, Gary. (1999). "Should You Clone Your Dog?: An Animal Rights Perspective on Somacloning." *Animal Welfare* 8: 407–420.

PHENOMENOLOGY

• • •

Phenomenology is an influential philosophical movement especially in relation to science and technology. It has developed critical studies of scientific rationality, artificial intelligence, electronic media, virtual reality, the Internet, and more. Leading contributors to the three waves of phenomenology are often drawn on in discussions of science, technology, and ethics: from the first wave of Edmund Husserl (1859–1938) through the second wave of Martin Heidegger (1889–1976) to the third wave of Maurice Merleau-Ponty (1908–1961). Even more prominent figures in debates about science, technology, and ethics discussions such as Hans Jonas (1903–1993), Emmanuel Lévinas (1905–1995), and Hannah Arendt (1906–1975) have also been strongly influenced by phenomenology, as have the critical assessments of science and technology to be found in later work by Albert Borgmann, Hubert L. Dreyfus, Andrew Feenberg, Michael Heim, Don Ihde, Langdon Winner, and others. Phenomenology nevertheless remains difficult to define, and its distinctive contributions not easy to pin down.

What Is Phenomenology?

It is difficult to define phenomenology in a way that will cover all its diverse traditions. In his monumental history of the phenomenological movement, not even Herbert Spiegelberg (1994) attempted a definitive formulation. In spite of this difficulty it is necessary to attempt some definition as a starting point—even if all phenomenologists do not accept it without qualification.

Initially, then, phenomenology may be described as an effort to disclose the transcendental features or presuppositions of the world as given in ongoing experience. Phenomenology takes as its basic concern our ongoing experiencing of the world within the unfolding horizon of temporality. Although the language of phenomenology often refers to "essences" in experience, it is not interested in some stable atemporal or a historical account of the world. For the phenomenologist essences do not stand outside of our ongoing existence. The transcendental horizon, the focus of its concern, is never divorced from the concrete experiences of everyday life. But at its foundation is the attempt to take the phenomena of human experience and subject them to deeper or broader examination than is done by the sciences, all of which, according to phenomenology, abstract from experience.

To extend this working definition, take the human experience of music and consider it phenomenologically. From the perspective of physics and physiology, music is constituted by a flux of waves of particular frequencies to which the inner ear may be sensitive. Indeed, once so analyzed, it is possible to create a technological device such as a tape recorder that is sensitive to these same sounds, and can even replay them on command. Human beings, however, when they hear sounds in everyday life never take them simply as a stream of sounds, rather they find themselves already listening to something particular—a cry for help, an automobile braking, construction noise, or a piece of music. Indeed it would take a very strange sort of attitude to hear sounds and take them as a flux of waves of particular frequencies. Listening is different than registering or recording; to listen is to already take sounds as this or that. In listening, the taking of sound as music implies an already existing sense of what music is, something that makes it possible for us to take these sounds as music rather than noise. Furthermore, in listening to music, this listening is informed by an ongoing sense (or unity) of movement, rhythm, tone, scale, style, and so forth. This ongoing active unity provides an active and ongoing framework that enables me, in the experience of listening (right now), to simultaneously "retain" the sounds I no longer hear (the past), and in anticipation to "fill in" the sounds I am not yet hearing, yet already anticipate (the future). As a phenomenological being I find myself listening to music, not merely recording sounds after the manner of a technological device. For phenomenologists the relevant question is: What is this ongoing framework that makes it possible for humans to listen to music rather than merely record sounds?

Even our encountering of mundane everyday objects takes as necessary an already existent sense or familiarity with the world. What makes it possible to encounter a chair—recognize it, see it, refer to it, use it as a chair rather than as a something else? Like sounds, we are always given it only in some one aspect. When we stand in "front" of it the "back" is not given to our senses as such. When we stand at the back the front is hidden from view. Yet when we approach the chair we do not take it as a confusing flux of sensation, but as that which it already is, a chair to sit or stand on.

What is it then that enables us to encounter music and chairs in their fullness even though we are always given, at any particular point in time, only some limited aspect of such phenomena? The answer of phenomenology is that it is the *transcendental horizon* that makes phenomena possible, where the transcendental is understood as "that which constitutes, and thereby renders the empirical possible" (Mohanty 1997, p. 52). In Don Ihde's words "phenomenology investigates the conditions of what makes things appear as such" (2003, p. 133). One could say that the transcendental is the background, or horizon, that makes the meaningful experience of the foreground possible. Yet insofar as such a formulation suggests a background that is somehow separate and "behind" that which appears in the foreground, it would be incorrect. Transcendental horizons or conditions are always and immediately already present in the very appearing as such—this is exactly what makes a horizon "disappear" or withdraw from our focal awareness. It is so evident that it simply does not come up as an issue.

It is this seemingly "forgotten" horizon that is the focus of phenomenology—indeed it is this horizon that phenomenologists want to call to our attention. All phenomenological "reductions" have as their purpose a "return" to this vital constitutive transcendental horizon. "Reduction" should be understood here in relation to its Latin root *re-ducere*, to lead back.

In place of the examples of listening to music or encountering a chair, we could also refer to engagements with such diverse phenomena as language, self, identity, sociality, and so forth. In the case of those phenomena known as science or technology we would attempt to provide an account of the transcendental horizons that constitute the scientific or technological and therefore render them possible in our everyday experience. What is it within our ongoing relation with the world that allows science or technology to show up as a way to structure that relationship? To this question phenomenologists have given many different and illuminating answers.

But what is the transcendental horizon? How and where do we find it? The answer differs from one phenomenologist to another. Husserl (1970, 1982, 1995) argued that it was the ongoing life of pure consciousness. For him the intentionality of consciousness allows things to appear as this or that thing. He thus proposed that we bracket out, or set aside, our normal everyday assumptions about the world—the natural attitude—and return to the life of pure consciousness.

By contrast, Roman Ingarden (1893–1970) maintained that the transcendental horizon is constituted by the *a priori* truths necessary for the factual world to be what we experience in ongoing experience. He proposed that we return to these truths, but also encouraged us to always ground ourselves in the real world as given in experience.

Heidegger (1962), in turn, argued that it is our always already immersion in the world of everyday life that is the transcendental or constituting horizon. For him active beings are always already busy in the world, and the world shows up precisely as that which it already is. We do not need a "bridge back to the world" from our concepts. We have never left the world of everyday life, and it is exactly this ongoing intimate relation with the world—our pre-ontological understanding of being as such—that is the very basis of all scientific knowledge. It is the "stuff" from which we construct all systems of logic, mathematics, and science.

Merleau-Ponty (1962) continues this discovery of transcendental horizons by focusing on the body, or more specifically on the always already embodiedness of our being. He calls on us to return to the already lived and situated body of our ongoing perception of the world. For him our scientific systems of orientation in time and space have their condition of possibility in our being a body—a lived body that is the ongoing horizon of orientation and meaning.

Despite their differences these phenomenologists all claim that the naturalistic empirical science (also referred to as objectivism or positivism) remains unreflective and uncritical of the importance of the ongoing constituting role of these various transcendental horizons. For example, scientists take the objects of their investigations—such as atoms, ozone layers, cultures, money, criminals, and so forth—as simply already given without considering the conditions that make it possible for them to encounter these phenomena as what they take them to be. In their emphasis on these already assumed objects of study the constituting horizons withdraw to be forgotten, thereby allowing them to move, in their analyses and arguments, way beyond the possibilities offered by the constitutive conditions of meaning. It is exactly the explication of these constitutive conditions or horizons of meaning that phenomenology seeks to call to our attention, in order to keep us from becoming lost in or misled by the abstractions of science and the powers of technology.

The Phenomenology of Everyday Encounters

To provide an illustration of the phenomenological approach it is useful to present in slightly more detail Heidegger's pivotal analysis of our everyday encounters. This presentation will then link to the work of Ihde, Borgmann, Dreyfus, and Lévinas.

For Heidegger the human encounter with things is fundamentally practical in orientation. We do not encounter chairs as chair objects—after the manner of designers or scientists—but as "possibilities for" sitting down or standing on or facing somebody, and so forth. Furthermore, the chair is a "possibility for" (what Heidegger called an "in-order-to") only within an already present referential whole including a multitude of possibility-for's. The transcendental horizon of meaning is the ongoing, unfolding referential whole in which every thing has its ongoing way of being that which it already is, while the whole draws on this very being to be the whole that it is. To describe this active and ongoing transcendental horizon of reference and meaning—in which the world and humans already implicate each other—Heidegger uses the notion of *being-in-the-world*, thus indicating the intimate relation between being and world. For Heidegger, any being whatsoever is a being only in an already assumed world—referential whole—that constitutes it as such. Heidegger argues that we humans-already-in-the-world (which he calls *Dasein*) exist in an ongoing structural openness toward the world in which the self and the world are always already a unity, a being-in-the-world (Heidegger 1962). Thus, we human beings (*Dasein*) have this unity as our ongoing way of being. That is why the world mostly makes sense rather than being mostly strange and unfamiliar.

Consider this example. Whenever we find ourselves or take note of ourselves, we do so already engaged in practical everyday activity in which things show up as "possibilities for" our practical intentions—as tools for this and that. When I switch on my laptop it already shows up as a possibility to write, communicate with my office, and so on. When we consider this world of practical activity we note that all the things we encounter already matter in some way or another—even if they matter only as useless, boring, or irrelevant. Heidegger claims that we, as *Dasein*, are always already "ahead" of ourselves—always already projected into the future as it were. In being ahead of ourselves things show up as this or that possibility-for. When we get up in the morning we already find ourselves acting in anticipation of the day ahead. When we get into our cars we already anticipate the journey. To put it rather abstractly, we are always and already projected as a necessary condition of that what we already are—as academics, politicians, managers, and so on. I did not so much decide to take up the project to write this entry as much as I found myself writing this entry as that which already made sense for an academic like me to do. Thus, as already projected beings, tools (opportunities) show up as tools (opportunities)—as possibilities-for. The world as possibilities-for shows up in particular ways to scientists (as scientists) that are different from that of artists (as artists) or managers (as managers).

This does not mean, however, that one can simply take the world any way one wants; the world—the scientific, art, or managerial world—is not simply of one's making. These tools are tools for this or that purpose only in as much as they already refer to other tools, which also already refer to them as their transcendental condition for being this or that tool. Here, "refer" is used in the sense of a necessary relation or reference for

the tool to be what it already is taken to be when taken up in practical activity. The laptop I am working on, to be taken up as a laptop, rather than a piece of assembled plastic and silicone, refers to application programs, which refer to operating systems, which refer to hardware, which refer to a power supply—all of which refer to suppliers, which refer to maintenance services, and so forth. Dreyfus (1991) calls this recursively defining, necessary nexus of relations, the tool or equipment whole.

When we take up these tools, as tools, however, we do not take them up for their own sake; we take them up within an already present reference to our projects. I do not simply bang on keys; I use the laptop to type, in order to write, do e-mail, surf the Internet, and so forth. Moreover, the writing of this entry already refers to the possibility of an encyclopedia, of which it would be a part. This encyclopedia already refers to editors, which refer to a potential audience, which refers to potential publishers, and so on. Furthermore, the writing of this entry also already refers to the publication of my work, which refers to a publication record, which refers to academic status, which refers to the possibility for promotion, and so forth. Heidegger (1962) calls this recursively defining and necessary nexus of projects, or for-the-sake-of relations, the involvement whole.

The equipment whole and the involvement whole refer to each other and sustain each other as an ongoing referential whole, horizon of meaning. Heidegger calls this referential whole "the world." We humans always already dwell in the world in which the world is mostly familiar (it is simply already there, "ready-to-hand" in Heidegger's terminology). Now sometimes the world "breaks down," and then we tend to encounter it as objects or events as such—it becomes occurrent or present-to-hand in Heidegger's terminology. When we type and the key gets stuck then we notice it "as a key"; otherwise we merely type. If it remains stuck the computer becomes occurrent "as a broken laptop." But as we start to take it apart, in an attempt to fix it, it recedes back into the background as something I am fixing.

The point of Heidegger's account is "that things show up for us or are encountered as what they are only against a background of familiarity, competence, and concern that carves out a system of related roles [recursively defining references] into which things fit. Equipmental things are the roles [recursively defining references] into which they are cast by skilled users of them, and skilled users *are* the practical roles [recursively defining references] into which they [become] cast themselves" (Hall 1993, p. 132). The phenomenological meaning of the world of science, technology, and ethics can be understood only within the always already defining referential whole, the world we are already "in"—or more correctly the world we always already are. Grasping this phenomenological foundation is essential to making sense of some of the authors most important for science, technology, and ethics.

Phenomenology in Science and Technology

Phenomenology has been used to analyze a number of aspects of science and technology in ways that have implications for ethics. What follows is a consideration of three major cases: artificial intelligence, consumer devices, and human–technology relationships.

DREYFUS ON ARTIFICIAL INTELLIGENCE. In critiquing artificial intelligence (AI) Dreyfus (1979, 1992) argues that the way skill development has become understood has been wrong. He argues, using the work of Heidegger, that the classical conception of skill development, going back as far as Plato, assumes that we start with the particular cases and then abstract from these to discover and internalize more and more sophisticated and general rules. Indeed, he argues, this is the model that the early artificial intelligence community uncritically adopted. In opposition to this view he argues, with Heidegger, that what we observe when we learn a new skill in everyday practice is in fact the opposite. We most often start with explicit rules or preformulated approaches and then move to a multiplicity of particular cases, as we become an expert. His argument draws directly on Heidegger's account of humans as beings that are always already in-the-world. As humans in-the-world we are already experts at going about everyday life, at dealing with the subtleties of every particular situation—that is why everyday life seems so obvious. Thus, the intricate expertise of everyday action is forgotten and taken for granted by AI.

As a way to critique the program of AI, Dreyfus provides an account of five stages of becoming an expert. A *novice* acts according to conscious and context-free rules and generally lacks a sense of the overall task and situational elements. The *advanced beginner* adds, through experience, situational aspects to the context-free rules to gain access to a more sophisticated understanding of the situation. The relationship between the situational aspects and the rules are learned through carefully chosen examples, as it is difficult to formalize them. The *competent person* will have learned to recognize a multiplicity of context-free rules and situational aspects. This may lead, however, to being overwhelmed because it becomes difficult to know what to include or exclude.

The competent individual learns to take a particular perspective on the situation, thereby reducing the complexity. Such "taking a stand," however, involves a certain level of risk taking that requires commitment and personal involvement. For the *proficient* most tasks are performed intuitively. As an involved actor the relevant situational aspects show up as part of the ongoing activity and need not be formalized. Nevertheless, a pause may still be required to think analytically about a relevant response. For the *expert* relevant situational aspects as well as appropriate actions emerge as part of the ongoing activity within which the expert is totally absorbed, involved, and committed. The task is performed intuitively, almost all the time. In the ongoing activity of the expert thousands of special cases are discriminated and dealt with appropriately.

With this phenomenological account of skill development in hand it is easy to see the problem for AI. Computing machines need some form of formal rules (a program) to operate. Any attempt to move from the formal to the particular, as described by Dreyfus above, will be limited by the ability of the programmer to formulate rules for such a shift—a shift forgotten by AI. Thus, what the computer lacks is an already there familiarity with the world that it can draw upon as the transcendental horizon of meaning to discern the relevant from the irrelevant in ongoing activity—that is, the computer is not a being-in-the-world in Heidegger's terms.

Dreyfus's critique pushed AI researchers into new ways of thinking. In particular it has led to the embodied cognition program of the Massachusetts Institute of Technology Artificial Intelligence Laboratory under the direction of Rodney Brooks. Nevertheless, even such programs of embodied cognition (or cog robots as they are called) would fail if AI cannot give an account of how a cog robot's own existence would be at stake—would matter. Without such a "stake"—without being ahead of itself—the cog robot would lack the fundamental transcendental horizon of intentionality and meaning, according to Heidegger. Phenomenology's call to a "return to the things themselves," to recover the supposed transcendental horizon of meaning, will continue to challenge the progress of AI. Moreover, it seems that many of our assumptions about the relationship between the technical and the social, even the supposition of such a relation itself, will continue to provide a multiplicity of opportunities for phenomenology to explore.

BORGMANN ON CONSUMER DEVICES. In thinking about our relationship with technology in modern contemporary life, Borgmann takes up the question of the possibility of a "free" relation with technology. He agrees with Heidegger that modern technology is a phenomenon that tends to "frame" our relationships with things, and ultimately ourselves and others, in a one-dimensional manner—the world as available resources for our projects. He argues that modern technology frames the world for us as "devices," and specifically as devices that hide the referentiality of the world—the worldhood of the world—upon which devices depend. Devices do not disclose the multiple conditions that are necessary for them to be what they are taken to be. Just the opposite is true: They try to hide the effort that is necessary for them to be available for use. Thus, a thermostat that we simply set at a comfortable temperature now replaces the process of chopping wood, building a fire, and maintaining it. Our relationship with the environment is reduced to, and disclosed to us as a control that we simply set to our liking. In this way devices de-world our relationship with things, in Heidegger's terminology. By relieving us of the burden of making and maintaining fires, our relationship with the world becomes disclosed in a new way—as one of disengagement. The world of things is not something to be engaged in, it is simply available for consumption.

Against such a disengaging relationship with things in the world, Borgmann argues for the importance of focal practices based on focal things. Focal things solicit our full and engaging presence. Compare, for example, the focal practice of preparing and enjoying a meal with friends or family to the solitary consumption of a fast-food meal. If one takes Borgmann's analysis seriously one might conclude that contemporary humans, being surrounded by devices, are doomed to increasingly relate to the world in a disengaged manner. Borgmann argues, however, that it is also possible to have a free relation with technology—even modern technology—if we imbed it in focal practices rather than use it, or accept it, as devices. Otherwise we will, as Heidegger (1977) argued, become the devices of our devices.

IHDE ON HUMAN–TECHNOLOGY RELATIONSHIPS.
Phenomenology does not function only as an approach to critique our relationship with technology. Ihde (1990) has used the resources of phenomenology to give a rich and subtle account of our relationship with technology. In thinking about the human–technology relationship Ihde characterizes four different *I–technology–world* relationships. The first type he calls "embodiment relations." In this case technology is taken into subjective perceptual experience of the world, thus transforming the subject's perceptual and bodily sense. In wearing my eyeglasses I not only see through them, they also become "see through." In functioning as that which they are,

they already withdraw into my own bodily sense of being a part of the ordinary way I experience my surrounding. He denotes this relationship as having the form (I–glasses)–world. This relationship, however, has a necessary "magnification/reduction structure" associated with it. Embodiment relations simultaneously magnify and amplify or reduce and place aside (screen out) what is experienced through them. The moon seen through a telescope is different from the moon perceived by the naked eye. The person at the other end of the telephone is brought to me across a great distance at the expense of being reduced to a voice.

The second type of human–technology relationship is what Ihde calls "hermeneutic." Here, the technology functions as an immediate referent to something beyond itself. Although I might fix my focus on a map, what I actually see—immediately and simultaneously—is not the map itself but rather the world it already refers to, the landscape suggested in the symbols. In this case the transparency of the technology is hermeneutic rather than perceptual. As I become skilled at reading maps they withdraw to become immediately and already the world itself. Ihde denotes this relationship as having the form I–(map–world).

The third type of human–technology relationship Ihde calls "alterity relations." In this case, technology is experienced as a being that is otherwise, different from myself—technology-as-other. Examples include things such as religious icons and intelligent robots (the Sony dog for example). In my interaction with these technologies they seem to exhibit a "life of their own," thus as I engage with them they tend to disengage me from the world of everyday life, hence their pervasiveness in activities such as play, art, and sport. Ihde denotes these as having the form I–technology–(world), indicating that the world withdraws into the background and technology emerges as a focal entity with which I momentarily engage—as I play with my robot dog for example.

Finally, Ihde recognizes a fourth type of human–technology relationship in which technology is not directly implicated in a conscious process of engagement on the part of the human. Ihde refers to these as "background relations." Examples include automatic central heating systems, traffic control systems, and so forth. These systems are "black-boxed" in such a way that we do not attend to them, yet we draw on them for our ongoing everyday existence. They withdraw as ongoing background conditions. Although he does not designate them as such, one might formalize these relations in the form: I–(technology)–world. These invisible background technologies can be powerful in configuring our world in particular ways, yet escape our scrutiny.

Ihde's phenomenological description of the human–technology relationship provides a useful way to give an account of many everyday relations of import to science, technology, and ethics. One can imagine a very interesting phenomenological analysis of the relationship between scientists and their instruments as done in the social study of science. Furthermore, the withdrawal of technology, into my body, into my perception, and into the background, has important political and ethical implications for its design and implementation, especially if one considers that every disclosure of the world "through" technology is also immediately a concealment of other possible disclosures. The car discloses possibilities for getting to places quickly, but also conceals, in its withdrawal, the resources (roads, fuel, clean air, etc.) necessary for it to be what it is—they act as devices in Borgmann's terminology. Indeed we often lose sight of the reduction/magnification structure as we simply use these technologies. As these technologies become more and more pervasive—almost a necessary condition of everyday life—it becomes more and more difficult to see that which has become concealed in their withdrawal. With Ihde's typology of I–technology–world relationships it might be possible to bring what has become concealed back to the foreground for critical attention and ethical reflection.

FURTHER CASES, AND LÉVINAS. There are many more authors that could be used to illustrate phenomenology's relevance and influence in the domain of science, technology, and ethics. For example there are Heim's studies of virtual reality (1993) and electronic writing (1999), or Richard Coyne's discussion of being in cyberspace (1995), Tony Fry's excellent essays on the televisual (1993), Terry Winograd and Fernando Flores's critique of the use of computers in organizations (1986), and many more. Nonetheless, it is the work of Lévinas (1969, 1991) that might serve as a final signpost on our phenomenological way. The reason for this is that Lévinas, although he starts within the phenomenological tradition, wants to turn our attention to the most basic encounter of all—that of the ethical.

Lévinas argues that Western philosophy, and phenomenology in particular, is a philosophy of what he calls the same, or the totality—a totality within which every otherness becomes "domesticated." By totality he means the expectation that all things will eventually "add up," will be accounted for; that somehow there is a larger whole or "system" in which everything will eventually find its place. For Lévinas this expectation already has its source in the ongoing synthesizing intentionality of consciousness itself. The transcendental horizon of

meaning, opened up by intentionality, is already colonized by our individual self-ish will to be. The gravity of our everyday existential project does not allow the other, as profoundly singular, to remain at the margins of our constituting horizon. Through our will to be—our always already projectedness in Heidegger's language—we have indeed already taken the place "in the sun" of the other. We, in our already in-the-worldness, are already guilty of violating the otherness of the other; we are already responsible, therefore we must respond. For Lévinas, taking up our responsibility for the other is the only possibility for transcending the self-ishness of the will to be.

Thus, what Lévinas points to is that although phenomenology provides a path back to the very constitutive possibilities of experience it also immediately implicates us as already responsible for violating the otherness of the other in these very possibilities. In our quest for meaning we find ourselves at the dawn of ethics, but we find ourselves already guilty. For Lévinas the ethical has always and already called into question the projects of science and technology. With Lévinas one might say that the success of science and technology has always come at the expense of masking the plight of the singular—the singular that is the incidental, idiosyncratic, and random error excluded from consideration in a world in which things always have to add up. Thus, for Lévinas, the most profound question is not the what or how of science and technology but the always already suffering of my neighbor, the specific one closest to me, that the projects of science and technology obscure even if they try to do what is right. Obviously Lévinas is not saying we should abandon science and technology. He is rather saying that we should allow the ethical, the singular other, to continually question and interrogate the already supposed legitimacy of science and technology. It is only in the currency of the singular, this individual here and now, that ethics has any possibilities.

Some Critical Comments

Phenomenology provides a variety of resources for examining relationships among science, technology, and ethics. But phenomenology also has limits. It is often criticized for essentialism and failures to provide rich accounts of the particular and the situated, such as those provided by social studies of science, as in the work of Bruno Latour and Steve Woolgar (1986), Michel Callon (1986), or Andrew Pickering (1995). Phenomenology does not appear able to explain why some technologies become accepted and used rather than others in the way social constructivist accounts do, as in the work of Wiebe E. Bijker and colleagues (1987). This tension between phenomenology and social constructivism permeates the work of Feenberg (1995), whose analysis of technology retains important phenomenological insights while working with findings from social studies of technology.

Indeed, other "post" phenomenology authors in science, technology, and ethics retain insights from phenomenology while trying to move beyond its limitations. Don Ihde (1993, 2003) suggests a post-phenomenology that is not centered on the subject but on embodiment. With the notion of "embodiment" he problematizes the ongoing interrelation between the active and perceiving body (or thing) and its environment of action (or use). Likewise, although Latour rejects phenomenology, he retains Heidggger's insight that a thing or tool is what it is within a referential whole. It is this perspective that makes it possible to conceive a thing as an "actant," and therefore constitute the "network" as a network. It would therefore seem reasonable to expect that phenomenology will remain important for those seeking to make sense of our relation with the phenomena of science, technology, and ethics.

LUCAS D. INTRONA

SEE ALSO *German Perspectives; Heidegger, Martin; Husserl, Edmund.*

BIBLIOGRAPHY

Bijker, Wiebe E.; Thomas P. Hughes; and Trevor J. Pinch, eds. (1987). *The Social Construction of Technological Systems: New Directions in the Sociology and History of Technology.* Cambridge, MA: MIT Press. This is a key text for social constructivism.

Borgmann, Albert. (1984). *Technology and the Character of Contemporary Life.* Chicago: University of Chicago Press. Borgmann's critique of technology in contemporary society.

Callon, Michel. (1986). "Some Elements of a Sociology of Translation: Domestication of the Scallops and the Fishermen of St. Brieuc Bay." In *Power, Action, and Belief: A New Sociology of Knowledge?* ed. John Law. London: Routledge and Kegan Paul. This is an important example of how science and technology becomes enrolled in political programs by different actors.

Coyne, Richard. (1995). *Designing Information Technology in the Postmodern Age: From Method to Metaphor.* Cambridge, MA: MIT Press. Applies phenomenology to the problem of information technology and design.

Dreyfus, Hubert L. (1979). *What Computers Can't Do: The Limits of Artificial Reason,* rev. edition. New York: Harper and Row. Dreyfus's devastating critique of artificial intelligence.

Dreyfus, Hubert L. (1991). *Being-in-the-World: A Commentary on Heidegger's "Being and Time," Division I.* Cambridge, MA: MIT Press. Good introduction to Heidegger's *Being and Time.*

Dreyfus, Hubert L. (1992). *What Computers Still Can't Do: A Critique of Artificial Reason.* Cambridge, MA: MIT Press.

Feenberg, Andrew. (1995). *Alternative Modernity: The Technical Turn in Philosophy and Social Theory.* Berkeley and Los Angeles: University of California Press.

Fry, Tony. (1993). "Switchings." In *RUA TV? Heidegger and the Televisual,* ed. Tony Fry. Sydney: Power Publications. Heideggarian account of our engagement with the televisual.

Hall, Harrison. (1993). "Intentionality and World: Division I of *Being and Time.*" In *The Cambridge Companion to Heidegger,* ed. Charles Guignon. Cambridge, UK: Cambridge University Press.

Heidegger, Martin. (1962). *Being and Time,* trans. J. Macquarrie and E. Robinson. Oxford: Blackwell. Heidegger's most influential work.

Heidegger, Martin. (1977). *The Question Concerning Technology, and Other Essays,* trans. William Lovitt. New York: Harper and Row. One of the key texts in the philosophy of technology.

Heim, Michael. (1993). *The Metaphysics of Virtual Reality.* New York: Oxford University Press. An excellent and accessible account of the metaphysical assumptions of virtual reality.

Heim, Michael. (1999). *Electric Language,* 2nd edition. New Haven, CT: Yale University Press. A phenomenologicaly informed account of writing technologies in general and of word processing in particular.

Husserl, Edmund. (1970). *The Crisis of European Sciences and Transcendental Phenomenology: An Introduction to Phenomenological Philosophy,* trans. David Carr. Evanston, IL: Northwestern University Press.

Husserl, Edmund. (1982). *Ideas: General Introduction to Pure Phenomenology,* trans. W.R. Boyce Gibson. London: Allen and Unwin; New York: Macmillan.

Husserl, Edmund. (1995). *Cartesian Meditations: An Introduction to Phenomenology,* trans. Dorian Cairns. Dordrecht, Netherlands: Kluwer Academic.

Ihde, Don. (1990). *Technology and the Lifeworld: From Garden to Earth.* Bloomington: Indiana University Press.

Ihde, Don. (1993). *Postphenomenology: Essays in the Postmodern Context.* Evanston, Illinois: Northwestern University Press. Ihde's development of a postphenomenology for studying technology.

Ihde, Don. (2003). "If Phenomenology Is an Albatross, Is Post-phenomenology Possible?" In *Chasing Technoscience: Matrix for Materiality,* ed. Don Ihde and Evan Selinger. Indianapolis: Indiana University Press.

Latour, Bruno, and Steve Woolgar. (1986). *Laboratory Life: The Construction of Scientific Facts.* Princeton, NJ: Princeton University Press.

Lévinas, Emmanuel. (1969). *Totality and Infinity,* trans. Alphonso Lingis. Pittsburgh: Duquesne University Press. Important work of the "early" Levinas.

Lévinas, Emmanuel. (1991). *Otherwise than Being; or, Beyond Essence*, trans. Alphonso Lingis. Dordrecht, Netherlands: Kluwer Academic.

Merleau-Ponty, Maurice. (1962). *Phenomenology of Perception*, trans. Colin Smith. New York: Humanities Press. An important work for understanding the "embodied" literature in studies of technology.

Mohanty, Jitendranath N. (1997). *Phenomenology: Between Essentialism and Transcendental Philosophy*. Evanston, IL: Northwestern University Press. A good summary of some of the important ideas of phenomenology.

Pickering, Andrew. (1995). *The Mangle of Practice: Time, Agency, and Science*. Chicago: University of Chicago Press. Shows the reciprocal interrelation between technology, humans, society, and scientific knowledge as produced in scientific and technological work.

Spiegelberg, Herbert. (1994). *The Phenomenological Movement: A Historical Introduction*, 3rd edition. Dordrecht, Netherlands: Kluwer Academic. A vast survey of most of the important authors in phenomenology.

Winograd, Terry, and Fernando Flores. (1986). *Understanding Computers and Cognition: A New Foundation for Design*. Norwood, NJ: Ablex Publishing. The application of Heidegger's work to the design of organizations and computer systems.

PLACE

• • •

Attention to the idea of place has grown steadily since the 1980s in the context of an increased focus upon physical localities and new ways of organizing knowledge. These developments have been in part a response to the forces of globalization. The world has become an economic and cultural commons: Companies such as Wal-Mart and Starbucks have extended their reach worldwide, as have Hollywood films, which now earn the majority of their revenues overseas. Moreover, modern science, technology, and economics have not just homogenized space; in many respects they have annihilated it. Air travel has become ubiquitous, and both information and human identity now mutate within the hyperreal environment of cyberspace. As a result, there is a growing feeling that people inhabit a "Geography of Nowhere" (Kunstler 1993) not only in the sense of the homogeneity of shopping malls and suburban tract homes but also in terms of the uniformity that results from the relentless drive toward individuation.

Attention to the distinctiveness of places from the perspective of architecture, geography, philosophy, or personal narrative thus may be seen as a reaction to contemporary social and economic unifications. More basically, however, the focus on place marks the recognition

of the irreducible fact that people attach themselves to and live out of particular landscapes, cultures, and bodily experiences (Tuan 1990). Abstractions—scientific, religious, or otherwise—must be complemented by the experience of lived, concrete existence. People construct their sense of identity through being born into or making a commitment to a landscape, a country, a culture, or a profession or "position." Attraction to place thus represents a persistent aspect of the human condition.

Recent attention to the concept of place also represents a response to deficiencies in the contemporary organization and use of knowledge. It is here that the term gains particular salience in relation to science, technology, and ethics. Environmental concerns offer especially good examples of place-based approaches to knowledge that strive to blend science, technology, and social concerns.

The Meaning of Place

The term *place* is used in several senses. It can identify a particular physical location such as the greater Yellowstone ecosystem, a mental location such as the Vietnam era, one's position in a social hierarchy, a field of research such as women's or Chicano studies, or a subject of controversy such as science wars. What all these senses share is participation in a part-whole relationship: A place is a delimitation within a larger geography, whether natural, cultural, or personal. However, to qualify as truly distinctive a location in some way must escape the terms that define all other spaces. This implies that there is something ineffable in the concept of place.

Place exists in an uneasy dialogue with the concept of space. In the early twenty-first century space is understood most commonly as a concept of physics, denoting the entirety of quantitative, mappable extension. In contrast, place typically is viewed as a psychological concept that highlights a person's subjective, affective response to a particular fragment of the world. This account, however, inverts the relation between place and space. Whatever powers science may have to describe extension, a person's initial experience of reality is always perspectival (Malpas 1999). Objective, mathematical accounts of space are derivative of people's embeddedness in particular places.

Place and Knowledge

Concerns with place have been particularly important in regard to the role of knowledge in society. Topical approaches to knowledge (from the Greek *topos*, meaning "place") challenge the traditional disciplinary manner of organizing and applying knowledge in terms of chemistry, history, and the like. This is not a trivial point: The way knowledge is organized determines the types of questions society asks as well as the way its questions are formulated. If, for instance, science and technology are viewed as forms of knowledge that are distinct from ethical and philosophical knowledge—or if ethics and philosophy are not considered "real" knowledge—it follows that the ethical consequences of the productions of scientists and engineers will be seen as quite distinct from their research.

Topical approaches to knowledge are part of a larger movement aimed at critiquing accepted practices within academia and other locations of knowledge production. In recent years interdisciplinary and transdisciplinary approaches to knowledge have become more prominent. Recent initiatives within Federal agencies such as the National Science Foundation (for example, funding for the IGERT, the Integrative Graduate Education and Research Traineeship program) highlight the increasing pressures on scientists and engineers from different disciplines to practice interdisciplinary collaboration in dealing with particular issues. Although place-based approaches are best seen as complementary to rather than in opposition to the disciplines, topical thinking nonetheless represents a new imperative, breaking through the logical space of disciplines to achieve a better purchase on human problems.

By its very nature a disciplinary approach to knowledge takes an analytic approach to its subject matter. The philosopher René Descartes (1596–1650) argued that people come to know a thing by breaking it into its smallest parts and thoroughly studying those parts. This is the analytic method. It is clear that this approach has a built-in bias in that it assumes that a problem can be subdivided into discrete units without a loss in understanding. Some issues, however, most notably environmental ones, are essentially holistic in nature. The Greater Yellowstone ecosystem, for instance, consists of something more than a series of "disciplines" (geologic, hydrologic, economic, and so on) worked on by different teams of professionals. As necessary as an understanding of these different systems is for improving the health of the ecosystem, environmental problems resist simple division into the categories of environmental science, economics, ethics, and the like. To address such problems effectively it is necessary to understand how those disciplines relate to and flow into one another at a particular location.

Topical thinking provides a means for tracing the ontological disruptions that occur when one attends to the holistic nature of a problem. Certainly a complex issue must be divided into pieces to understand its moving parts, but it also is necessary to retain a sense of the whole, seeking to understand the relation between and across the disciplines in a particular place. Otherwise one is left with a type of educated incoherence, with experts inhabiting their own privileged stances, largely failing to communicate with one another or with the public.

Place and the Environment

Environmental issues provide good examples for assessing the success of topical approaches to societal problems. This is the case in part because nature presents some of the most distinctive locations imaginable. Founding environmental documents such as the Endangered Species Act of 1973 reflect the importance people attach to the unique, as do attempts to eliminate "exotic" plants and animals to protect the distinctiveness of natural places.

Battles over issues such as the Arctic National Wildlife Refuge (ANWR) demonstrate the deeply interpretive nature of a topical approach to problems. Science has a reputation for providing objective information, but as Aristotle noted in the *Poetics*, there can be no science of the individual. When the U.S. Geological Survey attempted to determine oil reserves at the ANWR, its estimates varied from 4.3 to 11.8 billion barrels of oil, a range that was used to defend a variety of policy recommendations. Engineering questions were equally vexing, with some experts claiming that new engineering techniques could make drilling in the Arctic safe and others stating that the risks remained too great. Even the question of degrees of wilderness and natural beauty was debated: The ANWR was described as both pristine and having a long history of human modifications and as both stunningly beautiful and a boggy wasteland. Again, there is no science of the individual: Answers to the question of whether it is safe to drill in the ANWR cannot come through laboratory experiment or computer modeling, only through actual experiments.

Rather than seeing such results as repudiating claims about the usefulness of topical approaches to knowledge, one can view a topical approach as stripping the pretensions from types of knowledge that claim to escape the skein of interpretation. Science retains its claims to objectivity only by locking itself up in the laboratory. The clarity of disciplinary knowledge is bought at the cost of abstraction from the real world. Bringing knowledge into the field and to specific locations increases its relevance to people's lives.

ROBERT FRODEMAN

SEE ALSO *Environmental Ethics; Space.*

BIBLIOGRAPHY

Bachelard, Gaston. (1964). *The Poetics of Space.* Boston: Beacon Press. Classic treatment of human lived experience of space by a preeminent philosopher of science.

Casey, Edward. (1993). *The Fate of Place: A Philosophical History.* Berkeley: University of California Press. Traces the history of place and space from the Greeks to the postmodern theories of Foucault, Derrida, and Irigaray.

Kunstler, James Howard. (1993). *The Geography of Nowhere: The Rise and Decline of America's Man-Made Landscape.* New York: Simon & Schuster. A readable and provocative work that examines the role of the automobile in the destruction of American cities.

Light, Andrew, and Jonathan Smith, eds. (1999). *Philosophies of Place: Philosophy and Geography III.* Lanham, MD: Rowman and Littlefield. Includes the reflections of philosophers and geographers on the decline and attempted recreation of public spaces.

Malpas, J. E. (1999). *Place and Experience: A Philosophical Topography.* New York: Cambridge University Press.

Tuan, Yi-fu. (1990). *Topophilia: A Study of Environmental Perception, Attitudes, and Values.* New York: Columbia University Press. Influential book by a well-known geographer that explores the bonds between people and place.

INTERNET RESOURCE

Gross, Matthias, and Holgar Hoffmann-Riem. (2004). "Real-World Experiments: Strategies for Robust Ecological Design." Available from http://www.uni-bielefeld.de/iwt/realworld/index.html.

PLAGIARISM

• • •

Plagiarism is commonly defined as the unauthorized or unacknowledged appropriation of the words, graphic images, or ideas from another person. As such plagiarism can be a violation of intellectual property rights, although it is not in all cases illegal. It is in fact one of the most serious general issues in the practice of scientific scholarship, in part because its precise boundaries are not always easily determined and because concepts of plagiarism have evolved considerably over time. Opportunities for plagiarism and efforts to deal with it have also altered in conjunction with technological change.

Plagiarism overlaps yet is distinct from copyright infringement; the latter can occur with proper attribution, while plagiarism cannot, and the latter generally occurs with the failure to obtain permission to use copyrighted material (Lindey 1952). Further, while definitions of copyright infringement are based on statute, definitions of plagiarism—and their resultant interpretations—vary considerably across institutions (Myers 1998). According to a committee convened by the National Academies of Sciences and Engineering and the Institute of Medicine (1995), plagiarism and the falsification and fabrication of data or results constitute three forms of deceptive scientific misconduct that impede scientific progress as well as endanger foundational scientific norms. Data obtained from National Science Foundation and National Institutes of Health investigations into allegations of misconduct in the late 1980s and early 1990s suggest that plagiarism is the more common of these three (LaFollette 1992).

Historical Emergence

One interpretation of the evolving perspective on plagiarism holds that plagiarism was not discussed as an ethical issue until after the rise of individualism during and especially after the Renaissance. For instance, classical authors sometimes copied from each other without explicit acknowledgment, and occasionally attributed their own works to another author (a kind of reverse plagiarism) because they saw them as part of a tradition better represented by someone else. There are a number of pseudonymous works of Plato and Aristotle, and the first five books of the Bible, although attributed to Moses, were almost certainly written by someone else. In the area of graphic representation, all works of art from the studio of a master were commonly attributed to that master.

Such a view is nevertheless complicated by multiple factors, including classical understandings of originality, which significantly influenced sixteenth-century Italy and France (White 1935). Greek and Roman authors prized the imitation of previous works and considered established subject matter a common inheritance. Imitation was not synonymous with copying or piracy because classical authors often earned respect by identifying their esteemed sources. Further, established works were to be judiciously selected and reinterpreted via one's own experience specifically to expand and even surpass their prior treatment. To explain their notions of originality, classical writers such as Seneca and Plutarch used the metaphor of the bee, which draws nectar from many flowers yet transforms these into an altogether new creation. The classical notion of originality has influenced the modern scientific enterprise, as has the classical authors' general (though not universal) disdain for unattributed sources (White 1935).

The value placed on imitation also influenced literary and nonliterary texts written in early modern England (c. 1500–1800), a period in which views of plagiarism varied widely, from ethically venerable to venial or vile (Kewes 2003). This range of reactions stemmed in part from the complex interactions between plagiarism and "imitation, borrowing, adaptation, allusion, intertextuality, appropriation, copyright infringement," and other concepts (Kewes 2003, p. 2). Genre and intellectual context also mattered. For instance, some seventeenth-century religious figures did not acknowledge their debt to sources they copied or received inspiration from as they felt those covert sources could strengthen the power of their sermons. During the same period, Robert Boyle (1627–1691), often considered the father of modern chemistry, reprimanded those who had appropriated his experimental work without full attribution.

Technology shaped plagiarism perspectives considerably. General public disapproval of unacknowledged copying across contexts increased when printed texts (as opposed to handwritten ones) became more commonplace, in part because of publishers' desire to secure revenue. Moreover, after 1750 the value on imitation declined not coincidentally with an increased emphasis on originality (Kewes 2003) and the growth of individualism.

Contemporary Issues

Plagiarism is also a contemporary cultural issue. Because originality and individualism are viewed through cultural norms, perceptions of plagiarism vary widely across cultures and often contrast with the predominant Western view, especially among cultures that emphasize the community over the individual. Thus, in certain cultures "using the words and ideas of others without attribution is considered a sign of deep respect as well as an indication of knowledge" (Lunsford 2004, p. 169). By contrast, normative views in Europe and North America are embedded in the very origin of the word *plagiarism*, which derives from the Greek *plagios*, meaning crooked or treacherous, from which comes the Latin *plagiarius*, meaning kidnapper. Because the vast majority of scientific gatekeeping and production originates in Europe and North America, it is these notions of plagiarism and intellectual property that have been widely disseminated across cultures. Whether this dissemination constitutes linguistic and cultural hegemony has been debated (e.g., see Myers 1998; Scollon 1995).

Beyond its cultural aspects, plagiarism in the early twenty-first century is complicated by the difficulties in identification as well as the role of technology. Although identifying plagiarism may seem straightforward, context matters. For instance, paraphrased or summarized common knowledge does not require attribution, yet what is considered common knowledge varies among audiences. Further, whether the smallest unit of plagiarism should be the paragraph, sentence, or the phrase is open to debate (e.g., see St. Onge 1993). Also, if plagiarism is a deceptive form of scientific misconduct, it is important—yet sometimes complicated—to determine whether an act was malicious or unintentional.

Identifying plagiarism can also be intricate because scientific and technological works are frequently collaborative creations, with shaping influences from colleagues, coauthors, peer reviewers, journal referees, and editors. Although guidelines to distinguish various contribution levels help determine who should be referenced, acknowledged, or listed as a coauthor, these guidelines are far from universal (see NAS, NAE, and IOM 1995). Some researchers would define a given contribution as worthy of coauthorship, whereas others would classify the same contribution as worthy only of acknowledgment (see Buzzelli 1993).

Technological changes have created additional opportunities for plagiarism as well as its detection. For instance, multisite research collaborations involve significant electronic information sharing, which increases the amount and availability of information that can be plagiarized. Additionally, individuals raised with free Internet music, software, and other information have grown accustomed to easily accessible information, which may engender attitudes regarding plagiarism that differ from the previous few generations. The same information technologies that can facilitate plagiarism, however, are being used to expose plagiarists; Internet-based plagiarism prevention and detection services compare electronically submitted texts against massive information databases to identify unoriginal material.

Perhaps in part because contexts, definitions, and interpretations of plagiarism vary, consequences for plagiarizing also vary. Generally, cases of proven plagiarism can involve demotion, job loss, and varying degrees and types of loss of respect and ostracism from one's own field. Marcel C. LaFollette (1992) tells of a former director of the National Institute of Mental Health whose early-career plagiarism was detected, leading to his resignation from his academic positions and, several months later, reinstatement based on the merits of his overall career contributions. In another case, an award-winning malaria researcher at the Harvard School of Public Health was accused of plagiarizing portions of a National Institutes of Health (NIH) grant. After an investigation by both Harvard and the federal government's Office of Research Integrity, the accused assistant professor resigned and became ineligible to apply for federal funding for three years (Glenn 2004). A Spanish journal of micropaleontology stopped accepting manuscripts indefinitely from a researcher who for twenty years had allegedly plagiarized pictures of diverse organisms from other publications (Bosch 2004).

Consequences become particularly complex with plagiarism accusations between colleagues of different ranks. When a graduate student at Arizona State University accused his acclaimed mentor in plant biology of copying part of his work, the mentor stated that such practices were common in science and that he was justified because the graduate student was part of his research team. One-third of the mentor's article was reportedly taken directly from the student's work, which itself had appeared in an earlier publication. Shortly after the student contacted the editor who published his mentor's work, the student indicated that he experienced exclusion from major research projects. The mentor is a member of the National Academy of Sciences, formerly served on the editorial board of the journal *Science*, and was appointed by President George W. Bush to the President's Council on Science and Technology (Bartlett and Smallwood 2004b). In addition to the issue of power inequity between accuser and accused, rendering judgment in this case may have been complicated by the possibility that the university committee charged with the investigation perceived the reputation of the mentor (their colleague) as inextricably tied to the university's reputation. The consequences of plagiarism have become of increasing interest in an era in which a significant portion of scientific research is supported by public funds (Miller and Hersen 1992). Whether plagiarism is common or rare in science and technology research is a topic of debate (see LaFollette 1992, Miller and Hersen 1992, Bartlett and Smallwood 2004a).

Closely related to the issue of consequences are the mechanisms for addressing plagiarism allegations. Controversy exists over whether plagiarism cases should be handled by government agencies, university or other presses, professional societies, the legal profession, academic institutions, or some combination of these, and many are handled according to the specific attributes of the case. Because of the potential enormity of legal expenses, some professional societies have expressly refused any involvement in prosecuting plagiarism cases, and universities may also be wary of the costs of legal action (see Glenn 2004).

JON A. LEYDENS

SEE ALSO *Misconduct in Science*.

BIBLIOGRAPHY

Bartlett, Thomas, and Scott Smallwood. (2004a). "Four Academic Plagiarists You've Never Heard Of: How Many More Are Out There?" *Chronicle of Higher Education* 51(14): A8.

Bartlett, Thomas, and Scott Smallwood. (2004b). "Mentor vs. Protégé." *Chronicle of Higher Education* 51(14): A14.

Buzzelli, Donald E. (1993). "Plagiarism in Science: The Experience of NSF." *Perspectives on the Professions* 13(1): 6–7.

Glenn, David. (2004). "Judge or Judge Not?" *Chronicle of Higher Education* 51(14): A16.

Kewes, Paulina, ed. (2003). *Plagiarism in Early Modern England*. Basingstoke, Hampshire, UK: Palgrave Macmillan.

LaFollette, Marcel C. (1992). *Stealing into Print: Fraud, Plagiarism, and Misconduct in Scientific Publishing*. Berkeley: University of California Press.

Lindey, Alexander. (1952). *Plagiarism and Originality*. New York: Harper.

Lunsford, Andrea A. (2004). *The Everyday Writer*, 3rd edition. New York: Bedford/St. Martin's.

Miller, David J., and Michel Hersen, eds. (1992). *Research Fraud in the Behavioral and Biomedical Sciences*. New York: Wiley.

Myers, Sharon. (1998). "Questioning Author(ity): ESL/EFL, Science, and Teaching about Plagiarism." *Teaching English as a Second or Foreign Language* 3(2): 1–21.

National Academy of Sciences (NAS), National Academy of Engineering (NAE), and Institute of Medicine (IOM). Committee on Science, Engineering, and Public Policy. (1995). *On Being a Scientist: Responsible Conduct in Research*, 2nd edition. Washington, DC: National Academy Press.

Scollon, Ron. (1995). "Plagiarism and Ideology: Identity in Intercultural Discourse." *Language in Society* 24(1): 1–28.

St. Onge, K. R. (1993). "The Threshold of Plagiarism." *Perspectives on the Professions* 13(1): 2–3.

White, Harold Ogden. (1935). *Plagiarism and Imitation during the English Renaissance: A Study in Critical Distinctions*. Cambridge, MA: Harvard University Press.

INTERNET RESOURCE

Bosch, Xavier. (2004). "Plagiarism in Paleontology." *Scientist* September 22. Available from http://www.the-scientist.com.

PLANNING ETHICS

• • •

Planning is both a profession and a discipline that has at its foundation questions of how to best develop land, social programs, housing, parks, health services, and other aspects of human settlements. Planning ethics is focused on terms such as *best* as it appears in this characterization of planning, where ethics, or moral philosophy, provides a means of analyzing normative ways of responding to planning challenges.

Planning began largely as a community-led process focusing on aesthetics, safety, and health concerns at the neighborhood level. As planning became professionalized in North America in the late-1800s and early-1900s, urban design, economic vitality and order, beauty, and efficiency became prominent considerations. Planning issues later expanded to include environmental conservation and preservation, energy consumption, empowerment (including public participation), and heritage conservation (Hodge 1998, Krueckeberg 1994).

Historically the planning profession has evolved from an almost exclusive focus on the technical aspects of developing and conserving land to concern with a more holistic view of urban areas and regions. It has changed its disciplinary base from emphasizing engineering and architecture to striving for balance among the natural, physical, and social sciences. In addition, planning processes have shifted from focusing on technical, value-neutral expertise to addressing communicative processes, value-laden and normative analyses, and facilitation/mediation. Planning is thus often described as an art as well as a science (see for instance Canadian Institute of Planners 2004). While debates regarding these shifts are clear and progressive in academic circles, it is fair to say that society continues to view planners largely as technical experts in land development and, to a lesser extent, social and health programming.

Planning, Science, and Technology

Planning, science, and technology are connected in multiple ways. The use of science and technology by planning and planners is clear in the form of mapping techniques such as Geographical Information Systems (GIS), ecological theories, analytic and computing techniques, computer aided design, and others. Indeed one of the central criticisms of planning as an autonomous profession is the fact that it borrows methods, techniques, and tools from the social, natural, and physical sciences as well as the arts. This calls into question its independence as a field of inquiry, but is also often regarded as a strength in terms of underlining the inter- and multi-disciplinary nature of planning.

In addition to the use by planners of science and technology, technological advances have been linked to changing urban forms and activity patterns. Wireless communication, for example, calls into question the shape(s) of cities, transportation flows, and employment locations. Such phenomena have altered the perceptions and analyses of planners who help to shape these areas.

Conversely planning contributes to science and technology by demonstrating the effects of scientific theories and technological advances *on the ground* and can thus play a role in their refinement. Planners must, at the end of the day, develop a plan or make a recommendation, while scientists often study a given problem for an unlimited period of time. In this way, science and technology as used in planning has an immediacy that may lead to the adoption, adaptation, or abandonment of a given development.

Planning Ethics, Science, and Technology

Ethical aspects of planning, science, and technology may be discussed in terms of research as well as professional practice. Planning academics conduct research that contributes to the development of the field; planning practitioners also conduct research but their work is typically limited to issues with which they must deal in their everyday work. The ethical issues in both activities are similar, although particulars change. The following discussion includes examples from both fields of endeavor.

Ethics is used here as a synonym for moral philosophy; it does not replace other terms such as values, beliefs, morality, and morals. Instead it connotes a way of studying and addressing moral problems utilizing ethical theories and rigorous analysis. Ethical theories such as utilitarianism, Kantian thought, communitarianism, and rights-based morality, among others from sub-fields such as feminist ethics and environmental ethics, are used to help explore normative issues in planning and arrive at viable solutions.

Planning ethics, as part of professional ethics and, more generally, applied ethics, has been discussed in terms of five separate aspects of the field (Wachs 1985, Hendler 1995): everyday behavior; plans and policies; administrative discretion; the normative intent of the planning endeavor (planning theory); and planning techniques. Each category of ideas and action includes reference to issues of science and technology.

EVERYDAY BEHAVIOR. Everyday behavior refers to the actions of planners in the day-to-day context of their work. Conflict of interest is a typical ethical issue here. Should a land use planner accept a gift in return for expediting a development proposal? Should social planners bias a new service or program in ways that would help their family members? While such issues are commonly discussed in terms of planning ethics, they do not exhaust the field. As Joan Tronto argues, professional, including planning, ethics should be "about more than teaching [planners] that it is wrong to lie, to cheat or to steal (Tronto 1993, p. 134)."

Planners' behavior often includes their use of science and technology and the ethical aspects of that use. Most behaviors are linked to techniques and assessments that determine the efficacy of plans and policies. Some, however, include ethical issues that pertain directly to routine professional etiquette. For example, a bribe or a conflict of interest may pertain to the massaging of data (facilitated by such things as the sheer size of data sets that can be manipulated by computer programs), not only to the approval of a development proposal. Other concerns are equity, treatment of vulnerable populations (publics as well as colleagues), and relations with other professionals such as engineers and computer technologists. Included are such issues as sharing information with publics; for example, is communication via web sites and other means that require access to technology ethical when the target group is economically disadvantaged and may not have such access?

PLANS AND POLICIES. Plans and policies are inherently normative in that they allocate or reallocate resources among groups and individuals in a community or region. It is this normative content of plans and policies, as well as programs and projects, that is most strongly linked to ethics. Ethical theories and perspectives provide a conceptual basis for normative decisions in that a rights-based view of ethics, for example, directs plans and policies differently than would a utilitarian ethical theory. A plan that includes the provision of a transit route, or a park, in a particular neighborhood means that certain amenities will be part of this plan for particular people but not others. Such a distribution of *benefits* and *costs* is subject to ethical assessment.

Analyses of costs and benefits, as well as assessments of other ethical aspects of a plan or policy (such as social justice concerns), rest partly on the shoulders of science and technology. For instance, ecologists, environmental scientists, and other scientists, who study such things as carrying capacity, ecological stress, and environmental assessment, can determine whether a plan will result in the demise of a species or valued natural area. Similarly a transportation plan may be analyzed with regard to its ethical implications but one must understand the technical aspects of pollution generation and abatement, economic considerations, not to mention safety-related concerns pertaining to the materials used in road construction, the physical integrity of bridges, and traffic moderation, in order to conduct a rigorous ethical analysis. Further given the dearth of developable land in most urban centers, remediation of brownfield areas (lands on which polluting uses have occurred, thus necessitating corrective action) has become popular and the safety, cost, and efficiency of such action is subject to ethical, as well as scientific, analysis.

ADMINISTRATIVE DISCRETION. Administrative discretion pertains to the fact that planning roles are diverse and often ambiguous. This means that planners are often able to choose the role they wish to assume at any given time, where roles may vary from technician to mediator to advocate. This discretion gives rise to ethical considerations in that the selection of one role over another has implications for planners in their work. These implications pertain to clients, colleagues, employers, and publics in that all must know what to expect from the planners with whom they are working.

Planners may select roles that have more or less to do with science and technology. If they assume a role in which such expertise is required, they must ethically ensure that they have the necessary knowledge to act in this capacity. Most professional codes address this by referring to professional competence as a requirement for accepting, or carrying out, professional tasks. In addition to questions of competence, however, it is also possible that science and technology both broaden and restrict the role choices available. That is, some roles may be restricted when they require skills that are beyond the technical capacities of most planners. Conversely role choice might be broadened if such things as communication technologies make it easier for planners to collaborate with other experts, thus leading to more teamwork in planning.

PLANNING THEORY. Ethical aspects of planning theory pertain to the fundamental questions of why the planning profession should morally exist and how it is justified. Upholding individual rights, striving for maximum benefits for the greatest number of people, maintaining ecological integrity, ending oppression, and building community are all possible moral goals of the planning field (Beatley 1994, Hendler 1995, Howe 1994, Wachs 1985).

Science and technology enter into planning theory by indicating what is possible or feasible. It makes little sense to strive toward a goal that is, in fact, not physically achievable. Information provided by scientists indicates to planners what goals are reasonable in the face of available scientific knowledge. More specifically, an ethical analysis can suggest a way of life for society and, hence, to planners (for example, sustainable development). A scientific analysis can provide options as to how to achieve this goal and appropriate technology can assist in its implementation (solar energy, for example).

PLANNING TECHNIQUES. Planners use many analytic techniques ranging from statistical methods to economic forecasts to qualitative approaches. These are in addition to the methods inherent to each natural, physical, or social science that, together, make up the toolkit for most planning professionals. These techniques are connected to ethical ideas in that most make normative assumptions about their subject matter and such assumptions may be subject to ethical scrutiny. For example, assessing what is of sufficient value to count in a quantitative assessment of a particular development is an ethical, not a technical, question.

It is surprising to many scientists, as well as planners, to find that their methods and analyses are value-laden. This view of science and knowledge in general is consistent with a post-positivist perspective of the world in which it is recognized that experiencing the world from the perspective of a blank slate is not possible. Scientific knowledge is generated by people who perceive the world through particular lenses or filters. The best that science can do is to be as transparent as possible about this fact and its possible effect on decision making. The inherent subjectivity of knowledge is thus inescapable but this fact does not preclude critical assessment.

The choice of scientific techniques thus becomes subject to ethical inquiry, as do the data generated by such techniques or methods. Risk assessment, for instance, rests on definitions of risk, allocation of weights and probabilities to these risks, and normative conclusions as to what level of risk is appropriate. As already suggested, the same holds true for cost-benefit analysis, as well as environmental assessment. Forecasting methods and population projections, typically used in transportation planning, health planning, land use, and social planning are well-known for their often implicit value bases. More generally, certain methods can be linked to particular ethical theories; cost-benefit analysis, for instance, has been shown to be consistent

with themes in utilitarianism—an ethical theory that holds that preferred actions should result in greater aggregate benefits. While entirely legitimate as one moral argument, the use of a method that rests solely on this sort of theoretical base can be problematic insofar as it neglects other values such as individual rights, community, and more. Analyses based on one restricted method may be criticized for ignoring other equally legitimate moral positions. Similarly computing techniques, such as mapping large quantities of data in order to show distributions of such things as literacy, poverty, and illness are useful in helping planners to distribute needed services. However in amalgamating these data (in true utilitarian fashion), minority populations and their needs are often left out, given the emphasis of utilitarianism on "the greatest good."

Related to these methods is the issue of norms and standards; many plans and planning processes include the use of quantitative guidelines such as X amount of parkland for Y number of people (for example, one acre of neighborhood park area per 1000 population). The efficacy and usefulness of such standards is a legitimate ethical question, especially because most are conventions that were developed in very different historical and socio-political contexts, often with little in the way of logical or empirical justification. Also of relevance here are such things as allowable, tradeable pollution levels that enable planners to plan differently than they would if there were standards that were cast in stone. Planners concerned about air pollution, for example, would need to write their development plans in a way that incorporated more in the way of uncertainty if industries within their jurisdiction were allowed to generate more or less pollution by trading their emission allowances with other industries and still staying within their permitted limits.

In a more positive vein, advances in computer technology enable public participation—a longstanding tenet of good (hence, ethical) planning, given its contemporary normative emphases on democracy, empowerment and diverse interpretations of 'the' public interest. Such technology, through video conferencing, instant messaging, listservs, and chat rooms, provides potentially accessible means for discussion of planning issues and, perhaps, arrival at consensus on such issues. This applies especially to remote areas in which residents are not concentrated in a single geographic locale. Issues of equity, however, arise in ensuring that the populations most sought after in terms of their participation are indeed those able to access the technology needed to have a voice in the planning issue at hand.

Planning Ethics, Science, and Technology in Practice

Within these five categories of planning ethics, the consideration of planning techniques displays the deepest connections to science and technology. Yet as also indicated, each aspect of planning includes reference to issues of science and technology and accompanying ethical concerns. Scientific and technological developments change the face of planning and of planning ethics by altering the analytic tools and descriptive information available to planners making decisions that will impact people's lives.

All of these themes are manifested in the professional codes of planning organizations. Such codes are vehicles for ethical analysis and direction in that they present practical guidance for planners facing ethical problems, while also providing a vision of what the profession should be trying to accomplish. Developments in science and technology, however, are often poorly addressed in professional codes; such developments occur at a pace that is difficult to maintain in terms of revising and adopting a code of ethics or a code of conduct (see, for example, Canadian Institute of Planners 2004, American Institute of Planners 1991). For example, fast-moving advances in computer technology, which facilitate the fraudulent manipulation of information and which can be adapted, with increasing ease, to circumvent safeguards, should be an important consideration in professional codes. However, because professional organizations revise their codes sporadically at best, practitioners are left to extrapolate solutions for emergent and rapidly changing problems from dated principles. Similarly, the positive contributions made by science and technology to planning and planners should be addressed in professional codes as an example of good professional practice. For example, and as suggested above, facilitating public participation with the use of computer technology could be cited as ethically appropriate in the sections of codes that deal with planners' responsibilities to various publics.

Whether codes of ethics and conduct will keep pace with the challenges provided by scientific and technological developments remains to be seen. That the work of planners rests on science and technology is clear; what is less clear is how and whether science and technology can assist planning and planners in addressing their basic ethical concerns or whether they simply add their own ethical issues to the mix. Either way, discussions in planning ethics mirror, and contribute to, fundamental debates in ethics, science, and technology.

The interdisciplinary and applied nature of the planning field is a strength in this regard in that its analyses are far-reaching and pragmatic. The outcomes of the ethical decisions of planners, in their use of science and technology, become part of the lives of ordinary people in cities and regions. They thus become subject to scrutiny by all. Subsequent accountability by those accorded the status of *professional*, with all of the ethical implications of this label, necessarily follows.

SUE HENDLER

SEE ALSO *Management.*

BIBLIOGRAPHY

American Institute of Certified Planners. (1991). *AICP Code of Ethics and Professional Conduct.* Washington, DC: American Planning Association.

Beatley, Timothy. (1994). *Ethical Land Use.* Baltimore, MD: John Hopkins University Press.

Canadian Institute of Planners. (2004). *Statement of Values and Code of Professional Practice.* Ottawa: Author.

Frankena, William K. (1973). *Ethics,* 2nd edition. Englewood Cliffs, NJ: Prentice-Hall.

Hendler, Sue, ed. (1995). *Planning Ethics. A Reader in Planning Theory, Practice and Education.* New Brunswick, NJ: Center for Urban Policy Research. This is a collection of essays focused on the integration of ethics and planning. A wide variety of ethical theories and planning issues, including planning education, is represented.

Hodge, Gerald. (1998). *Planning Canadian Communities,* 3rd edition. Toronto: International Thomson Publishing.

Howe, Elizabeth. (1994). *Acting on Ethics in City Planning.* New Brunswick, NJ: Center for Urban Policy Research.

Krueckeberg, Donald, ed. (1994). *The American Planner: Biographies and Recollections,* 2nd edition. New Brunswick, NJ: Center for Urban Policy Research. This book represents a biographical approach to planning history in which notable men (and some women) are introduced to readers with emphasis on their contributions to planning cities and regions in the United States.

So, Frank; Israel Stollman; Frank Beal; and David Arnold. (1979). *The Practice of Local Government Planning.* Washington, DC: International City Management Association.

Tronto, Joan. (1993). *Moral Boundaries: A Political Argument for an Ethic of Care.* New York: Routledge.

Wachs, Martin, ed. (1985). *Ethics in Planning.* New Brunswick, NJ: Center for Urban Policy Research. As the first book-length discussion of ethics and planning, this collection includes discussions of ethical aspects of planning organizations, policymaking, administration and environment.

PLASTICS

• • •

Technologies have world-shaping powers. They have fundamentally changed ways of thinking as much as they influenced social practices. Plastics form a striking case.

Human beings are surrounded by plastics, in their computers, clothes, cars, kitchens, and beds, on their noses, and often in their bodies, in the form of hearing aids, hip replacements, and heart valves. In the early-twentieth century they were an odd curiosity; a century later a world without plastics is unthinkable and unlivable. They have permeated every conceivable practice and in most of these made themselves indispensable. It would, for example, be impossible to have twenty-first century supermarkets without plastic packaging, because the supermarket system is dependent on lightweight, airproof, and pre-packaged goods. In fact the transition from the traditional grocery store to the supermarket system was strongly encouraged by the emerging availability of plastic packaging materials in the 1950s and 1960s.

Plastics Science and Technology

The noun *plastics* is derived from the Latin *plasticus*, itself rooted in the Greek *plasein* meaning to mold; by connotation plastics are thus pliable, malleable, and adaptable. In scientific language, plastics are called polymers, a large and divergent group of materials with a wide range of properties. Their shared characteristic is that they consist of synthetically produced macromolecules, molecules about 1,000 to 100,000 times larger than, for example, the molecules of water or sugar. In a broad sense, synthetic rubbers and resins may also be called plastics. Some macromolecular materials (such as rubbers and resins) are found in nature, but the revolutionary thing about plastics is that they can be synthesized in the laboratory.

Launched in 1868 with the synthesis of celluloid by the American inventor John Wesley Hyatt (1937–1920), polymer synthesis was followed around the turn of the nineteenth century into casei formaldehyde (synthetic horn) and fenolformaldehyde, better known as Bakelite. These more or less accidental findings preceded the scientific understanding of macromolecular structures, which were first elucidated by the German chemist and Nobel Prize winner Hermann Staudinger (1881–1965) and his students in the 1920s. Chemistry thus opened the door to a riot in plastics design. New types that turned out to be especially successful included polyethylene (PE), polypropylene (PP), polyvinyl chloride (PVC), nylon, polystyrene (PS), and the synthetic rubber styrene butadiene rubber (SBR).

Cultural History

On top of this scientific and technological history is another even more exciting one concerning the public image and social embedding of plastics. As an exemplary case of the cultural response to controversial technology the history offers rich material for philosophical and ethical reflection. In all European and North American countries, the public appreciation of plastics exhibits a whimsical pattern, filled with opposite emotions and paradoxes, soaked with utopian and dystopian fantasies.

One peculiarity in the cultural history of plastics is that appreciation was out of step with development. Parallel to dramatic advances in quality and numbers of applications, the image of plastics deteriorated rapidly. The same qualities that were initially praised—such as their cheapness, lightness, unnaturalness, durability, moldability, imitative properties, ability to be mass produced, and resistance to wear and tear—subsequently became the basis of criticisms.

Jeffrey Meikle's *American Plastic* (1995) offers an excellent overview of this cultural transformation. The book is a gold mine of facts, stories, and opinions on plastics during the twentieth century, focusing on the United States but with some foreign perspectives as well. As an historian, Meikle does not articulate nor theorize the patterns of extreme and opposite public reactions, which call for philosophical interpretation. The public reactions from fascination to abomination cannot be explained by any simple irrationality or gut feelings on the part of the public, as is often claimed. Rather the ambiguous position of plastics in the cultural scheme is part of a deep-seated nature-culture dichotomy.

In the beginning, plastics were warmly embraced by scientists and the nonscientific public alike. Until World War II plastics existed mostly in chemical labs. Insofar as they were, like Bakelite, commercially produced, their quality and functions were rather poor, yet dreams of their potential were sky-high. Inventors and promoters portrayed plastics as unnatural or even supernatural substances. Plastics thus began with a positive reputation.

For the first time in history, human beings had been able to produce a raw material artificially. This was the general sentiment. Previously raw materials were products of nature that required human processing. Plastics were looked upon as unique exceptions to this rule, miraculous substances just waiting for human use. Their alleged unnaturalness gave rise to a widespread euphoria, their development considered a triumph of humans over nature.

At the end of World War I, Edwin Slosson, a journalist and director of the Science News Service, portrayed plastics chemists as *agents of applied democracy*. Rare and expensive materials, such as ebony and precious metals, which formerly had been "confined to the selfish enjoyment of the rich," were now "within the reach of every one" thanks to the imitative qualities of plastics. For Slosson "a state of democratic luxury" based on synthetic chemistry was at hand (Slosson 1919, p. 132–135). Fulfilling the ancient alchemists dream of transforming dirt into gold, chemists would gradually "substitute for the natural world an artificial world, molded nearer to its heart's desire" (Meikle 1995, p. 69).

Near the beginning of World War II, the applied chemists Victor Emmanuel Yarsley and Edward Gordon Couzens announced *The Expanding Age of Plastics* that would created a world brighter and clear than any previously known, "a world free from moth and rust and full of color" (Yarsley and Couzens 1941, p. 57). In such a world, *Plastic Man* would live in an abundance of safe, hygenic, strong, soft, and light objects, "a world in which man, like a magician makes what he wants for almost every need, out of what is beneath him and around him: coal, water, and air" (Yarsley and Couzens 1941, p. 68). Indeed, because of scarcities in traditional raw materials during World War II, war production of plastic or synthetic substitutes laid the base for postwar mass utilization.

But during the war the best plastics were reserved for the military and consumer plastics were often of inferior quality. As historian Meikle notes, U.S. civilians were faced with "shower heads of cellulose acetate that softened in hot water, with laminated products that separated when wet or stressed, with small moldings so devoid of resin that they shattered when dropped" (Meikle 1995, p. 166). Initial enthusiasm turned into ambivalence, as plastics came to connote inferior substitutes for real materials. When the war ended, the people felt free to demand *genuine* not *artificial* materials.

Yet postwar plastics were a booming business. Already in 1946, the average American used 3.5 kilos of plastics per year. Between 1950 and 1974, world production grew by an average 16 percent annually. Compared to other materials, plastics were the most expansive sector in many economies. At the same time, a growing call for "real," natural materials emerged. The quality of

artificialness and unnaturalness now had become the essence of plastics supposed flaw. Plastics started to symbolize a fake, cheap, materialist world that would lead to human alienation, cultural decay, and loss of control over technology.

An early sign of this kind of discomfort was expressed by the young biologist and journalist Rachel Carson (the future author of *Silent Spring* [1962]) in a women's magazine in 1947: "The witchery the chemist performs, turns them first into something unearthly, that gives you the creeps. You feel, when you go into a chemical plant where plastics are made, that maybe man has something quite unruly by the tail" (Carson 1947, p. 127). Roland Barthes, the French literary critic, voiced a similar distrust after he saw a large exposition on plastics in Paris. After his visit, Barthes feared that the whole world would become plasticized, even life. "Even one has already begun to produce plastic aortas," he wrote with disgust (Barthes 1957). But meanwhile Barthes supposed that living materials would not be imitated adequately. Plastics would remain inferior to natural materials, he declared, ignorant of the high-quality biomedical materials that would follow.

Although science, technology, and industry worked to overcome the inferior qualities of consumer plastics—and were remarkably successful in doing so—the nadir in public image was yet to come. This occurred in the 1960s and 1970s as environmental concerns turned plastics, along with nuclear radiation, into central emblems of self-destructiveness in high-tech society. According to novelist Norman Mailer, for instance, plastics were spreading through the country "like the metastases of cancer cells" (Meikle 1995, p. 177). In this climate most viewers of the film *The Graduate* (1967) immediately recognized its praise of plastics as a cynical joke, as a metaphor for the phony, banal and materialist world the protagonist has entered. The unsolicited career advice given to the new college graduate Benjamin Braddock (played by Dustin Hoffman) is simple: "Plastics. There is a great future in plastics." The words came to reflect dense cultural irony, because, of course, the future of plastics was the problem of waste.

An early spokesman of the plastics waste problem was the American biologist and environmentalist Barry Commoner. According to Commoner, the strength of plastics was also their essential flaw, an inability to degrade when discarded as waste: Only "human beings are uniquely capable of producing materials not found in nature [such as] is synthetic plastic, which unlike natural materials is not degraded by biological decay.

It therefore persists as rubbish or is burned—in both cases causing pollution" (Commoner 1971, p. 127). Not being biodegradable had lost its meaning of triumph over nature; on the contrary, it made that plastics were perceived as a permanent threat to nature, and the durability of plastic became a permanent threat to nature.

Then in the 1980s and 1990s, the public response to plastics shifted again. The issues of acid rain and greenhouse gases replaced the emblematic status of plastics as a source of environmental problems (Hajer 1995). Instead of condemning all plastics wholesale, even strict environmentalists began to distinguish different types associated with different degrees of environmental burden. Several organizational and technical strategies emerged to cope with plastics waste—from recycling to decomposing polymer materials into oil-like products and the development of biopolymers that degraded in sunlight. The plastics waste problem was not solved, but with technological and organizational fixes it became manageable.

Toward an Anthropological Ethics

How can one account for the fierce and contradictory emotions and changes in perception about plastics during the last century? They cannot be explained by the improving qualities of the material. Neither can they be explained by the dimension of plastics waste risks in comparison with other environmental risks. Explaining the whimsical pattern of public fascination and disgust about plastics by appealing to the emotional approach of the public—as chemists and spokespersons of the plastics industry were apt to do in reaction to environmental criticism—is unsatisfactory as well.

A richer understanding calls for taking into account fundamental, cultural assumptions toward new technologies. Technologies must be appropriated in order to make them fit into people's lives and practices. During the appropriation process both technologies and existing social orders often have to shift and adjust to one another. Plastics are ambiguous substances that did not always fit into existing cultural, symbolic categories. Under such circumstances erratic reactions are common.

In her pioneering work on impurity ideas in traditional societies, the British anthropologist Mary Douglas (1966) has described how border-crossing phenomena that do not fit into the cultural orders cause extreme reactions both of fascination and fear. Such a dual reaction is especially strong when something fits into two categories that were previously considered to be

mutually exclusive such as the human and animal, organism and machine, or nature and culture. The Nuer Tribe in Africa, for example experienced malformed babies as ambivalent beings, crossing the border between man and animal. Therefore they were treated as hippopotamus babies and put across the river. In the case of plastics, it is the nature–culture dichotomy that is decisive for its experienced ambivalence.

From the beginning, plastics were unlike natural raw materials, because they were artificially synthesized and therefore products of culture. This led to the interpretation of plastics as a miracle. Then in the climate of increasing environmental concern the nondegradability of plastics turned the miracle into monster. The coping strategies can be understood as attempts to put plastics in an acceptable cultural category. Product recycling brings the waste back into culture, while biodegradation makes *nature* out of it again. Although the waste problem is not solved, plastics have been culturally domesticated. They have become ethically accepted.

MARTIJNTJE W. SMITS

SEE ALSO *Artificiality.*

BIBLIOGRAPHY

Barthes, Roland. (1972). *Mythologies,* trans. Annette Lavers. New York: Hill and Wang. Original from Paris: Éditions du Seuil (1957).

Carson, Rachel. (1947). "Plastic Age." *Colliers* 120: 22, 49–50.

Commoner, Barry. (1971). *The Closing Circle; Nature, Man and Technology.* New York: Knopf.

Douglas, Mary. (1966). *Purity and Danger: An Analysis of the Concepts of Pollution and Taboo.* London: Routledge.

Hajer, Maarten A. (1995). *The Politics of Environmental Discourse: Ecological Modernization and the Policy Process.* Oxford: Clarendon Press; New York: Oxford University Press.

Meikle, Jeffrey L. (1995). *American Plastic: A Cultural History.* New Brunswick, NJ: Rutgers University Press.

Mossman, Susan, ed. (1997). *Early Plastics, Perspectives 1850–1950.* London: Leicester University Press; Washington, DC: Science Museum. Paperback edition from New York: Continuum (2000).

Slosson, Edwin E. (1919). *Creative Chemistry.* New York: Century.

Smits, Martijntje. (2002). *Monsterbezwering. De Culturele Domesticatie Van Nieuwe Technologie* [Taming monsters: the cultural domestication of new technology]. Amsterdam: Boom.

Sparke, Penny, ed. (1990). *The Plastics Age: From Modernity to Post-modernity.* London: Victoria and Albert Museum.

Published in the United States as *The Plastics Age: From Bakelite to Beanbags and Beyond*. Woodstock, NY: Overlook Press (1993).

Yarsley, Victor Emmanuel, and Edward G. Couzens. (1941). "The Expanding Age of Plastics." *Science Digest* 10(December): 57–59.

PLATO

• • •

Plato (428–347 B.C.E.), born in Athens, was a philosopher and founder of a school, the Academy. He was a student of Socrates and the teacher of Aristotle. Apart from a few letters, Plato's writing consists entirely of dialogues. These philosophical dramas display a mastery of composition, character, and action that rank him among the best of ancient poets. The range of philosophical problems treated in the dialogues and the quality of the treatment make this one of the most important bodies of work in the history of Western philosophy.

The chief character in most of the dialogues is Socrates, Plato himself never speaking. This raises two questions: First, to what extent does the Platonic Socrates correspond to the historical Socrates? And second, because Plato is silent, how can scholars determine what his views were? The standard answer is that Socrates or his occasional stand-in is always the mouthpiece of Plato, but that only the earlier dialogues present the authentic Socrates. There is no strong evidence for either conclusion. In this entry, the Socrates referred to is the character as he appears in Plato's dialogues.

Socrates in the Early Dialogues

Plato's early dialogues present the reader with the Socrates who *brought philosophy down from the heavens*. Pre-Socratic philosophers had been largely preoccupied with the study of the heavens and the earth, and especially with the phenomena of change and generation. Socrates apparently turned away from natural science to investigate the moral and political opinions of his fellow citizens. His habit of questioning them eventually resulted in his indictment, trial, and execution by the city of Athens. Plato uses this background both to mount a political defense of Socrates and to explain the kind of wisdom Socrates laid claim to.

In the *Apology*, the presentation of his defense, Socrates explains himself. According to the Oracle at Delphi, no one was wiser than Socrates. This astounded Socrates, for his philosophical investigations had convinced him that he knew nothing at all. He decided to

Plato, 428–348 B.C. The Greek philosopher founded the Academy, one of the great philosophical schools of antiquity. His thought had enormous impact on the development of Western philosophy.

test the oracle by interrogating those who were reputed to be wise. The politicians, he discovered, neither knew nor produced anything of value. The poets composed beautiful works, but could not explain how they did so or what their compositions meant. The artisans by contrast both produced useful things and understood what they were doing. Because of this, however, they supposed themselves wise about beauty, justice, and virtue, when in fact they were ignorant of such things. Socrates concluded that he was indeed wise in this one thing: He alone knew the full extent of his own ignorance.

But how can this meager knowledge, which Socrates calls *human wisdom*, be of any use? First, Socratic questioning can teach fellow citizens humility by showing them that they do not know what they think they know. What do they think they know? They know that power and wealth are the most valuable things. By undermining *these* opinions, Socrates was in effect urging Athenians to care about their own souls more than their property and the city's virtue more than its power. Small wonder they killed him for it.

In the *Euthyphro*, Socrates encounters a young man who is prosecuting his own father, an act that amounts to a radical assault on Greek familial morality. Euthyphro's boldness turns out to be supported by a hubristic confidence in his own understanding of piety. Socrates' relentless questioning demolishes that confidence, with the apparent result that Euthyphro drops his suit. So even if the only knowledge that is available is knowledge of one's own ignorance, philosophy can still be useful to the city by encouraging political moderation.

The utility of Socratic questioning is not limited to undermining bad arguments. The *Crito* provides a more positive account. In the absence of knowledge, one is left with opinions; Socrates, however, draws a distinction between the opinions of the many and those of the expert. If someone wants medical advice, that person does not put the matter to a vote but consults a doctor. If Socrates wants to know whether to accept his sentence or escape from jail, he will not be swayed by popular opinion but will turn to the expert in moral and political matters, presumably himself.

An opinion is never more than a guess; but an art or expertise consists of a set of educated guesses, informed by a long practice of questioning the evidence and alternatives. Expertise differs from philosophy insofar as it does not aim at comprehensive knowledge of the whole of things. In a theoretical sense the expert does not necessarily know anything, but in practical matters knows what he or she is talking about and so can be relied upon. If Plato's early dialogues were all there were to go on, one would conclude that Socrates was a political scientist and ethicist, and that these were the limits of Plato's ambitions.

This picture is substantially modified in later dialogues. Whereas in the *Apology* Socrates strenuously denies that he has anything to do with the physical sciences, in the *Phaedo* he confesses that, as a young man, he had a wonderful enthusiasm for physics, cosmology, and biology. But he came to believe that the reductionism of Greek science blinded its practitioners to the true nature of the phenomena they studied. Anaxagoras, for example, would explain the fact that Socrates is in jail by the position of his bones and muscles while ignoring the most important cause: the fact that the philosopher had concluded that he was obligated to accept his sentence. Without that last reason, Socrates exclaims, those bones and muscles would long gone from their prison.

This approach, applied to nature, obviously anthropomorphizes it. Socrates supposes that to explain the moon or the stars one must explain why it is best for them to be as they are. Perhaps the expert can get by with good guesswork, but real knowledge requires a consonance between human understanding and the world it seeks to understand. The mind looks for motives and justifications, and seeks answers in general ideas such as beauty and the good. What would have to be true of the world for such knowledge to be possible?

The Theory of Ideas

Socrates' most famous innovation was his theory of ideas. According to this principle the ideas by which human beings conceive of ordinary things are more real than the things themselves. Thus bigness is more real than a big tree, and unity and multiplicity more real than one person or the parts into which that person may be divided. Visible, tangible things are conceptually messy: relatively large and small; many and one at the same time; young and beautiful then, yet old and ugly now. But the idea of beauty is never ugly nor does the idea of one ever admit of division. That alone is real that simply is what it is, without contradiction, everywhere, and always.

Consider what happens when one approaches a mature oak tree from a distance. At first the tree appears so small that a person can cover it with one hand. Up close it is so large that it fills the horizon. But the tree cannot be both larger *and* smaller than an individual, nor does it really change as one approaches it. It is not what the eyes see but what the mind apprehends that is real. In the case of ordinary things, the true object is invisible, and what is visible is less than true.

Now compare the painting of a table with the fabrication of a table. The artist fashions an image of an image, at least twice removed from reality and bereft of dimension and substance. The artisan produces an actual table. He does so because he looks beyond any particular object to the idea or set of ideas that constitute the universal table. Just as images of a tree draw their reality from some object that is always, somehow, behind them, so human apprehension of various objects as one kind of thing—a tree or a table—draws on objects that are yet more universal and more real. It is in fact the ideas that generate reality, rather than vice versa.

Socrates' theory solves an impressive range of problems. It explains how human beings are able to perceive unities behind the otherwise chaotic manifold of sense impressions. It is also the basis of a theory of knowledge. Opinions are nothing more than temporal/spatial perspectives on things, and are therefore more or less unreliable. Knowledge is a grasp of ideas that never changes, for which reason it cannot fail.

This theory of knowledge in turn explains Socrates' moral perfection. How is it possible that Socrates alone seems never to succumb to temptation? Most people are guided by opinions about justice, and so are subject to changes in perspective. When one is owed money, justice means always paying debts. When one's luck changes, justice requires the forgiveness of debts. The philosopher by contrast is guided by the idea of justice. He is therefore perfectly steadfast in all circumstances. Even when confronted with his own imminent execution, Socrates says and does the same things in the same calm manner as he did before.

Political Philosophy

The theory of ideas is also the basis of Socrates' moral and political philosophy in the *Republic*. In that dialogue Socrates describes an ideal form of government consisting of three distinct classes. Philosophers rule, supported by a class of warriors called guardians. Both in turn are supported by a class of producers. The mores of the guardian class are shockingly radical. They practice a communism not only of property but of sex and reproduction, with no individual knowing who his or her own children are. Moreover women receive the same military training as men. In addition to these unprecedented social arrangements, the guardians' exposure to poetry, music, and religious teaching is tightly censored by the rulers.

The primary object of all these innovations is to prevent faction. The philosophers can rule because they alone are guided by the ideas of justice, moderation, and the good, and hence are incorruptible. A philosopher will never choose what is really bad because it looked good at the time. Because the guardians are not philosophers, their opinions about what is honorable and just must be scrupulously regulated by the ruling class, and private interest must be suppressed.

Socrates' ideal republic has been scathingly rebuked as both fantastic and totalitarian. But these criticisms forget the context. Its sole purpose is to provide a model of justice in the human soul. Like the republic, the individual soul is composed of distinct parts. If not, how could someone desire drink or revenge and at the same time want to resist such desires? In the well-ordered soul, intelligence governs the passions and the passions in turn discipline the appetites. When each part of the soul confines itself to its proper work, justice exists. By contrast, when passions or appetites take command of a person, injustice prevails.

The moral argument in the *Republic* seems to depend on the theory of ideas; however, in the *Gorgias* Socrates is able to derive much the same ethics from even the most jaundiced of moral opinions. Socrates' most frequent and persistent opponents in the dialogues were the sophists and orators. These men claimed to possess an art of persuasion whereby they could move an assembly or a jury to any conclusion someone might desire. Acquire that technology, either by learning it at a fee or by hiring one of its practitioners, and all the powers of state are at one's disposal. Even better, one may do whatever one desires without fear of prosecution. The young orator Polus knows exactly what the payoff is, the power to murder with impunity.

Socrates argues that sophism and oratory are not arts at all, but examples of flattery. An art, or *technē*, must be informed by some more or less correct notion of what is good for body or soul. Thus the arts of gymnastics and medicine aim to perfect the body and repair it, respectively, whereas the arts of politics and justice do the same for the soul. But just as cosmetics and gourmet cooking deliver what looks good, even if the person wearing the makeup or the food used to prepare the meal is in fact unhealthy, so sophism and oratory cater to vanity while doing harm rather than good.

The sophists held that the ends at which all human actions aim are unproblematic. Everyone wants the same things: wealth, reputation, beautiful lovers, and, occasionally, revenge. If one could rule other human beings, one could obtain an unlimited supply of these things and so be perfectly happy. Socrates argues that these ends are in fact problematic and may as easily bring ruin as happiness. No power is any good unless people know how to use it to get what is good for them; and that is not ruling over others but ruling oneself.

Platonism and Technology

Socrates' presentation of wisdom as expertise seems perfectly compatible with the development of technology. But the presentation was so overwhelmed by his theory of forms that it is almost invisible in the history of Platonism. There are good reasons to suppose that Socrates would have been at best indifferent to technological progress. He himself was so moderate in his appetites that he could live comfortably in *ten thousand-fold poverty*. In the *Republic* he suggests that the only city that is really natural is the city of sows, where human beings live very simple lives without any need for the arts and sciences.

The theory of forms provides powerful existential consolation, as the perfection of ideas is always available to the trained mind without need to modify the tangible world. During the Renaissance, Aristotle, whose philosophy was more oriented to practice, was popular whenever events seemed to be going well.

When foreigners invaded and governments collapsed, scholars turned back to reading Plato. A philosophy of consolation does little to encourage political or technological innovation.

KENNETH C. BLANCHARD JR.

SEE ALSO *Aristotle and Aristotelianism; Evolutionary Ethics; Evolution-Creationism Debate; Social Engineering; Virtue Ethics.*

BIBLIOGRAPHY

Guthrie, William K. C. (1975). *A History of Greek Philosophy*, Vols. 4 and 5. Cambridge, UK: Cambridge University Press. Guthrie's history is the best general work on Greek philosophy. Newcomers to the topic will find it especially useful.

Kraut, Richard, ed. (1992). *The Cambridge Companion to Plato*. Cambridge, UK: Cambridge University Press. A good collection of essays on various aspects of Plato's thought.

Plato. (1980). *The Laws of Plato*, trans. Thomas L. Pangle. New York: Basic Books. Very similar to Bloom's translation of *The Republic*.

Plato. (1991). *The Republic of Plato*, trans. Allan Bloom. New York: Basic Books. A very fine translation of this work, using consistent English terms and phrases for the same Greek terms. It also includes a challenging but very thorough essay.

Plato. (1998). *Gorgias*, trans. James H. Nichols Jr. Ithaca, NY: Cornell University Press. The best translation of one of Plato's most important ethical dialogues.

Plato and Aristophanes. (1984). *Four Texts on Socrates*, trans. Thomas G. West, and Grace Starry West. Ithaca, NY: Cornell University Press. An excellent translation of the most popular of Plato's short dialogues, with notes and brief commentary.

Rosen, Stanley. (1993). "*Techne* and the Origins of Modernity." In *Technology in the Western Tradition*, ed. Arthur M. Melzer et al. Cornell, NY: Cornell University Press. A profound treatment of the philosophical meaning of technology.

Strauss, Leo. (1963). "Plato." In *History of Political Philosophy*, ed. Leo Strauss and Joseph Cropsey. Chicago: Rand McNally. One of the best essays on Platonic political philosophy.

PLAYING GOD

• • •

The phrase, *playing God*, appears to be one a theologian might use. But in contemporary parlance it has taken on secular significance. It refers to the powers that science, engineering, and technology confer on human beings to understand and to control the natural world.

Celebration and Criticism

The playing God metaphor has been used in both celebratory and critical contexts. In celebration, H. G. Wells's novel *Men Like Gods* (1923) describes an advanced human civilization in which people lead the *life of demigods*, very free, strongly individualized ... a practical communism." Indeed the communist movement sometimes described itself as realizing previously thwarted divinelike possibilities in human nature. Inventor R. Buckminster Fuller proclaimed the advent of *No More Second-Hand God* (1963) through science and technology. The psychologist Erich Fromm, in his book *You Shall Be as Gods* (1966), argued the need to assume responsibilities for many new powers that were once attributed to supernatural entities. And the alternative culture Whole Earth Catalog (1968) declared on its cover, "We are gods and might as well get used to it."

Among the followers of Ayn Rand, playing god has been declared a virtue. Science fiction writers sometimes describe themselves as playing god. And for Kevin Kelly (1999), *nerd theology* involves repeatedly playing god, as in a learning game.

More commonly, however, playing God has served as a metaphor for criticizing the human exercise of excessive scientific and technological powers. Early Romantic writers implicitly criticized human aspirations to play God insofar as they mourned the loss of a sense of the sacred in the wake of scientific and technological progress. In the contexts of both celebration and criticism, there are, nevertheless, three overlapping meanings that can be discerned.

Three Meanings

The first meaning is associated with basic scientific research wherein human beings *learn God's awesome secrets*. Some research elicits a sense of awe and wonder over the complexity and majesty of the natural world that the human mind can apprehend. Science is like a light shining down into the previously dark and secretive caverns of natural mystery, revealing what had been hidden. The revelatory power of science leads human beings to believe they are gaining godlike powers. Few would argue against continuing the investigation because *learning for learning's sake* remains the morality of scientific knowledge.

The second meaning of playing God arises primarily within the field of medicine where doctors seem to have gained the *power over life and death*. In a medical emergency, the patient feels helpless, totally dependent upon the scientific training and personal skills of the attending physicians. Doctors, and the scientific training they received in medical school, stand between the patient and death. Similarly large-scale research programs dedicated to finding cures for cancer or HIV/AIDS provide society with hope in the face of helplessness. Here playing God takes on a redemptive or salvational component. The genre of jokes about doctors who think of themselves as gods reflects the wider anxiety over powerlessness plus human dependence upon doctors and their skills.

Two assumptions are at work in the medical meaning of playing God. First is the assumption that decisions regarding life and death are the prerogative of God. The second follows from the first: When a human being has the power of life and death, society places that person in a godlike role. This elicits a second anxiety; namely, worry that the person in the godlike role will succumb to the temptation of pride, or hubris. The concept of hubris articulates the more inchoate fear that human beings will presume too much, overreach themselves, violate some divinely appointed limit, and reap destruction. Anxiety over hubris marks the overlapping transition from the second to the third meaning of the phrase playing God.

To alter life and influence human evolution is the third meaning of playing God. Here science and technology team up so that understanding leads to control. Control over nature places human beings where only God belongs, and humans are challenged by the choice between good and evil. In atomic physics, the discovery of how a nuclear chain reaction works led to both nuclear medicine and weapons of mass destruction with the attendant threat of self-extinction. Taming nature by pesticide use in order to increase food production threatens the life-sustaining potency of the planet. The Human Genome Project has enhanced understanding of DNA, confronting society with unavoidable decisions: The choices made to alter or avoid altering the human genetic code may affect the evolutionary future of the human race and perhaps even human nature itself. If DNA is the essence of a human being, then people take the ability to change their very nature into their own hands when they modify it. To alter what has evolved borders on creating a new human nature; this is a reminder of humankind's godlike powers and the awesome responsibility imposed by those powers. The human race of tomorrow will be the result of scientific and technological decisions made in the present. The scientific community becomes a microcosm of the entire human community. The fear is that if scientists give into the temptation of hubris, evil will result.

The God in Question

A close look shows that the God of playing God is not the God of the Bible but divinized nature. Nature has absorbed the qualities of sacredness; science and technology risk profaning the sacred.

Contemporary fear of playing God connotes the ancient Greek myth of Prometheus. While creating the world, the sky-god Zeus was in a cranky mood. The Olympian decided to withhold fire from Earth's inhabitants, leaving the nascent human race to live in relentless cold and darkness. The Titan Prometheus, whose name means *to think ahead*, saw the value of fire to warm homes. He anticipated how fire could separate humanity from the beasts by making it possible to forge tools. Prometheus craftily snuck into the heavens where the gods dwelt and where the sun was kept. He lit his torch from the fires of the sun and carried the heavenly gift back to earth.

The gods were outraged that their stronghold had been penetrated and robbed. Zeus was particularly angry over Prometheus's impertinence and exacted a merciless punishment on the rebel. Zeus chained Prometheus to a rock where an eagle could feast on the Titan's liver all day long. The head of the pantheon cursed the future-oriented Prometheus: "Forever shall the intolerable present grind you down." The moral of the story is this: Pride or hubris that leads humans to overestimate themselves and enter the realm of the sacred will precipitate vengeful destruction. The Bible provides a variant: "Pride goes before destruction" (*Prov.* 16:18).

In early-twenty-first-century culture, dominated by Western science, Zeus no longer plays the role of the sacred. Nature does. Nature strikes back in the Frankenstein legend and the more contemporary, geneticized version of it described in Michael Crichton's novel *Jurassic Park* (1990) and the films adapted from it. The theme has become common: A mad scientist exploits a new discovery and crosses the line between life and death; nature strikes back with consequent chaos and destruction.

Theological articulations of caution in the face of human pride mirror the wider culture. In a 1980 task force report, *Human Life and the New Genetics*, the Council of Churches of Christ issued a warning: "Human beings have an ability to do Godlike things: to exercise creativity, to direct and redirect processes of nature. But the warnings also imply that these powers may be used rashly, that it may be better for people to remember that they are creatures and not gods." A United Methodist Church Genetic Science Task Force Report to the 1992 General Conference stated similarly, "The image of God, in which humanity is created, confers both power and responsibility to use power as God does: neither by coercion nor tyranny, but by love. Failure to accept limits by rejecting or ignoring accountability to God and interdependency with the whole of creation is the essence of sin" (United Methodist Church 2000, Internet site). In sum, humans can sin through science by failing to recognize limits and, thereby, violate the sacred.

Although the proscription against playing God can be applied to many fields of science, it is found most often in the field of genetics because DNA has garnered cultural reverence. The human genome has become tacitly identified with the essence of what is human. A person's individuality, identity, and dignity are associated with his or her DNA. Therefore if humans have the hubris to intervene in the human genome, they risk violating something sacred. This tacit belief is called the *gene myth* as well as *the strong genetic principle* or *genetic essentialism*. This myth is an interpretive framework that includes the assumed sacredness of the human genome and the fear of Promethean pride.

Theological anthropology questions the gene myth, doubting the equation of DNA with human essence or human personhood. In 2002 the National Council of Churches of Singapore issued *A Christian Response to the Life Sciences* that stated, among other things, "It is a fallacy of genetic determinism to equate the genetic makeup of a person with the person" (National Council of Churches 2002, p. 81). Such anthropology combats the gene myth and opens the door to ethical approval of cautious genetic engineering.

Contemplating careful employment of genetic technology to alter human DNA leads to concern over the distinction between therapy and enhancement. At first glance, therapy seems ethically warranted, whereas enhancement seems Promethean and dangerous. *Gene therapy* is the directed genetic change of human somatic cells to treat a genetic disease or defect in a living person. With 4,000 to 6,000 human diseases traceable to genetic predispositions—cystic fibrosis, Huntington's disease, Alzheimer's, and many cancers among them—the prospects of gene-based therapies are raising hopes for dramatic medical advances. Few if any cite ethical reasons to prohibit somatic cell therapy via gene manipulation.

Human genetic enhancement is the use of genetic knowledge and technology to do more than heal disease. Enhancement seeks to bring about improvements in the capacities of living persons, in embryos, or in future generations. Enhancement might be accomplished in one of two ways, either through genetic selection during screening or through directed genetic change. Genetic selection may take place at the gamete stage, or more commonly by means of embryo selection during preimplantation genetic diagnosis (PGD) following in vitro fertilization (IVF). Genetic changes could be introduced into early embryos, thereby influencing a living individual, or by altering the germ line, thereby influencing future generations.

Modest forms of enhancement are becoming possible. For example, introduction of the gene IGF-1 (insulin growth factor) into muscle cells results in increased muscle strength as well as health. Such procedure is quite valuable as a therapy; yet, it lends itself to enhancement as well. For those who daydream of so-called *designer babies*, the list of traits to be enhanced would likely include increased height or intelligence as well as preferred eye or hair color. Concerns raised by both secular and religious ethicists focus on economic justice—that is, wealthy families are more likely to take advantage of genetic enhancement services leading to a gap between the *genrich* and the *genpoor*.

Serious concerns have been raised over germ line intervention for purposes of both therapy and enhancement. *Germ line intervention* is gene selection or gene change in the gametes, which in turn would influence the genomes of future generations. Because the mutant form of the gene that predisposes for cystic fibrosis has been located on chromosome 4, researchers can devise a plan to select out that gene and spare future generations the suffering caused by a debilitating disease. This would constitute germ line alteration for therapeutic motives. In principle scientists could select or even engineer genetic predispositions to favorable traits in the same manner. This would constitute germ line alteration for enhancement motives.

Both of these scenarios are risky, and for the same reason. Too much remains unknown about gene function. It is probable that gene expression works in delicate systems, so it is rare that a single gene is responsible for a single phenotypical expression. If one or two genes are removed or engineered, scientists may unknowingly upset an entire system of gene interaction that could lead to unfortunate consequences. The proscription against playing God serves here as a warning to avoid rushing in prematurely with what appears to be an improvement but could turn out to be a disaster. Ethicists often advise that scientists and researchers proceed with caution—the precautionary principle—until the scope of knowledge is adequate to cover all possible contingencies.

Note that the precautionary principle does not rely upon the tacit belief that DNA is sacred. Rather it relies upon a principle of prudence that respects the complexity of the natural world and the finite limits of human knowledge.

TED PETERS

SEE ALSO *Death and Dying; Fetal Research; Frankenstein; Gene Therapy; Genethics; Genetic Counseling; In Vitro Fertilization and Genetic Screening; Prometheus.*

BIBLIOGRAPHY

Bruce, Donald, and Ann Bruce, eds. (1999). *Engineering Genesis: The Ethics of Genetic Engineering in Non-Human Species.* London: Earthscan Publications. A collection of essays on the ethical implications of plant genetics and human genetics.

Chapman, Audrey R. (1999). *Unprecedented Choices: Religious Ethics at the Frontiers of Genetic Science.* Minneapolis, MN: Fortress Press. An overview of the challenges posed to public consciousness by the genetic revolution.

Council of the Churches of Christ. (1980). *Human Life and the New Genetics.* New York: Author.

Dutney, Andrew. (2001). *Playing God: Ethics and Faith.* San Francisco: HarperCollins. A warning to be cautious in genetic research based on religious convictions.

Kelly, Kevin. (1999). "Nerd Theology." *Technology in Society* 21(4): 387–392.

National Council of Churches of Singapore. (2002). *A Christian Response to the Life Sciences.* Singapore: Genesis Books. A brief and brilliant analysis of the science of cloning and stem cell research from a theological perspective.

Peters, Ted. (2002). *Playing God? Genetic Determinism and Human Freedom,* 2nd edition. London, New York: Routledge. A theological and ethical analysis of the gene myth, genetic determinism, the so-called "gay gene," cloning, stem cells, and genetic patenting.

Peters, Ted, ed. (1998). *Genetics: Issues of Social Justice.* Cleveland, OH: Pilgrim Press. A collection of essays by scientists and theologians analyzing the human genome project.

INTERNET RESOURCE

United Methodist Church. (2000). "Book of Resolutions of the United Methodist Church 2000: Developments in Genetic Science." Available at http://www.electronic-church.org/Genetics/UMC-Book_of_Resolutions_Genetic_Science_and_Cloning.pdf.

POLANYI, KARL

• • •

Karl Polanyi (1886–1964) was born in Vienna on October 25 of Hungarian parents and became a leading economic historian of the twentieth century. His understanding of the Industrial Revolution as dependent on a disembedding of the economy from the broader culture offers an important perspective on globalization and suggestive insights relevant to relationships between science, technology, and ethics. After studies in Budapest, work as a lawyer, radical political activity, service in World War I where he was imprisoned on the Eastern front, and postwar convalescence and work as a journalist, he immigrated first to Great Britain (1933) and then to the United States and Canada (1940s), where he taught first at Bennington College and then at Columbia University. Because of past involvement with Marxist radicalism, his wife, Ilona Duczynska, was denied the right to live in the United States and Polanyi was forced to live in Canada and commute to New York. He died in Pickering, Ontario, on April 23. He was survived by his younger brother, the scientist and philosopher Michael Polanyi.

The Great Transformation

Polanyi's *The Great Transformation* (1944) has been recognized as a central contribution to economic sociology. The basic argument of this analysis of the Industrial Revolution is that capitalism is historically unique in its separation of economic relationships from other social interactions. All previous human economies were embedded in the sense of being integrated into familial, kinship, social, religious, and other interactions and obligations. The great transformation was not simply the development of new sources of power (steam), machines, and systems of production (division of labor), but the disembedding of production and market distribution from all other modes of interaction.

One key feature of the disembedding process was turning land, labor, and capital into what Polayni calls *fictitious commodities*. In reality neither land (nature) nor labor (people)—and only to a limited extent capital (whether liquid or fixed)—can ever have their price freely determined by market relations in the same way as industrial products. The self-regulating market as conceived by neoclassical economics nevertheless requires such an assumption. What Polanyi's analysis seeks to demonstrate is the fictitious character of these assumptions, both in relation to previous historical practices and as revealed in the failures of market economy in the early twentieth century.

For Polayni the great transformation of his concern was actually two quite different historical events: the collapse of nineteenth century civilization associated with World War I and the creation of the self-regulating market economy through the collaboration of industrialists, neoclassical economists, and liberal politicians. In the first sense his diagnosis of the great transformation was precisely the opposite of that of his contemporary Friedrich von Hayek in *The Road to Serfdom* (1944). For von Hayek the collapse that terminated the nineteenth century was caused by a failure to extend the market system to its logical conclusion and more fully remove state regulation of the economy. For Polanyi the reactions of communism, fascism, and Keynesian economics were legitimate efforts to reaffirm the proper subordination of industrialist economics to society and culture.

Polanyi's argument has been subject to criticisms by both anthropologists and economists, each raising essentially the same question: Does Polanyi not romanticize premodern economic orders? Is there really any alternative to the market economy, which is a natural historical development? Following *The Great Transformation* Polanyi undertook extensive studies of premodern economic practices in order to further substantiate his claims about the historical uniqueness of neoeconomic assumptions. One of the more influential results of this research was the collaborative publication of *Trade and Market in the Early Empires: Economies in History and Theory* (1957).

Application and Assessment

From Polayni's perspective the market economy is a historical anomaly. Although forms of trade and exchange can be found in all human societies, economic exchange had never previously been so independent of all other relations. The pattern found in modern economies is, of course, also that exhibited in analogous ways in science and technology: the development of autonomous communities of practitioners operating according to sets of rules that apply only to quite limited aspects of human behavior (as in the practice of the scientific method). Under such conditions rationalist ethics is forced to play a more important role in criticizing and moderating disembedded behaviors (economic, scientific, and technological) than ever before—while at the same time disembedding creates conditions that make ethics ever more ineffectual. Ethics is thus forced to adapt policy as its handmaid in order to overcome its own impotence.

But is it not the case that Polanyi was fundamentally mistaken, if not about the past then about the collapse of the free market system that supported the civilization of the long nineteenth century? As his daughter Kari Polanyi Levitt admits, "Polanyi was certainly premature in dismissing 'market economy' and 'market society' from the stage of history" (McRobbie and Polanyi Levitt 2000, p. 10). From the end of the Cold War and into the beginning of the twenty-first century, neoliberalism reemerged with the forces of globalization stronger than ever before. But this world was also one in which, as Polanyi Levitt notes, "disasters of famines, wars, new diseases and environmental degradation threaten the destruction of the social, cultural and ecological fabric which sustains life on earth." Under such conditions, is it not possible that Polanyi's "analysis of the dangers inherent in the elevation of 'the economic instance' over all other aspects of human endeavor" deserves continuing consideration? (McRobbie and Polanyi Levitt 2000, p. 10).

CARL MITCHAM

SEE ALSO *Governance of Science; Merton, Robert K.; Science Policy.*

BIBLIOGRAPHY

McRobbie, Kenneth, and Kari Polanyi Levitt, eds. (2000). *Karl Polanyi in Vienna: The Contemporary Significance of "The Great Transformation."* Montreal: Black Rose Books. Thirty papers from a conference in Vienna on the fiftieth anniversary of the publication of *The Great Transformation.*

Polanyi, Karl. (1944). *The Great Transformation.* New York: Rinehart. Reprint, Boston: Beacon Press, 1957. Second edition, Boston: Beacon Press, 2001.

Polanyi, Karl. (1968). *Primitive, Archaic, and Modern Economies: Essays of Karl Polanyi,* ed. George Dalton. Garden City, NY: Doubleday Anchor.

Polanyi, Karl; Conrad M. Arensberg; and Harry W. Pearson, eds. (1957). *Trade and Market in the Early Empires: Economies in History and Theory.* Glencoe, IL: Free Press.

Stanfield, J. Ron. (1986). *The Economic Thought of Karl Polanyi: Lives and Livelihood.* New York: St. Martin's Press. The last chapter connects Polanyi's economic history to an assessment of industrial technology and the alternative technology response present in E. F. Schumacher, Wendell Berry, and others.

POLANYI, MICHAEL

• • •

A physician and physical chemist who became a philosopher in middle age, Michael Polanyi (1891–1976) was born in Budapest, Hungary on March 12, the youngest child in a liberal Jewish family that provided a broad humanistic education. After medical training and completing a dissertation in chemistry, Polanyi rose to be an eminent physical chemist (publishing more than 200 scientific papers in his career) in Berlin; in 1933, he fled Nazi Germany and took a position in Great Britain at Manchester University. He was elected Senior Research Fellow at Merton College, Oxford, in 1959. Polanyi died in Northhampton on February 22.

Science and Society

From the 1930s forward, Polanyi often wrote about the governance of science and the fragile relation between science and society. Marxist-influenced politics and philosophical discussions about the nature and justification of science challenged Polanyi to probe such issues. He found that most of the ideas about science and society put forth by Western scientists and philosophers of science were as inadequate as the Marxist ideas. In response, Polanyi early argued that freedom was a prerequisite for establishing a community of inquiry in which individuals pursued truth and openly stated their findings. He further criticized centrally planned scientific research, arguing that opportunities for individual initiative are critical and that civil liberties and a democratic society provide important foundations for science.

Polanyi's dissatisfaction with philosophical accounts of science thus led him gradually to shift his interest from scientific research to philosophy. By the mid-1940s, Polanyi began to put together his own comprehensive philosophical account in *Science, Faith, and Society* (1946).

Personal Knowledge

Personal Knowledge, published in 1958 and based on his 1951–1952 Gifford Lectures, is a much broader articulation of his philosophical stance. Later publications refine and extend the framework of this book. *The Tacit Dimension* (1966) is particularly important because it reflects the way in which Polanyi's earlier emphasis on commitment was recast and enriched by working out an account of the structure of tacit knowing.

In Polanyi's sometimes dense texts, his constructive philosophy is bound up with searching criticisms of much modern philosophy. In his early formulations in the syllabus for his 1951 Gifford Lectures, Polanyi summarized his constructive philosophical project as setting forth a "fiduciary philosophy" that overcame the "restrictions of objectivism" and rehabilitated "overt belief" (Papers of Michael Polanyi, Box 33, Folder 1,

University of Chicago Library). All knowledge is based in belief, but this does not mean knowledge claims are necessarily without warrant. For Polanyi, a fully impersonal, objective knowledge is a false and destructive ideal embraced by modern western philosophy and science. In *Personal Knowledge,* he argues that doubt, celebrated since Descartes, is not heuristic and, in fact, is parasitical on belief. Knowledge must be understood in terms of the activity of a skillful and committed knower immersed in a community with a living tradition. Polanyi is a fallibilist and a metaphysical realist who argues for what he terms "personal knowledge," which is subject-grounded but not merely subjective. Truth claims are set forth with universal intent.

Polanyi's early interest in the administration of science led him to work out an epistemology of science that focuses on the person and discovery. His epistemology recasts ideas of Gestalt psychology in order to emphasize the active shaping of comprehension and the commitment of the knower. After *Personal Knowledge,* Polanyi came to better understand what he early called the "fiduciary" element in knowledge as he continued to explore the importance of the inarticulate. His later theory of tacit knowing claims all knowing involves an integration of subsidiarily or skillfully known elements to produce a focal comprehension. Thus, knowing has a from-to structure: It moves from subsidiaries or tacitly known particulars to a focus. Thought dwells, Polanyi argues, in its subsidiaries, and those subsidiaries function like parts of one's body that a person dwells in and skillfully coordinates in order to achieve certain objectives.

Some of Polanyi's ideas about science, and more generally about human knowing and the problems of modern society, parallel ideas developed by other thinkers in the mid-twentieth century. Several mid-century philosophers of science, such as Polanyi, backed away from narrow empirical approaches and took new interest in the practices of scientists and the history of science; philosophers such as Maurice Merleau-Ponty (1908–1961) wrote about the body and perception in ways that complement Polanyi's views.

Assessment

Polanyi's major constructive philosophical contribution is his theory of tacit knowing, which holds that all knowledge is grounded in tacitly held elements; a knower always relies on such unspecifiable elements to achieve focal awareness. This claim is a major break with the theory of knowledge developed in the modern philosophical tradition. Many of Polanyi's broader philosophical ideas about persons, communities, and the human project of exploring the universe are novel views that grow out of his new approach to the problem of knowledge. Taken as a whole, Polanyi offers a comprehensive philosophical vision that weaves together an epistemology, a philosophy of life, and an evolutionary cosmology.

Polanyi was deeply disturbed by what he regarded as the nihilistic tenor of modern culture; he aimed to restore confidence in the human capacity to discover meaning. His philosophical ideas are also sometimes linked with what is now called postmodern thought. But while he sharply criticized some elements of the Enlightenment tradition, Polanyi also affirmed Enlightenment values such as truth-seeking as necessary and worthy ideals. Polanyi was committed to the reliability of natural science, although he did not contend that only scientific knowledge was possible and important.

PHIL MULLINS

SEE ALSO *Liberalism; Participation.*

BIBLIOGRAPHY

Gelwick, Richard. (1977). *The Way of Discovery: An Introduction to the Thought of Michael Polanyi.* Oxford, UK: Oxford University Press. A clear basic introduction to Polanyi's philosophical ideas.

Grene, Marjorie. (1977). "Tacit Knowing: Grounds for a Revolution in Philosophy." *Journal of the British Society for Phenomenology* 8(3): 164–171. A brief account of the development and importance Polanyi's philosophical ideas by one of his major collaborators.

Polanyi, Michael. (1959). *The Study of Man.* Chicago: University of Chicago Press. Extends the account of *Personal Knowledge* to treat questions about historical knowledge.

Polanyi, Michael. (1964). *Science, Faith and Society.* Chicago: University of Chicago Press. First published in 1946, this is Polanyi's best early discussion of science, and includes a 1963 introduction by Polanyi that links early and later ideas.

Polanyi, Michael. (1966). *The Tacit Dimension.* Garden City, NY: Doubleday. The clearest statement of Polanyi's theory of tacit knowing.

Polanyi, Michael. (1969). *Knowing and Being: Essays by Michael Polanyi.* Chicago: University of Chicago Press. A collection edited and introduced by Marjorie Grene of fourteen Polanyi essays from the 1960s.

Polanyi, Michael. (1974). *Personal Knowledge Towards a Post Critical Philosophy.* Chicago: University of Chicago Press. First published in 1958, this is Polanyi's most comprehensive philosophical statement, which grew out of his 1951 and 1952 Gifford Lectures.

Polanyi, Michael, and Harry Prosch. (1975). *Meaning.* Chicago: University of Chicago Press. A late book, written with a collaborator, that attempts to extends Polanyi's epistemological framework to treat meaning in art, myth, and religion.

Polanyi, Michael. (1997). *Science, Economics and Philosophy: Selected Papers of Michael Polanyi,* ed. and introduced by R. T. Allen. New Brunswick and London: Transaction. Includes twenty-five Polanyi essays written from 1917 to 1972 as well as an annotated bibliography of Polanyi's publications on society, economics, and philosophy.

Polanyi, Michael. (1998). *The Logic of Liberty Reflections and Rejoinders.* Indianapolis, IN: Liberty Fund. Essays on science and politics, originally published in 1951.

Prosch, Harry. (1986). *Michael Polanyi: A Critical Exposition.* Albany: SUNY Press. An interpretation of Polanyi's philosophical work by one of his collaborators.

INTERNET RESOURCE

Polanyi Society. Available from http://www.mwsc.edu/orgs/polanyi/. Selected short essays by Michael Polanyi and other information.

POLICE

• • •

As members of a social institution which, like the military, is a legitimate employer of force in the service of the state, the police must adhere to strict standards of ethical conduct. The rapid pace of scientific and technological change has affected this ethically guided police work in two ways: The detective resources and enforcement powers at their disposal are altered by changes in science and technology; the powers available to illegitimate users of force, those whom the police are charged with opposing, are also altered. At several different levels, the law enforcement institution has adapted to these changes, which have brought both increased opportunity for improved service as well as challenges and controversies.

Police Ethics

Law enforcement officers represent the epitome of society insofar as they daily risk their lives to protect and serve the public and uphold the laws of the state. Their position of authority and their ability to legitimately use force in various contingencies, however, means that they must uphold the strictest of ethical standards in order not to abuse their power. Although it is unrealistic to hold law enforcement officers (or any human being) to standards of perfection, both citizens and the state expect the police to uphold certain values and norms. Although constitutional and other laws (for example, the U.S. Miranda rights of persons accused of crimes established in 1966) play a role in ensuring the ethical conduct of police, these measures are also supplemented by various codes of ethics drafted by professional law enforcement associations.

Two of the most important overarching codes are the Code of Ethics adopted in 1966 by the American Federation of Police and the Law Enforcement Code of Ethics adopted in 1957 (revised in 1989) by the International Association of Chiefs of Police (IACP 1992). The latter is often used as a model by individual police departments in crafting their own codes of conduct and ethics, which then serve as oaths taken by new officers. Another important document on the international level is the United Nations Code of Conduct for Law Enforcement Officials, adopted by the General Assembly in 1979. A key distinguishing feature of this code is the broad directive for police to protect human dignity and uphold human rights. *Criminal Justice Ethics,* a semiannual journal published by the Institute of Criminal Justice Ethics, and *Ethics Roll Call*, a quarterly journal published by the Center for Law Enforcement Ethics, serve as key resources for facilitating ongoing discussions in the field of law enforcement ethics, especially as changes in science and technology raise new questions about proper conduct.

Most codes of conduct and codes of ethics for police uphold certain general principles in order to prevent misconduct and abuse of power. These principles include: the duty to uphold the law and loyalty to the constitution; personal integrity, honesty, and honor; responsibility to know the law and understand the limits of one's power; and responsibility to use the least amount of force necessary to achieve the proper end. These principles (in addition to laws) are designed to guard against police deviance, or behavior inconsistent with norms and values. This can include misconduct (e.g., excessive or discriminatory use/non-use of force), corruption (forbidden acts involving misuse of office for gain), and favoritism (unfair treatment of friends or relatives).

One noteworthy point is the scarcity of references to the proper use of science and technology in most codes of ethics. As new technologies emerge and become available for both police and criminals (for example, improved surveillance mechanisms or more deadly weapons), so too do new ethical dilemmas that may or may not be adequately resolved by interpreting the general principles found in police codes of ethics.

Various forms of police deviance often have been exacerbated by inadequate accountability mechanisms. During the last half of the twentieth century, however, this was improved thanks in part to developments in communications and surveillance technologies that allow watchdog groups to monitor police behavior and record and share their findings. For the most part, the increased public scrutiny of police activities has helped to reinforce ethical conduct, but it can also interfere with police operations and unfairly stigmatize officers. One source of deviance is an incentive structure that attaches promotions to number and rate of arrests. This can distract officers from their principle duties of protecting and serving the public. Another common ethical problem is the tribal system of values that can evolve within such tight-knit communities as police departments. The "blue code of silence" sometimes leads to the cover-up of corruption and abuses of power.

Science and Technology in Police Work

Advances in science and technology have both improved the capabilities of law enforcement officers to perform their duties and raised several challenges and controversies. Transportation provides one example of how radically these developments have altered police work. Although foot and horseback patrols still play key roles in law enforcement, the introduction of police cars (first used in Akron, Ohio, in 1899 and popularized in the 1930s) has dramatically increased officer mobility. Now helicopters and motorboats complement ever more powerful police cars equipped with video cameras, laptop computers capable of accessing information systems, and global positioning systems (GPS). In addition to transportation, other areas of major scientific and technological change include identification and crime solving, computers and communication, monitoring and surveillance, and protection and control.

The discovery and utilization of deoxyribonucleic acid (DNA) has greatly improved the science of identifying people, which involves determining where they have been, what they did, and how they did it. With only a strand of hair, flake of dandruff, or drop of saliva, police laboratories are now able to positively identify individuals. Despite some debate on this issue, in 1996 the National Academy of Sciences determined there is no reason to question the reliability of DNA evidence. The creation of crime laboratories and advances in forensic science (e.g., the microscopic comparison of fibers, bullets, and other tangible evidence) has made identification of hard evidence a powerful means of detection. Fingerprinting, first widely used in the 1920s, is another technique that has vastly improved the ability of police to identify criminals.

In 1967, the Federal Bureau of Investigation (FBI) created the National Crime Information Center

A police officer checks the tracking system on a helicopter. This system is another example of new policing technologies. (*AP/Wide World Photos.*)

(NCIC), the first nation-wide computer filing system. This helped spark the large-scale computerization of police departments in the United States in the 1970s. Integrated networks of computer databases allow different police departments and different sectors of law enforcement to rapidly share information. Improvements in information and communications technologies can enhance the effectiveness of identification technologies by building national databases of license plates, fingerprints, and even DNA. The 911 system, two-way radios, cell phones, and satellite phones have also increased the ability of police to respond to the public's needs.

Advances in computer and information technology have also improved police monitoring and surveillance. For example, police in the City of London utilize an information technology scheme called Police Informant Management System (PIMS), which allows them to target specific criminal activities and manage informants more effectively. Other monitoring technologies used by the police include camera systems and GPS. Police surveillance work focuses on specific individuals, places, or vehicles deemed suspicious. This more covert work can involve the recording and monitoring of telephone or in-person conversations as well as electronic correspondence. Three notable technologies are the Echelon surveillance program (used to monitor electronic correspondence), the FaceTrac system that "reads" faces, and the Digital Angel tracking chip, which—disguised as jewelry or implanted under the skin—can track the wearer anywhere in the world. For all the advantages these techniques confer on police, there is still no replacement for the proper training of officers to infiltrate criminal groups.

Finally, advances in protective equipment (such as bulletproof vests and helmets) and less-than-lethal technologies have greatly improved police work. Especially during the 1960s, there were many attempts to develop riot control technologies and use-of-force alternatives to guns and the standard side-handle baton. Tried and largely abandoned technologies included rubber, plastic, and wooden bullets, dart and tranquilizer guns, an electrified water jet, and strobe lights (Seaskate 1998). Taser guns, which shoot two wire-controlled darts into the victim and deliver a 50,000-volt shock, bean bag rounds for crowd control, and pepper spray have been more widely employed. Another major development is the RoadSpike, a strip of remote-controlled retractable

A police officer uses a computer in his squad car. The use of technology has led to many improvements in law enforcement methods. (© DiMaggio/Kalish/Corbis.)

spikes that allows police to more safely and effectively stop fleeing vehicles while minimizing unintended damage to others.

Police departments have often been slow in reaping the advantages made possible through progress in science and technology (Seaskate 1998). For example, even with massive federal funding, computerization happened slowly and unevenly, since it took a long time for many departments to figure out how information technologies like records management systems and computerized crime mapping could be usefully implemented. Furthermore, many technologies have been adopted from the private sector, but the police also have needs for specialized technologies, which are more difficult to develop and apply. In the United States, the Office of Science and Technology of the National Institute of Justice is responsible for determining and supplying the special technology needs of the nation's police force and for fostering technology research and development.

The use of emerging science and technology by the police can also raise controversies. Cyrille Fijnaut and Gary T. Marx (1995) have argued that the increased use of technical means by the police is one manifestation of the growing technicization of social control (enforcing norms by preventing violations). They suggest that the increased use of technology in social control can cause more problems than it solves. Other controversies stem primarily from specific technologies. Enhanced monitoring and surveillance abilities have raised privacy issues. The reliability of fingerprinting has been questioned (see Cho 2002), especially following one case of wrongful arrest and one of wrongful conviction in 2004 based on faulty interpretation of the fingerprint evidence. Taser guns have also sparked controversy as many concede that they can save lives but that officers may use them too early or too often. More than forty deaths have been linked to Taser guns.

Another important impact of science and technology has been the increasing specialization of law enforcement. Police negotiators, special weapons experts, and tactics teams are often relied on in various circumstances. Deferring to experts is usually the best way of handling a critical situation, but only when time allows. In moments of urgency that require quick judgment and action, this strategy can turn into passive policing possibly to the extent of cowardice. The most horrific example is the Columbine High School tragedy (April 20, 1999), where the first responding police officers, knowing there were children being killed inside, failed to enter the school building. These duty-bound officers, supplied with firearms, body armor, and the color of law, chose to wait for the SWAT team rather than risk their own lives in an attempt to save the students. In the aftermath and on nationwide news, the Jefferson County sheriff stated he did not order his men into the school building because he did not want them hurt.

Criminal Adoption of New Technologies

Much of the same technology used by the police to counteract crime has also been adopted by criminals. Computerization and wireless communications are radically altering some forms of crime. For example, drug trafficking organizations often surpass the communications abilities of law enforcement, and even street-level dealers have access to state-of-the-art communications technologies. Electronic correspondence, the Internet, and cellular communications have made illegal transactions of all kinds more difficult to trace (Seaskate 1998). These technologies also allow terrorist cells to be extremely mobile and highly networked. The development of police technology in the future will be largely set within the context of this evolving technology race with criminals.

Police are forced to deal with new and more sophisticated criminal acts while maintaining their traditional roles of handling traffic, mediating domestic disputes, and providing a range of public services. In order to do so, they must devote a substantial portion of their time to continuing education. Law enforcement officers must attend refresher courses, mandated use-of-force training sessions, and other compendious schools just to keep up with court decisions and novel tricks and tactics being created by the criminal mind. This is in addition to the constant development of hardware, software, and scientific means of detecting criminal activity, which criminals in turn work hard to elude, often through the use of technology.

Unlike the police, however, criminals seldom invest in scientific research or are able to use science to develop new technologies. They are more limited to the

creative adaptation of existing technologies, after the manner of the creative consumer analyzed by Michel de Certeau (1984).

Assessment

It is important that the increased reliance on science and technology does not compromise the ethical standards of law enforcement officers. In order to avoid such a possibility, police departments and professional law enforcement societies should make any necessary updates to their codes of ethics. For example, given increased surveillance capabilities and powers (like those under the USA Patriot Act of 2001), police need to ensure that their conduct strikes the right balance between protection of civil rights (like the right to privacy) and the physical protection of citizens from harm. All such changes in science and technology are rapidly altering the context of police work, and law enforcement officers are continually challenged to find the proper use of new technologies to achieve the goals of protecting the public and upholding the law. The rational use of technology and force by the police requires active democratic involvement and citizen partnerships with the police in order to avoid the rise of a modern police state (see Stevens 2000, Wolfe and Zelman 2001).

Technology also plays a role in globalizing criminal activities. Transportation and communication technologies especially enable criminal and terrorist networks to operate and coordinate actions that span the globe. One possible response to this trend is a more central role for Interpol. Created in 1923, Interpol is the world's largest police organization with 182 member countries. It supports all organizations that combat international crime, and it facilitates and coordinates cross-border police cooperation. In this latter function, Interpol is dependent on communication and information technologies that allow multiple agencies to track criminal activities that cross political boundaries. Given the vital importance of technology for Interpol's mission, it may need to strengthen its budget to support research and development specifically targeted to its needs.

In fact, as science and technology become integral parts of police work, it is important that all governments establish rational bureaucratic structures capable of securing the necessary resources to develop and disseminate novel technologies and improved scientific practices. Furthermore, given the increased technological capabilities of criminals and terrorists, it is essential that police and other first responders are adequately trained and equipped to handle contingencies from hostage situations to attacks using weapons of mass destruction. These requirements became especially apparent for the United Sates in the aftermath of the terrorist attacks on September 11, 2001. One of the responses was the creation of the Department of Homeland Security (DHS) in 2002. A central element of the DHS is the Science and Technology Directorate, which works to counter terrorist threats by improving current technological capabilities and developing new technologies. This marks another step in the effort to coordinate and fund federal efforts and encourage industry in the task of providing police with the proper technologies to fulfill their vital mission. However, and in contrast, all of the science, technology and/or modern, crime-detecting gee-whiz gizmos are of no value if police conduct condones anything other than strict compliance to the highest of ethical standards.

CHUCK KLEIN
ADAM BRIGGLE

SEE ALSO *Forensic Science; Science, Technology, and Law; Security.*

BIBLIOGRAPHY

Barker, Thomas and Carter, David L. (1994). *Police Deviance*, 3rd edition. Cincinnati, OH: Anderson. Describes and analyzes many issues of police deviance and misconduct and offers recommendations for mitigation.

Cho, Adrian. (2002). "Fingerprinting Doesn't Hold Up as a Science in Court." *Science* 295(5554): 418. Reports that the U.S. Supreme Court found that fingerprinting does not uphold three of the four standards for federal rules of scientific evidence established in *Daubert v. Merrell Dow Pharmaceuticals*.

de Certeau, Michel. (1984). *The Practice of Everyday Life*, trans. Steven Rendall. Berkley: University of California Press. Marks a turning point in the study of culture from a focus on products and producers to consumers and their creative uses of products.

Fijnaut, Cyrille, and Gary T. Marx, eds. (1995). *Undercover: Police Surveillance in Comparative Perspective*. The Hague: Kluwer Law International. Argues from police surveillance to the larger thesis that social control is changing worldwide due to the functional needs of the modern state and common cultural beliefs.

Hansen, David A. (1973). *Police Ethics*. Springfield, IL: Charles C. Thomas.

Heffernan, William C., and Timothy Stroup, eds. (1985). *Police Ethics: Hard Choices in Law Enforcement*. New York: John Jay Press.

International Association of Chiefs of Police (IACP). (1992). "The Evolution of the Law Enforcement Code of Ethics." *Police Chief* 59(January): 14–17.

Stevens, Richard W. (2000). "What is a Police State?" In *Police State Polices Alert* Newsletter (Winter) Hartford, WI: Concerned Citizens Opposed to Police States.

Toffler, Alvin, and Heidi Toffler. (1993). *War and Anti-War: Survival at the Dawn of the Twenty-first Century*. New York: Little, Brown. Shows how technology alters the nature of conflict and argues that technology and innovation can be the wellsprings of a lasting peace.

Wolfe, Claire, and Aaron Zelman. (2001). *The State vs. The People*. Hartford, WI: Mazel Freedom Press. Defines and differentiates three different types of police-states (traditional, totalitarian, and modern authoritarian) and argues that the United States may be moving toward a form of modern authoritarian police state that grows organically out of fear about crime and terrorism.

INTERNET RESOURCES

Seaskate, Inc. (1988). *The Evolution and Development of Police Technology*. A technical report prepared for The National Committee on Criminal Justice Technology at the National Institute of Justice. Available from http://inventors.about.com/gi/dynamic/offsite.htm?site= http://www.nlectc.org/txtfiles/policetech.html. Contains a timeline of police technology and recommendations for federal decision makers on how to improve the technological capabilities of U.S. police.

POLICY

SEE *Science Policy.*

POLITICAL ECONOMY

• • •

Political economy is commonly defined as that branch of social science dealing with the production and distribution of wealth. The political economy of science and technology would thus focus on the production and distribution of scientific knowledge and technological capabilities that affect "who gets what." Although students of political economy sometimes claim to be objective, ethical issues are intrinsic to the subject.

Technology associated with the industrial revolution stimulated the pioneering political economic inquiries of Adam Smith (1723–1790) and David Ricardo (1772–1823). Smith and Ricardo were particularly interested in public policies that would maximize wealth creation. With the integration of science into the industrial value chain during the second industrial revolution of the late nineteenth century, it too became a subject of political economic scholarship.

From Ethics to Political Economy

The word *ethics* typically connotes issues of personal choice. In the context of science and technology, one might associate it with whether or not to use extraordinary means to extend life or to conceive a child. Yet society makes collective choices about science and technology as well, and these choices have profound moral implications. Many *extraordinary means* in medicine, for instance, emerged from research and development (R&D) projects that were supported directly by government funding or were subsidized indirectly through other policy measures.

In the absence of complete and unquestioned unanimity within a polity, collective choices involve the exercise of power. Persuasive or coercive authorities extract and redeploy resources or, equally important, determine how those who hold resources may use them. The U.S. government, to continue the example, spends nearly $30 billion per year on biomedical R&D. Its regulations, especially those of the Food and Drug Administration (FDA), further shape the flow of private biomedical R&D funding.

The prospect of action by the authorities induces the mobilization of interests. Individuals and organizations with a material, ideological, bureaucratic, or other stake in whether and how power is used seek influence. The potential recipients of biomedical R&D funding lobby governmental officials; think tanks advocate changes in the regulatory process; and groups representing patients work to enlarge the shares of the R&D pie devoted to the diseases with which they are most concerned.

Political economy embraces all of these activities: the intertwined exercise of public power and exertion of private influence to shape the allocation and use of societal resources. In the contemporary political economy of science and technology, money is the resource that is most visibly at stake, but it is not the only one. Property rights, access to markets, and skilled people are also very important.

Centralization and Decentralization

Technological innovation is an ancient and, some would argue, characteristically human process. The political economy of technology is nearly as old. Douglass C. North (1994), for instance, ascribes the invention of agriculture to the assertion of property rights over land. Agricultural production stimulated ancient industries such as metalworking only after centralized empires were established.

Yet highly centralized political economies, such as empires and communist systems, have fostered technological innovation only intermittently. They are vulnerable to bureaucratic ossification and the whims of leadership. During the Middle Ages, for instance, the Chinese Empire developed arts such as textile production and shipbuilding to a level that astonished European visitors. Then fifteenth-century emperors put an end to these endeavors, going so far as to impose the death penalty on any subject who dared to build a three-masted ship.

Capitalism has proven the most technologically fecund of all the great political economic systems, in large part because decision making about how technologically-relevant resources are used is largely decentralized. Competition among producers leads to experimentation with new ways of making things and with the making of new things, experimentation that is enabled by property rights and mediated by market prices (Rosenberg and Birdzell 1986). The results of these experiments are judged by a multitude of end users who, through their buying decisions, feed both resources and information back to the innovation system.

One must take care not to exaggerate the degree of decentralization. Capitalist enterprises are embedded in a larger framework of social institutions that depend on collective authority, albeit an authority that is circumscribed by constitution and culture. These institutions vary dramatically over time and across political jurisdictions, coevolving with the economic system and in response to military and other external challenges. The delicate balance of public and private power, of centralized control and decentralized experimentation, is a core theme of the political economy of science and technology.

Intellectual Property Rights

Intellectual property rights (IPR) exemplify the delicate balance. Patents, copyrights, and other forms of IPR allow holders to use the coercive power of the state to prevent others from using specific bits of knowledge for defined purposes for limited periods of time. This control over potential competition is designed to induce the substantial additional investment that is usually required to convert the protected knowledge into a commercially viable product or process. In the absence of IPR protection, potential innovators might be deterred by the prospect of rapid imitation. Yet very broad, very long, or very rigid IPR protection may be an equally powerful constraint on innovation, inhibiting cumulation and competition.

This basic theory of IPR has been articulated by Kenneth J. Arrow (1962), F. M. Scherer, and other economists, but it provides little practical guidance for setting the balance. This is left to political and legal processes. The historic contrast between Germany and the United States is striking in this regard. The German government has generally been much more tolerant of cooperation among rights-holders, building on the medieval guild tradition of exclusive control over the arts of production. The United States has often struck down such arrangements, not only when they take the form of contractual agreements, such as patent pools, but even when they result from single firms amassing market-dominating positions. Antitrust law has often been used to compel the licensing of intellectual property.

The political economy of intellectual property has become increasingly complex and contested as science and technology have grown in economic importance and the capacity to produce them have diffused globally. The pharmaceutical industry, for example, is more dependent than any other on patents. Pharmaceutical firms, not surprisingly, have lobbied and litigated to expand the scope and duration of IPR, with great success during the last decades of the twentieth century. New kinds of inventions, especially in biotechnology, have gained protection in the United States, and legislators, administrators, and judges have generally treated rights-holders more favorably than in the preceding decades.

Pharmaceutical firms were also at the forefront of an advocacy push that extended Euro-American principles of IPR protection to much of the rest of the international community through the agreement on trade-related aspects of intellectual property rights (TRIPS) within the World Trade Organization framework. TRIPS, however, seems to many actors and observers to have tipped the delicate balance too far in the direction of rights-holders. In response, a global movement has emerged to secure low-cost access to patented medicines for the treatment of diseases that are widespread in developing countries, such as tuberculosis and AIDS. Invoking the ethical principle that current human needs ought to be valued more than future corporate profits, this movement has for the moment stemmed the drift of international policy in favor of stronger IPR.

Trade

The association of IPR with the international trade regime is a new development in the political economy of science and technology. Traditional regulation of trade in goods, though, has long been understood to be a potentially powerful factor bearing on science and technology and the distribution of the benefits and costs associated with them. Indeed Adam Smith, one of the progenitors of the concept of political economy, argued in *The Wealth of Nations* (1776) that larger markets facilitate occupational specialization, which in turn fosters the development of science and technology. Among the specialized occupations to which Smith attributed economic significance was science itself: "philosophers or men of speculation, whose trade it is not to do anything, but to observe everything; and who, by that account, are often capable of combining together the powers of the most distant and dissimilar objects" (Penguin Classics ed. 1986, p. 115; or Book 1, Chapter 1).

The nineteenth-century German political economist Friedrich List (1789–1846) disputed the association that Smith made between *the extent of the market* and the development of scientific and technological capabilities. List argued that free trade allowed those who already had such capabilities to deepen them and reduced the odds that those who did not have them would acquire them. List's arguments have been cast in modern form by the theories of the *developmental state* and *strategic trade*. By striking a careful and dynamic balance between trade protection and openness to the world market, clever and powerful governments could—at least in principle and under particular circumstances—induce the creation of domestic high-technology industries that would not have flourished otherwise. The great inspiration for and proving ground of these theories has been East Asia, where first Japan and more recently the *four tigers* of Hong Kong, Singapore, South Korea, and Taiwan, joined the ranks of global high-technology powers.

An even greater test of these theories looms ahead as other developing countries, especially China and India, with more than a third of world population, seek to follow suit. China and India have both aggressively sought foreign direct investment since the 1980s, especially in areas such as semiconductor manufacturing and software development. They have also opened domestic markets to sales by foreign high-technology firms, but usually conditionally, using the leverage of market access to secure benefits from foreign firms for their own *infant* high-technology industries.

Whether these infants will mature into healthy adults that help to raise living standards in previously impoverished countries remains to be seen. Their growth could be stunted by, among other things, inept governance, capture of policy-making by narrow interests, or aggressive protectionist reactions in developed countries. The aspirations of billions of people for a better life hang in part on whether world trade policy-makers can steer effectively between the perpetually inequitable Scylla of unregulated trade and the stifling Charybdis of ratcheting protectionism.

Human Resources

The effectiveness of strategic trade policy depends not only on the intelligence and agility with which it is implemented, but also on the capacity of an economy to absorb ideas from abroad and generate new ones. Access to the richest scientific literature and the best blueprints, even in the context of cleverly protected markets, is no guarantee that domestic enterprises will move to the cutting edge of global competition. Tacit knowledge, which cannot be written down but is acquired through experience in doing science or operating technological systems, is another necessary ingredient in the development of scientific and technological capabilities. The people who have such knowledge, or have the capacity and incentive to acquire it, are thus critical resources in the political economy of science and technology.

Karl Marx (1818–1883), who put science and technology at the center of his pioneering political economic analysis, claimed to the contrary that technological innovation under capitalism merely displaced human capabilities. This process of alienation, as he called it, would ultimately motivate revolutionary upheaval as workers came to recognize their interest in controlling the means of production. The threat of technological displacement has occasionally prompted workers to exercise their collective power, albeit never to the point of overthrowing governments. Trade unions have fought to have a voice in the process of technological change in the workplace. Labor victories in such contests have sometimes led to slowdowns in the pace of innovation, but (contrary to Marxist expectations) have also often allowed enterprises to tap more effectively into the expertise of workers and even accelerate the pace of change.

More important, the Marxist focus on particular labor processes ignores the broader transformation of the economy brought about by the development of science- and information-based industries that began to appear in the waning years of Marx's life. Even if technology displaces and *deskills* workers in older industries, the growth of newer industries that rely more heavily on *knowledge workers* more than counterbalances those losses in the long run. Such industrial transitions do not occur solely as a result of shifts in private investment. Public investments are typically critical catalysts as well.

While the balance between worker voice and capitalist flexibility is important for the political economy of science and technology, the balance between current consumption and future-oriented public and private investments may be even more so, as suggested by the work of Robert Solow (1957), Paul Romer (1990), and others.

Universal public education at the primary and secondary levels, for example, seems to be a prerequisite for the development of a knowledge economy. The United States and Germany surpassed the United Kingdom in science and technology during the nineteenth century in part because they were willing to impose taxes (and break down social barriers) to provide education. The more recent East Asian *development miracle* similarly rests on a strong educational base.

Private investment enters the balance more forcefully at higher levels of education. University and graduate students may be able to recoup the costs of education through future earnings, even if they borrow funds to pay tuition. Responsibility for such an investment will tend to encourage diligence and attune students to the likely needs of future employers. Yet information about the future is sufficiently uncertain and the spillover benefits to society of a highly-trained workforce sufficiently great that significant public subsidies to higher education are justifiable. The U.S. university system has more private elements than most, but its rise to world leadership in the twentieth century coincided with an infusion of resources from taxpayers to students, such as scholarship grants, tuition loan guarantees, and publicly funded research assistantships.

The migration of highly skilled people complicates the political economy of science and technology. The immediate social benefits of graduates who emigrate spill over to their new neighbors, not those who paid for their education. The threat of a *brain drain* may prompt preventive or compensatory measures, such as controls on movement or exit taxes. In the longer run and under particular conditions, emigrees may nonetheless pay back the investment made by their places of origin by creating channels through which knowledge flows. Taiwanese *astronauts* in Silicon Valley, for instance, have helped to make their home country a global center for the information technology industry.

R&D Funding

Higher education is increasingly joined at the hip with scientific research in the institution of the research university. Involvement in research conveys tacit knowledge to students even as they produce formal knowledge, such as publications and patents, in conjunction with their professors and other researchers. The benefits of formal knowledge spill over even more easily than those of tacit knowledge. Indeed the academic scientific community has a distinctive political economy in which collective rewards in the form of prestige flow to individuals whose work has spilled over most broadly. This system discourages scientists from trying to appropriate the financial benefits that flow from an idea by keeping it secret or gaining IPR protection for it, because prestige can only be gained through widespread, low-cost diffusion of ideas.

Of course, as union organizers at Harvard once put it, "You can't eat prestige." Fortunately for scientists, material rewards tend to correlate with prestige, although less systematically than licensing fees correlate with intellectual property holdings. Private patrons inspired by the scientific spirit and the desire to bathe in reflected glory were a particularly important source of sustenance for scientists in the early-modern era. Private patronage continues in the early twenty-first century, but it is overshadowed by government and corporate support underlain by baser motives. Where the *communist* (as Robert Merton [1973] characterized it) or shared knowledge political economy of science meets the capitalist political economy of science and technology, sparks often fly.

The standard economic theory behind government funding of R&D carries forward the tradition of the noble patron: The financial burden of R&D with benefits that accrue to all in society should be shouldered by all. R&D that benefits only a few should be funded privately by those few. Economic research by Richard R. Nelson (1959) and Edwin Mansfield (1977), among others, suggests that many opportunities for socially valuable R&D go unrealized. Because the constituency for diffuse future benefits is usually weak, political processes tend to favor other uses of societal resources. In U.S. politics, a more specific and urgent mission, such as national defense or public health, must typically be marshaled to win significant government R&D funding, although those who manage and disburse these funds have often seen fit to support projects highly regarded by scientists but with only a distant relation to the stated mission.

That political forces impede the achievement of the socially optimal level of public investment presents no challenge to economic theory. A deeper problem is that prospective public and private benefits are more difficult to distinguish in practice than in principle; in fact some public benefits may be impossible to obtain unless people get rich providing them. The division of labor between the public and private sectors is not nearly so clean as the conventional categories of *basic research*, *applied research*, and *development* imply.

The biotechnology industry is the most prominent case in point. Publicly funded science underlies the industry, and publicly funded scientists routinely start firms to capitalize on their findings, often with investments from their own universities. Large pharmaceutical firms are major funders of academic researchers and entrepreneurial start-ups as well, making deals that may impose restriction on the free exchange of ideas in order to preserve the funder's pecuniary interest. At this flash point between the communist and capitalist political economies, hot debates have erupted over the rules that govern public funding as well as the norms that regulate the behavior of scientists and research universities.

As with property rights, access to markets, and human resources, the diffusion of scientific and technological capabilities globally has complicated efforts to find a workable balance in the allocation of R&D funding. Spillovers that accrue across borders, whether in the public or private sector, weaken incentives for governments to make public R&D investments. Collective action on behalf of the global public good is a tortuous process in the absence of a global authority capable of levying taxes. The largest multinational corporations have globalized their R&D infrastructures, drawing on brainpower from Barcelona to Bangalore to Beijing to Boston. But these firms do not yet form a cohesive constituency that lobbies for global public goods, nor should one expect that if and when they do their interests will coincide with the greatest good for the most people or any other broad ethical principle.

Creative Destruction

At any point in history, people who seek "to promote the progress of science and the useful arts" (U.S. Constitution, Article 1, Section 8) depend on access to ideas and materials to do their work. Access to these resources has never been free and unencumbered, but is instead conditioned by public power and private influence. Marx imagined an end-state to history in which all people would engage in creative work, but this utopia is, at best, far in the future. *Real existing socialism*, as the people's republics of the twentieth century were sometimes referred to, was far less efficient in its allocation of technologically-relevant resources than its capitalist competitor. It was also far less fair in allocating the costs and benefits associated with scientific research and technological innovation.

Capitalism, to borrow from Winston Churchill, is the worst political economy of science and technology, except for all the others. Critical resources, including property rights, access to markets, highly-skilled people, and R&D funding, are allocated through a messy mixture of market exchange and state action. The appropriate division of labor between the two mechanisms is clarified only somewhat by theory, and even these partial insights are honored in the breach. Some people get extraordinarily rich, and others are displaced, injured, or otherwise left out. The process of *creative destruction*, as Joseph Schumpeter (1950) famously labeled it, is intrinsically disruptive.

The political economy of science and technology is itself a continual work in progress. Globalization is forcing public authorities and private actors to reconsider priorities and rethink routines that were previously taken for granted. In this moment of transition may lie opportunities to nudge the system in more ethically satisfying directions.

DAVID M. HART

SEE ALSO *Capitalism; Economics and Ethics; Marx, Karl; Science, Technology, and Law; Smith, Adam; Socialism.*

BIBLIOGRAPHY

Arrow, Kenneth J. (1962). "Economic Welfare and the Allocation of Resources for Invention." In *The Rate and Direction of Inventive Activity: Economic and Social Factors.* Princeton, NJ: Princeton University Press. Fundamental neoclassical economic model of science.

Harhoff, Dietmar, Francis Narin, Frederic M. Scherer, and Katrin Vopel. (1999). "Citation Frequency and the Value of Patented Inventions." *Review of Economics and Statistics* 81: 511–515. Demonstrates the highly skewed distribution of value of inventions.

Mansfield, Edwin, et al. (1977). "Social and Private Rates of Return from Industrial Innovations." *Quarterly Journal of Economics* 91: 221–240. Shows how benefits spill over from innovations.

Merton, Robert K. (1973). "The Normative Structure of Science." In *Sociology of Science.* Chicago: University of Chicago Press. Originally published in 1942. Exposition of the value system of academic science.

Nelson, Richard R. (1959). "The Simple Economics of Basic Scientific Research." *Journal of Political Economy* 67: 297–306. Economic rationale for public support of science.

North, Douglass C. (1994). "Economic Performance Through Time." *American Economic Review* 84(3): 359–367. Links innovation process to institutional incentives.

Rosenberg, Nathan, and L. E. Birdzell. (1986). *How the West Grew Rich: The Economic Transformation of the Industrial World.* New York: Basic. Economic history from late middle ages centered on science and technology.

Romer, Paul M. (1990). "Endogenous Technological Change." *Journal of Political Economy* 98: S71–S102. Economic theory of increasing returns to knowledge.

Schumpeter, Joseph A. (1950). *Capitalism, Socialism, and Democracy*, Harper Colophon edition. New York: Harper and Row. "Creative destruction" as the core feature of capitalism.

Solow, Robert M. (1957). "Technical Change and the Aggregate Production Function." *Review of Economics and Statistics* 39: 312–320. Econometrically demonstrates importance of technological change in economic growth.

POLITICAL RISK ASSESSMENT

• • •

Support for scientific research and technological development, especially in developing countries, requires interstate and cross-border participation. Such development and technology transfer issues are subject not only to ethical evaluations but also to political risk assessments. The degree to which international investment projects, public and private, are attracted to or successful in many parts of the world is increasingly dependent not simply on technical but on social and political factors.

It has been argued that any engineering project worth $100 million or more is no longer a technical project, but a political enterprise. All political enterprises embody risks. Political risk—also known as country risk or sovereign risk—is most often defined as those conditions that a country can create *at home* that might undermine investment climate and cause investors to incur losses. Political risk also involves exposing a business to conditions *abroad* created by extra- and supranational political changes, policy decisions, social situations, inter-market relations of two or more regions, and global financial market oscillations, over which the country may or may not have control.

Political Risk Types

Developed countries as well as developing states generate political risks. More likely, such developed countries as the United States, Japan, and France can offer political risks of *regulatory* excesses, while developing countries such as Indonesia, Peru, and South Africa can offer *structural* risks such as regime instability, out-of-sync economic policies, and ethno-religious-cultural imbalances in development due to the monopoly of political power and economic wealth by a single dominant ethnic or religious group. Examples of regulatory risks are excessive environmental rules, market restrictions to favor or protect a certain domestic economic group (such as in the United States and Britain), or manipulating free market rules to promote national champion firms (such as in Japan, Germany, and France). Cases of structural political risks in the developing countries above all rest on the lack of the rule of law, an impartial court, the protection of private property, the sanctity of contracts, and transparency, as well as out-of-control corruption, excessive subsidies to state-chosen firms, and favored access to power and wealth by a state-favored ethnic or religious group. Developed countries engender fewer risks than developing countries, due to the better-developed legal institutions, norms, and practices for business. Also, multiethnic states tend to create more structural risks than countries with a single ethnicity. And a country with intractable economic and financial difficulties, whether developed or developing, runs greater risks for investors than a country with a prudently managed economy. In order to produce a carefully weighed assessment, risk variables must be quantitatively evaluated.

All countries generate and nurture five kinds of risks:

(1) political instability that can lead to regime change;

(2) macroeconomic and financial imbalances that can lead to a severe malfunctioning of the economy;

(3) social, cultural, and environmental risk that can affect human development;

(4) global linkages facilitate a country's integration into the global economy but insufficient ties can deny access to external capital, technology, resources, and markets, thus increasing the country's risks; and

(5) business environment risk, which allows the country to achieve the level of competitiveness against its neighbors.

Each of the five compartments or shells is self-defining and self-contained, while one or two inferior performances in the five shells can undermine the soundness of the other three or four, thus increasing the overall risk of a single entity, setting off a *contagion effect*. Conversely, two or three well-calibrated shells can lower the risks of the lesser shells, benefiting from a *free ride effect*. In brief, they are collectively interlinked and mutually reinforcing. A country framed in five well-balanced and well-reinforced shells offers little or no risk. And a country fraught with ethnic, racial, and religious strife as well as chronic economic crises and illiberal democratic practices will suffer from high political risk and discourage investors.

Political Risk Assessment Users

Avid users of political risk assessments are governments, global businesses, and increasingly nongovernmental organizations. Each needs to know the political, economic, sociocultural, and environmental conditions of a given country in which it seeks to successfully operate for profit, forge security and diplomatic alliances, cement friendly bilateral trade and financial relationships, or expand the participation of civil society in political processes. A visiting head of government needs to know about the strengths and weaknesses of the host country as well as his or her counterpart, while a global corporation must realistically assess the country's political risks before it commits millions of dollars to an investment project. A transnational nongovernmental organization needs to choose a right local partner in order to effect its global agenda, whether it be environmental, religious, scientific, developmental, or ethnocentric. Correctly assessing risks can increase the success of a state or corporate policy.

What constitutes a high risk for one country may be no risk for another. The United States may not be welcomed to certain countries due to historic policy differences, but Canada or Switzerland can watch over American interests. What is a risk for a bank may pose no risk for a mining or oil company. The U.S. foreign and defense policy in the post-9/11 era has increased risk for American businesses, due to escalating anti-Americanism around the world. A U.S. bank may not be welcomed in Sudan, a poor Muslin country that views with resentment Washington policies toward Islamic nations. But Sudan will welcome a U.S. oil company for its advanced technology and global market reach. Conversely, a Chinese firm can engage in a joint venture with IBM to access U.S. technology while reducing the political risks of hyperregulation and export control by the U.S. government, which considers China both a trading partner and security rival in the Asia Pacific.

Political risk is a dynamic phenomenon. Hence, political risk assessment requires a constant monitoring of all five categories of risks and fashioning of mitigation strategies. Multinational and global companies have come to manage their cash flows in a basket of currencies (dollars, euros, and yen) to mitigate the risk of the sudden devaluation and revaluation of a single currency, often a reflection of a country's fragile state of economy and unstable politics. In the contemporary globalized economy, sound assessment of political risk can save a company millions from regulatory or structural risks or can generate windfall profits, while a country can reduce security risk by engaging potential rivals in expanded trade and investment activities.

Some risks are *interstate*, others are *regional*, and still others are *global*. Before the days of regional free trading systems, such as the European Union, Mercado Comun del Sur (MERCOSUR), and the North American Free Trade Agreement (NAFTA), minor members wielded little influence in the global arena—politically, economically, and diplomatically. Today Portugal, Uruguay, and Mexico can wield more. To avoid an unpleasant showdown in bilateral relations, the United States often resorts to its formal and informal veto power in multilateral organizations such as the United Nations, the World Bank, and the International Monetary Fund to reject funding requests from less cooperative countries, thus reducing confrontational risks.

Political risk is therefore an outcome of policy choice; it can increase or decrease as the state chooses how to devise and implement its domestic and external policies. To maintain low political risk can lead to immeasurable loss of a country's independence, autonomy, and even sovereignty. In return, this can allow a country access to international capital, market, technology, and skilled labor. In the age of globalization, to insist on keeping independence, autonomy, and sovereignty can increase political risk and therefore be costly in both economic and political terms.

EUL-SOO PANG

SEE ALSO *Globalism and Globalization; Modernization; Risk Ethics; Science Policy.*

BIBLIOGRAPHY

Borner, Silvio; Aymo Brunetti; and Beatrice Weder. (1995). *Political Credibility and Economic Development.* New York: St. Martin's Press.

Moran, Theodore H., ed. (1998). *Managing International Political Risk.* Malden MA: Blackwell.

Pang, Eul-Soo, and Halim Saad. (2001). "Globalization, Regime Changes, and Political Risk." *Journal of Diplomacy and Foreign Relations* 3(2): 1–20.

Samuels II, Barbara C. (1990). *Managing Risk in Developing Countries: National Demands and Multinational Response.* Princeton NJ: Princeton University Press.

POLLUTION

• • •

Most often used in regard to the natural environment, the term *pollute* means to make foul or unclean, degrade ecological and/or human health, contaminate or defile, and, in a religious sense, render ceremonially impure or desecrate. The verb pollute derives from the Middle English *polute*, and this from the Latin *pollūt(us)*, the past participle of *polluere*, which meant to soil, defile. Pollution generally denotes an undesirable condition, where there is too much of something (the pollutant or contaminant) in a natural or other beneficial system. It is, then, not an objectively determined state of affairs. Rather decisions about pollution require both science (for example, identification, monitoring, and classification) and ethics and politics (such as debate about what is undesirable, what is acceptable, who should monitor pollutants, and who should be held accountable). Although pollution is, in a sense, an unavoidable by-product of human (and nonhuman) activity, it was not until the Industrial Revolution that it regularly occurred on a large-scale and became a public policy issue.

Measures have been taken to curb pollution, especially the public health activities in the 1800s and then again with the rise of the contemporary environmental movement in the 1960s. Largely due to political and economic incentives and advances in technology, many pollutants are declining. In several regions, however, pollution remains a serious problem threatening both human and environmental health. Pollution has long been seen as the most visible and costly reminder of a downside to the technological mastery of nature. The use of technologies to prevent and diminish pollution, however, may eventually eliminate this particular cause for technological pessimism.

Classifying and Describing Pollution

Environmental pollution can be either point source (such as emissions from factory smokestacks) or non-point source (for example, fertilizers and oil washed from lawns and parking lots into streams). It can occur suddenly, as in the massive radioactive plume released from a nuclear power plant in Chernobyl in 1986 or the 1.26 million barrels of oil spilled into Prince William Sound, Alaska, by the *Exxon Valdez* in 1989. However pollution usually stems from long-term emissions, as in the accumulation of carbon dioxide (CO_2) in the atmosphere resulting from fossil fuel combustion. Although most pollution is anthropogenic (human caused), some forms are naturally occurring. One example is radon gas that leaks from rocks into buildings and accounts for roughly 55 percent of the total radiation dose received by an average person in the United States. Anthropogenic and nonanthropogenic sources of pollution can also combine to produce deleterious environmental and health effects. One example is the combination of human-produced chlorofluorocarbons (CFCs) and naturally occurring ultraviolet radiation in the stratosphere, which initiates a reaction that depletes ozone. The methane gas produced by cows (and other farm animals) accounts for roughly 20 percent of all such global emissions. Although this is a natural source of pollution, the vast quantity of these animals would not exist without humans.

Any compound can be considered a pollutant if it is judged to exist either in excessive quantities or in the wrong place. For example, ozone in the stratosphere is regarded as beneficial, whereas ozone in the troposphere (the lowest layer of the atmosphere) is regarded as a pollutant because it contributes to smog, which causes harmful ecological, human health, and aesthetic effects. This is akin to exotic species, which can be considered pollutants, because they are located outside the boundaries of the area in which they evolved.

Pollution can be classified by economic sector (such as residential, industry, agriculture, transportation, and others), which can be helpful for regional governments implementing pollution reduction policies, but the relative contribution of each source varies markedly depending on the composition of regional economies. A more universalizable classificatory scheme groups pollutants according to the reservoir in which they are found: water, soil, air, and space. More detailed cataloging can then be carried out. Water pollution, for example, is typically sorted by type, including biological/pathogens, sedimentation, nutrients, toxic synthetic chemicals, and heat/cold. Water quality indicators include hardness (a measure of dissolved minerals), pH, temperature, dissolved oxygen, turbidity, and smell. The U.S. Environmental Protection Agency (EPA) has set national air quality standards based on six common (referred to as *criteria*) air pollutants: ozone, nitrogen dioxide, sulfur dioxide, particulate matter, carbon monoxide, and lead. Pollution in space (scraps from old satellites, rockets, or other spacecraft) is generally classified by size. The vast majority of the hundreds of thousands of human-produced particles in space are between one and ten centimeters in diameter, but even these small pieces have caused massive damage to satellites. Not included in the above list, but worthy of mention, is indoor air pollution (especially caused by smoking and the use of

FIGURE 1

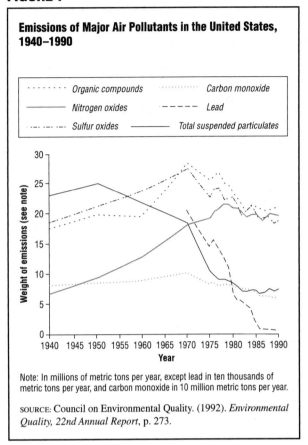

Emissions of Major Air Pollutants in the United States, 1940–1990

Legend:
- Organic compounds
- Nitrogen oxides
- Sulfur oxides
- Carbon monoxide
- Lead
- Total suspended particulates

Note: In millions of metric tons per year, except lead in ten thousands of metric tons per year, and carbon monoxide in 10 million metric tons per year.

SOURCE: Council on Environmental Quality. (1992). *Environmental Quality, 22nd Annual Report*, p. 273.

wood or coal-burning stoves), which poses grave health risks. Noise and even drugs can be considered pollutants insofar as they can have deleterious effects on the well-being of humans and other animals.

The severity of a pollutant depends upon its chemical nature, concentration, and persistence. One important equation used by environmental scientists to understand pollution derives from biogeochemical cycling: *Residence time = Reservoir size/Sum of all fluxes (in or out of the reservoir)*. A reservoir is simply any "compartment" that can serve as a storage place for pollutants. Examples include the ocean, atmosphere, and biosphere. Reservoirs can be defined more precisely depending on the pollutant or other compound of interest. For example, scientists interested in particulate organic carbon may choose to focus only on the upper ocean (where the majority of carbon is located). Flux refers to the rate at which the pollutant (or other compound of interest) moves in and out of the reservoir. Residence time is how long the pollutant stays in the reservoir of interest. CO_2, for instance, has a long residence time in the atmosphere, such that even if all emissions were immediately stopped, CO_2 levels would drop only very slowly. Sulfur dioxide (SO_2), by contrast, is water soluble, and because water has a relatively short residence time in the atmosphere (less than five days), SO_2 will quickly precipitate out.

This explains why CO_2 emissions pose problems on a global scale (because it stays in the atmosphere long enough to thoroughly mix around the globe), whereas SO_2 emissions pose problems on a regional scale (because it will precipitate out of the atmosphere somewhere within five days downwind of the source). The equation also helps explain why groundwater pollution is so much more difficult to clean up than surface water pollution. Groundwater aquifers have very low fluxes, meaning residence times for pollutants are quite high. Surface water systems, for the most part, have high fluxes, meaning that pollutants can be quickly flushed through the system.

Ethics and Deciding How Much is *Too Much*

Classifying and describing the behavior of pollutants in natural systems still leaves many questions unanswered, including: How much pollution should be allowed? What should count as pollution? How should societies determine the relative values of risks to people's health and other matters of concern (for example, ecosystem integrity and aesthetics), or, how should they determine when there is too much of something, thus turning simple presence into pollution?

Pollution may result in injustice, because its effects can be disproportionately suffered by the poor. For example, poor people can often only afford to live in neighborhoods that are crowded with polluting industries, yet they seldom have the resources to challenge polluters in the court system (the Erin Brockovich case is an exception to this rule). Similarly global climate change resulting from CO_2 emissions (significantly produced by wealthy nations) may have the most devastating impacts on poor nations unable to adapt to rising sea levels and other effects. Welfare economists conceptualize pollution as a problem of establishing the proper costs so that its effects are fairly distributed.

Despite these injustices and the more general detrimental effects of pollution, several economic theorists and philosophers have made a strong case that the proper reaction is not to eliminate pollution, but rather find the optimal amount of pollution. Julian Simon (1981) and his successor Bjørn Lomborg (1998) argue that economics correctly views pollution as a trade-off between cost and cleanliness. This has two main implications. First, the goal is not pristine, pollution-free environments, but rather an environment that is optimally clean in the sense that, at this point, citizens would rather pay for some other service or good than more pollution abatement (this is the willingness-to-pay criterion for determining optimal pollution). Second, measuring the goal of optimal pollution would best be accomplished by some metric of human welfare such as life expectancy.

Both Lomborg and Simon argue that pollution does not undermine human well-being in the long run. Although air pollution continues to worsen in developing nations, they are just making the same trade-offs that developed nations did during the Industrial Revolution. Indeed, as Lomborg (1998) states, "the environment and economic prosperity are not opposing entities: without adequate environmental protection, growth is undermined, but environmental protection is unaffordable without growth" (p. 210). Thus following the path of developed nations, as the developing world achieves higher levels of income, it will choose and be able to afford an ever cleaner environment.

William Baxter (1974) echoes Simon and Lomborg by contending that only humans should count in the calculus of determining optimal pollution. This does not mean that other species will be wantonly destroyed, he maintains, because humans both depend on them and enjoy them for aesthetic and recreational reasons. It does mean, however, that the claim "DDT use is damaging penguin populations" does not automatically mean that people must stop the use of DDT. In order for this result to follow, Baxter claims that it must be shown that the well-being of people would be less impaired by discontinuing the use of DDT than by harming penguins. This conclusion is rejected by theorists such as Aldo Leopold (1949), who argued that humans must take the *integrity, stability, and beauty* of the biotic community directly into account when making decisions that impact the environment. Indeed perspectives and values play an enormous role in how one perceives pollution and the state of the planet. For example, the controversies aroused by the works of Simon and Lomborg show that measuring pollution is as much a political as a technical endeavor.

From a traditional economic standpoint, pollution can be classified as an externality, that is, an unintended and unaccounted for spillover effect on an unconsenting third party. A good example is industrial activities in the Midwestern United States leading to acid rain in the Northeastern United States and Canada. This definition logically leads to attempts to fix market failures (instances where not all costs are appropriately taken into account). Thus environmental economists attempt to quantify the costs of pollution and integrate them into market transactions. Many models use the willingness-to-pay criterion or cost-benefit analyses to establish these costs. The philosopher Mark Sagoff (1988) argued this form of economics and its narrow notion of *physical spillover* would not rule out many projects or policies that might seem appalling, for example, the attempted conversion of Mineral King Valley in California into a Disney resort. Such a narrow notion leaves no room for many aesthetic and ethical values.

This led economists, especially since the 1970s, to replace the notion of physical spillover with that of transaction or bargaining cost in evaluating the efficiency of a project or a policy concerning pollution. The focus was thereby widened to cover any unpriced benefit or cost (that is, anything a person might be willing to pay for) even if markets do not typically price it correctly. However, as Sagoff also argues, if the wider, more recent notion of transaction or bargaining cost is used, then economic calculations establish policy goals, in the process reducing factual, moral, and aesthetic judgments to mere preferences. But economists have replied that these economic analyses are assessment tools, not decision-making mechanisms. Whatever method of analysis, policymakers need information and tools that will allow them to examine more explicitly and precisely, whether quantitatively or qualitatively, what those affected value about their programs, and how the value of these programs can be assessed. Even if cost-benefit analysis and willingness-to-pay are inadequate, the question remains: How to value?

Sagoff argues that traditional economic methods for determining optimal pollution are insufficient because they place individuals in the role of mere consumers or bidders. Instead he claims each individual should play the role of citizen or trustee of one's own and others' health and well-being. On this view, questions should aim to determine not what individuals would be willing to pay for their health and well-being, but what they would exchange these things for. That is, a willingness-to-sell criterion should be used. This implies that citizens have property rights to an unpolluted environment, thus assigning them the role of sellers and not mere bidders. As such, they may be unwilling to sell those rights or willing to sell them only at a much higher price than they would have been willing to pay for them. One question is whether this leaves room for consent and respect of property rights.

Smokestacks from a factory in Pittsburgh, Pennsylvania, belch black smoke into the atmosphere. (© Bettmann/Corbis.)

Solving Pollution Problems

The natural reaction to pollution problems by polities has been to use command-and-control style regulations and legislation. Indeed, around 300 C.E. local Roman magistrates passed laws regulating certain sources of air pollution in York, England, and in 1272 Edward I banned the use of *sea coal*, while parliament ordered punishment by torture and hanging of people who sold and burned the outlawed coal. The rise in environmental consciousness in the United States in the early 1970s saw the continuation of this trend as government legislation and agencies multiplied to prevent and decrease pollution. Some examples include the National Environmental Policy Act, 1969; the creation of the EPA, 1970; Clean Air Act Amendments, 1970 and 1977; the Clean Water Act, 1972 and amended in 1977; the Endangered Species Act, 1973; and the Toxic Substances Control Act, 1976.

Pollution does not respect political borders, however, and transboundary issues have increasingly required international cooperation in the development of pollution regulations. One notable example is the UN Framework Convention on Climate Change (Framework Convention), formally established in March

1994, which is a constitutive body specifying rules for making decisions about global climate change. Its major outcome is the 1997 Kyoto Protocol, which attempted to prescribe legally binding targets and timetables for emissions reductions. The most successful example of international cooperation to control pollution is the 1987 Montreal Protocol on Substances that Deplete the Ozone Layer.

Although government approaches to pollution problems often result in important successes, they also betray the fact that there are governmental failures just as there are market failures. Several reasons for these failures exist. On the international level, bodies such as the Framework Convention often lack political power. Bureaucrats, like all people, are self-interested, and when governmental structures are not designed to link authority with responsibility for program outcomes, "decision makers have few incentives to consider the full social costs of their actions" (Baden and Stroup 1981, p. v). Furthermore decision makers have only a limited capacity to comprehend complex social and environmental interactions, which can constrain their ability to make wise regulatory decisions.

One response has been to improve the structure of government, but another reaction has been to improve the structure of markets by implementing what Terry Anderson and Donald Leal term *Free Market Environmentalism* (1991). The underlying philosophy of this regulatory approach is that markets and environmental concerns can be made compatible by internalizing costs and establishing the proper incentives. They write, "Instead of intentions, good resource stewardship depends on how well social institutions harness self-interest through individual incentives" (p. 4). Examples of utilizing market mechanisms for pollution abatement include green taxes, marketable emissions permits (for example, cap-and-trade systems), and the elimination of harmful government subsidies. Command-and-control and free market regulatory strategies can often be used in conjunction to achieve desired outcomes. One example is cost-effectiveness analysis, where courts or legislatures establish goals, but economists utilize cost-benefit analyses to establish the cheapest ways of attaining those independently set goals within the market.

Technological innovations have been a major force in pollution prevention and abatement as industry has been either forced to comply with regulations or more subtly incentivized to increase efficiencies and reduce pollution outputs. For example, smokestack scrubbers and catalytic converters in automobiles mitigate pollu-

tion problems originally caused by the technologies of electricity generation and automobile transportation. Although instances of technological fixes to pollution problems abound (as well as technological devices to monitor pollution), it is also true that technologies continually present novel pollution problems. This holds for the thousands of novel synthetic chemicals produced every year (of which very little is known about possible long-range health impacts) as well as potential future scenarios such as the emergence of grey goo, unrestrained nanobot replication, that could potentially wreak havoc on human and environmental health (see Joy 2000). Such devastating possibilities (not to mention the realities of disasters such as the deadly poison leaked from the Union Carbide insecticide plant in Bhopal, India in 1984) cause some to argue for the relinquishment of potentially harmful technologies or even the abandonment of industrial capitalism and the modern way of life (see for instance Bradford 2001). Others claim that society must develop defensive technologies in an arms race to stay ahead of destructive technologies. For example, Ray Kurzweil (2003) envisions blue goo, police nanobots that combat the bad nanobots, as the solution to potential unrestrained nanotechnology self-replication.

For the most part, society has come a long way from the 1952 killer smog in London, which caused an estimated 4,000 deaths in a three-day period. As Lomborg (1998) asserts, London has not been as clean as it is now since 1585. Systems thinking is also catching on in the form of industrial ecology, material flows assessments, and product life-cycle analyses. Yet all is not well. Developing nations are at least temporarily experiencing high levels of pollution as they begin to industrialize. Poor peoples, even in developed nations, continue to suffer disproportionate hazards from pollution. Radioactive waste and CO_2 emissions remain long-term issues with potentially disastrous outcomes. In both of these cases, it has become apparent that the political challenges of altering behavior, making trade-offs between competing goods, and finding common ground in contexts marked by a plurality of values is even more daunting than the technical challenges presented by pollution. Work is needed in crafting flexible, democratic mechanisms for deciding optimal levels of pollution.

A. PABLO IANNONE

SEE ALSO *Automobiles; Aviation Regulatory Agencies; Conservation and Preservation; Dams; Ecology; Environmental Ethics; Environmental Regulatory Agencies; Global*

Climate Change; Three Gorges Dam; United Nations Environmental Program; Waste; Water.

BIBLIOGRAPHY

Anderson, Terry, and Donald Leal. (1991). *Free Market Environmentalism.* San Francisco: Westview Press. The first comprehensive treatment of the idea that free markets can achieve environmental goals.

Baden, John, and Richard Stroup. (1981). *Bureaucracy vs. Environment: The Environmental Costs of Bureaucratic Governance.* Ann Arbor: University of Michigan Press. Outlines the failures of bureaucratic solutions to environmental problems and offers suggestions for reforms based on aligning incentives with valued outcomes.

Baxter, William F. (1974). *People or Penguins: The Case for Optimal Pollution.* New York: Columbia University Press. Argues that a pollution-free society cannot exist without harming people.

Bradford, George. (2001). "We All Live in Bhopal." In *Environmental Ethics: Readings in Theory and Application,* 3rd edition, ed. Louis P. Pojman. Belmont, CA: Wadsworth.

Joy, Bill. (2000). "Why the Future Doesn't Need Us." *Wired* 8(4): 238–262. Also available from http://www.wired.com/wired/archive/8.04/joy.html. A pessimistic outlook on the impending loss of human control as genetics, nanotechnology, and robotics develop the ability to self-replicate.

Kurzweil, Ray. (2003). "Promise and Peril." In *Living with the Genie: Essays on Technology and the Quest for Human Mastery,* ed. Alan Lightman, Daniel Sarewitz, and Christina Desser. London: Island Press. Examines mid- and long-term hopes and dangers presented by technologies in the field of genetics, nanotechnology, and robotics.

Leopold, Aldo. (1949). *A Sand County Almanac: And Sketches Here and There.* London: Oxford University Press. Articulates the Land Ethic, which has inspired much thinking in nonanthropocentric or biocentric and systems-thinking in environmental ethics.

Lomborg, Bjørn. (1998). *The Skeptical Environmentalist: Measuring the Real Sate of the World.* Cambridge, UK: Cambridge University Press. A well-documented account of the positive trends in human welfare, including a section devoted to the decline in pollution problems.

Sagoff, Mark. (1988). *The Economy of the Earth.* Cambridge, UK: Cambridge University Press.

Simon, Julian. (1981). *The Ultimate Resource.* Princeton, NJ: Princeton University Press. Attacks the neo-Malthusians by arguing that human ingenuity effectively makes resources unlimited in the long run.

POLYGRAPH

• • •

The polygraph or so-called *lie detector* measures physiological responses to stress experienced by a subject during the course of an interrogation. The instrument monitors three physiological states: (a) cardio-vascular responses manifested by changes in blood pressure and pulse rate; (b) galvanic skin resistance that lowers as perspiration increases; and (c) breathing patterns that respond to changes in tension. Changes in any of these patterns can be detected as the subject experiences emotional reactions. The theory behind the polygraph assumes that people encounter measurable physiological changes in the act of deception. The heartbeat increases, blood pressure goes up, breathing rhythms change, and perspiration increases. All of these reactions are recorded on a moving chart for analysis by a trained polygraph technician.

The physiological connection with deception was assumed in the eighteenth century. English novelist Daniel Defoe suggested that "Guilt always carries fear around with it, there is a tremor in the blood of a thief, that, if attended to, would effectually discover him" (Gale 1988, p. 158). In 1915 Harvard psychologist William Marston devised an instrument to monitor the blood pressure of a subject under interrogation. In 1921 medical student John Larson came up with the first true polygraph, adding a measure of respiration along with blood pressure. In the 1930s, Leonarde Keeler integrated Larson's instrument with measurement of electrical skin conductivity into a single machine (Block 1977). Keeler's instrument remains in controversial use in the early twenty-first century in forensic and employment practice.

Supporters of the polygraph claim that it "is one of the most accurate means available to determine truth and deception" (American Polygraph Association 2002, Internet site). But polygraph credibility has yet to become accepted by the scientific community. A major study by the National Academy of Sciences (NAS) in 2002 found that while polygraph data is reliable, it lacks validity. *Reliability* is a measure of consistency, suggesting that the results are the same across different times, places, subjects, and conditions. *Validity* is a measure of appropriateness, suggesting that the test actually measures what it purports to measure. The NAS study found that if there were ten spies among 10,000 government employees, the lie detector would catch eight of them, but 1,598 loyal staff workers would also be falsely accused of deception. If the polygraph tests were adjusted to a much lower sensitivity, only forty-one people would be wrongly accused, but eight of the ten spies would escape detection (Moore et. al 2002). In other words, the polygraph is highly prone to type ii errors or false positives.

Because of such problems, use of the polygraph is practiced only at the fringes of legal and forensic practice, but it is in active use. The polygraph is utilized more for its utilitarian value to extract information than for its ability to measure truth or lies (Lykken 1984). Armed with a deceptively *scientific* instrument, an investigator may be perceived as able to *read the mind* of a subject. The ethical use of lie detection has been rationalized for its ability to extract information, even though the instrument cannot accurately discriminate between truth and lies. In this sense, Immanuel Kant's categorical imperative yields to John Stuart Mill's utilitarian ethic. The end of truth justifies for the modern detective the means of lying. Technical deception is practiced as a means of extracting reluctant truths.

MARTIN RYDER

SEE ALSO *Biometrics; Crime; Justice; Police.*

BIBLIOGRAPHY

Block, Eugene G. (1977). *Lie Detectors: Their History and Use.* New York: David McKay.

Gale, Anthony, ed. (1988). *The Polygraph Test—Lies, Truth and Science.* London: Sage.

Lykken, David Thoreson. (1984). "Polygraphic Interrogation." *Nature* 307: 681–684. Professor of psychology at the University of Minnesota and past president of the Society for Psychophysiological Research, David Lykken has been a leading critic of polygraph industry-sponsored research studies that claim high levels of accuracy.

Marston, William. (1924). "A Theory of Emotions and Affection Based upon Systolic Blood Pressure Studies." *American Journal of Psychology* 35: 469–506.

Moore, Mark H.; Carol V. Petrie; and Anthony A. Braga, eds. (2002). *The Polygraph and Lie Detection..*Washington, DC: National Academy Press. This publication from the National Academy of Sciences is a report of the Committee to Review the Scientific Evidence on the Polygraph, chaired by Stephen E. Fienberg, professor of statistics at Carnegie Mellon University. The committee's charge was to review research on polygraph examinations with specific focus on the reliability and validity of polygraph test results.

INTERNET RESOURCE

American Polygraph Association. "What is a Polygraph?" Available from http://www.polygraphplace.com/docs/information.shtml#polygraph. The American Polygraph Association represents the position and the interests of the polygraph industry.

POPPER, KARL

• • •

Karl Raimund Popper (1902–1994) was a philosopher of science and politics best known for advancing falsifiability as the criterion for distinguishing science from non-science and for a defense of what he termed the *open society*. Born in Vienna on July 28, Popper received his Ph.D. in philosophy from the University of Vienna in 1928. After teaching secondary school from 1930 to 1936, he fled the rise of Nazism and the impending *Anschluss* by emigrating to New Zealand, where he lectured in philosophy at Canterbury University College. In 1946 he moved to England, and three years later became professor at the London School of Economics, which he developed into a leading center for philosophy of science. He was knighted by Queen Elizabeth II in 1965 and elected fellow of the Royal Society in 1976. Popper remained active as a writer and lecturer until his death in Croydon, Surrey, on September 17.

Philosophy of Science

Popper's philosophy of science emerged in the context of Vienna Circle logical positivism, which held that scientific *and therefore all meaningful statements* are of two kinds, with their truth or falsity accordingly verifiable in one of two ways. Analytic statements (for example, Triangles are three-sided plane figures) are true or false simply on the basis of their conceptual and logical structure; synthetic (empirical) statements (such as The tree is green) are *verifiable* insofar as they can be *tested* by positive sense experience. Any statement that did not fit into one of these categories could not be counted as part of science and was considered cognitively meaningless.

Like the logical positivists, Popper was interested in distinguishing science from nonscience, but rejected its verification theory of meaning. Like others, he wanted to assess the theories of physics, Marxism, and psychoanalysis scientifically, but recognized that for abstract or general synthetic statements in physics (for example, The electron has a negative charge or F = ma) as much as in Marixism or psychoanalysis, it was often difficult to specify their direct derivation from sense experience. But upon hearing a lecture on the theory of relativity by Albert Einstein, Popper, then 17 years old, recognized a unique epistemic feature of Einstein's work, namely, that his theory clearly made some unexpected predictions that, if not observed, would falsify it. This contrasted with the theories of Karl Marx and Sigmund

Freud, which, despite many positive confirmations, were not subject to any straightforward falsification.

Thus in his first book, *Logik der Forschung* (The logic of scientific discovery) (1934), Popper argued that no number of positive confirmations at the level of empirical observation could establish a theory as true or probably true, although a single genuine counterinstance could refute a theory. This asymmetry between verification and falsification—one that could never be definitive, the other that could—provided the basis for a clear demarcation between science and nonscience, and became central to Popper's philosophical analysis of scientific rationality. While recognizing the meaningfulness of nonscientific statements in ethics and metaphysics in ways that logical empiricists refused to do, Popper nevertheless emphatically rejected Marxist and psychoanalytic theories as pseudoscience because he found them nonfalsifiable.

Without verification through confirmation, however, it was difficult to explain how scientific knowledge can accumulate or grow. But, for Popper, a "theory is comprehensible and reasonable only in its relation to a given problem-situation" (Popper 1963, p. 139). His proposed metric of scientific progress was that "the best tentative theories (and all theories are tentative) are those which give rise to the deepest and most unexpected problems" (Popper 1972, p. 286). Thus a rationally acceptable theory is one that can withstand criticisms as a proposed answer to questions posed by a problem-situation shared by members of a scientific discipline.

In short, Popper's response to issues regarding the growth of knowledge was this: Because a theory may be false, the appropriate rational response is to look for its weaknesses in order to get rid of them. Science progresses by the conjecture of bold (more general and falsifiable) theories proposed as solutions to the problems identified in prior theories. This analysis was a major influence on subsequent work in the philosophy of science, especially the turn toward philosophical analyses of the history of science by Thomas Kuhn and others.

Closed and Open Societies

Popper's problem-solving model led him to develop an evolutionary epistemology that accepted true theories and useful technologies as dual aims of science, while denying that either truth or utility can ever be determined definitively. In his effort to lay out the framework in which this evolutionary problem solving takes place,

Karl Popper, 1902–1994. The Austrian philosopher offered an original analysis of scientific research that he also applied to research in history and philosophy. *(Hulton Archive/Getty Images.)*

Popper developed a three-world ontology. World 1 is constituted by physical objects, world 2 by subjective experience, and world 3 by objective experience that presents in science, art, ethics, and politics. Popper argued that the world of science, which bears on world 1, evolves in ways analogous to organic evolution.

Popper further contrasted the growth of theory, which tended toward unifying explanations, and technology, which advanced through increased differentiation and specialization. This distinction enabled Popper to extend his critical thinking on theory and praxis to technics, and to balance the judgment that "the critique of technology ... is urgently necessary," often from the outside, with the insight that it would be dogmatic and irresponsible "to attack science and technology as a whole, when they alone permit the necessary corrections to be made" (Popper 1999, p. 101).

This ability to criticize science and its applications is, for Popper, the central feature of an open society where knowledge is freely available to all. Liberal democracy protects the identity and agency of individuals and allows for the peaceful removal of leaders. It is founded on critical rationalism, in that individuals are free to critique systems of thought and work incrementally through democratic processes toward better conditions.

"This is why rationalism is closely linked to the political demand for practical social engineering—piecemeal engineering, of course—in the humanitarian sense, to the demand for the rationalization of society, to planning for freedom, and to the control of freedom by reason. Such societal goals are not governed by *science*, or by a Platonic, pseudorational authority, but by Socratic reason that is aware of its limitations, and that therefore respects others and does not aspire to coerce anyone—not even into happiness." (Popper 1962, vol. 2, p. 238).

Popper believed that society is no more or less than the aggregate of individuals, and that history is indeterminate because it is driven by the consequences of individual choices rather than intrinsic laws. Thus the link between Popper's philosophy of science and social philosophy is fallibalism. Just as scientific progress is made by subjecting theories to critical scrutiny, so too the open society can be sustained only if individuals are free to critically evaluate government decisions and technological change and to modify each in light of such evaluation. Just as in scientific communities, differences in the open society should be resolved by critical discussion rather than force.

By championing the open society, Popper was primarily refuting the dangerous presuppositions at the heart of closed (totalitarian or authoritarian) societies rather than defending a libertarian ideology. As he argued in both *The Open Society and its Enemies* (1945) and *The Poverty of Historicism* (1961), the closed society is predicated on the related postulates of holism and historicism. Holism is the belief that societies are greater than the sum of their members and that society inexorably influences individuals to shape the course of history. Historicism, in Popper's usage, is the belief that history develops according to certain intrinsic principles toward a determinate end. The most significant implication of historicism is that a scientific method can be used to study history and formulate theories to predict social and political developments.

Popper believed historicism to be theoretically erroneous and socially dangerous. History, he contended, is unavoidably indeterminate and not amenable to predictive theories that can lead to falsifiable claims. Yet the view of history as the unfolding of an internal and knowable logic inevitably leads to totalitarian, centralized regimes. These governments feel justified in carrying out massive social engineering programs in order to fulfill a logic of history. Popper's position is that science must be demarcated from nonscience not only to guarantee the growth of knowledge, but also to guard against a tyrannical regime and the authority it could derive from an erroneous interpretation of history as scientific. For Popper, "The fact that we predict eclipses does not, therefore, provide a valid reason for expecting that we can predict revolutions" (Popper 1963, p. 340). Popper's political philosophy shows that the theoretical task of demarcating and limiting the sphere of science and its influence on human affairs is just as ethically important as the physical and political restraint of dangerous technologies.

Popper also derides the absurdity of a "scientific ethics" that would construct "a code of norms upon a scientific basis, so that we need only look up the index of the code if we are faced with a difficult moral decision" (Popper 1962, p. 237). Setting up scientific criteria of ethics relieves human beings of responsibility and therefore all ethical concerns. Thus scientific ethics (which includes ethical naturalism and its attempt to define human nature or the good) is actually an escape from the urgent problems of the moral life. The escape from personal responsibility is compounded and made more dangerous by the tendancy of tyrants to utilize some concept of scientific ethics (i.e., a knowable, natural law) to develop sociological laws and enforce programs of social engineering based on them. For Popper, then, it is crucial for the open society that moral laws remain distinct from natural laws. Only in this way will human choice, freedom, and rationality be entitled to enter the political realm.

Assessment and Extension

Popper's work has been a major stimulus for ongoing discussions regarding the philosophy of science and political philosophy. Popper's students Imre Lakatos (1922–1974) and Paul Feyerabend (1924–1994) became leading philosophers of science. The former defended Popper's critical and cumulative rationalism against the challenges of Kuhn's historically discontinuous paradigms by interpreting paradigms as research programs. The latter repudiated Popper's critical rationalism in the name of an epistemological anarchism that, he argued, was an extension of Popper's own creative openness. In political philosophy, Popper's historical interpretations of Plato, Hegel, and Marx have been hotly contested, but his overall influence has been salutory in its promotion of democracy and the critical assessment of technology.

One interpreter, Paul Levinson, has sought to bridge Popper's philosophy of science and political philosophy by means of the philosophy of technology. For Levinson, Popper's world 3 is too limited. In Levinson's technomaterialist reformulation of Popper's three-world ontology, the human mind (world 2) acting in and on the material world (world 1) forges technology (world 3). Technology thus "enjoys a unique ontological status commensurate with its unique role in the universe: with the execption of humans themselves, nothing is as special ... or as different from all other things" (Levinson 1988, p. 80). The practical criticism and revision of technology is for Levinson a material parallel to critical rationalism in science.

DOMINIC BALESTRA

SEE ALSO Incrementalism; Liberalism; Social Engineering.

BIBLIOGRAPHY

Levinson, Paul. (1988). Mind at Large: Knowing in the Technological Age. Greenwich, CT: JAI Press.

Popper, Karl. (1934). Logik der Forschung [The logic of scientific discovery]. Vienna: Julius Springer Verlag. English translation, The Logic of Scientific Discovery. London: Hutchinson, 1959.

Popper, Karl. (1961). The Poverty of Historicism, 2nd edition. London: Routledge. Revised book version of original articles from 1944.

Popper, Karl. (1962). The Open Society and Its Enemies, 2 vols. Originally published London, 1945.

Popper, Karl. (1963). Conjectures and Refutations: The Growth of Scientific Knowledge. London: Routledge. The basic statement of Popper's principle of falsifiability.

Popper, Karl. (1972). Objective Knowledge: An Evolutionary Approach. Oxford: Clarendon Press.

Popper, Karl. (1999). All Life is Problem Solving. London: Routledge. An autobiography.

POPULAR CULTURE

• • •

The term popular culture, often shortened to pop culture, crystallized around the middle of the twentieth century in recognition of the definitive emergence in European and especially North American society of mass-produced and -consumed cultural goods (including novels, recorded music, radio programs, motion pictures, and advertisements). Popular culture products are usually created by people who do not classify themselves as artists, and they are accepted by people who do not think of themselves as exercising aesthetic judgments. Other, more pejorative terms that have been used to refer to this phenomenon are mass culture (José Ortega y Gasset and others) and the culture industry (Theodor W. Adorno). The term was fashioned after the pop art ("popular art") movement that emerged in the late 1950s—a movement that saw artists appropriate images and commodities from consumerist culture as their subject matter. One of the most famous pop artists was the American Andy Warhol (1928?–1987), who created paintings and silk-screen prints of commonplace objects, such as soup cans, and pictures of celebrities, such as the actress Marilyn Monroe. Pop culture involves the representation of any aspect of consumerist society, not just visual, emphasizing the powerful impact of consumerism and materialism on contemporary life. Pop culture rejects both the supremacy of the "high art" of the past and the pretensions of avant-garde intellectualist trends of the present. It is highly appealing for this very reason. It bestows on common people the assurance that artistic texts are for mass consumption, not just for an elite class of cognoscenti. It is thus populist, popular, and public.

"High," "Low," and "Pop" Culture

Culture is a system of shared meanings. The Estonian semiotician Yuri M. Lotman (1922–1993) used the term semiosphere to encapsulate that very fact and to emphasize that the ways in which people come to understand the world is through the semiotic filters of the language, music, myths, rituals, and other codes that they acquire in cultural context (Lotman 1990).

The adjectives high, low, and popular have been used with culture to differentiate between levels of representation within the semiosphere. "High" culture implies a level considered to have a superior value, socially and aesthetically, than other levels, which are said to have a "lower" value. Traditionally, the high and low levels were associated with class distinctions—high culture was associated with the church and the aristocracy in Western Europe; low culture with "common folk." "Pop culture" emerged in the twentieth century, obliterating this distinction. As John Storey (2003) argues, the idea of pop culture replaced that of "folk" culture, becoming a target of autonomous academic study in the late 1950s when the French semiotician Roland Barthes (1915–1980) showed the importance of studying such things as wrestling and blockbuster movies in terms of how they generate cultural meanings. By the early twenty-first century, the study of pop culture had become a flourishing interdisciplinary area of investigation that had several important journals, including the *Journal of Popular Culture* (founded in 1967).

As Jean Baudrillard (1998) has emphasized, pop culture engages the masses, rather than the cognoscenti, because it takes the material of everyday life and gives it expression and meaning. Everything from comic books to fashion shows have mass appeal because they emanate from within the culture, not from sponsors or authority figures. As such, the makers of pop culture make little or no distinction between art and recreation, distraction and engagement.

The spread of pop culture as a kind of mainstream culture has been brought about by developments in cheap technology. The rise of music as a mass art, for instance, was made possible by the advent of recording and radio broadcasting technologies at the start of the twentieth century. Records and radio made music available to large audiences, converting it from an art for the elite to a commodity for one and all. The late-twentieth-century advent of satellite technology is responsible for the spread and appeal of pop culture throughout the globe. Satellite television, for example, is often cited as bringing about the disintegration of the former soviet system in Europe, as people became attracted to images of consumerist delights by simply tuning into American television programs. The Canadian communications theorist Marshall McLuhan (1911–1980) went so far as to claim that the diffusion of pop culture images through electronic media has brought about a type of "global culture" that strangely unites people in a kind of "global village" (McLuhan 1964). Clearly, the pop culture distraction factory has had an impact on the world far greater than that of the material it communicates.

Pop Culture as a Mythological System

Barthes (1957) claimed that a large part of the emotional allure of pop culture is due to the fact that it is based on the recycling of deeply entrenched mythical meanings. To distinguish between the original myths and their pop culture versions, Barthes designated the latter *mythologies*. In early Hollywood westerns, for instance, the mythic struggle of good versus evil manifested itself in various symbolic and representational forms—heroes wore white hats and villains black ones; heroes were honest and truthful, villains dishonest and cowardly; and so on. The Superman character of comic

book and cinematic fame, to cite another example, is a perfect example of a recycled hero, possessing all the characteristics of his mythic predecessors but in modern guise—he comes from another world (the planet Krypton) in order to help humanity overcome its weaknesses; he has superhuman powers; but he has a tragic flaw (exposure to the fictitious substance known as kryptonite takes away his power). Barthes claimed that pop culture is an overarching "mythological system." And because of this it imbues its own representations and spectacles with an unconsciously felt cogency.

As a consequence, Barthes argued, pop culture has had a profound impact on modern-day ethics. In the historical development of ethics, three principal standards of conduct have been proposed as the highest good: happiness or pleasure; duty, virtue, or obligation; and perfection, the fullest harmonious development of human potential. In traditional cultures, these standards were established through religious and philosophical traditions. In pop culture, they are shaped by spectacles, performances, and especially media representations. Ethical issues that are showcased on television, for example, are felt as being more significant and historically meaningful to society than those that are not. Television imbues them with significance and salience.

The power of the media to affect the interpretation of ethical behavior has inevitably led people to stage events for the cameras. The social critic Walter Truett Anderson (1990) calls these appropriately "pseudoevents," because they are never spontaneous, but planned for the sole purpose of playing to pop culture's huge audiences. Most pseudoevents are intended to be self-fulfilling prophecies. The media are thus the vehicles through which people come to grips with issues of lifestyle, ethics, and morality. The understanding of them, however, is fragmentary and ephemeral because the images of media are constantly in flux. The only constant in pop culture is, in fact, constant change. With few exceptions, most pop culture products and styles come and go quickly. Thus, while it has great appeal, pop culture has also had a powerful negative impact on traditional approaches to ethics.

Summary

Pop culture has become virtually mainstream culture, having obliterated the distinction between high, low, and folk culture. It has become a powerful force in modern-day society because it has great emotional appeal and because of its built-in tendency for constant change. The comic-book art of Charles Schulz (1922–2000) is a case in point. His comic strip *Peanuts*, which was originally titled *Li'l Folks*, debuted in 1950, appealing to mass audiences. Through the strip Schultz dealt with some of the most profound religious and philosophical themes of human history in a way that was unique and aesthetically powerful.

The movie *Amadeus* is another case-in-point. This 1984 work directed by Milos Forman (b. 1932) became a pop culture phenomenon in the decade of the 1980s. It is based on the 1979 play by British playwright Peter Shaffer (b. 1926) about the eighteenth-century rivalry between Austrian composer Wolfgang Amadeus Mozart and Italian composer Antonio Salieri. The play plumbs the meaning of art, genius, and the important role of music in the spiritual life of human beings. The film captures these themes visually and acoustically by juxtaposing the emotionally powerful music of Mozart against the backdrop of dramatized events in his life and the truly splendid commentaries of Salieri, who guides the audience through the musical repertoire with remarkable insight and perspicacity. Forman's camera shots, close-ups, angle shots, tracking shots (which capture horizontal movement), and zooming actions allows the viewer to literally see Mozart's moods (his passions, his tragedies, and so forth) on his face as he conducts or plays his music, as well as those of his commentator Salieri (his envy, his deep understanding of Mozart's art) as he speaks to his confessor. In effect, Mozart became a pop culture hero, so to speak, through the power of cinema.

MARCEL DANESI

SEE ALSO *Consumerism; Critical Social Theory; Entertainment; Information Ethics; Movies; Music; Robot Toys; Technocomics; Television.*

BIBLIOGRAPHY

Anderson, Walter Truett. (1990). *Reality Isn't What It Used to Be: Theatrical Politics, Ready-to-Wear Religion, Global Myths, Primitive Chic, and Other Wonders of the Postmodern World.* San Francisco: Harper and Row.

Barthes, Roland. (1957). *Mythologies.* Paris: Seuil. A critical analysis of the mythological structure of pop culture performances, from wrestling matches to blockbuster movies.

Baudrillard, Jean. (1998). *The Consumer Society.* Thousand Oaks, CA: Sage. An acerbic critique of the image-making techniques of consumerist culture and their effect on human cultural development.

Lotman, Yuri M. (1990). *Universe of the Mind: A Semiotic Theory of Culture.* Bloomington: Indiana University Press. A groundbreaking study of the semiosphere, showing that human psychic life is governed by sign-making tendencies that are tied to social context in the same way that human biological life is governed by organic tendencies that are tied to physical context.

Ortega y Gasset, José. (1932). *The Revolt of the Masses.* New York: Norton. Originally published in Spanish, 1929.

McLuhan, Marshall. (1964). *Understanding Media.* New York: McGraw-Hill. The classic study of the effects of technology and media on cultural evolution. The basic idea presented here is that media are extensions of sensory processes and thus felt to be emotionally powerful.

Storey, John. (2003). *Inventing Popular Culture: From Folklore to Globalization.* Malden, MA: Blackwell. In-depth analysis of the ideological structure of pop culture and its manifestations in the political and social spheres.

POPULATION

• • •

Population often is said to be the biggest problem facing the world. However, precisely what this problem is varies, depending on whether the issue is meeting basic human needs, the stress placed on the natural environment by increased consumption, changes in family structures, or demographic transitions within nations. Population is at once a conceptual, scientific, technological, ethical, and political issue.

Definitions

The simple definition of population as the total number of persons in a geographic area indicates the relativity of population to sometimes arbitrary boundaries. Other relevant factors in population studies are fertility, mortality, and mobility; empirical studies of those factors are often difficult to pursue and are subject to contentious interpretative frameworks. Scientific theories of population growth and its relationship to social stability or economic development often rely on intuitive or "commonsense" views that have not proved reliable. The influence of technologies on population growth or delimitation similarly is lacking in specificity.

Indicative of the complexity of this issue, the entry "Population Ethics" in the third edition of the *Encyclopedia of Bioethics* is the largest single composite, with more companion pieces than any other entry. Under the general title there are three entries on the elements of population ethics, an analysis of normative approaches, and eight entries describing the perspectives of different religious traditions. In this entry a brief review of how population became an issue is followed by an overview analysis of major ethical assessments that emphasize science and technology.

FIGURE 1

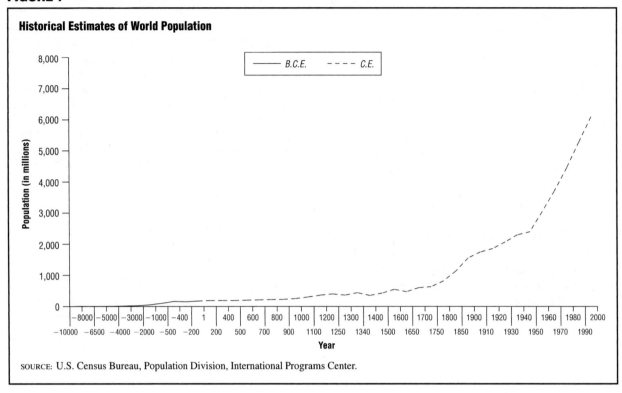

Historical Estimates of World Population

SOURCE: U.S. Census Bureau, Population Division, International Programs Center.

Population Issues

What is experienced directly is not population but people. Before the modern period there were only informal political-philosophical discussions of how different numbers of people in a state can affect its character, and Christian traditions sometimes highlight the biblical injunction to "be fruitful and multiply" (Genesis 1:27). For population to become a subject of debate and inquiry the modern techniques of political economics had to be brought to bear on issues related to both aggregate production and consumption, a process that began in seventeenth-century England and reached its first peak in the work of the economist Thomas Robert Malthus (1766–1834) (Glass 1973).

Before Malthus most early modern population theorists argued for the simple stimulation of population growth (the political philosophers Baron de Montesquieu [1689–1755] and Jean-Jacques Rousseau [1712–1778]) or predicted that in the near future, because of good health and long life, human commitments to procreation would be moderated in favor of more liberal pursuits (the political philosophers William Godwin [1756–1836] and the Marquis de Condorcet [1743–1794]). Malthus attacked both views in his *Essay on the*

Principle of Population, which appeared anonymously in 1798 and was revised with acknowledged authorship in 1803 and given the subtitle "Or a View of Its Past and Present Effects on Human Happiness; with an Inquiry into Our Prospects Respecting the Removal or Mitigation of the Evils Which It Occasions." Malthus continued to revise this work, with five more editions appearing during his lifetime.

It was Malthus who formulated what may be considered the classical form of the population problem. Malthus's argument was that population, by increasing when unchecked at a geometric rate (2, 4, 8, 16, etc.), outruns food supply, which grows only at an arithmetic rate (2, 4, 6, 8, etc.). What is known as a Malthusian catastrophe occurs when this happens and starvation forces some of the population back to a subsistence level. For Malthus this catastrophe can be prevented only through self-restraint or technology, meaning contraception or abortion. In later editions of his *Essay* Malthus further noted that increased wealth was correlated with reductions not only in mortality but also in fertility; this suggested that technological development could meliorate the problem more indirectly. However, Malthus did not foresee the ways in which advances in science and technology might alter growth in the food supply.

The central problem for the classical Malthusian view may be described as the extent to which human population increase becomes unchecked through scientific progress. For thousands of years the human population remained relatively stable, checked largely by the vicissitudes of nature. Over the course of the agricultural revolution (roughly 10,000 to 5000 B.C.E.) the human population rose to about 150 million worldwide. Only very slowly, over the next 1,500 years, did it increase to over 300 million. However, by 1700 world population had risen to 600 million, by 1800 to 900 million, and by 1900 to 1.6 billion (see Figure 1). These dramatic increases resulted from decreases in infant and adult mortality brought about by advances in public health technology and medicine as well as in scientific agriculture. In 2000 world population reached more than 6 billion. Food production was able to keep up with population growth as a result of radical developments in agriculture, from the industrialization of agriculture to the Green Revolution.

A second form of the population problem arose in the 1960s in association with the environmental movement. The first population problem was based on doubts that people could extract enough from the earth to support themselves. The second population problem arose from the concern that they would be so successful that they would alter the character of the natural world. The first problem focused primarily on whether humans would be able to sustain themselves, and the second on whether the earth was sustainable in the face of human abilities, through science and technology, to transform the world. The possibility that destruction of the earth might rebound on humans was, of course, a supporting worry.

Central to articulating the second form of the population problem, and thus playing a role similar to that of Malthus in regard to the first form, was the Club of Rome's study *The Limits to Growth* (Meadows, Meadows, Randers, and Behrens 1972). According to the limits to growth argument, which has been argued in equally dramatic fashion in Paul Ehrlich's *The Population Bomb* (1968) and Garrett Hardin's *Living within Limits* (1993), high-affluence industrial societies cannot indefinitely expand the exploitation of inherently limited natural sources such as oil and fresh water or pour wastes into inherently limited ecological sinks such as the oceans and the atmosphere. At some point the resources will run out, the ecological sinks will be full or destroyed, and the societies based on their consumption and pollution will collapse.

In response to this limitationist argument, Julian Simon and other expansionists have argued that science and technology are capable of expanding the resource base indefinitely and transforming pollution into raw materials that can be used for further productive activity. Simon's argument in *The Ultimate Resource* (1980) is that population itself and the human ingenuity manifested in the individuals who make up a free society are a more important resource than is any combination of minerals or vegetables on the planet. Under conditions of economic and political liberty human beings, through science and technology, can create the resources necessary for their indefinite expansion.

Population Ethics

According to Donald Warwick, "Those stating that there is a population problem base their assertions on three elements: perceived threats to social, moral, or political values; factual evidence; and theories explaining how population creates the conditions that threaten values" (Warwick 2004, p. 2035). For Warwick the primary need in population ethics is to distinguish these values, evidence, and theories and carefully adjudicate their interactions. Population ethics depends on an ethics of analyzing population issues that would eschew quick ideological appeals and emotional rhetoric. Those who argue for particular interpretations of population as a problem should state their values, sources of evidence, and theories explicitly. Conclusions and policy recommendations should follow from the careful analytic interrelation of those different elements.

With regard to overarching values Warwick further proposes respect for four fundamental rights: the rights to life, freedom, welfare, and fairness. As for evidence, many people would argue that scientific knowledge should trump other ideas about what should count as data. Theories about the relationships between values and evidence remain fundamentally problematic in relation to population, as they are in many other areas. What is important is to acknowledge the problematics even when conclusions and policy recommendations cannot be avoided because failure to reach a conclusion or make a recommendation will function as a conclusion or recommendation.

Against this background it is nevertheless useful to highlight at least three basic ethical arguments regarding population: what may, for want of better terms, be called limitationist ethics, libertarian ethics, and management ethics. The first two grow out of the limitationist and expansionist interpretations of the problem. The third is more the consensus view of the international development community.

LIMITATIONIST ETHICS. Garrett Hardin (1974) devised the term *lifeboat ethics*, suggesting that because the planet has limited resources, humanity should be thought of as having been cast adrift in space like survivors in a lifeboat. If there are too many passengers, the lifeboat will run out of supplies for those passengers. Using this logic, Hardin states that providing food aid to countries in crisis does not address the problems that created the need for such aid. Hardin's limitationist thesis is that people in poor countries should be allowed to starve because the net result of helping them would be negative for the planet as a whole. In his opinion, extensive food aid would court disaster. Another version of limitationist ethics would argue for limitations on consumption and the practice of voluntary simplicity in the lifestyles of people in wealthy countries.

LIBERTARIAN ETHICS. Julian Simon (1980) argued that the population problem has been fundamentally misperceived. Population growth is a good thing as long as you allow individuals freedom of choice, and grant them the economic and political freedoms to be creative in their uses of science and technology. "Human beings," he wrote, "are not just more mouths to feed, but are productive and inventive minds that help find creative solutions to man's problems, thus leaving us better off over the long run." For Libertarians like Simon, population is not the cause of our problems but the generator of solutions to all of our problems. The more people addressing problems the quicker they will be solved. Thus, population growth is a resource and not a threat to our future. Arne Naess referred to this view as the Cornucopian position.

MANAGEMENT ETHICS. Between the limitationist and libertarian positions is the Management ethics viewpoint. Proponents of this position, like the World Bank, are not as pessimistic as Hardin and the limitationists, nor as optimistic as Simon and the libertarians. The view that population is more of a two edge sword. Managed properly it can be a resource, a boon to the world. Left uncontrolled, it can have disastrous effects.

In a radically different take on the issue Barbara Duden (1992) questions the concept of population as a variable for economic problem solving. For Duden population is such an abstract concept that it creates situations in which human beings are deprived of their humanity as they are transformed into statistics to be manipulated by others. The problem of population is not its role in issues involving environmental resources (the limitationist perspective) or in fostering major misperceptions of problems (the libertarian perspective) but the tendency to lose sight of people in their existential reality as models are created to manage problems.

FRANZ ALLEN FOLTZ

SEE ALSO *Consumerism; Chinese Perspectives: Population Policy; Social Darwinism.*

BIBLIOGRAPHY

Duden, Barbara. (1992). "Population." In *The Development Dictionary: A Guide to Knowledge as Power*, ed. Wolfgang Sachs. London: Zed Books.

Ehrlich, Paul R. (1968). *The Population Bomb*. New York: Ballantine. Revised edition, 1971. Rewritten, with Anne H. Ehrlich, as *The Population Explosion* (New York: Simon & Schuster, 1990).

Glass, David Victor. (1973). *The Population Controversy: A Collective Reprint of Material Concerning the 18th Century Controversy on the Trend of Population in England and Wales. Compiled With an Introduction By D. V. Glass*. Farnborough, UK: Gregg International.

Hardin, Garrett. (1974). "Living on a Lifeboat." *BioScience* 24(10): 561–68.

Hardin, Garrett. (1993). *Living within Limits: Ecology, Economics, and Population Taboos*. New York: Oxford University Press.

Meadows, Donella H.; Dennis L. Meadows; Jürgen Randers; and Williams H. Behrens. (1972). *The Limits to Growth: A Report for the Club of Rome's Project on the Predicament of Mankind*. New York: Universe Books. This book has been revised or extended as *Beyond the Limits: Confronting Global Collapse, Envisioning a Sustainable Future* (Post Mills, VT: Chelsea Green, 1992) and *The Limits to Growth: The 30-Year Update* (White River Junction, VT: Chelsea Green, 2004).

Simon, Julian L. (1980). *The Ultimate Resource*. Princeton, NJ: Princeton University Press. Revised and enlarged as *The Ultimate Resource 2* (Princeton, NJ: Princeton University Press, 1996).

Simon, Julian, and Herman Kahn. (1984). *The Resourceful Earth*. New York: Basil Blackwell.

POPULATION POLICY IN CHINA

• • •

The People's Republic of China (PRC) has the largest population in the world. At the end of 2002, the population in China (excluding Hong Kong, Macao, and Taiwan) was 1.284 billion, and the birthrate was 12.86 births per year per 1,000 population, which results in a doubling every fifty-five years.

Historical Background

The large Chinese population is a result of historical factors. Before 1900 China had a predominately agricultural economy dependent primarily on manual labor, with a standard of living closely tied to the number of working children in a family. Traditionally, having many children brought higher welfare and happiness. As a result, China had a high birthrate.

In the twentieth century, with the gradual improvement of medicine, people's health improved, and as a result, the death rate decreased continuously, from 20 deaths per year per 1,000 population in 1945 to 9.5 in 1965. Since 1980 the death rate has remained constant at close to 6. Because of the huge population base, the number of people in China rapidly increased from 601.9 million in 1953 to 1.0318 billion in 1982. At the same time, employment shifted from agriculture to industry. If China had not instituted family planning policies, a great portion of resources would have had to go to supporting a now nonproductive segment of the population (children), slowing the pace of social development, which would be unfair to present and future generations.

Because high population growth strains societal resources in education, employment, and medical care, as well as other areas, the Chinese government implemented a policy of family planning that considers the interaction of science, technology, economics, and society. For instance, improvements in technology should increase the quality of life, advances in medicine will allow people to live longer lives, but too rapid a decline in the birthrate would mean that younger generations would eventually have to support too large an elderly population.

Policy Guidelines and Their Development

The PRC has adopted the following family planning policies: It encourages late marriage and late, fewer, but healthier babies. It seeks to avoid genetic and other birth defects, which are a disproportionately large drain on societal resources. It advocates a "one couple, one child" policy. It encourages rural couples who have a need for more children to space them properly. The government also provides strong support for family planning policies to raise the level of health among women and children. In 1981 the government established the State Family Planning Commission—now the State Family Planning and Population Commission—which seeks to provide a service-oriented approach to family planning.

Chinese family-planning policy is tailored to meet the practical living needs of people in different regions of the country. Provinces and autonomous regions decide specific family planning measures and regulations for minorities in accord with local conditions. China is also making strides in getting citizens to understand and accept its family planning policies. To this end some politicians and scholars have made great contributions. For example, in 1957 Ma Yinchu, a renowned economist, became a pioneer advocate of family planning when he presented to the National People's Congress his new population theory, in which he recommended controlling population size so as not to impede economic development. Yet Ma was ahead of his time, for he was soon criticized as a representative figure of erroneous idea. He was not able to publish his *New Population Theory* until 1979. In the early 1970s Premier Zhou Enlai also overcame diverse difficulties to promote stable family planning.

Since 1980 many academic societies for research on population and family planning policy have been established. In 1980 the Chinese Academy of Social Sciences created the Institute of Population Research. In 1981 the China Population Society was founded. Institutes for research on the population were in turn set up at Beijing University, Renmin University of China, and Xiamen University. These efforts of the government and research institutes have led to many publications. The government started publishing the *China Population Statistics Yearbook* in 1985 and the *China Population Paper* in 1988. In the late 1990s several important academic publications appeared, including the *Encyclopedia of Chinese Family Planning* (Peng Peiyun 1997). Subsequently, scholars made efforts to relate China's population policy to issues of sustainable development (Qin, Zhang, and Niu 2002), and a number of authors reflected on the importance of limiting the population not just for social development but also for preserving the quality of the environment (Li Shuhua 2003, Peng Keshan 1994, Zhou Yi 2003).

As a result of this research, the significance of family planning policy in the development of science, technology, economics, and society was now generally well recognized and accepted by the early 2000s. The implementation of a family planning policy has effectively controlled the rapid expansion of the population in the PRC, improved the quality of life and health, and made possible the greater development of science, technology, and society.

The Ethics of Population Control

Chinese population policy has been very controversial outside of China. The most common criticism is that the policy deprives people of their right to bear children and to decide for themselves how many children they will have. Another criticism is that because of a traditional desire for male children, the one-child policy encourages parents to abort or abandon female offspring. Within the historical and social context of China, however, the implementation of the "one couple, one child" policy during the 1980s represented a major shift from the much more coercive practices of the Cultural Revolution (1966–1976). Moreover, under some circumstances, Chinese policymakers argue, concerns for the common good should outweigh individual freedoms. Finally, as Margaret Pabst Battin (2004) has argued, although the Chinese policy may be "the most coercive population-limitation policy in any country, it is also the most fair" (p. 2095). Unlike the population-limitation policies of India, for instance, the Chinese policy applies equally to all groups.

WANG QIAN
QIAN QIAN

SEE ALSO *Chinese Perspectives; Eugenics.*

BIBLIOGRAPHY

Banister, Judith. (1987). *China's Changing Population.* Stanford, CA: Stanford University Press.

Battin, Margaret Pabst. (2004). "Population Policies, Strategies for Fertility Control In." In *Encyclopedia of Bioethics,* 3rd edition, ed. Stephen G. Post. New York: Macmillan Reference.

Conly, Shanti R., and Sharon L. Camp. (1992). *China's Family Planning Program: Challenging the Myths.* Washington, DC: Population Crisis Committee.

Li Shuhua. (2003). "Renkou kongzhi yu huanjing lunli" [Population control and environmental ethics]. *Nankai xuebao (Zhexue shehuikexue ban)* 2: 97–103.

Ma Yinchu. (1979). *Xin renkou lun* [A new population theory]. Beijing: Beijing chubanshe.

Peng Keshan. (1994). "Kongzhi renkou zengzhang, gaishan shengtai huanjing" [Controlling population growth, improving the ecological environment]. *Kexue xue yu kexue jishu guanli* 12: 10–14.

Peng Peiyun. (1997). *Zhongguo jihua shengyu quanshu* [Encyclopedia of Chinese family planning]. Beijing: Zhongguo renkou chubanshe.

Peng Xizhe and Zhigang Guo, eds. (2000). *The Changing Population of China.* Oxford: Blackwell.

People's Republic of China. State Council. (1994). *China's Agenda 21: White Paper on China's Population, Environment, Development in the Twenty-First Century.* Beijing: China Environmental Science Press.

Poston, Dudley L., Jr., and Baochang Gu. (1995). "Socioeconomic Development, Family Planning, and Fertility in China." In *Developing Areas: A Book of Readings and Research,* ed. Vijayan K. Pillai and Lyle W. Shannon. Oxford: Berg.

Qin Dahe, Zhang Kunmin, and Niu Wenyuan. (2002). *Zhongguo renkou ziyuan, huanjing yu ke chixu fazhan* [Chinese population resources, environments, and sustainable development]. Beijing: Xinhua chubanshe.

Scharping, Thomas. (2002). *Birth Control in China, 1949–2000: Population Policy and Demographic Development.* London: Routledge.

Sung Chien. (1985). *Population Control in China: Theory and Applications.* New York: Praeger.

Tang Re-Feng. (2004). "Chinese Population Policy." In *Bioethics: Asian Perspectives,* ed. Ren-Zong Qiu. Dordrecht, Netherlands: Kluwer Academic.

Wang, Gabe T. (1999). *China's Population: Problems, Thoughts, and Policies.* Brookfield, VT: Ashgate.

Zhou Yi. (2003). "Renkou yu huanjing ke chixu fazhan" [The sustainable development of population and environment]. *Guizhou shifan daxue xuebao (Shehui kexue ban)* 2: 23–31.

POSITIVE EUGENICS

SEE *Eugenics.*

POSITIVISM

SEE *Comte, Auguste.*

POSTCOLONIALISM

SEE *Colonialism and Postcolonialism.*

POSTHUMANISM

• • •

The posthumanist (sometimes called transhumanist) views human dignity as a matter of seizing the opportunity to modify and *enhance* human nature in ways that include the deceleration or arresting of aging, genetic engineering, the bodily introduction of nanotechnology and cybernetics, reproductive cloning, and even the *downloading of mind* into immortalizing computers. The anti-posthumanist responds that human dignity lies chiefly in accepting the existing contours of human nature as a gift, and that biotechnological efforts to recreate human nature according to inevitably arrogant and short-sighted images of *perfectability* should be greeted with severe skepticism. The debate between posthumanists and their critics over *the future of human nature* is rhetorically sharp; any resolutions can emerge only from inclusive discourse, with significant consensus on specific technologies of human modification arrived at only in the full light of disparate ethical self-understandings of the meaning of humanness both secular and sacred (Habermas 2003).

Radical vs. Qualified Posthumanism

The posthumanist, it is argued, has the superficial enthusiasm of the adolescent convert to some new image of the human, yet has little or no insight into the human condition or the narrative of history. Rather than free humans of biological constraints in a misplaced effort to transcend humanness by technology, the anti-posthumanist urges, to quote Leon Kass's 1985 publication title, "a more natural science."

But many posthumanists are deeply reflective. The 1974 Nobel Laureate in Medicine, Christian de Duve (2002), thoughtfully urges pursuing the goal of a *superorganism* as humans *reshape life*, and raises the question "After us, what?" De Duve warns against fearing the consequences of genetic engineering, or the seduction of a *return to nature* philosophy. De Duve contends that before even thinking of genetically modifying humans, society should focus on improving the chances of all its members to realize the potential they are born with (through suitable economic, social and family conditions). Fears should be focused on resource exhaustion and catastrophic epidemics. Nevertheless future generations will increasingly interfere with the human genome, he argues, and hopefully the decisions will not be left to a powerful bureaucracy, although a *genetic supermarket* using the individual choices of parents is not likely to exert more favorable effects on the gene pool.

Posthumanism as Technological Millennialism

Posthumanists embrace decelerated and even arrested aging, but only as part of a larger vision to re-engineer human nature, and thereby to create biologically and technologically superior human beings, as the narrative history of posthumanism by N. Katherine Hayles (1999) makes clear. Genetics, nanotechnology, cybernetics, and computer technologies are all part of the posthuman vision, including the downloading of synaptic connections in the brain to form a computerized human mind freed of mortal flesh, and thereby immortalized (Noble 1997). This last scenario of immortalized minds liberated from any biological substrate makes the biogerontological goal of *prolongevity* appear conservative.

Posthumanists do not believe that biology should in any sense be destiny, and seek a new sort of entity for whom human nature has been more or less overcome (Hook 2003). They urge humans to take human nature into their own re-creative hands as the next great step in evolution, achieving a post-modern morphological freedom. Their argument begins with the claim that, within the boundaries of technology, humans have always been reinventing themselves through applied technologies. Where should the lines be drawn? Besides as the Princeton University physicist Freeman Dyson writes, "the artificial improvement of human beings will come, one way or another, whether we like it or not," as scientific understanding increases, for such improvement has always been viewed as a "liberation from past constraints" (Dyson 1997, p. 76).

What is natural and what is unnatural, anyway? Homo sapiens long ago embarked on the human phase of evolution through technological prowess, and in the future lies nothing more monumental than increased novelty. At one time the very idea of human beings trying to fly was deemed heretical hubris in the light of eternity—*sub specie aeternitatis*. It would be a repetition of this error to argue that redesigning human nature runs afoul of the precautionary appeal to the complexities of evolution—*sub specie evolutionis?* Should people not set aside trepidation and with confidence rethink themselves in the light of human creativity? The postmodernists have paved the way by purportedly demonstrating that there is no essential aspect to human nature, and *vive le difference*. So it is that Gregory Stock (2002) introduces the idea of *superbiology* as human beings take full control of their own biology in turning toward perfection.

Technological Millennialism as Secularized Religion

David F. Noble (1997) has argued with some plausibility that the roots of this posthumanist project lie in Western European religion, and especially in the ninth century, when the *useful arts* came to be associated with the concept of human redemption. As a result, there exists a *religion of technology* that promotes the uncritical and irrational affirmation of unregulated technological advance. In essence technological advance is always deemed good. Noble hopes people can free themselves from the religion of technology, from which they seek deliverance, through learning to think and act rationally toward humane goals.

Millennialist religion is certainly relevant to the posthumanist vision. As Gerald J. Gruman has pointed out, the modern concern with enhancing longevity "stems from the decline since the Renaissance of faith in supernatural salvation from death; concern with the worth of individual identity and experience shifted from an otherworldly realm to the *here and now*, with intensification of earthly expectations" (Gruman 1966, p. 88).

With the transition to a this-worldly millennialist human horizon, a powerful current of thought emerged in which the goal of significantly extending the length of human life through biomedical science was affirmed. Gruman termed the concept prolongevity as "a subsidiary variant of meliorism, the belief that human effort should be applied to improving the world" (Gruman 1966, p. 89). Carl L. Becker, in his classic work, *The Heavenly City of the Eighteenth-Century Philosophers* (1932), had similarly interpreted the great ideas of the Enlightenment and the merging goals of science as based on a secularization of the medieval idea of otherworldly salvation, resulting in an advance toward a heaven on earth.

Indeed, Francis Bacon (1561–1626), a founder of the scientific method, in his millennialist and utopian essay "The New Atlantis" (1627), set in motion a biological mandate for boldness that included both the making of new species or *chimeras*, organ replacement, and the *Water of Paradise* that would allow the possibility to "indeed live very long" (Bacon 1996, p. 481). Three centuries before Francis Bacon, the English theologian Roger Bacon (c.1220–1292) argued that in the future the 900-year-long lives of the antediluvian patriarchs would be restored alchemically. Like many Western European religious thinkers, both Bacons saw death as the unnatural result of Adam's fall into sin. These dreams of embodied near-immortality could only emerge against a theological background that more or less endorses them. There are various other cultural and historical influences at work besides religion, but the initial conceptual context for a scientific assault on aging itself is a religious one (Barash 1983).

The modern goals of anti-aging research and technology, then, are historically emergent, at least in part, from a pre-modern religious drama of hope and salvation, Renaissance science transferred the task of achieving immortality from heaven to earth in the spirit of millennial hopes. The economy of salvation presented by the Italian poet Dante Alighieri was replaced by the here and now. There is a vibrant millennialist enthusiasm in the responsible biogerontologists, who have proclaimed aging itself to be surmountable to degrees through human ingenuity.

The Anti-Posthumanist Appeal

For every utopian there is a dystopian. Should individuals, viewing their own prospects for deceleration of aging, pursue such anti-aging treatments when and if they actually become available? Perhaps yes, if this assures one that diseases for which old age is the overwhelmingly significant risk factor can be avoided. But there is an important school of thought that cautions against the development of treatments to slow aging.

Individuals, when confronted with the availability of deceleration, ought to reflect carefully about the choice at hand, raising every question of relevance to themselves and to humanity. One of the wiser minds of the last century, Hans Jonas (1903–1993), an intellectual inspiration for contemporary anti-posthumanists, articulated these questions quite thoroughly. He wrote in 1985 that "a practical hope is held out by certain advances in cell biology to prolong, perhaps indefinitely extend, the span of life by counteracting biochemical processes of aging" (Jonas 1985, p. 18). How desirable would this power to slow or arrest aging be for the individual and for the species? Do people want to tamper with the delicate biological "balance of death and procreation" (Jonas 1985, p. 18), and preempt the place of youth? Would the species gain or lose? Jonas, by merely raising these questions, meant to cast significant doubt on the anti-aging enterprise. "Perhaps," he wrote, "a nonnegotiable limit to our expected time is necessary for each of us as the incentive to number our days and make them count" (p. 19). Jonas's later essays raising many of these same questions were published posthumously in 1996.

Many of the these issues are echoed in the writings of Leon Kass. Kass for the most part accepts biotechnological progress within a therapeutic mode; his issue is chiefly with efforts to enhance and improve upon the givenness of human nature. He draws on the technological dystopians, such as Aldous Huxley, as well as on the writings of C. S. Lewis (1898–1963). An early anti-posthumanist, Lewis wrote *The Abolition of Man* (1944) to defend a natural law tradition: What is, is good, and people should live within their God-given limits. He cautioned against a world in which one class of enhanced human beings would dominate and oppress the other. One might ask, then, if those freed from the decline of aging would become the superior and elite humans, while those who age would be deemed inferior.

In a creative essay, "L'Chaim and Its Limits: Why Not Immortality?" (2001) Kass argues against prolongevity in ways mostly raised by Jonas. He asserts, for example, that the gradual descent into aged frailty weans people from attachment to life and renders death more acceptable. He contends that numbered days encourage a creative depth in human nature—a depth that escaped so many of the immortal Greek gods and goddesses, whose often debauched and purposeless behavior made Plato wish to ban them from the ideal Republic. In addition, says Kass, a preoccupation with the continuance of life is a distraction from that which is best for the human soul. Finally Kass writes that in a world transformed by anti-aging research, youth will be displaced rather than elevated, and the parental investment in the young will give way to *my* perpetuation; and that in such a new world people will grow bored and tired of life, having *been there* and *done that*. These assertions are all thoughtful, creative, and appropriately cautionary, because the implications of slowing or arresting aging itself are obviously monumental and mixed. Responsibility to future generations precludes clinging to youthfulness. There is wisdom in simply accepting the fact that humans evolved for reproductive success rather than for long-lived lives Without such wisdom will people lose sight of their deepest creative motives? Possibly.

Another leading anti-posthumanist, Francis Fukuyama challenges those who would march society into a posthuman future, characterized by cybernetics, nanotechnology, genetic enhancement, reproductive cloning, life span extension, and new forms of behavior control. Undoubtedly the ambitions of posthumanists to create a new posthuman who is no longer human are arrogant, pretentious, and lacking in fundamental appreciation for natural human dignity. Fukuyama is also drawn to the dystopian genre and sees much more bad than good in efforts to significantly modify human nature. He argues powerfully that the anti-aging technologies of the future will disrupt all the delicate demographic balances between the young and the old, and exacerbate the gap between the haves and the have-nots. The concerns raised by political scientists such as Fukuyama are ones that the individual decision maker ought certainly to have in mind.

Conclusion

The anti-posthumanists often appeal to nature and character as morally valuable categories. They understand the proper human attitude toward evolved nature as one of humility, awe, and appreciation. Clearly the emerging technological power to control nature does not always constitute progress. The anti-posthumanist exhorts us to work with human nature to get the best out of it, rather than to seek cavalier domination in an effort to recreate what is already good. Better to accept natural limits, or so, anyway, is the spirit of anti-posthumanism. The perfectibility of humankind lies not in

modifying the human vessel, but in developing the treasures within, such as compassion, virtue, and dignity.

In summary the natural law traditions represented by anti-posthumanists exhort people to live more or less according to nature, and warn that efforts to depart from that will result in new evils more perilous than the old ones. How can society presume that the brave new world will be a better world? Should not the burden of proof be on the proponents of radical change? What right have people in the early 2000s to impose their own arbitrary images of human enhancement on future generations?

Posthumanist beliefs in the inevitability and desirability of transforming human nature see human beings as essentially technological beings who now have the opportunity to redirect the technological powers that they have been exercising on the nonhuman world onto human nature itself. Just as humans have made the world better through technological mastery, so will they be able to do with human nature, in the first instance by prolonging human life as it currently exists but then ultimately by transforming human life. Such a posthumanist future is the natural outcome of all previous human history and the specific form that a respect for human dignity takes in the twenty-first century.

By contrast, anti-posthumanists suggest that the proper human attitude toward evolved nature is one of humility, awe, and appreciation. Just as past technological manipulations of nonhuman nature have not always been beneficial, so the emerging technological power to control human nature does not always constitute progress.

STEPHEN G. POST

SEE ALSO *Aging and Regenerative Medicine; Artificiality; Bioethics; Cybernetics; Cyborgs; Dignity; Freedom; Future Generations; Human Cloning; Human Nature; Nanoethics; Utopia and Dystopia.*

BIBLIOGRAPHY

Bacon, Francis. (1996). *The New Atlantis.* In *Francis Bacon: A Critical Edition of the Major Works,* ed. Brian Vickers. Oxford: Oxford University Press.

Barash, David P. (1983). *Aging: An Exploration.* Seattle: University of Washington Press.

Becker, Carl L. 2003 (1932). *The Heavenly City of the Eighteenth-Century Philosophers.* New Haven, CT: Yale University Press.

de Duve, Christian. (2002). *Life Evolving: Molecules, Mind, and Meaning*. Oxford: Oxford University Press.

Dyson, Freeman J. (1997). *Imagined Worlds*. Cambridge, MA: Harvard University Press.

Gruman, Gerald J. (1966). "A History of Ideas About the Prolongation of Life: The Evolution of Prolongevity Hypotheses to 1800." *Transactions of the American Philosophical Society* 56 (Part 9). Philadelphia: American Philosophical Association.

Habermas, Jurgen. (2003). *The Future of Human Nature*. Cambridge, England: Polity Press.

Hayles, N. Katherine. (1999). *How We Become Posthuman: Virtual Bodies in Cybernetics, Literature and Informatics*. Chicago: University of Chicago Press.

Hook, C. C. (2003). "Transhumanism & Posthumanism." In *The Encyclopedia of Bioethics*, 3rd edition, ed. Stephen G. Post. New York: Macmillan Reference.

Jonas, Hans. (1985). *The Imperative of Responsibility: In Search of an Ethics for the Technological Age*. Chicago: University of Chicago Press.

Jonas, Hans. (1996). *Mortality and Morality: A Search for the Good after Auschwitz*, ed. L. Vogel. Evanston, IL: Northwestern University Press.

Kass, Leon R. (1985). *Toward a More Natural Science: Biology and Human Affairs*. New York: The Free Press.

Kass, Leon R. (2001). "L'Chaim and Its Limits: Why Not Immortality?" *First Things* 113(May): 17–24.

Lewis, C. S. (1996 [1944]). *The Abolition of Man*. New York: Simon and Schuster.

Noble, David F. (1997). *The Religion of Technology: The Divinity of Man and the Spirit of Invention*. New York: Penguin.

Pepperell, Robert. (2003). *The Posthuman Condition: Consciousness Beyond The Brain*. Bristol, CT: Intellect.

Stock, Gregory. (2002). *Redesigning Humans: Our Inevitable Genetic Future*.

INTERNET RESOURCE

Kass, Leon R. (2003). "Beyond Therapy: Biotechnology and the Pursuit of Human Improvement." President's Council on Bioethics. Available from http://www.bioethics.gov/background/kasspaper.html.

POSTMODERNISM

• • •

A movement in the arts and humanities known as *postmodernism* gained a foothold in Western society in the 1980s and 1990s. The term was coined originally by architects in the early 1970s to designate an architectural style that aimed to break away from the dominant modernist style, characterized by indistinct boxlike skyscrapers, apartment complexes, and government buildings that had degenerated into a sterile and monotonous structural formula. Postmodern architects called for greater individuality, complexity, and eccentricity in design, along with the use of symbols with historical value. Shortly after its introduction into architecture, the term started to catch on more broadly, adopted by many in other arts and the humanities.

Philosophical Roots

Postmodernism became fashionable as the articulation of a continuing cultural reaction against "scientific modernism" that initially emerged in Europe during the Romantic period. The origin of scientific modernism is generally traced to the scientific revolution and the Enlightenment, also known as the "Age of Reason." Enlightenment philosophers believed that scientific reason was the best method for discovering truth and that science could eventually solve all the mysteries of life. In the early nineteenth century, the dizzying growth of technology and the constantly increasing belief that science would triumph over religion further entrenched scientific modernism into Western culture. By the end of the century, Friedrich Nietzsche's famous assertion that "God is dead" encapsulated the radical worldview of modernity. This modernist triumph was manifest in architecture and design. Buildings were constructed with new industrial materials such as steel and concrete, and many consumer goods were given a streamlined design. (Modernism in literature, however, was more ambiguous. It both imitated science and technology in some areas, as with experimentation in form and adapting techniques influenced by cinematic montage, while often criticizing science and technology in its content, as in T. S. Eliot's *The Waste Land*.)

Actually, Nietzsche's assertion signaled at the same time the beginning of a reaction against modernism itself. By the early decades of the twentieth century, artists and composers en masse started to express this very reaction through new and unorthodox forms of representation—forms that came to have wide appeal, no matter how different from tradition. When architects rejected the sterile formulas of modernist style, their coinage of the term *postmodern* (literally "after the modern") caught on widely, because it expressed what had, in effect, been happening in the content of other arts for a considerable period of time.

In postmodernism, nothing is for certain. Even science and mathematics are perceived to be constructs of human invention, as subject to human vagary as are the arts. The essence of the postmodern perspective is irony. This is why it is often described as a "deconstructive" approach to knowledge and representation. As the sociologist Zygmunt Bauman has perceptively remarked, postmodernism constitutes "a state of mind marked above all by its all-deriding, all-eroding, all-dissolving *destructiveness*" (1992, pp. vii–viii). By the early twenty-first century, postmodernism had become a topic of study under various academic rubrics, from semiotics and philosophy, to popular culture studies. Among those who are considered to provide significant critical frameworks for any discussion of postmodernism are Jean-François Lyotard (1984), Frederic Jameson (1991), and Jean Baudrillard (1998).

Postmodernity versus Postmodernism

By the early 1980s Western society itself was being labeled increasingly as being "postmodern." For this reason, a distinction emerged between *postmodernity* and *postmodernism*. The former was coined to refer to the social tendency to view absolute systems of truth (such as religious ones) with skepticism, and the latter to any representational technique that exemplifies this tendency. An often-cited example of the latter is Godfrey Reggio's brilliant 1983 film *Koyaanisqatsi*. The movie shows how fragmented the postmodern world is through a series of discontinuous, narrativeless images of cars on freeways, atomic blasts, litter on urban streets, people shopping in malls, housing complexes, buildings being demolished, and so on, all of which mirror the world's spiritual fragmentation. The collage of images paints a turgid, gloomy world populated by countless cars, decaying buildings, and crowds bustling aimlessly about. Reggio incorporates the mesmerizing music of Philip Glass (b. 1937), which reflects the images tonally. Glass's slow rhythms tire viewers with their heaviness, and his fast tempi—which accompany a demented chorus of singers chanting in the background—assault viewers' senses.

Implicit in Reggio's movie is the view that technology has been a destructive force in Western society, rather than constructively—a postmodern theme that runs through many contemporary movies such as *The Matrix* (1999). The struggle of humanity against its technological machinery is seen by postmodernists as part of the contemporary human condition, as is its struggle against deviance and abnormality, portrayed in such postmodern movies as *Blade Runner* (1982), directed by Ridley Scott and *Blue Velvet* (1986), directed by David Lynch.

Ultimately, the aim of postmodernism is to critique the contemporary world and its overreliance on scientific approaches to human behavior, such as psychology.

As a critical movement, therefore, it has had an important impact on how people perceive science and all kinds of approaches based on reason and logic. In postmodern representations, human beings are typically portrayed as fulfilling no particular purpose for being alive, and life is depicted as a meaningless collage of actions on a relentless course leading to death and a return to nothingness. But this bleak portrait of the human condition somehow forces a person to think about that very condition, paradoxically stimulating a profound reevaluation of the meaning of life.

Summary

Postmodern ideas have been destabilizing the rationalistic and logocentric (language-influenced) worldview that took shape in the Renaissance. As a cultural movement, postmodernism has made people more inclined to question belief systems in every domain of society, including the scientific one. (Scientists have entered the postmodern debate, either supporting the basic principles of postmodern ideology or rejecting it outright. The principle of indeterminacy in physics, for example, is based on an implicit postmodern tenet—namely, that the observer's interpretation of a physical phenomenon cannot be eliminated from the observation itself. Physics became unconsciously postmodern when it transformed itself into a study of quantum phenomena which entail participation of the observer in the observed.) The main reaction against postmodernism is the age-old one against the concept of relativism—that all truths are constructed—vs. the notion of an objective world where truth can be discovered by reason alone.

This does not mean, however, that postmodernity is devoid of ethics or a sense of truth and reality. As mentioned, postmodern artists ask the fundamental questions of life: What is a human being? What is real? Is there any meaning to existence? It is true, however, that they approach these questions in ways that are radically different from previous ethical traditions. Postmodern discourse has had a great impact on modern-day society, influencing the ways in which people perceive such issues as right and wrong, real and unreal, and so on. But the postmodern way of seeing things seems to be losing its social grip during the first decade of the twenty-first century. Like all ideological and intellectual movements of the past, postmodernism has probably run its course, as new social and intellectual trends now embrace a reinvigorated sense of purpose beyond the purely ironic.

MARCEL DANESI

SEE ALSO *Lyotard, Jean-François; Semiotics: Language and Culture.*

BIBLIOGRAPHY

Baudrillard, Jean. (1998). *The Consumer Society.* Thousand Oaks, CA: Sage. An acerbic critique of the image-making techniques of consumerist culture and their effect on human cultural development, inducing a generic postmodern worldview in society at large.

Bauman, Zygmunt. (1992). *Intimations of Postmodernity.* London: Routledge. A classic study of the ironic and destructive fabric of postmodernity.

Jameson, Frederic. (1991). *Postmodernism; or, The Cultural Logic of Late Capitalism.* Durham, NC: Duke University Press. The most-quoted critique of capitalism from the standpoint of postmodern theory, deconstructing the social texts that capitalist culture produces on a regular basis.

Lyotard, Jean-François. (1984). *The Postmodern Condition: A Report on Knowledge,* trans. Geoff Bennington and Brian Massumi. Minneapolis: University of Minnesota Press. An in-depth study of postmodernism, postmodernity and its consequences on social processes. It is both a praise of postmodern thought and an implicit warning of its overall destabilizing effects.

POVERTY

• • •

The elimination of world poverty, along with such concerns as the protection of the biosphere and the maintenance of peace, is generally counted among the global challenges facing humankind in the twenty-first century. In the mainstream account, poverty is the state of individuals who lack sufficient money or material possessions required for a dignified life. It usually implies living under the constant threat of starvation, sickness, and social exclusion. Global poverty is intolerable for societies oriented toward the achievement of material affluence and freedom; its eradication is therefore an ethical imperative for world economic policy. The role of science and technology, however, is a contested terrain; whether they are part of the problem or the solution depends on how poverty is understood and acted upon.

Disputed Definitions

Global poverty, understood as a category that comprises nations with low income, is a statistical construct. It is based on the comparison of aggregate national income figures, an operation that was first performed in the early 1940s. As societies are ranked according to a single quantitative scale, each nation is assigned a position in the hierarchy of income, which, below a certain poverty line, may be classified as poor. Likewise, global poverty as a category for classifying households worldwide is based on the ranking of household incomes, which, below a certain poverty line (for example, one or two dollars a day) are defined as poor. In both cases, the multidimensional diversity of living conditions on the globe is thus reduced to a unidimensional difference between income levels. Such a model of the world, while providing order and orientation, rests on a belief in the primacy of economic success over any other civilizational achievement. It had emerged during the rise of national economies in Europe and the United States; its projection upon the rest of the world triggered the rise of the development epoch after World War II (Sachs 1992).

Since the 1970s, however, the income definition of poverty has been recurrently contested, reflecting profound disagreements about the socially good and desirable. On a first level, measurements of objective poverty disregard subjective poverty. Yet how people perceive themselves is an important dimension of poverty. Calling people poor who do not think of themselves in this way may be misleading, offensive, or both.

On a second level, indicators that focus on absolute income fail to account for the relative nature of poverty, the experience of which varies according to context. As a general standard of living rises, a given amount of income may buy less well-being, because consumption items, such as automobiles, that were once viewed as luxuries may have become necessities, or because activities, such as child care, which were once available free of charge, may have come to involve expenditures. So the modernization of poverty tends to offset the income gains lifting people above the poverty line.

On a third level, income indicators ignore the importance of nonmarket goods and services for well-being. Two households that are equally poor in monetary terms may have quite different levels of well-being depending on access to community networks, environmental assets, and public services. In other words, common wealth is an important source of well-being; neglecting it renders any statement about poverty contestable.

Finally, on a fourth level, because income indicators are usually based on household measurements, they ignore gender inequality within households. But income is seldom equally distributed among family members;

increases in household income tend to favor men over women.

Given the limitations of income as a measurement of poverty, social indicators have been put forth to capture information about a broader range of living conditions (Kanbur and Squire 2001). This approach, which has been particularly promoted by the United Nations Development Programme (UNDP), views income merely as an instrument for achieving desired outcomes. Money matters, but not alone; well-being may be only loosely correlated with income. In this perspective, the poor are seen as deprived of basic capabilities, such as education, health care, longevity, economic opportunities, and legal entitlements, that would permit them to lead the kind of life they value (Sen 1999). How capabilities are shaped depends only partly on household income; variables of age, gender, availability of public goods, market opportunities, and legal security may be equally important.

UNDP's human poverty index, for instance, concentrates on three aspects of human deprivation: longevity, literacy, and living standard. Longevity is measured by premature deaths, literacy by the percentage of adults who are literate, and living standard by the percentage of the population with access to health services and safe water, and the percentage of malnourished children. As it turns out, national income is not necessarily correlated to quality of life. For example, despite their rather low levels of income, the people of China, Sri Lanka, or Kerala, India, enjoy enormously higher levels of life expectancy than do much richer populations of Gabon, Brazil, Namibia, or, for that matter, African Americans in the United States (Sen 1999).

Furthermore, given the limitations of quantitative measurements in general, efforts have been made to represent conditions of poverty through the voices of the poor themselves, using participatory and qualitative research methods (Narayan et al. 2000). How do poor people view poverty and well-being? Again, the picture of poverty shifts; many poor are not primarily concerned about lack of money or services, but lack of security and political voice. Poverty is associated with a state of vulnerability, both as precariousness in the economic sense and as powerlessness in the political sense. Having secure livelihoods is perceived as more important than maximizing income, just as having voice and influence is seen as more relevant than the delivery of services. Such accounts of lived poverty suggest conceptual implications: Poverty results from a lack of power rather than lack of income. It is the outcome of social relationships that are structured in a way in which benefits accrue consistently to one group and costs to another. As aspirations for wealth and power are acted out in society, some groups of people are unable to gain access to life-supporting assets, be they productive, environmental, or cultural, whereas others succeed in securing conditions for stable, productive lives. Poverty can thus be defined as relative powerlessness; its mitigation calls for basic-rights rather than basic-needs strategies. Poverty, in the early twenty-first century, has turned from an issue of economic growth into an issue of human rights.

Contentious Strategies and the Relation to Science and Technology

It is commonplace to call for poverty alleviation, but opinions divide sharply as to how and by whom. Looking back at decades of conflict, a growth-based perspective may be distinguished from a people-based perspective. In the first perspective, poverty alleviation is seen as the collateral benefit of aggregate economic growth, spurred by world market integration and accompanied by redistributional policies. Investors, transnational companies, and planners figure highly as agents of development. In the second perspective, overcoming poverty calls for stronger rights of the poor to land, capital, culture, and participation. The poor themselves are seen as actors capable of shaping their lives, yet constrained by a lack of entitlements and political leverage.

Both perspectives differ also in terms of time and direction. Growth strategies trust in the trickle-down effect, which is expected to eventually spread the benefits of growth throughout society down to the poorest strata. The social and environmental costs of growth in the present are regarded as the price for benefits in the future. In contrast, in the people-based perspective, growth often fails to trickle down; consequently, there is no point in sacrificing human lives and natural resources in the present for speculative gains in the future. Instead, it is regarded as crucial to empower the poor for a dignified life here and now.

As to the direction of poverty alleviation, the growth perspective aims for higher purchasing power, without taking into account nonmonetary sources of well-being. It therefore tends to confuse frugality and destitution, lumping both together under the rubric of poverty (Rahnema 2003). A people-based perspective, however, considers communities that are poor in money capital, yet rich in natural and social capital, as a base of livelihood to build on. But as dams displaced people or cash crops replaced subsistence crops, livelihood economies have time and again been squashed in favor of the money economy. As a consequence, growth, in the name of poverty eradication, has often turned frugality into wealth for a few and destitution for many.

Similar lines of conflict pervade the use of science and technology in poverty alleviation. In a growth perspective, technology appears as crucial factor for raising the productivity of national economies, in particular through infrastructure investments in areas such as transportation, energy, and communications. Predominantly science-based, capital-intensive, and centrally controlled technological systems are expected to deliver growth and are viewed as the royal road to reducing poverty (ADB/OECD 2002).

Any technology, however, has an impact on the structure of social relationships; it allows some to capture the benefits and condemns others to carry the costs. To the extent that dams or highways, hybrid seeds or water supply systems, boost opportunities for the well-off and powerful while shifting additional burdens onto the poor and powerless, technologies have helped deepen poverty. Against this background, people-based strategies attempt to disseminate human-scale technologies that are designed to enhance the power of the weak (ITDG ET.AL 2003). Low-input agriculture, micro-power systems, rainwater collection, and hand-driven radios are examples of alternative technologies that are comparatively low in investment costs, are operated decentrally, and empower the poor in their daily activities. Whether or not technology is up to the ethical challenge of relieving the burden of poverty thus depends in the last instance on the degree of agency they give to the poor. Insofar as technologies enable the poor to broaden their scope of action at low financial, environmental and social cost, they may serve as stepping stones out of powerlessness.

WOLFGANG SACHS

SEE ALSO *Development Ethics; Digital Divide; Green Revolution.*

BIBLIOGRAPHY

Asian Development Bank, and Development Centre of the Organisation for Economic Co-operation and Development (ADB/OECD). (2002). *Technology and Poverty Reduction in Asia and the Pacific.* Paris: Author.

Intermediate Technology and Development Group (ITDG)/UNESCO/Television Trust for the Environment. (2003). VHS and booklet, *Small Is Working: Technology for Poverty Reduction.* London: Author.

Kanbur, Ravi, and Lyn Squire. (2001). "The Evolution of Thinking about Poverty: Exploring the Interactions." In *Frontiers of Development Economics: The Future in Perspective*, ed. Gerald M. Meier and Joseph E. Stiglitz. New York: Oxford University Press.

Narayan, Deepa; Robert Chambers; Meera K. Shah; and Patti Petesch. (2000). *Voices of the Poor*, Vol. 2: *Crying Out for Change*. New York: Oxford University Press.

Rahnema, Majid. (2003). *Quand la misère chasse la pauvreté* [When misery drives out poverty]. Paris: Fayard/Actes Sud.

Sachs, Wolfgang, ed. (1992). *The Development Dictionary.* London: Zed.

Sen, Amartya. (1999). *Development as Freedom.* New York: Knopf.

POWER SYSTEMS

• • •

Power systems represent the class of technologies used to generate electricity. The cost-effective generation, distribution, and use of electricity since the early twentieth century have changed ways of life in developed and developing countries alike. Electricity has made possible a special kind of economic and technological development, including conveniences at home and increased productivity at work. It is what drives modern society and is the foundation upon which the digital age is being built.

However electricity production has also had major impacts on the biosphere. The production of electricity, which is generated primarily from carbon-based fuels, has contributed largely to the increase in greenhouse gases. Electricity production is also responsible for acid rain and smog precursors, as well as mercury and other toxic air pollutants. In addition, two-thirds of electricity production globally is from nonrenewable resources. Thus an electrified society places future generations at risk by both destroying the biospheric services on which they depend for survival, and depleting natural resources. If one accepts a moral obligation for the health and happiness of future generations, the current power system model should be reconsidered.

Production Developments

A curiosity before the 1880s, electricity entered the mainstream in 1882 when Thomas A. Edison began generating and distributing direct current (DC) electricity from his Pearl Street station in New York City. This was soon followed by a host of other generation and distribution systems, most notably by George Westinghouse, who in 1895 began to produce alternating current (AC) electricity from a power plant at Niagara Falls. Soon after AC electricity became the dominant form used in homes and businesses. From 1900 through the 1930s, new appliances, such as vacuum cleaners, washing machines, refrigerators, radios, and televisions, found their way into U.S. and European homes; the Electric Age had begun.

The first electric power plants burned coal or wood to produce steam to power electric generators. In the United States and Europe fossil fuels dominated power markets throughout the first half of the twentieth century. For example, in the United States in 1950, production of electricity from coal, oil, and gas was 46 percent, 10 percent, and 13 percent respectively, while hydroelectric dams produced about 30 percent of the total.

In the 1960s, nuclear power was harnessed to generate the heat needed to power steam turbines. In the early twenty-first century in some countries, such as France, nuclear power contributes a majority of electricity. In the United States, fossil fuels still dominate, as shown in Tables 1 and 2.

The electrification of the industrial world is almost complete. However many people in the developing world still live without electricity. As these nations develop, electricity will surely play an increasingly important role. Even with advancements in energy efficiency and flattened population growth projections, one would expect significant increases in electricity consumption in the developing world throughout the twenty-first century.

Electrification of highly populated developing countries, particularly China and India, may have significant global repercussions. If the electrification of these countries occurs using fossil fuels such as coal, there are grave concerns about the impacts on greenhouse gas emissions and global warming. Similarly, if these countries move toward a nuclear future, concerns about weapons proliferation, safety, and nuclear waste management present additional challenges.

Future Assessment

Those in the developed world cannot expect developing countries to forego electrification as they move along the development path. However technical solutions may exist to limit the global problems associated with electricity production. One such solution is renewable energy. Although new hydroelectric dam sites are becoming scarce, electricity opportunities from wind, solar, and biomass are increasing. Wind power is now competitive with fossil fuels in many areas, while solar

TABLE 1

U.S. Fuel Use for Electricity, 2002	
Fuel Type	Percentage
Coal	50.2%
Petroleum	2.3%
Natural Gas	17.9%
Nuclear	20.3%
Hydro	6.9%
Other Renewables	2.2%

*Percentages may not add to 100% due to rounding.

SOURCE: U.S. Energy Information Administration.

technologies are currently cost effective in some remote locations or niche applications. Issues such as energy storage and delivery currently plague these technologies, but with appropriate technological advancements and economic assistance, the developing world may be able to achieve a future that has eluded the industrialized world—carbon-free electrification.

Another energy source that looks promising and would support a renewable electric future is hydrogen. Because hydrogen can be produced from the electrolysis of water, one could envision a system whereby electricity produced from a renewable resource, such as solar photovoltaics, could be used to generate hydrogen. This hydrogen could be stored and transported, and ultimately used in fuel cells to produce electricity where needed. However, despite recent media attention on hydrogen and the so-called hydrogen economy, the current state of technology and costs suggest that hydrogen will not become a genuine competitor to fossil fuels before the mid-twenty-first century.

The development of carbon sequestration technologies may provide another solution to biospheric problems posed by carbon-based power systems. These technologies are able to capture carbon emissions from power plants and transform or store this carbon to prevent atmospheric discharge and greenhouse gas buildup in the atmosphere. Considerable research is also being invested in other ways to reduce carbon emissions from coal-fired power plants.

Ethical Issues

The ethical implications of power production rest on the seriousness with which society holds its responsibilities to future generations. Since energy markets cannot adequately internalize the costs of fossil-fuel power generation (both in terms of current and future environmental externalities), many argue that government poli-

TABLE 2

World Electricity Production by Fuel, 2000

(Billion Kilowatt-hours)

Region	Thermal	Hydro	Nuclear	Geothermal and Other	Total
North America	2,997.1	657.6	830.4	99.0	4,584.0
Central & South America	204.1	545.0	10.9	17.4	777.4
Western Europe	1,365.4	557.5	849.4	74.8	2,847.1
Eastern Europe & Former U.S.S.R.	1,043.7	253.5	265.7	3.9	1,566.9
Middle East	425.3	13.8	0.0	0.0	439.1
Africa	333.7	69.8	13.0	0.4	416.9
Asia & Oceania	2,949.2	528.7	464.7	43.1	3,985.7
World Total	**9,318.4**	**2,625.8**	**2,434.2**	**238.7**	**14,617.0**

SOURCE: U.S. Energy Information Administration.

cies and international agreements are needed. However there is still uncertainty surrounding the distributional impacts of global warming (across space and time). This uncertainty has been used to thwart regulatory actions aimed at curbing carbon emissions from fossil-fuel power plants. Assuming continued uncertainty about the long-term impacts of greenhouse gas emissions, one would expect governments to be slow to take action in the near term. The development of marketable, cost-effective competitors to fossil fuels will likely be needed to displace early-twenty-first-century power systems. Thus far such technologies do not seem imminent.

JAMES J. WINEBRAKE

SEE ALSO *Alternative Energy; Alternative Technology.*

BIBLIOGRAPHY

Smil, Vaclav. (2003). *Energy at the Crossroads.* Cambridge, MA: MIT Press.

Winebrake, James J., ed. (2004). *Alternate Energy: Assessment and Implementation Reference Book.* Lilburn, GA: Fairmont Press.

INTERNET RESOURCES

Energy Information Administration. U.S. Department of Energy. Available from http://www.eia.doe.gov.

IEEE. (2005). "How Electricity Came to Be." Available at the IEEE Virtual Museum http://www.ieee-virtual-museum.org/.

PRAGMATISM

• • •

"Pragmatic" seems to have been used for the first time in the modern Western philosophical tradition by Immanuel Kant (1724–1804); for him, there was some connection with ethics, but little with science or technology in the modern sense. In the early twentieth century, pragmatics turns up as a third subdivision of a formal semantics triad (see Morris 1938), but it has only a remote connection to science by way of mathematics, and none to ethics or technology.

In most introductory accounts, "pragmatism" as a term for a philosophical approach is usually taken to be synonymous with a "pragmatic theory of truth." We can be sure about something if it has practical or real-world consequences. Enemies of philosophical pragmatism even caricature this as meaning that the test of the truth of a statement—even about ethics—is whether or not it works. Such characterizations are unfair to the nuanced thought of philosophers who have been willing to call themselves pragmatists—as has occurred during two periods: in the period of "classical American philosophy" (see Stuhr 2000) at the beginning of the twentieth century, and again at the end of the twentieth century.

Classical Pragmatists

In the early twentieth century, a number of philosophical pragmatisms sprang up, for example, that of Giovanni Papini (1881–1956) in Italy and Edouard Le Roy (1870–1954) (see Stebbing 1914) in France; but the best known of these was the school (in the loose sense) of American pragmatism, beginning with Charles Sanders Peirce (1839–1914) (who chose to call his approach "pragmaticism") and William James (1842–1910). But the best known of the American pragmatists was John Dewey (1859–1952), whose ideas were closely paralleled by those of his friend and colleague, George Herbert Mead (1863–1931). (On the classical pragmatists, their relationships with one another and with the term, see Menand 2001.)

The basic move of the classical pragmatists was to seek to replace what may be termed the epistemological account of knowledge as justified true beliefs (a definition that can be traced back to Plato [427–347 B.C.E.]) with an analysis of beliefs in terms of relationships to human action. Traditional epistemologies sought to identify the foundations of knowledge in some special cognitive activity or method. Peirce, however, adapting the suggestion of Alexander Bain (1818–1903)—a Scottish philosopher and friend of John Stuart Mill (1806–1873)—argued that beliefs are more properly interpreted as habits of acting than as representations of reality, and so not in need of some special foundations. All pragmatists reject both conceptual reference (concepts are true if they refer to real things) and coherence (concepts are true if they fit together logically) theories of knowledge prominent in empiricism and rationalism, respectively, in favor of some interpretation of inquiry that unites theoretical and practical knowledge as grounded in forms of learning to operate more effectively in the world. Such an approach easily ties knowing into science and technology, and in some instances to ethics, although this happens in different ways in different pragmatisms.

In this respect, the pragmatism of Dewey and Mead exhibited a special relationship to science, technology, and ethics. For Mead and Dewey, ethics is not a theoretical discipline but simply social problem solving using "the scientific method." What this meant for them was applying expert knowledge from *all* the sciences—from the natural sciences and engineering to sociology or social psychology—in democratic efforts of particular communities to solve urgent social problems. The communities in question ran the spectrum from families and technical communities all the way up to the world community, in the former League of Nations. As one of the best of recent interpreters of Dewey, Larry Hickman (2001, p. 51; see also Hickman 1990), puts the matter, Dewey thought that it is possible "to articulate a general method of intelligence that takes into account successful inquiry in many different areas of human activity," including various sciences, the arts, politics, jurispru-

dence, and so on. Yet while contemporary science-based technologies have made major contributions to this general method of intelligence, they are only one of many sources. In this way, Hickman thinks, Dewey avoids the charge that he favored scientism. Mead's version of the same general approach can be seen in the title of his "Scientific Method and the Moral Sciences" (1964).

Both Mead and Dewey had lifelong contacts with colleagues in the science departments of their universities and kept abreast of developments in the sciences, perhaps especially in physiological psychology but also in physics and biology and other fields. (On this aspect of Dewey's work, see Dalton 2002.) As for technology, both Mead and Dewey were highly critical of the then-new corporations, with their research and development laboratories. The problem was that the corporations were so often involved in what amounted to private wars to break the power of the new labor unions. For Dewey, "We must wrest our general culture from an industrialized civilization" in which science "is ultimately a reflex of the social conditions under which science is applied [in industry] so as to reach only a pecuniary fruition" (Dewey 1930, pp. 133–134). Mead's work with progressive reformers—for example, trying to mediate the struggle between strikebreaking corporate managers and the unions in Chicago in the early twentieth century—can be seen in Andrew Feffer's 1993 work *The Chicago Pragmatists and American Progressivism.*

Others among the early American pragmatists in some cases had similar views, but there were also many differences. James, for example, credited Peirce with the originating idea of American pragmatism, referring to Peirce's famous "pragmatic maxim": "A conception can have no logical effect or import differing from that of a second conception except, so far as . . . it might conceivably modify our practical conduct" (from his "Lectures on Pragmatism," 1903). But Peirce was primarily a scientist, mathematician, logician, and philosopher of science, not a social reformer.

James accepted the pragmatic maxim, which he rendered this way: "Grant an idea or belief to be true," then, "what concrete difference will its being true make in one's actual life?" (James 1907; see also James 1909). He was certainly pro-science enough to have founded the experimental psychology program at Harvard, but he also dabbled in spiritualist theories in ways that alienated other experimental psychologists. Moreover, though James was progressive in a patrician sort of way, he seems never to have given a thought to union organizing, and watched the "Chicago school's" activism in, at best, a detached sort of way (see McDermott 1967 and Gale 1999).

In summary, among these early American pragmatists, Peirce was primarily a philosopher of science interested in doing away with any certainty-seeking foundationalism of a Cartesian sort. James was the suave elder statesman, interested in pushing science, especially evolution, as a new cultural force, while maintaining a place for a liberal religion in this new culture. Mead and Dewey pushed pragmatism in the direction of progressive social reform, including a critique of the newly-powerful science-based corporations, basing their reforms on "the scientific method." This meant primarily a respect for expertise of all kinds, as long as it was combined with a democratic citizen activism aimed at challenging old verities while working out new and better social arrangements. Dewey was explicit that the only contribution of theoretical philosophizing in the traditional sense (however important on other grounds) was in "divesting ourselves of the intellectual habits we take on and wear when we assimilate the culture of our own time and place" (1925, p. 40; the view is best represented in *The Quest for Certainty*, 1929, and *Reconstruction in Philosophy*, 1920).

John Stuhr (2000) places the early pragmatists within a tradition of "classical American philosophy," and in doing so he adds context. Their works appeared among and were related to writings of significant American women writers (Jane Addams [1860–1935]), American idealists and personalists (Borden Parker Bowne [1847–1910]), African-American philosophers (Alain Locke [1886–1954]), and non-pragmatist naturalists. (Stuhr, p. 695, cites John Herman Randall, Jr. [1899–1980], as an example.) A similar, equally controversial contextualizing, appears in Cornel West's *The American Evasion of Philosophy* (1989).

Late Twentieth Century Pragmatism

Joseph Margolis (2002) characterizes all the above as the "early" American pragmatism, with which he contrasts the "revival" of pragmatism in American analytic philosophy after about 1980. The main representatives of this revival are Willard Van Orman Quine (1908–2000), Donald Davidson (1917–2003), Hilary Putnam (born 1926), and Richard Rorty (born 1931). In the revived version of American pragmatism, the focus is not on Mead and Dewey's "meliorizing" progressivism, with its suspicion of large science-based corporations, but on quarrels over different versions of epistemology. With the exception of Rorty, who wants

his pragmatism (he says it is more literary than philosophical) to join in leftist causes (1998), none of the revived pragmatists have much interest in ethics, less in technology, and an interest in science that is reducible to a scientistic model of human knowing—or opposition to it.

Margolis's is the best summary of these disputes, which he characterizes as involving two challenges to pragmatism: naturalism and postmodernism. The primary debate is between "pragmatism" and "naturalizing"—especially several debates between Rorty (claiming to speak for Quine and Davidson as well as himself) and Putnam. The conflict has to do with how to safeguard a "true" pragmatism from relapsing into a Cartesian quest for a guaranteed foundation of knowledge in science.

To summarize the account, at some cost to nuances, Margolis argues that although they call themselves pragmatists, Quine, Davidson, and Putnam are all concerned with essentially epistemological issues, and that they approach these in ways that are ultimately unfaithful to pragmatist inspirations. Insofar as Quine and others attempt to understand knowing in naturalistic or scientific terms, and turn epistemology into an empirical examination of cognition, they tend to put forth a new kind of foundationalism, which was just what the original pragmatists were at pains to avoid. The late-twentieth-century epistemological pragmatists tend toward realism rather than instrumentalism: that is, they want to defend a view of scientific knowledge as providing a privileged view of the world rather than the process of science as a privileged means or method for living in the world.

Margolis's account of the challenge of postmodernism and its manifestation in Rorty is easier to state. Postmodernism rejects not just science as a privileged form of knowing but science as a privileged method for living. Rorty's postmodernism is thus incompatible with classical pragmatism and its reliance on (but not idolization of) science specifically and expertise generally. For classical pragmatism, the need for the democratic governance of expertise does not reject or deny its benefits.

In the end, Margolis derives his own version of pragmatism from the failures of naturalism and postmodernism. This version places constructivism at the center of pragmatism. In Margolis's words, "questions of knowledge, objectivity, truth, confirmation, and legitimation are constructed in accord with our interpretive conceptual schemes." Thus, "though we do not construct the actual world, what we posit (constructively) as the independent world is epistemically dependent on our mediating conceptual schemes" (p. 22).

Future Prospects

At the end of his analysis, Margolis confesses doubt as to whether even his constructive pragmatism, with its unique combination of the best in pragmatism with the best in recent European philosophy, will succeed in the twenty-first century. (On European, especially German, interest in pragmatism, see Aboulafia, Bookman, and Kemp 2002.) Instead, Margolis fears that the naturalizers will continue to dominate analytical philosophy, especially in neo-Darwinism viewed as the best reductive model of the cultural world; in extreme linguistic views (originating with Noam Chomsky); and in a computational analysis of every form of human perception and intelligence. But Margolis still has hope, though he says at the end that pragmatists have little more than their original intuition to rely on. Other pragmatists would argue that pragmatisms are not based on mere intuition; that pragmatists have good arguments, for example, against reductionism.

All of this epistemological nitpicking among recent pragmatists would leave the earlier pragmatists shaking their heads. Mead and Dewey, and probably also James and Peirce, thought they had good reason to reject any epistemology based on foundationalist projects; such epistemologies are simply inconsistent with their scientific and progressive project (Palmer 2002). Mead (1934, p. 94), as one example, rejected all epistemology as "riff-raff"; and he pointed out, one by one, how all traditional epistemologies (traditional at the time of his writing) depended on individualist assumptions that are incompatible with a view of science as a social undertaking, dependent on a world taken for granted within particular science communities (Mead 1964).

Moreover, the earlier pragmatists have their nonanalytic followers; examples include Larry Hickman (1990, 2001) on technology; Sharyn Clough (2003) on feminist science studies; and Glenn McGee (1997), providing a pragmatic ethics of genetic engineering.

Still, it is true that even the earlier version of American pragmatism has difficulties to face—in addition to Margolis's claim that it is analytically naïve and unsophisticated. Some criticisms of a Deweyan philosophy of technology have been collected in Paul Durbin's special issue of *Techne* (2003), and they come from Heideggerians and neo-Heideggerians, from critical theorists and neo-Marxists, among others. In the end, it should be obvious that even the best philosophical version of pragmatism will continue to have its detractors.

PAUL T. DURBIN

SEE ALSO *Democracy; Dewey, John; Expertise; Pierce, Charles Sanders; Science Policy.*

BIBLIOGRAPHY

Aboulafia, Mitchell; Myrna Bookman; and Cathy Kemp, eds. (2002). *Habermas and Pragmatism.* London: Routledge.

Clough, Sharyn. (2003). *Beyond Epistemology: A Pragmatist Approach to Feminist Science Studies.* Lanham, MD: Rowman & Littlefield.

Dalton, Thomas C. (2002). *Becoming John Dewey: Dilemmas of a Philosopher and Naturalist.* Bloomington: Indiana University Press.

Davidson, Donald. (1986). "A Coherence Theory of Truth and Knowledge." In *Truth and Interpretation: Perspectives on the Philosophy of Donald Davidson,* ed. Ernest LePore. Oxford, UK: Blackwell.

Dewey, John. (2003 [1920]). *Reconstruction in Philosophy,* enlarged edition. Mineola, NY: Dover.

Dewey, John. (1925). *Experience and Nature.* Chicago: Open Court.

Dewey, John. (1929). *The Quest for Certainty.* New York: Minton, Balch.

Dewey, John. (1930). *Individualism, Old and New.* New York: Minton, Balch.

Feffer, Andrew. (1993). *The Chicago Pragmatists and American Progressivism.* Ithaca, NY: Cornell University Press.

Gale, Richard M. (1999). *The Divided Self of William James.* Cambridge, UK: Cambridge University Press.

Hickman, Larry A. (1990). *John Dewey's Pragmatic Technology.* Bloomington: Indiana University Press.

Hickman, Larry A. (2001). *Philosophical Tools for Technological Culture: Putting Pragmatism to Work.* Bloomington: Indiana University Press.

James, William. (1907). *Pragmatism: A New Name for Some Old Ways of Thinking.* New York: Longmans, Green.

James, William. (1909). *The Meaning of Truth: A Sequel to "Pragmatism."* New York: Longmans, Green.

Margolis, Joseph. (2002). *Reinventing Pragmatism: American Philosophy at the End of the Twentieth Century.* Ithaca, NY: Cornell University Press.

McDermott, John, ed. (1967). *The Writings of William James: A Comprehensive Edition.* New York: Random House.

McGee, Glenn. (1997). *The Perfect Baby: A Pragmatic Approach to Genetics.* Lanham, MD: Rowman & Littlefield.

Mead, George Herbert. (1938). *The Philosophy of the Act.* Chicago: University of Chicago Press.

Mead, George Herbert. (1964). "Scientific Method and Individual Thinker." In *Selected Writings,* ed. Andrew J. Reck. Chicago: University of Chicago Press.

Mead, George Herbert. (1964). "Scientific Method and the Moral Sciences." In *Selected Writings,* ed. Andrew J. Reck. Chicago: University of Chicago Press.

Menand, Louis. (2001). *The Metaphysical Club.* New York: Farrar, Straus and Giroux.

Morris, Charles W. (1938). *Foundations of the Theory of Signs*. Chicago: University of Chicago Press.

Palmer, L. M. (2002). "Vico and Pragmatism: New Variations on Vichian Themes." *Transactions of the Charles S. Peirce Society* XXXVIII(3): 433–440.

Papini, Giovanni. (1913). *Pragmatismo*. Milan.

Peirce, Charles Sanders. (1903). "Lectures on Pragmatism." In *Collected Papers of Charles Sanders Peirce*, ed. C. Hartshorne and P. Weiss. Cambridge, MA: Harvard University Press.

Putnam, Hilary. (1980). *Reason, Truth, and History*. Cambridge, UK: Cambridge University Press.

Putnam, Hilary. (1994). "Sense, Nonsense, and the Senses: An Inquiry into the Powers of the Human Mind." *Journal of Philosophy* 91(9): 445–517.

Quine, W. V. (1969). "Epistemology Naturalized." In *Ontological Relativity, and Other Essays*. New York: Columbia University Press.

Rorty, Richard. (1998). *Achieving Our Country: Leftist Thought in Twentieth-Century America*. Cambridge, MA: Harvard University Press.

Rorty, Richard. 1986. "Pragmatism, Davidson, and Truth." In *Truth and Interpretation: Perspectives on the Philosophy of Donald Davidson*, ed. Ernest LePore. Oxford, UK: Blackwell.

Stebbing, L. Susan. (1914). *Pragmatism and French Voluntarism*. Cambridge, UK: Cambridge University Press.

Stuhr, John J., ed. (2000). *Pragmatism and Classical American Philosophy: Essential Readings and Interpretive Essays*, 2nd edition. New York: Oxford University Press.

West, Cornel. (1989). *The American Evasion of Philosophy: A Genealogy of Pragmatism*. Madison: University of Wisconsin Press.

INTERNET RESOURCES

Durbin, Paul T., ed. (2003). *Techne* 7(1). Available from http://spt.org. Special author-meets-critics issue on the Deweyan philosophy of technology of Larry Hickman.

Hickman, Larry A. (2003). "Revisiting Philosophical Tools for Technological Culture." *Techne* 7(1): 74–93. Available from http://spt.org.

PRAXIOLOGY

• • •

Praxiology, occasionally *praxeology* and rarely *praxæology*, is from the Greek *praxis* meaning goal-directed action, and *logos* in the sense of *knowledge* or *information*. Apparently having stipulative origins in French, namely, *praxéologie* (Mitcham), the lexical term praxiology was introduced by Tadeuz Kotarbiński (1886–1981) in 1965. Polish philosopher and co-founder, with Jan Łukasiewicz and Stanislaw Leśniewski of the Warsaw Center of Logical Research (Warsaw Circle), Kotarbiński used praxiology to reference an area in the philosophy of action that was distinguished from other such areas by its focus on efficient action. With adaptations to engineering, business, law, and more, and with discussions relating efficient action to mathematics, the natural sciences, technology, and ethics, praxiology has developed along three major lines: Kotarbińskian, analytic, and synthetic.

Kotarbińskian praxiology, also traditional or classical praxiology, begins with a practical situation said to be complex and exigent, and with a wish to change it to some prescribed future situation. The process of changing a practical situation is subjected to nine value foci called the Es (Collen): efficiency, effectiveness, efficacy, ethicality, economy, educability, executability, evaluability, and expendability. Inasmuch as some Es are factual in nature, for example, efficiency, praxiological inquiries in such areas have been referred to as sciences. Although some Es are more qualitative in nature than others, for example, ethicality, no evaluative hierarchy among the Es exists. Thus economics can compete with ethics in praxiological decision making. The remaining lines of praxiological development focus on one or another phase in the process of change.

Analytic praxiology including pragmatic praxiology refers to an analysis of a situation, specifically, a prediction—based on knowledge of its component parts and their connections—of its response to prescribed stimuli or service conditions. The name *pragmatic praxiology* derives from the centrality given to the prediction of consequences in the theories of pragmatism crafted by Immanuel Kant and Charles Peirce (Ryan et al. 2002). The main question is epistemological: What do humans know will result from what they do? The task of responding to this question often falls to the sciences. Historically significant contributions to analytic praxiology may well be found in the histories of systems analysis and cybernetics. (Mitcham 1994).

Synthetic praxiology including design praxiology extends the task of analytic praxiology from creating knowledge about consequences of action to the making of plans for action. A design is a choice from a portfolio of possible future situations; it is a choice based on analyses of these situations and the processes required to realize them. The main question is methodological: How do humans change the world to realize their wishes? Historically significant contributions to synthetic praxiology surely lie among the works of Wojciech W. Gasparski on design and Henryk Skolimowski on the ethics of design ends, but they may also be found in the histories of operations research and management science. (Mitcham 1994).

Kotarbińskian, analytic, synthetic, and other praxiologies comprise a general praxiology spawning applications to the professions. Because of its transdisciplinary aspirations, taxonomic issues arise where such applications, or special praxiologies, meet the academic disciplines of professional education. Would a praxiology of law correspond to jurisprudence? Would theology be a praxiology for organized religion? Where does praxiology of education fit into philosophy of education? If management science is rightly called management technology, would praxiology be its philosophical aspect? (Bunge 1999) Is military science a praxiology?

The transdisciplinary mode is but one of four modes by which praxiology might engage another learned discipline. In the cross-disciplinary mode the tools and methods of praxiology are used to inquire into another discipline. For example, instead of attempting to prove that engineering is a case of praxiology, one might demonstrate that engineering possesses praxiological properties or natures. In the multidisciplinary mode, tools and methods of praxiology are brought together with those of other disciplines. Remaining intact and distinct, these disciplines join to produce novel subdisciplines. For example, when Ludwig von Mises made praxiology the method of the Austrian School of Economics, he crafted the subdiscipline that can be called praxiological economics. In the interdisciplinary mode, tools and methods of praxiology may likewise be brought together with those of other disciplines, but they would not remain intact. Rather essentials of each would be organized into coherent wholes or novel disciplines displaying principles that disagree with principles of their parent disciplines. For example, chemical engineering, which possesses nonscientific principles, namely Koen's (2003) heuristics, is to a degree the result of an interdisciplinary engagement of praxiology with chemistry.

At about the same time that Kotarbiński was working out praxiology, John Dewey (1859–1952) was working out his naturalism. Both of their transdisciplinary ideas began with practical situations. Dewey worked within a Cartesian framework developing cognitive abilities to make change, which loosened the grip that classical education had on education. In the cross-disciplinary mode with education, Dewey emphasized the needs of the individual to advance the ideals of a capitalistic democracy, and gave ethics primacy. Kotarbiński worked within a Marxist framework developing the human will to make change. Putting ethics in the Es

with economics, Kotarbiński emphasized the needs of the state. In the United States, the technocracy movement of the 1930s, which advocated a dictatorship of engineers (Layton 1971), and the communist scare in the early-1930s, which was followed by McCarthyism in the 1950s, were not favorable to praxiology. In Poland, Nazi oppression and subsequent communistic regimes virtually cut off international scholarly communications. These social factors left the STS movement, which was underway in the United States by the early 1970s, to independently develop many ideas discussed in praxiology. In 1978, Karol Wojtyla became Pope John Paul II, the first Polish Supreme Pontiff, and interest in Polish scholarship increased. By 1978 though, STS gained currency with an attendant lessening of the importance of the theory of praxiology.

TAFT H. BROOME, JR.

SEE ALSO *Efficiency.*

BIBLIOGRAPHY

Alexandre, Victor, and Wojciech W. Gasparski, eds. (2004). "French and Other Perspectives in Praxiology." In *Praxiology: The International Annual of Practical Philosophy and Methodology*, Vol. 12. New Brunswick, NJ: Transaction Publishers. Discusses work by scholars from France, Finland, Great Britain, Poland, Portugal, Spain, and the United States. Topics include cooperative actions; university education, and corporate governance in Central and Eastern Europe; innovation in Spain; information systems; and fuzzy logic.

Bunge, Mario. (1999). "Ethics and Praxiology as Technologies." Virginia Polytechnic and State University. Available from http://scholar.lib.vt.edu/ejournals/SPT/v4n4/bunge.html. Argues that the intersection of ethics and praxiology is of a sufficient technological nature to comprise philosophical technology; modifies Kotarbińskian praxiology by elevating the status of ethics in the Es to primacy.

Collen, Arne, and Wojciech W. Gasparski, eds. (2003). "Systemic Change Through Praxis and Inquiry." In *Praxiology: The International Annual of Practical Philosophy and Methodology*, Vol. 11. New Brunswick, NJ: Transaction Publishers. Topics include change as a systemic idea from a research methodologist's point of view; prevalence of hierarchy and control; praxiology in research; the research process as means of systemic change; the Es of praxiological inquiry; and traditional, pragmatic, and design praxiologies.

Dewey, John. (1958). *Experience and Nature.* New York: Dover. Dewey's 1925 naturalism as a metaphor for praxiology.

Koen, Billy Vaughn. (2003). *Discussion of the Method: Conducting the Engineer's Approach to Problem Solving.* New York: Oxford University Press. The engineering method of change as a universal method of creating utopia; the most innovative aspect is Koen's notion of engineering heuristics.

Kotarbiński, Tadeuz. (1965). *An Introduction to the Sciences of Efficient Action*, trans. Olgierd Wojtasiewicz. New York: Pergamon Press. Work began in Poland in 1920; the seminal work on praxiology.

Layton, Edwin T. (1971). *The Revolt of the Engineers*. Cleveland, OH: Case Western University Press. A historical account of the development of U.S. engineering professionalism from 1900 to 1940.

Mises, Ludwig von. (1996). *Human Action: A Treatise on Economics*, 4th revised edition. San Francisco: Fox & Wilkes. Originally published in 1949, Mises's cross-disciplinary application of praxiology to economics for the Austrian School of Economics.

Mitcham, Carl. (1994). *Thinking Through Technology: The Path Between Engineering and Philosophy*. Chicago: Chicago University Press. A critical introduction to the philosophy of technology offering alternatives to the treatment of technology as magic.

Ryan, Leo V., and Wojciech W. Gasparski, eds. (2002) "Praxiology and Pragmatism." In *Praxiology: The International Annual of Practical Philosophy and Methodology*, Vol. 10. New Brunswick, NJ: Transaction Publishers. Presents a 1972 Kotarbinski study of practicality as well as addresses the relevance of pragmatism to management, business ethics, law, and pragmatic inquiry. Includes a model for teaching pragmatism and a review of pragmatism in Europe.

PRECAUTIONARY PRINCIPLE

• • •

The precautionary principle was introduced into environmental politics in response to a perception that existing policies did not provide adequate protection to the environment. The most prominent formulation was adopted as Principle 15 of the Rio Declaration from the 1992 United Nations Conference on Environment and Development: "In order to protect the environment, the precautionary approach shall be widely applied by States according to their capabilities. Where there are threats of serious or irreversible damage, lack of full scientific certainty shall not be used as a reason for postponing cost-effective measures to prevent environmental degradation" (United Nations 1992). The principle has important implications to the interpretation of science and the regulation of technology, and is an expression of values in relation to the environment.

Terminology

It is important to distinguish between the precautionary principle, a precautionary approach, and precautionary action. The precautionary principle is a framework for thinking that provides foresight in situations characterized by uncertainty, ignorance, and ambiguity, and where there are potentially large pros and cons for both regulatory action and inaction. As a principle, it may have legal standing with implications for applications in the international arena. In the European Union, precaution is interpreted as such a principle with legal standing, and was officially adopted as such in the Maastricht Treaty of 1992.

A *precautionary approach* is a way of doing things along the same lines of thought, but has no legal standing. In international trade disputes, the United States tends to interpret the precautionary principle as an approach and not a principle with legal standing. A *precautionary action* is simply a measure taken to implement the thought behind the principle, or it may be an isolated action taken for other or related reasons.

History

Histories of the precautionary principle and of precautionary actions are different. Precautionary actions are known from before the term was invented. Examples can be drawn from legislation in both the European Union and the United States. However, precaution as a principle dates back to German legislation on air pollution from 1976, where the principle was called "Vorsorgeprinzip," where the German word *Vorsorge* means "care" as much as "precaution" (Boehmer-Christensen 1994). The difference is subtle but significant, because "care" is a positive expression of responsibility and prudence, while "caution" has a connotation of "not daring" and "risk aversion," frequently used in a derogatory way implying that one is too cautious. The German legislation of 1976 introduced many measures related to duty ethics: BAT (Best Available Technology), ALARA (As Low As Reasonably Achievable), LCA (Life Cycle Analysis), and the concept of cleaner production. The common feature of these approaches is that one has an obligation to do the best from the perspective of reason, prudence, and environmental sustainability.

The precautionary principle was given many interpretations at various international conferences (such as the North Sea conferences in 1984 and 1987, and the Bergen Conference on Sustainable Development in 1990) building up to the Rio conference in 1992. The

European Union adopted the principle at the constitutional level in the Maastricht Treaty (1992) by a simple statement in Article 174: Community policy "shall be based on the precautionary principle... ." A commission communication provided an interpretation of the precautionary principle (European Commission 2000), and at the end of the same year this interpretation was endorsed in the Nice Treaty.

In the United States, Kenneth Foster, Paolo Vecchia, and Michael Repacholi (2000) published in *Science* a policy commentary on the E.U. interpretation. This commentary argued that under the practical interpretations adopted by the European Union, the precautionary principle was not in conflict with the weight of evidence analysis approach more typically employed by scientists and health administrators in the United States. Retrospective historical analyses (Harremoës, Gee, MacGarvin, et al. 2002) and contemporary case studies (Tickner 2003) have tended to support this assessment. Andrea Saltelli and Silvio Funtowicz (2003) and others also have explored options for operationalizing basic intuitions involved in the precautionary principle.

Basic Interpretations

The overall impression is that the precautionary principle is a response to societies strongly influenced by *positivism*, which tends to regard scientific and technological development as a priori beneficial. The background is an increasing awareness of the potentially detrimental effects of scientific and technological development. Accordingly, the precautionary principle may be interpreted in several ways and is subject to intensive debate in scientific, technical, social, legal, and political terms. There are two extreme misinterpretations of the principle.

One misinterpretation considers the precautionary principle a one-sided argument for the elimination of all adverse effects on health and environment. To demand absolute proof of safety before undertaking any action is, of course, not realistic, and this view is derogatorily referred to as the *risk averse* interpretation. Were one to apply the precautionary principle in this sense to the precautionary principle, the principle itself would have to be rejected.

A counter-misinterpretation considers any use of the precautionary principle as an unwarranted, costly, unjustified approach to environmental protection, especially in comparison to existing approaches of risk assessment and management and the effect of tort liabi-

lity law. The precautionary principle is considered a threat to the foundation of technological progress because it would halt innovation and development.

Both interpretations are typical of the polarized debate. The more common or balanced interpretation is that the precautionary principle may be applied when uncertainties are so great that it is impossible to predict the impact of technological development with any degree of accuracy, there are good grounds to suspect danger, and yet policy decisions need to be made. More specifically, in cases of significant uncertainty, when there are both sufficient scientific grounds for suspecting that a new development may have a potential for causing large scale, serious, or irreversible harms, the precautionary principle simply judges that it is more prudent to err on the side of safety. This is obviously an extension of the Hippocratic principle to avoid doing harm, to minimize risk when attempting to do good.

Of course, possible harms can always be considered from more than one perspective. Where environmentalists may see potential harms to the environment from the introduction of a new technology, economists may envision potential harms to the economy from blocking introduction of the same technology. Indeed, taking economic development as the status quo rather than the natural world, economists and business leaders can easily appeal to the precautionary principle to limit technological regulation, claiming that false alarms cause more harm than failures to identify and act on potential dangers. In the face of arguments to this effect by, for example, Bjørn Lomborg (2001) and others, it has been argued that on balance "the evidence indicates that we are receiving substantial benefits from our response to environmental alarms" (Pascala, Bulte, List et al. 2003, p. 1188).

Normative, Balance of Proof, and Risk Perspectives

Further specification of the balanced or moderate interpretation is nevertheless required and has taken at least three forms. The normative specification calls for a number of concerns to be addressed in all cases where there are reasonable possibilities of large-scale, serious, or irreversible harms. This specification has been developed in response to a tendency to disregard a number of prudent concerns. It is important, for instance, to identify and constructively account for uncertainty and ignorance, to assure interdisciplinary perspectives, to evaluate a range of options, to take full account of the values and perspectives of all stakeholders, to assure regulatory independence, and to act on reasonable grounds for concern. In the case of potentially large-scale, ser-

TABLE 1

Type I and Type II Errors

Experimental Results	Worldly Reality	
	− (not harmful)	+ (harmful)
+ (harmful)	False positive Type I error	True
− (not harmful)	True	False negative Type II error

SOURCE: Courtesy of Poul Harremoës.

ious, or irreversible harms, it is further appropriate to choose robust solutions that are adaptable to changing circumstances, because initial decisions are necessarily going to be taken under significant uncertainty, ignorance, or ambiguity (European Environmental Agency 2001).

The balance of proof specification is based on consideration of the risk of a mistaken decision. It can be assumed that no procedure for identification and documentation of environmental harmlessness is certain. There will always be some degree of uncertainty, because of the statistical uncertainties associated with practical experiments and cognitive uncertainties regarding cause-effect relationships.

The distinction between Type I and Type II errors is relevant here as indicating two possible ways in which laboratory results can differ from real-world phenomena. As they occur in a court of law, the two errors are those of convicting an innocent person (Type I) and failing to convict a guilty person (Type II). In the first instance, laboratory experiments could reject a presumption of environmental safety regarding some new technology (guilty verdict), when in fact it is safe (or innocent). This type of mistake or error is known as a *false positive*. In the second instance, laboratory results could fail to reject a presumption about the environmental safety of a new technology (judge the technology not guilty), when in fact it is unsafe (guilty). This is known as a *false negative*. See Table 1. As Kristin Shrader-Frechette has argued, the dangers of Type I errors are risks to industry (and thus economic risks to the public), whereas the danger of Type II errors pose risks to the environment (and thus health risks to the public).

Insofar as the legal system places its emphasis on avoiding Type I errors (false convictions), it is necessarily subject to Type II errors (false acquittals). In a similar manner, insofar as science is more concerned to avoid false assertions (that X causes harm when it does

not) than false denials (that X is does not cause harm when it does), because it is denials that can be falsified by experiment whereas assertions can never be fully confirmed by experiment, then science may be said to have a bias toward letting guilty technologies go free. From this perspective, the precautionary principle promotes shifting the balance of proof from concern with avoiding false convictions to avoiding false acquittals. In medicine, too, physicians have traditionally been concerned first and foremost with avoiding treatments that might harm.

In the regulation of developments that may be harmful to humans or the environment, the standard with respect to choice of acceptable types and levels of error may be thus reasonably quite different from those acceptable in science or in criminal courts. Levels of proof may be graded as follows: "vague, circumstantial, substantial, beyond reasonable doubt, certain." The required level of proof must be determined in relation to the potential harm and the claimed benefits of the activity in question. Cases of potential large-scale, serious, or irreversible harms may justify setting the level of proof at a lower level than "beyond reasonable doubt." In the European Union, "reasonable grounds for concern" is suggested as level of proof for invocation of the precautionary principle with regard to the regulation of chemicals and technological activities (European Commission 2000, p. 9).

These issues are also important to the question of who shall carry the burden of proof. Should the producer, manufacturer, or importer, on the one side? The government, on another side? Or the public, by means to liability suits, on still another side? In many cases, society has adopted the principle of prior approval (positive lists) before placing on the market certain products, such as drugs, pesticides, and food additives. Accordingly, the precautionary principle incorporates a proposed reversal of the burden of proof from the public to the proponent of any development that has a potential for large-scale, serious, or irreversible harms.

This is highly controversial, because the free market economy tends to be based on the principle that any economic activities are permissible as long as they are legal and subject to tort liability for the recuperation of damages. Opponents to the precautionary principle consider any restriction of this economic liberty as detrimental to technical and economic development. In the case of development of new chemicals and genetically modified organisms (GMOs), industry tends to consider such developments potentially so beneficial to society that industry should not have to bear the costs of a greater burden of proof; instead, liability should be invested in society as a whole.

The risk specification involves a comprehensive risk assessment in accord with the standards established in this field. The normative and balance of proof specifications may be included in the process. But a technical risk assessment is assumed to be more scientific and objective, involving as it does hazard identification, dose-response assessment, exposure assessment, and risk characterization (Environmental Protection Agency 1997, Lewalle 1999).

Social studies of science have, however, provided grounds for questioning the complete objectivity of such procedures, which are always undertaken by human beings with their own interests and perspectives. This is why the normative approach explicitly insists that a wide range of stakeholder values and perspectives be considered from the beginning even in framing the issue.

Subsequent risk management involves risk evaluation, emission and exposure control, and risk monitoring, plus risk communication. Risk evaluation and the regulation of emissions and exposures is where political and ethical choices come most obviously into play. What are the values and perspectives to be considered, and what is an acceptable risk? The European Commission communication on the precautionary principle, for instance, explicitly states that "the protection of public health should undoubtedly be given greater weight than economic considerations" (2000, p. 19).

Supplementary Principles

Invoking the precautionary principle nevertheless requires other principles to be considered. The European Commission has named five of these.

- *Proportionality*. Any decision is required to be proportional—that is, even a preliminary invocation of the precautionary principle must consider the balance between the pros and cons of a precautionary action, accounting for all aspects known at the time. In proportionality all concerns may count, not only consequences, but also deontological concerns, like duties, rights, and considerations of justice.

- *Non-discrimination*. Invocation of the precautionary principle means that comparable situations should not be treated differently, unless there are objective grounds for doing so.

- *Consistency*. Measures should be similar to previously adopted measures in similar circumstances.

- *Pros and cons* of action versus lack of action. Even in a provisional invocation of the precautionary principle, an analysis should be made of the factors pointing in favor versus against action or no-action.

- *Scientific development*. It is an essential part of the invocation of the precautionary principle to initiate research and monitoring in order to reduce the uncertainty and ignorance that cause the invocation.

Precautionary Principle Implementation

The means by which the precautionary principle should be implemented also have been the focus of much debate. Implementation must be related to other significant developments associated with risk assessment and management, and principles of good governance.

The tendency is to employ *participatory, discursive, and adaptable* procedures. The participatory processes require the participation at an early stage of all relevant stakeholders, as well as an ongoing discourse with stakeholders for the duration of the project. Adaptive procedures are the logical consequence of the fact that uncertainty and ignorance prevail in the decision making. Accordingly, it has to be publicly admitted that any decision could be false and susceptible to change in the light of new information obtained from research and monitoring.

Concrete regulatory actions can take many forms, from initiation of research and monitoring in order to decrease uncertainty and ignorance, to outright ban of the activity in question. Consider, for example, the case of endocrine disrupters.

Endocrine disrupters are natural hormones, which may be discharged in large quantities (such as the female hormone, estrogen, large amounts of which are discharged in wastewater, in large part due to increased excretion of residues from use of contraceptive pills), or hormone-like, artificial substances with a similar effect (such as Tributyltin [TBT], which is used in antirust paint on boats).

It has been demonstrated that increased concentrations of endocrine disrupters may cause sexual disturbances called *imposex* in fish and invertebrates in the aquatic environment (European Environment Agency 2001, p. 135–143). The first reaction was to increase research, because the evidence was insufficient to justify regulatory actions. With increasing evidence of serious effects, however, TBT has been banned for use on pleasure boats. Measures aimed at paints on commercial ships are forthcoming.

In the case of release of estrogen with wastewater, the question is whether scientifically-based suspicions of serious harms are sufficient to invoke the precautionary principle and demand either a ban of contraceptive pills or, more likely, to demand that wastewater treatment include the removal of endocrine disrupters before water is discharged into the environment. The key question is whether to invoke the precautionary principle immediately or wait for the results of a larger and more reliable risk assessment, which may be time consuming due to a need for more research.

Legal Status

Ambiguities remain regarding interpretation, application, and implementation. Ultimately such ambiguities will be reduced by case law precedence built up through court decisions. Several judgments already point in this direction.

Internationally, the Agreement on Sanitary and Photosanitary Measures and the Agreement on Technical Barriers to Trade have been brought before the World Trade Organization (1997). In Europe, an influential case is that of antimicrobial growth promoters brought before the European Court of Justice (1999).

A European Commission ban on certain antibiotics as growth promoters in animal production was upheld by the court with reference to the precautionary principle because of scientifically-based indications that widespread use of antibiotics might adversely affect the bacterial resistance to related antibiotics for humans (European Environment Agency 2001). However, the judgment also outlines the severe limitations and formal requirements associated with invoking the precautionary principle.

Ethics

The precautionary principle is not a scientific principle. It is an ethical principle in the sense that it makes a statement regarding values and the proper procedures for governance and due process. It is prudent to take action in spite of lack of complete scientific evidence when there are significant uncertainties, recognized ignorance, and ambiguity, combined with scientifically-based suspicions of large scale, serious, or irreversible harms. This is a deontological principle in the sense that

it prescribes an approach to prudent action in response to the awareness of the situation.

However, for the precautionary principle to be invoked there must first have been a preliminary risk assessment combined with a preliminary cost-benefit or cost-effectiveness assessment. In ethical terms, what must happen first is a preliminary utilitarian appraisal, the uncertainty of which may provide the justification for invoking the precautionary principle at the time of decision making. The challenge is to "avoid paralysis by analysis" (European Environmental Agency 2001, p. 181).

POUL HARREMOËS

SEE ALSO *Science Policy; Uncertainty.*

BIBLIOGRAPHY

Boehmer-Christensen, S. (1994). "The Precautionary Principle in Germany: Enabling Government." In *Interpreting the Precautionary Principle,* ed. Tim O'Riordan and James Cameron. London: Earthscan.

Cranor, C. F. (1999). "Asymmetric Information, The Precautionary Principle, and Burdens of Proof." In *Protecting Public Health and the Environment, Implementing the Precautionary Principle,* ed. Carolyn Raffensberger and Joel Tickner. Washington, DC: Island Press.

European Court of Justice. (1999). *Pfizer v. EC* (T 13/99) and *Alpharma v. EC* (T 70/99).

Foster, Kenneth R.; Paolo Vecchia; and Michael H. Repacholi. (2000). "Science and the Precautionary Principle," *Science* 288(5468): 979, 981.

Harremoës, Poul; David Gee; Malcom MacGarvin; et al. (2002). *The Precautionary Principle in the 20th Century: Late Lessons from Early Warnings.* London: Earthscan.

Harremoës, Poul. (2003). "Ethical Aspects of Scientific Incertitude in Environmental Analysis and Decision Making." *Journal of Cleaner Production* 11: 705–712.

Lomborg, Bjørn. (2001). *The Skeptical Environmentalist: Measuring the Real State of the World.* Cambridge, UK: Cambridge University Press.

Lewalle P. (1999). "Risk Assessment Terminology: Methodological Considerations and Provisional Results." *Terminology Standardization and Harmonization* II(1–4): 1–28.

Pascala, S.W.; E. Bulte; J.A. List; and S.A. Levin. (2003). "False Alarm over Environmental False Alarms." *Science* 301: 1187–1188.

Popper, Karl R. (2000). *Conjectures and Refutations: The Growth of Scientific Knowledge,* 5th rev. edition. London: Routledge and Kegan Paul.

Saltelli, Andrea, and Silvio Funtowicz. (2003). "The Precautionary Principle: Implications for Risk Management Strategies." *European Journal of Oncology* 2: 67–80.

Shrader-Frechette, Kristin S. (1991). *Risk and Rationality: Philosophical Foundations for Populist Reforms*. Berkeley: University of California Press.

Tickner, Joel A., ed. (2003). *Precaution, Environmental Science, and Preventive Public Policy*. Washington, DC: Island Press. A collection of 24 analyses, many of them contemporary case studies.

INTERNET RESOURCES

Environmental Protection Agency. (1997). "Risk Assessment and Risk Management in Regulatory Decision-Making." *The Presidential/Congressional Commission on Risk Assessment and Risk Management, Final Report* Volume 2, Glossary. Available from http://www.riskworld.com.

European Commission. (2000). "Communication from the Commission on the Precautionary Principle." Available from http://europa.eu.int/comm/.

European Environment Agency. (2001). "Late Lessons from Early Warnings: The Precautionary Principle 1986–2000." *Environmental Issues Series* 22. Available from http://reports.eea.eu.int.

European Union. (1992). Treaty of the European Union (The Maastricht Treaty). Available from http://www.europa.eu.int.

United Nations. (1992). "Rio Declaration on Environment and Development, United Nations, A/CONF.151/26, vol. I." Available from http://www.un.org/documents.

World Trade Organization. (1997). "Agreement on the Application of Sanitary and Phytosanitary Measures." Available from http://www.wto.org.

PREDICTION

• • •

Prediction is a central concept in science and politics, with important ethical implications. Its meanings, however, differ in important if subtle ways in these different realms, which can cause confusion about the appropriate relationship between science and society. Distinguishing clearly between the meanings is crucial for understanding and prescribing an appropriate role for science in political and ethical decision making.

Prediction as Confirmation

One way of looking at science, championed especially by the influential twentieth-century philosopher Karl Popper, is to view knowledge acquisition as a process of first making, and then testing, falsifiable hypotheses (Popper 1992). According to this view, the first of these two activities generates predictions about the consequences of the hypotheses, and the second confirms or refutes the predictions. Perhaps the emblematic example of this type of prediction is how Albert Einstein's

general theory of relativity, published in 1916, predicted that the path traveled by light would be bent by the force of gravity, and was later confirmed by Arthur Eddington's 1919 experiment in which this bending was observed during a solar eclipse. From this perspective, science is in its essence a prediction-generating-and-testing activity.

But the notion of prediction as inherent in science itself rests on a particular meaning of the word and a particular notion of what counts as science. Whereas common usage of the word prediction refers to the foretelling of future events, philosophers of science have viewed prediction as the process of deducing consequences from hypotheses independent of any sense of time. Indeed hypotheses that are temporally dependent for their correctness are said to have very little predictive power, because they are only true under limited circumstances.

The power of science, from this perspective, lies in its ability to make highly general, experimentally testable predictions about natural phenomena *independent of time or other contexts external to the phenomena*—light should always be bent by gravity. The temporal or locational power of a prediction (if one drops a paperweight from a particular desk at a particular time, one can predict that it will accelerate toward the center of the earth at thirty-two feet per second per second) is trivial compared to the more general, explanatory power of Isaac Newton's gravitational law (the attraction exerted by gravity between any two bodies is directly proportional to the masses of the bodies and inversely proportional to the square of the distance between them). The power of such generality is especially on display when scientific principles inform technological innovation, because the application of invariant laws ensures the uniform behavior of engineered devices from airplane wings and barometers to electronic circuits and clock pendulums.

Yet this view of science is problematic because it dismisses as nonscientific those disciplines that are not grounded in using experiments to test falsifiable predictions, such as most social sciences, paleontology and geology, system-level biology, ecology, and even some branches of the quantitative physical sciences dealing with complex, non-linear systems. Notably these are also the disciplines of science that most directly seek to understand the complexity of human experience and natural systems. More subtly, but of equal importance, generality in scientific prediction is almost always achieved through careful control of experimental conditions, or stripping away contextual *complications*. Newtonian mechanics, for example, operates as predicted in vacuums, in frictionless environments, and on rigid bodies, conditions that are often not met in the real world. Generality, as Nancy Cartwright (1983) argues, is often achieved at the expense of reality. Thus if science is in its essence a prediction-generating activity, where prediction means *logical inference,* then science can have only a limited capacity to inform about how the real world works.

Prediction as Foretelling

Yet it is in this real world that people make decisions about how to act. Indeed turning now to prediction in its more conventional sense, human decision making can be understood as an inherently predictive activity. Human action at every scale from the most mundane to the most ambitious aims at connecting the actions that one takes to some set of desired or expected outcomes. While such fields as philosophy, psychology, and economics have confronted the problem of how humans can make better decisions in light of existing knowledge and experience, it is only in the past several decades that science and technology have begun to offer the credible promise of actually predicting future events as an aid to decision making. Simultaneous advances in computer power, data acquisition technologies, and mathematical modeling are now being applied to predicting everything from economic trends and election results to the spread of diseases and the behavior of the ocean-climate system. This promise of a scientifically legitimate predictive capability has proven extremely attractive to decision makers interested in problems as diverse as selecting appropriate crops to plant in a specific region, assigning rates to insurance policies, and negotiating international environmental treaties. To serve such interests, each year billions of dollars are spent on science and technology aimed at improving predictive capabilities.

Prediction of future events is most familiar—and successful—in the area of short-term weather. Indeed there is a crucial connection between the familiarity and the success of weather predictions. Weather forecasters, who make upward of 10 million forecasts each year in the United States alone, are able to constantly test and refine their predictive skills because they can compare their predictions to the weather conditions that actually occur (*realizations*). At the same time, decision makers who use weather forecasts (everyone from individuals deciding whether to carry an umbrella to military generals deciding how to deploy their forces) are able to develop judgment about the reliability of weather forecasts based on their own multiple experi-

ences, and integrate this judgment into their decision processes. Moreover the production of weather predictions is linked to their use by a dynamic and sophisticated enterprise, including the mass media and companies that sell weather predictions, whose goal is to communicate the forecasts to those who might benefit from them. Finally many decisions based on weather predictions—for example, whether to evacuate a town, mobilize snow plows, or ground a fleet of airplanes—entail significant costs, and if predictions turn out to be wrong, those issuing them may be held accountable for their mistakes.

One or more of these attributes—the ability to test predictions, gain experience in their use, communicate them effectively, and hold people accountable—are missing from most other areas of scientific prediction. The consequences of this distinction are especially important in public policy, for example, where decision makers turn to scientists to predict future costs of a public program, the health consequences of particular levels of chemicals in the environment, the future prospects of an endangered species, the regional climate impacts that can be expected as a result of greenhouse gas emissions, or the behavior of buried nuclear waste. Such predictive challenges are characterized by the fact that the event or condition to be predicted plays out over decades or even centuries, and may represent a temporally and spatially unique set of conditions within an open system. The learning necessary to improve predictive accuracy in such cases is thus very difficult to acquire, because (a) predictions cannot be compared to actual outcomes; (b) causes of error in predictions are contingent on specific conditions and thus cannot be generalized; or (c) in many cases, both.

Obstacles to Prediction

A well-known example of the first kind of difficulty are predictions of long-term climate change, which are the source of considerable scientific and political debate, yet cannot be confirmed in the time frame within which policy decisions about climate change will have to be made. Representative of the second problem was the inability of economic models to predict the change in the relationship between energy consumption and economic growth in the United States that occurred after the Arab oil embargoes of the 1970s. Economists had understood economic growth to be tightly coupled to rising energy use, and thus predicted that the embargo would cripple the U.S. economy. Actual events showed that existing energy technologies could be mobilized to significantly boost energy efficiency, and thus economic activity, in the absence of increased consumption (Schurr 1984). While this insight is important and revealed why past predictions were wrong, it is unlikely to add much to the ability to predict future economic growth trends, because such trends are influenced by innumerable variables of which energy use is only one.

Moreover it is likely that accurate predictions of the behavior of some types—perhaps most types—of natural and social systems are impossible even in theory, due to their complexity, nonlinearity, and openness. Frequentist approaches to prediction, which rely on probabilistic characterizations of past system behavior to predict future behavior, founder on the fact that, for an open system, there is no reason to think that past behavior (even if it has been correctly characterized) will continue unchanged into the future. Deterministic approaches to prediction seek to avoid the pitfalls of frequentist strategies by using first principles (described mathematically) to ascertain causal relations between past, present, and future conditions. Yet determinism has to confront the practical reality that choices must always be made about which aspects of the system are worth characterizing, and which are not. For an open system, such choices are always made on the basis of incomplete knowledge. For example, long-term behavior of complex systems are often dependent on small variations in initial conditions, which means that knowledge of present conditions would have to be characterized with complete accuracy to insure accurate predictions, while errors in this characterization would tend to compound over time. This is why weather forecasts, which depend on knowing the present state of the atmosphere and then projecting future behavior, are accurate to a maximum of about two weeks.

Alternatives to Prediction

In this light, one may well ask the following: Given the limitations, what good is all this science aimed at predicting complex systems?; and If the ability to predict the future is really so limited, how are people going to be able to make successful decisions? These questions are not unrelated.

In considering the first question, the important point is that insight about how complex systems behave may be valuable for reasons other than an ability to predict the future. Charles Darwin's theory of natural selection is one of the most powerful, influential, and enlightening theories of modern science, yet it is predictive in neither the explanatory sense (in that it is not easily falsifiable) nor the temporal sense (in that it can reveal little about how species will evolve in the future). Yet it

offers enormous insight into how the natural world works, insight that can enhance understanding and appreciation of the interconnectedness of all things and inform decision making in light of this awareness. Similarly research on ecosystems, the climate system, social systems, and the connections among such systems can help explain causal relations among various system components, characterize past and present conditions, act as an alert to impending problems, and point toward potential solutions. But it is a very different thing to ask science to elucidate the general relations between greenhouse gas emissions and climate behavior (a difficult enough task), than it is to demand that science accurately predict how these relations will unfold on a regional level through the twenty-first century.

From this perspective it is important to recognize that, while decisions always carry with them some expectation of what the future will look like after the decision is made, good decisions—those that move in the direction desired by the decision makers—do not depend on accurate predictions. Numerous strategies exist for making effective decisions in the presence of scientific insight but the absence of accurate predictions.

One approach is prevention. For example, past experience shows that many areas of California are subject to earthquakes, and this knowledge has been sufficient to guide activities, such as better construction practices, that can reduce loss of life and property from earthquakes, without needing to predict them. Another approach is trial and error informed by understanding and monitoring. For example, the Federal Reserve Board modulates macroeconomic behavior in the United States by making small, incremental changes in interest rates and then seeing how those changes affect economic performance. Similarly, biologists and natural resource managers have increasingly been drawn to an adaptive, incremental approach for managing fragile ecosystems. The role of science here is to assess current conditions, suggest plausible cause-and-effect relations for guiding decisions, and then monitor the effects of actions taken to manage the system. This allows learning both from success and error, and it keeps the costs of errors relatively small, because decisions are incremental.

A third approach is to adopt hedging strategies for addressing future risks whose probabilities cannot be accurately predicted. For example, a *no-regrets* approach to global warming could mandate the adoption of energy efficient technologies whose lifetime costs are about the same as less-efficient technologies, and at the same time introduce reforms in land-use practices and insurance coverage that would reduce exposure to future climate events, whether or not they are caused by global warming. A fourth strategy is to introduce redundancy into the system, for example by combining geologic and engineering containment strategies for isolating nuclear waste, rather than depending on only one approach.

Political and Ethical Implications

Rather than making decisions in anticipation of a particular, predicted future, these sorts of approaches aim at building resilience into a system, a quality that allows for desired outcomes to be attained under a variety of plausible futures. Yet these approaches also demand that political commitments to action be made under conditions of uncertainty. This demand is not inherently problematic—indeed all decisions are made under conditions of uncertainty—but the rising expectation that science can provide accurate predictions may undercut the political motivation to actually take action, especially if such action entails political risk. The short-term benefits for both politicians and scientists of the predictive approach are clear: Politicians can avoid making tough decisions yet point to research as a step in the right direction, while scientists receive more funding to develop more accurate predictions. As a result, however, political discourse can shift from a discussion about the values and ethics that should inform action, to an endless debate about the technical merits of contesting scientific predictions. This dynamic is on stark display in a number of high-profile environmental controversies. The elusive promise of accurate scientific predictions may not only delay necessary action, but undermine the vitality of democratic debate.

DANIEL SAREWITZ

SEE ALSO *Global Climate Change; Incrementalism.*

BIBLIOGRAPHY

Cartwright, Nancy. (1983). *How the Laws of Physics Lie.* Oxford:, UK Oxford University Press.

Popper, Karl. (1992 [1935]). *The Logic of Scientific Discovery.* New York: Routledge. Hugely influential philosophical treatment of how reliable scientific knowledge is created.

Sarewitz, Daniel R.; Roger A. Pielke Jr.; and Radford Byerly Jr., eds. (2000). *Prediction: Science, Decision Making, and the Future of Nature.* Covelo, CA: Island Press. Contains expanded treatment of many of the arguments and examples in this article.

Schurr, Sam H. (1984). "Energy Use, Technological Change, and Productive Efficiency: An Economic-Historical Interpretation." *Annual Review of Energy* 9: 409–425.

PRESERVATION

SEE *Conservation and Preservation.*

PRESIDENT'S COUNCIL ON BIOETHICS

• • •

Since the 1970s, many governmentally sponsored advisory committees have been formed to offer advice about the ethical and political issues arising from biomedical research and biotechnology. In the United States, one of the most prominent of these is the President's Council on Bioethics (Council), which was established by President George W. Bush in November 2001. The work of the Council illustrates how hard it is to deliberate about the ethical issues provoked by modern science and technology in a political arena of partisan conflict and moral diversity. This is particularly difficult when the ethical and political discussion is influenced by the controversy over abortion and the moral status of human embryos. And yet despite these difficulties, the Council stands out as an attempt to promote a Socratic discussion in political debates about the ethical implications of science and technology.

Creation of the Council

On August 9, 2001, Bush gave a nationally televised speech on stem cell research. Stem cells are found in embryos, and have the power to grow into all of the specialized cells of the body (liver cells, muscle cells, brain cells, and so on). Some scientists believe that stem cells could be used to repair or replace the damaged cells that cause human diseases and disabilities such as Alzheimer's disease, Parkinson's disease, diabetes, and spinal cord injuries. But extracting stem cells from human embryos destroys the embryos, and Bush and others believe this is unethical because it means killing potential human life. This creates a conflict between the moral good in relieving suffering through medical research and the moral good in respecting potential human life. The political background for this controversy is the debate over abortion. Bush and many of his conservative supporters regard abortion as murder because they think that as soon as a human egg is fertilized, there exists a human being with a right to life.

The key issue in Bush's stem cell speech was whether federal funding should be provided to support human embryonic stem cell research. His decision was to allow such funding only for those stem cell lines that had been extracted before August 9, 2001. This would allow funding to support the research, but it would not promote future destruction of human embryos. He ended his speech by announcing that he would appoint a presidential council under the chairmanship of Leon Kass to study the ethical and political issues surrounding such biomedical research.

Kass received a bachelor's degree in biology and medical degree from the University of Chicago; he received a Ph.D. in biochemistry from Harvard University. After working at the National Institutes of Health (NIH) and the National Academy of Science (NAS), he taught for four years as a tutor at St. John's College (Annapolis, Maryland). Kass then returned to the University of Chicago as a professor in the Committee on Social Thought. At St. John's and the University of Chicago, he taught seminars on classic texts of philosophy, literature, and theology.

Kass was influenced by Leo Strauss, who was also a teacher at the University of Chicago and St. John's. Strauss and his students sought to revive the ancient philosophic wisdom of Socrates, Plato, and Aristotle and the ancient theological wisdom of the Bible. In promoting these traditions, the Straussians were critical of modern traditions of thought beginning with political philosophers such as Niccolò Machiavelli (1469–1527), Francis Bacon (1561–1626), and René Descartes (1596–1650). They were particularly skeptical about the philosophical project of Bacon and Descartes, which promoted a new science that would allow human beings to conquer nature. The Straussians feared that this scientific conquest of nature would become a willful quest for power unconstrained by moral or religious limits. When Kass expressed similar skepticism about modern science and technology in his published writings, he won the respect of U.S. political and religious conservatives who shared his suspicion that science was subverting moral and religious traditions. His writings warning against the dehumanizing effects of biotechnology attracted the attention of Bush's conservative advisors, which led to Kass's appointment as chair of the Council.

In consultation with Kass, Bush appointed seventeen other people to the Council: Elizabeth Blackburn, a professor of biochemistry at the University of California-San Francisco; Stephen Carter, a professor of law at Yale University Law School; Rebecca Dresser, a professor of law at Washington University School of Law;

Daniel Foster, a professor of medicine at the Southwestern Medical School of the University of Texas; Francis Fukuyama, a professor of international studies at Johns Hopkins University; Michael Gazzaniga, a professor of neuroscience at Dartmouth College; Robert George, a professor of politics at Princeton University; Mary Ann Glendon, a professor of law at Harvard University Law School; Alfonso Gomez-Lobo, a professor of philosophy at Georgetown University; William Hurlburt, a physician and a professor in the Program in Human Biology at Stanford University; Charles Krauthammer, a syndicated columnist for the *Washington Post*; Paul McHugh, a professor of psychiatry at the Johns Hopkins University Hospital; William May, a professor emeritus of theology at Southern Methodist University; Gilbert Meilaender, a professor of theology at Valparaiso University; Janet Rowley, a professor of medicine at the University of Chicago Medical School; Michael Sandel, a professor of government at Harvard University; and James Q. Wilson, a professor emeritus of management at the University of California–Los Angeles and a former professor of government at Harvard University.

Criticism, Conflict, and the Work Begins

Bush's critics thought the Council was biased because it included so many political and religious conservatives (such as Fukuyama, George, Glendon, Krauthammer, Meilaender, and Wilson), who would generally agree with Kass and Bush. But it soon became clear that there was genuine disagreement on the Council, and that some members of the Council (such as Blackburn, Gazzaniga, and Rowley) were strong proponents of biotechnology who rejected Kass's moral criticisms of science.

Scholars of bioethics complained that the Council had no members who were professional bioethicists. This was a deliberate move by Kass. At the first meeting of the Council, Kass indicated that he would lead the Council away from the methods and topics that dominate bioethics as a professional field of academic expertise. "This is a council on bioethics, not a council of bioethicists," he explained at the January 17, 2002, meeting. "We come to the domain of bioethics not as experts but as thoughtful human beings who recognize the supreme importance of the issues that arise at the many junctions between biology, biotechnology and life as humanly lived." He stated that the Council was not required to reach complete agreement, and he quoted from the president's executive order creating the Council: "The council shall be guided by the need to articulate fully the complex and often competing moral positions on any given issue and may, therefore, choose to proceed by offering a variety of views on a particular issue rather than attempt to reach a single consensus position." Kass doubted that complete agreement was likely in any event, because if the Council engaged in serious discussions of the competing human goods at stake in biomedical research and technology, disagreement would surely arise as different people would weigh those various human goods in different ways. For example, some might give more weight to the human good of respect for potential human life and less weight to the human good of relieving human suffering, while others might do the opposite. What was important, Kass insisted, was that every serious point of view be considered as part of a deliberative debate that would probably not reach consensus.

Kass was guiding the Council towards a tradition of ethical and political inquiry that goes back to Socrates, Plato, and Aristotle in ancient Greece. In it, thoughtful people work through the great questions of human life by debating the meaning of the human good—often using classic texts that illuminate fundamental alternatives—without expecting to reach final agreement on the answers to those questions. Kass had been initiated into that Socratic tradition during his years as a student and a teacher at the University of Chicago and St. John's College.

In the first meeting of the Council, Kass led the members in a discussion of a short story by Nathaniel Hawthorne—"The Birth-Mark" (1844)—about a scientist who unintentionally kills his beautiful wife while trying to surgically remove a slight birthmark on her cheek. The clear lesson of the story was that the scientific quest for perfection and power could be destructive in its lack of respect for human beings with all the imperfections of mortal creatures. Reporters and others at this first meeting remarked on the serious—even philosophic—tone of the Council's discussions. It was clear that Kass would turn the discussions of the Council into something like a college seminar on science, technology, and the meaning of human nature. Transcripts of the Council's meetings were posted on its Internet site along with copies of its formal reports. All of this material was designed by Kass to stimulate interested citizens across the nation into serious reflection on the moral character of modern science and technology.

And yet, as must be the case for any committee appointed by the president, the intellectual discussion of the Council could not be separated from partisan political debate. This became clear when the Council released its first formal report, which was on human cloning. The Council debated both reproductive cloning (or cloning to produce children) and therapeutic cloning (or cloning for biomedical research). Bush had argued vigorously for a legal ban on all forms of human cloning. The Council was unanimous in recommending a total ban on reproductive cloning, but it was divided on therapeutic cloning. Cloning human embryos could have therapeutic value in producing human stem cells that would be genetically identical to those of patients who need such cells for restoring damaged tissue, thus avoiding the problem of immune rejection. Some Council members thought this sufficient reason to approve therapeutic cloning. But others who believed that human life begins at the moment at conception considered the destruction of embryos to be murder, and so rejected therapeutic cloning. Still other members who thought that embryos were less than fully human but still deserved deep respect also rejected therapeutic cloning. Kass feared that a lack of consensus for a complete ban on therapeutic cloning would embarrass the president. To avoid this, he convinced a majority of Council members to recommend a four-year moratorium on the process.

When the Council's report was released, some members who opposed Kass's position complained that he had put pressure on three *swing voters*—Dresser, Fukuyama, and McHugh—to agree to the moratorium recommendation. Four of the members who voted to recommend federal funding for embryonic stem cell research—Blackburn, Gazzaniga, Foster, and Rowley—published a statement criticizing the Council's recommendations.

Early in 2004, the two-year terms for the Council members expired. Bush reappointed fifteen of the eighteen members for another two-year term. Carter and May resigned voluntarily. But Blackburn was dismissed. Bush and Kass filled the three vacancies with people who were like-minded to them—Benjamin Carson, a neurosurgeon at Johns Hopkins University; Peter Lawler, a professor of political science at Berry College in Georgia; and Diana Schaub, a professor of political science at Loyola College in Maryland. Prior to their appointments, Lawler and Schaub publicly stated their agreement with Kass's intellectual stance on biotechnology; Schaub had been a student of Kass's at the University of Chicago. Blackburn wrote articles protesting that her dismissal was politically motivated because she had opposed the positions taken by Kass and Bush. Kass responded that politics was not involved at all. The controversy was widely reported in newspapers and science journals as an indication of Bush's effort to promote his political goals among his science advisers.

Politics and Religion

Many contend that the Council's work is distorted by political pressure. In response, Kass argues that critics have not read the Council's reports carefully enough to see how fair it is in surveying arguments on all sides of every debate. Kass notes that journalists concentrate all their attention on the political implications of the Council's recommendations rather than the intricate reasoning supporting those recommendations. To avoid this criticism, which started with the first report, Kass designed the subsequent reports as surveys of opposing positions on moral issues in biotechnology that offer few specific recommendations. The Council has issued reports on using biotechnology to enhance human life, stem cell research, and regulation of reproductive technologies. These reports clearly favor Kass's position that biotechnology might endanger moral values. Yet the reports always include arguments on the other side of the debate. This is Kass's way of promoting serious and fair-minded discussion of the deep moral questions raised by modern science and technology.

Nevertheless bioethicists such as George Annas criticize Kass for leading a "neoconservative bioethics council" that pursues "a narrow, embryo-centric agenda" (Annas and Elias 2004, p. 19). Although Annas concedes that the moral status of human embryos is an important issue, he cites many other important topics in bioethics such as access to healthcare, dangerous commercialization of science and medicine, pricing of drugs, and bioterrorism. Annas also charges that neoconservatives such as Kass have failed to embrace a global bioethics based on human rights because embryos do not have the same status as human beings in international codes of human rights, such as the "Universal Declaration of Human Rights."

In the presidential campaign of 2004, John Kerry criticized Bush for not funding embryonic stem cell research because of religious beliefs not shared by most people. Bush used this issue to win votes from conservative Christians identified with the "religious right." Although religion is rarely mentioned in the council's meetings and reports, some of the members of the council are motivated by religious objections to biotechnology. Kass has written a book on the Bible in which he interprets the Book of Genesis as condemning science and technology as part of the "humanist dream" of "the city of man," particularly as depicted in the biblical story of the Tower of Babel (Kass 2003, pp. 219, 242–243). For Kass, this is part of the Bible's general warning that all civilization expresses the impious pride of human beings.

Council reports show extraordinary intellectual and moral rigor in probing the political and ethical issues arising in modern biotechnology. This reflects Kass's deep understanding of how science and technology arose in the seventeenth century as a project of modern political philosophers to give human beings power over nature. And yet the reports also show how intractable the ethical debate becomes when it is entangled in abortion politics, and in the controversy over whether embryos should be treated as fully human with the same moral standing as children or adults.

LARRY ARNHART

SEE ALSO *Bioethics; Bioethics Committees and Commissions.*

BIBLIOGRAPHY

Annas, George J., and Sherman Elias. (2004). "Politics, Morals, and Embryos." *Nature* 431: 19–20.

Blackburn, Elizabeth H. (2004). "A 'Full Range' of Bioethical Views Just Got Narrower." *Washington Post*, March 7, p. B02.

Blackburn, Elizabeth H., and Janet Rowley. (2004). "Reason as Our Guide." *PloS Biology* 2(4): 1–3. Criticisms of two reports of the Council from two members.

Hall, Stephen S. (2002). "President's Bioethics Council Delivers." *Science* 297(5580): 322–324. Report on the controversy surrounding the cloning report of the Council.

Kass, Leon. (1985). *Towards a More Natural Science.* New York: Free Press.

Kass, Leon. (2002). *Life, Liberty, and the Defense of Dignity: The Challenge of Bioethics.* San Francisco: Encounter Books.

Kass, Leon. (2003). *The Beginning of Wisdom: Reading Genesis.* New York: Free Press.

Kass, Leon. (2004). "We Don't Play Politics With Science." *Washinton Post*, March 3, p. A27.

Kass, Leon, and James Q. Wilson. (1998). *The Ethics of Human Cloning.* Washington, DC: AEI Press.

Kristol, William, and Eric Cohen, eds. (2002). *The Future Is Now: America Confronts the New Genetics.* Lanham, MD: Rowman and Littlefield. Collection of writings showing the conservative fear of biotechnology.

Lawler, Peter. (2004). "Restless Souls." *New Atlantis* 4(Winter): 42–46.

President's Council on Bioethics. (2002). *Human Cloning and Human Dignity: The Report of the President's Council on Bioethics.* New York: PublicAffairs.

President's Council on Bioethics. (2003). *Beyond Therapy: Biotechnology and the Pursuit of Happiness.* Washington, DC: Dana Press.

President's Council on Bioethics. (2004a). *Monitoring Stem Cell Research.* Washington, DC: Government Printing Office.

President's Council on Bioethics. (2004b). *Reproduction and Responsibility: The Regulation of New Biotechnologies.* Washington, DC: Author.

Rowley, Janet D.; Elizabeth Blackburn; Michael Gazzaniga; and Daniel Foster. (2002). "Harmful Moratorium on Stem Cell Research." *Science* 297 (5589): 1957. Criticisms of the cloning report from four members of the Council.

Schaub, Diana. (2003). "Slavery Plus Abortion." *Public Interest* 150(Winter): 41–46.

Weiss, Rick. (2004). "Bush Ejects Two From Bioethics Council." *Washington Post*, February 28, p. A06.

INTERNET RESOURCE

President's Council on Bioethics. "Transcript of Meeting of January 17, 2002." Available from http://www.bioethics.gov/transcripts/jan02/jan17.html.

PREVENTIVE ENGINEERING

• • •

Preventive approaches for the engineering, management, and regulation of modern technology distinguish themselves from their conventional counterparts by using design and decision processes that obtain the desired results while preventing or minimizing undesired effects. The term *preventive engineering* was coined by the author in 1989, and has since become a term of some importance in Canada.

Through the beginning of the twenty-first century, societies tended to direct technological and economic growth by means of a kind of design and decision-making that may be compared to driving a car by concentrating on its performance as indicated by the instruments on the dashboard and only occasionally glancing out to see where it is heading. The result has been many preventable "collisions" with human life, society, and the biosphere. The metaphor is appropriate because engineers, managers, and regulators make decisions whose consequences fall mostly outside of their domains of expertise, where they cannot "see" them. This leaves them little choice but to concentrate on obtaining the maximum possible desired outputs from the requisite inputs and to measure success in terms of performance values (output/input ratios such as efficiency, productivity, profitability, cost-benefit ratios, and gross domestic product [GDP]).

The result of non-preventive engineering is a system in which problems are created in every domain of specialization and left to be dealt with by other specialists in whose domain of competence they fall. In this way, an "end-of-pipe" approach has become institutionalized, making it very difficult to get to the roots of any problem, due to an intellectual and professional division of labor in every contemporary university, corporation, and government. The consequence is a labyrinth of technology: a patchwork of compensations that merely shift problems from one place to another.

It is arguable that the costs of maintaining and expanding this labyrinth are substantially undercutting gross wealth production. According to some calculations, net wealth production has been declining for decades (Daly and Cobb, 1989). It is also estimated that more than 90 percent of what the current (non-preventive) system extracts from the biosphere does not end up in salable products (Allenby and Richards 1994). The primary product would appear to be waste. Similarly, according to socio-epidemiology, workplaces have become one of the primary sources of physical and mental illness (Karasek and Theorell 1990). Unfortunately, many economic, social, and environmental policies address symptoms as opposed to root problems.

Because it has become widely recognized that most of the social and environmental consequences of any engineered product, process, or system are determined during the design phase, the present system helps to design a future society and its relations with the biosphere by omission, paying only peripheral attention to undesired consequences. Preventive approaches can turn this situation around.

Preventive approaches grew out of a study attempting to determine the extent to which the current engineering system is preventive in its orientation (Vandenburg 2000). A typical North American undergraduate engineering curriculum was examined by asking two questions: (1) How much do future practitioners learn about the way technology interacts with human life, society, and the biosphere? (2) To what extent do they learn to use this knowledge in a negative feedback mode to adjust design and decision-making to ensure that the desired results are achieved while simultaneously preventing or greatly minimizing undesired results? These questions were converted into two research instruments to score each component of every course. It was found that in the technical core, little or no reference was made to society and the biosphere, and even when there was, little or no use was made of it in a negative feedback mode. In the complementary studies component of the curriculum, little reference was made to modern science and technology, even though few aspects of modern societies are imaginable without them. Hence, students encounter some disciplines that are full of technology and little else, and others full of everything else and little technology. It is no wonder that successful design courses have been almost non-existent. This non-preventive orientation was also found in the curricula of other professions. The research also showed that the situation changed very little over the last few decades of the twentieth century.

The second phase in the study examined whether the above situation changes significantly after graduation, when practitioners enter specific areas of application (Vandenburg 2000). Using the same research instruments, the latest methods and approaches were scored in the areas of materials and production, energy, work, the urban habitat, and computer-based systems. The results showed that, except for a small cluster of methods and approaches, the same non-preventive situation prevailed. This exceptional cluster was then compared with its conventional counterparts.

The author's study made apparent that conventional approaches separate the economy of technology from the ecology of technology because they generally take the form of a two-stage approach. The economy of technology strips away all contexts (human life, society, and the biosphere), leaving only the requisite inputs and the desired outputs of a technology. From the process of converting requisite inputs into desired outputs, participating specialists abstract those aspects that are coterminous with their domains of competence. Alternatives can no longer be assessed in terms of their meaning and value for human life, society, and the biosphere because these specialists have no such knowledge. Instead, they must be assessed in terms of their contribution to the performance of the process as measured by performance values. Such accounting of outputs and inputs is essential for the effective use of scarce resources. However, it is insufficient to ensure that greater outputs are not partly or wholly achieved at the expense of human life, society, and the biosphere. In a second stage, specialists deal with undesired effects only to ensure that these are within the acceptable limits set out by applicable regulations. The two-stage process assumes that the technical and economic optimum achieved in the first stage is not made sub-optimal by the second stage. The first stage is seen as creating wealth and the second as dealing with unavoidable costs. Conventional approaches are fundamentally non-preventive and non-precautionary in their

structure. They are based on the production of *gross* wealth, not on optimizing the creation of *net* wealth by subtracting social and environmental costs. Nor do they ask the question how increased wealth correlates with well-being.

In contrast, the methods and approaches receiving much higher scores in the author's study integrate these two stages by adjusting design and decision-making to obtain the desired results while preventing, as much as possible, the undesired ones. These come closer to the way one normally drives a car, by looking out the windows and occasionally glancing at the dashboard. They are equipped with negative feedback regarding their consequences, while conventional approaches are not.

From this comparative study emerged a prescription based on the concept of preventive approaches. In 2002, the Canada Foundation for Innovation recognized this concept as one of twenty-five important recent innovations.

WILLEM H. VANDERBURG

SEE ALSO *Engineering Ethics*.

BIBLIOGRAPHY

Allenby, Braden R., and Deanna J. Richards. (1994). "Introduction." In *The Greening of Industrial Ecosystems*, eds. Braden R. Allenby and Deanna J. Richards. Washington, DC: National Academy Press.

Daly, Herman E., and John B. Cobb, Jr. (1989). *For the Common Good: Redirecting the Economy Toward Community, the Environment and a Sustainable Future*. Boston: Beacon Press.

Downey, James, and Lois Claxton, eds. (2002). *Inno' Va - Tion: Essays By Leading Canadian Researchers*. Toronto: Key Porter Books.

Karasek, Robert, and ?öres Theorell. (1990). *Healthy Work: Stress, Productivity, and the Reconstruction of Working Life*. New York: Basic Books.

Vanderburg, Willem H. (2000). *The Labyrinth of Technology*. Toronto: University of Toronto Press.

PRISONER'S DILEMMA

• • •

The Prisoner's Dilemma is one of the simplest yet most widely applicable situations studied in game theory. The Prisoner's Dilemma was discovered by Melvin Dresher and Merrill Flood at the Rand Corporation in 1950, but its name comes from the following story, which was supplied shortly afterward by the Princeton mathematician

Harold Kuhn. The story and its analysis have been used in different ways to draw forth ethical implications.

The Basic Story

Two men are caught committing an illegal act. If neither one confesses, there is enough evidence to ensure that each man will get one year in jail. If both confess, each one gets five years in jail. However, if one confesses and the other does not, the man who does not confess gets ten years in jail but the confessor who incriminates his partner gets off free. This is a special case of the normal form game illustrated in matrix form in Figure 1.

The *normal form* specifies a *strategy set* for each player and a *payoff* for each player as a function of the choice of strategy by each player. In this matrix player 1, the row player, chooses a row, and player 2, the column player, chooses a column. In general the Prisoner's Dilemma requires that $T > R > P > S$. This means that if both players defect, each one receives P, whereas if they both cooperate, each one receives $R > P$. If one player cooperates and the other defects, the defector gets T, which is the largest of the four numbers, and the cooperator gets S, which is the smallest. In Kuhn's example cooperate means "not confess," defect means "confess," and the four payoffs are $T = 0$, $R = -1$, $P = -5$, and $S = -10$. As a mnemonic device one can say that T is the temptation to defect, R is the reward for mutual cooperation, P is the penalty for mutual defections, and S is the sucker's payoff for cooperating when one's partner is defecting.

Note that whatever player 1 does, the best response for player 2 is to confess. This is the case because $T > R$ ($0 > -1$ in our example), and so player 2 does better to confess if player 1 cooperates and $P > S$ ($-5 > -10$ in our example), and so player 2 does better to confess if player 1 confesses. Therefore, if both players are self-interested and rational (i.e., they maximize their payoffs), both players will defect, and so their payoffs will be (P, P), which is $(-5, -5)$ in this example.

The Prisoner's Dilemma also can be described in an *extensive form*, which involves displaying the various moves of the players, as well as the payoffs, using a *game tree*, as is shown in Figure 2.

Perhaps the most important application of the Prisoner's Dilemma is to increase one's understanding of the role of market competition in promoting efficiency, growth, and material wealth. Although traditional economic theory posits the ability of markets to "get the prices right" and thus achieve allocational efficiency, a more important effect of market competition is to subject producers in the same industry to Prisoner's

FIGURE 1

The Normal Form of the Prisoner's Dilemma

	Cooperate	Defect
Cooperate	R, R	S, T
Defect	T, S	P, P

SOURCE: Courtesy of Herbert Gintis.

Dilemma–like situations in which mutual defection means that each producer chooses to produce high quality at a low price.

Consider, for instance, an industry with two firms. If they cooperate, they will choose a common price that maximizes total profits (the so-called *monopoly price*) and split total sales. However, each has an incentive to undercut the other's price to increase its own profits by taking sales away from the other. Thus, each producer will "defect" by charging the competitive price no matter what the other producer does (Gintis 2000). In effect, market competition, at least when it is working properly, *disciplines* producers, forcing them to act in the public interest.

The Public Goods Game

When there are n players, the Prisoner's Dilemma is known as the Public Goods Game, which is described as follows. Suppose a team of n players can each contribute an amount b to the group at a cost $c < b$ to each contributor. Each player decides independently of the others whether to cooperate (contribute) or defect (not contribute). Suppose at the end of the game the n players share their proceeds equally. Then if m of the players cooperated, each cooperator will earn mb/n, whereas each defector will earn $mb/n - c$. To see whether it pays to cooperate, consider one of the players, say, player A, and assume that $m - 1$ other players cooperate. By cooperating, player A earns $mb/n - c$, whereas by defecting, player A earns $(m - 1)b/n$. Comparing these two quantities, one can see that cooperating pays off more than defecting does precisely when $b > nc$. That is, a self-interested player A will cooperate only if A's share of the b that A contributes to the group, which is b/n, is greater than A's cost c. If $n = 2$ and $b > c > b/2$, the Public Goods Game becomes a Prisoner's Dilemma in which $T > R > P > S$ becomes $b/2 > b - c > 0 > b/2 - c$.

The Public Goods game was made famous by Garrett Hardin (1915–2003), whose article "The Tragedy of the Commons" (1968) argued that all people have a collective interest in maintaining the natural environment, yet if all people are self-interested, each one will overexploit the environment, even though each one hopes that others will act to preserve the environment. For instance, if ten fishers share a lake, the number of tons of fish that can be harvested season after season (the so-called *sustainable yield* of the lake) may be 1,000, which is 100 tons per fisher. However, each individual fisher may prefer to take 200 tons even if this endangers the yields in future years. In this case, cooperate means "take 100 tons of fish" and defect means "take 200 tons of fish." A fisher who is self-interested will hope others cooperate but will defect no matter what the other fishers do.

Other examples of social situations that can be couched as *n*-player Prisoner's Dilemmas are (a) pollution, in which each firm hopes the others cooperate (refrain from polluting a river) but defects no matter what the others do; (b) population control, in which each family hopes the other families limit the number of children they bear but bears as many children as it can no matter what the others do; (c) community participation, in which all benefit when all contribute to community projects (schools, roads, public parks, and gardens) but each community member would rather stay home and let the others do the work; and (d) a situation in which a group of farmers share irrigated water; each gains from diverting a large amount of water from their common pool, but all benefit when the water is used in moderation.

Perhaps the most important aspect of the Prisoner's Dilemma is that empirical investigation shows that in real life communities have a variety of resources available to moderate the use of the commons in a reasonable way (Yamagishi 1986, Ostrom 1990). Both state control and privatization of common resources have been advocated, but neither the state nor the market has been uniformly successful in solving common pool resource problems. This is the case because state officials have priorities that often conflict with those of the local resource users and because privatization often concentrates power and wealth in the hands of the individual or group to which the common goods are assigned.

In contrast to the proposition of the tragedy of the commons argument, common pool problems sometimes are solved by voluntary organizations rather than by a coercive state. Among those cases are communal tenure in meadows and forests, irrigation communities and other water rights, and fisheries. These cases often involve local

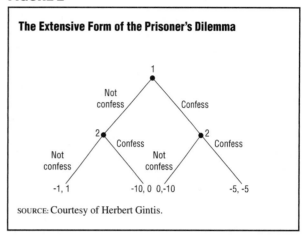

The Extensive Form of the Prisoner's Dilemma

SOURCE: Courtesy of Herbert Gintis.

self-organizing regimes that rely on implicit or explicit principles, norms, rules, and procedures rather than the command and control of a central authority.

If agents were truly self-interested, it is not clear how such self-organization could work effectively. However, the fact is that when people play the Prisoner's Dilemma in the laboratory for real money, they very often prefer to cooperate rather than defect as long as their partners cooperate as well (Kiyonari, Tanida, and Yamagishi 2000). Thus, people are generally not well described by the self-interest principle, a fact that has opened up a new research area in human behavior in recent years (Gintis, Bowles, Boyd, and Fehr 2004). This human tendency to cooperate lies at the root of self-organized solutions to common pool resource problems.

Ethical Implications

The Prisoner's Dilemma has important implications for ethical theory. It shows, for instance, that the philosopher Immanuel Kant's (1724–1804) categorical imperative is at best highly ambiguous and at worst fatally flawed. The categorical imperative states that one ought to "act according to that maxim which the actor would at the same time will to become a universal law" (*Critique of Practical Reason*, 1788). In the Prisoner's Dilemma each party would prefer that cooperating were a universal law because in that case the mutually desired outcome would be attained.

However, only in very special cases do players coordinate on the mutual cooperation outcome, and almost never does the duty to cooperate seem to be a defensible ethical commitment. For instance, producers in the same industry who cooperate on Kantian grounds would harm a market economy by colluding to maximize profits at the expense of the public. Similarly, if a person believes that his or her partner will defect, the first person nevertheless is obliged by the categorical imperative to cooperate. Although cooperating in this case may be a nice thing to do ("turn the other cheek"), it would be difficult to defend as a moral duty.

Of course, Kantian ethics is not the only ethical theory that is compromised by game theory in general and by the Prisoner's Dilemma in particular. Utilitarianism also suggests that people act to maximize the sum of utility. In the case of the Prisoner's dilemma this means that each player should cooperate no matter what the other player does. This also lacks plausibility as a general ethical principle.

HERBERT GINTIS

SEE ALSO *Decision Theory; Game Theory; Rational Choice Theory*.

BIBLIOGRAPHY

Gintis, Herbert. (2000). *Game Theory Evolving*. Princeton, NJ: Princeton University Press.

Gintis, Herbert; Samuel Bowles; Robert Boyd; and Ernst Fehr. (2004). *Moral Sentiments and Material Interests: On the Foundations of Cooperation in Economic Life*. Cambridge, MA: MIT Press.

Hardin, Garrett. (1968). "The Tragedy of the Commons." *Science* 162: 1243–1248. Hardin's influential argument is expanded in John A. Baden and Douglas S. Noonan, eds., *Managing the Commons*, 2nd edition. (Bloomington: Indiana University Press, 1998), originally edited by Hardin and Baden (San Francisco, Freeman, 1977).

Kiyonari, Toko; Shigehito Tanida; and Toshio Yamagishi. (2000). "Social Exchange and Reciprocity: Confusion or a Heuristic?" *Evolution and Human Behavior* 21: 411–427.

Ostrom, Elinor. (1990). *Governing the Commons: The Evolution of Institutions for Collective Action*. Cambridge, UK: Cambridge University Press.

Poundstone, William. (1992). *Prisoner's Dilemma*. New York: Doubleday.

Yamagishi, Toshio. (1986). "The Provision of a Sanctioning System as a Public Good." *Journal of Personality and Social Psychology* 51: 110–116.

PRIVACY

• • •

Discussions about privacy are intertwined with the use of technology. The publication that began the debate about privacy in the Western world was occasioned by the introduction of the newspaper printing press and photography. Justices Warren and Brandeis wrote their article on privacy in the *Harvard Law Review* (Warren and Brandeis 1890) partly in protest against the intrusive activities of the journalists of those days. They argued that there is a "right to be left alone" based on a principle of "inviolate personality." Since the publication of that article the debate about privacy has been fueled by claims for the right of individuals to determine the extent to which others have access to them (Westin 1967) and claims for the right of society to know about individuals.

The Nature of Privacy Claims

Inspired by subsequent developments in U.S. law, a distinction can be made between (1) constitutional privacy or decisional privacy and (2) tort privacy or informational privacy (DeCew 1997). The first refers to the freedom to make one's own decisions without interference by others in regard to matters seen as intimate and personal, such as the decision to use contraceptives. The second is concerned with the interest of individuals in exercising control over access to information about themselves.

Statements about privacy can be either descriptive or normative, depending on whether they are used to describe the way people define situations and conditions of privacy and the way they value them or are used to indicate that there ought to be constraints on the use of information or information processing. Informational privacy in a normative sense refers typically to a nonabsolute moral right of persons to have direct or indirect control over access to (1) information about oneself, (2) situations in which others could acquire information about oneself, and (3) technology that can be used to process information about oneself.

Privacy Accounts

Functionalist accounts of privacy argue that privacy serves other values (such as security or autonomy) and that its importance therefore should be explained in terms of those other values. Reductionist accounts argue that privacy claims are really about something else, such as property. Intrinsicalist accounts argue that privacy is valuable in itself (Rössler 2004).

The scarcity account (Fried 1970, Rachels 1984) claims that privacy creates a scarcity of information that allows people to be selective in determining which information they share with whom. In this way one can distinguish between persons with whom one chooses to be close, not so close, or not close at all. On a utilitarian account (Posner 1981) privacy norms are valuable if and insofar as they support valuable social institutions, practices, or actions. Their justification is therefore utilitarian. The moral self-ownership account (Reiman 1984) observes that environments of intensive surveillance and monitoring, such as prisons and mental asylums, convey the message to inmates that they no longer belong to themselves but are owned by the institution. Privacy norms convey the opposite message to individuals: that they own themselves. Autonomy accounts (Benn 1984) emphasize that privacy provides individuals with the autonomy to decide to be unobserved and the discretion to choose to whom to disclose which facts about themselves. Spying and accessing information about persons preempt their autonomous decisions in this respect. A moral autonomy account (Kupfer 1987), in contrast, argues that privacy serves moral autonomy, a second-order autonomy or an autonomy of self-concept. Only when one has a certain amount of control over who has access to oneself can one live a full-fledged moral life in the sense that one feels free to experiment, make mistakes, and criticize oneself. The gaze of others compromises the strong evaluation perspective, which is essential for moral autonomy and for which human beings have a basic capacity. Intimacy accounts (Gerstein 1978, Inness 1992) highlight the importance of intimate relations in human lives. Intimacy seems possible only if information associated with certain types of activities and relations is not widely accessible. A human dignity account (Bloustein 1964) maintains that privacy expresses respect for human dignity and the integrity of a person. According to a property account (Thompson 1975), privacy claims are claims of ownership of personal information and should be rendered as such.

More recently a type of privacy account has been proposed that acknowledges that there is a cluster of related moral claims (cluster accounts) underlying appeals to privacy (DeCew 1997, Van den Hoven 1999, Nissenbaum 2004).

The following types of moral reasons for the protection of personal data and for providing direct or indirect control over access to those data can be distinguished.

1. Prevention of information-based harm. Unrestricted access by others to one's passwords, characteristics, and whereabouts can be used to harm the data subject in a variety of ways.

2. Informational inequality. Personal data have become commodities. Individuals are usually not in a good position to negotiate contracts about the use of their data and do not have the means to check whether partners live up to the terms of the contract. Data protection laws aim at establishing fair conditions for drafting contracts about personal data.

3. Informational injustice and discrimination. Personal information provided in one sphere or context (for example, health care) may change its meaning when used in another sphere or context (such as commercial transactions) and may lead to discrimination and disadvantages for the individual.

4. Encroachment on moral autonomy.

These formulations all provide good moral reasons for limiting and constraining access to personal data and providing individuals with control over their data.

Technology

Information and communication technology has introduced a vast array of possibilities for linking, coupling, and merging databases. Internet searches are logged and can be charted through the use of cookies and spyware. Telecommunications traffic and location data are used to fight crime and global terrorism. Transactional, logistical, and radiofrequency identification data and vehicle registration systems are used to streamline supply chains and improve traffic control. Biometrical data, identification data, and authentication data are used to authorize users and manage access. Profiling and data-mining techniques are used to extract the maximum amount of useful information from what is available (Tavani 2004).

Genetic information constitutes a special type of information about people. It is used not only in health care and health insurance but also in policing and forensics. Genetic information is perceived as constitutive of individual human beings.

Nanotechnology also gives rise to privacy concerns. Miniature recording devices provide almost limitless storage capacity. Ubiquitous software and new recording materials may allow almost anyone to capture data about almost anyone else everywhere and all the time, a state that has been referred to as nano-panopticism (Gutierrez 2004).

Neuroimaging techniques such as computerized axial tomography, positron emission tomography, and functional magnetic resonance imaging make it possible to visualize the inner working and structure of the brain. The images show rational thought, memory activity, and emotional activity in reaction to stimuli and can be used to show a panoply of individual characteristics, defects, malfunctions, and deviancies.

Law, Regulation, and Indirect Control over Access

Data protection laws are in force in almost all countries. The basic moral principle underlying these laws is the requirement of informed consent for processing by the data subject. Furthermore, processing of personal information requires that its purpose be specified, its use be limited, individuals be notified and allowed to correct inaccuracies, and the holder of the data be accountable to oversight authorities (Europa 2004). Because it is impossible to guarantee compliance of all types of data processing in all these areas and applications with these rules and laws in traditional ways, so-called privacy-enhancing technologies and identity management systems are expected to replace human oversight in many cases (Agre and Rotenberg 1997). The challenge with respect to privacy in the twenty-first century is to assure that technology is designed in such a way that it incorporates privacy requirements in the software, architecture, infrastructure, and work processes in a way that makes privacy violations unlikely to occur.

JEROEN VAN DEN HOVEN

SEE ALSO *Genethics; Geographic Information Systems; Information; Information Ethics; Information Society; Internet; Monitoring and Surveillance; Security; Sociological Ethics; Telephone.*

BIBLIOGRAPHY

Agre, Philip E., and Marc Rotenberg, eds. (1997). *Technology and Privacy: The New Landscape.* Cambridge, MA: MIT Press.

Benn, Stanley I. (1984). "Privacy, Freedom and Respect for Persons." In *Philosophical Dimensions of Privacy: An Anthology,* ed. Ferdinand David Schoeman. Cambridge, UK, and New York: Cambridge University Press.

Bloustein, E. (1964). "Privacy as an Aspect of Human Dignity: An Answer to Dean Prosser." *New York University Law Review* 39: 962–1007.

DeCew, Judith Wagner. (1997). *In Pursuit of Privacy: Law, Ethics, and the Rise of Technology.* Ithaca, NY: Cornell University Press.

Fried, Charles. (1970). *An Anatomy of Values.* Cambridge, MA: Harvard University Press.

Gerstein, Robert. (1978). "Intimacy and Privacy." *Ethics* 89: 76–81.

Inness, Julie C. (1992). *Privacy, Intimacy and Isolation.* New York: Oxford University Press.

Kupfer, Joseph. (1987). "Privacy, Autonomy and Self-Concept." *American Philosophical Quarterly* 24: 81–89.

Nissenbaum, Helen. (2004). "Privacy as Contextual Integrity." *Washington Law Review* 79: 101–139.

Posner, Richard A. (1981). *The Economics of Justice.* Cambridge, MA: Harvard University Press.

Rachels, James. (1984). "Why Privacy Is Important." In *Philosophical Dimensions of Privacy: An Anthology,* ed. Ferdinand David Schoeman. Cambridge, UK, and New York: Cambridge University Press.

Reiman, Jeffery H. (1984). "Privacy, Intimacy, and Personhood." In *Philosophical Dimensions of Privacy: An Anthology,* ed. Ferdinand David Schoeman. Cambridge, UK: Cambridge University Press.

Rössler, Beate, ed. (2004). *Privacies. Philosophical Evaluations.* Stanford, CA: Stanford University Press.

Tavani, Herman T. (2004). *Ethics and Technology: Ethical Issues in an Age of Information and Communication Technology.* New York: Wiley.

Thomson, Judith Jarvis. (1975). "The Right to Privacy." *Philosophy and Public Affairs* 4: 295–314.

Van den Hoven, M. Joren. (1999). "Privacy and the Varieties of Informational Wrongdoing." *Australian Journal of Professional and Applied Ethics* 1(1): 30–44.

Westin, Alan F. (1967). *Privacy and Freedom.* New York: Atheneum.

Warren, Samuel D., and Louis D. Brandeis. (1890). "The Right to Privacy." *Harvard Law Review* 4: 193–220.

INTERNET RESOURCES

Europa. (2004). Material on European Union data protection laws is available athttp://europa.eu.int/comm/internal_market/privacy/index_en.htm.

Gutierrez, Eva. (2004). "Privacy Implications of Nanotechnology." Available at http//:www.epic.org.

PROBABILITY

• • •

Basic Concepts of Mathematical Probability
History, Interpretation, and Application

BASIC CONCEPTS OF MATHEMATICAL PROBABILITY

Widely used in everyday life, the word *probability* has no simple definition. Probability relates to chance, a notion with deep roots in antiquity, encountered in the works of philosophers and poets, reflected in widespread games of chance and the practice of sortilege, resolving uncertainty by the casting of lots. The mathematical theory of probability, the study of laws that govern random variation, originated in the seventeenth century and has grown into a vigorous branch of modern mathematics. As the foundation of statistical inference it has transformed science and is at the basis of much of modern technology. It has exercised significant influence in ethics and politics, although not always with full appreciation of either its strengths or its limitations.

Thousands of scientists, engineers, economists, and other professionals use the methods of probability and statistics in their work, aided by readily available computer software packages. But there is no strong consensus on the nature of chance in the universe, nor on the best way to make inferences from probability, so the subject continues to be of lively interest to philosophers. It is also part of daily experience—the weather, traffic conditions, sports, the lottery, the stock market, insurance, to name just a few—about which everyone has opinions.

The use of probability in science and technology is often quite technical, involving elaborate models and advanced mathematics that are beyond the understanding of nonspecialists. High-profile controversies may hinge on oversimplification by advocates and the media, unexplored biases, or a lack of appreciation of the extent of uncertainty in scientific results. Yet policy decisions based on such flawed evidence may have far-reaching economic and social consequences. Awareness of the role of probability is thus essential for judging the quality of empirical evidence, and this implies a moral responsibility for citizens of a democratic society.

Although many different techniques of the theory of probability are now in use, they all share a set of basic concepts. It is possible to express these concepts without advanced mathematics, but the concepts themselves are deep, and the results often counterintuitive. Insight may thus require persistent pondering. This entry presents the basic concepts in concise form, using only elementary mathematics. Further details and many applications are found in a wide range of introductory textbooks, written on various levels of mathematical abstraction.

A Simple Example

Consider as a first example the probability that a newborn child is a boy. One approach would be to use the theoretical model shown in Figure 1. According to Mendelian genetics, sex is determined by whether the sperm carries the father's X or Y chromosome; the egg has one of the mother's two X chromosomes. In a cell division called meiosis the twenty-three pairs of human chromosomes segregate to form two haploid (unpaired complement) cells called gametes, each containing twenty-two autosomes and one sex chromosome. In fertilization the male gamete (spermatozoan) fuses with a female gamete (ovum) to form a zygote, a diploid (double complement) cell with one set of chromosomes from each parent, its sex determined by the father. Assuming

FIGURE 1-2

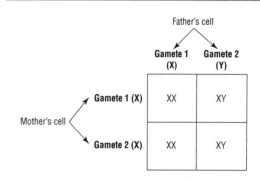

Figure 1: Probability That Newborn Child Is a Boy
THEORY (MENDELIAN GENETICS)

Probability (male) = 2/4 = 1/2 = 0.5

In a cell division called meiosis the 23 pairs of human chromosomes segregate to form two haploid cells called gametes, each containing 22 autosomes and one sex chromosome (X and Y for males, and X and X for females). In fertilization a male gamete (spermatozoan) fuses with a female gamete (ovum) to form a zygote, a diploid cell with one set of chromosomes from each parent, its sex determined by the father.

SOURCE: Courtesy of Valerie Miké.

Figure 2: Probability That Newborn Child Is a Boy
OBSERVED RELATIVE FREQUENCY

US National Center for Health Statistics
Data for United States 1991–2000

Probability (male) = .5118

Range in annual proportions: .5113–.5123
Average number of births per year: 3.98 million

SOURCE: Courtesy of Valerie Miké. Based on data in *Statistical Abstract of the United States* (1999: Table no. 93, and 2002: Table no. 68). Available from http://www.census.gov.

that the four possible outcomes are equally likely, two of them being XY, the probability that the child is male, written as Probability (male), can be defined as 2/4 = 1/2 = .5.

A second approach would look at the observed relative frequency of boys among the newborn, such as shown in Figure 2. In the ten-year period from 1991 to 2000 there were approximately 39,761,000 registered births in the United States. Of these 20,348,000 were boys, with a relative frequency of .5118. The annual proportions ranged between .5113 and .5123. One could say that the probability of a newborn child being male is .5118, or approximately .51.

FIGURE 3

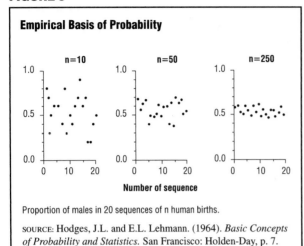

Empirical Basis of Probability

Proportion of males in 20 sequences of n human births.

SOURCE: Hodges, J.L. and E.L. Lehmann. (1964). *Basic Concepts of Probability and Statistics*. San Francisco: Holden-Day, p. 7.

Which answer then is correct? Most people would agree that the empirical result, based on such a large sample, has to override the model. In fact, the excess of boys among newborns has been observed throughout the world for centuries. The theoretical model is thus not an entirely correct representation of reality.

But what about the sex ratio in smaller samples? Figure 3 presents an experiment based on actual hospital records. The three graphs show the proportion of boys in 20 sequences each of 10, 50, and 250 consecutive births. Note that there is great variation in the sequences of 10, less for 50, and by 250 the proportions settle just above .5. Any one study yields only a single point, and the result from a small sample could be way off. For example, a researcher seeking to establish the proportion of boys among the newborn from a sequence of 10 could come up with a result of .2 or .9! In this example the approximate answer is already known, but in general this is not the case. The use of sample sizes too small to yield meaningful results is a serious problem in practical applications, as is the employment of inadequate theoretical models.

Two Definitions of Probability

Figures 1 and 2 illustrate two ways of defining the notion of probability in a mathematical context.

CLASSICAL DEFINITION. If there is a finite number of possible outcomes of an experiment, all equally likely and mutually exclusive, then the probability of an event is the number of outcomes favorable to the event, divided by the total number of possible outcomes.

This is the case shown in Figure 1, where the probability that a newborn infant is male is given as 2/4 = .5. Customary examples include tossing an unbiased coin or throwing a balanced die. Most situations, however, do not involve equally likely outcomes. Nor does this definition explain what probability is, it just states how to assign a numeric value to this *primitive* idea in certain simple cases.

STATISTICAL DEFINITION. The probability of an event denotes the relative frequency of occurrence of that event in the long run.

In Figure 2, the probability of a newborn infant being male is estimated to be about .51. This is also called the *frequentist* definition and is the one in common use. But it is not a fully satisfactory definition. What does "in the long run" mean? And what about situations in which the experiment cannot be repeated indefinitely under identical conditions, even in principle?

The Axiomatic Approach

A mathematically precise approach is provided by a third definition, the so-called axiomatic definition of probability, which incorporates the other two and is the foundation of the modern theory of probability. It begins with some abstract terms and then defines a few basic axioms on which an elaborate logical structure can be built using the mathematical theories of sets and measure. Probability is a number between zero and one, but nothing is specified about how to assign it. Assignment may be based on a model or on experimental data. Developments are valid if they follow from the axioms, as in other branches of mathematics, independently of any correspondence to phenomena of the physical world.

SAMPLE SPACE AND EVENTS. The framework for any probabilistic study is a *sample space*, often denoted by the letter S, a set whose *elements* represent the possible outcomes of an *experiment*. Subsets of S are called *events*, denoted by A, B, C, and so on. Consider an example of a finite sample space, and let S be the records of 100 consecutive births in a large urban hospital. Events are subsets of these records, defined by some characteristic of the newborn, such as sex, race, or birthweight. Assume further that this sample space of 100 births includes 51 boys, 9 of the infants were of low birthweight (LBW, defined as ≤ 2,500 grams), and 20 of the mothers smoked (actually, admitted to smoking) during their pregnancy; 3 of these mothers had LBW babies.

Hospital data of this type can be used, for example, to assess the relationship between smoking and low birthweight, important for the development of public health measures to lower the incidence of LBW. In a formal statistical design called a *case-control study*, a set of LBW babies is closely matched with controls of normal weight, to determine the proportion in each group whose mother smoked. Based on extensive data obtained from hundreds of hospital patients, this was the research method that led to the discovery that smoking is a cause of lung cancer. The case presented here is artificially simple, introduced to illustrate the abstract concepts that form the basis of mathematical probability.

THE ALGEBRA OF EVENTS. The relationships among events in a sample space can be represented by a Venn diagram, such as Figure 4. Let A = LBW babies, and let B = babies whose mother smoked. The event that A does not occur may be denoted by A' ("A prime" or "not A"), consisting of the 91 babies of normal birthweight; A and A' are called *complementary* events. The event that both A and B occur, the *intersection* of A and B, is denoted by $A \cap B$ ("A intersection B"), or simply AB, the set of 3 LBW babies whose mother smoked. The event that either A or B occurs (inclusive or), the *union* of A and B, is denoted by $A \cup B$ ("A union B"), the set of 26 babies who were LBW or their mother smoked, or both. Two events M and F are *mutually exclusive* if the occurrence of one precludes the occurrence of the other. Their intersection MF is the *null set* or *impossible event*, denoted by ϕ (the lower case Greek letter *phi*), where $\phi = S'$, consisting of none of the experimental outcomes. For example, if M and F are the sets of male and female newborns, respectively, then (setting aside the complications of intersexuality) their intersection is an impossible event.

THE AXIOMS OF PROBABILITY. The probability of an event A, denoted $P(A)$, is a number that satisfies the following three axioms:

Axiom 1: $0 \leq P(A) \leq 1$ for all events A in S

Axiom 2: $P(S) = 1$

Axiom 3: $P(A \cup B) = P(A) + P(B)$, if $AB = \phi$.

Stating the axioms in words, the probability of any event A in the sample space S is a number between zero and one, and the probability of the entire sample space is one (because by definition S contains all events). Furthermore, if two events are mutually exclusive (only one of them can occur), then the probability of their union (one or the other occurs) is the sum of their probabilities. These axioms are sufficient for a theory of finite

FIGURE 4-5

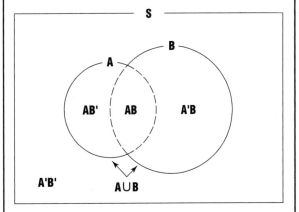

Figure 4: Events in a Sample Space

Venn diagram showing events in a sample space, the basis of the axiomatic approach to probability.

SOURCE: Courtesy of Valerie Miké.

Figure 5: Pascal's Triangle

```
                1   1
              1   2   1
            1   3   3   1
          1   4   6   4   1
        1   5  10  10   5   1
      1   6  15  20  15   6   1
    1   7  21  35  35  21   7   1
  1   8  28  56  70  56  28   8   1
1   9  36  84 126 126  84  36   9   1
↓   •   •   •   •   •   •   •   • ↓
```

Each number other than one is the sum of the two directly above it. The nth row represents the binomial coefficients C(n, r).

SOURCE: Courtesy of Valerie Miké.

sample spaces, and Axiom 3 can be generalized to more than two mutually exclusive events. Treatment of infinite sample spaces requires more advanced mathematics.

ELEMENTARY THEOREMS. The following results are immediate consequences of the axioms.

Theorem 1: $P(\phi) = 0$

Theorem 2: $P(A') = 1 - P(A)$

Theorem 3: $P(A \cup B) = P(A) + P(B) - P(AB)$.

The first two theorems state that the probability of the impossible event is zero, and the probability of "not A" is one minus the probability of A. Also called the *addi-*

tion theorem, the third statement means that elements that are in both sets should not be counted twice; the probability of overlapping events must be subtracted. In the hospital example, assuming that individual records are equally likely to be selected, so that the classical definition applies, $P(A) = 9/100 = .09$, $P(B) = 20/100 = .2$, and $P(AB) = 3/100 = .03$. Then the probability that a baby selected at random is either LBW or its mother smoked or both is $P(A \cup B) = .09 + .20 - .03 = .26$.

Conditional Probability and Independence

The two related concepts of conditional probability and independence are among the most important in probability theory as well as its applications. It is often of great interest to know whether the occurrence of an event affects the probability of some other event.

CONDITIONAL PROBABILITY. If $P(B) > 0$, the *conditional* probability of an event A given that an event B has occurred is defined as

$$P(A|B) = \frac{P(AB)}{P(B)}, \qquad (1)$$

that is, the probability of A given B is equal to the probability of AB, divided by the probability of B. For example, consider the conditional probability that a baby selected from the sample of 100 is LBW given that its mother smoked. Then $P(A|B) = .03/.20 = .15$. For nonsmoking mothers, represented by B', the probability of a LBW child is

$$P(A|B') = P(AB')/P(B') = .06/.80 = .075.$$

Rearranging equation (1), and also interchanging the events, assuming $P(A) > 0$, yields the *multiplication theorem* of probability:

$$P(AB) = P(A|B)P(B)$$
$$\text{and } P(AB) = P(A)P(B|A).$$

These relationships, obtained from the definition of conditional probability, lead to the definition of independence.

INDEPENDENCE. Two events A and B are said to be *independent* if the occurrence of one has no effect on the probability of occurrence of the other. More precisely, $P(A|B) = P(A)$ and $P(B|A) = P(B)$, if $P(A) > 0$ and $P(B) > 0$. The events A and B are defined to be independent if

$$P(AB) = P(A)P(B).$$

For example, one would expect a mother's smoking status to have no effect on the sex of her child. So selecting a hospital record at random, the probability of obtaining a boy born to a smoker would be the product of the probabilities, or $(.51)(.20) = .10$.

Assuming the independence of events is a common situation in applications. A prototype model is that of tossing a fair coin, with probability of heads $P(H) = .5$. Then the probability of two heads is $P(HH) = .5 \times .5 = .25$, of three heads is $P(HHH) = .5^3 = .125$, and the probability of n consecutive heads is $(.5)^n$. It follows from Theorem 2 that the probability of at least one tails, or equivalently, the probability of not all heads, is one minus the probability of all heads.

Taking a more real-life (although still oversimplified) example, consider the safety engineering of a space shuttle consisting of 1,000 parts, each of which can fail independently and cause destruction of the shuttle in flight. If each part has reliability of .99999, that is, its chance of failure is one in 100,000 launches, is that a sufficient safety margin for the shuttle? Application of the results above yields

$$P(\text{at least one component failure}) =$$
$$1 - (.99999)^{1,000} = .01,$$

that is, on average one in a hundred shuttle missions will fail, a somewhat counterintuitive result and an unacceptably high risk. With a component failure rate of one in 10,000, the chance of shuttle failure would be one in ten. Achievement of a failure rate of only one in a million per individual parts would be needed to lower the probability of a tragic launch to .001, one in a thousand.

BAYES'S THEOREM. The definition of conditional probability yields formulas that are useful in many applications, and one of these has become known as Bayes's theorem.

Given two sets A and B in a sample space S, with $P(A) > 0$ and $P(B) > 0$, Bayes's theorem can be written in its simplest form as

$$P(A|B) = \frac{P(A)P(B|A)}{P(A)P(B|A) + P(A')P(B|A')}. \qquad (2)$$

Here $P(A)$ is called the *prior probability* of A and $P(A|B)$ the *posterior probability*. Using the definition of conditional probability, the equation shows how to go from the known (or assumed) probability of an event A to estimating its probability given that the event B has occurred. Formula (2) can be generalized to n mutually exclusive events A_k that are *jointly exhaustive* (that is, one of them must occur and their union is S), and $P(A_k) > 0$, for any $k = 1, 2, \ldots, n$,

$$P(A_k|B) = \frac{P(A_k)P(B|A_k)}{P(A_1)P(B|A_1) + \ldots + P(A_n)P(B|A_n)}.$$

Bayes's theorem is sometimes referred to as a formula for finding the conditional probabilities of causes. As a somewhat oversimplified example in medicine, it may be used to diagnose (by selecting the highest posterior probability) which of n diseases A_k a patient has, given a particular set of symptoms B, when the prior probability of each disease in the general population is known, as is the probability of this set of symptoms for each of the candidate diseases. The use of conditional probabilities in medical diagnosis has been extensively developed in the field of biostatistics.

Bayes's theorem is also referred to as a formula for revising probabilities as new information becomes available. It is basic to a mode of induction called *Bayesian inference*, where, in contrast to classical or frequentist inference, previous information about a scientific problem is combined with new results to update the evidence. This approach pertains to an alternative, *subjective* interpretation of probability, in which the prior probability may be a *personal* assessment of the truth of the hypothesis of interest.

Random Variables and Probability Distributions

Research studies generally seek some quantitative information. In the present mathematical framework, these are numeric values associated with each element of the sample space, and the outcome is determined by the selection of elements in the experiment. The concepts involved are rather abstract. They are needed to connect the intuitive notion of probability with established mathematical entities on which standard operations can be performed to develop a mathematical theory.

RANDOM VARIABLE. The numeric quantity or code associated with each element of a sample space is called a *random variable*, usually denoted by the capital letters X, Y, and so on. Many different random variables can be assigned to the same sample space, depending on the aims of the study. A random variable may be *discrete* or *continuous*. The number of values assumed by a discrete random variable is finite or denumerably infinite (meaning that it can be put in one-to-one correspondence with the positive integers). A special case is the *binary* random variable, which has two outcomes (coded 1 and 0: heads/tails, success/failure, boy/girl). A continuous

FIGURE 6

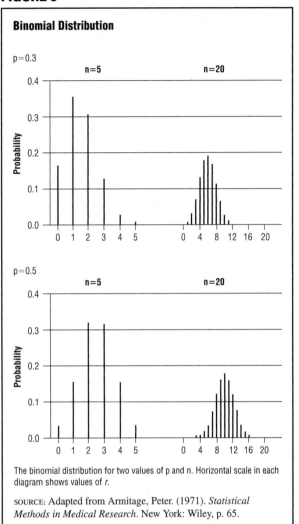

Binomial Distribution

The binomial distribution for two values of p and n. Horizontal scale in each diagram shows values of r.

SOURCE: Adapted from Armitage, Peter. (1971). *Statistical Methods in Medical Research*. New York: Wiley, p. 65.

random variable assumes values along a continuum (e.g., temperature, height, weight). The random variables associated with each baby in the sample space S of 100 hospital records include sex, race, birthweight, and mother's smoking status.

PROBABILITY DISTRIBUTION. The set of probabilities of the possible values of a random variable is called the *probability distribution* of the random variable. The sum of the probabilities is one, because it includes the entire sample space, and $P(S) = 1$. In the simplest case of only two possible outcomes, such as the sex of a newborn child, the distribution consists of $P(\text{male}) = .51$ and $P(\text{female}) = .49$.

PARAMETERS OF A DISTRIBUTION. Parameters are constants that specify the location (central value) and shape of a distribution, often denoted by Greek letters. The most frequently used *location parameter* is the *mean*, also called the *expected value* or *expectation* of X, E(X), denoted by μ (lower case *mu*). Others are the *median* and the *mode*. E(X) is the weighted average of all possible outcomes of a random variable, weighted by the probabilities of the respective outcomes. An important parameter that specifies the *spread* of a distribution is the *variance* of the random variable X, Var(X), defined as $E(X-\mu)^2$ and denoted by σ^2 (lower case *sigma square*). It is the expected value of squared deviations of the outcomes from the mean, always positive because the deviations are squared. The square root of the variance, or σ, is called the *standard deviation* of X, which expresses the spread of the distribution in the same units as the random variable. These concepts are illustrated below for two basic probability distributions, one discrete and the other continuous. When Greek letters are used, it is assumed that the parameters are known. In statistical applications their values are usually estimated from the data. The variance is important as a measure of how widely the observations fluctuate about their mean value, with a small variance providing a more precise estimate of the unknown "true" mean μ.

BINOMIAL DISTRIBUTION. Independent repetition of a *Bernoulli trial*, an experiment with a binary outcome (success/failure) and the same probability p of success, n times yields the *binomial distribution*, specified by the parameters n and p. The random variable X, defined as the number of successes in n trials, can have any value r between 0 and n, with probability

$$P(X = r) = C(n,r)p^r(1 - p)^{n-r}, \qquad (3)$$

where $C(n,r)$, the *binomial coefficient*, is the combination of n things taken r at a time, given by the formula

$$C(n,r) = \binom{n}{r} = \frac{n!}{r!(n - r)!}. \qquad (4)$$

(The symbol $n!$ is called "n factorial," the product of integers from 1 to n; $0! = 1$. For example, $3! = 1 \times 2 \times 3 = 6$.) Equation (3) is called the *probability function* of the binomial random variable. While random variables are generally denoted by capital letters, the values they assume are shown in lower case letters. (Elementary textbooks, however, do not always make this distinction.) For the binomial distribution $E(X) = np$ and $Var(X) = np(1-p)$.

Returning to the hospital example, assume that 30 of the 100 infants belong to a minority race, and five records are selected at random. Then X, the number of minority babies selected, could be 0, 1, ..., 5. The probability that there is no minority baby among the five is

$$P(X = 0) = C(5,0)(.3)^0(.7)^5 = .17.$$

$C(5, 0) = 1$, because there is only one outcome in which all five babies are white. To obtain the entire distribution, $C(5, r)$ needs to be calculated for the other values of r using formula (4). The binomial coefficients $C(n, r)$ can also be read off *Pascal's triangle*, shown in Figure 5. $C(5, r)$ is the fifth row, yielding the coefficients 1, 5, 10, 10, 5, 1. Applying these to equation (3) for all values of r, with $n = 5$ and $p = .3$, results in the binomial distribution shown in Figure 6, top row, left. The distribution for 20 babies is shown alongside, with expected value

$$E(X) = np = (20)(.3) = 6.$$

This means that on average one can expect 6 babies of a random sample of 20 to belong to a minority group. The second row in Figure 6 shows the binomial distribution for $p = .5$ and $n = 5$ and 20, respectively.

NORMAL DISTRIBUTION. It is seen that for $n = 20$ the distribution looks bell-shaped, and is *symmetric* even for the case $p = .3$, which is *skewed* for $n = 5$. In fact, it can be shown that the binomial distribution is closely approximated by the *normal distribution*, shown in Figure 7. The formula for the normal curve is

$$f(x) = \frac{1}{\sqrt{2\pi}\sigma}e^{-(x-\mu)^2/2\sigma^2},$$

the most famous equation of probability theory. To be read as "f of x," the symbol stands for "function of x," its numerical values obtained by computing the expression on the right for different values x of the random variable X. The distribution is completely determined by the parameters μ and σ, but also involves the mathematical constants $\pi = 3.142$ and $e = 2.718$, the base of the natural logarithm. Curves A and B have different means μ (4 and 8), but the same spread σ (1.0); B and C have the same mean μ (8), but different spreads σ (1.0 and .5). It can be seen that for each of these normal distributions most of the outcomes (actually about 95 percent) are within 2 standard deviations of the mean.

The normal random variable is continuous and can take on any value between minus and plus infinity. For continuous distributions $f(x)$ is called the *probability density function* of the random variable, which describes the shape of the curve. But for a continuum one can speak of the probability of the random variable X only for an interval of values x between two points; it is given by the corresponding area under the curve, obtained by integral calculus. The total area under the curve is one, by definition, as it includes all possible outcomes. The normal distribution plays a central role in statistics, because many variables in nature are normally distributed and also because it provides an excellent approximation to other distributions.

Two Basic Principles of Probability Theory

The most fundamental aspect of mathematical probability can be observed empirically as a fact of nature, and also proved with rigor. This phenomenon can be expressed in the form of two principles. They are given here in their simplest versions, to convey the essential result.

LAW OF LARGE NUMBERS. This laws hold that, in the long run, the relative frequency of occurrence of an event approaches its probability. It is illustrated by the empirical results of Figure 3. Stated more precisely: As the number of observations increases, the relative frequency of an event is within an arbitrarily small interval around the true probability, with a probability that tends to one. The *law of large numbers* connects observed relative frequency with the mathematical concept of probability, and has been proved with increasingly refined bounds on the true probability. A more general formulation pertains to the sample mean approaching the true mean, or expected value. If the occurrence of an event is denoted by 1 and its nonoccurrence by 0, then the relative frequency is the mean of the observations, which approaches the expected value p.

CENTRAL LIMIT THEOREM. This theorem states that, in general, for very large values of *n*, the sample mean has an approximate normal distribution. The theorem can be proved with great precision for a variety of conditions, without specifying the shape of the underlying distribution. Figure 6 suggests the result for the binomial distribution. A striking example is given in Figure 8, which shows the distribution of averages of 5 digits, selected at random from the integers between 0 and 9. This discrete random variable has a *uniform distribution*, where each outcome has the same probability .10. Yet the normal approximation is quite good already for this small sample size. The *central limit theorem* is a powerful tool for assessing the state of nature in a wide range of

FIGURE 7-8

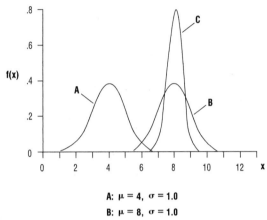

Figure 7: Normal Distribution

$$f(x) = \frac{1}{\sqrt{2\pi}\,\sigma}\, e^{-\frac{(x-\mu)^2}{2\sigma^2}}$$

μ = mean = median = mode
σ^2 = variance; σ = standard deviation

A: $\mu = 4$, $\sigma = 1.0$
B: $\mu = 8$, $\sigma = 1.0$
C: $\mu = 8$, $\sigma = 0.5$

The normal distribution for different values of μ and σ. The formula includes the constants $\pi = 3.142$ and $e = 2.718$, the base of the natural logarithm.

SOURCE: Courtesy of Valerie Miké.

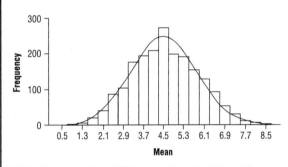

Figure 8: Central Limit Theorem

Distribution of means from 2,000 samples of 5 random digits with the approximating normal distribution.

SOURCE: Armitage, Peter. (1971). *Statistical Methods in Medical Research*. New York: Wiley, p. 87.

circumstances, with measures of uncertainty provided by the normal distribution.

Concluding Remarks

The concepts discussed here form the basis of the mathematical theory of probability, which—unlike the interpretation of probability—is not subject to controversy. The interested newcomer has a wide choice of textbooks as guides in further pursuit of the subject. The main criterion of selection should be comfort with the level of abstraction and the style of presentation: neither too terse nor too wordy. The purpose of symbol in mathematics is the unambiguous and universal expression of concepts. The use of symbol is an indispensable, welcome shorthand for those who understand; it should never be a hindrance to understanding.

Many ethical issues in science and technology require greater insight on the part of the public and call for better education concerning the extent of related uncertainties. But how does one promote understanding of a deep and complex notion such as chance and its myriad manifestations in everyday life? For the mathematical approach a good way is to start early: Encourage the young to play numbers games, to work on puzzles exploring the different ways things can happen, to confront logical paradoxes, and to savor the joy of insight—the aha! experience. Doing mathematics because it is fun enhances intuition and develops the habit of critical thinking, helping the child to grow into a self-confident adult always in search of understanding. But when is it too late? To play mathematical games the only requirement is to be young at heart.

VALERIE MIKÉ

SEE ALSO *Risk: Overview; Statistics; Uncertainty.*

BIBLIOGRAPHY

Anderson, David R., Dennis J. Sweeney, and Thomas A. Williams. (1994). *Introduction to Statistics: Concepts and Applications*, 3rd edition. Minneapolis/St. Paul: West Publishing. Contains several chapters on probability, with many examples; no calculus required.

Armitage, Peter. (1971). *Statistical Methods in Medical Research*. New York: Wiley. Includes a concise summary of the basic concepts of probability theory. Fourth edition, cowritten with Geoffrey Berry and J. N. S. Matthews published in 2001, Malden, MA: Blackwell Science.

Edwards, A. W. F. (2002). *Pascal's Arithmetical Triangle: The Story of a Mathematical Idea*. Baltimore: Johns Hopkins University Press.

Feller, William. (1950, 1966). *An Introduction to Probability Theory and Its Applications*. 2 vols. New York: Wiley. A classic text of probability; Vol. 1 requires only elementary mathematics.

Gardner, Martin. (1978). *Aha! Aha! Insight*. New York: Scientific American. A popular collection of mathematical games, one of a series by this author, illustrated with cartoons; includes a chapter on combinatorics, a basic component of probability.

Gnedenko, Boris V. (1962). *The Theory of Probability*, trans. B. D. Seckler. New York: Chelsea Publishing. Translation of *Kurs teorii veroyatnostei*. A classic text on an advanced level by a leading Russian mathematician of the twentieth century.

Hacking, Ian. (2001). *An Introduction to Probability and Inductive Logic*. Cambridge, UK: Cambridge University Press. Introductory textbook for students of philosophy, with many examples from everyday life.

Hodges, J. L., Jr., and E. L. Lehmann. (1964). *Basic Concepts of Probability and Statistics*, 2nd edition. San Francisco: Holden-Day. Textbook in a more mathematical context, but does not require calculus. Second edition published 1970.

Kolmogorov, Andrei N. (1956). *Foundations of the Theory of Probability*, 2nd English edition, trans. Nathan Morrison. New York: Chelsea Publishing. Translation of *Grundbegriffe der Wahrscheinlichkeitsrechnung* (1933). The original work on the axiomatic basis of probability theory.

Kotz, Samuel; Norman L. Johnson; and Campbell B. Read, eds. (1982–1999). *Encyclopedia of Statistical Sciences*. 9 vols. plus supp. and 3 update vols. New York: Wiley.

Kruskal, William H., and Judith M. Tanur, eds. (1978). *International Encyclopedia of Statistics*. 2 vols. New York: Free Press.

Laplace, Pierre-Simon de. (1812). *Théorie analytique des probabilités* [Analytic theory of probability]. Paris: Courcier. First comprehensive treatment of mathematical probability.

Laplace, Pierre-Simon de. (1951). *A Philosophical Essay on Probabilities*, trans., from the 6th French edition, Frederick Wilson Truscott and Frederick Lincoln Emory. New York: Dover. Translation of *Essaie philosophique sur les probabilités*, 1819. Addressed to the general public, included as the introduction to the third edition (1820) of the work listed above.

Riordan, John. (2002 [1958]). *An Introduction to Combinatorial Analysis*. New York: Dover. A classic of combinatorial analysis, a branch of mathematics basic to probability.

Strait, Peggy Tang. (1989). *A First Course in Probability and Statistics with Applications*, 2nd edition. San Diego, CA: Harcourt Brace Jovanovich. A careful, thorough presentation of mathematical concepts and techniques for the beginner, with hundreds of examples from a wide range of applications.

Weaver, Warren. (1963). *Lady Luck: The Theory of Probability*. Garden City, NY: Anchor Books. A witty, engaging introduction to probability, addressed to the general reader.

HISTORY, INTERPRETATION, AND APPLICATION

It is often said that something is "probably the case" or "probably not the case." The word *probable* comes from the Latin *probabilis*, meaning commendable, which itself derives from *probare*, to prove. Indeed, the English *probable* and *provable* have the same etymologic origin. The scientific study of probability takes the everyday notions of recommending and approving and gives them strict definitions and systematic analysis, something that narrows their focus while enhancing their power to inform. Insight into related matters is essential in advanced technological societies where experts regularly give technical advice to a public that must then decide whether or not to accept it. This may involve the development of new government policies or actions to be taken by individuals, such as submitting to a new medical treatment.

But there are other complex issues to consider. It is generally understood that probability has something to do with chance, a concept of enduring fascination throughout history. While philosophers explore alternative interpretations of probability that lead to different modes of induction in science, there remains the enigma of the role of chance in the world. Given the theories of quantum physics and evolutionary biology proclaiming a universe of chance, how do these impact the fundamental questions of philosophy that sooner or later confront every thinking person: Who am I? Why am I here? How should I live my life?

Reflecting in search of insight, it is important to distinguish between what is science and what is philosophy, and to differentiate between the speculations of philosophers—traditionally fraught with controversy—and the daily activities of practicing scientists. There is a need to understand the role of probability in science and technology, as well as its relation to the perennial questions of human existence. After a brief sketch of the history of probability, the present entry offers some thoughts on this vast and profound subject, concluding with a discussion of the applications of probability at the start of the twenty-first century.

Highlights of History

This quick survey of the history of probability is presented in two sections, beginning with the evolution of mathematical concepts and then turning to their use in philosophical speculation.

THE RISE OF MATHEMATICAL PROBABILITY. There are earlier records of mathematics applied to games of chance, but the beginning of the theory of probability is generally identified with the 1654 correspondence between the two French mathematicians Blaise Pascal (1623–1662) and Pierre de Fermat (1601–1665) concerning the so-called problem of points in gambling. The question was how to divide the stakes between two players who part before completing the game. To arrive at the solution, Pascal introduced the binomial distribu-

tion for $p = .5$ and found the coefficients by means of the arithmetical triangle, a curious numeric structure now named after him. In 1657 the Dutch mathematician Christiaan Huygens (1629–1695) published his monograph *De Ratiociniis in Ludo Aleae* (Reasoning on games of chance), the first printed mathematical treatment of games of chance. In these games equally likely outcomes, such as the six faces of a balanced die, were the assumption that led to the classical definition of probability. The first major work devoted to probability theory was *Ars Conjectandi* (The art of conjecturing) by the Swiss mathematician Jakob (Jacques) Bernoulli (1654–1705), published in 1713. It contained the first form of the law of large numbers.

About this time in England, attention focused on the by then established systematic recording of births and deaths and related practical issues of insurance and annuities. Relative frequency was applied to mortality data by the merchant John Graunt (1620–1674), whose *Natural and Political Observations … Made upon the Bills of Mortality* (1662) marked the beginning of actuarial science. The stability of observed ratios suggested the second, the statistical or frequentist, definition of probability. William Petty (1623–1687), physician and mathematician, coined the term *political arithmetic* in his quantitative analysis of social phenomena that would become the foundation of modern economics. Also working in England, the French mathematician Abraham de Moivre (1667–1754) wrote *The Doctrine of Chances; or, A Method of Calculating the Probabilities of Events in Play* (1718, 1738, 1756), another landmark in the history of probability. The second and third editions of the book include his discovery of the normal curve as the limit of the binomial distribution.

Important advances were made in the first part of the nineteenth century. The normal distribution, applied to measurement variations in astronomy, was studied by the French mathematician Pierre-Simon de Laplace (1749–1827), author of the first comprehensive work on probability, *Théorie analytique des probabilités* (1812; Analytic theory of probability). Laplace discovered and proved the earliest general form of the central limit theorem. The normal curve is also called the Gaussian distribution, after the German mathematician Carl Friedrich Gauss (1777–1855), who developed it as the law of errors of observations, in conjunction with the principle of least squares, in which it plays a key role. Least squares, a method for combining observations to estimate parameters by minimizing the squared deviations of the observations from expected values involving the parameters, became a basic tool in astronomy, geodesy, and a wide range of other areas.

Probability came to be used for the analysis of variation in itself, not as errors to be eliminated, in the social sciences and in physics and biology. The intense study of heredity triggered by Charles Darwin's (1809–1882) theory of evolution, spearheaded by his cousin Francis Galton (1822–1911), would lead to the new field of mathematical statistics around the turn of the twentieth century. The axiomatic foundation of the modern theory of probability was the work of the Russian mathematician Andrei N. Kolmogorov (1903–1987), published in 1933.

PROBABILITY AND PHILOSOPHY. The notion of probability dates back to antiquity, and beyond games of chance to questions of philosophy, of permanence and change, of truth and uncertainty, of knowledge and belief. The revival of interest in the thought of the ancients during the Renaissance brought about an interplay of intellectual currents with scientific discoveries that energized a renewed search for explanation and meaning. The role of chance was at the core of developments from the start.

Pascal posed a challenge to skeptics of his day in the famous "Wager" of his *Pensées*, published posthumously in 1670, in which the question of God's existence was to be answered as if by the toss of a coin at the end of life. Presenting arguments for betting that God exists, Pascal developed basic elements of decision theory concerning courses of action in the face of uncertainty.

The work of Isaac Newton (1642–1727), his universal law of gravitation and his synthesis of cause and effect explained by laws of physics in a fully determined universe, launched the era of modern science. Since then, reports of scientific advances have been at the forefront of public consciousness, dominant factors to be integrated into any cohesive worldview. Newton's system involved his concept of an omnipresent deity who maintains the motion of heavenly bodies, and this led to a lively natural theology (part of philosophy, as it does not have recourse to Revelation) in the eighteenth century. In contrast to the observed regularity of planetary orbits there was variability in human affairs, but here the stable patterns of long-run frequencies also seemed to imply design and purpose. The constant excess of males among the newborn was a recurring example.

In 1710 John Arbuthnot (1667–1735), physician and scholar, published an influential essay titled "An Argument for Divine Providence, Taken from the Constant Regularity Observed in the Births of Both Sexes."

He found that in the eighty-two consecutive years on record more boys than girls had been born in London. He reasoned that because boys were at greater risk of dying young as a result of their duties in the world, there was a need in a monogamous society for more boys to be born, and this was wisely arranged by Providence. His article contained the earliest example of a test of a statistical hypothesis, concluding that the observed result would be highly unlikely if in fact the true probability of a boy was one-half.

De Moivre aimed to show that probability had more consequential objects than the frivolous pastime of gambling, and in the second and third editions of *The Doctrine of Chances* argued for its serious mission in proving the existence of God. While chance produces irregularities, he wrote, it is evident that these are governed by laws according to which events happen, and the laws serve to preserve the order of the universe. We are thus led "to the acknowledgment of the great MAKER and GOVERNOUR of all; *Himself all-wise, all-powerful, and good*" (1756, p. 252).

One of the most famous documents in the history of science is "An Essay towards Solving a Problem in the Doctrine of Chances," by Thomas Bayes (1702–1761), an English clergyman also interested in probability. It is the first expression in precise, quantitative terms of one of the chief modes of inductive inference. The essay contains what is now called Bayes's theorem and is central to approaches known as Bayesian inference. The manuscript was published posthumously in 1763, with an introduction by the Reverend Richard Price (1723–1791). In delineating the importance of Bayes's achievement, Price suggested that his method of using the probabilities of observed events to compare the plausibility of hypotheses that could explain them is a stronger argument for an intelligent cause than the appeal to laws obtained from chance events proposed by de Moivre. More generally, as asserted by Price and explored by modern scholarship, Bayes's method in a sense evades the problem of direct induction posed by the Scottish philosopher David Hume (1711–1776), who rejected the very possibility of inductive reasoning. A Bayesian does not claim to justify any set of beliefs as uniquely rational. But having a belief structure that satisfies the axioms of probability, one's earlier *personal probability* (*degree of belief*) can be updated by new evidence in a coherent, reasonable manner. Bayes's method, the argument goes, provides a uniquely rational way to learn from experience.

In Germany, using results from England as well as his own extensive collection of data, Johann Peter

Süssmilch (1707–1767), military chaplain and mathematician, wrote the first analytic theory of population, *Die göttliche Ordnung in den Veränderungen des menschlichen Geschlechts, aus der Geburt, dem Tode, und der Fortpflanzung desselben erwiesen* (1741; The divine order in the fluctuations of the human race, shown by the births, deaths, and propagation of the same). Through his pioneering work in demography Süssmilch sought to discern in the detected patterns of population trends, in this natural order, the eternal laws of God.

As the use of probability expanded in the nineteenth century, so did philosophical concern with the problem of chance in a deterministic universe, with questions of causality, proof, natural law, free will. Speculation entered a new phase with the theory of evolution, when chance assumed a dominant role, to be enhanced by quantum theory in the early twentieth century. The debate continues with renewed vigor, in the light of new developments in cosmology, evolutionary biology, and other related disciplines.

Interpretation: A Commentary

The following discussion of various aspects of probability does not aim to be comprehensive or exhaustive. Rather, it offers some comments to stimulate thought and further exploration of this deep, complex subject.

OBJECTIVE VERSUS SUBJECTIVE PROBABILITY. Probability has a dual nature, recognized since its emergence in the seventeenth century. It may be *aleatory* (frequentist, from "dicing") or *epistemic* (pertaining to knowledge), also called *objective* or *subjective* probability. Objective probability takes a sort of Platonic view, assuming the existence of idealized states, represented by a mathematical model and estimated by observed relative frequency. Subjective probability is degree of belief, and it involves personal judgment.

Both interpretations are common in everyday use. The probability that a newborn child is a boy, which is .5 according to Mendelian genetics and .51 as observed relative frequency, provides two examples of objective or frequency-type probability. The subjective or belief-type may refer to any statements expressing some belief or opinion. It can be illustrated by the high-profile Terri Schiavo case of early 2005. A severely brain-damaged woman, on artificial nutrition and hydration for years, had her feeding tube removed by court order at the request of her husband but against the strong objections of her parents. There were many conflicting reports in the media concerning important aspects of the case, so that no one not directly involved could possibly know

the facts for sure. In the absence of a living will, a key factor was the husband's claim, challenged by others, that prior to being stricken fifteen years earlier the young woman had clearly stated her wishes not to be kept alive under these circumstances. The diverse opinions expressed in public and private debates were examples of subjective probability, not determined by objective information, but reflecting the division in American society on a host of related issues.

The precise interpretation of probability in science has been of special concern to philosophers. The theory of subjective probability is the theory of coherence of a body of opinion, guided by its conformance to the axioms of probability that both types must obey, with probability as a number between zero and one. There are several approaches of subjective probability, explained and illustrated with simple examples in Ian Hacking's 2001 textbook *An Introduction to Probability and Inductive Logic*.

The subjective probability of a proposition may be defined as the value to the user of a unit benefit contingent on the truth of the proposition. The concept of personal value or *utility* is central to decision theory in economics and the behavioral sciences. But in general statistical inference, the two interpretations of probability are in direct opposition, with no resolution likely in the foreseeable future. The subjective approach, usually called *Bayesian*, involves combining one's *prior probability*, based on a qualitative assessment of the situation, with new information to obtain the *posterior probability*. A key controversial issue is the subjective choice of the prior probability. Critics of objective probability counter that relative frequency itself is subjective, because it depends on the denominator used, and what about situations in which long-run repeated experimentation under identical conditions is not possible, even in principle? And so it goes. But any approach of logic has its intrinsic limitations. There are no right or wrong answers to the debates of philosophers; probability and chance are among the primitive concepts always open to analysis, such as knowledge, cause, and truth.

Some points to remember: Unless otherwise indicated in the title of a published report, the "default" method of analysis is based on objective probability and the classical (Neyman-Pearson) theory of statistical inference. From the viewpoint of communicating scientific results to the public, often in media sound bites, objective probability seems to be the more suitable method. In any case, under many conditions the results are similar. But discoveries are not made by formula. Creative scientists know what is happening in their own field and entertain ideas in the context of their own views. Out of this may emerge something new after years of search and many blind alleys. Ethical concerns pertain to violation of the codes of research conduct and false reporting of results, whatever the claimed method of confirmation.

CHANCE AT THE HEART OF REALITY? From the great Aristotelian synthesis of antiquity to the late nineteenth century, physical determinism with strict causality was a basic assumption of science and philosophy. Chance was taken as a measure of ignorance, a lack of knowledge of the complex interaction of unknown causes. This changed with the theory of evolution, involving random mutation and natural selection, and was followed in the early twentieth century by the discovery of quantum mechanics and indeterminism at the fundamental level. According to Heisenberg's uncertainty principle, the position and momentum of elementary particles can be considered together only in terms of probabilities. These theories endow chance with a distinct identity, as an explanatory principle of effects without a cause.

Is chance then an intrinsic part of nature, a feature of reality? That was the Copenhagen interpretation of quantum theory, accepted by the majority of physicists, although it never became unanimous. Albert Einstein expressed his opposition in the famous statement: "God does not play dice with the universe." An alternative view is to differentiate between interaction in nature and the level of measurability in physics (Jaki 1986). But the acceptance of chance in quantum mechanics does not imply a lawless universe; the probabilities of the different states can be precisely measured, and on a macroscopic scale nature appears to follow deterministic laws. There is also the concept of contingent order: Events that may be random still obey a larger law; an example would be random mutation in biology, within the structure of Mendelian genetics.

Again, some points to consider: Training in physics at the doctoral level is required to appreciate the implications of quantum mechanics. The subject has no intuitive meaning for nonspecialists, and there is continued disagreement among physicists. Speculation on the nature of reality belongs to philosophy, even if done by physicists. Intrinsic to the intellectual motivation of working scientists is a philosophy of realism, the belief in an external world of order that is accessible to human inquiry. In this context chance remains a measure of uncertainty, and that is the relevant interpretation for the applied sciences and technology.

OBSERVING RANDOMNESS. The word *random* cannot be defined precisely; one can say only what it is not. In textbooks of probability and statistics it is generally an undefined term, like *point* in geometry. The random numbers generated by computer and used in many research applications are in fact produced by given rules and as such are not random; *pseudorandom* is the proper technical term. There is much ongoing research on the concept of randomness. The simplest common example of a random experiment, the flipping of a coin, has been analyzed in terms of Newton's laws of physics, with upward velocity and rate of spin of the coin determining the outcome. Similar analyses hold for dice and roulette wheels.

Chaos theory has found that very little complexity in a deterministic system is needed to bring about highly complex phenomena, often unpredictably "chaotic" behavior. Almost imperceptible differences in the initial conditions can result in widely diverging outcomes. First noted in a computer simulation of a weather system, this has become known as the "butterfly effect," the image of a butterfly flapping its wings causing a hurricane somewhere across the globe. The phenomenon has been observed in a variety of fields, and the theory being developed has application in a wide range of disciplines, including hydrodynamics, biology, physiology, psychology, economics, ecology, and engineering. The important observation is that even many phenomena that are adequately covered by deterministic theories of classical physics prove to be chaotic, suggesting that there are real limitations on what can be learned about physical systems.

Clearly here scientific determinism does not imply epistemological determinism (meaning that results can be established with certainty). The phenomena appear random and need to be addressed in terms of probabilities. These discoveries should teach caution in expectations for the claimed effects of various aggressively promoted economic and social policies for giant systems such as the United States and other nations.

FREE WILL AND THE LAWS OF PROBABILITY. As a simple example, consider a local telephone calling region where the length of a call does not affect its cost. Residents can call anyone in the region they wish, at any time they wish, and talk as long as they wish, for one unit charge per call. Then the probability distribution of call durations for any given time period will be an *exponential distribution*. The number of calls arriving at an exchange during a fixed time interval will follow a *Poisson distribution*, with higher means for busy periods of telephone traffic. These precisely defined laws make possible the efficient design of communications systems.

From the engineering viewpoint the calls, initiated by the free will of large numbers of individuals, are random, following known probability laws with parameters that are estimated from observations.

PURPOSE IN THE UNIVERSE? The evolution controversy is often presented to the public as the conflict between two diametrically opposed fundamentalist views: Strict Darwinism, according to which chance variation and natural selection are sufficient to explain the origin of all life on Earth, and so-called creationism, which accepts a literal interpretation of the Book of Genesis of the Old Testament. In fact the situation is more complex.

Some evolutionary biologists hold that further structures beyond strict Darwinism are needed to account for the complexity of living systems. They are naturalists, whose explorations use the latest scientific advances to seek better explanations in the natural order. Many mainstream believers accept the fact of evolution, and those interested in science also question the mechanism of evolution. They are creationists in the sense that they believe in Creation, but they seek to learn what science has to say about how the world came into being. They believe that there is purpose in the universe, and see no problem with considering intelligent design as one of the explanatory hypotheses. Because the aim is to understand all of life and human experience, they do not think it rational to exclude any viable hypotheses.

Working along these lines are the American researchers Michael J. Behe, William A. Dembski, and Stephen C. Meyer, who argue that the complex specified information found in the universe, including irreducibly complex biochemical systems, cannot be the product of chance mechanisms and thus provides evidence of intelligent design (Behe, Dembski, and Meyer 2000). In cosmology the big bang theory of the origin of the universe and the anthropic principle concerning conditions necessary for the existence of life may be used in speculations of natural theology. Any emerging results that show consistency of science with the tenets of belief should be discussed openly, along with everything else. Submit it all to the test of time.

THE RELEVANCE OF PASCAL. The work of Pascal, of enduring interest for 300 years, was the subject of books by two prominent thinkers of the twentieth century—the Hungarian mathematician Alfréd Rényi (1921–1970) and the Italian-German theologian and philosopher of religion Romano Guardini (1885–1968), who held the philosophy chair "Christliche Weltanschauung" (Christian worldview) at the University of Munich.

Letters on Probability (Rényi 1972) is a series of four fictitious letters by Pascal to Fermat, assumed to be part of the lost correspondence between the two mathematicians. Addressed to the general reader, it is a witty and charming exploration of the notion of chance and probability, in the cultural context of the seventeenth century that shows the timelessness of the subject. In the last letter Pascal reports on a dialogue he had with a friend concerning the merits of objective and subjective probability. They discussed *De rerum natura* (On the nature of things), by the Roman poet-philosopher Lucretius (fl. first century B.C.E.), in which he described the Greek atomistic philosophy of Democritus (c. 460–c. 370 B.C.E.) and Epicurus (341–270 B.C.E.); they wondered what the ancients might have meant by chance and random events. In its images of whirling atoms the poem conveys a striking picture of Brownian motion. Pascal is here an advocate of objective probability, reflecting the views of the author.

Pascal for Our Time (Guardini 1966) is a biography placing an immensely gifted believer at the point in the history of ideas when the scientific consciousness of the modern age had fully emerged, but that of the previous era had not yet faded. Pascal is presented as a human being who—simultaneously endowed with keen insight in science, psychology, and philosophy—seeks with reflection to justify his existence at every moment. Guardini shows Pascal's relevance at the intellectual and cultural watershed reached by the twentieth century.

For Pascal thinking was the basis of morality, and a reasoned search the way to proceed to find meaning. Human longing far surpasses what this life has to offer: "Man infinitely transcends man" (Pascal 1995, #131; the numbering refers to the fragments in this edition of the *Pensées*). A totally committed search is the only option of reason. But the search is feebleminded if it stops before reaching the absolute limits of reason: "Reason's last step is the recognition that there are an infinite number of things which are beyond it. It is merely feeble if it does not go as far as to realize that" (#188). Faith offers more knowledge, but it has to be consistent with the evidence of sense experience: "Faith certainly tells us what the senses do not, but not the contrary of what they see; it is above, not against them" (#185).

The ultimate limits of human reason, perceived by Pascal, were established in the twentieth century with Kurt Gödel's incompleteness theorem in mathematics. The search Pascal so strongly urged was taken up by the natural theologians, among others, and it continues into the twenty-first century. And for thoughtful believers there still cannot be a conflict between faith and science.

THE ETHICS OF EVIDENCE. The comments shared above fit into a proposed framework for dealing with uncertainty, the *Ethics of Evidence* (Miké 2000). The Ethics of Evidence calls for developing and using the best evidence for decision-making in human affairs, while recognizing that there will always be uncertainty—scientific as well as existential uncertainty. It calls for synthesis of the findings of all relevant fields, and taking personal responsibility for committed action. Philosophical questions such as the nature of reality and purpose in the universe cannot be decided by the latest findings of a particular science. The French philosopher Étienne Gilson (1884–1978) argued in his book *The Unity of Philosophical Experience* (1999 [1937]) that this age has been going through the last phase of the current cycle of twenty-five centuries of Western philosophy. A new philosophical synthesis is needed, with a first principle that integrates the accumulating insights of science and other disciplines.

Application of Probability

Since the 1960s much historical scholarship has focused on what Gerd Gigerenzer and colleagues (1989) aptly described as *The Empire of Chance: How Probability Changed Science and Everyday Life*. There are encyclopedias devoted to the subject, with probability as an integral component of the field of statistics. Probability is the basis of theories of *sampling, estimation* of parameters, *hypothesis testing,* and other modes of *inference,* in a multitude of complex designs for the simultaneous study of variables of interest.

Reminiscent of the beginnings with games of chance, the Hungarian mathematician John von Neumann (1903–1957) published a seminal essay in 1928 on the theory of games of strategy, opening up entirely new paths for mathematical economics. He collaborated with the Austrian economist Oskar Morgenstern (1902–1977), by then both in the United States, on their classic work *Theory of Games and Economic Behavior* (1944). The theory of games provides models for economic and social phenomena, including political and military contexts, in which participants strive for their own advantage but do not control or know the probability distribution of all the variables on which the outcome of their acts depends. An important extension is noncooperative game theory, which excludes binding agreements and is based on the concept of Nash equilibrium, used to make predictions about the outcome of strategic interaction.

It is named after its originator, the American mathematician John F. Nash (b. 1928). Game theory is inference in the form of decision-making.

More generally, there are *stochastic processes,* in what is called the probability theory of movement; these are systems that pass through a succession of states, usually over time, as distinct from deterministic systems in which a constant mechanism generates data that are assumed to be independent. Examples of these include epidemic theory, study of complex networks, finance theory, genetic epidemiology, hydrology, and the foundations of quantum theory.

Ethical aspects of probability pertain to knowing and using the proper techniques to clarify and help resolve problems in science and technology, with close attention to remaining uncertainties. If mechanisms of action are fully understood, as in many engineering systems, careful design and built-in redundancies will result in reliable performance within specified probabilities. But in most areas of interest, such as medical, social, and economic phenomena, the number of variables is large and the mechanisms often unknown or at best poorly understood. Thus only a selection of potentially relevant factors can be studied in any one tentative model, amid vast uncertainties. Misuse of such limited results makes the public vulnerable to manipulation by state, market, and a multitude of interest groups. It seems impossible to overstate the importance of awareness and education concerning these issues.

VALERIE MIKÉ

SEE ALSO *Pascal, Blaise; Risk: Overview; Statistics; Uncertainty.*

BIBLIOGRAPHY

Arbuthnot, John. (1710). "An Argument for Divine Providence, Taken from the Constant Regularity Observed in the Births of Both Sexes." *Philosophical Transactions of the Royal Society of London* 27: 186–190. Reprinted in *Studies in the History of Statistics and Probability,* Vol. 2., ed. Maurice G. Kendall and R. L. Plackett. New York: Macmillan, 1977.

Bayes, Thomas. (1763). "An Essay towards Solving a Problem in the Doctrine, of Chances." *Philosophical Transactions of the Royal Society of London* 53: 370–418. Reprinted in *Studies in the History of Statistics and Probability,* Vol. 1, ed. Egon S. Pearson and Maurice G. Kendall. London: Griffin, 1970.

Behe, Michael J.; William A. Dembski; and Stephen C. Meyer. (2000). *Science and Evidence for Design in the Universe.* San Francisco: Ignatius Press. Authors are trained in biochemistry, mathematics, and philosophy.

Bernoulli, Jacques. (1713). *Ars Conjectandi* [The art of conjecturing]. Basel: Impensis Thurnisiorum.

David, Florence N. (1962). *Games, Gods, and Gambling: The Origins and History of Probability and Statistical Ideas from the Earliest Times to the Newtonian Era*. London: Charles Griffin. Illustrated story of the prehistory of probability and its early development. Assessment of Pascal's contribution questioned by other scholars, such as Rényi (1972).

Edwards, A. W. F. (2002). *Pascal's Arithmetical Triangle: The Story of a Mathematical Idea*. Baltimore: Johns Hopkins University Press.

Eisenhart, Churchill, and Allan Birnbaum. (1967). "Tercentennials of Arbuthnot and de Moivre." *American Statistician* 21(3): 22–29.

Gigerenzer, Gerd; Zeno Swijtink; Theodore Porter; et al. (1989). *The Empire of Chance: How Probability Changed Science and Everyday Life*. Cambridge, UK: Cambridge University Press. Summary of a two-volume work by a team of historians and philosophers of science, written for a general audience.

Gilson, Étienne. (1999 [1937]). *The Unity of Philosophical Experience*. San Francisco: Ignatius Press. Analysis of the history of Western philosophy with a proposed new philosophical synthesis.

Graunt, John. (1662). *Natural and Political Observations Mentioned in a Following Index, and Made upon the Bills of Mortality*. London. Reprinted in *Natural and Political Observations Made upon the Bills of Mortality*, ed. Walter F. Willcox. Baltimore: Johns Hopkins University Press, 1939.

Guardini, Romano. (1966). *Pascal for Our Time*, trans. Brian Thompson. New York: Herder and Herder. Translation of *Christliches Bewußtsein: Versuche über Pascal*, 1935.

Hacking, Ian. (1975). *The Emergence of Probability: A Philosophical Study of Early Ideas about Probability, Induction, and Statistical Inference*. London: Cambridge University Press. Includes speculation on the dual nature of probability.

Hacking, Ian. (1990). *The Taming of Chance*. Cambridge, UK: Cambridge University Press. Continuation of 1975 work (see above), exploring the development of probability to the beginning of the twentieth century.

Hacking, Ian. (2001). *An Introduction to Probability and Inductive Logic*. Cambridge, UK: Cambridge University Press. Introductory textbook for students of philosophy, with many examples.

Huygens, Christiaan. (1657). *De Ratiociniis in Ludo Aleae* [Reasoning on games of chance]. In *Exercitationum Mathematicarum Libri Quinque* [Five books of mathematical exercises], ed. Frans van Schooten. Leiden, Netherlands: Johannis Elsevirii.

Jaki, Stanley L. (1986). "Chance or Reality: Interaction in Nature versus Measurement in Physics." In his *Chance or Reality and Other Essays*. Lanham, MD: University Press of America. Analysis of the controversy over the interpretation of quantum mechanics by a noted historian of science.

Kolmogorov, Andrei N. (1956). *Foundations of the Theory of Probability*, 2nd English edition, trans. Nathan Morrison. New York: Chelsea Publishing. Translation of *Grundbegriffe der Wahrscheinlichkeitsrechnung*, 1933. The original work on the axiomatic basis of probability theory.

Kotz, Samuel; Norman L. Johnson; and Campbell B. Read, eds. (1982–1999). *Encyclopedia of Statistical Sciences*. 9 vols. plus supp. and 3 update vols. New York: Wiley.

Kruskal, William H., and Judith M. Tanur, eds. (1978). *International Encyclopedia of Statistics*. 2 vols. New York: Free Press.

Laplace, Pierre-Simon de. (1812). *Théorie analytique des probabilités* [Analytic theory of probability]. Paris: Courcier. First systematic treatment of probability theory.

Laplace, Pierre-Simon de. (1951). *A Philosophical Essay on Probabilities*, trans., from the 6th French edition, Frederick Wilson Truscott and Frederick Lincoln Emory. New York: Dover. Translation of *Essaie philosophique sur les probabilités*, 1819. Addressed to the general public, included as the introduction to the third edition (1820) of the work listed above.

Miké, Valerie. (2000). "Seeking the Truth in a World of Chance." *Technology in Society* 22(3): 353–360. Discusses the work of Pascal in a contemporary cultural context.

Moivre, Abraham de. (1718). *The Doctrine of Chances; or, A Method of Calculating the Probabilities of Events in Play*. London: W. Pearson. 2nd edition, London: Woodfall, 1738. 3rd edition, London: Millar, 1756. Reprinted: New York: Chelsea Publishing, 1967; Providence, RI: American Mathematical Society, 2000.

Pascal, Blaise. (1995). *Pensées* [Thoughts], trans. A. J. Krailsheimer. London: Penguin. Originally published in French, 1670. Fine modern English translation, with an introduction by the translator.

Peterson, Ivars. (1990). *Islands of Truth: A Mathematical Mystery Cruise*. New York: Freeman. One in a series of richly illustrated books by a science writer on new ideas in mathematics, addressed to the lay reader; includes chaos theory.

Rényi, Alfréd. (1972). *Letters on Probability*, trans. László Vekerdi. Detroit, MI: Wayne State University Press. Translation of *Levelek a valószínűségről*, 1969. Incisive and witty exploration of the notion of probability, in the form of fictitious letters assumed to be part of a lost correspondence between Pascal and Fermat. Written for the general reader.

Stigler, Stephen M. (1990). *The History of Statistics: The Measurement of Uncertainty before 1900*. Cambridge, MA: Harvard University Press, Belknap Press. A comprehensive history, tracing the interplay of mathematical concepts with the needs of several applied sciences that gave rise to the field of statistics.

Süssmilch, Johann Peter. (1741). *Die göttliche Ordnung in den Veränderungen des menschlichen Geschlechts, aus der Geburt, dem Tode, und der Fortpflanzung desselben erwiesen*. [The divine order in the fluctuations of the human race ...]. Later enlarged ed. reprinted, Augsburg, Germany: Verlag-Cromm, 1988. First analytic theory of population by a founder of modern demography.

von Neumann, John, and Oskar Morgenstern. (1944). *Theory of Games and Economic Behavior*. Princeton, NJ: Princeton University Press.

PRODUCT SAFETY AND LIABILITY

• • •

As people become increasingly dependent on the use of engineered products, product safety and liability become issues of worldwide importance. In many countries, however, there are no strong traditions promoting safety standards in the technical design and testing of consumer products, nor are there methods of legal redress when such standards are not met. The ethics of product safety and liability is thus reasonably addressed by treating the United States as a leading case study, with the inclusion of some supplementary references to related developments in other countries. It is also necessary to acknowledge the role of product safety standards in relation to global trade practices.

U.S. Perspective

According to figures from the Internet site of the U.S. Consumer Product Safety Commission (CPSC), consumer products are annually responsible for more than 22,000 deaths and 29 million injuries (more than two deaths and 3,000 injuries per hour) at a total annual cost (including property damage) of more than $700 billion. Although the magnitude of these numbers may be subject to argument, they support the contention that product-related injuries were the primary factor in deaths of people from ages one to thirty-six, exceeding deaths from cancer and heart disease (Andre and Velasquez 1991). Staggering as such numbers are, product safety has significantly *increased* over the past three decades: "The CPSC's work to ensure the safety of consumer products—such as toys, cribs, power tools, cigarette lighters, and household chemicals—contributed significantly to the 30 percent decline in the rate of deaths and injuries associated with consumer products over the past 30 years" (CPSC).

Just as Rachel Carson's *Silent Spring* (1962) marked the beginning of the modern popular environmental movement, the publication of Ralph Nader's *Unsafe at Any Speed* (1965), which documented the neglect of safety features in the design of the Chevrolet Corvair and other U.S. automobiles, launched the contemporary consumer product safety movement. Nader influenced a number of federal laws concerned with public health and safety, including the National Traffic and Motor Vehicle Safety Act (1966), the Consumer Product Safety Act (1972), and the Freedom of Information Act (1966), as well as numerous not for profit consumer rights organizations.

In the intervening years, a succession of highly publicized product safety cases has fueled public interest in the topic including those concerning the Ford Pinto (1970s), the Dalkon Shield Intrauterine Device (1970s–1980s), the Bjork/Shiley heart valve (1979–1986), the Therac-25 radiation therapy machine (1985–1987), the Ford/Firestone tire recalls (2000), the health risks attributed to smoking, numerous airline crashes, and, perhaps the most spectacular product failure of all, the space shuttle *Challenger* (1986). Product safety is now promoted by many governmental and nongovernmental organizations including national product safety testing and certification organizations such as the Underwriters Laboratories (founded 1894) in the United States; the International Organization for Standardization (ISO, founded 1946); consumer groups such as the Consumers Union (founded 1936), publishers of the popular magazine *Consumer Reports*; and socially conscious investment groups such as the Calvert Fund (created 1990). The Worldwide System for Conformity Testing and Certification of Electrical Equipment (IECEE), maintained by The International Electrotechnical Commission (IEC) (founded 1906), includes a code of ethics for product safety certification programs.

Product Safety, Liability, and Engineering Ethics

During the same period as the Carson and Nader books, professional engineering societies began to take more seriously the role of engineers and the engineering profession as stewards of product safety. All contemporary codes of engineering ethics state that engineers have a responsibility to protect *the public safety, health, and welfare,* and most codes state that this duty should be held *paramount.*

The notion that safety is of primary importance in engineering is also fundamental to nearly all academic treatments of engineering ethics (Herkert 2000). A key concept is the notion of *professional responsibility,* which many ethicists characterize as a type of moral responsibility arising from special knowledge possessed by an individual (Whitbeck 1998). Philosopher Mike Martin and engineer Roland Schinzinger argue that professional responsibility in engineering involves "the creation of useful and safe technological products while respecting the autonomy of clients and the public, especially in matters of risk-taking" (Martin and Schinzinger 1996, p. 42).

Yet while product safety is central to discussions of engineering ethics, the closely related legal concept of product liability is often ignored, or even attacked by engineering professionals and others. "Developing from the Industrial Revolution, U.S. product liability law is derived from case law and restatements of law anchored in contract and tort. It is based on the belief that consumers need protection from business and that business should bear the costs of harms inflicted on consumers" (Product Liability Lawyer Resource Center Internet site). Over time, the legal standard regarding product liability has evolved from the doctrine of *let the buyer beware,* to a legal theory requiring a determination of negligence on the part of the manufacturer, to the modern legal standard of strict liability (liability imposed without fault). Product liability claims can be based on manufacturing defects, design defects, and information defects (lack of appropriate warnings).

Judgments in product liability cases can include both compensatory (reimbursement for costs) and punitive damages; large judgments have often been the focus of attention in the controversy over product liability, especially in cases when the judgment may seem out of proportion to the harm. In one notorious case, a jury awarded a woman nearly $3 million for burns she received when she spilled coffee purchased at a McDonald's drive-up window.

Critics of current product liability law, including many professional engineering societies, call for rollbacks often approaching the old *let the buyer beware* policies. For example, in 1996 Congress passed legislation that would have severely limited the effect of product liability litigation by placing a cap on punitive damages and enacting stricter requirements for holding manufacturers liable. President Bill Clinton vetoed the bill; however, the debate over product liability reform continued.

The proponents of product liability reform argue that the current system unjustly rewards plaintiffs and stifles technological innovation, resulting in a lack of competitiveness on the part of U.S. manufacturers and decreased product safety. Supporters of the current system counter that it generally works as intended in discouraging the manufacture of defective products and compensating people injured by such defects (Hunziker and Jones 1994). To some the debate over product liability reform is a classic business/consumer conflict. A *New York Times* editorial (1996), for example, described proposed legislation as "The Anti-Consumer Act of 1996." Despite the arguments of both sides, the evidence is mixed concerning whether product liability rewards result in improvements in product safety (Hunziker and Jones 1994).

Engineers and engineering societies have tended to side with the proponents of product liability reform (Herkert 2001, 2003). A vice president of engineering of a major U.S. automobile company, for example, has argued that product liability restricts engineering practice by inhibiting innovation, discouraging critical evaluation of safety features, and preventing implementation of new or improved designs (Castaing 1994). The 1998 position statement on product liability of IEEE-USA, a unit of the Institute of Electrical and Electronics Engineers (IEEE) concerned with professional issues in the United States, calls for stringent limits on product liability including holding the manufacturer blameless when existing standards are met, adequate warnings are provided, or the product is misused or altered by the user. Other engineering societies, such as ASME International (formerly the American Society of Mechanical Engineers) have also actively supported product liability reform (ASME International 2001).

Given the primary responsibility of engineers for public safety, health, and welfare stated in the codes of ethics, it is surprising that the product liability issue has not drawn more attention from the perspective of engineering ethics (Herkert 2001, 2003). There is little, if any, evidence, however, to suggest that engineering societies promoting changes in the product liability system have considered the effect that decreasing the impact of product liability would have from the point of view of engineering ethics. On the whole, the engineering community has paid little attention' to the ethical implications of product liability. For example, a major study of product liability and innovation by the National Academy of Engineering (Hunziker and Jones 1994), which considered such issues as corporate practice, insurance, regulation, and the role of scientific and technical information in the courtroom, touched only briefly on ethics (in a chapter on the need to address public risk perceptions) (Fischhoff and Merz 1994). Even the ethics literature is equivocal on the issue of product liability. For example, one well-known essay on engineering responsibility in the Ford Pinto case advocated stronger regulation and fines and imprisonment for corporate officials to achieve desired levels of safety, giving only passing notice to the role of product liability litigation (DeGeorge 1981).

One aspect of product liability and calls for its reform that can be readily identified as an ethical issue is the notion of *standard of care* (Kardon 1999). Though usually considered in a legal context, the standard of care in engineering design is also important in considering the ethical responsibilities of engineers. Many discussions of product liability turn on the concept of standard of care. Examples include such classic engineering ethics cases as the Turkish Airlines DC-10 disaster, where some blamed the luggage handlers for failing to secure the poorly designed cargo door, and the McDonald's coffee case, where public (and engineering) opinion generally held the product's *user* responsible for the accident. In such attitudes there is an assumption that the user should be held to a standard of care in use of a product equivalent to the standard of care applied to designers and manufacturers in its creation.

The McDonald's Coffee Case

Observers often tend to blame the victim in accidents of this kind. Such cases, however, are rarely that clear cut, as Howard Twiggs notes when commenting on the McDonald's case:

> That case demonstrates how well our system works. Unfortunately, headlines and misrepresentations by civil justice's opponents misshaped public opinion about [the] case against McDonald's. The public was led to believe that a woman driving a car was holding a cup of McDonald's coffee between her knees, spilled it, burned herself, and hired a trial lawyer who conned a jury into awarding her $2.86 million. (Twiggs 1997, p. 9)

Included among the facts of the case as cited by Twiggs to buttress his point were the following:

- The accident occurred in a parked car.
- The coffee was served scalding hot (180°–190° F), which can cause third-degree burns in seven seconds; this is 40–50 degrees hotter than normal coffee service. The victim suffered third-degree burns over 6 percent of her body.
- McDonald's had earlier reports of more than 700 people, including infants, being burned by its coffee.
- The victim attempted to settle out of court for $20,000 in medical bills.
- The jury awarded $200,000 for actual damages, which they reduced to $160,000 because they found the victim partly at fault.
- The jury based its award of $2.7 million in punitive damages on two days of coffee sales by McDonald's.

- The trial judge reduced the punitive damages to three times actual damages ($480,000) and ordered postverdict mediation where the case was settled.
- Despite telling the jury at trial that they would not do so, McDonald's immediately stopped selling coffee at this temperature.

Lessons for Engineering Design

On the face of it, the assumption that the victim is to blame in such instances undermines the notion that professionals have ethical responsibilities that go beyond those of nonprofessionals. A counter example more in tune with notions of professional responsibility would be an engineering designer who attempts to foresee preventable harm to users by anticipating common forms of product misuse, a doctrine sometimes applied in legal rulings concerning standard of care (Kardon 1999).

Roger Boisjoly (1998), the renowned whistle-blowing engineer in the *Challenger* case, argues that design engineers do have the obligation to anticipate product safety problems, even in so-called instances of product misuse. Following his blacklisting in the aerospace industry, Boisjoly became a consultant specializing in forensic engineering. As a forensic engineer, he became involved with product safety cases that included defective trigger lock switches on handheld drills, unstable step stools, and tipping problems in common household stoves; in most cases the products had met applicable regulatory standards. Boisjoly testified in two cases involving stove-tipping accidents; in one an adult and in the other a child leaned on open oven doors and were scalded with hot food being prepared on the stove's burners. Similar to the McDonald's case, the manufacturers had been provided ample evidence of the defect by prior complaints and litigation. As part of his investigation, Boisjoly, in about two weeks, designed an inexpensive collapsible door hinge that solved the problem. As Boisjoly demonstrates, ensuring product safety involves more than meeting engineering standards and avoiding liability—an engineer's professional obligation to protect public safety includes anticipating safety hazards and where possible designing the hazards out of the system.

International Issues

While political concerns over product safety and liability in the United States continue to focus on the relative responsibility of manufacturers and consumers, additional issues are prevalent in the rest of the world. In Europe debate is centered on needed harmonization of product safety standards both within the European Community and with respect to other nations, most notably the United States. Such concerns are primarily motivated by a desire to lower trade barriers but they also have important product safety implications because safety issues and standards can vary from country to country (Mader and Krøigaard 1999). In the developing world, as in so many other aspects of technological development, the outlook for product safety is much worse. An article calling for establishment of a consumer product safety commission of India points out safety and health problems with the entire range of consumer products, including unprocessed or improperly packaged food, unsafe rail transport, and dangerous toys and other hazards that lack child-proofing (Desikan 1999). Such inequities will continue in the absence of enforcement of national product safety standards and until fair and effective international standards are developed and recognized.

JOSEPH R. HERKERT

SEE ALSO *Engineering Ethics; Ford Pinto Case.*

BIBLIOGRAPHY

Carson, Rachel. (1962). *Silent Spring.* Boston: Houghton Mifflin. Biologist's celebrated and controversial account of the ecological impacts of pesticides.

Castaing, François J. (1994). "The Effects of Product Liability on Automotive Engineering Practice." In *Product Liability and Innovation*, ed. Janet R. Hunziker and Trevor O. Jones. Washington, DC: National Academy Press.

De George, Richard T. (1981). "Ethical Responsibilities of Engineers in Large Organizations: The Pinto Case." *Business and Professional Ethics Journal* 1: 1–14. Classic essay on moral responsibilities of would-be whistleblowers that uses the Ford Pinto's questionable gas tank design as a case study.

Editorial. (1996). "The Anti-Consumer Act of 1996." *New York Times*, March 21, Section A, p. 24.

Fischhoff, Baruch, and Jon F. Merz. (1994). "The Inconvenient Public: Behavioral Research Approaches to Reducing Product Liability Risks." In *Product Liability and Innovation*, ed. Janet. R. Hunziker and Trevor. O. Jones. Washington, DC: National Academy Press.

Herkert, Joseph R. (2000). "Engineering Ethics Education in the USA: Content, Pedagogy, and Curriculum." *European Journal of Engineering Education* 25(4): 303–313.

Herkert, Joseph R. (2001). "Future Directions in Engineering Ethics Research: Microethics, Macroethics and the Role of Professional Societies." *Science and Engineering Ethics* 7(3): 403–414.

Herkert, Joseph R. (2003). "Professional Societies, Microethics, and Macroethics: Product Liability as an Ethical Issue in Engineering Design." *International Journal of Engineering Education* 19(1): 163–167.

Hunziker, Janet R., and Trevor O. Jones, eds. (1994). *Product Liability and Innovation.* Washington, DC: National Academy Press. Report of a committee of the National Academy of Engineering on the impact of the U.S. product liability system on technological innovation.

Martin, Mike W., and Roland Schinzinger. (1996). *Ethics in Engineering,* 3rd edition. New York: McGraw-Hill. Classic text on engineering ethics co-authored by an engineer and a philosopher.

Nader, Ralph. (1965). *Unsafe at Any Speed: The Designed-in Dangers of the American Automobile.* New York: Grossman. The consumer advocate's expose of unsafe automobile designs, especially General Motors' Corvair.

Twiggs, Howard. (1997). "How Civil Justice Saved Me From Getting Burned." *Trial Magazine* June: 9. Also available from http://www.atla.org/secrecy/data/twiggs.aspx.

Whitbeck, Caroline. (1998). *Ethics in Engineering Practice and Research.* Cambridge, England: Cambridge University Press. Engineering ethics text unique in its attention to both professional ethics and research ethics.

INTERNET RESOURCES

American Society of Mechanical Engineers International. "Public Policy Agenda: 2003-2004." Available from http://www.asme.org/gric/Agenda/PPA2003-2004StateIssues.html.

Andre, Claire, and Manuel Velasquez. (1991). "Who Should Pay? The Product Liability Debate." *Issues in Ethics* 4(1). Available from http://www.scu.edu/ethics/publications/iie/v4n1/pay.html.

Boisjoly, Roger. (1998) "Professionalism." IEEE-USA. Available from http://www.ieeeusa.org/PACE/LIBRARY/boisjoily.html. The noted whistleblower from the space-shuttle Challenger case discusses professionalism in engineering design drawing on his experiences as a design engineer and as a forensic engineer.

Desikan, R. (1999) "Product Safety: A Long Way to Go." *The Hindu.* Available from http://www.hinduonnet.com/folio/fo9910/99100180.htm. Special *Consumer* issue, October 31, 1999.

IEEE-USA. "Tort Law And Product Liability Reform." Available from http://www.ieeeusa.org/forum/POSITIONS/liability.html.

Kardon, Joshua B. (1999). "The Structural Engineer's Standard of Care." Online Ethics Center for Engineering and Science. Available from http://onlineethics.org/cases/kardon.html.

Mader, Donald A., and Søren Krøigaard. (1999). "Achieving Harmonization of Product Safety Standards." Compliance Engineering. Available from http://www.ce-mag.com/archive/1999/novdec/guesteditorial.html.

Product Liability Lawyer Resource Center. "Product Liability Laws Evolution." Product Liability Lawyer.com. Available from http://www.productliabilitylawyer.com/evolutionOfProductLiability.cfm.

U.S. Consumer Product Safety Commission (CPSC). "CPSC Overview." Available from http://www.cpsc.gov/about/about.html.

Virginia Department of Public Health. "Product Safety." Available from http://www.vahealth.org/civp/product/.

Worldwide System for Conformity Testing and Certification of Electrical Equipment (IECEE). "Code of Ethics Applicable to a Product Safety Certification Organization." Available from http://www.iecee.org/cbscheme/html/chcode.htm.

PROFESSIONAL ENGINEERING ORGANIZATIONS

• • •

Professional engineering organizations are the primary channels by which engineers working in particular technical disciplines, or otherwise possessing common interests, share technical knowledge, regulate professional practice, influence public policy, and maintain the traditions and reputation of the profession. These organizations, as well as the profession of engineering itself, are of relatively recent origin, arising during the Industrial Revolution. In contrast, the primary object with which engineering is concerned—technology—is of ancient origin.

Historical Background

Throughout the history of civilization, humans have been engaged in developing and adjusting to changed circumstances for technological development. Construction, shipbuilding, irrigation, mining, metallurgy, and military fortification are prominent examples of technologies with extensive histories. Prior to the eighteenth century, the bulk of knowledge and practices in these areas was largely uncodified, slow to spread between geographic regions, and passed from one generation to another mainly through apprenticeship.

During certain periods, the artisans and tradespeople who plied these skills organized themselves for mutual benefit. In the late Roman and Byzantine periods, such organizations were called *collegia*, and in medieval times, *guilds*. Among the purposes these organizations served, were the regulation of prices, product quality, and entry into the craft. But with the coming of the Scientific and Industrial Revolutions, the status of guilds diminished as the pace of technological development accelerated and the expansion of trade routes increased the availability of imported goods.

By the late-eighteenth century, developments such as the advent of steam power, the increased complexity of military ordnance, the rise of canal building, and the genesis of mechanized production had begun to cause significant changes in society, and the need for a more formal means of acquiring and transmitting technical training began to grow. One leader in the creation of technical schools was France, first for military engineers, and then for engineers engaged in civilian projects. This model for technical education, which relied heavily upon mathematics, spread to other parts of continental Europe by the early-nineteenth century, and to England and the United States in the following decades.

Although England lagged France in developing technical schools, it was at the forefront of the Industrial Revolution by virtue of industrious, self-made engineers such as John Smeaton (1724–1792), who is widely considered to be the founder of the civil engineering profession. In 1771 he formed the Society of Civil Engineers, which was later renamed the Smeatonian Society. The meetings of this society were generally informal, and membership was not necessarily restricted to engineers; rather it also included those who had business or political interests in the engineering of public works.

In 1818 the Institution of Civil Engineers (ICE) was founded in England and is considered to be the earliest of the modern professional engineering societies. Its membership was restricted to practicing engineers and meetings were expressly for the purpose of exchanging technical information. Although the ICE grew slowly during its first couple of decades, these two characteristics formed the basic blueprint for subsequent societies, the next one of which was the Institution of Civil Engineers of Ireland formed in 1835. The Swiss Society of Engineers and Architects, followed in 1837, and then in 1847 the British Institution of Mechanical Engineers and the Royal Institution of Engineers in the Netherlands were formed. Between 1850 and 1900, no fewer than thirty additional professional engineering societies began operating in Europe, Scandinavia, North America, South America, South Africa, and Japan. Subsequently the number and types of professional engineering societies grew rapidly such that by the start of the twenty-first century hundreds of organizations existed worldwide.

Diversity of Technical Disciplines

The first main differentiation among types of professional engineering societies occurred along disciplinary lines. The original term civil engineering was meant to distinguish engineers engaged in the building of public works from military engineers. By the mid-nineteenth century, the rise of steam power, railroads, and mechanized production led to a divergence between mechanical engineering and civil engineering. By the latter part of the 1800s, societies had formed for mining engineering, electrical engineering, marine engineering, and sanitary engineering. In the United States, five organizations have become known as the *founder societies*. These are the American Society of Civil Engineers (ASCE, formed in 1852), the American Institute of Mining, Metallurgical, and Petroleum Engineers (AIME, formed in 1871), the American Society of Mechanical Engineers (ASME, formed in 1880), the Institute of Electrical and Electronics Engineers (IEEE, formed in 1963 from the merger of the American Institute of Electrical Engineers [AIEE, formed in 1884] with the Institute of Radio Engineers [IRE, formed in 1912]), and the American Institute of Chemical Engineers (AIChE, formed in1908). In 1904 the then existing four ancestor organizations formed a meta-organization known as the United Engineering Society (UES) in an effort to unify the engineering profession, but it failed to thrive. In 1979 the American Association of Engineering Societies (AAES) was founded with a similar goal. However the continued emergence of new and dissimilar engineering disciplines (e.g., automotive, aerospace, industrial, nuclear, computer, and biomedical), along with the increasing diversity of knowledge within each discipline, has proved to be a powerfully fragmenting force within the profession, and has generally thwarted attempts at unification. Thus the proliferation of professional engineering organizations accelerated through the twentieth century, paralleling the expanding scope of science and technology.

For this type of society, one organized around a particular technical discipline, the primary purposes are typically (a) to foster the presentation, discussion, and dissemination of the latest technical information and practices relevant to the discipline and its associated industry; (b) to provide a mechanism for overseeing the development of technical codes and standards relating to safety and uniformity in that industry; and (c) to promote the reputation and welfare of both the profession and the industry. In support of these main functions, societies frequently take on additional roles, such as supporting educational programs, lobbying political bodies, establishing professional ethics codes, documenting the history of the discipline, and offering various career development and continuing education benefits to members.

The technical engineering societies span a broad spectrum with respect to size, scope of activities, and focus of mission. Some tend to have close ties with particular industries, and engage in very practical activities that serve to promote and support those industries. Others maintain more independence, and pursue a broader agenda of technical and professional development activities. Overall these technically-oriented engineering societies, via research journals, conference proceedings, and trade magazines, are responsible for the bulk of engineering technical publication worldwide.

The technical societies are also instrumental for the development of technical codes and standards, which either serve to facilitate the compatibility of products and services across an industry, or which become incorporated in laws prescribing safe engineering practices. For example since its inception ASME has been engaged in the work of standardizing the specifications for such items as screw threads and pipe fittings, and in developing safety codes for the design of boilers and pressure vessels, explosions of which had been a serious safety hazard throughout the 1800s. The IEEE has been responsible for developing codes and standards on topics ranging from electrical insulation to digital communications protocols. What in the United States have been the purview of non-governmental organizations have in Europe, however, often been the responsibility of a government ministry.

Regulation of Professional Practice

The traditional focus of the discipline-specific engineering societies—developing a particular body of technical knowledge and overseeing its application in related industries—has proved to be a powerful organizing principle that is relatively loose and inclusive, largely transcending geographic boundaries, employment status, and political climate. In contrast there is another organizing principle that is more parochial, more exclusive, and more entwined with political and legal affairs. This organizing principle, which has given rise to a different type of professional engineering organization, is the idea that the title *engineer*, and *the practice* of engineering, ought to be controlled, either through a legislated process for *licensure*, or otherwise formalized procedures for *registration*. The organizations that have developed around this idea are the various state, provincial, and national societies and boards that oversee and promote professional licensure or registration.

In the United States the first law regarding the licensing of engineers was enacted in Wyoming in 1907 in response to disputes over property and water rights caused by incompetent surveyors. Other states also enacted engineering licensure laws following negative events, such as the St. Francis Dam collapse in California in 1928 and a school boiler explosion in Texas in 1937, both of which resulted in hundreds of lives lost. By 1950 all states had licensing laws. In 1934 the National Society of Professional Engineers (NSPE) was founded in the United States with the mission of promoting "the competent, ethical, and professional practice of engineering," mainly through the endorsement of licensure, which is a requirement for NSPE membership. In addition each state has its own NSPE affiliate organization, many of which, such as the Ohio Society for Professional Engineers (formed in 1878), pre-date the NSPE itself. Because licensing laws are enacted at the state level, these state-level organizations lobby state legislatures to maintain and improve the laws, and work with the state boards that oversee their enforcement. Licensure generally requires an education from an accredited institution, passage of qualifying examinations, and a specified number of years of probationary engineering experience.

Notwithstanding these developments, in the United States licensure has remained a difficult issue for the engineering profession. Most state licensing laws restrict the use of the *Professional Engineer* title and the offering of *engineering services* to the public. These requirements for licensure have had the biggest effects on civil engineers engaged in the design and construction of public works, and on consulting engineers. However the majority of engineers are employed by companies to do internal product design and development, product testing, technical sales, or project management. These engineers are exempt from licensure, with the result that less than 20 percent of engineers are licensed in the United States. NSPE and its state affiliates have struggled to convince more engineers of the benefits of licensure to both the individual and the profession.

While licensing laws affect only a small minority of engineers in the United States, legal constraints on engineering practice are even less strict in many other countries. In the United Kingdom, for example, neither the title of engineer nor the practice of engineering are restricted. There is, however, a voluntary engineering registration system that confers the title *Chartered Engineer* upon qualified applicants. This registration process is governed by the Engineering Council (UK), which is an independent, royal-chartered organization comprising most of the discipline-specific engineering societies in Great Britain as corporate members. In continental Europe, a few countries, notably Germany, Italy, Austria, and Luxembourg, place a significant degree of legal restriction on engineering practice, while in most other countries the constraints are more lax, or else nonexistent. The *European Federation of National Engineering Associations* (FEANI) serves to coordinate engineering registration qualifications between European nations to allow engineers the freedom to practice across international borders. FEANI confers the title EUR ING (European Engineer) to qualified applicants. In a related international effort, the Engineers Mobility Forum (EMF), together with the Engineer Coordinating Committee of the Asia-Pacific Economic Cooperation (APEC), comprising national engineering organizations from many countries in Oceania, Asia, Africa, North America, and Europe, have created the *International Registry of Professional Engineers* to facilitate comity in engineering qualifications between countries.

The overriding concern of these engineering professional organizations is to protect the reputation, professional status, and economic interests of the engineering profession by ensuring that engineers, regardless of technical specialty, are certified competent in their practice. In addition these organizations seek to influence political bodies to generate legislation and international agreements protective of the professional status of engineering and conducive to profitable engineering practice. One hallmark of this category of professional organization is the emphasis on the promulgation of codes of ethical conduct for engineers. Though details of the ethical codes vary from organization to organization, the codes generally emanate from a few central canons that are somewhat universal. These include holding public safety and welfare of paramount importance, performing work only in areas of competence, making public statements in an objective and truthful manner, and maintaining the interests and confidentiality of clients and employers. In areas where engineering practice is restricted by licensure laws, elements of these ethical codes are generally incorporated into the legal code. Most of the discipline-specific professional organizations have also adopted their own similar codes of ethics that members are expected to uphold.

Other Engineering Organizations

In addition to organizations devoted to technical interests or professional status, there are various other types of special purpose engineering professional organizations. Some of these are aimed at developing a supportive community of interest with respect to race, culture, or gender, such as the Society of Women Engineers, the National Society of Black Engineers, and the Society of Hispanic Professional Engineers. Engineers Without

Borders is an international humanitarian network that seeks to assist disadvantaged communities worldwide and to promote responsible and sustainable engineering. Other organizations are devoted to promoting quality and innovation in engineering education. These include the American Society for Engineering Education, the International Network for Engineering Education and Research, and the European Society for Engineering Education. Many countries have established national advisory organizations, comprising some of the most highly respected engineers, for the purpose of assisting government on matters of public policy related to technology. Examples include the Royal Academy of Engineering in Great Britain, the National Academy of Technologies of France, and the National Academy of Engineering in the United States.

Conclusion

The engineering profession is broad in scope, encompassing topics from nuts and bolts to satellite communications, and from deep-sea oil exploration to medical implants. It is heterogeneous in constitution, with practitioners running the gamut from independent consultants to employees of large, multinational corporations, and performing job functions from detailed component design to company CEO. Perhaps because of the diverse nature of the profession, there is a corresponding profusion in the number and types of engineering professional organizations, each seeking to meet the professional needs of some portion of the engineering community.

BYRON P. NEWBERRY

SEE ALSO *Association for Computing Machinery; Engineering Ethics; Federation of American Scientists; Institute of Electrical and Electronics Engineers; Institute of Professional Engineers New Zealand; Nongovernmental Organizations; Research Integrity; Union of Concerned Scientists.*

BIBLIOGRAPHY

Armytage, W. H. G. (1966). *A Social History of Engineering.* Cambridge, MA: MIT Press. A look at the relationship of engineering and society from antiquity to the twentieth century

Buchanan, Robert A. (1989). *The Engineers: A History of the Engineering Profession in Britain, 1750–1914.* London: Jessica Kingsley Publishers.

Layton, Edwin T. 1986 (1971). *The Revolt of the Engineers: Social Responsibility and the American Engineering Profession.* Cleveland, OH: Case Western Reserve University Press. The best single history of U.S. engineering to 1940.

Reader, William J. (1987). *A History of the Institution of Electrical Engineers, 1871–1971.* London: Peregrinus. A company history, not very critical.

Sinclair, Bruce. (1980). *A Centennial History of the American Society of Mechanical Engineers.* Toronto: University of Toronto Press. A history of the organization and a look at some of the tensions arising from the dual professional-employee roles of engineers.

Watson, Garth. (1988). *The Civils: The Story of the Institution of Civil Engineers.* London: Thomas Telford.

Watson, Garth. (1989). *The Smeatonians: The Society of Civil Engineers.* London: Thomas Telford.

PROFESSIONAL ETHICS

SEE *Profession and Professionalism.*

PROFESSION AND PROFESSIONALISM

• • •

Engineering is generally considered a profession, but science, or at least some of the sciences, are sometimes counted as professions and sometimes distinguished from them. Often, a dispute about the professional status of a science begins when someone proposes it have a code of ethics. What is a profession? What has professional status to do with ethics? What distinction, if any, exists between the professional status of engineering and science? Why should the professional status of either matter?

Four Senses of "Profession"

In ordinary usage, *profession* has at least four senses. First, *profession* can be a mere synonym for *vocation* (or *calling*), that is, any *useful* activity to which one devotes (and perhaps feels called to devote) much of one's life. (If the activity were not useful, it would be a hobby rather than a vocation.) *Profession* in this sense has no necessary relation to income. Even a *gentleman*—in the now outdated sense describing someone rich enough to live comfortably without working—might have such a profession. Max Weber's "Science as a Vocation" (1901) explains how a now-bureaucratized professoriate can still be a vocation in this sense. Weber never uses the term *profession.*

Second, *profession* can be a synonym for *occupation,* that is, any typically full-time activity (defined by function or discipline) by which practitioners generally earn a living. In this sense, one may, without irony, speak of a professional thief or professional athlete. The opposite of *professional* (in this sense) is *amateur* (one who engages in the activity for love rather than money) or *dilettante* (one who lacks the seriousness of those who must live by such work). This is the sense of *profession* from which *professionalism* derives. To exhibit professionalism is to exhibit the knowledge, skill, or judgment characteristic of someone who makes a good living in the occupation. Both engineers and scientists are now generally professionals in this sense, though science still seems to have more room than engineering for amateurs and dilettantes.

Third, *profession* can refer to any occupation one may openly admit to or profess, that is, an honest occupation: While athletics can be a profession in this sense, neither thieving nor being a gentleman can. Thieving cannot because it is not honest; being a gentleman (in its outdated sense) cannot because, though an honest way of life, it is not an occupation. Occupation seems to be the (primary) sense of *profession* in Émile Durkheim's seminal work on professions (written about the same time as Weber's work on vocation).

These three senses of *profession* are alike in having obvious synonyms. If *profession* had only these senses, it would, being redundant, seem destined to disappear from use. Its increasing popularity suggests that these three senses derive from a fourth, the primary sense and the source of the term's popularity. *Profession* in this fourth sense is a special kind of honest occupation. There are at least two competing approaches to defining it: the sociological and the philosophical.

Sociological Definitions

The sociological approach to defining *profession* has its origin in the social sciences. Its language tends to be statistical; the definition does not purport to state necessary or sufficient conditions for an occupation to be a profession, but merely what is true of "*most* professions," "the *most* important professions," or the like. Generally, sociological definitions understand a profession to be any honest occupation whose practitioners have high social status, high income, advanced education, important social function, or some combination of these or other features easy for the social sciences to measure.

Sociological definitions differ a good deal. Some emphasize public service, (individual) autonomy, (group) self-regulation, dangerous knowledge, having a code of ethics, or the like, while others do not. What explains the great variety of sociological definitions? Part of the explanation is that, being statistical, such definitions are not threatened by a few counter-examples. But that is only part of the explanation. Another factor is that when the counter-examples grow more numerous than the professions fitting the definition, defenders can distinguish between *true professions, fully developed professions,* or *paradigms* and those not fitting the definition (*pseudo-professions, less well developed professions,* or *quasi-professions*). The only professions that appear on every sociological list of true, fully developed, paradigmatic professions are law and medicine. When evidence suggests that even these do not fit the definition, sociologists can retreat again, claiming that their definition states an *ideal type* that actual professions only approximate. When asked why this ideal type is chosen over another, sociologists generally explain the choice in terms of a theory of society they accept (Marxist, Weberian, Durkheimian, or the like). Sociological definitions seem to derive from theory, not evidence. The way professions understand themselves plays a surprisingly small part in the sociological approach.

For most sociological definitions, little distinguishes contemporary professions from what used to be called the *liberal professions* (those few honest *vocations* requiring a university degree in most of early modern Europe). Carpentry cannot be a profession (in the sociological sense) because both the social status and education of carpenters are too low. Science is a profession in this sense because scientists have relatively high status, high income, advanced education, and important social functions. Technical managers also form a profession in this sense because they too tend to have high income, high status, advanced education, and an important social function. According to most sociological definitions, Europe and the Americas have had professions for many centuries.

Philosophical Definitions

The philosophical approach to defining *profession* attempts to state necessary and sufficient conditions. A philosophical definition is therefore much more sensitive to counter-example than sociological definitions are. Philosophical definitions may be developed in one of (at least) two ways: the Cartesian or the Socratic.

The Cartesian way tries to make sense of the contents of one person's mind. One develops a definition by asking oneself what one means by a certain term, setting out that meaning in a definition, testing the definition by counter-examples and other considerations, revising whenever a counter-example or other consideration seems to reveal a flaw, and continuing that process until one has put one's beliefs in good order.

In contrast, the Socratic way seeks common ground between one or more philosophers and *practitioners* (those who normally use the term in question and are therefore expert in its use). A Socratic definition begins with the definition a practitioner offers. A philosopher responds with counter-examples or other criticism, inviting practitioners to revise. Often the philosopher will help by suggesting possible revisions. Once the practitioners seem satisfied with the revised definition, the philosopher again responds with counter-examples or other criticism. And so the process continues until everyone is satisfied with the result. Instead of the private monologue of the Cartesian, there is a public conversation. But neither the Cartesian nor the Socratic approach is empirical (in the way the sociological approach at least claims to be). They are equally analyses of concepts. They differ primarily in how they understand concepts. For the Cartesian, concepts are more or less private; for the Socratic, they are a public practice.

What follows is a Socratic definition: "A profession is a number of individuals in the same occupation voluntarily organized to earn a living by openly serving a certain moral ideal in a morally permissible way beyond what law, market, and morality would otherwise require."

According to this definition, the members of a would-be profession must have an occupation. Mere gentlemen cannot form a profession. Hence, members of the traditional liberal professions (clergy, physicians, and lawyers) could not form a profession until quite recently—until, that is, they ceased to be gentlemen, began to work for a living, and recognized that change in circumstance. That seems to be well after 1800. Most professions are much younger than the function they perform or the discipline they exploit.

The members of the would-be profession must not only have an occupation, they must *share* it. So, for example, chemists and chemical engineers cannot form one profession because they are trained in different academic departments, learn different skills, and generally do different work. They belong to different occupations.

Ethics and Professions

According to the Socratic definition above, each profession is designed to serve a certain moral ideal, that is, to contribute to a state of affairs everyone (all rational persons at their rational best) can recognize as good. So, physicians have organized to cure the sick, comfort the dying, and protect the healthy from disease; engineers, to help produce and maintain safe and useful objects; and so on. But a profession does not just organize to serve a certain moral ideal; it organizes to serve it *in a certain way*, that is, according to standards beyond what law, market, and morality would otherwise require. A would-be profession, then, must set *special* (morally permissible) standards. Otherwise it would remain nothing more than an honest occupation. Among its special standards may be a certain minimum of education, character, or skill, but inevitably some of the standards will concern conduct. These standards of conduct will be ethical (as distinct from moral): they will govern the conduct of all members of the *group* simply because they are members of that group (and not, as ordinary moral standards do, just because they are moral agents).

These special standards will, if effective, be ethical in another sense as well. They will be *morally* binding on members of the profession (and only them). The members of a profession must pursue their profession openly; that is, engineers must declare themselves to be engineers, chemists must declare themselves to be chemists, and so on. The members of a (would-be) profession must declare themselves to be members of that profession in order to earn their living by that profession. They cannot be hired as such-and-such (say, an engineer) unless they let people know that is what they are. If their profession has a good reputation for what it does, the declaration of membership will aid them in earning a living. People will seek their help. If, however, the profession has a bad reputation, their declaration of membership ("I am a tinker") will be a disadvantage. People will shun their help. The profession's special way of pursuing its moral ideal is what distinguishes its members from others in the same occupation, and from what the members would be but for their profession.

Of course, the declaration of membership must be true. Those who declare membership in a profession to which they do not belong are mere charlatans, quacks, impostors, or the like. How membership is determined may vary a good deal from one profession to another. Some professions have only a set curriculum to assure minimum knowledge. (Graduate with the appropriate degree and one is a chemist.) Other professions have only a test. (Pass the examination and, however one learned the discipline, one is an actuary.) And other professions have a more complex standard. (So, for example, to be a physician, one must graduate with a certain degree, work under supervision for a time, and pass certain examinations.) What all professions share are special standards distinguishing members from others. Whatever their origin, these standards, once accepted in practice, constitute the *professional organization*. The professional organization (that is, the profession) is distinct from any technical, scientific, or mutual-aid society members of a profession may form.

The members of a profession, being free to declare membership or not, will generally declare membership if, but only if, the declaration benefits them overall—that is, serves some purpose of their own at what seems reasonable cost. The purpose may be high-minded, self-interested, or even selfish. Whatever the purpose of individuals, their membership in a profession identifies them as engaged in pursuing the profession's moral ideal according to the morally permissible special standards the profession has adopted. *Occupations* can be "value free" (that is, have no special commitments); *professions* cannot.

Where members of a profession declare their membership voluntarily ("I am an architect"), they are part of a voluntary, morally permissible, cooperative practice. They are in position to have the benefits of the practice, employment as members of that profession, because the employer sought such-and-such and they (truthfully) declared their membership. They will also be in position to take advantage of the practice by doing less than the standards of practice require, even though the expectation that they would do what the standards require as declared members of the profession is part of what won them employment. If cheating consists in violating the rules of a voluntary, morally permissible, cooperative practice (that is, taking unfair advantage of the practice), then every member of a profession is in a position to cheat. Because cheating is morally wrong, every member of a profession has a moral obligation, all else equal, to do as the profession's special standards require.

A profession's ethics imposes moral obligations on members of that profession. These obligations may, and generally do, vary from profession to profession (and, within a single profession, may also vary over time). These obligations appear in a range of documents, including standards of education, admission, practice, and discipline. A code of ethics is the most general of these documents, the one concerned with the practice of the profession as such.

Status and Profession

According to the Socratic definition above, an occupation's status as a profession is (more or less) independent of license, state-imposed monopoly, and other special legal intervention. Such special legal interventions are characteristic of bureaucracy rather than profession. In principle, professions are not the creatures of law; and, even in practice, some professions (such as Certified Computer Professionals) do without license, monopoly, and other legal protection against market pressures, except for protection of their designation (such as "CCP") analogous to that the law gives to trademarks to protect the consumer from counterfeits.

An occupation's status as a profession is, according to this definition, also more or less independent of its social status, income, and other social indexes of profession. There is, for example, no profession of technical managers, even though technical managers have relatively high social status, income, and education and important social functions. What technical managers lack is a common moral ideal beyond law, market, and ordinary morality—and common standards, including a code of ethics, settling how that ideal should be pursued. There is, in contrast, certainly a profession of nursing, though nurses typically earn much less than technical managers and have much lower social status. The only high status a profession entitles one to is being regarded as more reliable or trustworthy in what one does for a living than one would (probably) be if that way of earning a living were not organized as a profession. This high status is deserved only insofar as the profession continues to meet the special standards it has set for itself. An occupation should become a profession in this fourth sense if, but only if, it is willing to assume the burdens that generate that high status. The current popularity of the terms *professional* and *professionalism* is evidence that, on the whole, the professions have been handling that burden pretty well.

MICHAEL DAVIS

SEE ALSO *Codes of Ethics; Durkheim, Émile; Professional Engineering Organizations.*

BIBLIOGRAPHY

Abbot, Andrew. (1988). *The System of Professions: An Essay on the Division of Expert Labor.* Chicago: Chicago University Press. An important example of a sociological approach to professions.

Bayles, Michael. (1981). *Professional Ethics.* Belmont, CA: Wadsworth. A widely used text.

Camenisch, Paul. (1983). *Grounding Professional Ethics in a Pluralistic Society.* New York: Haven Publishing. Eclectic approach to explaining the moral status of professional obligations. Contrasts nicely with Bayles' consequentialism.

Chalk, Rosemary. (1980). *AAAS Professional Ethics Project: Professional Ethics Activities in the Scientific and Engineering Societies*. Washington, DC: American Association for the Advancement of Science. A collection and analysis of codes.

Davis, Michael. (1998). *Thinking Like an Engineer*. New York: Oxford University Press. A detailed argument for understanding engineering as a profession and understanding that status as central to how engineers actually work.

Davis, Michael. (2002). *Profession, Code, and Ethics*. Aldershot, UK: Ashgate. A defense of the Socratic definition of profession. Includes a chapter on whether scientists have any professional obligations (and several chapters on engineering).

Durkheim, Émile. (1957). *Professional Ethics and Civic Morals*, trans. Cornelia Brookfield. London: Routledge. Lectures given in the 1890s but not published until 1947. Seminal work in sociology of professions although it is actually only about occupations.

Freidson, Eliot. (2001). *Professionalism: The Third Logic*. Chicago: University of Chicago Press. A sociologist arguing that professions are "third logic," distinct both from free market and bureaucratic regulation. Full of interesting examples and important insights.

Gewirth, Alan. (1986). "Professional Ethics: The Separatist Thesis." *Ethics* 96: 282–300. Classic refutation of the claim that professional status to some degree frees professionals from the constraints of ordinary morality.

Goldman, Alan. (1980). *The Moral Foundations of Professional Responsibility*. Totowa, NJ: Littlefield Adams. The classic defense of the claim that some professions are to some degree free of ordinary morality.

Koehn, Daryl. (1994). *The Ground of Professional Ethics*. London: Routledge. An attempt to understand professions as founded on a "covenant"; a worthy example of the Cartesian approach.

Kultgen, John. (1988). *Ethics and Professionalism*. Philadelphia: University of Pennsylvania Press. A philosopher attempting to understand professions using the sociological approach.

Larson, Magali Sarfatti. (1977). *The Rise of Professionalism: A Sociological Analysis*. Berkeley: University of California Press. A good example of the Marxist approach to professions. Dated but still important.

Weber, Max. (1958 [1901]). "Science as a Vocation." *From Max Weber: Essays in Sociology*, ed. H. H. Gerth and C. Wright Mills. New York: Oxford University Press. First published in 1901. A seminal work in the sociology of professions (though it is about vocations, not professions).

PROGRESS

• • •

The idea of progress is unique to the cultural tradition of Western Europe and from its birth has had a strong association with ethical issues raised by new knowledge and technological innovation. Although there are allusions to it in the twelfth and thirteenth centuries, the concept first appeared in its modern sense in the transition from the Middle Ages to the Renaissance. The idea was introduced by the early humanists in the context of their invention of the division of history into three periods: a classical age, encompassing the cultures of Greece and Rome from about 600 B.C.E. to 400 C.E.; a culturally dark "middle age" from about 400 to 1300; and their own age, self-proclaimed as a renaissance, or rebirth, of cultural excellence that began in the fourteenth century. In the seventeenth and eighteenth centuries, progress was explicitly coupled to the primacy of objective reason in human affairs and the promise of technological progress became an explicit dogma of the eighteenth century Enlightenment. In the nineteenth and twentieth centuries, progress became the mantra of industrial capitalism, proclaiming the blessing it conferred on society even as the reality of progress came under attack, first by the Romantics, then by philosophers and intellectuals more broadly, and finally by social and political activists.

Defining Progress

What the word *progress* means has thus changed significantly since the mid-fourteenth century. Common to all definitions, however, is the claim that *something* is better than it had been and promises to get better still in the future. What that something is, is what has changed over time. For the humanists, the something was *high culture*—literature, poetry, painting, sculpture, and architecture—and, perhaps surprisingly for humanists to be proud of, technology. All of these, they argued, were better in the fifteenth and sixteenth centuries than they had been and they promised to keep getting better. In the seventeenth and eighteenth centuries, the definition of progress, though it looked to the growing power of modern science as evidence, widened to an identification of progress with intellectual and social reform, and thus with the claim that the subject of progress was the human condition itself, which not only could be, but in fact was being improved by the efforts of human beings themselves. Through initiative, courage, reason, and inventiveness, it was argued, individuals were improving the world in which they found themselves and in the process making people better as people.

In the nineteenth and twentieth centuries, the idea of progress became increasingly complex and controversial. For one thing, the claim that art and literature were progressing fell out of favor. They changed, of course, but many dismissed any judgment that impressionism was better than Renaissance painting or that Yeats was a better poet than Milton. Cultural forms change but do not move toward an ultimate perfection, nor do there exist objective criteria for judging across these forms. Meanwhile contemporary science and technology in effect co-opted the idea of progress, claiming improvement as self-evident. And even as the ideal of human progress shaped nineteenth- and twentieth-century social and political reform movements—liberalism, socialism, and communism—increasingly strident challenges were raised against the claim that the human condition and human beings had improved in any essential way.

The bitterness of the criticism of progress in the late-twentieth century was in part the legacy of two murderous world wars, in part the failure of many social and political reform movements to effect lasting improvements in the quality of life when they achieved power, and in part a response to the emergence of environmental, social, and personal problems linked to applications of increasingly powerful scientific theories and technological innovations. Relevant, too, was the historicism and relativism of much twentieth-century social science and philosophy, according to which there were no universal, objective, and hence value-neutral criteria for judging whether a change of any sort was an unqualified improvement. In the realm of technology, there are objective criteria for comparing and evaluating changes because artifacts are means to ends defined by their makers. Given the intended purpose of a camera, for example, one model can be said to be better or worse than another. But because the notion of purpose or end in relation to nature was abandoned in modern science, there is no basis in science or in technology for judging the value of the ends to be served by technologies and therefore no basis for judging that changes to natural entities are improvements. This isolation of ends from means creates an ethical gulf between technical knowledge and its applications that was only fully appreciated in the second half of the twentieth century, a gulf that further undermined claims of progress even in science and technology.

Progress as Threat and Ideal

From its introduction by the humanists, progress was a profoundly new and a profoundly secular idea, and the claim of real and promised improvement that it made was extraordinarily bold. The idea of progress challenged what had been a deeply rooted belief in pre-modern Western culture, inherited from antiquity, that the golden age of humankind lay in the past and that the aging of the Earth entailed decay for it and its inhabitants, analogous to the aging of individual living organisms. Furthermore the idea of progress implies a directionality to history and to time that contrasts sharply with the cyclical conceptions of time and of history dominant in antiquity. Finally the idea of progress implies an activist role for humans in defining their well-being and in causing it, in the present and for the future.

Judaism and Christianity, through their respective messianic and salvational doctrines, had already introduced an anticlassical directionality to history and time, but this directionality was the culmination of a divine plan and in the hands of God; it was not open to calculated, self-interested human intervention. Attributing value to improving the human cultural or material condition in a Christian context posed a direct challenge to transcendent religious values, and the claim that humans could by their own efforts make themselves better posed an even greater threat. The broad public appeal of and occasional resistance to the ideology of progress, first in Europe and then globally, thus reveals a great deal about these societies and their deepest values.

In the fourteenth century, long before the first hints of modern science or modern philosophy, the idea of progress had already emerged in Western Europe, tentatively in the context of the twelfth- and thirteenth-century university movement, but clearly in the writings of the poet Petrarch, heir to Dante and father of humanism. The humanists are inaccurately depicted as worshipping Greek and Roman literary culture and seeking to reconstruct it imitatively. Petrarch's conception of a Renaissance was not the rebirth of antique ways of living and writing in the manner of a Williamsburg, Virginia. It was a rebirth of the style standard set in antiquity, after a long *dark age* during which this standard, especially in literature, art, and manners, was debased. As a start, then, but only as a start, the humanists sought by emulation first to recover, then to master, and ultimately to improve upon, what the Ancients had achieved—to use ancient texts as stepping-stones to still greater accomplishment. Bees, Petrarch noted, take pollen from flowers but transform it into honey, which is better than pollen. This is the humanist conception of progress: to take the pollen of stylistic excellence from ancient art and transform it into the honey of still greater art.

The idea of progress is expressed clearly enough here for it to have become an issue by the end of the fifteenth century. With the invention of increasingly powerful gunpowder-based weaponry; of printing by movable metal type followed by the rapid growth of a vigorous international printed book industry; of central vanishing point perspective and the flowering of Renaissance art and sculpture; of new, more complex forms of musical harmony and composition; of new, more powerful types of machinery; and with the voyages of discovery east to India and west to the Americas, culminating in Magellan's circumnavigation of the globe in 1525, all enabled by new techniques of mapmaking and navigation, defenders of progress argued that the ancients had been far surpassed by the *moderns*. There followed, throughout the sixteenth century and into the seventeenth, set piece entertainments, popular in courts across Western Europe and in many books and essays, called the Battle of the Ancients and the Moderns in which the claim that *we* were superior to ancient predecessors was defended against the argument that the ancients were superior in quality, as human beings, in spite of subsequent superficial technological superiority.

By the 1660s, the idea of progress was no longer open for debate. Joseph Glanville's *Plus Ultra* (1668) was a paean to the new experimental philosophy, enabling humankind to *go further*, to exceed all limitations previously set by ignorance and superstition (and religion!) on what people can know and achieve. While the engine of progress in the fifteenth and sixteenth centuries had been identified with inventiveness or creativity, especially in art and technology, with the seventeenth-century rise of modern science and philosophy, the engine of progress became reason, especially as exemplified in science and mathematics. This identification of progress with reason became a central dogma of modernism: that through the exercise of reason human beings can improve life on Earth without limit. In both modern philosophy, whether rationalist or empiricist, and in modern science, reason subsumes inventiveness and shifts the focus of progress from art and technology to understanding, with technological innovation merely a fruit or byproduct of understanding.

It is this version of the idea of progress that is at the heart of the eighteenth-century Enlightenment and expressed in Thomas Paine's *Age of Reason* (1795). It is the justification for the republican *experiment* that created the United States and inspired the French revolution; that without kings, history, or God, the exercise of reason alone can create better societies than have ever existed, societies in which people will be happier, healthier, more prosperous, longer-lived, and more productive, for themselves and for others. The clear expectation that basing action on reason would produce better people is articulated in the Marquis de Condorcet's 1793 "Sketch for a Historical Depiction of the Progress of the Human Mind" (*L'esprit humaine*), written, ironically and tragically, on the eve of Condorcet's imprisonment by agents of the very Revolution whose ideals he proclaimed.

Progress Under Attack

The case for the rationalist interpretation of progress was based on the manifest superiority of modern science over ancient, medieval, and Renaissance science, of modern philosophy—René Descartes, Benedict de Spinoza, Gottfried Wilhelm Leibniz, John Locke, and Immanuel Kant—over ancient, medieval, and Renaissance philosophy, and on the continually increasing power of technology, especially after the invention in the late-eighteenth century of mass production machinery and the steam engine. But the Romantic poets, novelists, and playwrights—among them Samuel Taylor Coleridge, William Blake, and William Wordsworth in England, and Novalis and Heinrich Wilhelm Kleist in Germany—rejected the hegemony of reason in human affairs, the capacity of reason to serve as an engine of truly human progress, and even the possibility of a happy ending to human history by creating an earthly, secular version of Paradise. With the spread of the Industrial Revolution and the *dark Satanic mills* (as Blake called them) that were its progeny, of the railroads with their noise and pollution, with the growing, poverty-ridden urban proletariat, the case for social progress weakened.

Progress within science and in technology, however, could hardly be gainsaid. Scientific theories clearly kept getting better in terms of explanatory power, prediction, control, and revelation of hitherto unknown aspects of reality. New inventions—steam-powered factories, ships, and railroads; the telegraph; synthetic dyes; electricity; the telephone; the automobile; and flight—gave people unprecedented capabilities and poured out in seemingly endless profusion. But the note that had been sounded in the sixteenth-century Battle of the Ancient and the Moderns was sounded again: Does any of this scientific and technological progress mean social or human progress? Does it make people better? Is the human condition in fact better than it was before, or is it merely different? Again every improvement entails a change, but not every change entails an improvement!

On what grounds can people judge which changes are improvements? How can they tell which capabilities provided by technological innovations are worth adopting? To whom or to what do people turn to learn how to apply knowledge or implement innovations and set goals, for which particular technologies can provide helpful means? In the absence of goals, means become ends in themselves. Neither technology nor science can help to identify which ends to pursue with their aid: technology because it is purely a means, and science because value-neutrality is central to the methodology of modern science.

The equation of progress with the application of value-neutral reason became increasingly problematic in the course of the nineteenth century. Echoing the earlier Romantic poets, philosophers from Arthur Schopenhauer and Søren Kierkegaard to Friedrich Nietzsche and Henri-Louis Bergson formulated criticisms of reason that undermined its capacity to serve as the engine of human or social progress. By the end of World War I, the claim that through science and reason Western societies and their inhabitants had improved rang hollow. This feeling was intensified by the global slaughter of World War II, a war in which the most advanced forms of value-neutral rationality, science, and technology were proudly allied to the value-laden nonrationality of politics.

The Price of Progress

In the course of the twentieth century, then, it became clear that the price of modern science and science-based technology was that the ties between knowledge and action were sundered. Even as the rate of development of theories in the sciences and the pace of technological innovation accelerated, driven by massive public and corporate funding and by the creation of reinforcing social institutions, even as science and technology became the dominant agents of social change and became inextricably entangled with personal and social life and values, the ethical divide separating knowledge and action widened. It seemed that progress could be defined unequivocally with respect to scientific theory change and technological innovation, but claims that social and personal life style changes were progressive were highly equivocal. Suddenly the ethical implications of science and technology became central issues for society, but there existed no conceptual tools, comparable in power to those available to scientists and engineers, for grappling with these issues, nor did the average person have the political and economic power to challenge the institutions that exploited science and technology.

In fact even the confidence that progress could be defined objectively with respect to scientific theory change and technological action was severely shaken in the 1960s. Technological change can be evaluated objectively but only with respect to parameters that incorporate arbitrary value judgments: A high speed Internet connection is better than a slower speed con-

nection if the values of speed and of being connected to the Internet at all are accepted as givens. These values, of course, cannot be judged objectively. An analogous challenge was raised with respect to science, because from its beginning modern science had as its primary objectives discovering the nature of things, revealing the hidden causes of why things happen, and disclosing reality. In the nineteenth century, questions were raised about the relation between increasingly abstract mathematical physical models of nature and what was *really* out there, but the prevailing view remained that scientific theories changed because newer theories were truer to reality than older ones. To be sure, quantum theory raised more serious questions about the relation between physics and reality than had been asked in the nineteenth century; and the Copenhagen Interpretation of quantum mechanics invented by Niels Bohr and Werner Heisenberg argued that physics could not provide a picture of reality, only an empirically satisfactory account of experience.

It was only in the 1960s, however, that a broad consensus grew among intellectuals, challenging the progressive and objective character of scientific knowledge. People had no real access to the new realities that scientists claimed to be encountering and thus no way to know whether such advances truly constituted progress. This consensus was precipitated by the debate over Thomas Kuhn's *The Structure of Scientific Revolutions* (1962), which led to a broad historical, philosophical, and social scientific critique of the concept of objectivity and for many scholars a rejection of the possibility of objective knowledge. This in turn triggered the so-called *Science Wars* of the 1980s and 1990s in which the objectivity of scientific knowledge and the progressive character of scientific theory change were defended by physical and life scientists. But even if the objectivity of scientific knowledge were conceded, bridging the ethical gulf between value-neutral knowledge and its applications remains an issue in the early-twenty-first century.

STEVEN L. GOLDMAN

SEE ALSO *Change and Development; Development Ethics; Theodicy; Wells, H. G.; Wittgenstein, Ludwig.*

BIBLIOGRAPHY

Becker, Carl. (1993). *The Heavenly City of the Eighteenth Century Philosophers.* New Haven, CT: Yale. An important early twentieth century historian's analysis of the goals of enlightenment social and political reformers. Very well written.

Burtt, Edwin A. (2003). *The Metaphysical Foundations of Modern Science*. New York: Dover. A classic pre-1960s study of the ideas and values underlying the seventeenth century scientific revolution.

Goldman, Steven L., ed. (1989). *Science, Technology and Social Progress*. Bethlehem, PA: Lehigh University Press. A collection of thought-provoking essays by scientists, historians and philosophers on the question of whether modern science and technology have improved the quality of human life.

Hacking, Ian. (2000). *The Social Construction of What?* Cambridge, MA: Harvard University Press. A clearly written, balanced, and insightful account of the "science wars" of the late twentieth century.

Kelly, Donald R. (1991). *Renaissance Humanism*. Boston: Twayne. Excellent introduction to the distinctive ideas and values of key humanist thinkers and why they are important to understanding the riser of modernity.

Kitcher, Philip. (2001). *Science, Truth and Democracy*. Oxford: Oxford University Press. Excellent, deeply thought-provoking discussion of bridging the gap between technical knowledge and its applications in a democratic society.

Kuhn, Thomas. (1962). *The Structure of Scientific Revolutions*. Chicago: University Of Chicago Press. This is the book—short, sharply focused and clear—that precipitated the challenge to the claims that scientific knowledge was truly objective.

PROMETHEUS

• • •

The Punishment of Prometheus, as depicted on a Laconian cup, c. 555 B.C.E. (© Scala/Art Resource, NY.)

In ancient Greek mythology the hero Prometheus (meaning forethought) rose up to the heavens to light a torch from the Sun's fire, then brought it back to Earth for humankind. This fire, stolen from the sun god Helios, transformed humankind into something superior to other living beings. As retribution, Zeus sentenced Prometheus to be chained to a rock while an eagle forever gnawed at his liver; Hercules killed the eagle and freed him. Zeus's divine justice included a ruse for Prometheus's brother Epimetheus (meaning afterthought). He received the gift of an all-good, incomparably beautiful wife, Pandora, who came accompanied by a box that was never to be opened. Pandora could not resist the temptation and opened the box, releasing upon humankind a manifold of miseries and evils—along with hope.

In Greek literature the story of Prometheus can be found in three sources: Hesiod's *Theogony* and *Works and Days* (eighth century B.C.E.) and Aeschylus's *Prometheus Bound* (fifth century B.C.E.). (Aeschylus's drama is the only extant part of a trilogy that began with *Pro-metheus Fire-Carrier* and concluded with *Prometheus Unbound*.) Plato's *Protagoras* also provides a version of the myth in which Prometheus steals *technai* (technics) from Hephaestus and Athena, after which Zeus commands Hermes to give human beings a sense of justice and shame so that they might live with their new abilities (*Protagoras* 320d-322d). Plato further has Prometheus mentioned as a giver of problematic gifts in the *Gorgias* (523d-e), the *Politicus* (also known as *Statesman* (274a), and the *Philebus* (16e). After Plato, however, it is significant that Prometheus does not have a prominent place in Greek or Roman or even medieval European literature.

In modern culture, however, Prometheus plays a more significant and somewhat altered role. As Karl Kerényi (1963), among others, notes, he often represents a creative rebellion against the limitations of the human condition, for which he is unjustly punished. Although humanity pays for its productive creations, Prometheus is to be admired for his courage and the heroic self-sacrifice that accompanies technological progress. At the same time, new discoveries, driven by hope springing eternal, repeatedly bring forth negative unintended consequences. In counterpoint to such a Promethean fate, Ivan Illich (1972) presented the image of *Epimethean Man*, who in retrospect learns to practice what, in the early-twenty-first century, is called the "precautionary principle."

Among the many modern reflections on the Prometheus story are the short lyric poem of the same name by Johann Wolfgang von Goethe (1774) and the poetic play, *Prometheus Unbound*, by Percy Bysshe Shelley (1819). The Dirck van Baburen painting *Prometheus Being*

Chained by Vulcan (1623) is representative of a novel visual interest. Ludwig van Beethoven's Geschöpfe des Prometheus (ballet, opus 43, 1801) and Erocia (third symphony, opus 55, 1801) both reveal the composer's personal sense of confrontation with Promethean struggles. The best-known modern adaptation is, however, Mary Shelley's Frankenstein, or The Modern Prometheus (1816).

More recently Carl Orff's opera Prometheus (1968), Richard Schechner's performance work The Prometheus Project (1985), and Tony Harrison's film Prometheus (1998) all link the story to technology, although in different ways. Orff's music has been described as anticipating technomusic. Schechner's performance employs projected images to connect Hiroshima and pornography. In Harrison's film, miners from a closed colliery pit are melted down and made into a golden statue of Prometheus, which is then trucked by Hermes across Europe from Dresden to Auschwitz and eventually to Greece. Allegorically, Hermes, the messenger god in mythology, returns the current age to the immortality of ancient Greece; so too each epoch age revives the original impulse of the promethean myth and this recurrent hope: Carrying the human torch back to its source, like an Olympian returning home, connotes carrying on with humanity, its eternal re-emergence rising from human ashes and senseless destruction to rebirth, with glories restored and horrors transcended.

Finally the extent to which the Prometheus story may serve as a continuing vehicle for reflections on issues related to science, technology, and ethics is indicated by simply noting the titles of the following books: John M. Ziman's Prometheus Bound: Science in a Dynamic Steady State (1994); Thomas Parke Hughes's Rescuing Prometheus: Four Monumental Projects that Changed the Modern World (1998); Norman Levitt's Prometheus Bedeviled: Science and the Contradictions of Contemporary Culture (1999); Darin Barney's Prometheus Wired: The Hope for Democracy in the Age of Network Technology (2000); Arthur Mitzman's Prometheus Revisited: The Quest for Global Justice in the Twenty-first Century (2003); and William Newman's Promethean Ambitions: Alchemy and the Quest to Perfect Nature (2004).

MARY LENZI
CARL MITCHAM

SEE ALSO Faust; Frankenstein; Playing God.

BIBLIOGRAPHY

Illich, Ivan. (1972). Deschooling Society. New York: Harper and Row.

Kerényi, Karl. (1963). Prometheus: Archetypal Image of Human Existence, trans. Ralph Manheim. New York: Pantheon. First German publication, 1946. A Jungian commentary that references Goethe's poetry more than Aeschylus's drama.

PROPERTY

• • •

Property is defined as that which is owned, including both tangible things and the right to engage in certain actions. In the physical and social sciences one can speak of a property of a thing or an object in describing a characteristic of that thing. Here property is restricted to the right or authority to determine how a resource is used. Society designates who holds a resource and how it is used through governmental enforcement of laws or through social custom and tradition. Property rights determine not only who is allowed to use a resource but how exclusive the use is, who has the ability to preclude use, and how property may or may not be acquired and exchanged. Property therefore helps define the relationship between individuals and between groups of individuals.

Property and property rights depend on the answers to two fundamental questions: In a just society, what criteria should be used to distribute resources? and What types of property rights structures should be recognized? The philosopher John Locke's (1632–1704) concept of natural rights provides a starting point: Individuals have property rights to themselves and their labor a priori. In other words, independent of institutional, legal, cultural, or social constraints imposed by others, persons have rights to themselves and the products of their labor as long as they do not impede others from exercising the same rights. The idea of natural rights is both intuitive and morally appealing yet is insufficient because in the course of human interactions people tend to impede others from realizing their natural rights.

Property, Technology, and Science

Technology plays a central role in the use and protection of property, and in many instances property rights depend on access to and control of technology. In the extreme case wilderness real estate may be accessed only with a helicopter, but even a computer cannot be used without electricity. Not only the effectiveness but also the security and transferability of property rest on the ability to monitor and enforce property rights. Technologies for monitoring and securing physical property can take the form of locks, fences, security cameras, tamperproof devices, and alarms. In the absence of monitoring

TABLE 1

Property: Compatibility and Exclusivity

	Exclusive	Non-exclusive
Incompatible	Pure private goods: apples, TVs, automobiles	Common property resources: fisheries, roads, groundwater
Compatible	Public goods: concerts, internet access	Pure public goods: national defense, clean air

SOURCE: Adapted from Perloff (2001), p. 628.

or enforcement technologies even well-defined property rights lack meaning.

The control of technology not only determines who has access to property but also may influence capital flows to support research and development. A communications network may facilitate connections between some inventors and venture capitalists and exclude others. At the same time, when research and development are funded primarily by private interests, this can create a risk of subordinating scientific inquiry to profit making. Similarly, easy technological access to venture capital or other types of investments can distort scientific interests.

Scientific research and technological development themselves depend on secure property rights. Without secure intellectual property rights incentives to engage in research and development are lowered because the rewards may not accrue to those who produce those goods.

Characteristics of Property

Property has two main characteristics that determine how well it functions, that is, how easily it can be transferred and how well related rights can be monitored or enforced. The two characteristics are incompatibility and exclusiveness. A good or service is incompatible if consumption by one person precludes consumption by another person. If one person eats an apple, another person cannot eat it. Exclusivity means that the owner of a good should receive all of the benefits and costs of ownership including the ability to exclude consumption of the good for those who do not pay. Owners of a movie theater can exclude customers who do not pay. Owners of a drive-in are less able to exclude non-paying customers. The degree to which something is incompatible and exclusive determines the degree to which a good is private or public. The matrix (Table 1), which is adapted from the work of Jeffrey Perloff (2001, p. 628), summarizes the possible combinations of incompatibility and exclusiveness.

A pure private good is one that is both incompatible and exclusive, such as many consumer goods. With pure private goods there are no restrictions on the property right to use or exchange the good. The value of the good is determined by the ability to exchange the good with others on mutually agreeable terms. With private goods the market price reflects the value of the property right to the good or service in its best use. However, the table does not convey the extent to which most goods and services are hybrids somewhere in the middle. Most goods have degrees of compatibility (that is, they are nonrival) or exclusivity.

Private Goods

The basis of neoclassical economics is private property. In fact, the economic historian Donald McCloskey (1985) defines modern economics as the science of property such that property itself is defined not merely as a thing but as a social relation. If everything owned and exchanged is costless, the property right to the thing being exchanged belongs to the person who values it the most. If the thing has no value, no one will bother with it, and hence there is no need to define the relationship between the thing and anyone who would possess it. The value of property depends on its scarcity. If more than one person desires a thing, property rights to that thing define the relation not only of the owner to the thing but of the owner to anyone else who may value the thing and of nonowners to the thing. The effectiveness of property as a social relation therefore depends on the definition of this social relation and its transferability. For economists well-functioning markets for pure private goods depend on clearly defined property rights.

In regard to the question of how property historically has been defined or how it can be clearly defined economists resort to the tautological argument that rights become well defined when it is in someone's interest to do so. This answer leads to inequitable income distributions. To understand this one can use Allan Schmid's argument about capitalization and the role of property (Schmid 1987). Schmid contends that the property right to exchange facilitates capitalization, or the conversion of future values into present values. In other words one person can consume today by trading his or her future production for someone else's current consumption. The ability to exchange the present for the future provides incentives to innovate and to invest in scientific research and technological development. If property is not transferable over time, the individual producer will have to wait until production is finished in order to consume. In this case there is neither borrowing nor lending.

The problem here is that the way property rights are defined affects the rewards given to innovators. If markets are competitive, the benefits of transferable property rights and their concomitant technological advances will be shared by everyone in society over time. However, perfectly competitive markets rarely exist outside economic theory textbooks. Market power in imperfect markets, which are the norm, means that some individuals have easier access to credit and capital. This typically results in capital markets that provide instant wealth to innovators and an astounding degree of income inequality. The stock market magnifies this inequality and has not always produced capital for new investment; instead it often provides power, both political and economic, that strengthens property holders' interests at the expense of those who do not have access to capital.

Public Goods

At the other end of the spectrum a pure public good such as clean air is both compatible (nonrival) and nonexclusive. The less exclusive a good is, the more difficult it is to monitor and enforce property rights to that good. In the case of a pure public good property belongs to everyone. Because the benefits of ownership accrue to everyone but the costs accrue to no one in particular, the provision of a pure public good depends on someone bearing the cost of production. In many cases with high costs of provision, such as national defense, the government must step in and pay by collecting tax dollars from those who benefit. The share of benefits may not reflect the proportional tax share borne by each taxpayer accurately.

When property is publicly owned or when the acquisition and exchange of property is not well defined, determining a just distribution of resources and deciding who controls access become an ethical minefield. The decision about who gets to decide and how questions of allocation and distribution are decided is complex. Locke's idea of natural rights to oneself and one's labor is difficult if not impossible to extend to communal property. The political philosopher Karl Marx's (1818–1883) version of socialism was an attempt to make collective decisions about the production and distribution of both common and private property. For Marx property could not be appropriated (literally "made one's own"). If profits were realized from collective production, they would be shared equally among all people according to each person's needs. However, Kenneth Arrow's impossibility theorem challenges the possibility of a common social choice.

Between Pure Public and Pure Private Goods

Between pure public goods and pure private goods lies the murky continuum of fuzzy property rights. Some resources are rival but nonexclusive. Groundwater can be accessed by anyone with a pump, but once removed from the ground, it is typically the property of the person who owns the pump. The classic example of a common property resource is the commons, or town pasture, where individuals were allowed to graze their animals without cost. In this case each individual can graze additional animals on the commons without any additional cost to the individual but with a cumulative detrimental cost to the commons. The net effect of each individual's rational actions when property rights are absent and individuals are free to use the resource without cost or at a cost that does not reflect the true value of use is complete degradation of the resource, or what Garrett Hardin (1968) called the tragedy of the commons.

Intellectual property, or ideas, innovations, and inventions, also lies between pure private and pure public goods. Intellectual property rights such as patents and copyrights are some of the most difficult rights to qualify. Intellectual property is both intangible and compatible (nonrival). Once intellectual property is produced, anyone can enjoy it at zero or very low additional cost. Thus, the cost of developing the first unit is often great and the incentive to produce it does not exist unless the producer can charge more than one person for use of the property and recoup the cost of research, development, and normal operating costs. Without the incentive to innovate, new ideas and consequently new technology are slow to develop, especially if other individuals can duplicate the intellectual property easily.

The response of societies to this quandary is to grant patents or copyrights, which are property rights and as such allow individuals to earn an economic profit. Without property rights innovators and entrepreneurs are not willing or able to invest in research and development because they do not have the requisite capital for the endeavor or because they will not reap rewards commensurate with their efforts.

This raises the question of whether science itself is a public or a private good. By granting patents the government essentially places science in the private realm and grants corporations monopoly power over goods that may have a public nature. With scientific research an alternative to patents would be to make all scientific research and development publicly funded. That might allow science to remain independent of corporate power and better serve the public interest. Detractors of this idea believe that without incentives individuals will not develop new ideas. They also believe that without property rights ideas will not be secure. However, less cynical thinkers maintain that the pursuit of science is not necessarily motivated by financial reward. Creativity does not follow a schedule and does not answer to the auditor.

The way property is defined, appropriated, and exchanged is one of the most frequently discussed topics in economic and political philosophy. The challenge to society is to define property rights clearly and in a manner that allows transparent monitoring and enforcement of those rights and to recognize that property rights to some types of entities can lead to gross inequality. Meeting this challenge may lead to greater investment in technology and better-informed choices for individuals and society. Sometimes, however, the nature of the way people interact interferes with clear and effective property rights.

WILLARD DELAVAN

SEE ALSO *Intellectual Property*.

BIBLIOGRAPHY

Christman, John. (1994). *The Myth of Property*. New York: Oxford University Press. A contrarian view of the capitalist idea of property.

Demsetz, Harold. (1967). "Toward a Theory of Property Rights." *American Economic Review* 57(2): 347–359. A classic early article with a formal neoclassical approach to property.

Hardin, Garrett. (1968). "The Tragedy of the Commons." *Science* 162: 1243–1248. The defining article on common property resources.

McCloskey, Donald. (1985). *The Applied Theory of Price*. New York: Macmillan. One of the best-written and most provocative intermediate microeconomic theory textbooks; by a prolific, enigmatic economic historian.

Perloff, Jeffrey M. (2001). *Microeconomics*, 2nd edition. Boston: Addison Wesley Longman.

Schmid, Allan A. (1987). *Property, Power, and Public Choice*, 2nd edition. Westport, CT: Praeger. A profound look at the institution of property.

PROSTHETICS

• • •

In a narrow sense, prosthetics is a branch of medicine, specifically of surgery, concerned with the replacement of missing body parts (upper and lower limbs, and parts thereof) after amputation. It is related to orthotics, a branch of medicine that deals with the support of weak or ineffective joints or muscles using supportive braces and splints. In dentistry, prosthetics or prosthodontics is that branch concerned with the replacement of missing teeth and other oral structures. In this narrow sense, a prosthesis is a replacement artificial limb or tooth. In a broader sense, *prosthesis* is the name for any artifact used to restore bodily functions, and *prosthetics* is the field concerned with the development and fitting of artificial body parts, which is the sense at issue here.

Approaches to Prosthetics

Prostheses in this broad sense are an important focus of the relatively new field of bioengineering, or biomedical engineering, which is concerned with the application of engineering techniques to medicine and the biomedical sciences. Bioengineering is itself a broad field, with applications ranging from molecular imaging tools to medical radiation devices. The development of prosthetic techniques and devices is only one of its interests.

Several areas in bioengineering have special relevance to prosthetics. Rehabilitation engineering is an area concerned with ameliorating the impairments of individuals with disabilities. It includes prosthetics and orthotics as defined at the beginning of this entry, but also addresses other disabilities, specifically sensory and speech impairments. It does not address functional impairments in internal organs, however. Other relevant areas include tissue engineering, which involves the repair or replacement of organic cells, tissues, or organs with laboratory-grown biological substitutes; biomaterials engineering, which aims to develop synthetic or natural materials that can replace or augment tissues, organs, or bodily functions; biomechanics, which studies the human musculoskeletal system and its mechanical aspects and includes artificial limb and joint design; cardiovascular engineering, which studies the cardiovascular and blood system and develops techniques and systems for diagnosis, intervention, therapy, and replacement; and neural engineering, which studies the nervous system and develops means to repair or replace damaged and nonfunctioning nerves and sensory systems. Neuroprosthetics is a rapidly growing subfield of neural engineering that aims to develop devices or systems that communicate with nerves to restore functionality of the nervous system.

Although research in prosthetics and bioengineering is primarily aimed at restoring damaged human functions, there has been a growing interest in the augmentation of human functions. Human augmentation or enhancement is a relatively new field in bioengineering directed at developing prosthetic devices that augment normal function or prevent injury to function.

Together with artificial intelligence and robotics, bioengineering is the successor of bionics (a conflation of *biological electronics*), which emerged in the 1950s with the aim of using biological design principles to create novel technological devices and mechanical substitutes for the extension of biological organs. Bionics is specifically concerned with the development of bionic devices or bionic implants, which are electromechanical devices that do not merely replace a body part but also closely mimic or surpass the behavior of a replaced organ, and that are often able to communicate with the nervous system. To attain its aims, bionics relied on a feedback-control framework that was provided by cybernetics, the science of communication and control in animal and machine. Cybernetics has been partially superseded by systems theory, a field that studies the general principles underlying the organization of systems of any kind. Cybernetics has yielded the term *cyborg*, a conflation of *cybernetic organism*, meaning an organism that is part human, part machine. A cyborg is an individual whose biological functions are aided or controlled by technological devices, particularly by bionic implants.

A large number of human biological functions can be restored or improved with the aid of prostheses. The list of implants and related devices is extensive:

- artificial limbs, including robotic ones and ones with sensory feedback to the body

- artificial joints, hips, and vertebrae

- artificial muscles made of polymer

- artificial skin used to promote healing

- artificial bone used to help heal fractures and replace diseased bone

- bracing systems, cervical implants, and spinal cages to support the spine

- silicone or plastic implants to build bony structures of the face

- breast implants

- penile implants

- dental implants and false teeth

- speech synthesizers and artificial larynxes to restore speech

- retinal implants (experimental), intraocular lenses, and artificial corneas to restore vision

- cochlear implants that replace the inner ear and involve a microphone, speech processor, and wiring to the nervous system

- artificial nerves (experimental)

- cardiac pacemakers, defibrillators, artificial heart valves, and heart-assist pumps

- artificial hearts (experimental)

- artificial blood vessels and urological systems

- artificial blood (experimental)

- implanted drug-delivery systems (experimental)

- electrodes implanted in the brain to control seizures or tremor

- implanted chips to locate persons or to regulate devices in "intelligent environments"

- orgasmatrons (implants for women that produce orgasms; experimental)

- spinal neuroimplants with handheld remote control to block pain signals

- motor neural prostheses based on functional electrical stimulation systems, which stimulate motor nerves for movement, respiration, and bladder function

- artificial hippocampi in the brain (experimental)

Research is underway on bioartificial livers, kidneys, pancreases, lungs, and other organs, as well as on more advanced neural prostheses to restore functions of the brain and nervous system.

Anthropological Theories

Most philosophical and anthropological theories that refer to the notion of prosthesis are not so much concerned with understanding prosthetic technologies as normally defined but with an understanding of technology in general by means of the concept of prosthesis. Prosthesis is used as a metaphor to understand technology and its relation to human beings. In prosthetic theories of technology, which have been proposed since at least the late nineteenth century by a variety of different authors, it is claimed that there is no essential distinction between prosthetic and other technologies, because all technologies in some way aim to replace or augment aspects of human functioning. This view has been proposed by, among others, Marshall McLuhan (1911–1980), Henri Bergson (1859–1941), Arnold Gehlen (1904–1976), Ernst Kapp (1808–1896), and Lewis Mumford (1895–1990).

According to the prosthetic view of technology, every technological artifact or system extends the human organism in that it takes human faculties outside the body, thus amplifying already present abilities. The body is itself a toolbox that its owner uses to do things in the world. Technical artifacts serve to replace, extend, or augment tools in this organic toolbox. Weapons and tools such as bows, knives, and saws are extensions of human hands, nails, and teeth; clothing extends the heat control and protection functions of the skin; the wheel extends the mobility functions of the legs; bags extend the ability of the hands and arms to carry things; the radio and telephone extend hearing; television and photography extend the visual function; writing and print media extend human language and memory functions; and the computer extends a large variety of human cognitive functions. Prosthesis, in the narrow sense, is therefore only an instance of the general ability of technology to extend or replace functions of the human organism, and all technologies should be understood in terms of their relation to human functioning.

Even if this view is correct, it is recognized by many authors that all artifacts do not extend the human organism in the same way. Some technological artifacts have a symbiotic relation to the body, whereas others function independently. A relevant distinction seems to exist between artifacts that serve as direct extensions of human functioning by engaging in a symbiotic relationship with human limbs, senses, or other body parts, such as telescopes, glasses, hammers, and canes, and those artifacts that operate separately from the body and are themselves the object of interaction or perception, such as dinner plates, stereo systems, and computer screens. Phenomenologist Don Ihde (1990), drawing on the work of Maurice Merleau-Ponty (1908–1961), argues that humans are able to engage in embodiment relations with some artifacts, which are incorporated into the body schema or body image, meaning that they are integrated with the image that human beings have of their own sensorimotor abilities—an image that defines them as agents and separates them from a world that is to be engaged. (Other artifacts remain separate and subject to interpretative or hermeneutic relations.) Embodiment relations have found support in psychological studies of body schemas.

Cyborg Theories

Cyborg theory or cyborgology—the multidisciplinary study of cyborgs and their representation in popular culture—provides another perspective on prosthetics. Studies in cyborg theory tend to use the notion of the cyborg as a metaphor to understand aspects of contemporary—late modern or postmodern—relationships of technology to society, as well as to the human body and the self. In cyborg theory, the notion of cyborg refers to hybrid organisms in science fiction (e.g., *The Six Million Dollar Man*, *RoboCop*, *X-Men*, *Star Trek*'s The Borg), contemporary human beings with prostheses or implants, as well as (contemporary) human beings in general, who are all conceived as cyborgs in the sense of being inherently dependent on technology.

The advance of cyborg theory as an area of academic interest has been credited to Donna Haraway, in particular to her 1985 "Manifesto for Cyborgs." In this essay, Haraway presents the cyborg as a hybrid organism that disrupts essentialist presuppositions of modern thinking, with its black-and-white dichotomies of nature–culture, human–animal, organism–technology, man–woman, physical–nonphysical, and fact–fiction. Cyborgs have no preexisting nature or stable identity, and cut through oppositions because of their thoroughly hybrid character. Haraway holds that modernity is characterized by essentialism and binary ways of thinking that have the political effect of trapping beings into supposedly fixed identities and oppressing those beings (animals, women, blacks, etc.) who are on the wrong, inferior side of a binary opposition. She argues that the hybridization of humans and human societies, through the notion of the cyborg, can free those who are oppressed by blurring boundaries and constructing hybrid identities that are less vulnerable to the trappings of modernistic thinking.

According to Haraway and other authors such as N. Katherine Hayles (1999) and Chris Hables Gray (1995), this hybridization is already occurring on a large scale. Such hybridization is a consequence of the transition since World War II from an industrial to an information society, as a result of technological advances in biotechnology, information technology, and cybernetics. In the new world order that is ensuing, boundaries are constantly blurring, and linguistic categories and symbols increasingly reflect this fact. Many basic concepts, such as those of human nature, the body, consciousness, and reality, are shifting and taking on hybrid, informationalized meanings. In this postmodern, posthuman age, power relations morph, and new forms of freedom and resistance are made possible.

Sharing the positive outlook of cyborg theorists on the technological transformation of human nature, but otherwise quite distinct from it both politically and phi-

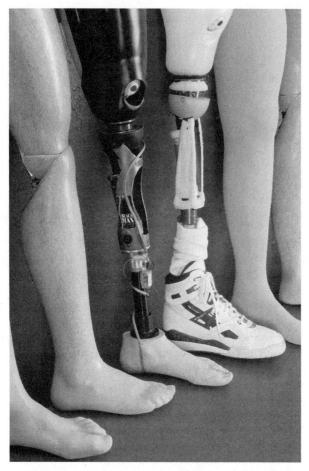

Various prosthetic legs. (© Roger Ressmeyer/Corbis.)

losophically, transhumanism is a recent school of thought or movement that advocates the progressive transformation of the human condition through technological means. Its early inspirational source was FM-2030 (formerly, F. M. Esfandiary) (1989), a futurist who wrote on the notion of the transhuman in the 1970s and 1980s, while its current main organizing body is the World Transhumanist Association, cofounded in 1998 by Nick Bostrom and David Pearce. Transhumanists want to move beyond humanism, which they commend for many of its values, such as its orientation toward reason and science, its commitment to and belief in progress, and its rejection of faith and worship, but which they fault for a belief in some fixed human nature. Transhumanists want to use modern technology to alter human nature in order to augment human bodily and cognitive abilities and extend human life. They see converging developments in genetic engineering, biomedical engineering, artificial intelligence, nanotechnology, and cognitive science as transcending human nature, thus leading humanity to a transhuman or posthuman

condition. They argue that this development should receive full support, because of its potential to enhance human autonomy and happiness and eliminate suffering and pain, and possibly even death.

Ethical Issues

The research, development, application, and use of prostheses and implants raise a number of ethical issues relating to health and safety, distributive justice, identity, privacy, autonomy, and accountability. Special ethical issues are raised by human augmentation or enhancement research.

HEALTH AND SAFETY. The functioning of a prosthesis for the remainder of someone's life cannot be predicted reliably on the basis of a few clinical trials with human subjects or tests with animals. There is a real risk, therefore, that people will be fitted with prostheses or implants that malfunction, have harmful side effects, or are even rejected by the autoimmune system. Negative experiences with silicone breast implants and artificial hearts have already shown the body's resistance to technological interventions. Ideally, prostheses would be tested over many years, decades even, and involve a large number of human subjects. But such extensive clinical trials and experimental uses are often considered too lengthy and costly and raise ethical issues by making guinea pigs out of human beings. Tests on animals often cannot serve as a substitute, while raising ethical issues of their own.

JUSTICE. The development of increasingly sophisticated prostheses and implants presents issues of distributive justice: Will there be a division between biological haves and have-nots? Will there be a division between those who receive no prosthesis or a low-quality or high-risk one and those who receive the best medical care? Do people have a moral right to a replacement part for a malfunctioning organ, when such parts exist? And will all be able to obtain implants that are attuned to their biological characteristics and lifestyle? In a 2003 incident in the United Kingdom, a black woman with an amputated foot was told that she would have to be fitted with a white prosthetic limb unless she paid an additional £3,000 (U.S.$ 5,500) for a black one. Although this is an obvious instance of discrimination, the situation is not always so clear. Who, for example, should pay the extra costs when a person has mild allergic reactions to a prosthesis and demands a much more expensive version that will not cause such reactions? Do producers have a duty to develop special prostheses for people whose biological features do not fit the norm, and should they be able to charge extra for those?

IDENTITY. Acquiring a prosthesis requires people to come to terms with the fact that a part of their body is artificial, and that they are dependent on a piece of technology for their biological functioning. This may be even more of an issue with bionic and neuroprosthetic implants, which may display or induce behaviors only partially controllable, with which one may thus find it hard to identify. Even more so, cognitive prostheses, which are neuroprostheses that aid cognitive function, may be developed in the future, and these may undermine identity even more directly as they directly interface with the mind. Some critics of prostheses have argued for the integrity of the human body, with all its defects and flaws, and worry that as humans increasingly become cyborgs, the essence of humanity will be lost. Social identity may be at issue as well. A particular controversy has arisen over cochlear implants; deaf advocates have argued that they may place children in between the deaf world and the hearing world, and that they may end up destroying the deaf community with its rich history and culture.

PRIVACY. Privacy issues are at stake when implants process or store information or emit identifying signals that can be registered from a distance. Implantable chips for tracking, already common in pets and livestock, are also being considered for children and adults, and they make it possible to trace individuals over long distances. Sensory and neuroprosthetic devices and prostheses equipped with biosensors process and sometimes store information about people's biological states, behaviors, and perceptions that may be accessed by third parties.

AUTONOMY. Prostheses can clearly enhance individual autonomy by restoring functions, but it has been argued that they can also reduce it. Having a prosthesis means being intrinsically dependent on technology. A prosthesis also creates dependence on others for maintenance, diagnosis, and testing. Bionic and neuroprosthetic implants may not even leave their wearer in complete control of their actions or thoughts.

ACCOUNTABILITY. Bionic and neuroprosthetic implants may raise issues of accountability, because the behavior or cognitive processes of their wearers will be determined in part by the workings of machines. If such individuals cause accidents or make bad decisions, who is to blame: they or their implants?

ETHICAL ASPECTS OF HUMAN AUGMENTATION. The field of human augmentation or enhancement raises a number of special ethical issues in addition to the ones already mentioned. Is it ever morally permissible to destroy or impair healthy human tissue or organs to fit an augmentation, considering that this destruction may be irreversible? Can an employer require an employee to have enhanced functions, or put a premium on the possession of such functions? Human augmentations is still a young field, and questions of this sort have mainly been raised in relation to cosmetic surgery, which can be understood as a special type of human augmentation with the purpose of enhancing aesthetic rather than functional qualities. Specifically, breast implants intended to create bigger breasts—as opposed to restoring breasts after a radical masectomy—have created controversy because they have been argued to be "unnatural" and to involve health and safety risks that cannot be justified by reference to their subjective aesthetic value. If certain augmentations become popular, there is also a risk that they will become accepted as the norm and people without them will be seen as cripples. To an extent, this is already happening with breast implants and other cosmetic surgery in some communities, but it may also happen with prostheses that enhance perceptual, motor or cognitive functions.

A large part of the debate on human augmentation, finally, has focused on military applications, specifically the possibility of creating supersoldiers. But should military research be devoted to the creation of a supersoldier, involving implants, steroids, amphetamines, genetically altered muscles, integrated weaponry, and lightning-fast artificial nerves?

Many parts of the human body can already be replaced by prosthetic devices, and revolutionary developments in bioengineering are rapidly expanding the reach of prosthetics. Biomedical engineers and medical specialists have a special, professional responsibility in dealing with the ethical issues that arise as a result, as they are primarily responsible for the development and fitting of prostheses. Many ethical issues also need to be addressed at the level of legislation and public policy. Special moral concerns are raised in the areas of human augmentation or enhancement and neuroprosthetics.

PHILIP BREY

SEE ALSO *Androids; Bioengineering Ethics; Cyborgs; Disability; Posthumanism; Therapy and Enhancement.*

BIBLIOGRAPHY

Brey, Philip. (2000). "Technology as Extension of Human Faculties." In *Metaphysics, Epistemology, and Technology*, ed. Carl Mitcham. London: Elsevier/JAI Press. A study of prosthetic theories of technology from Kapp to McLuhan and beyond.

FM-2030. (1989). *Are You a Transhuman? Monitoring and Stimulating Your Personal Rate of Growth in a Rapidly Changing World*. New York: Warner Books. Culminative statement of the transhumanist project by this transhumanist pioneer.

Gray, Chris Hables, ed. (1995). *The Cyborg Handbook*. New York: Routledge. Reader with important papers in cyborg theory as well as some historic papers in cybernetics, with a foreword by Gray.

Hayles, N. Katherine. (1999). *How We Became Posthuman: Virtual Bodies in Cybernetics, Literature, and Informatics*. Chicago: University of Chicago Press. A broad-ranging study of how cybernetics and computer science have fostered a disembodied, posthuman conception of human beings as information constructs.

Ihde, Don. (1990). *Technology and the Lifeworld: From Garden to Earth*. Bloomington: Indiana University Press. A phenomenological study of the relation between human beings and technology, including embodiment relations.

PSEUDOSCIENCE

• • •

The distinction between ideas and activities that represent science and those that represent nonscience is usually clear; no one confuses physics with art or chemistry with poetry. Nevertheless, there are ideas and activities related to bodies of knowledge that are not characterized clearly as science or nonscience and sometimes are claimed by their proponents to be science but are considered by most scientists to be pseudoscience. For example, the National Science Foundation (2002) conducted a poll on the different forms of pseudoscience accepted by Americans:

- Thirty percent believe that unidentified flying objects (UFOs) are space vehicles from other civilizations.
- Sixty percent believe in extrasensory perception (ESP).
- Forty percent think astrology is scientific.
- Thirty-two percent believe in lucky numbers.
- Seventy percent accept magnetic therapy as scientific.
- Eighty-eight percent agree that alternative medicine is a viable means of treating illness.

Most scientists reject these beliefs, which are variously called pseudoscience, voodoo science, junk science, crackpot science, or plain nonsense. However, from the perspective of those making the claims, what is being presented is more like a new aspect of science, an alternative science, prescience, or revolutionary science. In a culture in which science is given high status—indeed, this is said to be an age of science—one would expect political theories (scientific socialism), religions (Christian science, scientology, creation science), and even literature (science fiction) to try to associate themselves with science. Precisely for this reason attempts to define the boundaries of science and pseudoscience and to distinguish pseudoscience from mistaken science or not fully accepted science raise ethical as well as epistemological issues.

The Boundary Issues

Here one is faced with a "boundary problem": Where does one draw the boundary between science and pseudoscience and between science and nonscience? The problem is that it is not always or even usually clear where one should draw the line. Whether a claim should be put into the set labeled science or the one labeled pseudoscience depends on both the claim and the definition of the set. In this regard it is useful to expand the heuristic into three categories: normal science, pseudoscience, and borderlands science. The following are examples of claims that might best be classified in one of those three categories:

Normal science: heliocentrism, evolution, quantum mechanics, big bang cosmology, plate tectonics, neurophysiology of brain functions, punctuated equilibrium, sociobiology/evolutionary psychology, chaos and complexity theory, intelligence and intelligence testing

Pseudoscience: creationism, Holocaust revisionism, remote viewing, astrology, Bible code, alien abductions, Bigfoot, UFOs, Freudian psychoanalytic theory, recovered memories

Borderlands science: superstring theory, inflationary cosmology, theories of consciousness, grand theories of economics (objectivism, socialism, etc.), SETI, hypnosis, chiropractic, acupuncture, cryonics, omega point theory

Because these categories are provisional it is possible for them to be moved and reevaluated with changing evidence. Indeed, many normal science claims at one time were pseudoscience or borderlands science. SETI (the search for extraterrestrial intelligence), for example, is not pseudoscience because it does not claim to have found anything (or anyone) yet, is conducted by professional scientists who publish their findings in peer-reviewed journals, polices its own claims and does not hesitate to debunk the occasional signals found in the data, and fits well within the general understanding of the history and structure of the cosmos and the evolution of life. However, SETI is not normal science because its central theme has not surfaced as reality. UFOlogy, by contrast, is pseudoscience. Its proponents do not play by the rules of science, do not publish in peer-reviewed journals, ignore the 90 to 95 percent of sightings that are fully explicable, focus on anomalies, are not self-policing, and depend heavily on theorizing about government conspiracies and cover-ups, hidden spacecraft, and aliens holed up in secret caves in Nevada.

Similarly, superstring theory and inflationary cosmology are at the top of borderlands science, soon to be elevated into full-scale normal science or abandoned altogether, depending on the evidence that is starting to come in for these previously untested ideas. What makes them borderlands science instead of pseudoscience (or nonscience) is the fact that their practitioners are professional scientists who publish in peer-reviewed journals and are trying to devise ways to test their theories. By contrast, creationists who devise cosmologies that they think will fit biblical myths are typically not professional scientists, do not publish in peer-reviewed journals, and have no interest in testing their theories except against what they believe to be the divine words of God.

Theories of consciousness are borderlands science and psychoanalytic theories are pseudoscience because the former are being tested and are grounded in sound facts of neurophysiology whereas the latter have been tested, have failed the tests repeatedly, and are grounded in discredited nineteenth-century theories of the mind. Similarly, recovered memory theory is pseudoscience because it now is understood that memory is not like a videotape that one can rewind and play back and that the very process of "recovering" a memory contaminates that memory. Hypnosis, by contrast, is tapping into something else in the brain, and there may very well be sound scientific evidence in support of some of its claims; therefore, it remains in the borderlands of science.

Eliminating Pseudoscience

When one encounters a claim, there is no simple set of rules by which one can determine whether it is science and pseudoscience. However, there are a number of questions that can help illuminate its validity.

Signs of the Zodiac. Astrology is one of the most popular forms of pseudoscience. (© *Historical Picture Archive/Corbis.*)

1. How reliable is the source of the claim? All scientists make mistakes, but are the mistakes random, as one might expect from a normally reliable source, or are they directed toward supporting the claimant's preferred belief? Scientists' mistakes tend to be random; pseudoscientists' mistakes tend to be directional.

2. Does this source often make similar claims? Pseudoscientists have a habit of going well beyond the facts, and so when individuals make many extraordinary claims, they may be more than iconoclasts. What one is looking for here is a pattern of fringe thinking that consistently ignores or distorts data.

3. Have the claims been verified by another source? Typically pseudoscientists make statements that are unverified, or are verified by a source within their own belief circle. One must ask who is checking the claims and even who is checking the checkers.

4. How does the claim fit with what is known about how the world works? An extraordinary claim must be placed in a larger context to see how it fits.

Illustration depicting an alien abduction. Television shows like "The X-Files" have dramatized this pseudoscientific phenomenon. (© *Corbis*.)

When people claim that the pyramids and the Sphinx were built more than 10,000 years ago by an advanced race of humans, they are not presenting any context for that earlier civilization. Where are its works of art, weapons, clothing, tools, trash?

5. Has anyone made an effort to disprove the claim, or has only confirmatory evidence been sought? This is the confirmation bias, or the tendency to seek confirmatory evidence and reject or ignore disconfirmatory evidence. The confirmation bias is powerful and pervasive. This is why the scientific method, which emphasizes checking and rechecking, verification and replication, and especially attempts to falsify a claim, is critical.

6. Does the preponderance of evidence converge on the claimant's conclusion or a different one? The theory of evolution, for example, is proved through a convergence of evidence from a number of independent lines of inquiry. No single fossil or piece of biological or paleontological evidence has the word *evolution* written on it; instead there is a convergence from tens of thousands of evidentiary bits that adds up to a story of the evolution of life. Creationists conveniently ignore this convergence, focusing instead on trivial anomalies or currently unexplained phenomena in the history of life.

7. Is the claimant employing the accepted rules of reason and tools of research, or have those rules and tools been abandoned in favor of others that lead to the desired conclusion? UFOlogists exhibit this fallacy in their continued focus on a handful of unexplained atmospheric anomalies and visual misperceptions by eyewitnesses while ignoring the fact that the vast majority of UFO sightings are fully explicable.

8. Has the claimant provided a different explanation for the observed phenomena, or is it strictly a matter of denying the existing explanation? This is a classic debate strategy: Criticize one's opponent and never affirm what one believes in order to avoid criticism. This strategy is unacceptable in science.

9. If the claimant has proffered a new explanation, does it account for as many phenomena as does the old explanation? For a new theory to displace an old theory it must explain what the old theory did and then some.

10. Do the claimants' personal beliefs and biases drive the conclusions or vice versa? All scientists have social, political, and ideological beliefs that potentially could slant their interpretations of the data, but at some point, usually during the peer-review system, those biases and beliefs are rooted out or the paper or book is rejected for publication.

This final point reveals the ethical nature of science and the way it differs from pseudoscience. Whether the ethics comes from within the individual scientists or from the system of science is irrelevant. The point is that the system works to weed out error, bias, and fraud. Ethical issues arise when pseudoscience masquerades as science for political purposes, as occurs when biblical fundamentalists attempt to legislate their religious beliefs by calling them creation science and have them taught in public school science classes. Serious ethical concerns arise when quasi-scientific claims have health consequences, as do many of the claims of alternative and complementary medicine. The application of nonscientific or pseudoscientific treatments in place of scientifically proven medicine can be dangerous and even deadly.

Here too may be seen how the market, commercialism, and politics can also promote pseudoscience. The tobacco industry maintained that smoking does not cause cancer, for many years beyond when it was reasonable to do so because the evidence for the link between smoking and cancer was overwhelming. The Bush administration's insistence on more data on global warming before preventative measures should be taken is another example of politics overriding science, because virtually all environmental scientists agree that global warming is real.

Science may be flawed, but as Albert Einstein once observed: "One thing I have learned in a long life: that all our science, measured against reality, is primitive and childlike—and yet it is the most precious thing we have."

MICHAEL SHERMER

SEE ALSO *Misconduct in Science; Skepticism.*

BIBLIOGRAPHY

Gardner, Martin. (1996). *The New Age: Notes of a Fringe Watcher.* Buffalo, NY: Prometheus. A collection of essays from the most prolific skeptical author of the modern period, covering dozens of paranormal and pseudoscience claims made on the fringes of the New Age.

Gould, Stephen Jay. (1981). *The Mismeasure of Man.* New York: Norton. Debunks theories of racial differences in intelligence and shows how science can be influenced by cultural forces and also how it can be self-correcting.

Park, Robert L. (2000). *Voodoo Science: The Road from Foolishness to Fraud.* New York: Oxford University Press. The author, a physicist, analyzes cold fusion, perpetual motion machines, free energy, and how well-meaning scientists can go down the road to self-deception.

Sagan, Carl. (1995). *The Demon-Haunted World: Science as a Candle in the Dark.* New York: Random House. A collection of the astronomer's most powerful writings on UFOs, alien abductions, and all manner of pseudoscience and paranormal experiences.

Schick, Theodore, and Lewis Vaughn. (1996.) *How to Think about Weird Things: Critical Thinking for a New Age.* Mountain View, CA: Mayfield. The best textbook on pseudoscience, well organized for both teachers and students and widely adopted in colleges and universities.

Shermer, Michael, general ed. (2002). *The Skeptic Encyclopedia of Pseudoscience.* Santa Barbara, CA: ABC-CLIO. A two-volume compendium that includes various levels of analysis, from short encyclopedic entries to lengthy evaluations, as well as full-fledged investigations and pro-and-con debates.

Williams, William F., ed. (2000). *Encyclopedia of Pseudoscience.* New York: Facts on File. Includes introductions dealing with "Ethical Issues in Pseudoscience" (by Carl Mitcham) and the boundary problem (by William F.

Williams) as well as two others on the phenomenon of anomalies.

INTERNET RESOURCES

National Science Foundation. (2002). *Science Indicators Biennial Report.* The section on pseudoscience, "Science Fiction and Pseudoscience," is in Chapter 7, "Science and Technology: Public Attitudes and Public Understanding." Available from http://www.nsf.gov/sbe/srs/seind02/c7/c7h.htm.

PSYCHOLOGY

• • •

    Overview
    Humanistic Approaches

OVERVIEW

Psychology, defined broadly, is the study of individual behavior. *Individual* can refer to a human or an animal, and *behavior* can encompass anything an individual does, thinks, or feels. Because there are so many things that individuals do, think, and feel, psychology is divided into many subareas that each study a different aspect of individual behavior. For example, some psychologists study how individual behavior is affected by those with whom the individual interacts, others investigate how the brain works to produce thoughts and feelings, and still others study the causes of feeling and thought disorders such as depression and schizophrenia.

Historical Emergence

Psychology is a relatively new scientific field. Wilhelm Wundt (1832–1920) founded the first official psychology laboratory in 1879 at the University of Leipzig, Germany. Psychology has roots, however, in ancient philosophy. Many of its concerns—such as personality development, rationality, language acquisition and use, the structure of consciousness, and the mind–body connection have been addressed by philosophers. Plato (c. 428–347 B.C.E.) stressed the distinction between body and mind, and argued that knowledge depended on the rational soul. Aristotle (384–322 B.C.E.) argued for a unity of body and mind, and that knowledge has a base in sensory perception.

What makes psychology different from philosophy is its efforts to adapt the scientific method to the investigation of individual behavior. To some extent psychology constitutes an effort to place traditional ethics, which may also be defined as the study of human behavior, on scientific foundations. Historically these efforts have taken place in two different settings, the laboratory and the clinic. Early on, these settings gave rise to largely separate approaches that progressed in relative isolation from one another.

From the beginning experimental and clinical psychology expressed different ideals. The experimental division worked from the ideal of scientific curiosity. Its goal was to understand the normal or everyday, for example, attention or memory. In contrast, the clinical division worked from the ideal of helping people and understanding problems. Its goal was to understand the unusual or problematic, for example, depression or antisocial behavior.

Experimental Psychology

The first school of experimental psychology was structuralism, which emerged in Germany in the late nineteenth century. The pioneers of structuralism were physicists and physiologists who attempted to study sensations and perceptions as they would chemistry or biology, by measuring variables and examining how they interacted. Wundt, the founder of structuralism, had the goal of understanding and describing the contents of mind, that is, the basic elements of a person's immediate experience. The technique he developed, which his student Edward Tichener (1867–1927) championed in the United States, was called introspection. In introspection, trained scientists report their mental experience during rigorously controlled conditions. The structuralists were not interested in individual differences, and they did not believe in observing external things, only internal, mental events, so they did not come to much agreement.

In the United States, a second school of experimental psychology, called functionalism, emerged. American psychologists trained in Germany reinterpreted structuralism by emphasizing mental processes and their functions and applications. This approach, led by William James (1842–1910), was much more pragmatic, stressing the utility of mental functions such as attention or memory. The functionalists also argued that both the mental and physical (external) aspects of experience should be studied. Functionalism, however, lacked the scientific rigor of structuralism and instead was a more philosophical approach.

A third experimental movement, Gestalt psychology, emerged in Germany as another reaction to structuralism. The underlying principle was that the whole is different from the sum of the parts, that in breaking things apart into their components one loses the unified whole, or gestalt. Max Wertheimer (1880–1943), the founder of this movement, began with research on how humans can see movement in a series of static images. Although the Gestalt psychologists began with questions such as this based on sensation and perception, they broadened their perspective to ask how people interact with their environments and how this interaction organizes mental activity.

Clinical Psychology

Early clinical psychology was founded by Sigmund Freud (1856–1939), an Austrian physician who chose to study the mind rather than the body. He argued that unconscious processes could explain much of human behavior, including the development of personality and a variety of psychological disorders. Freud's theories dominated the clinical psychology landscape, as he was one of the first people to view mental illness as something to be treated and understood. Although his name is widely recognized, his theories are not well understood by the general public, and his approach had little in common with the experimental psychology of the same period. His technique, called psychoanalysis, was based on observation of individual patients, not on generating and testing predictions using the scientific method. With his practice and theories, however, Freud built a foundation for clinical psychology.

Becoming Scientific

Around 1900, a shift occurred in experimental psychology, namely, the behaviorist movement. Behaviorism arose as a reaction to the subjective nature of both early experimental and clinical psychology. An important early influence on behaviorism was Ivan Pavlov (1849–1936), a Russian physiologist who studied learning and relationships between a stimulus and a response. In a famous experiment, he trained dogs to associate a bell with food, so that they salivated in response to the bell even when the food was not present.

The behaviorist movement was largely defined by the work of John B. Watson (1878–1958). Watson criticized existing psychology research methods for being too subjective and not rigorous enough. He argued that psychology should focus on observable behavior rather than internal mental events. Behaviorism focused on the relationships between stimuli in the environment and behavioral responses. B. F. Skinner (1904–1990), a later but influential figure, extended early behaviorist principles to operant conditioning, or learning from rewards and punishments. Skinner also claimed that development could be explained in terms of behaviorist principles. For example, he argued that development of language was based on simple conditioning rules.

Behaviorism rejected many questions that were ethically relevant, for example, the nature of consciousness or how humans think and reason, because it claimed that these were not things open to scientific investigation. It created its own ethical dilemmas, however. Because behaviorists claimed that learning and conditioning rules could explain everything, people could be viewed as blank slates—anyone could become anything given the right circumstances. But this could portend a darker future in which the behavior of individuals could easily be shaped and controlled through conditioning.

Advances also occurred in clinical psychology because there remained a need for understanding and changing behavior in order to help individuals. Through a series of rejections and adaptations of Freud's theories, the humanist approach to clinical practice emerged. Important figures who modified Freud's work include Alfred Adler (1870–1937) and Carl Jung (1875–1961). Adler's theories were still considered psychoanalytic, but for him, social forces and creativity played an important role. He claimed that the individual tried to compensate for an inferiority felt in childhood, striving for perfection while moving through life. Neither Adler nor Jung were empirical psychologists; they were practitioners and theorists.

Further evolution of Freud's ideas, combined with influences from the existential movement in philosophy, which emphasized personal responsibility, led to the emergence of the humanist movement in clinical psychology. Important figures in this movement were Abraham H. Maslow (1908–1970) and Carl Rogers (1902–1987). Maslow described a hierarchy of needs: Individuals need to first meet their basic needs, such as those for food and safety, before they can meet higher human needs, such as those for belonging, knowledge, or beauty. Rogers advocated a new practice called client-centered therapy, in which the therapist and client (the person seeking help) have a personal relationship based on empathy. In practice, this focused on the process of better knowing one's self.

Ethics played a role in this shift from Freud's psychoanalysis to humanism. For humanists, it was important to recognize personal autonomy and potential, rather than to see individuals as victims of circumstances, unconscious powers, and unconscious thoughts or feelings.

Contemporary Psychology

In the early twenty-first century, experimental and clinical psychology translate into two types of professionals: research psychologists and practice psychologists. Research psychologists conduct experiments to study individual behavior in order to better understand it. Practice psychologists (who include counselors and therapists) use what is known about individual behavior to help individuals understand or change their behavior. In mainstream psychology, the distinction between research and practice is purely a functional distinction between the primary activities of the psychologists in each group. It is important to note that both groups work on and from the same body of knowledge. There still exist some approaches to practice that are based on philosophical or theological systems as opposed to empirical findings, but to the extent that there is no empirical evidence of their treatment efficacy they are not considered part of scientific psychology.

Modern psychology is a product of interactions between the clinical and experimental divisions. While the two divisions are not fully integrated, experimental data informs the practice of psychology, and insights from practice lead to new research in experimental psychology. In addition, psychology has been informed by other fields, including neuroscience, computer science, linguistics, and education. While many areas of specialization have formed, particularly within the academic research community, psychology is still interdisciplinary in that these specializations frequently interact. For example, neuroscientific research on how thoughts can affect mood can be used to develop methods for treating depression.

Ethics for Psychology

Psychologists face many ethical issues in their roles as research scientists and as clinical professionals. Many of these issues stem from the use of human and animal subjects in research, and the need to assure the safety and privacy of individuals seeking treatment. There are a variety of professional organizations for psychologists in each subspecialty area, and many of these organizations have developed codes of ethics. The primary code of ethics for professional psychologists, however, belongs to the American Psychological Association (APA), which is the largest professional association of psychologists worldwide, with 150,000 members as of 2005.

The APA has published ten revisions of its ethics code since it was first formulated in 1953. Unlike most professional codes of ethics, the APA code was developed pragmatically, based on a survey of ethical dilemmas encountered by APA members. The ninth revision, published in 1992, was the first time it included specific standards for academic scientists addressing teaching, training, supervision, research, and publishing. The tenth revision, published in 2002, eliminated language that appeared to allow use of the code to punish psychologists unfairly, increased protections for disempowered groups, and eliminated redundancy and vagueness. This tenth revision contains five general principles to guide the goals of research and practice, and ten standards for the conduct of psychologists.

The general principles included in the code are beneficence and nonmaleficence, fidelity and responsibility, integrity, justice, and respect for people's rights and dignity. The code has been criticized for not specifying an underlying ethical theory (e.g., utilitarianism, deontological ethics) to guide the evaluation of options and assist ethical decision-making. Further, the code lacks guidelines for valuing ethical principles in situations where conflicts arise. The ethics code of the Canadian Psychological Association has addressed this issue by providing a hierarchy that explicitly ranks the general principles it sets forth. The APA code also uses nontraditional ethical language, stating the principles and standards in terms of what psychologists "do" and "do not do," rather than in terms of what they "ought" or "should" do.

The ethical standards put forth in the APA code cover issues relevant to psychologists in their roles as scientists, teachers, and service providers of various types, and are enforced by the Ethics Committee of the APA according to its published rules and procedures. Detection of ethical violations are collected passively, in response to complaints, rather than actively (e.g., by auditing). Punishments for ethical violations can include expulsion from the APA and directives for corrective actions such as supervision, education, treatment, or probation. Other agencies and associations may also use the APA ethics code for assessing the behavior of psychologists.

Psychology for Ethics

In addition to following ethical principles in their professional work, psychologists can also use their expertise to contribute to ethical discussions in a number of ways. For instance, psychological research on moral development has investigated topics such as the development of moral reasoning over the lifespan, the nature of psychological components that are required for moral behavior to take place, and the contributions of social factors (e.g., persuasion, conformity, expectations) to moral discernment. The findings from these studies can be used

to help understand and assess culpability for moral infractions, and perhaps also provide direction for helping individuals decrease moral infractions. In a related vein, the emerging field of positive psychology is researching the causes and consequences of individual strengths and happiness, in order to help people develop positive traits such as resiliency and self-efficacy.

The results of research in psychology can also be used to inform specific ethical issues. Although research does not provide a basis for establishing standards for ethical behavior (called the "naturalistic fallacy"), it can provide information about the efficacy of certain means for bringing about desired ends. In many situations, psychology can provide information about the psychological consequences of various social policy alternatives, so that decisions can be based on available evidence. For example, in 2004 the APA filed amicus briefs on issues such as the juvenile death penalty and same-sex marriage, conveying research findings about brain development and decision-making ability in adolescents in the former, and research on relationship characteristics, parenting ability, and psychological benefits of marriage for both same-sex and heterosexual couples in the latter.

AMY SANTAMARIA
ELIZABETH J. MULLIGAN

SEE ALSO *Aristotle and Aristotelianism; Choice Behavior; Freud, Sigmund; Jung, Carl Gustav; Plato; Skinner, B. F.*

BIBLIOGRAPHY

American Psychological Association (APA). (2002). "Ethical Principles of Psychologists and Code of Conduct." *American Psychologist* 57(12): 1060–1073. The tenth revision of the code of ethics put forth by the APA, the primary professional organization for psychologists in the United States.

Bersoff, Donald N., ed. (2003). *Ethical Conflicts in Psychology*, 3rd edition. Washington, DC: American Psychological Association. Discusses the ethics codes of both the American and Canadian Psychological Associations and includes sections on specific ethical issues, as well as more general chapters on applying ethics and teaching ethics.

Brennan, James F. (2003). *History and Systems of Psychology*, 6th edition. Upper Saddle River, NJ: Prentice Hall.

DuBois, James M., ed. (1997). *Moral Issues in Psychology: Personalist Contributions to Selected Problems*. Lanham, MD: University Press of America. Discusses the relation of psychology to philosophy, including how psychology can contribute to moral issues.

O'Donohue, William T., and Kyle E. Ferguson, eds. (2003). *Handbook of Professional Ethics for Psychologists: Issues, Questions, and Controversies*. Thousand Oaks, CA: Sage.

Part I in particular provides an overview of the philosophical systems that might underlie an ethics code (by Andrew Lloyd and John Hansen), a discussion of the APA ethics code in the context of critical thinking about moral and ethical questions (Michael Lavin), and an overview of studies on the development of moral reasoning across the lifespan (Karl H. Hennig and Lawrence J. Walker).

HUMANISTIC APPROACHES

The history of psychology in the twentieth century is the history of a discipline struggling to balance values that seemed, more often than not, to exist in mutual tension. Some psychologists emphasized the necessity of empirical rigor in research, others promoted the development of individual emotional health and maturity, while still other mid-century thinkers would advance larger social and ethical concerns. Psychology's quest to establish itself as a science, combined with its historical emphasis on the connections between the human self and human well-being, produced a discipline of broad application and intense vitality, one uniquely suited to address the problems and opportunities of humankind in a technological age. Nowhere is this better illustrated than in the rise of what has become known as humanistic psychology.

Background

Efforts to limit psychological research to observable phenomena or behavior along began with reactions against the introspective psychological research program of Wilhelm Wundt (1832–1920). In the form of behaviorism, these efforts dominated psychological theory and practice between the two world wars. As conceived by such founders as Ivan Pavlov (1849–1936), John B. Watson (1878–1958), and B. F. Skinner (1904–1990), behaviorism aspired to be wholly objective. Watson insisted upon leaving consciousness and other metaphysical concerns aside for an experimental precision that could not be attained using "internal perception" or any other introspective methods. He articulated his fundamental complaint about previous psychological thought when he wrote, "Behaviorism claims that consciousness is neither a definite nor a usable concept. The behaviorist, who has been trained always as an experimentalist, hold, further, that belief in the existence of consciousness goes back to the ancient days of superstition and magic" (Watson 1924, p. 2). Behaviorism attempted to show that phenomena previously studied using introspective methodologies could be examined much more effectively from a perspective of stimulus and response; only those observations verifiable in more than one instance by more than one observer would be allowed to qualify as scientific.

Behaviorism has had lasting effects on the discipline and practice of psychology, including the development of highly objective experimental standards, new statistical methods, and behavior therapies. During the middle decades of the twentieth century, Skinner took Watson's ideas to their ultimate objective extreme, concentrating on the larger goal of predicting and controlling a broad range of human behavior. Much of this work was understandably focused on education and pedagogy, and Skinner and his colleagues often articulated an idealistic quest for positive techniques to solve human problems and improve society.

As psychology honed its experimental methods and techniques, its expanding scientific powers nevertheless brought ethical concerns to the foreground. Some of the most famous and influential studies in the field of behaviorist psychology, while revealing new insights into human consciousness and behavior, also highlighted the need for ethical standards in research practices. For instance, Watson and Rosalie Rayner's 1920 "Little Albert" study conditioned an eleven-month-old child to fear a white rat by pairing its presentation with a loud and startling noise—a fear that the young child generalized to similar animals and objects, and from which he was never deconditioned. Stanley Milgram's elaborate 1965 obedience experiments led subjects to falsely believe that they were carrying out orders to administer extremely severe electric shocks to another person. In the Stanford prison experiment in 1971, Philip Zimbardo assigned subjects to either a prisoner or guard role for a two-week simulation; the growing intensity of the situation and the subjects' increasing absorption into their roles, however, forced Zimbardo to halt the experiment after six days.

While each of these studies provided new discoveries in conditioning, obedience, roles, and attitudes, their effects on human subjects also provided strong arguments for reforms in experimental ethics. Over time, psychology established stringent ethical guidelines for informed consent, debriefing practices, and weighing potential deceptions or risks to subjects (including risks to animals as well as humans) against potential research benefits.

In contrast to the behaviorist attempt to eliminate consciousness by means of a methodological focus on overt behavior, Sigmund Freud (1856–1939) sought to downgrade consciousness through investigations into the power of the unconscious and its influence on behavior. Freud's psychoanalysis, however, developed

FIGURE 1

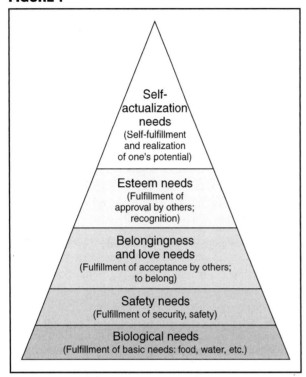

Pyramid of Abraham Maslow's hierarchy of needs.

primarily in a medical-clinical setting, drawing on clinical experience to formulate theories of human emotional abnormality and irrationality. Psychoanalysis thus developed a theory of human nature that highlighted the hidden complexities of the human psyche. But in the tradition of such thinkers as Plato, Augustine, or Jean-Jacques Rousseau, Freud was also concerned with human capability for self-understanding, the freedom that self-awareness can bring, and, more specifically, capacities to cope with life in more rational ways.

In psychoanalytic practice, too, ethical questions were brought to the fore. Like behaviorism, psychoanalytic theory challenged common conceptions of moral responsibility. Close relations between patient and psychoanalyst sometimes led to behaviors such as sexual relations that clearly violated social and traditional professional norms.

The Humanistic Movement

It was in reaction to both behaviorism and psychoanalysis that humanistic psychology began in the 1950s to develop its special approach to the study and treatment of human behavior—with a new ethical commitment. Among the precursors was Alfred Adler (1870–1937), who criticized Freud's emphasis on sexuality. Humanis-

tic psychology was also influenced by existentialist philosophy, with its focus on human struggles for meaning in a world characterized by scientific and technological dehumanization in such blatant forms as death camps and atomic bombs as well as in what existentialist philosophers from Søren Kierkegaard to Albert Camus saw as the more subtle forms of bourgeois culture.

Holocaust survivor Viktor Frankl (1905–1997), for example, in his book *Man's Search for Meaning* (1959), drew on his unique experiences to argue the power of human decision in the face of the most dehumanizing circumstances. But it was Carl Rogers (1902–1987), Abraham H. Maslow (1908–1970), and Rollo May (1909–1994) who most typified what Maslow himself termed the "third force" in psychology (the first force being behaviorism and the second, psychotherapy). Maslow in particular played a significant role in the development of the humanistic psychology movement, turning from his training in behaviorism to argue for a broader, more holistic version of human health. Maslow believed that no psychological theory could be truly complete unless it took into account complex human factors and motivations such as love and connection (Figure 1). Maslow's "self-actualization" theory of personality and his development of a human "hierarchy of needs" both stressed the universal human potential for achievement.

Even more representative, insofar as humanistic psychology brings its perspective to bear on science and technology, is the work of Erich Fromm (1900–1980). Like other third-force humanistic psychologists, Fromm sought to refocus the central ideas of Freudian psychoanalytic theory to address the moral, emotional, and spiritual crises of an increasingly violent and technology-oriented global society. He was less interested in simple human adaptability, techniques for the control of behavior, or strategies of coping than in nurturing humanity's basic ability to meet the challenges of a difficult transition into modernity, with its changing political systems and assumptions; various physical and spiritual displacements; astounding technological innovations in health, industry, and war; and, later in the century, the long Soviet–American nuclear standoff.

Forced to flee Germany after Adolf Hitler's election in 1933, Fromm was particularly concerned with the development of a "technetronic" society and its dehumanizing implications. He reserved his most incisive critiques for behaviorist strivings for absolute objectivity, arguing that behavioristic theories merely served the cerebral and technical prejudices of industrial society. Understanding, Fromm believed, should be different than "scientific" description. His criticism was often less than subtle: "[George] Orwell's *1984* will need much assistance from testing, conditioning, and smoothing-out psychologists in order to come true. It is of vital importance to distinguish between a psychology that understands and aims at the well-being of man and a psychology that studies man as an object, with the aim of making him more useful for the technological society" (Fromm 1968, p. 46).

Writing from a position similar to that of the existentialist thinkers, Fromm recognized that humankind had lost its traditional religious-ethical moorings, and he worried that the powerful attraction of technology and machinery was evidence that technological society had simply exchanged its religious faith (and humanistic values) for material and technical values: If something is *possible* (build the atom bomb, go to the moon), we should do it; production of *more* is preferred to production of better. People had lost, in that exchange of values, their capacity for deep emotional experiences, and with them their capacity to engage life with any sense of meaning. "Today," Fromm wrote in 1968,

> a widespread hopelessness exists with regard to the possibility of changing the course we have taken. This hopelessness is mainly unconscious, while consciously people are 'optimistic' and hope for further 'progress.' . . . [People] see that we have more and better machines than man had fifty years ago. . . . They believe that lack of direct political oppression is a manifestation of the achievement of personal freedom. (p. 5)

Aside from arguing for a reevaluation of technical values, Fromm advocated a reemergence of practical humanist perspectives, including altered forms of material consumption; an emphasis on social activity against what he perceived as a new and cancerous cultural passivity; changed attitudes about the place and capabilities of the worker in large organizations; more person-oriented, responsible, and imaginative bureaucratic systems; and spiritual renewal focused on faithful practices involving compassion instead of allegiance to ideology or code. Despite the existence of good reasons for pessimism, Fromm displayed the same hopefulness in his own attitudes that he argued would be necessary for the renewal of individuals and society: "The history of man shows precisely what you can do to man and at the same time what you *cannot* do. If man were infinitely malleable, there would have been no revolutions; there would have been no change because a culture would have succeeded in making man submit to its patterns without his resistance" (Fromm 1968, p. 62). One of Fromm's primary goals was to help initiate a resistance against the unimpeded development of a technological culture he believed had come to threaten humanity's connections to broader social and environmental contexts.

Ethics

Despite its inherently ethical orientation, humanistic psychology seldom explicitly couched its concerns in terms of "ethics." No doubt one reason is that both behaviorist and psychoanalytic thought had become over the course of decades extremely skeptical of ethical and moral language, so often used in order to advance destructive or manipulative ideologies, or simply to mask people from themselves. Humanistic psychologists nevertheless believed that to the extent that the inner life of human beings is taken seriously, and human nature conceived of as capable of freedom, people will be better equipped to examine the relationships that have been put at risk.

This fundamental commitment is clearly expressed in the Code of Ethical Principles of the UK Association of Humanistic Psychology Practitioners (UKAHPP). According to its first fundamental principle, "UKAHPP Members respect the dignity, worth and uniqueness of all individuals. They are committed to the promotion and protection of basic human rights, the integrity of the individual and the promotion of human growth, development and welfare. They affirm the self-determination, personal power and self-responsibility of the client." Note, in the last sentence, how the language of "patient" is rejected in favor of "client." More than any other group of psychologists, humanistic psychologists see themselves as working with and for others rather than as being superior to them. In this respect humanistic psychology presents a challenge for all scientists to reconsider the ways in which they conceive themselves as distinct or separate from the larger nonscientific public.

GLENN R. WILLIS
MARGARET M. BRACE

SEE ALSO *Choice Behavior; Freud, Sigmund; Jung, Carl Gustav; Skinner, B. F.*

BIBLIOGRAPHY

Becker, Ernest. (1973). *The Denial of Death.* New York: Free Press.

Blass, Thomas. (2004). *The Man Who Shocked the World: The Life and Legacy of Stanley Milgram.* New York: Basic.

Fromm, Erich. (1968). *The Revolution of Hope: Toward a Humanized Technology.* New York: Harper and Row.

Fromm, Erich. (1973). *The Anatomy of Human Destructiveness.* New York: Holt, Rinehart and Winston.

Homans, Peter. (1989). *The Ability to Mourn: Disillusionment and the Social Origins of Psychoanalysis.* Chicago: University of Chicago Press.

Watson, John B. (1998 [1924]). *Behaviorism.* New Brunswick, NJ: Transaction Publishers.

INTERNET RESOURCE

UK Association of Humanistic Psychology Practitioners (UKAHPP). "UKAHPP Code of Ethical Principles." Available from http://www.ahpp.org/ethical/code_of_ethics.htm.

PSYCHOPHARMACOLOGY

· · ·

Psychopharmacology is defined as the use of drugs to modify mental or behavioral performance. In general, psychopharmacology is used in the treatment of biologically based mental illnesses, although there has been increased interest in using drugs to enhance performance in healthy individuals.

Psychopharmacology assumes a strong mind-brain connection, if not a complete reduction of mind to brain. However, early theories about the relationship of brain chemistry to behavior were weak and post hoc. Most drugs were discovered accidentally and adopted because of their effects on symptoms (Valenstein 1998). Only later were theories of the ways drugs act on the brain developed, followed by theories of how mental states are related to brain chemistry. Although the mechanism of action on neuronal receptors has been elucidated for many drugs, the mechanisms by which drugs influence behavior at the whole-brain level are poorly understood. There are no definitive biological markers for diagnosing mental illness, and thus diagnosis relies on a clinical judgment of whether symptoms are present.

Although theories that mental illnesses result from a specific underlying chemical imbalance are not well substantiated, they have encouraged afflicted persons to seek treatment by reducing some of the stigma associated with psychological theories of mental illness. The discovery of drugs that alleviate some of the most debilitating symptoms of mental illness also allowed deinstitutionalization to occur, resulting in much more effective, community-based treatment programs

Psychopharmacologic Agents

The major classes of psychopharmacologic agents (Schatzberg and Nemeroff 1998) are antipsychotics (also known as neuroleptics), antidepressants, anxiolytics (antianxiety agents), and mood stabilizers. In addition, cognitive enhancing drugs are receiving increasing interest and use.

ANTIPSYCHOTICS. Antipsychotics, which are used primarily for the treatment of schizophrenia, reduce symptoms such as paranoia, visual and auditory hallucinations, and delusions. The first drugs of this type— phenothiazines— initially were produced as synthetic dyes. Later research in the 1940s showed that they act on the central nervous system as antihistamines. When administered to patients with allergies, they produced side effects that included decreased muscle tone, reduced nausea, and mild elation. That led to their use to relax patients before surgery, treat Parkinson's disease, and calm agitated and manic patients. In manic patients these drugs were also found to reduce the psychotic symptoms associated with the disorder. Antipsychotics are the most successful variety of psychopharmaceutical agents, reducing symptoms in 90 percent of patients in the acute phase of the disorder. Long-term use, however, may result in negative side effects that include tardive dyskinesia, which is characterized by involuntary motor movements such as those seen in Parkinson's patients, and neuroleptic malignant syndrome, which is a potentially fatal side effect. In contrast to the typical antipsychotics, the development of atypical antipsychotics has focused on not only reducing psychotic symptoms but also on improving negative symptoms (for example, loss of motivation, social withdrawal, and affective flattening) associated with schizophrenia and reducing adverse side effects.

ANTIDEPRESSANTS. Antidepressant pharmaceuticals also were discovered fortuitously. After World War II chemical companies had a surplus of the rocket fuel hydrazine, which they began modifying in an attempt to find new compounds with properties that might be useful for medical purposes. In the course of testing one of the new compounds against tuberculosis it was found to cause euphoria as a side effect. A derivative of hydrazine was synthesized as iproniazid, and after animal testing showed that the drug increased alertness, it became a treatment for depression.

There are three primary classes of antidepressants: selective serotonin reuptake inhibitors (SSRIs), monoamine oxidase inhibitors (MAOIs), and tricyclics. Common SSRIs include fluoxitine (Prozac) and paroxitine

(Paxil), MAOIs are exemplified by phenylzine (Nardil) and isocarboxazid (Marplan), and tricyclics include amitriptyline (Elavil) and imipramine hydrochloride (Tofranil).

All these drugs are considered equally effective in reducing depressive symptoms. The typical response rate of patients with uncomplicated unipolar depression to antidepressants is about 65 percent, compared with a 30 percent response rate on placebo. In addition to reducing current symptoms of depression these drugs appear to reduce the 50 percent relapse rate of major depressive episodes by 50 percent over the course of one year. The observed response rate with pharmacologic agents has been found to be identical to that of cognitive-behavioral psychotherapy. Treatment for depression that combines pharmacologic treatment and psychotherapy is more effective than is either modality alone (Burns 1980).

Side effect profiles or counterindications usually influence which drugs are prescribed more frequently. Side effects of SSRIs include headache, tremor, nausea, diarrhea, insomnia, agitation, nervousness, and sexual dysfunction. More important, SSRIs were found to increase suicidal behavior among adolescents, prompting the U.S. Food and Drug Administration (FDA) to mandate "black box" warnings to that effect on prescription bottles. Side effects of MAOIs include weight gain, orthostatic hypotension (drop in blood pressure when standing up quickly), delayed ejaculation, insomnia, and cholinergic side effects such as blurred vision, constipation, dry mouth, speeded heart rate, and urinary retention; MAOIs also require a diet that avoids foods with tyramine as hypertension and cerebral hemorrhage or death (rare) may occur. Side effects of tricyclics include weight gain, sexual dysfunction, cholinergic side effects, and sedation.

ANXIOLYTICS. The first drugs for treating anxiety were discovered in 1945 during testing of drugs designed to combat infectious bacteria. Those early drugs were found to be extremely habit-forming and to produce drowsiness, and ultimately were replaced by a class of antianxiety drugs called benzodiazepines (for example, alprazolam [Xanax] and diazepam [Valium]) that were discovered after an unexpected chemical reaction occurred in a compound that originally had been developed for use as a dye.

Benzodiazepines are very effective in reducing anxiety. In fact, since their introduction in the 1960s anxiolytics have been referred to as "happy pills." Eighty-two percent of patients on alprazolam show improvement compared with 42 percent on placebo. These drugs are also effective in reducing the occurrence of panic attacks. Benzodiazepines are among the most frequently prescribed drugs; however, they are habit-forming, and many of the 7 million prescriptions written yearly in the United States are in response to simple stresses of everyday life rather than debilitating conditions.

MOOD STABILIZERS. Mood-stabilizing drugs are used in the treatment of bipolar (manic-depressive) illnesses, in which patients suffer from recurrent cycles of depressive moods followed by manic periods. Lithium is the primary treatment for manic-depressive illness. Its effectiveness was discovered in the course of testing the hypothesis that uric acid would increase mania. Uric acid was difficult to work with because it was not easily soluble, and so lithium urate was used instead and surprisingly reduced mania. The FDA approved lithium treatment for mania in 1970 after a double-blind study showed that all the manic patients on lithium remained well, whereas half the patients who were switched from lithium to a placebo relapsed.

Additional double-blind, placebo-controlled studies have found that 70 to 80 percent of patients show improvement on lithium. Lithium reduces the intensity of manic and depressive episodes and decreases the overall number of episodes. Major side effects include excessive thirst and volume of urine, memory problems, tremor, and weight gain. In addition, high doses can lead to endocrine and renal complications.

COGNITIVE ENHANCERS. Cognitive-enhancing drugs are designed to improve cognitive functions such as memory and attention. Pharmacological agents to improve memory function are particularly important in slowing the memory loss observed in patients with Alzheimer's disease. The focus of this research has been on drugs that influence the brain systems involved in learning and memory. More specifically, agents are being developed to help patients retain memories that may be lost as individuals age. In clinical trials, donepezil (Aricept), rivastigmine tartrate (Exelon), and galantamine (Reminyl) all have been shown to reduce cognitive decline in the early stages of Alzheimer's. However, the measures of performance used have been very general tests of cognitive functioning, and so the exact cognitive function that is affected by these drugs is unclear. Side effects of these medications usually occur at higher doses and include gastrointestinal problems, dizziness, and headaches.

Drugs that enhance attentional functioning have been developed for the treatment of attention deficit hyperactivity disorder (ADHD). These drugs help patients maintain attention on a task over an extended period and reduce impulsive motor behaviors. Treatment for ADHD has consisted primarily of psychostimulants, including methylphenidate (Ritalin), d-amphetamine (Dexedrin), and a mixture of amphetamine salts (Adderall). Although the idea of giving stimulants to reduce hyperactivity is counterintuitive, these drugs have been shown to reduce symptoms in 70 to 80 percent of children with ADHD and are much more effective than are psychological treatments. These drugs reduce psychomotor activity and restlessness and increase a patient's ability to pay attention. Some of the more frequently reported side effects of psychostimulants include weight loss, social withdrawal, irritability, and insomnia.

With the development of these drugs interest has increased in the possibility of creating cognitive enhancers for healthy adults. Early research investigating the effects of Alzheimer's drugs on memory in healthy adults showed little to no improvement in memory function. As a result of that failure pharmaceutical companies began to focus on drugs that influence the formation of new memories and the retention of memories. However, data showing clinical effectiveness of these drugs were not available in the first years of the twenty-first century. In contrast to memory enhancement, much research suggests that healthy individuals who take methylphenidate and other psychostimulants show improvements in working memory and sustained attention. Conflicting research not only failed to replicate those improvements but found impairments in other cognitive functions. Before these drugs are prescribed for cognitive enhancement, their effects on healthy adults must be confirmed in controlled clinical trials.

Ethical Considerations

The ethical issues surrounding psychopharmacology can be grouped into four categories: research on psychopharmacologic agents, use in clinical treatment of illness, use for performance enhancement, and prophylactic or preventive use.

RESEARCH. Ethical issues in psychopharmacological research are largely the same as those in research in general and include issues related to the ethical treatment of animals, the informed consent of participants, and the appropriate use of placebos in control groups (Roberts and Krystal 2003). Because of the high commercial value of these products conflicts of interest among scientists are also important.

CLINICAL TREATMENT. Ethical issues that arise in the treatment of mental illness include informed consent (whether treatment is taking place inside or outside a research study), weighing the risks of side effects against the benefits of treatment (particularly the increased risk of suicide among adolescents taking certain antidepressants and the risks to the fetus or child of a pregnant or nursing mother receiving treatment), and access to treatment (for example, whether financial ability should determine which patients get access to newer, more expensive antipsychotics and which get cheaper, older, and less effective generic medications).

However, there are also "big picture" issues concerning what is viewed as an illness and when treatment should be directed at the individual rather than the environment. With most prescriptions for antidepressants and anxiolytics being written by general physicians rather than mental health professionals, drugs often are prescribed for dispositional characteristics or problems that are not biological in nature (for example, for persons dealing with stressful life events, grieving from a loss, or pessimistic by nature) even though psychological interventions designed to enhance coping skills could be more effective. Further, environmental change may be more effective than individual interventions in reducing the prevalence of some mental illnesses.

For example, with suicide as a leading cause of death among college students, perhaps it would be more efficacious to think about rampant depression as a problem stemming from the environment rather than from the individual. This changes the focus of treatment to modifications of the environment, (such as transition programs, peer support resources, and so on) as opposed to treating the many individuals who are suffering as a result of that environment. This amounts to taking a human factors approach to society and asking how to take what is known about cognitive strengths and limitations and use it to redesign cultural institutions to maximize productivity and benefit the individual while minimizing stress.

COGNITIVE ENHANCEMENT AND PROPHYLAXIS. Ethical questions concerning cognitive performance enhancement and prophylactic use have begun to be addressed. (President's Council on Bioethics 2003). The use of enhancing drugs when deficits are present (for example, for patients with Alzheimer's or ADHD) is subject to the same ethical questions as is the use of other psychopharmaceuticals for the treatment of illness. New ethical concerns arise when drugs are used to enhance performance in patients in whom no deficits are present (Farah, Illes, Cook-Deegan, et al. 2004) or to prevent illness when no signs of illness are present. For example, the potential risks of taking a drug are much more important in risk-benefit calculations when the patient's quality of life is high in the absence of the drug.

Questions of access and coercion also arise: With the use of Ritalin as a study aid reportedly on the rise among high school and college students, does this constitute an unfair advantage to the users? Will nonusers feel pressured to use the drugs in order to compete with users? Will those who cannot afford the drugs be left behind? In addition, general questions of what it means to be a person have been asked: To what extent is character built by coping with the limitations and imperfections present in oneself and others? Should individuals be free to experiment on themselves with such drugs? Will achieving human perfection make people happier?

Finally, with the contemporary emphasis on genetic contributions to psychiatric disorders, individuals eventually may be able to take drugs prophylactically based on genetic tests that assign an increased probability of developing a disorder. This raises concerns about whether this information should be supplied to individuals and how they will interpret the risks. It also prompts questions about how likely an outcome should be before information is given or action is taken. In light of the relative lack of understanding of brain system dysfunctioning in disorders, prophylactic use of drugs likely will become an increasingly serious issue.

ELIZABETH J. MULLIGAN
ERIC D. CLAUS

SEE ALSO *Emotion; Neuroethics; Psychology.*

BIBLIOGRAPHY

Burns, David D. (1980). *Feeling Good: The New Mood Therapy.* New York: Avon. Provides a comprehensive overview of pharmacologic and cognitive-behavioral treatments for depression.

Farah, Martha J.; Judy Illes; Robert Cook-Deegan; et al. (2004). "Neurocognitive Enhancement: What Can We Do and What Should We Do?" *Nature Reviews Neuroscience* 5: 421–425. A response to the President's Council on Bioethics (2003). Presents an argument in support of the development and use of cognitive enhancing drugs.

President's Council on Bioethics. (2003). *Beyond Therapy: Biotechnology and the Pursuit of Happiness.* Washington, DC: Author.

Roberts, Laura Weiss, and John Krystal, eds. (2003). Special issue on psychopharmacology and ethics. *Psychopharmacology* 171(1): 1–119.

Schatzberg, Alan F., and Charles B. Nemeroff, eds. (1998). *The American Psychiatric Press Textbook of Psychopharmacology,* 2nd edition. Washington, DC: American Psychiatric Press.

Valenstein, Elliot S. (1998). *Blaming the Brain: The Truth about Drugs and Mental Health.* New York: Free Press. Presents the historical development of drugs used to treat mental illness and critically analyzes the role of the pharmaceutical industry in promoting biological theories of mental illness.

PUBLIC POLICY CENTERS

• • •

Policy centers or *think tanks* (as they are often called) are an influential, diverse part of the U.S. not-for-profit sector. Those that contribute to discussions of science, technology, and ethics include organizations such as the liberal progressive Institute for Philosophy and Public Policy at the University of Maryland and the culturally conservative Ethics and Public Policy Center in Washington, DC. (Bioethics centers, which also contribute to these discussions, constitute a special category of policy centers and are considered in a separate article.)

Historical Background

Policy centers have grown in number and significance since the foundation of the Carnegie Endowment for International Peace in 1910 and the Institute for Government Research (IGR) in 1916, the first private organizations dedicated to analyzing public policy issues at the international and national levels, respectively. Subsequently IGR founder Robert Somers Brookings (1850–1932) established two supporting organizations: the Institute of Economics and a graduate school bearing his name. The Brookings Institution was formed when these three groups merged in 1927.

Both the Carnegie Endowment (with a staff of 100 and operating expenses of more than $19 million) and the Brookings Institution (with a staff of 275 and expenses of about $40 million) are still going strong, and have been joined by roughly 100 active think tanks in the Washington, DC, area. These include a number of additional policy centers that have expanded since their rather humble beginnings—among them, the Heritage Foundation (with more than 200 staff and more than $34 million in revenue); American Enterprise Institute (with 60 resident scholars, more than 100 adjunct scholars, and more than $18 million in revenues); the Urban Institute (including ten major policy centers with large staffs and operating expenses of more than $77 million); the Cato Institute (with 90 full-time staff, 60 adjunct scholars, 16 fellows, and revenues of roughly $13 million); and the Institute for Policy Studies (with a staff of 30 and expenses of roughly $1.5 million). (Staff and budgetary information is available from the Internet site of each organization, except Cato, obtained from an annual report.)

The expansion in both the numbers and influence of these organizations provides testament to the increasing complexity in government policy making and the growing demand for specialized knowledge and advice. Politicians and bureaucrats who make and implement policy often rely on outside experts to translate academic research and dialogue into predigested, understandable information and recommendations.

The term *think tank* originated in the United States during World War II to describe the secure environment where military and civilian experts developed military strategy. Subsequently the term was applied to contractors (such as the Rand Corporation) that worked closely with the military on both long-term strategy and short-term consulting. During the 1960s and 1970s, the use of the term was further expanded—first to include organizations focusing on international affairs, and then more broadly to cover organizations working on domestic political, economic, or social issues (McGann 2002).

The Role of Policy Centers in the United States

Think tanks inhabit the world of *nongovernmental organizations*—the third sector—and their success is primarily evaluated in terms of influence on the political process and the media. Think tanks operating in Washington, DC, at the beginning of the twenty-first century represent divergent points of view (for example, liberal, conservative, or libertarian) and cover a wide range of subject matter (from international relations to the environment, bioethics to economics) (Ricci 1994). Some specialize in one issue or field—for instance, the Pew Center on Global Climate Change, the Ethics and Public Policy Center. They are also diverse in their activities, roles, and sources of funding. As a result, neatly defining and categorizing think tanks is not an easy task. Nonetheless think tanks generally conduct policy research and analysis, and provide advice. In the United States, think tanks do any or all of the following:

- Serve as incubators for ideas that may later inform policy making;
- Provide a public forum for the exchange of ideas and debate;

- Provide advice to policymakers and offer expertise to the media;
- Advocate for particular positions—often crossing the line from think tanks to *do tanks*. (McGann 2002)

The influence of think tanks in Washington is considerable. While modern-day politicians often publicly eschew the *policy elite*, the variety and complexity of issues public officials confront often results in their reliance on such experts—if not directly, then indirectly (Smith 1991). Policymakers' staffs and outside stakeholders to whom they turn for advice routinely rely on publications and briefings by policy center staffs. Recent offerings by well-established think tanks such as Brookings and AEI include seminars on topics as diverse as ocean policy, Chinese labor issues, post-election Iraq, global warming, and the science of happiness.

In addition to being ubiquitous as *pundits* on television news programs and roundtables, think tank fellows and researchers often rank high in surveys and journal articles as individuals with the greatest influence on Washington, DC, policymakers (Ricci 1994). Over the years, think tanks have provided an important forum for independent research and strategic thinking that has informed important public policy debates.

Policy Centers with a Purpose

Ethical issues flow from the influence of policy centers on the process of governing. While campaign finance receives a great deal of public scrutiny, the influence of special interests on policy centers, which in turn influence elected officials, is often ignored. In addition to think tanks that may have a certain thrust (some would say bias) in approaching a wide sweep of policy issues, or that develop deep expertise in a specific subject area, a number of think tanks have been established to promote or attack certain policy proposals. Corporate interests financially support some of these and a central mission of such policy centers is to promote their sponsors' agenda. While financial support is sometimes acknowledged, such information is often not provided on the web sites of these centers, in their meeting materials, or in their publications. However most think tanks are established as not-for-profits. In order to maintain 501(c)(3) nonprofit status, lobbying must represent only a fraction of total expenditures for the organization and financial records must be disclosed (although not necessarily in a widely accessible manner).

The creation of for-profit ventures that merge lobbying, think tank, and journalism functions has further complicated the scene. Staff in some Washington, DC, area think tanks are as likely to come from Capitol Hill offices or the field of journalism as they are from the halls of academia. They use their skills and contacts to actively lobby for policy positions espoused by their clients. Such *journo-lobbying* has been called "an attempt to dominate the entire intellectual environment in which officials make policy decisions ... funding everything from think tanks to issue ads to phony grassroots pressure groups" (Confessore 2003, from Internet site). Blurring the spectrum of journalism and think tanks and lobbying raises obvious concerns about real or apparent conflicts of interests.

Analytical work published by various policy centers can range from rigorously researched, documented, and peer-reviewed books that serve an important role in elevating the policy debate to brief issue papers or even just press releases or short articles with little or no supporting analysis. With the advent of the Internet and email, centers can develop and widely disseminate *fact sheets* in minutes. Questions regarding the expertise of researchers, rigor and review of work product, and independence of analysis cast doubt upon the intellectual integrity of some think tanks. Because early twenty-first century think tanks weigh in on so many issues of scientific, social, and economic significance, the danger of an independent-sounding think tank fronting for specific private-sector interests under the guise of objective research and analysis provides reason to be concerned. Some articles in the popular press have revealed strategies to do just that (Confessore 2003, Cushman 1998).

Benefits of Policy Centers

In spite of concerns about think tanks with a specific corporate agenda, many play a valuable role where they conduct genuinely objective research and provide analyses critical to informing government policy making. Their publications and workshops often provide a rich resource for those wanting to understand complex technical, economic, and scientific issues and how they relate to questions of policy. Whether affiliated with universities (for example, the Institute for Philosophy and Public Policy at the University of Maryland) or independent, they provide a rich research environment for scholars. They allow research staffs the luxury of delving deeply into important topics regardless of the current political climate or government sponsorship, thus providing important and stable intellectual capital.

VICKI ARROYO

SEE ALSO *Bioethics Centers; Science Policy.*

BIBLIOGRAPHY

Cushman, John H., Jr. (1998). "Industrial Group Plans to Battle Climate Treaty." *New York Times* April 26, sec. 1, p. 1.

McGann, James G., and R. Kent Weaver, eds. (2002). *Think Tanks and Civil Societies: Catalysts for Ideas and Action.* New Brunswick, NJ: Transaction Publishers.

Ricci, David M. (1994). *The Transformation of American Politics: The New Washington and the Rise of Think Tanks.* New Haven, CT: Yale University Press.

Smith, James A. (1991). *Idea Brokers: Think Tanks and the Rise of the New Policy Elite.* New York: Free Press.

INTERNET RESOURCES

American Enterprise Institute. Available from www.aei.org.

Brookings Institution. Available from www.brookings.org.

Carnegie Endowment for International Peace. Available from www.carnegieendowment.org.

Cato Institute. Available from www.cato.org.

Confessore, Nicholas. (2003). "How James Glassman Reinvented Journalism—As Lobbying." *Washington Monthly,* December. Available from http://www.washingtonmonthly.com/features/2003/0312.confessore.html

Heritage Foundation. Available from www.heritage.org.

Institute for Policy Studies. Available from www.ips-dc.org.

Urban Institute. Available from www.urban.org.

PUBLIC RELATIONS
SEE *Advertising, Marketing, and Public Relations.*

PUBLIC UNDERSTANDING OF SCIENCE

• • •

Concern for the public understanding of science constitutes a field of teaching and research focused on the communication of science and technology to the nonscientific public. As science, technology, and society become increasingly intertwined, public communication concerning science and technology is of ever more obvious importance to relations between science, technology, and ethics.

Basic Issues

Strong belief in the social importance of scientific and technological knowledge is part of the professional heritage of scientists and engineers. But to a significant portion of the general population, regardless of educational level, many scientific and technological developments remain mysterious. Such mysteriousness arose originally both from the unique powers of science as well as the specialization of scientific knowledge. It easily degenerates into either excessive faith in or mistrust of scientific-technological developments, attitudes that in turn become a challenge for relations between scientific-technological knowledge and the public. This is especially true because even in the presence of irrational and easily manipulated faith and fears, the enormous powers of science and technology call for control by democratic decisions, the ultimate intelligence of which sometimes depends on a measure of scientific and technological literacy. The public understanding and communication of science have become topics of increasing concern since the 1960s as public attitudes to science became more ambivalent than the overweening optimism that reigned immediately after World War II.

For a variety of reasons, increased public understanding of science has been seen as preferable to a strict separation between science and the public. These reasons include: benefits to science, economic growth, national power and influence, participation by individuals in democratic societies, increased work skills, skills for public policymakers faced with issues that have scientific and technological dimensions, and intellectual, aesthetic, and moral benefits (Thomas and Durant 1987). There is mild consensus that the front-end loaded approach to science education needs supplemental adult education. But contention remains on such factors as how to conceive of "the public" (Miller 1983), how to measure "understanding," and what specific responsibilities are apportioned to scientists, engineers, and members of the public. A few even contend that increased public understanding may damage science and technology policy decisions (Trachtman 1981). Others fear the movement will foster a flattening scientism, or that it is solely motivated by scientists' wish for more public money.

The public communication of science and technology includes, in its widest sense, all of the means, manners, and sites that promote an interaction among science, technology, and the public. The media play an important role in the diffusion of scientific-technological information and in the analysis of the results, limits, benefits, and risks of technoscience. Popularization opens science and technology communication to new voices, to new information generators, and to new critics. But despite a growing acceptance of such activities by the scientific community since the 1980s, popularization is still rarely encouraged or rewarded by academic institutions. But simple linear, one-way, hierarchical models of communication processes are slowly being replaced with more nuanced representations of complex interchanges between scientists and various publics (Gregory and Miller 1998).

Scientists and engineers have been transformed, intentionally or not, into communicators, active participants in public debates, and spokespersons of scientific-technological knowledge. Some professional codes of ethics reflect the nature of new responsibilities brought on by these roles. In some senses there is a clear distinction between the roles of researcher and communicator. Technoscientific communicators must be able to set their knowledge in novel contexts, using different jargon, and often on short timescales, and be more aware of ethical, legal, and societal implications. But in another sense, both roles require respect for others, awareness of personal biases, and the formation of reasonable arguments. Programs for training technoscientists to communicate about their work in a clear and effective way are growing.

Increasingly researchers contend that the communication of scientific-technological knowledge should not be an attempt to achieve the exclusive goal of gaining the *confidence* of the public in scientific-technological matters. Rather, the main goal should be to make the public *participants* in these matters. David Layton and others (1993) argue that the lack of public understanding of science is often conceptualized in terms of a paternalist "deficit model" in which passive lay consumers of knowledge have cognitive gaps (i.e., ignorance) that need to be filled by the producers of expert objective knowledge. They propose an "interactive model" that rejects the objectivity of expert knowledge, the passivity of nonexpert consumers, and the homogeneity of the public. Science is interpreted as an interactive partner that should be responsive to diverse, context-dependent societal demands—where credibility is more important than objectivity. Many agree that this contextual and interactive approach is an improvement over the deficit model, but it is important to recognize and accommodate the knowledge asymmetries that necessarily remain between experts and the public (Miller 2000).

These newer models capture the continuous process of mutual and reciprocal construction between various technoscientific and societal communities. The process is a dynamic one of negotiating the meaning and worth of scientific-technological knowledge involving different actors. The social context and networks of people influence, in turn, the manners of perceiving this knowledge.

Most policy issues in complex, modern societies reveal the attributes of "post-normal science" (Funtowicz and Ravetz 1993) characterized by uncertainty, because there is no consensus concerning values, there are many conflicts even about the facts of the matter, and it is necessary to make urgent decisions. Post-normal science provides, in this sense, a fairly coherent explication of the necessity for greater participation in political-scientific processes. This also means that, in order for the public to gain a clear understanding of the potential and limitations of science, an inclusive dialogue will move much of the backstage scientific disagreements into the forefront (Miller 2000). Clearly the resultant understanding will not be a noncritical appreciation or acceptance.

Research Programs

The first public understanding of science research program emerged in the United States in the wake of the Soviet launch of *Sputnik I* (1957) and fears that U.S. students were not learning sufficient science. The Physical Science Study Committee (PSSC) at Harvard University, headed by Gerald Holton, F. James Rutherford, and Fletcher Watson, spearheaded development of new, more engaging physics curricula for both high schools and colleges that focused on the practice of science and included a measure of the history and philosophy of science. The National Science Board followed this work with the commencement in 1972 of the biennial "Science Indicators" surveys to gauge knowledge of and attitudes about science. In the 1980s, broader science education reforms were initiated. One example is the American Association for the Advancement of Science (AAAS) Project 2061, which began in 1985 (the most recent year in which Halley's comet appeared) and constitutes a long-term initiative to advance literacy in science, mathematics, and technology so that by 2061 (when Halley's comet makes its next appearance) fundamental change will have been achieved. By the 1990s the term *public understanding of science* had largely been replaced in the United States by concerns for *scientific literacy* and to some extent *technological literacy*. It was also argued that science, technology, and society (STS) education had an important role to play in developing such literacy in the non-scientific public.

Other public understanding of science research programs appeared in Europe. In the United Kingdom, especially, promoting the public understanding of science has been a major activity that traces its lineage back to the creations of the Royal Institution (1799) and the British Association for the Advancement of Science (1831). For instance, according to its charter, the Royal Institution—which is not to be confused with the Royal Society—was founded for "diffusing the knowledge, and facilitating the general introduction, of useful mechanical inventions and improvements; and for teaching, by courses of philosophical lectures and experiments, the application of science to the common purposes of life." It was at the Royal Institution that Michael Faraday in 1826 initiated the Friday Evening Discourses (for adults) and his famous Christmas Lectures on science (for young people).

The more proximate origin, however, was a decision of the Royal Society in 1985 to establish a working party to examine the extent and nature of the public understanding of science and its adequacy for an advanced democracy. The resulting Bodmer Report (1985) led to establishment of the standing Committee on the Public Understanding of Science (COPUS) and a continuing series of reports and initiatives. A 1993 white paper titled "Realising Our Potential" further confirmed the commitment of the United Kingdom to the public understanding and communication of science.

In February 2000 a select committee of the House of Lords published a report titled *Science and Society* that reflected recent changes in the "deficit model" interpretation of the science communication problem and the associated belief this could be remedied by more scientific-technological knowledge. This report reconceptualized the relationship between science and society in a way that emphasized contextual and interactive approaches. It led to proposals to replace "Public Understanding of Science" with "Public Engagement with Science and Technology" (PEST)—and in 2003 to a reorganization of COPUS as a national umbrella organization. A similar contextual and audience-centered approach arose slightly earlier from research performed in the United States (Lewenstein 1992).

The European Union has conducted two major studies that centered on determining the level of knowledge and attitudes of the population. Is the public knowledge of science increasing? Not much, to judge from the Eurobarometer 1992 and 2001 surveys in which interviewers used comparable tests. Although nearly half of all Europeans (45.3%) declared in the 2001 survey (European Commission 2002), "I am interested in science and technology," one in two of them also believe that they are not well informed. In 2001 the European Commission established a "Science and Society" program to promote scientific education and culture structured in thirty-eight actions. It underlined the importance of improving the channels of communication. These efforts are also bolstered by the European Collaborative for Science, Industry, and Technology Exhibitions, which include 300 member institutions and attract over 30 million visitors annually.

NICANOR URSUA
TRANSLATED BY JAMES A. LYNCH

BIBLIOGRAPHY

Bodmer, Walter. (1985). *The Public Understanding of Science*. London: Royal Society. Known as the Bodmer Report.

Cope, Edward Meredith. (1970 [1877]). *The Rhetoric of Aristotle, with a Commentary*. Vol. II. Revised and edited by John Edwin Sandys. Dubuque, IA: William C. Brown.

Dickson, David. (2000). "Science and Its Public: The Need for a 'Third Way.'" *Social Studies of Science* 30(6): 917–923.

European Commission. (2002). "Europeans, Science, and Technology: Survey Findings." *RTD Info: Magazine for European Research*, Special Issue, March 2002. Also available from http://europa.eu.int/comm/research/rtdinfo/previous_en.html.

European Commission. (2002). "Talking Science." *RTD Info: Magazine for European Research*, no. 34. Special Issue, September 2002. Also available from http://europa.eu.int/comm/research/rtdinfo/previous_en.html.

Funtowicz, Silvio, and Jerome R. Ravetz. (1993). "Science for the Post-normal Age." *Futures* 25(7): 739–755.

Gregory, Jane, and Steve Miller. (1998). *Science in Public: Communication, Culture, and Credibility*. New York: Plenum Press.

Hilgartner, Stephen. (1990). "The Dominant View of Popularization: Conceptual Problems, Political Uses." *Social Studies of Sciences* 20(3): 519–539.

House of Lords. Select Committee on Science and Technology. (2000). *Science and Society*. London: Her Majesty's Stationery Office.

Irwin, Alan. (1994). "Science and Its Publics: Continuity and Change in the Risk Society." *Social Studies of Science* 24: 168–184.

Irwin, Alan, and Brian Wynne, eds. (1996). *Misunderstanding Science? The Public Reconstruction of Science and Technology*. Cambridge, UK: Cambridge University Press.

Layton, David; Edgar Jenkins; Sally Macgill; and Angela Davey. (1993). *Inarticulate Science? Perspectives on the Public Understanding of Science and Some Implications for Science Education*. Driffield, East Yorkshire, UK: Studies in Education.

Lewenstein, Bruce V., ed. (1992). *When Science Meets the Public*. Washington, DC: American Association for the Advancement of Science. Fourteen papers derived from a conference of the AAAS Committee on Public Understanding of Science and Technology.

Miller, Jon D. (1983). *The American People and Science Policy*. New York: Pergamon Press.

Montgomery, Scott L. (2003). *The Chicago Guide to Communicating Science*. Chicago: University of Chicago Press.

Thomas, Geoffrey, and John Durant. (1987). "Why Should We Promote the Public Understanding of Science?" *Scientific Literacy Papers* 1: 1–14.

Trachtman, L. E. (1981). "The Public Understanding of Science Effort: A Critique." *Science, Technology, and Human Values* 6(36): 10–15.

Wynne, Brian. (1992). "Misunderstood Misunderstandings: Social Identities and Public Uptake of Science." *Public Understanding of Science* 1(3): 281–304.

Wynne, Brian. (1992). "Public Understanding of Science Research: New Horizons or Hall of Mirrors?" *Public Understanding of Science* 1(1): 37–43.

Wynne, Brian. (1995). "The Public Understanding of Science." In *Handbook of Science and Technology Studies*, ed. Sheila Jasanoff, Gerald E. Markle, James C. Petersen, and Trevor Pinch. Thousand Oaks, CA: Sage.

INTERNET RESOURCES

European Commission. (2001). "Science and Society: An Action Plan." Available from http://europa.eu.int/comm/research/science-society/action-plan/action-plan_en.html.

Miller, Steve. (2000). "Public Understanding of Science at the Crossroads." Available from http://www.bshs.org.uk/conf/2000sciencecommpapers/miller.doc.

Rademakers, Lisa. (2000). "Discovering a Code of Ethics for Science Journalism." Available from http://www1.stpt.usf.edu/peec/Rademakers.pdf.

PUGWASH CONFERENCES

• • •

In 1995 the Pugwash Conferences and one of its co-founders, the physicist Sir Joseph Rotblat, shared the Nobel Peace Prize in recognition of their decades-long work to reduce the threat of nuclear war and seek the abolition of nuclear weapons. As announced by the Norwegian Nobel Committee, Pugwash and its then president, Joseph Rotblat, were being recognized "for their efforts to diminish the part played by nuclear arms in international politics and in the longer run to eliminate such arms. It is the Committee's hope that the award of the Nobel Peace Prize for 1995 to Rotblat and to Pugwash will encourage world leaders to intensify their efforts to rid the world of nuclear weapons" (Norwegian Nobel Committee Communique, 13 October 1995).

The purpose of the Pugwash Conferences is to bring together, from around the world, influential scientists, scholars, and public figures concerned with reducing the danger of armed conflict and seeking cooperative solutions for global problems, especially those at the intersection of science, technology, and security. Meeting in private as individuals, rather than as representatives of governments or institutions, Pugwash participants exchange views and explore alternative approaches to arms control and tension reduction with a combination of candor, continuity, and flexibility not often possible in official diplomatic meetings. Because of the stature of many of the Pugwash participants in their own countries, insights from Pugwash discussions tend to penetrate quickly to the appropriate levels of official policy-making.

Origins and Organization

The Pugwash Conferences take their name from the small fishing village of Pugwash, Nova Scotia, Canada, site of the first meeting in 1957, which was attended by twenty-two eminent scientists from the United States, Soviet Union, Europe, Japan, Canada, and Australia. The stimulus for this first Pugwash meeting was the "Manifesto" issued in 1955 by Bertrand Russell and Albert Einstein, and also signed by Max Born, Percy Bridgman, Leopold Infeld, Frederic Joliot-Curie, Herman Muller, Linus Pauling, Cecil Powell, Joseph Rotblat, and Hideki Yukawa, which called upon scientists of all political persuasions to assemble to discuss the threat posed to civilization by the advent of thermonuclear weapons. American philanthropist Cyrus Eaton hosted the 1957 meeting at Thinkers' Lodge in Pugwash, his birthplace, and Mr. Eaton continued to provide crucial support for Pugwash in its early years.

From that beginning evolved both a continuing series of meetings at locations all over the world—with a growing number and diversity of participants—and a decentralized organizational structure to coordinate and finance this activity. Pugwash convenes between eight and twelve meetings per year, consisting of the large annual conference, attended by 150 to 250 people, and the more frequent workshops and study group meetings, which focus on specific issues and typically involve twenty to fifty participants.

Although very loosely structured—anyone who attends a Pugwash Conference becomes a member—the organization has been presided over since its inception by a series of distinguished scientists. Among the presidents, besides Rotblat, have been Nobel Laureate in chemistry Dorothy Hodgkin and Sir Michael Atiya, both from the United Kingdom, and Professor M. S. Swaminathan of India. Since 2002 the Secretary General has been Professor Paolo Cotta-Ramusino, who is a professor of mathematical physics at the University of Milan, and the executive director has been Dr. Jeffrey Boutwell of the United States (former associate executive officer at the American Academy of Arts and Sciences). A twenty-eight-member council, which generally meets once per year, and a six-member executive committee provide formal governance for Pugwash. Council members are elected every five years at the Quinquennial Conferences, held since 1962, which approve the long-term goals and bylaws of Pugwash. Marie Muller, professor of international politics at the University of Pretoria, is chair of the Pugwash Council. Pugwash has four small permanent offices, in Rome, London, Geneva, and Washington, DC, which help coordinate activities with more than fifty national Pugwash Groups around the world.

Evolution of the Pugwash Agenda

During the height of the Cold War, when few official channels existed between the Soviet Union/Eastern Europe, and the United States and Western Europe, Pugwash helped create unofficial lines of communication among scientists and policy makers, which in turn contributed to laying the groundwork for some of the most important arms control treaties of the period, including the Partial Test Ban Treaty of 1963, the Non-Proliferation Treaty of 1968, the Anti-Ballistic Missile Treaty of 1972 and SALT I accords, the Biological Weapons Convention of 1972, and the Chemical Weapons Convention of 1993. Despite subsequent trends of generally improving international relations and the emergence of a much wider array of unofficial channels of communication, Pugwash meetings play an important role in bringing together key scientists, analysts, and policy advisers for sustained, in-depth discussions of crucial arms-control issues, particularly in the areas of nuclear, chemical, and biological weapons.

In the early-twenty-first century, the Pugwash Workshops on Nuclear Weapons focused on bringing together scientists and policy makers from areas of regional tension such as South Asia, the Korean Peninsula, and the Middle East to discuss ways of reducing the threat posed by nuclear and other weapons of mass destruction in those regions.

The Pugwash Chemical and Biological Warfare Workshops, which began in 1959, meet twice per year, involving scientists and other technical experts, official negotiators, and industry representatives to explore means of strengthening the international prohibitions on the development and deployment of chemical and biological weapons (CBW) as well as possible CBW terrorist threats.

The Pugwash Workshops on Energy, the Environment, and the Social Responsibility of Scientists capitalize on the global network of Pugwash scientists to hold meetings and consultations on the major scientific and technological issues facing the international community. The workshops cover issues such as global climate change and future world energy needs as well as more specific topics, such as two workshops held in Cuba on public health and medical research. The Pugwash Conferences also have as one of its major goals the promulgation of ethical norms for the scientific community, which was the subject of a workshop in Paris, France, in June 2003.

While Pugwash findings reach the policy community most directly through the participation of members of that community in Pugwash meetings and through the personal contacts of other participants with policy makers, additional means of disseminating policy analysis include the *Pugwash Newsletter* (published twice per year), *Pugwash Occasional Papers* and *Issue Briefs*, and the Pugwash website. Some Pugwash publications include *Nuclear Terrorism: The Danger of Highly Enriched Uranium* (2002) and *U.S.-Cuban Medical Cooperation: Effects of the U.S. Embargo* (2001), and others more generally focused on global perspectives regarding issues of humanitarian intervention and the ramifications of missile defenses for nuclear stability.

Complementing Pugwash is an international Student/Young Pugwash movement, inaugurated in 1979. This is a global network of national groups with their own agendas and goals. Although organizationally separate from the Pugwash Conferences, International Student/Young Pugwash helps introduce students and younger scientists and scholars to the principles and objectives of Pugwash.

Founded on the principle of the individual responsibility of scientists for their work, the Pugwash Conferences have worked toward the twin goals of abolishing nuclear weapons and the peaceful settlement of international disputes since 1957. Emerging challenges in science, technology, and international politics of the twenty-first century make those principles and goals more relevant than ever.

JEFFREY BOUTWELL

SEE ALSO *International Relations; Rotblat, Joseph.*

BIBLIOGRAPHY

INTERNET RESOURCES

Pugwash Conferences on Science and World Affairs. Available from www.pugwash.org. Website for the Pugwash Conferences.

International Student/Young Pugwash. Available from www.student-pugwash.org. Website for the ISYP.

PURE AND APPLIED

• • •

The terms *pure science* and *applied science* began to appear in British usage some time after 1840, and were regularly used by American scientists from about 1880 through the 1930s, when pure science began to be replaced by *basic* or *fundamental science* (Kline 1995). While there is no firm consensus on how applied science differs from either pure science on the one hand, or engineering and technology on the other, distinctions made between pure and applied science are relevant to ethics because of the presence of widely held beliefs that pure science is more or less ethically innocent or neutral, and that any ethically troubling matters arise only when science is applied to practical matters.

Motives and Content

One generally recognized basis for distinguishing pure from applied science is the motives or aims of scientists: If one is engaged in science in order to increase one's understanding of the world, one is doing pure science, whereas if one is doing science in order to solve problems regarding human activity, one is doing applied science. A similar approach, more sociological, is to distinguish pure and applied science according to the setting and source of the aims directing scientific activity: Pure science is academic science, and applied science is science in commercial firms or on government projects. Scientists in academia have the freedom, within broad limits, to pursue their own aims, investigating whatever matters strike their curiosity, for however long it might take. Traditionally, their findings are their own property. Scientists working for industry or government are not at liberty to choose their own aims. They work on projects of others' choosing, and face strict limits of time and resources. Their findings belong to their employers.

So science is pure to the extent that its aims are internal to scientific practice (truth, demonstration), with minimal intrusion of external aims (money, status, social welfare). In contrast, applied science refers to science *applied* to external aims, typically in commercial or governmental projects.

While most scholars recognize that applied and pure science have different motives or aims, some maintain that practical motives of control and use cannot be the defining feature of applied science, because on this conception science conducted with a practical aim, engineering, and technology are all applied science. Yet the consensus from recent scholarship is that neither engineering nor technology is accurately characterized simply as applied science, because both involve forms of knowledge and skill that are not derivable from scientific theory or experiment. While engineering and technology employ science among their elements, they are distinguished from applied science by their cognitive content.

Considering cognitive content suggests that there is a second sense of the term applied science. There exist what are called *the applied sciences,* as the term is used, for example, in descriptions of university schools or programs. Here applied science is distinguished from basic science, a distinction based on content. Science is basic if it enhances human understanding of the class of entities with which it is concerned. Applied science refers to the sciences that start from the theories, models, and methods of basic science and use them to understand those material properties and processes that show promise of enabling the synthesis of new materials or creation of new energy-generating or transforming processes. For example, optoelectronics and electroceramics are applied sciences based particularly on the physical theories of thermodynamics and kinetics.

There is considerable overlap between these distinctions between applied science (content) and science applied (motive), because the applied sciences are ultimately motivated by practical aims of control and use. Yet making this distinction allows one to more accurately represent cases of, on the one hand, pure applied science (for example, physicists, typically in academic settings, studying the electrical properties of ceramic materials, having as their primary motive the production of knowledge) and, on the other, basic science done with a practical intent (for example, scientists employed by biotech firms who work on characterizing fundamental molecular mechanisms).

Ethical Implications

The difference in aims of pure science and science applied to practical matters suggests an important difference in the norms appropriate to these practices, specifically a difference in norms regarding proper procedure under conditions of uncertainty, when one does not know or cannot predict the outcome of some course of action.

In pure science, it is considered preferable to limit false positives (claims of an effect when none is present—also known as Type I errors) rather than false negatives (claims of no effect when an effect is present—Type II errors). That is, it is seen as worse to accept a falsehood (Type I error) than to reject a truth (Type II error). An epistemological value judgment of this sort is usually seen as healthy, cautious skepticism, a virtue when doing science.

Kristin Shrader-Frechette (1990) argues, however, that this approach is not the most rational one when applying science, at least in situations of uncertainty. In the applications of science in situations of uncertain outcomes, two types of errors are relevant: one may accept and develop an application that proves to be on balance harmful, or one may reject the development of an application that is on balance beneficial. When scientific rationality is used to evaluate situations with these kinds of possible outcomes, the result is a preference for erring in accepting developments that might be harmful, rather than for erring in rejecting developments that might prove harmless. If science is seen as seeking to maximize truth, it would seem to be most rational to push forward with the development of knowledge, or its applications, on the grounds that error, whether conceptual or practical, will be more likely discovered and then dealt with, thus further maximizing truth, whereas failure to go forward with an investigation means that the truth in that domain will not come out.

But the aim of science applied to practical matters is not the maximization of truth. If it is to be seen as the maximization of something, it is the maximization of welfare, and once welfare is a concern then rationality demands a consideration of values other than purely epistemological ones.

If one takes a consequentialist utilitarian perspective, concern focuses not only on the probability of a hypothesis being true but also on the likely consequences following from a hypothesis. Practical errors arising in the application of science can adversely affect large numbers of people. If the situation is one of genuine uncertainty, meaning that it is not possible to assign probabilities to various outcomes, and some outcomes are worse than others, it can be argued that the most rational strategy is to act as if the worst consequence that could happen will happen, and thus seek to minimize the possibility of the worst-case scenario. That is, in a situation in which it is not possible to assign probabilities to either possible beneficial consequences or possible disastrous consequences, then it is better to forego possible benefits, if doing so prevents possible disasters.

If one takes a deontological perspective such as that of Immanuel Kant (1724–1804), matters of the social and legal obligation, informed consent, and the voluntariness of risk become relevant in deciding whether to apply some scientific knowledge. Shrader-Frechete concludes that, while the proper procedural norms in pure science are strictly epistemological, the proper procedural norms for applying science to practical matters are both epistemological and ethical.

Apart from consideration of the different procedural norms of pure science and science applied, some conclusions can be drawn about the general relevance to ethics of the distinctions between pure science and science applied, and basic science and applied science.

For duty-based ethical perspectives such as Kant's, and virtue-based moral perspectives with their focus on character, the distinction of pure science versus its applications, based as it is on motives for action, will have moral significance. For example, respect for the autonomy of persons would support the moral permissibility of all basic science, regardless of what might be done with the resulting knowledge. In contrast, utilitarian and other consequentialist approaches focus on foreseeable consequences rather than motives, and the pure/applied distinction will have little importance. If it can be foreseen that the knowledge gained from some basic science will most likely produce more harm than good, the motives of the scientists are beside the point: Such knowledge should not be gained, at least not in the referenced context. Those doing pure science have an obligation to consider not only *how* they should proceed but also *whether* they should proceed.

With respect to the basic/applied distinction regarding content, those for whom consequences determine the rightness of actions will not concern themselves with whether those consequences result from basic or applied science. For nonconsequentialists, pure applied science, like basic science, would always seem to be permissible, while the morality of the practical application of applied science will depend on whether those involved act upon their obligations toward others.

Beyond Science

It remains to be considered whether the previous analysis might be relevant in other areas in which the pure/applied distinction is used. Certainly it is common to speak of pure and applied ethics, pure and applied art—and, on rare occasions, distinctions may even be drawn between pure and applied engineering or technology.

With regard to ethics the pure/applied distinction can, as in science, be drawn on the basis of motives or content. With reference to motives, people pursue ethical reflection in the pure sense simply as a topic of interest in its own right, or in the applied sense when they do so in order to lead better lives. As with science, the sociological context of the former would probably be the university, of the latter a clinical or other practical setting. (In some interpretations, pursuit of the former itself leads to a better life.) With reference to content, ethics can be basic in the sense of engaged with fundamental insight into theories and principles or applied in the sense of making particular decisions. Whether and to what extent the further analysis of the different epistemological and ethical assessments of Type I and Type II errors applies remains an open question. Nevertheless, with regard to pure/applied art, it can be suggested that parallel reflections would be relevant.

With regard to engineering and technology and the pure/applied distinction, issues become more problematic. In part this is because of the application factor that is already built into these disciplines. As one observer has described it, "Pure technology is the building of machines for their own sake and for the pride or pleasure of accomplishment" (Daedalus 1970, p. 38). Samuel C. Florman (1976) refers to something similar when he analyzes "the existential pleasures of engineering." Any pure engineering or pure technology, pursued for its own sake, is nevertheless something more closely engaged with the world, and thus more directly subject to ethical assessment, than pure or basic science. It is difficult to imagine engineering or technology ever being as pure or basic in an ethically relevant sense as pure or basic science.

RUSSELL J. WOODRUFF

SEE ALSO *Neutrality in Science and Technology.*

BIBLIOGRAPHY

Bunge, Mario. (1966). "Technology as Applied Science." *Technology and Culture* 7(3): 329–347.

Daedalus of New Scientist [pseud.]. (1970). "Pure Technology." *Technology Review* 72(8): 38–45.

Feibleman, James K. (1972). "Pure Science, Applied Science, Technology: An Attempt at Definitions." In *Philosophy and Technology: Readings in the Philosophical Problems of Technology,* ed. Carl Mitcham and Robert Mackey. New York: Free Press.

Florman, Samuel C. (1976). *The Existential Pleasures of Engineering.* New York: St. Martin's Press. 2nd edition, 1994.

Kline, Ronald. (1995). "Construing Technology as Applied Science: Public Rhetoric of Scientists and Engineers in the United States, 1880–1945." *Isis* 86(2): 194–221.

Niiniluoto, Ilkka. (1993). "The Aim and Structure of Applied Research." *Erkenntnis* 38(1): 1–21.

Ravetz, Jerome R. (1971). *Scientific Knowledge and Its Social Problems.* Oxford: Clarendon Press.

Shrader-Frechette, Kristin. (1990). "Island Biogeography, Species-Area Curves, and Statistical Errors: Applied Biology and Scientific Rationality." *Proceedings of the Biennial Meeting of the Philosophy of Science Association* 1990, vol. 1: 447–456.

Q

QUALITATIVE RESEARCH
• • •

Since the seventeenth century modern science has emphasized the strengths of quantitatively based experimentation and research. The success of quantitative research in the so-called hard sciences, especially physics and chemistry, stimulated attempts to extend quantitative work into the social or human sciences, where its application was somewhat problematic. A countermovement with ethical dimensions developed during the nineteenth century as increased attempts at exploration and colonization resulted in efforts to document "native" cultures in qualitative ways; that countermovement contributed to the formalization of methods in anthropology. In the twentieth century qualitative methods were adopted in sociology; many of the applied disciplines, such as nursing, education, and business; and human and rural ecology, geography, and engineering. By the 1970s qualitative research and qualitative inquiry had become the rubrics of a reformist movement in the social sciences, with professional associations, journals, and basic reference works appearing into the twenty-first century.

Basics

Many distinct qualitative research methods were developed and formalized, including ethnography, phenomenology (as a method), conversational or discourse analysis, narrative inquiry, grounded theory, participant observation, and ethology. Those methods were complemented by research designs and analytic strategies that allowed data of different levels and types to be accessed, such as focus groups, case studies, and action research. Qualitative research is used in micro and macro descriptions, concept and theory development, and evaluation, all of which often combine or overlap and add to the complexity of methods. There are also different perspectives or schools of thought on qualitative research, such as Marxism, phenomenology, ethnomethodology, cultural theory, symbolic interactionism, feminism, critical theory, and structuralism. These theoretical underpinnings provide a lens that focuses an inquiry on particular purposes, agendas, and goals so that a researcher may choose to conduct, for example, a critical ethnography or formulate a feminist-grounded theory.

Transcending such differences among schools of qualitative inquiry, all qualitative research exhibits seven basic characteristics. The most important are (1) thick description, or rich and relevant descriptions of the social, cultural, linguistic, and material contexts in which people live; (2) the presentation of the perspective of the people being studied (the emic, or natives', point of view); and (3) the use of relatively small and purposefully selected (rather than large and randomly selected) samples. Qualitative inquiry also involves (4) the inductive development of explanation, concepts, and theory; (5) reliance on observational and interview data; (6) the use of textual data involving content and thematic analysis (rather than numerical data and statistical analysis); and (7) techniques of verification that assess the trustworthiness of data, replication, and saturation.

Contributions

What does qualitative inquiry contribute to knowledge? Using microanalytic inquiry, qualitative researchers explore, document, evaluate, and diagnose mechanisms and individual, group, or organizational behavior for purposes such as investigating problems (e.g., drug errors); processes of teaching, learning, or care giving; naturally occurring interactions between individuals and groups; and behavioral indexes (e.g., expressions of pain) and situations (e.g., drug trafficking).

Qualitative researchers also explore the subjective subjectively. They are concerned with perceptions, beliefs, and values and with the responses and experiences of people. Qualitative researchers look for norms and for exceptions to both obvious and less recognized patterns of behaviors. That research illuminates, explicates, and interprets to provide understanding. This knowledge allows the recognition of humanity in oneself and in others, leading to the ability to care for and teach people, run organizations and programs, and identify practices and develop policy. Qualitative inquiry provides the information, substance, rationale, and interventions needed for the optimal funding of social programs.

Qualitative researchers develop pertinent and useful concepts and valid theories. "Knowing what is actually happening" essentially removes subjectivity and enables action, providing organizing systems and paradigms and thus facilitating efficient, effective, and cohesive approaches to, for instance, health care and education.

Issues and Ethics

Qualitative research arose in the nineteenth century as a form of ethical resistance to what was seen as an unwarranted extension of quantitative methods. That challenge has been revived by attempts by what is known as the Cochrane Collaboration (a group that supports and publishes meta analysis of research, usually clinical drug trials, and evaluates the research using criteria recommended by Archie Cochrane, that support experimental design). Qualitative data is dismissed as "anecdotal" and is valued least to promote quantitative criteria for evidence in the assessment of healthcare interventions, in which efficacy, evaluation, and certainty are valued above context-based and applied knowledge. That approach devalues the contribution of qualitative inquiry. Moreover, the valuation of science for objective knowledge, experimental design, and hard data and measurement has devalued qualitative inquiry in universities and funding agencies, making qualitative inquiry a lower priority in curricula and in the agendas of funding agencies.

Recent efforts to strengthen qualitative inquiry, along with an increasing awareness of the limits of quantitative inquiry and its complementary relationship with quantitative inquiry, have led to increasing interest in mixed-method design, especially research designs that combine qualitative and quantitative inquiry. However, the underlying debate about the rigor of qualitative inquiry continues to constitute an ongoing challenge to qualitative researchers. Are qualitative findings rigorous enough to stand on their own, or should qualitative theories be tested quantitatively? Can qualitative results be generalized?

Despite criticisms, qualitative research is considered a powerful tool for eliciting the meaning of situations and for making sense of the complexity of life as it is lived and communicating that complexity. In the 1990s the art-based qualitative movement used techniques from the theater, the presentation and dissemination of qualitative findings, and the elicitation of qualitative data that reveals the implicit. Qualitative results also may be represented in the form of poetics and even as art installations in efforts to facilitate understanding of the worldview of the other.

In qualitative research ethics also comes into play. Issues of consent are paramount, dealing with subjects not only agreeing to participate in a qualitative study but to remain in that study over time. Such consent is considered ongoing, and the onus is on the researcher to ensure that participants are fully cognizant of the nature of a project. Because the quality of the data is dependent on the relationship with the participant (the establishment of trust) and because of the intimate nature of the topics qualitative researchers study protection of a participant's privacy by providing anonymity and confidentiality is important. The paradox here is that in the process of concealing identities the altering and/or removal of identifiers changes the data and creates the risk of impairing validity. However, this protection of the rights of the individual is one of the hallmarks of qualitative inquiry. It is this, along with its interest in patterns of human behavior, that distinguishes qualitative inquiry from journalism.

JANICE M. MORSE

BIBLIOGRAPHY

Denzin, Norman K., and Yvonna S. Lincoln, eds. (2000). *Handbook of Qualitative Research*, 2nd edition. Thousand Oaks, CA: Sage. A manual describing the strategies for conducting most of the qualitative methods.

Denzin, Norman K., and Yvonna S. Lincoln, eds. (2002). *The Qualitative Inquiry Reader*. Thousand Oaks, CA: Sage. Reprints of seminal articles in qualitative inquiry.

Morse, Janice M., ed. (1994). *Critical Issues in Qualitative Research Methods*. Thousand Oaks, CA: Sage. Addresses important issues in qualitative research, with authors' discussion of each chapter.

Morse, J. M. and Richards, L. (2002). *Readme First for a User's Guide to Qualitative Methods*. Thousand Oaks, CA: Sage. A basic text and guide to the qualitative methods literature.

Strauss, Anselm, and Juliet M. Corbin. (1998). *Basics of Qualitative Research: Techniques and Procedures for Developing Grounded Theory*, 2nd edition. Thousand Oaks, CA: Sage. A text describing the method of grounded theory.

INTERNET RESOURCES

International Journals of Qualitative Methods. Available at http://www.ualberta.ca/~ijqm/. An open access journal that specializes in qualitative methods.

The Qualitative Report. Available at www.nova.edu/ssss/QR/practice.html. An online journal that also provides links to web pages, papers and other texts, other journals, and course syllabuses.

R

RACE

• • •

Race, at a most basic level, is a system for classifying people by various forms of similarity and difference. Race is a culturally, socially, and scientifically defined concept whose meaning—depending on the period in history, geographic location, and the scientific or technological context—has changed over time. Race is a fluid concept. The meaning of race has evolved from a term describing livestock lineage to a tool used in medical diagnoses. The ethical implications of race in relation to science and technology depend on the ways in which it is deployed and by whom. In this regard, race can be used to make informed scientific and technological decisions, or it can be used to reinforce cultural stereotypes and regimes of discrimination.

Origins of Race

Prior to the sixteenth century, the current connotations of race did not exist. The most common use of the term *race* was in reference to the domestication of livestock. A "racial stock" was a group of animals bred for a specific purpose. In the sixteenth century, this animal husbandry term migrated and began to be used to describe peoples. Race became a way to explain differentiations within "human stock." Europeans were the first to use the terms *race* and *stock* to delineate between different human groups. Customs and regional origins, as well as religious values and beliefs, determined the degree of difference. The characteristics attributed to races and stocks were similar to those now attributed to culture. Race did not carry powerful biological overtones. Soon, however, it became a way of evaluating and differentiat-

ing between those considered to be civilized and those deemed to be uncivilized.

Indeed, for the Enlightenment philosophes and scientists of the seventeenth and eighteenth centuries, what was most important was the human race as a whole and the prospects for its progressive advancement. Enlightenment science advocated at least two propositions that severely limited the use of race as a justification for social discrimination. First, Enlightenment anthropologists were monogenists rather than polygenists; that is, they believed that human beings were created only once. As confirmed by the ability of all human beings to interbreed, all human beings were one species, and variations were the results of varieties within the species, not differences between species. Second, for the Enlightenment, environment and education were considered much more important than heredity. When the Baron de Montesquieu in his *Spirit of the Laws* (1748) argued that human differentiation was caused by environmental and historical factors, the corollary was that such differentiations were of secondary importance and could be overcome by means of education. On the basis of such views, France's Constituent Assembly abolished slavery in 1791 shortly after the beginning of the French Revolution, and the British abolished the slave trade in 1821.

Over the course of the eighteenth century, however, the understanding of race changed from a difference based on geographic boundaries and cultural heritage to one based on physical differences that could be easily categorized into human "types." This perception of race had its roots in the tenth edition of Carolus Linnaeus's *Systema Naturae* (1758). In this volume

Linnaeus brought together perceptions of cultural and physical characteristics to describe race, a formulation that marked the emergence of a racialized discourse within Western science. Linnaeus argued that four "races" existed with specific physical features, emotional temperaments, and intellectual abilities: Homo americanus—reddish, choleric, erect, tenacious, content, free, and ruled by custom; Homo europaeus—white, ruddy, muscular, stern, haughty, stingy, and ruled by opinion; Home asiaticus—yellow, melancholic, inflexible, light, inventive, and ruled by rites; Homo afer—black, phlegmatic, indulgent, cunning, slow, negligent, and ruled by caprice. In differentiating species into subspecies based on elements that are common to the entire species, Linnaeus linked elements such as skin color directly to perceived behavioral propensities and eventually to biological variation.

Such a system of classification became increasingly used to distinguish not human variation but different species. Distinctions made at the subspecies level enabled value judgments to be made about superiority, inferiority, domination, and subserviency, based on physical attributes. As the Enlightenment commitment to the primacy of environment over heredity faded, this solidified perceptions that the characteristics displayed by each subspecies were immutable. Based on common characteristics, race evolved, from an indicator of similarity and difference, to a system of classification, and finally to a concept that imbedded cultural and physical characteristics into individual biological makeup. By the nineteenth century race as a biological and scientific concept had been firmly instantiated within scientific studies undertaken by natural philosophers Georges Cuvier (1812) and Charles Darwin (1859).

Racialization of Science

The nineteenth century also saw the racialization of science. Racialization is a social process by which beliefs about race become instruments of social categorization, cultural classification, political judgments, and economic decisions. New scientific work emerged to validate the underlying implications within Linnaeus's system of classification. Louis Agassiz (1850), Pierre Paul Broca (1861), and Samuel George Morton (1839), as well as others, endeavored to produce scientific evidence confirming their beliefs that white Europeans were at the top of the racial hierarchy. Researchers used the now discredited sciences of *polygeny*, that racial groups had different origins and were different species; *phrenology*, the study of the shape and protuberances of the skull to reveal character and mental capacity; and

craniometry, the measurement of the skull to determine its characteristics as related to sex, race, or body type, to separate and differentiate races. According to Audrey Smedley, author of the 1993 book *Race in North America*, the reconceptualization of race in the nineteenth century created "a social mechanism for concretizing and rigidifying a universal ranking system that gave Europeans what they thought was a perpetual dominance over indigenous people of the New World, Africa, and Asia" (pp. 303–304). The hierarchy soon became understood as the natural order of things.

The scientifically supported perceived difference in races produced a Western ideological position of global superiority. The racialization process created an environment in which nonwhite peoples were viewed as socially, culturally, and intellectually inferior. It produced a scientific rationality that sustained this belief structure. The ways in which political and racial ideologies influenced science is well illustrated in the work of the French scientist Paul Broca (1824–1880). When Broca's craniometric studies produced results suggesting that Germans possessed larger brains than the French, he adjusted his data for body size, in order to show that German brains constituted a smaller percentage of overall body mass than French brains did. In like manner, when Broca found that people of African heritage had larger cranial nerves than Europeans, this clearly meant that cranial nerves did not contribute to intellectual activity of the brain. It is these processes of racialization in science that justified beliefs in racial superiority and inferiority, which in turn enabled racism to flourish. The racism was masked by religious authorities, and the racialized scientific truths of eugenics and Social Darwinism further reinforced the misperception of racial difference that reverberates to the present day.

By the late nineteenth century, racial difference became the dominant lens through which the Western world perceived racial and ethnic otherness. This perspective directly influenced the scientific and technical opportunities for those who were not white. In the United States, science codified the social attitudes about black inferiority and became the dominant obstacle inhibiting blacks, as well as other nonwhite persons, from engaging in scientific and technical work. Those who were able to partially overcome the barriers created by a tradition of racialization and contribute to science and engineering were regularly dismissed as exceptions or marginalized for what was assumed to be substandard work by substandard humans. By the beginning of the twentieth century, it was widely held in scientific and technical communities that people of African descent had contributed nothing worthwhile to the scientific and technical development of the modern world.

At the 1913 annual meeting of the American Association for the Advancement of Science, James McKeen Cattell, at the time the owner and editor of the journal *Science,* confirmed this opinion. In a speech titled "Science, Education, and Democracy," he argued that while there was a need for more educational opportunities for Negroes, it was clearly understood that "[t]here is not a single mulatto who has done creditable scientific work" (Cattell 1914, p. 154). This statement—which repeats equally negative judgments found in both David Hume's essay "Of National Characters" (1753) and Immanuel Kant's "On the Different Races of Man" (1775)—overlooks the highly regarded work by the agricultural chemist George Washington Carver (c. 1864–1943), the physician Rebecca Cole (1846–1922), the developmental biologist Ernest Everett Just (1883–1941), and the inventor Granville T. Woods (1856–1910). Nevertheless, their racial identification made their scientific and technical careers difficult at best.

Scientific Criticism of Race

During the early twentieth century scientists also began to challenge the conceptions of race developed in nineteenth-century science. For many it became an ethical issue when research began to reveal that many scientists altered their data to fit the valued racial hierarchy of the day. The foremost critic of scientific racism was the eminent anthropologist Franz Boas (1940). Boas applied a scientific rigor to counteract the social and racialized rigor of the nineteenth-century racial science. He recalculated data, exposed the inaccuracies, and provided evidence that would argue strongly against the racialization of science. His work indicated that many scientists molded their data to fit a worldview that aimed to maintain and strengthen a racial hierarchy that located Europeans at the top. By deploying the power of genetics and biology, he was able to begin breaking the hold that racialized assumptions about human variation had in science. But the perception had been so deeply imbedded in scientific practice that it would take decades to destabilize it. It is in this regard that the U.S. Public Health Service could conduct a forty-year experiment, known as the Tuskegee Syphilis Study (1932–1972), on 399 black men in the late stages of syphilis (Jones 1993).

The rise of Nazism represented a new wrinkle in the tradition of racialized science. What distinguishes the Nazi agenda from other historical genocidal efforts was its reliance on science. For instance, in 1934 the Nazi deputy party leader, Rudolph Hess, spoke of National Socialism as applied biology. Nazi racial purification, based on a racialized biomedical vision, escalated from forced sterilization to holocaust (Lifton 1986).

The claims of inherent racial inferiority during the reign of Nazism and the subsequent Holocaust provided an important impetus for the United Nations to produce a public statement challenging the scientific basis of race. The United Nations contended that such wholesale disregard for human life was made possible by the continued propagation of racial inequality. To reconstitute the ways in which race had been constructed, the United Nations Educational, Scientific and Cultural Organization (UNESCO) convened a panel of social and natural scientists and charged them with producing a definitive statement on racial difference. The panel produced two statements: Statement on Race (1950) and Statement on the Nature of Race and Race Differences (1951). Primarily written by Ashley Montagu, a student of Boas, the statements declared that race had no scientific basis and called for an end to racial thinking in scientific and political thought. Within the next two decades UNESCO would release two more statements: Statement on the Biological Aspects of Race (1964) and Statement on Race and Racial Prejudice (1967). Although important, these statements did not immediately influence social policy and the public attitudes that had been ingrained about race.

Continuing Issues

Scientifically the importance of race diminished over the latter part on the twentieth century. Race reemerged, however, with the organization of an international research project to determine the DNA sequence of the human genome. The Human Genome Project (HGP) began in 1990, and researchers produced a complete map in 2003. One of the major goals of the HGP was to find and elucidate the function of human genes. Some of the most promising and troubling outcomes of the HGP in the context of race have to do with genetic therapy. Genetic researchers contend that the human genome consists of chromosome units or haplotype blocks. Haplotype maps (HapMaps) can possibly provide a simple way for genetic researchers to quickly and efficiently search for genetic variations related to common diseases and drug responses.

The danger is that this research might re-ensconce the biological concept of race within scientific practice and knowledge production. It is already common practice for physicians to base clinical decisions on a patient's perceived race. The positive potential of Hap-Maps could be overshadowed by the manipulation of genetic data to support racialized stereotypes, renew claims of genetic differentiation between races, and add biological authority to ethnic stereotypes. These pitfalls arise when genetic data become the basis on which racially specific drugs or treatments are designed. In 2003 the U.S. Food and Drug Administration proposed guidelines that would require all new drugs be evaluated for their effects on different racial groups. In the contemporary world, the genetic origins of race reappear much more quickly than they are eliminated.

The connections between biology and race are far from settled. In thinking about the future ethical implications of this relationship, it is necessary to consider what function the multiple manifestations of race will serve within social, cultural, scientific, medical, and technological practices, as well as the ways in which researchers will deploy race within the conflicting and overlapping realms. As a result, race will continue to be one of multiple issues and concepts that will determine on what terms we as a society will engage each other humanely.

RAYVON FOUCHÉ

SEE ALSO Class; Eugenics; Feminist Ethics; Genocide; Holocaust; Human Rights; IQ Debate; Nazi Medicine; Social Darwinism; Tuskegee Experiment.

BIBLIOGRAPHY

Agassiz, Louis. (1850). "The Diversity of Origin of the Human Races." Christian Examiner 49: 110–145. A statement on polygeny and the disparate origins of human races.

Boaz, Franz. (1940). Race, Language, and Culture. New York: Macmillan. A collection of the most important essays written by Franz Boas on the science of anthropology.

Broca, Paul. (1861). "Sur le volume et la forme du cerveau suivant les individus et suivant les races" [On the volume and form of the brains of individuals and races]. Bulletin de la Société d'anthropologie de Paris 2: 139–207, 301–321, 441–446. Argues that intelligence can be related to anthropometic differences.

Cattell, J. McKeen. (1914). "Science, Education, and Democracy." Science 39(996): 154. Speech presented to the American Association for the Advancement of Science (AAAS) stating minorities have not contributed to science.

Cuvier, Georges. (1812). Recherches sur les ossemens fossiles de quadrupèdes [Research on the fossil bones of quadrupeds], Vol. 1. Paris: Deterville. A collection of anatomical studies that is one of the early works of vertebrate paleontology.

Darwin, Charles. (1859). On the Origin of Species by Means of Natural Selection. London: John Murray. Contends that species change though an evolutionary process of natural selection.

De Montesquieu, Baron. (2002 [1748]). Spirit of the Laws. Amherst, NY: Prometheus Books. A comparative study of three types of government: republic, monarchy, and despotism.

Dubow, Saul. (1995). Scientific Racism in Modern South Africa. Cambridge, UK: Cambridge University Press. Examines the relationship between science and colonial power in south Africa, Zimbabwe, Mozambique, and Mauritius.

Duster, Troy. (2003). Backdoor to Eugenics, 2nd edition. New York: Routledge. Discusses the ethical struggles and social implications of new genetic technologies.

Fouché, Rayvon. (2003). Black Inventors in the Age of Segregation: Granville T. Woods, Lewis H. Latimer, and Shelby J. Davidson. Baltimore: Johns Hopkins University Press. Reconeptualizes what it means to be an African American inventor.

Gould, Stephen J. (1996). The Mismeasure of Man, rev. edition. New York: Norton. Examines the history, politics, and power of science and the way biology has been deployed to construction racial difference.

Graves, Joseph L., Jr. (2001). The Emperor's New Clothes: Biological Theories of Race at the Millennium. New Brunswick, NJ: Rutgers University Press. Uses the scientific method to argue that races do not exist as a biological category.

Harding, Sandra, ed. (1993). The "Racial" Economy of Science: Toward a Democratic Future. Bloomington: Indiana University Press. A series of essay that address the ways that aspects of science are racially constructed.

Jones, James H. (1993). Bad Blood: The Tuskegee Syphilis Experiment, rev. edition. New York: Free Press. Details the history and affects of a government sponsored experiment on African American syphilitics.

King, James C. (1981). The Biology of Race, rev. edition. Berkeley and Los Angeles: University of California Press. Contends that race is a social, not biological, concept.

Kuttner, Robert E., ed. (1967). Race and Modern Science: A Collection of Essays by Biologists, Anthropologists, Sociologists, and Psychologists. New York: Social Science Press.

Lifton, Robert Jay. (1986). The Nazi Doctors: Medical Killing and the Psychology of Genocide. New York: Basic. Describes the conflicting roles that German doctors played in Nazi genocide.

Linnaeus, Carolus. (1758). Systema Naturae [System of nature], 10th edition. Stockholm: Laurentii Salvii Homiae.

Manning, Kenneth R. (1983). Black Apollo of Science: The Life of Ernest Everett Just. New York: Oxford University Press.

McMurry, Linda O. (1981). George Washington Carver: Scientist and Symbol. New York: Oxford University Press.

Morton, Samuel George. (1839). Crania Americana; or, a Comparative View of the Skulls of Various Aboriginal Nations of North and South America. Philadelphia: Dobson. Claims to find different cranial capacities between races.

Smedley, Audrey. (1993). *Race in North America: Origin and Evolution of a Worldview.* Boulder, CO: Westview Press. Traces the evolution of race over three centuries as it transitions from a folk category to a concept used to define superiority and inferiority.

Stepan, Nancy Leys. (1991). *The Hour of Eugenics: Race, Gender, and Nation in Latin America.* Ithaca, NY: Cornell University Press. A comparative study examining how eugenics was taken up by scientist and social reformers in Mexico, Brazil and Argentina.

RADIATION

• • •

Radiation is everywhere. Life would not exist on Earth without radiation from the sun. Additionally, many important technological activities are based on radiation, such as radio and telecommunications. Another type of radiation is used for producing X-ray images in industrial and medical applications. Radiation is also emitted as a side effect from various technological activities. Some types of radiation are known to be harmful to human beings and need to be carefully managed. Other types are not believed to be dangerous, but are a source of worry among the general public. An example is possible radiation risks from power lines, cellular phones, and cellular base stations, which since the 1980s have received considerable media attention.

Protection of humans and the environment from the harmful effects of radiation is called radiation protection. The field of radiation protection evaluates scientific knowledge of adverse health effects from radiation and influences legislation and regulations for protection. The field is complex and involves intricate ethical problems. Lauriston S. Taylor, one of the pioneers of radiation protection during the early 1900s, once said, "Radiation protection is not only a matter for science. It is a problem of philosophy, morality and the utmost wisdom" (1980, p. 854).

It is important to distinguish between ionizing and nonionizing radiation. The biological effects of the two types of radiation are very different, as are therefore the methods of protection. Radiation is ionizing if the energy of the radiation suffices to remove an electron from an atom to create an ion. Conversely, if the energy does not suffice to create ions it is called nonionizing.

Nonionizing Radiation

The most important types of nonionizing radiation are electromagnetic and consist of electric and magnetic waves propagating at the speed of light. Electromagnetic radiation comes from both natural and technological sources and has different properties depending on the frequency of the electromagnetic waves. Low-frequency electromagnetic fields and radio waves come from electric appliances, power lines, radio and television broadcasting, and natural sources such as thunderstorms. Microwaves are used in microwave ovens, radar, and telecommunications. Infrared radiation, visible light, and ultraviolet radiation are emitted from the sun, artificial light, and other technical applications. Electromagnetic radiation with frequencies above visible light has enough energy to change chemical bonds and cause ionizations. Ultraviolet radiation lies on the borderline between nonionizing and ionizing radiation, but is usually considered nonionizing.

The biological effects of nonionizing electromagnetic radiation depend on the frequency and the intensity of the radiation. Low-frequency electromagnetic fields and radio waves pass through human bodies without any apparent effects, but can induce electrical currents and stimulate human nerve cells at high intensities. Microwaves cannot penetrate far into human bodies, but high intensities can cause heating of tissue and burn injuries to the skin. Infrared radiation and visible light can produce surface heating and cause harm to the eye in high intensities. Ultraviolet radiation cannot penetrate the skin, but is known to cause skin cancers.

Claims that low-frequency electromagnetic fields and microwaves can cause cancer are controversial. These types of radiation have insufficient energy to damage the DNA directly, and no other mechanism is known through which they could cause cancer. The prevailing scientific view is that these types of radiation are unlikely to cause cancer. Other effects, such as reduced fertility, memory loss, and fatigue, have been reported, but there is no consistent evidence for these kinds of adverse health effects.

International and national recommendations on exposure limits for nonionizing radiation are based on guidelines from the International Commission on Non-Ionizing Radiation Protection (ICNIRP). The ICNIRP is a nongovernmental organization officially recognized by the World Health Organization (WHO), the International Labour Organization (ILO), and the European Union (EU). The ICNIRP recommends exposure limits for different types of nonionizing radiation. The exposure limits are set with a margin of safety to the level at which health effects occur. The ICNIRP guidelines are based on scientifically verified health effects of nonionizing radiation. Potential, but not proven, hazards are not used as a basis for the limits.

The most important ethical issue regarding nonionizing radiation concerns how to deal with potential health hazards that are scientifically controversial. Examples include the possible risks of radiation from power lines and cellular base stations. Typical exposure levels in these cases are substantially lower than the exposure limits recommended by the ICNIRP, but they do introduce new exposures into society and, in the case of cell phones, such exposures are centered around sensitive parts of the human body. Thus, some countries have, in addition to the recommended exposure limits, adopted precautionary strategies for managing possible hazards from nonionizing radiation. These strategies include the use of prudent avoidance and the precautionary principle.

Prudent avoidance can be defined as a general reduction of needless exposure. This means taking simple, easily achievable, low-cost measures, even in the absence of a demonstrable health hazard. *Prudent* refers to expenditures and does not include any requirement for assessment of the potential health benefits of adopted measures. In practice, this means that the location of new facilities can be influenced by prudent considerations, but need not be modified, because this would involve higher costs. Prudent avoidance can also take the form of voluntary measures, for example, to recommend that manufacturers of mobile phones minimize radiation exposure to the head.

The precautionary principle is not a single, well-defined principle, but the basic idea is that measures against a possible hazard ought to be taken even if evidence for the existence of the hazard does not suffice to be treated as a scientific fact. It is usually thought that the application of the precautionary principle should be science-based and should reference plausible explanations for possible mechanisms for hazards. A common further requirement is that precautionary measures should be temporary and subject to review when further knowledge is gathered. Because scientific evidence and plausible mechanisms are missing for possible risks of low levels of nonionizing electromagnetic radiation, it has been argued that the precautionary principle is inappropriate for these types of radiation.

Adopting precautionary approaches are not unproblematic. What level of precaution should be taken, and what should be the basis for the decision? The WHO has argued that precautionary approaches regarding nonionizing electromagnetic radiation should be adopted with care, and under the condition that scientific assessments of risk and science-based exposure limits are not undermined by arbitrary precautionary approaches.

Ionizing Radiation

Radiation is ionizing if it has enough energy to ionize atoms and molecules. There are two types of ionizing radiation: high-frequency electromagnetic radiation and particle radiation. Examples of ionizing electromagnetic radiation include gamma rays and X rays. Most particle radiation is ionizing. Common types of particle radiation are alpha (helium nuclei), beta (electrons), neutron, and proton radiation.

Ionizing radiation originates from both nonhuman and human sources. Nonhuman or natural sources of ionizing radiation are cosmic rays and naturally occurring radioactive substances in Earth's crust, the human body, air, water, and food. The level of natural exposure varies around the globe, and cosmic radiation is more intense at higher altitudes. The total exposure from all natural sources is called natural background radiation. The natural background radiation is by far the greatest contributor to human exposure to ionizing radiation.

Some human activities can enhance the exposure from natural sources. Examples include radon gas from the soil that concentrates in buildings, mining, and the combustion of fossil fuels that contain radioactive substances. Aircraft passengers and crew are subject to higher levels of cosmic radiation at flight altitudes. Environmental contamination by radioactive residues come from atmospheric nuclear weapons tests (performed between 1945 and 1980), the Chernobyl accident (1986), and the operation of nuclear power plants. These activities contribute only a small fraction of the global average exposure to ionizing radiation.

The largest human-made exposures to ionizing radiation stem from medical procedures. Medical exposures include diagnostic exposures (such as X-ray examinations) and therapeutic exposures (as in tumor treatment). Occupational exposure to ionizing radiation affects workers in industry, medicine, and research. The level of occupational exposure is generally similar to that of the average natural exposure. A few percent of workers are exposed to radiation levels several times greater than the average natural exposure. A comparison between the average exposures from different sources of ionizing radiation is listed in Table 1.

The biological effects of ionizing radiation are generally well known. Ionizing radiation can cause cell death and acute harm to organs if sufficient numbers of cells are damaged. Another type of damage occurs in cells that are modified. This may lead to inheritable genetic changes and the development of cancer, which may manifest itself decades after exposure. Acute effects

TABLE 1

Annual Average per Person Effective Doses of Ionizing Radiation in Year 2000 from Natural and Human-made Sources

Source	Worldwide annual per person effective dose (mSv)	Range or Trend of Exposure
Natural background	2.4	Typically ranges from 1–10 mSv, depending on circumstances at particular locations, with sizeable population also at 10–20 mSv.
Diagnostic medical examinations	0.4	Ranges fron 0.04–1.0 mSv at lowest and highest levels of health care.
Atmospheric nuclear testing	0.005	Has decreased from a maximum of 0.15 mSv in 1963. Higher in northern hemisphere and lower in southern hemihere.
Chernobyl accident	0.002	Has decreased from a maximum of 0.04 mSv in 1986 (average in northern hemisphere). Higher at locations nearer to accident site.
Nuclear power production	0.0002	Has increased with expansion of program but decreased with improved practice.

SOURCE: UNSCEAR (2000).

Range or trend of exposures from the different sources: Natural background typically ranges from 1–10 mSv, with sizable population also at 10–20 mSv. Diagnostic medical examinations ranges from 0.04–1.0 mSv at lowest and highest levels of health care. Atmospheric nuclear testing has decreased from a maximum of 0.15 mSv in 1963. Chernobyl accident has decreased from a maximum of 0.04 mSv in 1986 (average in northern hemisphere). Higher at locations nearer accident site. Nuclear power production has increased with expansion of programme but decreased with improved practice.

occur if the radiation dose is substantial (as in accidents), while it is believed that cancer and hereditary effects may be caused by the modification of a single cell. As the dose increases, the probability of these effects also increases.

The effects and penetration of ionizing radiation depend on the type of radiation. Exposure from ionizing radiation is therefore quantified by the effective dose, which is a measure that takes the type of radiation into account. The unit for the effective dose is the sievert (Sv). One sievert is a very large dose, and it is common to express the effective dose in millisieverts instead (1 mSv = 0.001 Sv). Sometimes the unit rem is used instead (1 rem = 0.01 Sv).

Epidemiological data argue for a linear relation between the dose and the cancer risk from ionizing radiation for intermediate dose levels. A linear dose–effect relation means that an increase in dose implies a corresponding increase in effect. Because of statistical limitations, the dose–effect relation cannot be determined for low doses. Therefore, the risks of low-dose ionizing radiation must be estimated based on knowledge of biological mechanisms that cause or inhibit cancer and inheritable defects. The dose–effect relation for low doses is important, because the exposure to the public or in normal work situations are in ranges where the risk is uncertain (below 50 mSv).

It is especially important to know if there is a threshold for the dose–effect relation for ionizing radiation. If there is no threshold, there is a (small) risk associated with even very low exposure levels. The pre-

vailing scientific consensus, represented by the United Nations Scientific Committee on the Effects of Atomic Radiation (UNSCEAR), is that a threshold is unlikely and that a linear dose–effect relationship for small doses is consistent with current knowledge about the mechanisms by which ionizing radiation causes harmful effects. This view is challenged by those who believe that there is a threshold (and thus no risk) for very low doses of ionizing radiation. Some even argue for a positive effect called hormesis at very low levels.

Setting Standards

Radiation protection from ionizing radiation is generally the same all over the world, because of the profound influence of the International Commission on Radiological Protection (ICRP), a nongovernmental organization whose recommendations are used by both national radiation protection authorities and international organizations as a basis for more detailed guidelines. The ICRP works under the assumption that the risk of cancer and hereditary effects from low doses of ionizing radiation is without a threshold and that the dose–effect relation is linear—the so-called linear, no threshold assumption. This approach to the risks of low-dose ionizing radiation can be seen as precautionary, although the assumption is supported by scientific knowledge.

The 1990 ICRP recommendations are based on a system of three principles: justification, optimization, and dose limitation. The justification principle states that no additional dose should be tolerated unless there is an associated benefit to the exposed individuals or to society that outweighs the detriment. Though the principle may seem obvious, its application gives rise to complex ethical issues. The concepts of benefit and detriment are difficult to define, and calculations are often associated with great uncertainties and errors. Other ethical issues include how the benefit for society can be weighed against the detriment to individuals, issues of free and informed consent, and who should make the decisions (for example, stakeholders or experts).

According to the optimization principle, total exposure should be kept as low as reasonably achievable (or ALARA), with economic and social factors taken into account. (Based on the acronym, this principle is sometimes called the ALARA principle.) What is reasonable depends on economic considerations, which means that doses need not be lowered further if the economic cost would be too high. The principle is thus a trade-off between economics and protection. Cost–benefit analysis has often been applied for optimization of protection, although the ICRP stresses that it is only one possible method.

The optimization principle does not consider the distribution of doses among individuals. A strict application of the principle may thus, at least in theory, lead to a situation in which a few individuals are exposed to substantially higher doses than others. The optimization principle can be seen as utilitarian or consequentialist, focusing on total rather than individual effects.

The dose-limitation principle requires that individual doses not exceed unacceptable levels. This principle can be seen as deontological, because it implies a duty to protect individuals from undue harm. In many cases, the optimization principle and the dose-limitation principle coincide, but there can be cases in which the two principles conflict. In the ICRP system such conflicts are resolved by first applying the dose-limitation principle and after that the optimization principle, deontology before utility.

Under the common assumption that cancer and hereditary effects do not have a threshold, a dose limit (above zero) cannot yield a completely safe level. The dose limits should, according to the ICRP, be regarded as the boundary to unacceptable doses, and protection should essentially be due to the optimization principle. As a dose limit cannot yield a wholly safe dose, a decision on a dose limit will always involve value judgments and ethical considerations. What is acceptable or not is a complex ethical issue, and judgments are not necessarily the same in all contexts.

The dose limits recommended by the ICRP are 1 mSv per year for the public and 20 mSv per year for occupational exposure. A special question regarding dose limits is why it is acceptable for workers to be exposed to higher risks than the public. This is an ethically problematic issue, not just for radiation protection. Arguments that have been used are that the limit for the general public concerns exposure for the whole life and not just the working life, and that the public includes children and other more susceptible individuals. Workers may also be informed of their exposure levels and thus voluntarily accept them, whereas the public has no alternative.

An important concept in radiation protection from ionizing radiation is the collective dose. The collective dose is defined as the mean dose for each individual in an exposed population multiplied by the number of individuals. There has been considerable controversy over what influence the value of the collective dose should have. Considerable collective doses can arise from exposure to large populations even if the dose to each individual is very low. This may be the case in global contamination from radioactive substances (such as in atmospheric nuclear weapons tests) or in contamination that stretches very far into the future. If the risk of cancer from ionizing radiation is proportional to the dose and without a threshold, it follows that the expected number of cancer cases is proportional to the collective dose. In spite of this, it has been argued that small individual doses should not pose a problem even if the collective dose is great.

Arguments to the effect that "risks ought to be disregarded if they are sufficiently small" are called de minimis arguments. Common arguments for calling risks de minimis are that they are trivial compared to other risks humans accept, that they are trivial in comparison to natural risks, or that they have to be disregarded in order to avoid the allocation of unreasonably large economic resources to investigate or manage them. It has often been claimed that risks with a probability on the order of magnitude of one in a million or smaller are de minimis. Nevertheless, such a general de minimis level is ethically problematic because it would allow many small risks that in combination may yield a large risk for an individual. Furthermore, many small risks to many people may also yield a large total effect. For example, exposing each of ten million persons to an independent risk of death of one per million yields ten expected fatalities. Also, the mathematical "law of large numbers" yields that the actual outcome will be around ten fatalities.

Another ethical problem in radiation protection arises from the long-term management of radioactive waste. Radioactive materials may be dangerous for hundreds of thousands of years, and mistakes made now may affect future generations. This problem is not exclusive to radioactive waste, because many other technological activities have consequences reaching far into future; examples include emissions that may lead to global climate change and damage to the ozone layer. The discussion regarding radioactive waste is nevertheless important, because many countries have not made final decisions for long-term management of the radioactive waste from nuclear reactors and/or nuclear weapons. The problem of distant future effects poses intriguing ethical problems. What is the moral status of future, nonexisting individuals and what duties do persons today have toward them? The International Atomic Energy Agency (IAEA) is of the opinion that radioactive waste should be managed in such a way that predicted impacts on the health of future generations will not be greater than today and that no undue burden is imposed on future generations.

PER WIKMAN

SEE ALSO Chernobyl; Hormesis; International Commission on Radiological Protection; Regulatory Toxicology.

BIBLIOGRAPHY

International Atomic Energy Agency (IAEA). (1996). *Radiation Protection and the Safety of Radiation Sources.* Vienna: Author. The foundation document of the IAEA system of protection.

International Commission on Non-Ionizing Radiation Protection (ICNIRP). (1998). "Guidelines for Limiting Exposure to Time-Varying Electric, Magnetic, and Electromagnetic Fields (Up to 300 GHz)." *Health Physics* 74(4): 494–522. Describes the rationale for the ICNIRP guidelines for limiting exposure to electromagnetic fields.

International Commission on Non-Ionizing Radiation Protection (ICNIRP). (2001). "Review of the Epidemiologic Literature on EMF and Health." *Environmental Health Perspectives* 109(suppl. 6): 911–933. An authoritative overview of the knowledge base for the risks from electromagnetic fields.

International Commission on Non-Ionizing Radiation Protection (ICNIRP). (2002). "General Approach to Protection against Non-Ionizing Radiation." *Health Physics* 82(4): 540–548. The foundation document of the ICNIRP system of protection.

International Commission on Radiological Protection (ICRP). (1992). *1990 Recommendations of the International Commission on Radiological Protection.* Oxford: Pergamon Press. The foundation document of the ICRP 1990 system

of protection that has had a profound influence on radiation protection internationally.

Lindell, Bo. (1985). *Concepts of Collective Dose in Radiological Protection: A Review for the Committee on Radiation Protection and Public Health of the OECD Nuclear Energy Agency.* Paris: Organisation for Economic Co-operation and Development, Nuclear Energy Agency. A detailed guide to the concept of collective dose, and related ethical issues.

Silini, Giovanni. (1992). "Ethical Issues in Radiation Protection: The 1992 Sievert Lecture." *Health Physics* 63(2): 139–148. Reviews ethical issues that have been considered in the development of radiation protection from ionizing radiation.

Sowby, David, and Jack Valentin. (2003). "Forty Years On: How Radiological Protection Has Evolved Internationally." *Journal of Radiological Protection* 23(2): 157–171. An overview of international organizations in radiation protection and the development of the ICRP recommendations since 1950.

Taylor, Lauriston S. (1980). "Some Nonscientific Influences on Radiation Protection Standards and Practice: The 1980 Sievert Lecture." *Health Physics* 39(6): 851–874. A classic text on more philosophical aspects of radiation protection by one of the first members of ICRP.

United Nations Scientific Committee on the Effects of Atomic Radiation (UNSCEAR). (2000). *Sources and Effects of Ionizing Radiation: UNSCEAR 2000 Report to the General Assembly, with Scientific Annexes.* 2 vols. New York: United Nations. An authoritative overview of the knowledge base for the risks from ionizing radiation.

Wikman, Per. (2004). "Trivial Risks and the New Radiation Protection System." *Journal of Radiological Protection* 24(1): 3–11. Examines ethical and philosophical aspects of a novel way of thinking in radiation protection.

RADIO

• • •

Radio includes a broad group of technologies that utilize electromagnetic radiation (also called radio waves) to transmit and/or receive information. Examples of radio technologies can be drawn from numerous industries, applications, and end users. A partial listing would include radio (and television) broadcasting, maritime communications, radio navigation, cellular telephony, satellite communications, numerous military applications, wireless computer networking, noncontact identification systems, military and meteorological radar, global positioning systems, and radio astronomy (see Figure 1).

What all these systems have in common is the conversion of electrical energy from one form into another, specifically, from electrical currents bound in conduc-

Figure 1: Part of the Very Large Array (VLA) radio telescope operated by the National Radio Astronomy Observation in Socorro, New Mexico. The VLA is capable of receiving extremely faint energy from extragalactic sources. (*JLM Visuals.*)

tive materials such as wires and cables into unbounded electromagnetic radiation that is free to propagate through space, the atmosphere, or another nonconducting medium. This is the process of radio transmission. Radio reception is the reverse process, in which incoming electromagnetic radiation is converted into electrical currents in the antennas, wires, and components of a radio receiver.

Historical Developments

The following material is a brief history of the development of radio technology with an emphasis on related ethical, political, and legal issues. This history draws on Christopher Sterling and John Michael Kittross's *Stay Tuned* (2002).

The background of radio was the earlier practical development of wired electronic signal transmission and reception, as in the telegraph (1830s and 1840s) and James Clerk Maxwell's electromagnetic theory (1860s), which was confirmed by Heinrich Hertz's laboratory experiments (1880s). It was his ability to draw on those previous achievements that enabled Guglielmo Marconi

(1874–1937) (see Figure 2) to transmit and receive the first wireless telegraph messages in 1895, an experiment that he followed up with wireless transmissions across the English Channel (1899) and the Atlantic (1901).

The rapid development of radio led in 1910 to the Wireless Ship Act in the United States, which required a radio and an operator on all oceangoing passenger vessels. Through World War I the U.S. Navy continued to control radio facilities, while the U.S. Congress debated the future government role in relation to the new technology. Shortly after the war, in 1921, thirty broadcasting stations went on the air, using only two frequencies or channels.

In 1922 President Herbert Hoover hosted the first radio conference, which called for government regulation of radio technology, limited advertising, and classification of radio stations by the services they provided. Two years later the British physicist Sir Edward Victor Appleton conducted the first experiment with radio range-finding equipment, reflecting radio waves off the ionosphere to determine its height. This was an important step in the development of radar.

Later in the 1920s President Calvin Coolidge signed the Radio Act of 1927, establishing the Federal Radio Commission (FRC). In that decade the National Association of Broadcasters issued a code of radio advertising and programming ethics.

In 1932 the engineer Karl Jansky discovered a strong source of radio noise that later was discovered to originate outside the solar system; this marked the beginning of radio astronomy. In 1934 the Federal Communications Commission (FCC) was established to replace the FRC. Later in the decade, in 1937, the first practical mobile radio, the DR38a transmitter-receiver, was developed.

During World War II both Axis and Allied engineers made significant advances in land, mobile, maritime, and airborne radio as well as radar. After the war, in 1948, scientists at Bell Laboratories demonstrated the potential uses of the transistor. Between 1945 and 1960 numerous television stations began broadcasting coast to coast, linked by microwave radios.

The year 1958 marked the invention of the integrated circuit. In the 1960s the concept of a broadband mobile telephone system was outlined. In 1969 the first frequency-resuing commercial cellular system was used on trains running from Washington to New York. By the 1980s analog cellular telephone use had become widespread. Digital cellular systems with increased capacity were introduced in the 1990s. Another significant development was the FCC auction of spectrum for the Personal Communications Services (PCS) band.

The Radio Frequency Spectrum as a Limited Natural Resource

The electromagnetic spectrum contains frequencies from below 1 Hertz (one cycle per second) to above 10^{25} Hz. However, a much smaller subset of those frequencies lend themselves to terrestrial radio systems. Although there is not universal agreement on the boundaries, the "radio spectrum" is the subset of the electromagnetic spectrum with frequencies from 100,000 Hz to 100 GHz (10^5 to 10^{11} Hz).

The lower end of the radio spectrum is less suited for most communications applications. The rate at which information can be transmitted (the data rate) becomes lower as the frequency decreases. This does not mean that low-frequency waves travel through space more slowly because all electromagnetic radiation travels at the speed of light. However, the theoretical rate of information transfer decreases with decreasing frequency. This gives rise to a lower limit to the frequency

band that can be used for most radio systems. Additionally, the ionosphere becomes opaque at lower frequencies, limiting some applications, although enhancing others.

At higher frequencies the entire atmosphere (not just the ionosphere) becomes opaque except for a few "windows" in which electromagnetic radiation is free to propagate without being absorbed significantly (see Figure 3). There is an optical window (the atmosphere is transparent to the frequencies human eyes can detect), and there is a radio window. Transmission of signals at frequencies above this window are absorbed or scattered rapidly by the atmosphere, similarly to the way fog limits visible frequencies. The opaque nature of the atmosphere at higher frequencies establishes an upper limit to the radio spectrum; thus, the radio spectrum is capped in its upper and lower ends. This means that the radio spectrum is a limited natural resource. Because of its immense importance and finite nature, the radio spectrum presents significant distributive justice issues.

Ethics, Politics, and Law

The ethical, political, and legal aspects of radio can be arranged in a four-fold taxonomy. Although there is significant overlap amongst the categories, they are useful in conceptualizing the major issues and highlighting the important ethical traditions pertaining to radio development and use.

First, there are issues surrounding the technological development of radio that pertain to topics in engineering ethics. For example, the use of radio for military applications and growing concerns about the health effects of electromagnetic frequencies present ethical challenges to engineers who are responsible for upholding the safety, health, and welfare of the public.

Second, radio content and use issues instantiate several aspects of broadcast journalism ethics as they place responsibilities on program directors, journalists, and radio managers. These obligations are traditionally formalized in codes of ethics such as the NAB code of radio advertising and program ethics and the Radio-Television News Directors Association (RTNDA) code of ethics, which states that electronic journalists ought to serve as trustees of the public reporting the truth with fairness, integrity, and independence.

Third, the broader cultural and societal impacts of radio raise issues explored in the philosophy of technology and the field of Science, Technology, and Society (STS) studies. Radio technologies reciprocally interact with various elements of culture to co-produce societal

Figure 2: Guglielmo Marconi, considered the father of radio. *(Hulton-Deutsch Collection/Corbis.)*

changes and personal life experiences. In the United States, for example, conservative talk radio programs have exerted massive influence over the political landscape and Christian programming has also come to dominate certain markets, which has influenced conceptions about religion in the public sphere. Such developments underscore the idea that radio is not a neutral medium, but rather an active agent that is used to selectively broadcast some voices and messages rather than others. It is a political and cultural force, albeit somewhat eclipsed by television. Interestingly, the rise of opinion and advocacy programs on radio seemed to foreshadow a general shift in media (furthered by the Internet and the "blogosphere") away from trust in a few supposedly neutral broadcast centers to a variegated spectrum of information streams.

Lastly, questions of how radio should be used and regulated raise fundamental issues from political philosophy such as distributive justice, the proper relationship between government and private enterprise, censorship and the proper limits to freedom of speech, and the concentration of corporate control over media.

As a common resource, it has been widely maintained that the radio spectrum must be centrally regulated to insure fairness and efficiency. For example, the International Telecommunication Union (ITU) is a regulatory body within the United Nations system that helps coordinate global telecommunications networks and services. Additionally, each country has its own national frequency allocation plan. In Germany, for example, each state exercises its own authority over radio broadcasting rather than a centralized federal entity. In the United States that plan is administered by the FCC. The FCC is an independent government agency, directly responsible to Congress, which plans, allocates, and monitors the use of the radio spectrum for nongovernment users. FCC rules pertaining to free speech and censorship tend to raise the most public controversy, especially those relating to indecency, obscenity, and profanity. These rules do not apply to satellite and cable broadcasting. The National Telecommunications and Information Administration (NTIA) is responsible for the allocation and assignment of frequencies for use by the federal government The national frequency allocation plan divides the spectrum into a

FIGURE 3

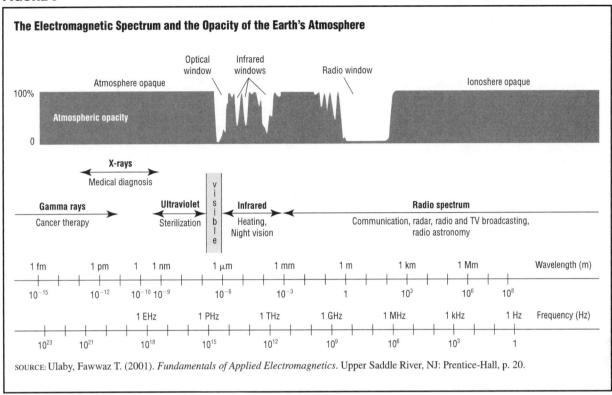

The Electromagnetic Spectrum and the Opacity of the Earth's Atmosphere

SOURCE: Ulaby, Fawwaz T. (2001). *Fundamentals of Applied Electromagnetics*. Upper Saddle River, NJ: Prentice-Hall, p. 20.

The electromagnetic spectrum and the opacity of the earth's atmosphere. The optical and radio windows are frequency bands transparent to electromagnetic radiation.

multitude of frequency bands, reserving bits of spectrum for different types of users and reducing channel interference. It plays a vital role in balancing the often conflicting needs of commercial, military, scientific, and educational uses.

Although some level of government regulation may be necessary, many advocate further deregulation in order to capture the benefits of market competition and avoid inefficiency, corruption, or other unethical practices by centralized bureaucrats. Others, however, fear that deregulation will lead to further corporate monopolization of local markets. In the United States, concerns are developing that the increased corporate consolidation of radio diminishes its locality, threatens the democratizing value of free and independent communication, homogenizes music play lists, and undermines journalistic quality.

Similar debates about the proper roles of private and public or community radio sparked the 1967 creation of the U.S. Public Broadcasting Act, which established the Corporation for Public Broadcasting (CPB). The CPB receives annual appropriations from Congress to support independent local stations and National Public Radio (NPR), which was established in 1970. Although this helps defend the independence, integrity, and diversity of radio journalism, it also raises accountability issues about the use of federal funds.

College and community listener sponsored radio stations also attempt to secure independence and diversity at the fringes of corporate media conglomerations. In 2000, the U.S. government began issuing licenses for low-power (below 100 watts) radio stations partially to provide another avenue for local communities (especially low-income and minority) to obtain diverse, community-oriented information. Most of these licenses have been obtained by rural communities and churches, and they have not had the expected impact on urban areas that are most dominated by commercial radio. Concerns have been raised that Christian stations are monopolizing these markets, thus producing the same drawbacks from consolidation. There is also some concern that these stations interfere with broadcasts from bigger stations. Furthermore, many low-power radio broadcasts still operate illegally as "pirate" stations. Some of these stations are switching to internet broadcasts in attempts to avoid federal lawsuits.

Current Trends

As more uses of radio technologies are conceived, developed, and marketed (e.g., cell phones and wireless internet connections) and as demand for existing uses continues to grow, the radio spectrum will become increasingly crowded. Interference among users will become increasingly difficult to avoid and solve. Modulation schemes that are more tolerant of interference such as spread spectrum–based technologies should see increased use, as should hardware-based solutions such as more sophisticated filtering. Spectral crowding also will result in the continued migration toward higher frequencies despite the greater atmospheric attenuation and other technological obstacles. Finally, both the general public and those involved in the technical industries will be forced to become more aware of the limits of the radio spectrum, the importance of coordination and regulation, issues involving radio interference, and spectral crowding.

J. BRIAN THOMAS

SEE ALSO *Advertising, Marketing, and Public Relations; Communication Ethics; Communication Systems; Entertainment; Networks.*

BIBLIOGRAPHY

Kobb, Bennett. (2001). *Wireless Spectrum Finder: Telecommunications, Government, and Scientific Radio Frequency Allocations in the U.S., 30 MHz–300 GHz,* 6th edition. New York: McGraw-Hill.

Kraus, John D. (1986). *Radio Astronomy,* 2nd edition. Powell, OH: Cygnus-Quasar.

Radio Astronomy: Observing The Invisible Universe. (1999). Video. Produced by Dick Young Productions, New York, NY, for the National Science Foundation, USA.

Reed, Dana George, ed. (2001). *The ARRL Handbook for Radio Amateurs.* Newington, CT: American Radio Relay League.

Sterling, Christopher H., and John Michael Kittross. (2002). *Stay Tuned: A History of American Broadcasting,* 3rd edition. Mahwah, NJ: Lawrence Erlbaum.

RAILROADS

• • •

Railroads use flanged wheels rolling over fixed rails for human transportation; the vehicles on these rails are commonly called *trains* because they are usually composed of a train of cars linked together. Trains have distinct characteristics that have called for specialized legal and policy regulation, and to some extent for the application of ethical principles.

Prior to the development of steam locomotion, early horse-drawn trains ran on tracks serving mines, where the ground was otherwise too uneven for wheeled vehicles. The first horse-drawn trains began operating at English coal mines in the 1630s. In 1758, the British Parliament established the Middleton Railway in Leeds; it began to adopt steam locomotives in 1812. The Middleton Railway claims to be the oldest railway in the world; however, at this time it carried only freight, not passengers. The first public steam-operated passenger railway was the Stockton & Darlington in England, which began operations in 1825. Commenting on railroad developments and aspirations at the time, the *English Quarterly Review* wrote: "What can be more palpably absurd and ridiculous than the prospects held out of locomotives traveling *twice as fast* as stagecoaches! We should as soon expect the people . . . to suffer themselves to be fired off on . . . [a] rocket, as to put themselves at the mercy of such a machine, going at such a rate" (Bianculli 2001, vol. I, p. 15).

The Nineteenth Century Experience

Early American railroads competed with canals, packet steamers, stagecoach lines, and turnpike companies for investment. Government did not immediately intervene on the side of the new technology; as late as 1856, the Erie Canal was subsidized by a tax on rail traffic. Local interests did not always want the railroad in the early years. Farmers tended to oppose them because the locomotives set fire to crops, scared livestock, and, most significantly, brought in cheap produce from elsewhere to compete with local products.

In February 1815 the New Jersey legislature passed the first railroad charter in the United States, authorizing a horse-drawn train to connect Trenton and New Brunswick. During the 1820s, almost every state granted railroad charters. John Stevens (1749–1838) built the first successful American steam locomotive in 1825, the same year the Stockton & Darlington began operation in Great Britain.

From the outset, an excitement for the technological possibilities was attached to railroad development that drove an unprecedented rush of development and adoption. Trains were seen as powerful tools and symbols of nation building. Just two years after the opening of the Stockton & Darlington, the Baltimore & Ohio was chartered as the first westward-bound railroad in the United States; and in 1831, President Andrew Jackson (1767–1845) in a message to Congress portrayed railroads as the binding force that would hold the most remote parts of the new nation together. A French

New York, New Haven, and Hartford diesel engine. (*Library of Congress.*)

observer remarked, "The American seems to consider the words democracy, liberalism, and railroads as synonymous terms" (Bianculli 2001, vol. I, p. 17). Jackson later became the first U.S. president to ride a steam-powered train.

In 1830 the Baltimore & Ohio began operations, pulled initially by horses and mules, switching to its steam locomotive, the "Tom Thumb," a few months later. A New York City to Washington line was in place by 1840, and a decade later, the country had 9,000 miles of track in service. Railroads permitted the development of urban centers not on rivers, and most railroad development was east-to-west, connecting rivers to each other instead of running parallel to them. However, most early railroads were short, local, and did not connect to one another.

Railways were the most capital-intensive enterprise the world had ever seen, far exceeding mills. They largely drove the development of the joint-stock company and therefore of modern Wall Street-style finance.

From scarcely twenty-five miles of public railroad worldwide in 1825, the mileage grew to over 160,000 miles in fifty years, with approximately one third of that being in the United States. As American eyes looked to the west, the railroads took on a new importance as the tool by which western lands would be secured to the Union and then controlled. In addition to other financial incentives, the federal government offered railroads ten to twenty square miles of adjoining land for every mile of track built. This resulted in the grant of 338,000 square miles to the railroads, which then realized additional profits developing or selling this land or leasing it out. In some cases, these land grants emboldened the railroads to lay track away from the nearest large towns, confident new towns would develop right alongside. In other cases, the railroads demanded subsidies from towns in order not to bypass them. When San Bernardino refused to pay the Southern Pacific, the railroad created the town of Colton, California, just five miles away.

A race began to finish the transcontinental railroad; the Union Pacific, originating at Omaha, Nebraska, headed west, while the Central Pacific, beginning in Sacramento, laid track east. The two competitors bickered over where the lines would meet; if the Ulysses S. Grant (1822–1885) administration had not intervened to force both roads to accept a meeting place in Utah, they would have ended up running parallel to one another for some 1,500 miles. The transcontinental railroad was completed in 1869.

From 1870 through about 1890, the railroads played a major role in the settlement of the west. In this twenty-year period, the Denver population increased from 5,000 to 107,000, while Minneapolis went from a town of 13,000 people to one of 164,000. But already by 1871, land grants were a fertile source of political scandal, with accusations that the railroads were charging exorbitant fees and foreclosing on tenants who could not pay.

The nineteenth-century railway was a major tool of nation-building and national identity. Canadian technology and media philosopher Harold Innis saw the railway as a bulwark of centralization, territorial expansion, nationalism, and state authority. Like the United States, Canada also was consolidated by the building of a transcontinental road, which reinforced the new nation's extremely tenuous control west of Ontario. "[T]he drive for railways embodied a sense of divine purpose, a mission to conquer the surrounding wilderness, that made the colonists, rather unexpectedly, less British and more American" (den Otter 1997, p. 12). For cultural historian Wolfgang Schivelbusch (1986), by forcing the creation of time zones to help schedule train traffic and turning journeys across great distances into well-ordered experiences, the railroad brought about the industrialization of time and space.

The Twentieth Century

From 1850 to about 1950, trains were the primary means of inland transport, but in the age of automobiles and airplanes there is some question as to whether trains are still needed. Unlike Europe, where the train has deep aesthetic, environmental, and cultural appeal, the United States flagged in its commitment to a national railway system. They are "of marginal utility and relevance to most people . . . more nostalgia than interest" (Perl 2002, p. 1). In the United States, those who defend the perpetuation of rail lines often do so on sentimental and historical grounds, though environmental arguments (that each train obviates the hydrocarbon emissions of a number of automobiles and trucks) are also applicable.

Trains were already perceived as a fading technology in the United States as early as the 1940s, as government aggressively supported the automobile by building highways everywhere.

In the face of competition from the car and later from the passenger airline, private American railroads in the 1950s began to close down passenger service while maintaining the more lucrative freight contracts. Although state railroad boards sometimes fought aggressively to preserve passenger service, regulatory responsibility shifted to the federal Interstate Commerce Commission, which agreed that the train was of declining utility. From 1958 to 1971, about 75 percent of passenger train mileage was abandoned by the railroads. But at the same time it became harder for them to compete with trucks and aviation in the freight business, and the railroad share of intercity freight declined from 68 percent in 1944 to 44 percent in 1960.

When automobiles and then airplanes first became prevalent, the railroads struggled to cover their fixed costs (track building and maintenance) out of a declining revenue. By contrast, automobile and aviation interests never became financially responsible for their entire infrastructure: Automobile manufacturers and trucking companies did not own the highways, airlines did not build airports. The infrastructure they require is paid for with public money, while the railroads had long been responsible for their own costs.

The Amtrak Corporation was founded in 1971 with $25.4 billion in federal subsidies and grants, as a response to the frightening bankruptcy of the Penn Central Railroad, which had been losing $375,000 a day on its passenger service. Amtrak took over passenger lines from twenty participating railroads, which were offered a choice of stock in Amtrak or a tax break. Only one tax-paying railroad chose the stock. At the time, the National Association of Railroad Passengers said that Amtrak was "operated by people who don't want it to succeed." Amtrak was also described as a "policy blocker," preventing more radical legislation (Perl 2002, p. 99). Amtrak has been a failure as a commercial entity, losing much more money than anyone anticipated. As of early 2005, the George W. Bush administration was proposing that Amtrak receive no further funding from the federal government.

Aesthetic and environmental considerations aside, trains only make sense if they provide speed and convenience equal to or greater than automobiles, at less cost than airplanes. Japan has succeeded in creating high-speed rail lines that connect directly to airports and travel more rapidly than cars. The trend at Amtrak has been the opposite. After debuting the Metroliner, which went from New York to Washington in under three hours, Amtrak has slowed this train down so that it is barely faster than the regular, less expensive service.

Anthony Perl (2002) notes that passenger railroads suffer from the perception that they should be profit-making entities rather than a national service. No one complains that New York subway fares only cover 71 percent of the cost of operating the system, while Amtrak is considered a failure for recouping 78 percent of its costs.

Public Service or Private Enterprise?

The question of whether trains should be a public service or private enterprise has played out most dramatically in Great Britain, where the nationalization of British Rail during the Thatcher era was based on the premise that "private = good, public = bad" (Murray 2001, p. 2). Andrew Murray describes the nationalization of British Rail as privatization run amok, a solution without a problem, since the entity that was replaced had a very high record of safety and reliability. It has been supplanted by a strange patchwork of several principal players and hundreds of subsidiary ones, with the tracks all owned by one entity, Railtrack, the rolling stock placed in separate leasing companies and leased back to franchisees, and maintenance and repair services sold to thirteen other companies that subcontract much of the work. The piece most visible to the public—the franchisee train operators, which include several of Britain's major bus companies and also Virgin Airways—own nothing except their trademarks.

The result has been a substantial increase in bureaucracy, decline in decisiveness and speed of decision-making, and a general lack of cooperation among the various entities. Examples include the fact that operators will no longer wait for connecting trains to arrive (they pay a fine if they start late, regardless of the reason); tickets on one line are not accepted on competing lines rolling over the same tracks, so if you miss your connection to London you often cannot go out on the next train without buying another ticket; substantial increases in overtime, and therefore in exhausted workers driving trains, as the lines make their declining base of experienced employees work harder, rather than hiring and training additional ones; and a terrible lack of interest in safety measures unless mandated by government. Some train crashes have resulted, with substantial loss of life and stories of safety systems switched off or malfunctioning. Murray is skeptical that these problems can be solved without re-nationalizing the railroads.

The history of trains, like that of dams and other nineteenth-century technologies, describes an arc from symbol of political and economic power to a nostalgia-supported technology left behind in a strictly technological competition with other interests and solutions. The future of trains will depend very much on the practicality of new technological innovations to make them compete effectively with automobiles, at prices that make sense. Without massive federal subsidies and a major change in governmental thinking, trains may not prevail for environmental or sentimental reasons alone.

JONATHAN WALLACE

SEE ALSO *Bay Area Rapid Transit Case; Roads and Highways; Ships*.

BIBLIOGRAPHY

Bianculli, Anthony J. (2001). *Trains and Technology: The American Railroad in the Nineteenth Century*, 3 vols. Newark: University of Delaware Press. A detailed technological history of nineteenth century steam locomotion.

den Otter, A.A. (1997). *The Philosophy of Railways: The Transcontinental Railway Idea in British North America*. Toronto: University of Toronto Press. Describes the social impact of Canadian railway building.

Innis, Harold Adams. (1971). *A History of the Canadian Pacific Railway*. Toronto: University of Toronto Press. Technology philosopher Harold Innis, a predecessor of Marshall McLuhan, analyzes the impact of the railroad on conceptions of nationhood, distance and time.

Murray, Andrew. (2001). *Off the Rails: Britain's Great Rail Crisis—Cause, Consequences, and Cure*. London: Verso. A sarcastic and lively analysis of the effects of railway privatization in Great Britain.

Perl, Anthony. (2002). *New Departures: Rethinking Rail Passenger Policy in the Twenty-first Century*. Lexington: University of Kentucky Press. An analysis of the results of the nationalization of U.S. passenger service via creation of the Amtrak Corporation.

Schivelbusch, Wolfgang. (1986). *The Railway Journey: The Industrialization of Time and Space in the Nineteenth Century*, trans. Anselm Hollo. Berkeley: University of California Press. Original German publication, 1977.

RAIN FOREST

• • •

The ethical and policy issues associated with rain forests are doubly related to technology and science: While technology has provided the tools for cutting down rain forests, science has produced knowledge about their importance that leads to the questioning of such practices.

If one compares maps of the world featuring maximum biodiversity, deserts, and desertification (for example, putting side by side Mittermeier et al., *Hotspots* [2000], p. 19; the *Encyclopedia of Deserts* [1999], inside cover; and the *World Atlas of Desertification* [1997], pp. 44–45), the most striking feature is the proximity of maximum and minimum biodiversity in well-defined bands that circle the globe—because of the heat of the sun at the Equator and related atmospheric and climate effects. That is, the areas that contain the highest levels of biological diversity are almost all endangered to a high degree as well.

Kathlyn Gay (2001) introduces her summary of worldwide research and activism on rain forests by describing tropical rain forests as those close to the Equator and characterized by a minimum of 80 to 120 inches of rainfall per year that make up 6 percent of the surface of the Earth. These are found in parts of Central and South America, Africa, Asia, and the United States, with the best-known being in Amazonia. Others are located in Papua New Guinea, the islands of Madagascar, Malaysia, Thailand, Mexico, Colombia, and Ecuador. Gay's book covers temperate rain forests as well, such as those in the Pacific Northwest of the United States and Canada. With respect to either kind, tropical or temperate, the reason for researcher and activist interest is the impact of forests on climate, including precipitation, soil, and the carbon cycle so necessary for terrestrial life. Decimation of the rain forests would have a lasting impact on world climate, and would also affect winds, rainfall, and heat patterns, especially in the rich equatorial band around the globe.

Deforestation as Problem

Deforestation is a particularly difficult issue in certain areas. The best-known problem area is the Amazon rain forest. Susanna Hecht and Alexander Cockburn (1989) claim that Amazon deforestation is based in the policies of post-World War II Brazilian military governments. In 1964 Brazil began a massive interior settlement program that promoted forest clearing for cattle ranching. Much of the clearing also took place near gold strikes, since cattle grazing allows "large amounts of land—and the mineral rights below it—to be claimed with minimal labour" (Gay 2001, p. 46). Clearing also undermined rubber tapping in the forests, stimulating the rubber tapper Chico Mendes (1944–1988) to highlight the manifold social and environmental problems being created by deforestation (Burch 1994). His murder helped stimulate creation of the World Rainforest Movement (founded 1986) that has criticized the UN Food and Agricultural Organization (FAO) and World Bank support for national forest clearing initiatives (World Rainforest Movement 1992).

Focus on the human dimension of deforestation is further emphasized in *Tropical Deforestation* (1996), which makes the sweeping claim that "government management of forests often results in deforestation, whereas local community management of forests is usually more likely to contribute to forest conservation" (Sponsel, Headland, and Bailey 1996, p. xx) This broad conclusion is based on anthropological studies that detail work in Mayan Mexico, Polynesia, India, Kenya and other areas of Africa, the Philippines and New Guinea, as well as Madagascar, the Amazon, and other areas of Central and South America.

A more extensive discussion of the problem is provided by Sing Chew (2001), who traces ecological tragedies from 3000 B.C.E. to the year 2000 C.E., under a series of imperial regimes. Chew argues that in every case, from ancient Mesopotamia through Greece and Rome to the Portuguese and Spanish Empires and later European imperialism, deforestation was a constant concomitant of political aggrandizement and empire building—along with the continuing rise in population.

Sustainable Possibilities

Few scholars challenge the link between government policies and deforestation. But some observers such as Bjørn Lomborg, while admitting that overexploitation may be taking place, nevertheless argue that the situation has been exaggerated. For instance, although in 1988 the Brazilian space agency announced that its satellites showed 7,000 fires destroying 2 percent of the Amazonian rain forest per year, subsequent corrections reduced this figure to 0.5 percent, and "in actual fact, overall Amazonian deforestation has only been about 14 percent" since humans arrived (Lomborg 2001, p. 114). Such figures raise important questions of scientific ethics and responsibility on many sides of this important issue.

A number of other scientists, especially environmental economists, argue that tree cutting—even timber harvesting on a large scale—can be managed sustainably. Eberhard Bruening, for example, maintains that it is possible "to mimic nature and utilize inherent ecosystem dynamics and indeterminism to improve self-sustainability and economic viability" (Bruening 1996, p. x). Bruening is not overly optimistic that current managers and their government supporters can do this, but he thinks matters could change if *community-oriented*

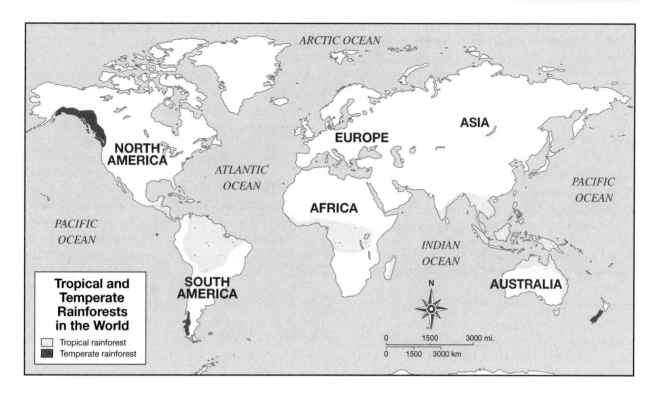

Tropical and Temperate Rainforests in the World

☐ Tropical rainforest
■ Temperate rainforest

forestry were initiated or expanded. (In Bruening's opinion, it has begun in some places including Sarawak in Southeast Asia.) Others emphasize forest-related activities that may prove more profitable than cutting trees in rain forests. For example, Douglas Southgate (1998) discusses ecotourism and its successes in Costa Rica, along with that country's genetic prospecting agreements, debt-for-nature swapping, and offers to serve as a *sink* for other countries in carbon-sequestration trading deals. (Activity in Nicaragua and Guatamala underlie similarly optimistic assessments of profitable alternatives, as described at length by Olman Segura-Bonilla [2000]).

In terms of science, technology, and ethics as related to rain forests (especially tropical rain forests), there is a broad consensus (represented here by Gay) that unethical forest management policies and practices have been implemented by governments since the Bronze Age. The science to support this claim, usually deforestation mapping from satellites, points to continuing tree cutting in spite of environmentalists' outrage—although the precise extent is contested. Indeed others argue that *sustainable management* (of tree cutting) is possible, even in tropical rain forests, provided that scientifically sound forest management practices are employed (Bruening 1996). Proponents of this theory also point to the ever-increasing demand for wood and wood products in the world economy, adding that rain

forests can be economically productive in other ways—some even as alternatives to deforestation.

PAUL T. DURBIN

SEE ALSO *Biodiversity; Deforestation and Desertification; Ecological Restoration; Ecology; Environmental Ethics; Environmentalism; Global Climate Change; Sierra Club; United Nations Environmental Program; Water.*

BIBLIOGRAPHY

Bruening, Eberhard F. (1996). *Conservation and Management of Tropical Rainforests: An Integrated Approach to Sustainability.* Wallingford, England: CAB International.

Burch, Joann Johnson. (1994). *Chico Mendes, Defender of the Rainforest.* Brookfield, CT: Millbrook.

Chew, Sing C. (2001). *World Ecological Degradation: Accumulation, Urbanization, and Deforestation 3000 B.C.–A.D. 2000.* Walnut Creek, CA: Alta Mira.

Gay, Kathlyn. (2001). *Rainforests of the World,* 2nd edition. Santa Barbara, CA: ABC-CLIO.

Hecht, Susanna, and Alexander Cockburn. (1989). *The Fate of the Forest: Developers, Destroyers and Defenders of the Amazon.* London: Verso.

Lomborg, Bjørn. (2001) *The Skeptical Environmentalist: Measuring the Real State of the World.* New York: Cambridge University Press.

Mares, Michael A., ed. (1999). *Encyclopedia of Deserts.* Norman: University of Oklahoma Press.

Middleton, Nick, and David Thomas, eds. (1997). *World Atlas of Desertification*, 2nd edition. London: Arnold.

Mittermeier, Russell A.; Norman Myers; Cristina Goettsch Mittermeier; and Patricio Robles Gil. (2000). *Hotspots*, CEMEX, Conservation International; and Chicago: University of Chicago Press.

Segura-Bonilla, Olman. (2000). *Sustainable Systems of Innovation: The Forest Sector in Central America*. Aalborg, Denmark: University of Aalborg.

Southgate, Douglas. (1998). *Tropical Forest Conservation: An Economic Assessment of the Alternatives in Latin America*. New York: Oxford University Press.

Sponsel, Leslie E.; Thomas N. Headland; and Robert C. Bailey. (1996). *Tropical Deforestation: The Human Dimension*. New York: Columbia University Press.

World Rainforest Movement. (1992). *Rainforest Destruction: Causes, Effects and False Solutions*. Penang, Malaysia: Author.

RAMSEY, PAUL

• • •

Theologians are often marginalized in public discussions about contemporary social, political, scientific, and technological issues in the United States. (Robert) Paul Ramsey (1913–1988) reminds us of an earlier era when particularly able American theologians were public intellectuals taken seriously by policy makers, the media, and members of the general public.

Life

Born the son of a Methodist minister in Mendenhall, Mississippi, on December 10, Ramsey would always maintain his Methodist connections but follow the path to a public pulpit as one of the leading ethicists of his generation. A 1935 graduate of Millsaps College in Jackson, Mississippi, he published his first essay that same year as a newly appointed teacher of history and social sciences at his alma mater. Departing in 1939 for Yale University, he graduated a year later with a bachelor of divinity degree and continued toward his Ph.D. As he studied under H. Richard Niebuhr (1894–1962), he moved away from the liberal idealistic theology he had acquired at Millsaps and adopted the theological realism of his mentor and the latter's equally well-known brother, Reinhold Niebuhr (1892–1971) at Union Theological Seminary in New York City.

After serving as an assistant professor of Christian ethics at Garrett Biblical Institute in Evanston, Illinois, and completing his Ph.D. at Yale, in 1944 he joined Princeton University where he was eventually (in 1957) appointed as the Harrington Spear Paine Professor of Religion. On retirement from Princeton, he continued work at the independent Center for Theological Inquiry until his death. He was elected a member of the Institute of Medicine (1972) for his pioneering contributions to bioethics, an unusual distinction for a theologian. He died on February 29 in Princeton, New Jersey. His papers reside in the Duke University Library.

Christian Bioethics

The crux of Ramsey's ethics is a focus on the Christian concept of agape as the chief determinant of human and institutional action. Contrary to Roman Catholic teaching, he rejected the relative autonomy of natural law and morality, aligning himself with deontological normative theories. He believed convictions as informed by theology provided the essential basis for all lasting deontological commitments. Ramsey was highly critical, however, of facile pronouncements by ecclesiastical bodies concerning social policy. He maintained throughout his life that theologically informed convictions can and should be expressed in the public arena, a position that, by the end of his professional career, would be strongly challenged on many fronts.

Approaching ethical decision making using the method of complex case studies, Ramsey specifically condemned the dropping of atomic bombs on Hiroshima and Nagasaki and the experiments on mentally disabled children at Willowbrook State School in Staten Island, New York. On the other hand, he upheld just war theory and believed that while any military action should be regretted, such action was often essential to prevent a greater evil. This led Ramsey to be a staunch proponent of the U.S. engagement in Vietnam and yet, consistent with his agape ethic, also to strenuously uphold the rights of persons to engage in sit-ins and other forms of nonviolent protest. He approved of the use of tactical nuclear weapons but did not believe that the mutually assured destruction (MAD) doctrine of U.S. Cold War policy was acceptable since it targeted innocent civilians living in cities for its chief deterrence potency.

A chief protagonist throughout Ramsey's life was Joseph Fletcher (1905–1991) and his *situation ethics*. While both adopted agape as their central frame of reference, they interpreted its import and action quite differently, with Fletcher arguing that one should always act in a situation to maximize happiness for the greatest number, a principle that Ramsey found highly problematic in actual applications—including those employed by Fletcher himself—due to its lack of consistent principles or rules. He believed that Fletcher's focus on individual acts would lead to a weakening of the very principle of love it was intended to realize. Charles Pinches and others have argued that Fletcher and Ramsey, despite their surface differences, are both principle monists.

Assessment

Ramsey's most lasting contributions have been in the arena of medical ethics; a fact signaled by the reissue of many of his works in this area and medical conferences devoted to his ethical approach. He was one of the first ethicists to explore difficult medical cases and use them to frame general policy approaches to such issues as abortion, euthanasia, organ transplants, artificial organs, and emergency room triage. He strongly argued against removal of the term *person* from decisions at the beginning and end of human life, since he recognized that only persons have rights. He maintained that the dying had a right to choose their own death without heroic interventions from medical personnel but rejected any concept of *death with dignity*, consistent with his theological views of death as the last human enemy to be overcome by Jesus Christ.

Despite his disagreements with aspects of it, he drew deeply on Roman Catholic moral tradition so fruitfully that scarcely any Protestant or Catholic ethicist working in the early-twenty-first century neglects the other tradition. At the same time, many of his arguments have been characterized as too focused on Christian theological content and concepts to serve as a useful language for broad public dialogue and not specific enough to be used exclusively by the Christian community to frame its own distinct positions. Many consider Ramsey to be the father of bioethics, although he would be aghast at how that discipline quickly jettisoned from the public sphere the very kind of theologically rich language he was trying to promote.

DENNIS W. CHEEK

SEE ALSO *Bioethics; Medical Ethics.*

BIBLIOGRAPHY

Long, Stephen D. (1993). *Tragedy, Tradition, Transfiguration: The Ethics of Paul Ramsey.* Boulder, CO: Westview Press. Still the most comprehensive study of his work.

McKenzie, Michael C. (2000). *Paul Ramsey's Ethics: The Power of 'Agape' in a Postmodern World.* New York: Praeger. An excellent discussion of the central role of agape in Ramsey's thought.

Ramsey, Paul. (1993 [1953]). *Basic Christian Ethics.* Louisville, KY: Westminster/John Knox Press. Reprint with new introduction of Scribner's original edition.

Ramsey, Paul. (1970). *Fabricated Man: The Ethics of Genetic Control.* New Haven, CT: Yale University Press.

Ramsey, Paul. (1978). *Ethics at the Edge of Life: Medical and Legal Intersections.* New Haven, CT: Yale University Press.

Ramsey, Paul. (2002). *The Patient as Person: Explorations in Medical Ethics,* 2nd edition. New Haven, CT: Yale University Press.

Vaux, Kenneth L., and Mark Stenberg, eds. (2003). *Covenants of Life: Contemporary Medical Ethics in Light of the Thought of Paul Ramsey.* New York: Springer-Verlag. A *festschrift* with eleven contributors and an interview with Ramsey shortly before his death.

Werpehowski, William. (2002). *American Protestant Ethics and the Legacy of H. Richard Niebuhr.* Washington, DC: Georgetown University Press. Explores the writings of Paul Ramsey, Stanley Hauerwas, James Gustafson, and Kathryn Tanner in relation to H. Richard Niebuhr, his brother Reinhold Niebuhr, and to one another.

Werpehowski, William, and Stephen D. Crocco, eds. (1994). *The Essential Paul Ramsey: A Collection.* New Haven, CT: Yale University Press. A superbly edited volume.

RAND, AYN

• • •

One of the twentieth century's best known novelists and philosophers, Ayn Rand (1905–1982), who was born in Saint Petersburg, Russia on February 2, and died in New York City on March 6, celebrated the individual in dramatic stories with unconventional characters and plots. The heroes of her four novels are engineers, scientists, architects, and industrialists. Her philosophy, which she called Objectivism, champions the rational productive individual.

In 1936, ten years after her arrival in the United States, Rand published her first novel, *We the Living.* Set in Russia shortly after the communist revolution of 1917, it tells the story of Kira Argounova, a young woman who wants to become an engineer and build bridges, and her struggle to live in a collectivist society at war with the individual.

Rand's second major publication, the novelette *Anthem,* published in 1938, is set in a bleak future in which freedom and individualism have been eliminated in the name of the common good. The achievements of the Industrial Revolution have been lost; people have been reduced to using candles. Against this background of decay one man defies society and rediscovers

Ayn Rand, 1905–1982. Rand began to form her philosophy of rational self-interest, which she called "objectivism," at an early age. This view became the basis for her immensely popular writings, which included *The Fountainhead* and *Atlas Shrugged*. (AP/Wide World Photos. Reproduced by permission.)

teries but a science whose purpose is to teach people how to think and live, a science as capable of certainty and proof as is physics or mathematics.

The central idea of Rand's philosophy is that reason is human being's means of survival. Only through a process of reasoning—cold, hard, scientific, logical thought—can an individual understand the world and thus survive and prosper in it. This is why the heroes in her novels are scientists, engineers, and businesspersons; they are rational thinkers.

Rand accordingly defended the power of reason: She argued that the testimony of the senses is unquestionably valid, that human concepts and language con connect one to the facts of reality, and that logic is the only method for reaching truth. She rejected all forms of mysticism and supernaturalism on the grounds that such doctrines defy reason and contradict the fundamental laws of reality.

In regard to ethics Rand advocated rational self-interest. The task of ethics, she argued, is to teach one the principles—the virtues—that one must practice to realize the values that sustain one's life. No outside power, whether society or an alleged god, has the right to demand that one sacrifice one's values and live for its sake. The good is to live one's own life and attain happiness. This is accomplished through a resolute commitment to the virtue of rationality. For Rand the moral and the practical are one.

In regard to political philosophy Rand argued that a proper social system must accord with the individual's nature as a rational being. Individuals in society must be free to live, think, produce and keep the results of their work, and pursue their own goals. They must have the rights to life, liberty, property, and the pursuit of happiness. The social system that results from the protection of individual rights, Rand taught, is laissez-faire capitalism. That system was approached in the freest countries in the nineteenth century, and Rand argued that the thought and productivity that capitalism unleashed made possible the ensuing unprecedented prosperity in those countries.

Rand was one of the twentieth century's champions of science and technology and the rational mind that creates them. She therefore was an opponent of ideological movements that praise more primitive lifestyles, such as the New Left and environmentalism. An increasingly industrialized society, Rand held, is the proper environment for a rational being. Although her thought, which challenged contemporary views, was largely ignored in academic circles during her lifetime, it is

individual thought, science, and technology, along with the importance of the self.

Rand's third novel, *The Fountainhead*, was published in 1943. Her first major commercial success, *The Fountainhead* is the story of Howard Roark, an innovative young architect who thinks and lives for himself and refuses to copy the designs of the past, and of the opposition he faces from a society that worships tradition and mindless conformity.

Rand's last novel, *Atlas Shrugged*, was published in 1957. Its focus is the heroic individuals who, like the titan of Greek mythology, carry the world on their shoulders: the scientists, inventors, and businesspersons who create the knowledge and technology that sustain human life. *Atlas Shrugged* describes how those "men of the mind," as Rand calls them, liberate themselves from a society that denounces them as evil.

In presenting her vision of the hero, Rand created a new philosophy, Objectivism, on which she elaborated in her later, nonfiction writings. She argued that the subject of philosophy is not a realm of nonsense or mys-

receiving growing attention from scholars in the early twenty-first century.

ONKAR GHATE

SEE ALSO *Freedom*.

BIBLIOGRAPHY

Mayhew, Robert, ed. (2004). *Essays on Ayn Rand's We The Living*. Oxford: Lexington Books. This is a collection of sixteen essays by thirteen different contributors on the history as well as literary and philosophical content of Rand's first novel. For the sake of disclosure, please note that I contributed one essay.

Peikoff, Leonard. (1993). *Objectivism: The Philosophy of Ayn Rand*. New York: Meridian. Presents the essentials of Rand's system of philosophy; written by her foremost student.

Rand, Ayn. (1943). *The Fountainhead*. New York: Signet.

Rand Ayn. (1957). *Atlas Shrugged*. New York: Signet.

Rand, Ayn. (1959). *We the Living*. New York: Signet.

Rand, Ayn. (1961). *For the New Intellectual: The Philosophy of Ayn Rand*. New York: Signet. This collection contains the key philosophical passages from Rand's novels as well as a lead essay written by her explaining how philosophy has shaped the course of Western history and why new thinkers are needed.

Rand, Ayn. (1964). *The Virtue of Selfishness: A New Concept of Egoism*. New York: Signet. A collection of essays, by Rand and a colleague, on her ethics of rational self-interest.

Rand, Ayn. (1995). *Anthem*. New York: Signet.

Rand, Ayn. (1999). *Return of the Primitive: The Anti-Industrial Revolution*, ed. Peter Schwartz. New York: Meridian. A collection of essays, by Rand and editor Peter Schwartz, on cultural trends in politics and education.

Smith, Tara. (2000). Viable Values: A Study of Life as the Root and Reward of Morality. Lanham, Maryland: Roman and Littlefield. A critical study of the foundations of Rand's ethics, written by a professor of philosophy at the University of Texas.

RATIONAL CHOICE THEORY

• • •

Rational choice theory is a tool for devising a scientific explanation of the way individuals make choices; it is based on the notion that individuals attempt to find the most effective method of attaining their personal goals. Rational choice theory is a fundamental instrument for understanding ethical behavior and is compatible with the idea that such behavior is rooted in the biology of human nature.

An Illuminating Example

Suppose a person has $10 to spend in a store that has goods X and Y at prices p_x and p_y. To determine how the person will spend the $10, an economist assumes that the person has a preferences function $u(x, y)$ that he or she maximizes subject to the income constraint $p_x x + p_y y = \$10$, where x is the amount of X purchased and y is the amount of Y purchased. The preference function $u(x, y)$ reflects exactly how the person values different "bundles" of X and Y. For instance, if $u(2, 5) = u(3, 1)$, it is known that if the person has two units of X and five units of Y and if one takes four units of Y from the person, one must give the person an additional unit of X to compensate for the loss.

The assumption that people maximize their preferences subject to the appropriate constraints has proved fruitful in economics. Maximization subject to constraints also is used widely in biology to predict, for instance, how a predator will allocate its time among various prey or how a bumblebee will decide which flower patches to harvest and which ones to ignore (Alcock 1993).

However useful this function is, it may not be true that humans and animals really have utility functions in a meaningful physiological sense. Rather, their choices are the product of extremely complex and poorly understood neurological and hormonal processes. Why, then, is maximization subject to constraints used so successfully? The answer is that a choice process need only satisfy three simple conditions to be represented by a utility function. It is said that an agent is rational when those conditions are satisfied.

Basic Conditions

By a *preference ordering* $\succeq$ on a set A it is meant that a relation such that $x \succeq y$ may be either true or false for various pairs x, y in A. In words one states $x \succeq y$ as "x is weakly preferred to y" (Kreps 1990).

The first condition on $\succeq$ is *completeness*, which means that for any two members of the set, one is weakly preferred to the other (for any x, y in A, either $x \succeq y$ or $y \succeq x$). Note that this implies that any member of A is weakly preferred to itself (for any x in A, $x \succeq x$). Generally, this is a very plausible condition, but it is possible to think of cases in which it will fail to hold. Note that it is necessary to have $x \succeq x$ by the completeness condition. This is why $\succeq$ is referred to as weak preference. One can define strong preference as $x \succ y$, meaning that "it is false that $y \succeq x$." One can use elementary logic to prove that if $\succeq$ satisfies the completeness condition, then $\succeq$ satisfies the following *exclusion condition*: If $x \succeq y$, then it is false that $y \succeq x$.

The second condition is *transitivity*, which states that if x is weakly preferred to y and y is weakly preferred to z, then x is weakly preferred to z. In symbols this is written $x \succeq y$ and $y \succeq z$ implies $x \succeq z$. It is hard to see how this condition could fail for anything one would be likely to call a preference ordering. In terms of strong preference the transitivity condition becomes "if x is strongly preferred to y and y is strongly preferred to z, then x is strongly preferred to z." Again, one can use elementary logic to show that weak preference transitivity implies strong preference transitivity.

The third condition is the *maximization condition*, which states that from any set an agent will choose an element that is weakly preferred to any other member of the set. This condition, which also is called the *independence of irrelevant alternatives*, seems completely unobjectionable, but one can think of cases in which it will fail to hold. For instance, suppose A is a big basket of crumpets of different sizes. When given any pair of crumpets to choose from, the agent chooses the smaller of the two or chooses randomly if they are the same size. This satisfies completeness and transitivity. But suppose that if given a choice among any number of crumpets, the agent always chooses the next to largest, perhaps because he or she does not want to seem greedy. Then the agent will always choose the smaller when choosing among two but will not do this when choosing among more than two.

When these three conditions are satisfied, along with a technical *continuity* condition, there always exists a utility function such that the agent behaves as if maximizing this utility function over the set A from which he or she is constrained to choose. *Rational choice theory* is the study of the behavior of agents who satisfy these conditions, who are called *rational actors*.

Background and Misconceptions

The origins of the rational actor model lie in nineteenth-century utilitarianism and particularly in the works of Jeremy Bentham (1748–1832) and Cesare Beccaria (1738–1794), who interpreted utility as happiness. In *Foundations of Economic Analysis* (1947) the economist Paul Samuelson (born 1915; winner of a Nobel Prize in economics in 1970) removed the hedonistic assumptions of utility maximization by arguing that utility maximization presupposes nothing more than the conditions listed above.

The rational actor model has been misrepresented by those who embrace it and thus has been misunderstood by those who do not. The most prominent misunderstanding is that rational actors are self-interested. For instance, if two rational agents bargain over the division of money they jointly earned, it is thought that rational action requires that each agent try to maximize his or her share. Similarly, it is thought that if a rational actor votes in an election, he or she must be motivated by self-interest and will vote for the candidate most likely to secure his or her personal gain.

Of course, if one considers the term *rational* in the broadest philosophical sense, there is nothing irrational about caring for others, believing in fairness, or making sacrifices for social ideals, and such personal goals do not contradict rational choice theory. For instance, suppose a man with $100 is considering how much to consume personally and how much to give to charity. Suppose he enjoys a tax break such that for each $1 he contributes to charity, he is obliged to pay only $p < 1$. Then that person can be treated as maximizing his utility for personal consumption x and contributions to charity y, say, $u(x, y)$, subject to the budget constraint $x + py = 100$. Clearly, it is perfectly rational for him to choose $y > 0$. Indeed, James Andreoni and John H. Miller (2002) have shown that people in fact behave as rational actors in making choices of this type. For instance, when the price p increases, individuals tend to lower the quantity q of contributions to charity.

A second misconception is that the rational choice model assumes that the choices people make are in their own interest, when in fact people often are slaves to passions that are distinctly self-harming. For instance, it often is held that people are deluded or irrational when they choose to smoke cigarettes, engage in unsafe sex, commit crimes in the face of extremely heavy penalties, or sacrifice their health to junk food consumption. It is not clear, however, that these behaviors in any way violate the principles of rational action.

Weakness of Will

Those behaviors have in common a certain *weakness of will*. Smokers may know that their habit will harm them in the long run but cannot bear to sacrifice the present urge to indulge in favor of a far-off reward of a healthful future. Similarly, a couple in the throes of sexual passion may appreciate the fact that they may regret their inadequate precautions in the future, but they cannot control their present urges. This is not irrational but rather time-inconsistent.

A very clear laboratory experiment illustrates this time inconsistency (Ainslie and Haslam 1992). If subjects are offered a choice between $10 today and $11 a week from today, many will take the $10 today. However, if the same subjects are offered $10 to be delivered a year from today or $11 to be delivered a year and a week from today, many of the same subjects who could not wait a week right now for an extra 10 percent prefer to wait a week for an extra 10 percent provided that the agreed on wait is in the future. This finding corresponds to the everyday notion that people are subject to temptation and failure of will, leading them to accept high long-term penalties for small short-term pleasures.

It is instructive to see exactly where the conditions for rational choice are violated in this example. Let x mean "$10 at some time t" and y mean "$11 at time $t + 7$," where time t is measured in days. Then the present-oriented subjects display x Y when $t = 0$ and Y x when $t = 365$. Thus, the exclusion condition for is violated, and because the completeness condition implies the exclusion condition, the completeness condition must be violated as well.

Despite first appearances time-inconsistent agents can be modeled as rational actors (Ahlbrecht and Weber 1995). To do that one simply insists that the distance between the time of choice and the time of delivery of the object chosen be included explicitly in the analysis. Thus, x_0 means $10 delivered immediately and x_{365} means $10 delivered a year from today, and similarly for y_7 and y_{372}. Then the observation that x_0 Y$_7$ and Y$_{372}$ x_{365} is not a contradiction.

Indeed, here is a simple utility function involving what is called *hyperbolic discounting* (Ainslie 1975). Let z_t mean the amount of money delivered t days from today. Then let the utility of z_t be $u(z_t) = z/(t + 1)$. The value of x_0 is thus $u(z_t) = u(10_0) = 10/1 = 10$ and the value of y_7 is $u(z_t) = u(11_7) = 11/8 = 1.375$, and so x_0 y$_7$. But $u(x_{365}) = 10/366 = 0.027$ whereas $u(Y_{372}) = 11/373 = 0.029$, and so Y$_{372}$ x_{365}.

Prospect Theory

Prospect theory represents a fundamental contribution to rational choice theory that first was proposed by Daniel Kahneman (born 1934; winner of a Nobel Prize in economics in 2003) and Amos Tversky (1937–1996). According to prospect theory, agents value alternatives with respect to a *status quo* position that represents their current situation. This status quo position serves as a reference point with respect to which gains and losses are evaluated.

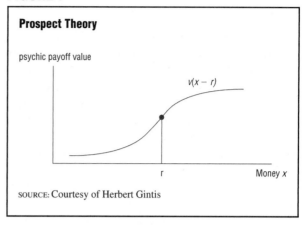

FIGURE 1

Prospect Theory

psychic payoff value

$v(x - r)$

r

Money x

SOURCE: Courtesy of Herbert Gintis

Suppose, for instance, an agent has utility function $v(x - r)$, where r is the status quo and x represents a change from the status quo. Prospect theory asserts that there is a "kink" in $v(x - r)$ such that the slope of $v(\cdot)$ is two to three times as great just to the left of $x = r$ as it is to the right, the curvature of $v(\cdot)$ is positive for positive values and negative for negative values, and the curvature goes to zero for large positive and negative values. In other words agents (a) are two to three times more sensitive to small losses than they are to small gains; (b) exhibit declining marginal utility over gains and declining absolute marginal utility over losses; and (c) are very insensitive to either large gains or large losses. This utility function is exhibited in Figure 1.

There are many regularities in experimental data on human behavior that do not fit prospect theory well (Kahneman and Tversky 2000). For instance, returns on equities (stocks) in the United States have exceeded the returns on bonds by about 8 percentage points averaged over the last 100 years. If this were due to risk aversion (concavity of utility function) alone, the average individual would be indifferent between a sure $51,209 and a lottery that paid $50,000 with probability 1/2 and a lottery that paid $100,000 with probability 1/2. This is, of course, implausible, as virtually everyone would choose the risky lottery in this situation. However, a loss aversion coefficient (the ratio of the slope of the utility function over losses at the kink to the slope over gains) of 2.25 is sufficient to explain this phenomenon. This loss aversion coefficient is very plausible from experiments. In a similar vein people tend to sell stocks when they are doing well but hold on to stocks when they are doing poorly. A similar phenomenon holds for housing sales: Homeowners are extremely averse to selling at a loss and will sustain operating, tax, and mortgage costs for long periods in the hope of obtaining a favorable selling price.

One of the earliest recognitions of loss aversion took the form of the so-called *ratchet effect* discovered by James Duesenberry (born 1918). Duesenberry noticed that over the business cycle, when times are good, people spend all their additional income, but when times start to go bad, people incur debt rather than curb their consumption. As a result there is a tendency for the savings ratio to decline over time. For instance, in one study unionized teachers consumed more when the next year's income was going to increase (through wage bargaining) but did not consume less when the next year's income was going to decrease. This behavior can be explained with a simple loss aversion model. A teacher's utility can be written $u(c(t) - r(t))$, where $c(t)$ is consumption in period t and $r(t)$ is the reference point (status quo point) in period t. Suppose the reference point changes as follows: $r(t + 1) = \alpha r(t) + (1 - \alpha)c(t)$, where α [0, 1] is an adjustment parameter ($\alpha = 1$ means no adjustment, and $\alpha = 0$ means complete adjustment to last year's consumption). Note that when consumption in one period rises, the reference point in the next period rises, and vice versa.

One curious implication of prospect theory is the *endowment effect*: By virtue of bhaving something, people tend to value it more than they are willing to pay for it if they do not have it. A common example is the rare wine effect: If a typical consumer wins a $200 bottle of wine in a contest, she will save it for a special occasion and drink it then. However, the consumer would never pay more than $20 for a bottle of wine and could have sold the prize wine if she desired to.

The status quo bias inherent in prospect theory leads to important *framing effects* that can distort effective decision making. In particular, when it is not clear what the appropriate reference point is, decision makers can exhibit inconsistency in their choices. Kahneman and Tversky (2000) give a dramatic example from health care policy. Suppose it is expected that there will be a flu epidemic in which 600 people are expected to die if nothing is done. If program A is adopted, 200 people will be saved, whereas if program B is adopted, there is a $\frac{1}{3}$ probability that 600 will be saved and a $\frac{2}{3}$ probability that no one will be saved. In one experiment, 72 percent of a sample of respondents preferred A to B. Suppose that if program C is adopted, 400 people will die, whereas if program D is adopted, there is a $\frac{1}{3}$ probability that nobody will die and a $\frac{2}{3}$ probability that 600 people will die. It was found that 78 percent of the respondents preferred D to C even though A and C are

equivalent and B and D are equivalent. Note that in the choice between A and B the alternatives involve gains whereas in the choice between C and D the alternatives involve losses, and people are loss-averse. The inconsistency stems from the fact that there is no natural reference point for the decision maker because the gains and losses are experienced by others, not by the decision maker himself or herself.

Why Rational Choice Theory Works

One important question remains: Why might one expect the conditions for rational choice to hold? The traditional answer is that humans are rational beings and the conditions for rational choice are the only conditions that satisfy the demands of reason. There are several problems with this justification. The most important is that the rational choice model often applies extremely well to nonhuman species, including insects and plants (Alcock 1993), whose mental apparatus falls far short of the capacity to exercise rational thought. Perhaps equally important, it is clear that humans often make choices that fail the test of right reason (e.g., weaknesses of will, including substance abuse, procrastination, and impulsive behavior), yet their choices do not violate the rational choice conditions.

A more contemporary explanation of the ubiquity of rational choice comes from evolutionary biology. Biologists define the fitness of an organism as its expected number of offspring, and the basic tenet of evolutionary biology is that fitness maximization is a precondition for evolutionary survival. If organisms maximized fitness directly, the conditions of rational choice would be directly satisfied because one could represent the organism's utility function as its fitness.

However, it is known that organisms, including humans, do not maximize fitness directly. For instance, moths fly into flames, few animals are capable of avoiding automobiles in the road, and humans voluntarily limit family size. In fact, biological fitness is a theoretical abstraction that is unknown to virtually every real-life organism. Rather than literally maximizing fitness, organisms have relatively simple preference orderings that are themselves subject to selection in accordance with their ability to promote fitness (Darwin 1872). One can expect preferences to satisfy the completeness condition because an organism must be able to make a choice in any situation it habitually faces or it will be outcompeted by another organism whose preference ordering can be used to make such a choice.

For similar evolutionary reasons one would expect the transitivity condition to hold in regard to choices that have some evolutionary meaning to the rational agent. Of course, unless the current environment of choice is the same as the historical environment in which the individual's preference system evolved, one would not expect an individual's choices to be fitness-maximizing or even welfare-improving. For instance, people in advanced technological societies have a tendency to obesity that can be explained by a weakness of will and a preference for high-calorie foods that may not be fitness-enhancing today but doubtless was at some times in the evolutionary history of the human species, which until about 10,000 years ago reflected the conditions of existence of small hunter-gatherer bands under constant threat of starvation.

Implications

Rational choice theory lies at the foundation of all behavioral science because natural selection strongly tends to select for preferences that satisfy the conditions of the rational actor model. Rational choice does not presuppose "reason," but it does presuppose adaptivity to an evolutionary environment. The fact that some behavioral disciplines, such as sociology, anthropology, and psychology, tend to ignore or reject the rational choice model through misunderstanding it arguably explains their relative immaturity and lack of unified principles in comparison with biology and economics, which tend to accept the principles of rational choice.

The most important implications of rational choice theory for ethics are as follows: (a) Weakness of will is not irrational and probably is an ineluctable dimension of the behavioral repertoire of humans; (b) because rational agents need not be selfish, it is not irrational to act altruistically and to care for others or to hate or act vindictively; and (c) what humans want and what they find ethically satisfying depend on their preference structures, which derive from an interaction between their species history and their personal histories. This argues for a behavioral ethics in which ethical principles are derived not from an appeal to introspection or reason but from the material conditions of the life of the human species.

HERBERT GINTIS

SEE ALSO *Choice Behavior; Decision Theory; Game Theory; Prisoner's Dilemma.*

BIBLIOGRAPHY

Ahlbrecht, Martin, and Martin Weber. (1995). "Hyperbolic Discounting Models in Prescriptive Theory of Intertemporal Choice." *Zeitschrift für Wirtschafts- und Sozialwissenschaften* 115: 535–568.

Ainslie, George. (1975). "Specious Reward: A Behavioral Theory of Impulsiveness and Impulse Control." *Psychological Bulletin* 82: 463–496.

Ainslie, George, and Nick Haslam. (1992). "Hyperbolic Discounting." In *Choice over Time*, eds. George Loewenstein and Jon Elster. New York: Russell Sage.

Alcock, John. (1993). *Animal Behavior: An Evolutionary Approach.* Sunderland, MA: Sinauer.

Andreoni, James, and John H. Miller. (2002). "Giving According to GARP: An Experimental Test of the Consistency of Preferences for Altruism." *Econometrica* 70(2): 737–753.

Darwin, Charles. (1872). *The Origin of Species by Means of Natural Selection*, 6th edition. London: John Murray.

Kahneman, Daniel, and Amos Tversky. (2000). *Choices, Values, and Frames.* Cambridge, UK: Cambridge University Press.

Kreps, David M. (1990). *A Course in Microeconomic Theory.* Princeton, NJ: Princeton University Press.

Samuelson, Paul. (1947). *The Foundations of Economic Analysis.* Cambridge: Harvard University Press.

Tversky, Amos, and Daniel Kanheman. (1990). "Rational Choice and the Framing of Decisions." In *The Limits of Rationality*, eds. Karen Schweers Cook and Margaret Levi. Chicago: University of Chicago Press.

RAWLS, JOHN

• • •

Bordley John Rawls (1921–2002) was born in Baltimore, Maryland, on February 21, educated in philosophy at Princeton University, and served in the military in the Pacific theater during World War II. He taught at Cornell University and at the Massachusetts Institute of Technology before becoming a professor at Harvard University where he taught philosophy for almost forty years. His theory of justice transformed twentieth-century political philosophy and has important implications for understanding the ethics of science and technology in terms of political governance and economics of the marketplace. He died in Cambridge, Massachusetts, on November 24.

Major Works

Rawls's major works include *A Theory of Justice* (1971), *The Law of Peoples* (1993), *Political Liberalism* (1993), and *Justice as Fairness: A Restatement* (2001). His

John Rawls, 1921–2002. The American philosopher was one of the most important political philosophers in the late 20th century. His *A Theory of Justice* developed principles of justice for a liberal society and challenged utilitarian political philosophy. (© *Jane Reed/Harvard University News Office.*)

writings have been widely distributed and translated into more than twenty languages.

Rawls developed his thought against the background of two existing philosophies: (a) utilitarianism, which employs the principle "the greatest good for the greatest number," and (b) emotivism, which claims moral and political judgments are basically personal or social preferences. Rawls finds both views inadequate, and in *A Theory of Justice* argues at length for a concept of "justice as fairness," which entails the economically "just distribution" of societal benefits and burdens through democratic procedures and institutions. Political procedures for advancing justice must run parallel to those of technological and economic progress.

In effect, Rawls revivifies theories of justice, rights, and international law that have their roots in Immanuel Kant (1724–1804) and social contract theory, as a broad response to totalitarianism and post–World War II inequities. Also, as a World War II veteran, Rawls authored "Fifty Years after Hiroshima," in which he argued against the use of the atomic bomb, and the employment of nuclear technology for nuclear weap-

onry. The crux of Rawls's argument may be found in a set of hypothetical conditions as follows. Imagine yourself in some "original position" in which you know that you are going to be placed in a complex world among persons with different abilities living in complex social institutional arrangements. At the same time you are prohibited by a "veil of ignorance" from knowing which abilities you might be given or which social institutions you will initially occupy.

In such a situation, Rawls argues, all persons, being both rational and self-interested, would choose to structure their social world around two principles of justice. The first, the "equal liberty principle," would establish equal basic rights and liberties for all. The second, the "difference principle," would defend inequalities on two conditions: (a) equality of opportunity (positions open to all having comparable prospects, talents, and abilities) and (b) economic and social inequalities distributed to benefit those disadvantaged by their social position. Rawls's argument is that when people do not know what abilities or benefits, or deficits and liabilities they might be given, such frame of mind affects the social order they would accept as just or fair. Moreover, such a well-ordered society based upon these principles will justly pair political democracy with economic capitalism.

Since the democratic-inspired revolutions of the eighteenth century, liberal philosophers have argued that rational individualism, republican democracy, and capitalism together could do more than any other systems to increase human rights, opportunities, and goods for more people. Historically, however, philosophers have also noted the recurring divide between rich and poor. In *Political Liberalism,* Rawls thus charges future progress, whether in government or business, in science or technology, with a moral imperative: Use political liberalism to promote justice, to ensure equal rights, and to acquire human rights as well as economic ones.

Rawls's principles of justice remain critical in evaluating these future problems and progress. Reminding his readers, in *The Law of Peoples,* that burdens accompany goods, and responsibilities come with liberties, Rawls analyzed who and what institutions will bear these responsibilities and duties to provide just and more equitable rights in a world in which people are actually situated, and materially advantaged or disadvantaged. Rawls directly formulated definitive tenets for law, rights, and duties that must be publicly instituted to address ongoing concerns and conflicts of minorities, pluralities, or the majority of global peoples. Therein, cosmopolitan individuals, technical experts, scientists, political leaders, and multinational corporations alike could find the principles, laws, and procedures in place to address fairly their worldly operations, disputes, and affairs.

Assessment

Rawls's work has inspired countless commentaries and critical replies in the United States and abroad. For instance, from its first publication in 1971 to its revised edition in 1999, *A Theory of Justice* has been challenged by communitarians and feminists. Both argue that *Theory* is too abstract and individualistic, despite its broad global outreach to diverse peoples, governments, and cultures. Arguably, Rawls draws heavily from Kant's rationalist, individualistic ethics and political philosophy of contractarian government, whereby citizens and their states jointly contract and consent (implicitly and explicitly) to institute and legitimate the just rule of their government.

Rawls has been criticized not only from the left (communitarians and feminists) but also from the right (libertarians and free-market theorists). Most notably, Robert Nozick (1938–2002), Rawls's well-renown Harvard colleague, was also his life-long critic, promulgating a counter theory known as the "entitlement theory" of social justice. In short, Nozick's theory extends another long-standing Western trend, libertarianism, which, like political liberalism, also originated in the eighteenth century, starting with Adam Smith's *The Wealth of Nations* (1776). Nozick wrote *Anarchy, State, and Utopia* (1974) in direct response as a critique of Rawls's *Theory of Justice*. Nozick thereby enlivened visions of justice based upon free-market capitalism and a minimalist state, in which the state serves solely to protect its members from violence and theft, and hence should possess no rights to interfere with one's property acquisition, use, and distribution, nor with any technological innovations and enterprises, unless fraud and unlawful force have been committed or contracts breached.

Rawls's political liberalism provides critical assurance that rational principles of justice and ethical government can control global capitalism, biotechnology, and engineering enterprises, so as to assure more of the world's people that liberties, goods, and opportunities can be more fairly distributed. Because Rawls rejects the premise that the powers and forces of right, possessed by people who are merely empowered and advantaged by circumstance or their societal position, can legitimately constitute justice, his *Theory* can test the progress made, and that still must be made, toward expanding global liberties and economic justice. In demonstrating "justice

as fairness," Rawls firmly reestablishes liberal political philosophy: In facing global pluralism—diverse beliefs, values, and bases for differing notions of good—politically just principles and powers for human rights-distribution are morally required to evaluate and improve the actual positions of individuals, states, and global peoples in working toward greater fairness.

MARY LENZI

SEE ALSO *Human Rights; Justice; Liberalism.*

BIBLIOGRAPHY

Freeman, Samuel, ed. (2003). *The Cambridge Companion to Rawls.* New York: Cambridge University Press.

Rawls, John. (1971). *A Theory of Justice.* Cambridge, MA: Harvard University Press, Belknap Press. Rev. edition, 1999.

Rawls, John. (1996). *Political Liberalism and the "Reply to Habermas."* New York: Columbia University Press. *Political Liberalism* originally published in 1993.

Rawls, John. (1999). *The Law of Peoples; with, "The Idea of Public Reason Revisited."* Cambridge, MA: Harvard University Press. Includes a major reworking of the essay "The Law of Peoples," originally published in 1993, along with the essay "The Idea of Public Reason Revisited," originally published in 1997.

Rawls, John. (2000). *Lectures on the History of Moral Philosophy*, ed. Barbara Herman. Cambridge, MA: Harvard University Press.

Rawls, John. (2001). *Justice as Fairness: A Restatement*, ed. Erin Kelly. Cambridge, MA: Harvard University Press.

REGENERATIVE MEDICINE

SEE *Aging and Regenerative Medicine.*

REGULATION AND REGULATORY AGENCIES

• • •

Regulation is a concept that is associated intimately with science, technology, and ethics. In the most general sense regulations control or direct human activities in accordance with a rule that has been promulgated. Neither sciences nor technologies could exist without internal processes of professional self-regulation. Biology includes research on the processes that regulate early embryonic development. The larger societies in which science and technology are embedded are dependent on forms of regulation that run the gamut from social to legal and governmental. Ethics is a form of regulation that often is seen as being more conscious or self-critical than social regulation and more broad than legal regulation.

The modern social construction of regulatory agencies as part of government was one attempt to respond to the complexity of advancing technological societies by "delegating legislation" that established appropriate institutional bodies to create and enforce "administrative laws" in specific areas of need such as water treatment, radio wave frequency allocation, and air traffic control. Reactions to the bureaucratic inefficiencies sometimes introduced by such agencies has led to countermovements for deregulation.

Historical Background and Modern Emergence

Regulations existed from the earliest periods of human history. Heads of tribes established rules that enabled closely related groups to live at peace within defined territories; rules of marriage, divorce, compensation for damage, bequests, and the status of slaves were set out in the Code of Hammurabi, which was carved in stone in Babylon in the 1700s B.C.E. Before that time the definitions of key weights and measures were established; for example, the mina (one-sixtieth of a talent) was a unit of payment that was mentioned specifically in the Code of Hammurabi. A talent, which might have been the weight a man could carry with comfort (about sixty pounds), had superseded the ox or cow as the unit of exchange.

In the era that preceded democratic governments the all-powerful prince was able to promulgate regal (regulatory) powers to modulate the behavior of his subjects according to his wishes. After the emergence of liberal democracies in the 1700s C.E. individuals and organizations within a society often were allowed to behave as they wished as long as they did not violate any of the rules and regulations crafted to ensure social order and the well-being of the society.

Those rules and regulations constitute a subset of the ethics of a society that are formulated and promulgated by those elected to a representative assembly. That assembly or body of lawmakers acts in place of the prince and therefore may be seen as an agent that regulates the affairs of the society. This is an example of the first level of the regulatory agency: the parliament or legislature.

In a democracy this type of regulatory agency involves the full complement of the members of the society who are eligible to vote and provides laws that have to be obeyed on pain of penalty when they are flouted. Those laws are upheld by an enforcement authority consisting of the police and if necessary the army that brings people suspected of lawbreaking before a judiciary where argument is presented with or without lawyers before a judge and a body of peers (the jury). If the guilt of the accused is established, punishment is meted out in the form of a fine, imprisonment, or another type of penalty.

The second level consists of religious authorities. In this case regulations or ethics are based on interpretations of sacred texts by clerics who have been given the authority to make such determinations by the head of the order or by the collective will of the congregants. The matters that are dealt with at this level are subject to compliance with laws of the state that override ecclesiastical regulation if there is a conflict. Thus, the way the church conducts its business and the messages the church promotes in helping members establish a workable relationship with the deity are an area of regulation for which this agency is fully responsible.

A third tier of regulation operates through groups of individuals who are selected by governmental departments and given authority by the issuance of specific laws to regulate the behavior of particular industries or service organizations. The first body of this type was set up in 1852 by the U.S. Congress as the Steamboat Inspection Service. That body was required to establish and maintain standards of design and production for the boilers that were used to power the paddles of steamboats plying the Mississippi River. Before that time explosions of those boilers resulted in the deaths of hundreds of passengers. Eventually that situation led to the establishment of a professional society, the American Society of Mechanical Engineers (ASME), that drew up codes of conduct to govern the education and practical training of the engineers involved in boiler design and construction along with specific codes that governed the construction of boilers that then were incorporated into local and state law.

In 1887 in the United States the Interstate Commerce Commission (ICC) was established to, among its other regulatory activities, prevent destabilizing competition in railway fares and set fare rates that would allow investment in new track and facilities as well as provisions for maintenance and safety measures without preventing the delivery of dividends to encourage further investment.

Other countries and international organizations established their own regulatory agencies. The United Nations (1945) and its subagencies, notably the World

Health Organization, the Food and Agricultural Organization, and the World Bank, were set up. In addition to a variety of international laws, those agencies provide regulations that control trade and the sustainable use of resources as well as the financial control of terrorism. The Treaty of Rome in 1957 established the European Union, which may issue directives whose power is binding on its members. There is also an International Organization for Standardization (1947) that has issued 14,000 international standards that enable world trade to proceed with confidence and a World Intellectual Property Organization that deals with regulations involving patents.

U.S. Regulatory Agencies

During the twentieth century some fifty regulatory agencies were established by the U.S. Congress. Some of the tasks undertaken by those bodies can be of major importance, for example, regulation of the quality of food and drugs through U.S. Food and Drug Administration regulations for pharmaceuticals and vaccines that often require manufacturers to test their products for safety, efficacy, and the consistency of their production process over a period of five to fifteen years at a cost of $500 million to $1 billion per product. Other tasks are trivial, including setting the when times a drawbridge may be raised or lowered.

Those agencies regulate financial operations (the Securities and Exchange Commission, established in 1933) and control the way people use their local environments (the Environmental Protection Agency, established in 1970). All aspects of the work environment are covered by the Occupational Safety and Health Administration (1970), and the Nuclear Regulatory Commission was set up in 1977 to supervise the development of civil nuclear installations. The development of the executive department of the Congress devoted to agricultural matters has spawned numerous regulatory agencies that oversee most aspects of agricultural practice. When it can be demonstrated that there is an overarching social need for regulation, members of Congress seem to be willing to provide the legal powers or instruments that give the agencies they create the tools to do their jobs.

Some of the functions that are served by American regulatory agencies include the following.

REGULATION OF COMPETITION. Although the liberal nature of the American democracy provides for the freedom of individuals and corporations to compete in attracting the attention of customers, corporations sometimes have colluded in setting prices or availabilities that have affected prices in ways that benefit corporations disproportionately. Such conglomerates have been disaggregated by law, and competition has to be active between the disaggregated entities that have been formed. For this reason the Standard Oil Company was broken up in 1911 and the Bell System's telephone monopoly was broken down to the AT&T company and the seven "Baby Bells" in 1982.

CONTROL OF COMPANY ACTIVITIES IN RELATION TO THE ENVIRONMENT. Most manufacturing companies acquire raw materials and convert them to final products, in the process producing solid, liquid, and gaseous wastes. At one time the disposal of that waste was a matter for company determination. Because there have been serious examples of wastes contaminating environments and damaging the health of local people (the Love Canal in New York State was so polluted that it took twenty years to clean up), regulations have been used to protect local residents and workers in the polluting factories.

PROVISION OF INFORMATION ABOUT PRODUCTS. The need to provide composition and calorific data on foods has turned supermarket shopping into an exercise in nutritional virtuosity. Additionally, data in advertisements have to comply with the realities of products and financial deals have to be expressed in ways that provide complete and comprehensible information to those about to take out loans or mortgages.

PROTECTION OF THE WEAK (CHILDREN) AND INFIRM. Regulations also may express the more basic virtues that are considered the hallmarks of a proud and independent society. These virtues include equality of opportunity; nondiscrimination on the basis of racial, ethnic, or religious affiliations; and the need to protect privacy on the street on in a column of data.

Criticisms

Any regulatory regimen is established at a cost. There is a burgeoning bureaucracy to deal with and costs in terms of time and trouble whenever a licence is required to make or do something. This may provide a hurdle for those who are innovating, who may be put off by the specifications they will have to meet to manufacture a product. There is also the consideration that regulations depart from the ideals of a liberal democracy that is premised on the least involvement of the state in the day-to-day activities of its citizens. In the United Kingdom the criticism that is leveled at the government as it seeks to advise and regulate the way people live, eat, and use mind-affecting drugs is that the government has become the "nanny" of the state.

A corollary of this situation is that regulations have to be devised to regulate the regulators. In the United States the Office of Management and Budget (OMB) was set up by a presidential executive order to determine the cost-effectiveness of the activities of the regulatory agencies that have been established by Congress.

People may live in a liberal society that purports to promote freedom of the individual and the corporation, yet they are biological organisms that need to have multiple levels of control to enable them to function. There are at least four levels of biochemical control of cellular function—environmental, enzymatic, energetic, and genetic—in addition to hormonal, neuronal, instinctive, subconscious, and conscious control systems. There are also social control systems, among which regulatory agencies are only one. There is little doubt that the application of a multitiered system of controls provides people with enhanced survival chances: Whether survival is always the only value is another issue.

R. E. SPIER

SEE ALSO *Aviation Regulatory Agencies; Environmental Regulatory Agencies; Foucault, Michel; Regulatory Toxicology; Science, Technology, and Law.*

BIBLIOGRAPHY

Unger, Stephen H. (1994). *Controlling Technology: Ethics and the Responsible Engineer*, 2nd edition. New York: Wiley.

INTERNET RESOURCES

Department of the Environment, Heritage, and Local Government. (2004). "Local Agenda 21." Available at http://www.environ.ie/DOEI/DOEIPol.nsf/wvNavView/Local+Agenda+21?OpenDocument&Lang=.

Environmental Protection Agency, Office of Site Remediation Enforcement. (2004). "The Lovel Canal." *Cleanup News.* Available at http://www.epa.gov/compliance/resourcesnewsletter/cleanup/cleanup3.pdf.

European Union. (2004). Available at http://europa.eu.int/index_en.htm.

International Organization for Standardization. (2004). Available at http://www.iso.org/iso/en/aboutiso/introduction/index.html#four.

Morris, Jane Anne. (1998). "Sheep in Wolf's Clothing." *By What Authority* 1(1). Available at http://www.poclad.org/bwa/fall98.htm.

Office of Management and Budget. (2003). "Report to Congress on the Costs and Benefits of Federal Regulations" Available at http://www.whitehouse.gov/omb/ inforeg/2003_cost-ben_final_rpt.pdf.

Office of Management and Budget, Office of Information and Regulatory Affairs. (2004). Available at http://whitehouse.gov/omb/inforeg/Chap1.htm.

United Nations. (2004). Available at http://www.un.org/english/.

United Nations Suppression of Terrorism Regulations. Available at http://www.fin.gc.ca/news01/01-082e.html.

REGULATORY TOXICOLOGY

• • •

Regulatory toxicology is the branch of toxicology (the study of adverse effects of chemicals) that uses scientific knowledge to develop regulations and other strategies for reducing and controlling exposure to dangerous chemicals.

The legal framework in this area is promulgated by governmental agencies. Examples of such agencies in the United States are the Food and Drug Administration (FDA), the Environmental Protection Agency (EPA), and the Occupational Safety and Health Administration (OSHA). Corresponding agencies exist in the European Union (EU) at the national or union level. The primary examples of authorizing legislation in the United States are the Food, Drug, and Cosmetic Act (1938), the Occupational Safety and Health Act (1970), the Clean Air Act (1970), the Federal Insecticide, Fungicide, and Rodenticide Act (1972), the Toxic Substances Control Act (1976), and the Clean Water Act (1977). Corresponding laws exist in the EU.

The Society of Toxicology (United States), EUROTOX (Europe), and the International Union of Toxicology (IUTOX) (global) are major professional organizations. The Society of Toxicology has published a code of ethics for toxicologists that requires its members to:

• Strive to conduct their work and themselves with objectivity and integrity.

• Hold as inviolate that credible science is fundamental to all toxicologic research.

• Seek to communicate information concerning health, safety, and toxicity in a timely and responsible manner, with due regard for the significance and credibility of the available data.

• Present their scientific statements or endorsements with full disclosure of whether or not factual supportive data are available.

• Abstain from professional judgments influenced by conflict of interest and, insofar as possible, avoid situations that imply a conflict of interest.

- Observe the spirit, as well as, the letter of law, regulations, and ethical standards with regard to the welfare of humans and animals involved in their experimental procedures.

- Practice high standards of occupational health and safety for the benefit of their co-workers and other personnel. (Society of Toxicology)

Toxicological Data and Assessment

Toxicity or adverse effects data are obtained either from experimental systems using animals or cell cultures, or from epidemiological studies of humans. The legally required testing differs among groups of chemical substances, ranging from no testing for many industrial chemicals to extensive requirements for pharmaceuticals.

A general problem is that the adverse effects of many chemicals, whether alone or in combination, are unknown. This is due to low data requirements, to statistical limitations in the available data, and to the cocktail effect or the interaction of chemicals. As a rough rule of thumb, epidemiological and experimental studies cannot reliably detect excess incidences of adverse effects of about 10 percent or smaller, and in many cases excess incidences of higher than 10 percent may go undetected. For relatively common types of disease, incidences are between 1 percent (leukemia) and 10 percent (breast cancer in Swedish women). Therefore even in the more sensitive studies, the limits of an observable excess lifetime risk are in the order of 1/100 or 1/1000, a level the public often considers unacceptable.

Once data are collected they are used to formulate toxicological assessments. Toxicological *health assessments* aim at identifying the potential adverse effects that a substance may cause in humans. This includes a description of the nature of these effects, their likelihood of occurrence, and their extent or severity.

The process of toxicological assessment is usually divided into four steps (National Research Council 1983, European Commission 2003). The first step of *hazard identification* aims at determining the inherent properties of a substance in order to identify the types of adverse effects to be included in further analysis.

The second step is *dose-response assessment*. The purpose of the dose-response assessment is to describe the relationship between the size of the dose and the response in the exposed. This is essential, because a high dose of a substance with low toxicity can be lethal, while a very low dose of a substance with high toxicity may be harmless. See Figure 1.

FIGURE 1

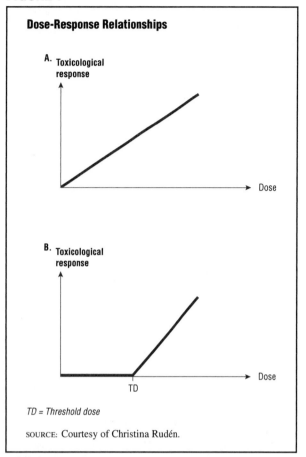

Diagram A shows a linear dose-response relationship increasing from zero exposure. In diagram B a threshold dose is indicated (denoted "TD").

The choice of a toxicological management strategy may depend on whether the dose-response relationship is considered to be linear from zero exposure or if a *threshold dose* is anticipated. A threshold dose is a dose under which no adverse effects are expected.

The lowest dose that has been shown to give rise to a statistically significant adverse effect compared to unexposed controls is the *Lowest Observed Adverse Effect Level* (LOAEL). The highest dose that has been administered without any observed statistically significant adverse effect is the *No Observed Adverse Effect Level* (NOAEL). A *benchmark dose* (BMD) is obtained by fitting a dose-response model to data, and from that model estimating a dose that corresponds to a predetermined change in the toxicological response investigated. The low-level change in response compared to background associated with the BMD is commonly termed the *benchmark response level* (BMR). Continuous dose-response data or incidence data may be used as a basis for these calculations. In the latter case, the BMD is

FIGURE 2

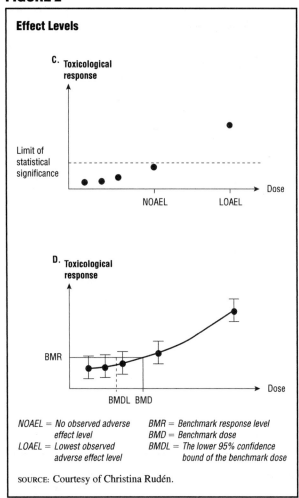

Effect Levels

NOAEL = No observed adverse effect level
LOAEL = Lowest observed adverse effect level

BMR = Benchmark response level
BMD = Benchmark dose
BMDL = The lower 95% confidence bound of the benchmark dose

SOURCE: Courtesy of Christina Rudén.

Diagram C shows the NOAEL/LOAEL approach, and diagram D shows the benchmark dose approach (BMD). The NOAEL/LOAEL is based on effect data for specific dose levels, while the BMD is obtained by curve-fitting of effect data.

generally defined as a 1 percent to 10 percent change in the incidence of the effect compared to background. In any case, the lower 95 percent confidence bound of the benchmark dose (the BMDL) is suggested as an alternative to NOAEL or LOAEL as a starting point for the determination of reference values for estimating acceptable exposure levels. See Figure 2.

The NOAEL, LOAEL, or BMDL should be defined for *critical effect*. Critical effect is the adverse effect that occurs at the lowest dose.

The third step is *exposure assessment*. This aims at determining the likelihood of exposure and estimates the magnitude and duration of the doses, as well as the potential exposure routes. Exposure assessment must be based on monitoring data and/or the use of theoretical exposure models.

The final step is *risk characterization*, which involves comparing the exposure data to the dose-response information in order to characterize the risk in qualitative and (if possible) quantitative terms.

Conclusive dose-response data are rarely available in humans, and therefore risk characterization often involves *extrapolation* from animal data to assess human risk. Absent contrary evidence, it is generally presumed that the effects seen in the test species under experimental conditions are relevant to humans. This presumption is supported by the fact that common test species are physiologically similar to humans.

In *environmental risk assessment* the same basic procedure applies. The outcome of hazard identification and dose-response assessment is the *Predicted No Effect Concentration* (PNEC), and exposure assessment estimates the *Predicted Environmental Concentration* (PEC). In the risk characterization process, the PEC/PNEC ratios are calculated. Extrapolation is made from experimental data (a limited number of single species) to the ecosystem (millions of species and multiple exposures interacting).

Extrapolation of data is hampered by scientific uncertainty. Resolving all uncertainties inherent in extrapolation would require testing on humans and/or an unreasonable number of animals. The presumptions used to overcome gaps of knowledge in assessment involve value judgments.

Toxicological Management

There are a number of possible risk management options in regulatory toxicology, ranging from public education to the banning of toxic substances. Two central systems are classification with labeling and exposure limits.

The *classification and labeling system* is an important part of international chemicals control because the classification process constitutes a background for further regulatory actions. According to the criteria for classification, substances (and preparations) are classified according to their inherent properties. Those fulfilling the criteria have to be provided with a warning label. Agenda 21, adopted at the United Nations Conference on Environment and Development in 1992, provided the international mandate to develop a globally harmonized system (GHS) for the classification and labeling of chemicals. The work was coordinated and managed under the auspices of the Interorganization Programme for the Sound Management of Chemicals (IOMC), administered by the World Health Organization (WHO). The aim is to have the GHS system fully implemented and operational by 2008.

Another major regulatory strategy is the setting of *exposure limits*. In the workplace such limits are called Occupational Exposure Limits (OEL), or Threshold Limit Values (TLV). Limits for exposure via food and drinking water are called Acceptable (or Tolerable) Daily Intake (ADI or TDI).

A health-based exposure limit is usually derived starting with either an experimentally estimated NOAEL/LOAEL, or a BMDL for the effect of concern. To overcome variability and other uncertainties, the experimental dose level is adjusted with an appropriate *uncertainty factor* to reach an exposure level assessed as not associated with adverse effects in humans.. The size of the uncertainty factor may vary from one to several thousands depending on the severity of the effect, the nature of the exposure, the exposed population, data-gaps, and uncertainties in the database.

Toxicological management is based on scientific evidence, but in the decision-making process nonscientific considerations are also taken into account. Examples of such considerations are the technical feasibility of the decision including availability of alternative technical processes, socioeconomic consequences, and value-based judgements of what health effects are acceptable.

CHRISTINA RUDÉN

SEE ALSO *Radiation; Regulation; Risk; Safety Engineering: Practices; Safety Factors.*

BIBLIOGRAPHY

Gad, Shayne C., ed. (2001). *Regulatory Toxicology*, 2nd edition. London: Taylor and Francis. A reference book of the requirements and regulations that both government agencies and non-government organizations promulgate for establishing the safety of a wide spectra of chemical products. It includes information on the United States, the European, and the Japanese systems, and is aimed for the full range of professionals in this field

National Research Council. (1983). *Risk Assessment in the Federal Government—Managing the Process.* Washington DC: National Academy Press. This volume evaluates past efforts to develop and use risk assessment guidelines, reviews the experience of regulatory agencies with different administrative arrangements for risk assessment, and evaluates various proposals to modify procedures. The book's conclusions and recommendations can be applied across the entire field of environmental health.

National Research Council, Committee on Risk Assessment of Hazardous Air Pollutants, Board on Environmental Studies and Toxicology, Commission on Life Sciences. (1994). *Science and Judgement in Risk Assessment.* Washington DC: National Academy Press. This report is aimed at a multidisciplinary audience with different levels of technical understanding. It addresses for instance the background of risk assessment and current practice at EPA, specific concerns in risk assessment, such as extrapolations, and cross-cutting issues that affect all parts of risk assessment. For example, how should uncertainty be handled?

Rudén, Christina. (2003). "Science and Transscience in Carcinogen Risk Assessment—The European Regulatory Process for Trichloroethylene." *Journal of Toxicology and Environmental Health*, part B: Critical Reviews 6(3): 257–277. In this article the European Union regulatory process for classification and labelling is described, it also reports an example of how hazard assessments are performed within this legislation.

Van Leeuwen, Cornelis Johannes: Josephus Ludovicus: and Maria Hermens, eds. (1995). *Risk Assessment of Chemicals: An Introduction.* Dordrecht, The Netherlands: Kluwer Academic Publishers. This book provides an introduction to risk assessment and management of chemicals, including background information on sources and emissions, distribution and fate processes, toxicology and ecotoxicology, and basic principles and methods for hazard and risk assessment within the legislative framework. It is intended for students and professionals within this field.

INTERNET RESOURCES

European Commission (2003). "Technical Guidance Document on Risk Assessment." Available from http://ecb.jrc.it. This is a set of technical guidance is issued by the European Commission as a help to carry out the risk assessments required within the EU legislations. It includes technical details for conducting hazard identification, dose - response assessment, exposure assessment and risk characterisation in relation to human health and the environment.

Society of Toxicology. Code of ethics. Available from http://www.toxicology.org/memberservices/aboutsot/ethics.html.

RELATIVISM

SEE *Pluralism: Social Pluralism.*

RELIABILITY

• • •

The term *reliability* can be used to indicate a virtue in a person, a feature of scientific knowledge, or the quality of a product, process, or system. Personal unreliability makes an individual difficult to trust. Unreliability in science calls the scientific enterprise into question. Lack of reliability in technology or engineering undermines utility and public confidence and perhaps commercial success. In all cases the pursuit of reliability is a conscious goal.

Scientific Reliability as Replication

Reliability in science takes its primary form as replicability. Research experiments and research must be performed and then communicated in such a way that they can be replicated by others or the results cannot become part of the edifice of science. Both replicability in principle and actual replication by diverse members of the scientific community are central to the processes of science that make the knowledge produced by science uniquely reliable and able to be trusted both within the community and by nonscientists.

Replication is easier to achieve in some scientific domains than in others, but when it fails, the science is judged unreliable. Historically replication was established first in physics and chemistry, and so in the physical sciences especially lack of replicability can become newsworthy. For example, the inability of other scientists to replicate the experiments on which Stanley Pons and Martin Fleischmann based their announcement of the discovery of cold fusion in 1989 doomed the credibility of their claims.

As Harry Collins and Trevor Pinch (1998) have shown in case studies, the replication of particular experiments often depends on the phenomenon of "golden hands." Not all experimenters are equally skilled at setting up and performing experiments, and subtle differences can be more relevant than it is possible to articulate clearly in the methods section of a research article.

In science another version of replicability is associated with peer review. Peer review procedures for scientific publication and for decision making about grants in effect depend on two or more persons coming to the same conclusion about the value of a report or proposal. Assessments must be replicated among independent professionals to support reliable decisions. Several evaluations of the peer review process in various disciplines have been performed (Peters and Ceci 1982). Many of those reports suggest that the system is unreliable because reviewers often fail to agree on the quality of a scientific article. Unreliability in this process undermines the internal quality controls of science, thus hampering progress. It also raises epistemological questions about the constitution of truth.

For instance, even if two reviewers judge a paper to be of high quality, both may be mistaken because they failed to spot a statistical error. In this sense reliability (agreement between reviewers) does not constitute validity (internal consistency or the absence of obvious errors of logic) (Wood, Roberts, and Howell 2004). However, on another level the negotiation of scientific claims within the scientific community is an integral part of determining what is true. Thus, in this sense reliability is a way of making or legitimating truth claims. These issues are made more complex by the role of editors in synthesizing disparate claims by reviewers and the question of whether reliability can be assessed by the metric of agreement between reviewers.

Another example of the issue of replicability in science is associated with the development of the *Diagnostic and Statistical Manual of Mental Disorders (DSM)* in psychiatry. Before this compendium of standardized descriptions of mental disorders was published, diagnoses of psychological illnesses lacked reliability. For example, if three physicians independently saw a patient with a psychological illness, it was unlikely that they would make the same diagnosis. Indeed, this remained the case through the publication of the original *DSM* in 1952 and *DSM-II* in 1968. It was only with the increasing detail and sophistication of *DSM-III*, published in 1980, that the psychiatric community began to achieve a significant measure of reliability in its diagnostic practices and psychiatry became more respected as a science.

This case suggests the connection between reliability and professionalization (the formation of a specialized academic discipline) because replicability was made possible only after a community of practitioners developed a shared conceptual language and a methodology that were sufficiently nuanced to communicate and establish likes as likes. Reliability as a way of establishing truth through replication thus is a product of both material reality and the way peers conceptualize the world and are able to replicate that conceptualization among themselves.

Functional Reliability in Engineering

Engineering or technological reliability is the probability that a product, process, or system will perform as intended or expected. Issues include the expected level of reliability, the cost-benefit trade-offs in improving reliability, and the consequences of failure. When these issues involve persons other than those inventing or tinkering with the relevant products, processes, or systems, with consequences for public safety, health, or welfare, ethical issues become prominent. Just as in science, reliability, in this case in the form of functional reliability, is a precondition for the integration of a particular technological device into the accepted or trusted edifice of the built environment.

Any technological product, process, or system is designed to perform one or more specified functions. In principle, the performance of the system can be defined mathematically and the demands placed on the system can be specified. Because uncertainties are associated with all aspects of systems in the real world, these descriptors should be defined in terms of uncertainties and reliability should be computed as the probability of intended performance. Because most systems have effects beyond their stated output (radiation, accidents, behavior modification, etc.), a comprehensive model must include all possible outcomes. Because complicated models all are based on extrapolations of the basic principles, the fundamental concepts are described in this entry.

The demands placed on a system include environmental and operational loads, which for simplicity will be designated here as a single demand, D. The capacity of the system to absorb those loads and perform its function is designated C, for capacity. The satisfactory operation of the system simply entails that the capacity be at least as large as the demand. This is expressed mathematically as $S = C - D \geq 0$ in which S represents satisfactory performance. In probability terms this becomes $P(S) = P(C - D) \geq 0$.

Each of these basic quantities can be described probabilistically by its probability function: $F_D(d)$ for demand and $F_C(c)$ for capacity. It is usually a safe assumption that the capacity (a function of the physical system) and the demand (a function of the operating environment) are statistically independent. In this case the reliability of the system is given by

$$P(S) = \int f_C(x) F_D(x)\, dx$$

in which $f_C(x)$ is the probability density function of the capacity (the derivative of the capacity probability function, $F_C(x)$) if the capacity is a continuous variable and otherwise is the probability mass function of the capacity (analogous to a histogram). In words the preceding equation indicates that one should assign a probability that the capacity is a particular value ($f_C(x)$) and then multiply by the probability that the demand is no greater than that value of capacity ($F_D(x)$). This process then is repeated for all possible values of the demand and the capacity, and the results are added (that is what the integration function does for continuous variables). The integrand of the equation above is shown in the Figure 1.

TIME DEPENDENCY. Most systems are not designed to be used just once but instead to perform over an intended period. In this case, the demand and the capacity become time-dependent variables and the probability of satisfactory performance is interpreted as being

FIGURE 1

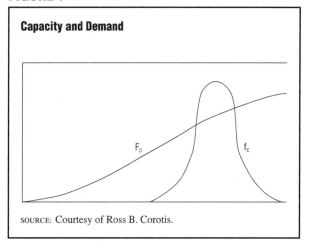

Capacity and Demand

F_D f_C

SOURCE: Courtesy of Ross B. Corotis.

The probability function of demand, F_D, is multiplied by the probability density function of capacity, f_C, and the resulting quantity is then integrated over all values to yield the reliability of the system.

over an intended design lifetime. The formulation of the previous section then is interpreted as being at a single point in time, and the results are integrated over the lifetime.

Most technology displays a characteristic failure curve that is relatively steep at the beginning of the design lifetime, during which time initial defects are discovered. The failure rate then decreases to a steady-state value that exists over most of the design lifetime of the technology. As the technology nears the end of its useful lifetime, the failure rate again rises as parts begin to wear out.

When failure is due to relatively rare events such as environmental hazards, unusual parts wear, and abnormal use, simplified time-dependent models can be developed on the basis of the independent occurrence of these unusual events. These models usually are based on the Poisson process model, which is the simplest among the time-dependent processes that are referred to as stochastic processes. The Poisson model assumes that the occurrence of each event is independent of the past history of performance of the technology.

Systems reliability adds another level to this analysis. A system is a technology that is composed of multiple parts. Usually it is necessary that the parts work together properly for the system as a whole to function as desired. Systems theory builds on the theory described above to consider multiple capacities and demands, and many theories and models have been developed to analyze the risks of systems (Haimes 1998). Because systems analysis can be complicated, formalized approaches such as decision tree analysis (Clemen 1996) and event tree and fault tree analysis have been developed (Page 1989). Approximate analyses use the concepts of systems reaching a discrete number of undesirable states that are referred to as limit states. One then evaluates the probability of reaching those states by using approximate analyses such as the first-order, second-moment (FOSM) method, in which the limit state is approximated by a straight line and the full probability descriptors of the demands and capacities are approximated by the first and second moments of the probability function, which usually are the average and the standard deviation (Melchers 1999).

Software reliability can be used to illustrate some of the issues mentioned here. Newly engineered software is notoriously unreliable. After in-house testing and even after beta (user) testing in the field or market, "patches" regularly have to be introduced as new problems arise. Sometimes those problems arise because of a lack of correctness in the underlying code, and at other times because of a lack of robustness in the overall design. Software engineers also can fail to appreciate the ways users may choose to utilize a particular piece of software, and hackers and others may try to exploit weaknesses in ways that undermine reliability. As software illustrates, the pursuit of functional reliability in engineering and technology is a never-ending quest with ethical implications.

Ethics of Reliability

Despite its ethical importance in science and technology reliability has been subject to little extended ethical analysis. With regard to persons, in which case the virtue of reliability manifests itself as trustworthiness, there has been more discussion. However, the following comments on the ethics of reliability in general are only preliminary observations.

First, as has been suggested in this entry, technological reliability is what makes engineered artifice the basis for improved material well-being. It is for this reason that a few technical professional ethics codes include the promotion of reliability as an explicit obligation. For example, in the Code of Ethics (developed 1948) of the American Society for Quality (founded in 1946), the third fundamental principle commits a member to promote "the safety and reliability of products for public use." However, although in some instances unreliability in products may be attributed to a failure of intention, in other cases it is caused by evolutionary changes in nature (e.g., the evolution of antibiotic-resis-

tant bacteria), economic change (as occurs when parts cease to be available for cars or other vehicles) or unintended consequences. Indeed, unintended consequences are one of the most common ways to conceptualize breakdowns in technological reliability as engineered devices bring about unexpected scenarios. This both raises questions about the degree to which reliability can be an ethical obligation and suggests the need for engineers to consider the wider ramifications of technology in their analyses of reliability and to build flexibility into their designs.

Another instance in which reliability has been adopted explicitly as an ethical concept related to technology occurred at a Poynter Journalism Values and Ethics in New Media Conference in 1997. That conference drafted an ethics code that included the following recommended "Online Reliability Statement":

> This site strives to provide accurate, reliable information to its users. We pledge to:
>
> Ensure information on our Web site has been edited to a standard equal to our print or broadcast standards.
>
> Notify our online users if newsworthy materials are posted from outside our site and may not have been edited or reviewed to meet our standards for reliability.
>
> Update all our databases for timeliness, accuracy and relevance.
>
> Warn users when they are leaving our site that they may be entering a site that has not embraced the content reliability protocol.

The idea here is that professional standards of reliability in the print media need to be transported consciously into a new technological media framework. Similar statements about the need for commitment to reliability in information delivery related in one way or another to technology have been discussed with regard to both medicine and computers.

With regard to science replicability generally is thought of as a self-regulating process that serves both as a method for epistemological quality control and as a way to prevent scientific misconduct, including fabrication, falsification, and plagiarism. Thus, it is a mechanism for nurturing trust within the scientific community. The dominant perception that scientists deal with absolute certainties often undermines public trust in science when scientists openly communicate uncertainties in their research or when a scientific finding of high public concern is disputed and eventually overturned ("In Science We Trust" 2001).

The notion of reliability as replicability also manifests a certain hierarchy of values or axiology in the pursuit of knowledge. Alvin Weinberg (1971) has noted that physics serves as the ideal science (of which other sciences are more or less distorted images) because of the universalizability and replicability of its findings. It most closely approximates deeply entrenched Western beliefs about truth as timeless and noncontextual. However, this ingrained cultural deference to this ideal of science can lead to misunderstandings of science and unrealistic expectations about its contributions to complex political decisions.

Questions also might be raised about the issue of reliability in ethics itself. The human sciences, including ethical inquiry, proceed by means of dialectical and hermeneutical processes that are different from the models of the engineering construction of reliable artifacts or the scientific construction of reliable knowledge claims. In the popular imagination ethical and other value claims often are treated as matters of religious commitment, subjective preference, or legalistic requirements. However, a more nuanced appreciation of the process of ethical argumentation can point to possibilities for reliability.

Substantive agreement and reliability can be found, for instance, in some common documents, such as the Universal Declaration of Human Rights of 1948. Procedural reliability is manifested in the democratic considerations of ethics and other values that also are able to proceed toward common interest solutions through reasonable argumentation, tolerance, compromise, and openness of mind, procedures not dissimilar to those involved in the pursuit of an always provisional scientific truth.

Thus, the test for reliability in ethics may not be replicability, but it also may not be as distant from the actual workings of science as is maintained by many people. Indeed, when it comes to practical affairs, the desirable trait for both science and ethics may not be replicability so much as something more akin to the functional reliability of technology. That is, reliable science and ethics, much like reliable technologies, help human beings navigate toward common goods within complex situations marked by uncertainties and pluralities.

ROSS B. COROTIS
CARL MITCHAM
ADAM BRIGGLE

SEE ALSO *Uncertainty*.

BIBLIOGRAPHY

Clemen, Robert. (1996). *Making Hard Decisions: An Introduction to Decision Analysis*. Pacific Grove, CA: Duxbury Press.

Collins, Harry, and Trevor Pinch. (1998). *The Golem: What You Should Know about Science*, 2nd rev. edition. New York: Cambridge University Press. This controversial book, which argues through a series of seven case studies for the influence of social factors in the assessment of experiments, originally was published in 1993.

Haimes, Yacov. (1998). *Risk Modeling, Assessment, and Management*. New York: Wiley.

"In Science We Trust." (2001). *Nature Medicine* 7(8): 871.

Melchers, Robert. (1999). *Structural Reliability Analysis and Prediction*. New York: Wiley.

Page, Lavon. (1989). *Probability for Engineering*. New York: Computer Science Press.

Peters, Douglas P., and Stephen J. Ceci. (1982). "Peer Review Practices of Psychological Journals: The Fate of Published Articles, Submitted Again." *Behavioral and Brain Sciences* 5: 187–255. Finds that eight of twelve articles resubmitted after they had been published were rejected by the same journal.

Weinberg, Alvin. (1971). "The Axiology of Science." *American Scientist* 58(6): 612–617.

Wood, Michael; Martyn Roberts; and Barbara Howell. (2004). "The Reliability of Peer Reviews of Papers on Information Systems." *Journal of Information Science* 30(1): 2–11.

INTERNET RESOURCES

American Society for Quality. "Code of Ethics." Available from http://www.asq.org. The 1948 code of ethics can be found on their site at http://www.asq.org/join/about/ethics.html

The Poynter Institute. "Online Content Reliability Guidelines." 1997. Includes the "Online Reliability Statement." Available from http://legacy.poynter.org/research/me/nme/me_samprot.htm#relability.

RESEARCH ETHICS: OVERVIEW

• • •

Research ethics is typically divided into two categories: those issues inherent in the practice of research, and those that arise in the application or use of research findings. In the United States, ethical practice has come to be known as the responsible conduct of research (RCR); outside the United States another common term is good scientific practice (GSP). Ethical issues associated with the application of research findings deal with their use in the support of legal, social, or economic policy as well as their technological applications (e.g., genetic engineering in therapy and agriculture, bioweapons development, and dam siting and construction).

Many entries in the *Encyclopedia of Science, Technology, and Ethics* cover different aspects of research ethics in more detail. Prime examples include the entries on "Responsible Conduct of Research" and "Scientific Integrity," the composite on "Misconduct in Science," and the series dealing with various aspects of genetics. The focus here is on a more synthetic overview that also highlights some points missing elsewhere.

Background

Both aspects of research ethics came to the forefront of public attention at the end of World War II and have developed more fully over the mid-twentieth century. Leading discussions have often but not always taken place in the United States.

RESEARCH PRACTICE. Initially ethical concerns regarding research practice emphasized the use of humans as research subjects. The revelation of Nazi atrocities at the close of World War II focused international attention on research that subjected individuals to high altitude experiments in low-pressure chambers, freezing due to exposure or submersion in ice water, starvation or seawater as their primary source of fluids, and infection with malaria, typhoid, streptococcus, and tetanus. Judges presiding over the trial of Nazi physicians drafted the Nuremberg Code (1946), which has since been followed by additional ethical codes most prominently the World Medical Association Declaration of Helsinki (1964; most recently revised in 2002). For further depth on these issues see the entries on "Nazi Medicine" and "Human Subjects Research."

In the early 1970s the U.S. Tuskegee Syphilis studies came to light (see "Tuskegee Experiment) and focused national attention on human subjects treatment in the United States. This research, carried out from 1932 to 1972, recruited disadvantaged, rural black males who had contracted syphilis to participate in the study of the course of untreated disease. Although no clearly effective treatment was initially available, when it became apparent that penicillin was effective, participants were not given this medication. When these studies were made known, the U.S. Congress mandated a commission to identify, develop, and articulate the ethical principles that underlie and must guide the acceptable use of human volunteers and subjects in biomedical research. The commission's work resulted in the Belmont Report (National Commission 1978) that serves as the foundational document for research involving humans in the United States.

In the 1980s other egregious examples of scientific misconduct were exposed, including the fabrication and falsification of data, and plagiarism (Broad and Wade 1982, LaFollette 1992). While these were not the first instances of misconduct in science—the Piltdown Man fraud was initiated in 1912—they raised serious concerns not only within but beyond the scientific community. Indeed the U.S. Congress began to demand more consistent oversight of the process of research funding which led to establishment of the Office of Scientific Integrity within the National Institutes of Health that ultimately became the Office of Research Integrity (ORI) in the Department of Health and Human Services.

Moreover, within the scientific community, it became clear that concerns regarding serious scientific misconduct were only the tip of the iceberg in the sense that the professional standards, expectations of colleagues, and ethical values of the research community with regard to many aspects of research practice were not clearly articulated nor widely understood. There was, and is, a wide range of accepted practices without much discussion of the underlying assumptions and wider implications that place those practices along the continuum of preferred, acceptable, discouraged, and prohibited practices. As a result, trainees and even more established researchers are not always clear about the acceptability of established or ongoing practices within the community.

For example, while plagiarism (the misrepresentation of the writings or ideas of another as one's own) is clearly deceptive and unacceptable, other publication practices can also be problematic. The practice of "honorary" authorship—that is, including in the list of authors individuals who have not made a clear and significant intellectual contribution to the published work—became increasingly widespread over the latter part of the twentieth century. The practice of adding names to the list of authors (sometimes without the knowledge or consent of the individual "honored") in exchange for a reagent, a strain of mice, laboratory space, or past tutelage not only tends to "dilute" the apparent contribution of other authors (depending on a reader's assumptions), but also to deny honorary authors any opportunity to make fully informed decisions about their associations with the work.

APPLICATION OF RESEARCH FINDINGS. The end of World War II also brought greater awareness of the ethical implications of the uses of science and technology. The use of the atomic bomb by the United States on Japan raised a host of questions regarding the social responsibility of scientists and engineers for the consequences of their work. The Manhattan Project reflected a national priority to devote all resources, including scientific expertise to winning the war. Yet those working on the project could only speculate on the immediate and long-term health and environmental effects of an atomic explosion. Moreover, as scientist J. Robert Oppenheimer mused, the science was so "technically sweet" that its appeal overrode concerns about the creation of an enormously destructive bomb so unlike the conventional weapons with which people were already familiar.

In the 1960s Rachel Carson and others called attention to the dangers of chemical pollutants in the environment, and reactions took place against some of the kinds of chemicals being used in many agricultural, industrial, and military activities. In the 1970s developments in molecular biology (specifically techniques with recombinant DNA) led researchers to convene a conference in Asilomar, California, to discuss the implications and potential hazards of genetic engineering. This is often identified as the first widespread, proactive effort on the part of the scientific community to acknowledge and address its social responsibility.

The discussion has become more nuanced and complex as the impact of human activity on the environment and on other species as well as other human populations has become more apparent. Whether in the construction of large engineering projects such as dams that dramatically alter the landscape, inundate archaeological treasures, and displace the local population, or in the oftentimes poorly executed use of genetically engineered crops in developing nations, or in many other technological applications, their larger ethical and social implications have become the focus of increasing examination, debate, and institutional reform.

The Responsible Conduct of Research and Good Scientific Practice

Progress in science depends on trust between scientists that results have been honestly presented. It also depends on members of society trusting the honesty and motives of scientists and the integrity of their results (European Science Foundation 2000). Fostering this trust requires clear and strong ethical principles to guide the conduct of scientific research. In the United States, ethical research practice is generally referred to as RCR or the responsible conduct of research. The ORI, the

U.S. federal agency primarily concerned with education in RCR, has identified nine core instructional areas in RCR (Office of Research Integrity 2005, Steneck 2004). Areas (1) through (5) deal with the actual conduct of research while areas (6) through (9) are associated with interactions between members of the scientific community.

1. Data Acquisition, Management, Sharing, and Ownership. This area focuses on the ways in which data are recorded, whether in notebooks or in other formats (such as electronic records, photographs, slides, etc.), and how and for how long they should be stored. It explores as well the question of who owns the data, who is responsible for storing them, and who has access to them. Issues of privacy and confidentiality of patient information as well as intellectual property issues and copyright laws are included.

2. Conflict of Interest and Commitments. Discussion of conflicting interests and commitments acknowledges the potential for interference in objective evaluation of research findings as a result of financial interests, obligations to other constituencies, personal and professional relationships, and other potential sources of conflict. It also considers strategies for managing such conflicts in order to prevent or control inappropriate bias in research design, data collection, and interpretation.

3. Human Subjects. Ethical treatment of human research subjects references the requirements of the Office of Human Research Protections (OHRP), which are based on the ethical principles outlined in the Belmont Report (National Commission 1978). These principles include especially (a) respect for persons as expressed in the requirement for informed consent to participate and protection of vulnerable populations such as children and those with limited mental capacity; (b) emphasis on beneficence that maximizes the potential benefits of the research and minimizes risks; and (c) attention to considerations of justice in the form of equitable distribution of the benefits and burdens of the research across populations. Adequate attention to patient privacy and the variety of potential harms including psychological, social, and, economic is essential.

4. Animal Welfare. Research involving animals emphasizes animal welfare in accordance with the regulations of the Office of Laboratory Animal Welfare (OLAW). Principles here emphasize respect for animals used in research (Russell and

Burch 1959) in the form of "the three Rs": reduction of the number of animals used, replacement of the use of animals with tissue or cell culture or computer models or with animals lower on the phylogenetic scale whenever appropriate and possible, and refinement of the research techniques to decrease or eliminate pain and stress.

5. Research Misconduct. Dealing with allegations of research misconduct is essential given its potential for derailing a research career. Definitions of scientific misconduct, including fabrication, falsification, and plagiarism as well as other serious deviations from accepted practice that may qualify as scientific misconduct, as distinguished from error, and protections for whistleblowers are important components of this topic.

6. Publication Practices and Responsible Authorship. Publication practices and responsible authorship examine the purpose of publication and how that is reflected in proper citation practice, criteria for authorship, multiple, duplicate and fragmentary publication, and the pressure to publish. This area also considers allocation of credit, the implications and assumptions reflected in the order of authors, and the responsibility of authorship.

7. Mentor/Trainee Responsibilities. The mentor/trainee relationship encompasses the responsibilities of both the mentor and the trainee, collaboration and competition, possible conflicts and potential challenges. It also covers the hierarchy of power and potential for the abuse of power in the relationship.

8. Peer Review. The tension between collaboration and competition is embodied in the peer review process for both publication and funding. In this area of RCR issues associated with competition, impartiality and confidentiality are explored along with the specifics of the structure and function of editorial and review boards and the *ad hoc* review process.

9. Collaborative Science. Not only does research build on the work of others, but more and more investigators from disparate fields work together. The collaborative nature of science requires that often implicit assumptions about common practices such as authorship and data sharing need to be made explicit in order to avoid disputes.

In Europe, the term of art for discussion of research ethics is GSP or good scientific practice (European Science Foundation 2000). However, unlike RCR,

which emphasizes guidelines for positive research behaviors, there is a tendency in other countries to emphasize the avoidance of negative behaviors. This means that despite the name (good scientific practices) discussion focuses on scientific misconduct. For instance, it the pursuit of GSP, the U.K. Office of Science and Technology (OST), the oversight body of the U.K. Research Councils, categorizes scientific misconduct into two broad groups. The first pertains to the fabrication and falsification of research results. The second category pertains to plagiarism, misquoting, or other misappropriation of the work of other researchers. The OST statement "Safeguarding Good Scientific Practice" (1998) stresses the need to avoid misconduct by means of self regulation of and by the research community, arguing that "Integrity cannot be prescribed" (Office of Science and Technology).

With the creation of the Danish Committee on Scientific Dishonesty in 1992, Denmark became the first European country to form a national body to handle cases of scientific dishonesty—again with the aim of promoting GSP. This has prompted similar practices in other Scandinavian countries (Vuckovic-Dekic 2000).

A serious case of scientific misconduct in Germany in 1998 sparked the creation of the international Commission on Professional Self Regulation in Science. This Commission was charged to explore causes of dishonesty in the science system, discuss preventive measures, examine the existing mechanisms of professional self regulation in science, and make recommendations on how to safeguard them. It published a report titled "Proposals for Safeguarding Good Scientific Practice," which advised relevant institutions (universities, research institutes, and funding organizations) to establish guidelines of scientific conduct, policies for handling allegations, and rules and norms of good practice (Commission on the Professional Self Regulation in Science 1998). Fearing over-regulation, the commission recommended that institutions retain authority for establishing misconduct policies (rather than establishing a centralized committee as in the United States and Denmark).

Ethical Issues in the Application of Research

The Enlightenment creed *Sapere aude!* (Dare to know!) symbolized the distinctively modern belief that scientific research is an ethical responsibility, indeed a moral obligation of the highest order. Ancient or premodern thinkers generally maintained that there were limits to the quest for knowledge, beyond which lay spiritual and physical dangers. Although there is a long tradition of critiques of this foundational modern commitment (e.g., Wolfgang von Goethe's *Faust* and Mary Shelly's *Frankenstein*), they have become more refined, extended, and institutionalized in the latter half of the twentieth century as science and technology began to profoundly alter both society and individual lives. The ramifications of various technological developments (e.g., atomic energy, genetic engineering) have demonstrated that unfettered research will not automatically bring unqualified goods to society.

Daniel Callahan (2003) has argued that there is a widespread assumption of the "research imperative," especially in the area of biomedicine and health care. Though a complex concept, it refers to the way in which research creates its own momentum and justification for gaining knowledge and developing technological responses to diverse medical conditions. It can pertain to the ethically dubious rationale of pursuing research goals that are hazardous or of doubtful human value, or the rationale that the ends of research justify the means (no matter how abhorrent). It can also pertain to the seemingly noble goal of relieving pain and suffering. Yet this commitment to medical progress has raised health care costs and distracted attention from the ultimate ends of individual happiness and the common good. Research, no matter how honorable the intent of those performing and supporting it, must be assessed within the context of other goods, rather than elevated as an overriding moral imperative (Jonas 1969, Rescher 1987).

As is considered in entries on "Science Policy" and "Governance of Science," the core assumption of the inherent value of research was operationalized in post-World War II U.S governmental policies for the funding of scientific research. What came to be known as the "linear model" of science-society relations posited that investments in "basic" research would automatically lead to societal benefits (Price 1965). However, the framers of this policy never specified how this "central alchemy" would occur, and they did not adequately address the need to mitigate negative consequences of scientific research (Holton 1979). The economic decline of the late 1970s and 1980s, the end of the cold war in the early 1990s, and the growing federal budget deficits of the same period combined to stimulate doubts about the identity of purpose between the scientific community and society (Mitcham and Frodeman 2004).

The very fact that societal resources are limited for the funding of scientific research has stimulated questions about what kind of science should be pursued. For instance, physicist and science administrator Alvin Weinberg argued in the 1960s that internal assessments of the quality of scientific projects and scientific researchers should be complemented by evaluation of scientific merit as judged by scientists in other disciplines, of technological merit, and of social merit. For Weinberg, because of the limited perspective of those within the community, "the most valid criteria for assessing scientific fields come from without rather than from within the scientific discipline that is being rated" (1967, p. 82).

Put simply, while the internal ethics of research asks: "How should we do science?" the external ethics of research takes up a suite of questions involving participants beyond the immediate scientific community and addressing more fundamental ends. As Daniel Sarewitz (1996) noted the pertinent questions are "What types of scientific knowledge should society choose to pursue? How should such choices be made and by whom? How should society apply this knowledge, once gained? How can "progress" in science and technology be defined and measured in the context of broader social and political goals?" (p. ix).

Myriad attempts have been made to reformulate the relationship between scientific research and political purposes, where the criteria for assessing science derive partially from without rather than from within a particular scientific discipline. Models include Philip Kitcher's ideal of "well-ordered science" (2001) and the concept of "use inspired basic research" put forward by Donald Stokes (1997). Such revised social contracts for science shift the focus from maximizing investments in research to devising mechanisms for directing research toward societal benefits; a shift from "how much?" to "toward what ends and why?" Legislation such as the 1993 U.S. Government Performance and Results Act (GPRA) reflects this focus on the social accountability of publicly funded science, as do technology assessment institutions and ethical, legal, and social implications research performed in conjunction with genome and nanotechnology research.

The prioritization of research projects is another important area in this regard, including the issue of how much money to allocate to the study of different diseases, which often raises ethical concerns about systematic discrimination. The effective use of scientific research and technologies in development policies intended to decrease poverty and improve the health of those in developing countries is a related topic. Diverse experiences with the Green Revolution, for example, show the importance of context in directing research toward common interests and away from negative outcomes such as ecological harms and the exacerbation of wealth disparities. Both of these topics raise the important issue of the role of various publics in guiding and informing scientific research and technological applications.

Although it is still largely true that "more money for more science is the commanding passion of the politics of science" (Greenberg 2001, p. 3), a number of critics and policy makers understand that more is not necessarily better. Scientific progress does not always equate to societal or personal progress in terms of goals such as safety, health, and happiness (Lightman, Sarewitz, and Desser 2003). The potential unintended physical harms that may result from scientific research have long been recognized and debated in terms of the roles of scientists and non-scientists in risk assessment. More recent developments, especially in bio- and nanotechnology research, and the growing specter of catastrophic terrorist attacks have lent a more urgent tone to questions about "subversive truths" and "forbidden knowledge" (e.g., Johnson 1996).

Limiting scientific research raises practical questions such as "Who should establish and administer controls?" and "At what level should the controls be imposed?" (Graham 1979). Some (e.g., McKibben 2003) have advocated the large scale relinquishment of whole sectors of research such as nanotechnology. Others, including the innovator Ray Kurzweil, argue for a more fine-grained relinquishment and the prioritizing of funding for research on defensive technologies to counteract potential misuses of science. This view holds that the optimal response to the potential for bioterrorism, for example, is to lessen restrictions on and increase funding for bioweapons research so that preventive measures and cures can be developed.

Discussion of the ethical implications of the use of scientific research is, at its core, about procedures for democratic decisions and the allocation of authority and voice among competing societal groups. This can be construed in broad terms ranging from criticisms of Western science as a dominant even hegemonic way of knowing that drowns out other voices, to defenses of science as an inherently democratizing force where truth speaks to power. These vague issues take on importance in concrete contexts that concern judgments about the appropriate degree of scientific freedom and autonomy within democratic societies. The most important area in which these issues arise is the use of scientific knowledge in formulating public policies.

Although bureaucratic political decision-making has come to rely heavily on scientific input, it is not obvious how the borders and interstices between science and policy should be managed. On the one hand, it seems appropriate that research undertaken by scientific advisory panels (as distinct from research in general) be somehow connected to the needs of decision makers. On the other hand, sound procedures for generating and assessing knowledge require a degree of independence from political (and corporate) pressures. Failure in the first instance leads to generation of irrelevant information and often delayed or uninformed action. Failure in the second case leads to conflicts of interest or the inappropriate distortion of scientific facts to support pre-existing political agendas (Lysenkoism is an extreme example) or corporate policies.

The latter instance is often couched in terms of the "politicization of science," which is a perennial theme in science-society relationships (e.g., Union of Concerned Scientists 2004). Yet in order to attain the democratic ideal of being responsive to the desires and fears of all citizens, the politicization of science in the sense of explicitly integrating it into the larger matrix of goods (and evaluating it from that standpoint) is proper. Scientific research can be "misused" when it is inappropriately mischaracterized (e.g., to over-hype the promise of research to justify funding) or delegitimized (Pielke 2004) and it is important to enforce ethical guidelines against these practices. However, the more common misuse of science that ranges from intentional to unconscious, is the practice of arguing moral or political stands through science (Longino, 1990). This can inhibit the ethical bases of disputes from being fully articulated and adjudicated, which often prevents science from playing an effective role in policy making (Sarewitz 2004).

Teaching Research Ethics

Science educators and researchers have generally believed their responsibility was to teach scientific concepts and laboratory techniques, and it was expected that professional values and ethical standards would be picked up by observing good examples. However, as a result of well-publicized and serious instances of scientific misconduct in the 1980s, the research community has become aware of the need to address the responsible conduct of research explicitly. Thus in 1989 the U.S. National Institutes of Health (NIH) began calling for formal instruction for NIH funded pre- and post-doctoral trainees in the responsible conduct and reporting of research (National Institutes of Health 1989). Moreover, in support of expanding the NIH requirement, both the report of the Commission on Research Integrity, "Integrity and Misconduct in Research" (1995) and the report of the international Commission on Professional Self Regulation in Science, "Proposals for Safeguarding Good Scientific Practice" (1998), highlighted the fact that education in RCR /GSP has been largely neglected worldwide and should be addressed for both trainees and senior scientists. In addition, recognition of the ethical implications of science and technology has led to the incorporation of these topics into many courses and programs aimed at teaching research and engineering ethics. It is widely appreciated that students need to understand that science and technology are not value free and that scientific information can be used for good or ill, misused or abused.

While it is widely believed that "by the time students enter graduate school, their values and ethical standards are so firmly established that they are difficult to change" (Swazey 1993, pp. 237–38) there is a solid body of evidence that supports the view that in fact adults *can* be taught to behave ethically through specific educational programs introduced at the undergraduate and postgraduate level (Rest et al. 1986; Bebeau et al. 1995). This is closely linked to the individual's reconceptualization of his or her professional role and relationship to society. Educational programs can affect awareness of moral problems and moral reasoning and judgment. Moreover, studies show that moral perception and judgment influence behavior.

There is some controversy regarding the emphasis of research ethics education, that is, whether to focus on the rules and regulations, expectations and standards of the research community, or to emphasize moral development. However in reality, teaching research ethics entails both communicating the standards and values of the community and promoting moral development through increased ethical sensitivity and ethical reasoning. Thus the goals of education in research ethics are to:

1. Increase awareness and knowledge of professional standards. Toward this end, professional standards and ethical values of scientific research and conventions are identified and clarified, as is the range of acceptable practices along the continuum of preferred, acceptable, discouraged, and prohibited. In the process, the assumptions that underlie accepted practices are examined and the immediate and long-term implications of these practices are assessed.

2. Increase awareness of ethical dimensions of science. This includes examination of the issues associated with both research practice and the application of research findings.

3. Provide experience in making and defending decisions about ethical issues. Case studies designed to illustrate common research practices and situations are generally used. Discussion of these cases invariably entails in-depth analysis of affected parties, points of conflict, implications of various courses of action, and examination of the expectations, needs and responsibilities of the different characters in the scenario.

4. Promote a sense of professional responsibility to be proactive in recognizing and addressing ethical issues associated with research.

A number of key characteristics of educational programs in research ethics have been identified (Bird 1999, Institute of Medicine 2002). These reflect principles of effective adult education as well as common sense. Programs that are *required* emphasize the view that ethical issues are inherent in research and that awareness of the ethical values and standards of the research community are an essential component of professional education. *Interactive* discussion of ethical issues and concerns raised by a realistic case provides participants with an opportunity to share their experience and solve problems in a context. This approach employs principles of learning science that have been identified through research on how people learn (Bransford et al. 1999). *Broad faculty involvement* in educational programs in research ethics demonstrates that this is valued by professionals across the discipline and incorporates a variety of experience and a range of perspectives with regard to accepted practices. Programs should *begin early* in research education (e.g., undergraduate science laboratory courses) and *continue* throughout college and graduate or other professional education. In so doing, individuals can reflect on their own experience, and their understanding and appreciation of ethical concerns and strategies for problem solving can evolve. When the various components of graduate education (i.e., courses, seminars, laboratory meetings, etc.) address ethical issues they *reinforce and complement* each other.

A variety of formats and strategies have been developed to teach research ethics. The most effective are case-based and integrate discussion of research ethics into all of the various elements of research education: as modules in core courses, stand-alone full semester or short courses on research ethics, departmental seminars, workshops, laboratory and research team meetings, one-on-one interactions between trainees and research supervisors, and computer-based instruction (Swazey and Bird 1997, Institute of Medicine 2002). Each approach has strengths and weaknesses.

Through explicit discussion of ethical issues associated with the practice of research and the application of research findings the research community acknowledges the complexity of the issues and the need to address them. Specifically addressing RCR reaffirms the responsibility of the research community for research integrity, individually and collectively, and the necessity of providing this information to its members. Identifying and examining the ethical issues associated with the application (or misapplication) of research findings emphasizes the responsibility of researchers and of citizens in general to examine and assess the ramifications of science and technology for society.

STEPHANIE J. BIRD
ADAM BRIGGLE

SEE ALSO *Accountability in Research; Animal Welfare; Chinese Perspectives: Research Ethics; Ethics: Overview; Misconduct in Science: Overview; Nazi Medicine; Science: Overview; Sociological Ethics.*

BIBLIOGRAPHY

Bebeau, Muriel J.; Kenneth D. Pimple; Karen M. T. Muskavitch; Sandra L. Borden, and David L. Smith. (1995). *Moral Reasoning in Scientific Research: Cases for Teaching and Assessment.* Bloomington, IN: Indiana University.

Bird, Stephanie J. (1999). "Including Ethics in Graduate Education in Scientific Research." In *Perspectives on Scholarly Misconduct in the Sciences,* John M. Braxton, ed. Columbus: Ohio State University Press.

Bransford, John D., A. L. Brown, and R. R. Cocking. (1999). *How People Learn: Brain, Mind, Experience, and School.* Committee on Developments in Science of Learning and Commission on Behavioral and Social Sciences and Education. Washington DC: National Academies Press.

Broad, W.J., and Nicholas Wade. (1982). *Betrayers of the Truth: Fraud and Deceit in the Halls of Science.* New York: Simon and Schuster.

Callahan, Daniel. (2003). *What Price Better Health?: Hazards of the Research Imperative.* Berkeley, CA: University of California Press.

Commission on Research Integrity (CORI). (1995). *Integrity and Misconduct in Research.* Washington, DC: U.S. Department of Health and Human Services, Public Health Services.

Graham, Loren R. (1979). "Concerns about Science and Attempts to Regulate Inquiry." In *Limits of Scientific Inquiry,* Gerald Holton and Robert S. Morison, eds. New York: W.W. Norton.

Greenberg, Daniel S. (2001). *Science, Money, and Politics: Political Triumph and the Ethical Erosion.* Chicago: University of Chicago Press. Seeks to explain the success of autonomous science embedded in U.S. politics from World War II through the beginning of the twenty-first century and argues that lobbying for money by scientists has corroded their integrity.

Holton, Gerald. (1979). "From the Endless Frontier to the Ideology of Limits." In *Limits of Scientific Inquiry,* Gerald Holton and Robert S. Morison, eds. New York: W.W. Norton.

Institute of Medicine and National Research Council (2002). *Integrity in Scientific Research: Creating and Environment that Promotes Responsible Conduct.* Washington, DC: National Academies Press.

Johnson, Deborah. (1996). "Forbidden Knowledge and Science as Professional Activity," *The Monist,* 79(2): 197–217.

Jonas, Hans. (1969). "Philosophical Reflections on Experimenting with Human Subjects," *Daedalus,* 98: 219–247.

Kitcher, Philip. (2001). *Science, Truth, and Democracy.* Oxford: Oxford University Press. Argues that epistemic values do not stand above or apart from practical social concerns and offers a new model for controlling and directing scientific inquiry.

LaFollette, Marcel C. (1992). *Stealing into Print: Fraud, Plagiarism, and Misconduct in Scientific Publishing.* Berkeley, CA: Oxford University Press.

Lightman, Alan; Daniel Sarewitz, and Christina Desser, eds. (2003). *Living with the Genie: Essays on Technology and the Quest for Human Mastery.* Washington, DC: Island Press. Collects sixteen essays on the central tension between the increasing pace of scientific and technological change and an immutable human core, stressing the importance of individual and public decisions in shaping the outcomes of this tension.

Longino, Helen E. (1990). *Science as Social Knowledge: Values and Objectivity in Scientific Inquiry.* Princeton, NJ: Princeton University Press.

McKibben, Bill. (2003). *Enough: Staying Human in an Engineered Age.* New York: Times Books. Advocates setting limits on the pursuit of knowledge and the quest for greater material wealth as they threaten to undermine the essence of being human.

Mitcham, Carl, and Robert Frodeman. (2004). "New Directions in the Philosophy of Science: Toward a Philosophy of Science Policy," *Philosophy Today.* 48(5 Supplement): 3–15.

National Commission for the Protection of Human Subjects (1978). *The Belmont Report: Ethical Principles and Guidelines for the Protection of Human Subjects of Research.* Washington, DC: Department of Health, Education, and Welfare, Government Printing Office, (OS) 78–0012.

National Institutes of Health. (1989). Reminder and Update: Requirement for Programs on the Responsible Conduct of Research in National Research Service Award Institutional Training Programs. NIH Guide for Grants and Contracts.

Price, Don K. (1965). *The Scientific Estate.* Cambridge, MA: Harvard University Press. Critiques the linear model of science policy that derived from Vannevar Bush's *Science—The Endless Frontier* (1945) and comments on fundamental issues in the relationship between science and politics.

Rescher, Nicholas. (1987). *Forbidden Knowledge and Other Essays on the Philosophy of Cognition*. Boston: D. Reidel. Argues that knowledge is only one good among others. Chapter one is titled "Forbidden Knowledge: Moral Limits of Scientific Research."

Rest, James R.; Muriel Bebeau, and J. Volker. (1986). "An Overview of the Psychology of Morality." In *Moral Development: Advances in Research and Theory*, James R. Rest, ed. New York: Praeger.

Russell, W. M. S. and Burch, R. L. (1959). *The Principles of Humane Animal Experimental Technique*. London: Methuen.

Sarewitz, Daniel. (1996). *Frontiers of Illusion: Science, Technology, and the Politics of Progress*. Philadelphia, PA: Temple University Press. Traces modern myths about science and its relation to society, outlines the problems they raise, and concludes with recommendations to form a new mythology.

Sarewitz, Daniel. (2004). "How Science Makes Environmental Controversies Worse," *Environmental Science and Policy* 7(5): 385–403. Case studies and explanations pertaining to the way in which some environmental conflicts become "scientized."

Steneck, Nicholas H. (2004). *Introduction to the Responsible Conduct of Research*. Washington, DC: Government Printing Office.

Stokes, Donald E. (1997). *Pasteur's Quadrant: Basic Science and Technological Innovation*. Washington, DC: Brookings Institution Press. Analyzes the goals of understanding and use and offers a model of use-inspired basic research to help both science and society.

Swazey, Judith P. (1993). "Teaching Ethics: Needs, Opportunities and Obstacles." In *Ethics, Values, and the Promise of Science*. Forum Proceedings, February 25-26, 1993. Research Triangle Park, NC: Sigma Xi, the Scientific Research Society.

Swazey, Judith P. and Bird, Stephanie J. (1997). "Teaching and Learning Research Ethics" In *Research Ethics: A Reader*, Deni Elliott and Judy E. Stern, eds. Dartmouth, NH: University Press of New England.

Vuckovic-Dekic, Lj. (2000). "Good Scientific Practice." In *Archive of Oncology*, 8 (Suppl. 1): 3–4.

Weinberg, Alvin M. (1967). *Reflections on Big Science*. Cambridge, MA: MIT Press.

INTERNET RESOURCES

Commission on Professional Self Regulation in Science. (1998). "Recommendations of the Commission on Professional Self Regulation in Science." Available from http://www.dfg.de/aktuelles_presse/reden_stellungnahmen/download/self_regulation_98.pdf.

European Science Foundation. (2000). "Good Scientific Practice in Research and Scholarship." Available from http://www.hrb.ie/storage/researchfunding/fundingpolicies/goodscientificpractice.pdf.

Office of Research Integrity. (2005). "Education—Responsible Conduct of Research." Available from http://ori.dhhs.gov/education/ed_rcr.shtml.

Office of Science and Technology. (1998). "Safeguarding Good Scientific Practice." Available from http://www.ost.gov.uk/research/councils/safe.htm#2.1.

Pielke, Roger, ed. (2004). "Report on the Misuse of Science in the Administrations of George H.W. Bush and William J. Clinton," available from http://sciencepolicy.colorado.edu/admin/publication_files/resourse-1429-ENVS%204800%20Report.pdf. Presents a taxonomy of the misuse of science and illustrates it through six case studies.

Union of Concerned Scientists. (2004). "Scientific Integrity in Policy Making: An Investigation into the Bush Administration's Misuse of Science," available from http://www.ucsusa.org/documents/RSI_final_fullreport.pdf. Illustrates through several vignettes the argument that individuals in the administration of President George W. Bush have suppressed or distorted research findings and undermined the quality and integrity of the appointment process.

RESEARCH INTEGRITY

• • •

Integrity (from the Latin *integritas*, meaning *whole* or *complete*) refers in ethics to adherence to a code or a usually high standard of conduct. Research integrity thus indicates doing research in accord with standards that properly inform and guide that activity—without deviance under any inappropriate influences. Integrity in this sense has close correlates with authenticity and accountability. Research integrity is also often considered the flip side of research misconduct. Whereas the topic of research misconduct concentrates on the definition, identification, adjudication, and consequences of malfeasance committed by scientists in the course of their research; research integrity concentrates on, as the Institute of Medicine's 2002 report, *Integrity in Scientific Research*, was subtitled: "creating an environment that promotes responsible conduct" of research (Institute of Medicine, p. x). Having received considerable public attention since the 1980s, however, research integrity is a contested issue both within the scientific community and between the community and its patrons.

Public and Professional Tensions

Part of the conflict over research integrity occurs over identifying the appropriate code or standard. Sociologist Robert K. Merton (1973) described four norms of science—communalism (or communism), universalism, disinterestedness, and organized skepticism—that are often cited as antecedent to codes to which scientists are supposed to adhere. But other scholars argue that such norms are not well recognized among all scientists

(Mitroff 1974), or that they are merely self-serving *vocabularies of justification* for scientific autonomy (Mulkay 1975), or that they might have served as guideposts historically but that they are being supplanted by counternorms that are more bureaucratic and commercially oriented (Ziman 1990).

Many professional societies have written or revised codes of ethics or guidelines for research integrity that encompass normative issues ranging from formal, regulatory definitions of research misconduct (for example fabrication, falsification, and plagiarism) to more subtle professional behavior such as authorship practices and mentorship. In the early-twenty-first century, professional bodies such as the Accrediting Board for Engineering and Technology (ABET) require training in ethics and research integrity for accredited undergraduate engineering programs. Scientific journals have also assumed an active role in defining integrity for their authors around topics such as credit for authorship, conflict of interest, and responsibility for corrections and retractions.

Research integrity is often connected not only with the attempt of the scientific community to encourage ethical behavior within its own ranks, but also with its attempt to maintain professional autonomy from public interference. As such, it is an aspect of the social contract for science in which the scientific community implicitly promised to maintain the integrity of its research in exchange for an unusual lack of oversight—despite public patronage. This tacit agreement was substantially reconfigured during the 1980s and 1990s, as both parties recognized that the promotion and assurance of research integrity must be a collaborative, rather than an autonomous, enterprise (Guston 2000).

The public patrons of research in liberal democracies have a special interest in research integrity not only because of the instrumental use of science and technology for public purposes (for example, only good science can lead to the promises of health, economic advancement, environmental quality, and military security, among others), but also because of the ideological support that good science offers the state by demonstrating its effectiveness and by reifying the concepts of representation and causality upon which representative government is based (Ezrahi 1990). In the United States, research integrity has become a pressing issue to the funding agencies and professional societies that mediate between public patrons and practicing scientists. A driving force for attention to research integrity was the promulgation of rules in 1990 by the National Institutes of Health (NIH) to require institutions participating in training grants to provide training in the responsible conduct of research. Such training often includes discussions not only of misconduct, but also of whistle-blowing, the protection of human and animal research subjects, the mentoring relationship, and the consequences of recently emergent economic relations in research including conflicts of interest and intellectual property rights. In 2000 the Office of Research Integrity (ORI) of the U.S. Public Health Service proposed more specific and broadly applicable rules for training in the responsible conduct of research, but as of 2004 these rules had not been implemented.

Because of the increasing recognition that the effects of research—for good or for ill—go beyond the scientific community, there is increasing attention as well to what some (particularly in engineering ethics) call *macroethics*, or the responsibility that scientists and engineers have to behave with integrity not just toward each other and toward their direct patrons but to society more broadly conceived (Herkert 2001). This agenda includes helping to craft private and public policies that make appropriate use of science and its products, assuring that the knowledge-based innovations to which they contribute are not only technically virtuous but socially benign, and even accepting greater involvement of nonscientists in some aspects of technical decision making. This agenda has historical roots, for example, in the characterization of activism by atomic physicists in nuclear weapons policy or molecular biologists in recombinant DNA policy as *scientific responsibility*.

Unresolved Questions

Despite increasing recognition of the importance of research integrity to both the scientific community and the broader society, and the consequent need for collaboration to assure it, several questions remain. One is whether the primary responsibility for assuring the integrity of research lies with individual researchers; research institutions such as universities, professional societies and the community of science; or public patrons of research. The Institute of Medicine (2002) concludes that research institutions should have the primary role, but that public patrons of research have an important oversight role and that individual integrity is still the backbone of the system.

A second question is, given the importance of some institutional role in research integrity, why so few exist. As one such institution, ORI—initially created to investigate allegations of research misconduct—has, in the early-twenty-first century, been changing its agenda toward encouraging training in research integrity and even sponsoring *research on research integrity*. The National Science Foundation (NSF) has also sponsored projects on research integrity, including the On-Line Ethics Center.

A third question is whether greater collaboration between science and society may legitimate an increasingly malign political interference, rather than a benign influence, on public science. The Waxman report, which issued from the U.S. House of Representatives, and a similar report from the Union of Concerned Scientists in 2004, for example, claim to document dozens of threats to research integrity from the intrusion of political agendas into scientific and technical decision making in the bureaucracy.

A fourth question, which makes the others all the more difficult to manage, is—as the Institute of Medicine (2002) concluded—how to create reliable ways to assess the overall integrity of the research environment, as well as the efficacy of any particular interventions (including educational ones). The lack of empirical evidence means that the scientific community can legitimately call for additional research on research integrity, but it also means that political demands for action may be met with less than satisfactory responses.

DAVID H. GUSTON

SEE ALSO *Accountability in Research; Ecological Integrity; Misconduct in Science: Overview; National Institutes of Health; Office of Research Integrity; Professional Engineering Organizations; Social Contract for Science.*

BIBLIOGRAPHY

Ezrahi, Yaron. (1990). *The Descent of Icarus: Science and the Transformation of Contemporary Democracy.* Cambridge, MA: Harvard University Press. Scholarly account of the co-dependence of scientific and democratic ideologies.

Guston, David H. (2000). *Between Politics and Science: Assuring the Integrity and the Productivity of Research.* New York: Cambridge University Press. Includes detailed political and institutional account of research integrity in U.S. in the 1980s and 1990s.

Herkert, Joseph R. (2001). "Future Directions in Engineering Ethics Research: Microethics, Macroethics and the Role of Professional Societies." *Science and Engineering Ethics* 7: 403–414. Responsible research also includes broader societal responsibilities.

Institute of Medicine. Committee on Assessing Integrity in Research Environments. (2002). *Integrity in Scientific Research: Creating an Environment That Promotes Responsible Conduct.* Washington, DC: National Academy Press. The scientific establishment focuses on the research environment to assure integrity.

Merton, Robert K. (1973). "The Normative Structure of Science." In *The Sociology of Science: Theoretical and Empirical Investigations,* ed. Norman Storer. Chicago: University of Chicago Press. Originally published in 1942. The *locus classicus* for the norms of the scientific community.

Mitroff, Irving I. (1974). *The Subjective Side of Science: A Philosophical Inquiry in the Psychology of the Apollo Moon Scientists.* Amsterdam: Elsevier. A non-Mertonian perspective on scientific norms.

Mulkay, Michael J. (1975). "Norms and Ideology in Science." *Social Science Information* 15: 637–656. Challenges Merton's norms.

Union of Concerned Scientists. (2004). *Scientific Integrity in Policymaking: An Investigation into the Bush Administration's Misuse of Science.* Cambridge, MA: Author. Independent group of scientists alleges that U.S. science has been politicized.

Ziman, John. (1990). "Research As a Career." In *The Research System in Transition,* eds. Susan Cozzens, Peter Healey, Arie Rip, and John Ziman. Boston: Kluwer Academic Publishers. Traditional Mertonian norms have been altered by new practices, including commercial relations, in science.

INTERNET RESOURCE

Waxman, Henry A. (2003). *Politics and Science in the Bush Administration.* Washington, DC: U.S. House of Representatives, Committee on Government Reform, Minority Staff. Available from www.house.gov/reform/min/politicsandscience/pdfs/pdf_politics_and_science_rep.pdf. Politicians allege that U.S. science has been politicized.

RESPONSIBILITY

• • •

Overview
Anglo-American Perspectives
German Perspectives

OVERVIEW

Ethical responsibility is one of the most commonly employed concepts in discussing the ethics of science and technology. Scientists have obligations for the "responsible conduct of research." The professional responsibility of engineers calls for attending to the public safety, health, and welfare consequences of their work. Entrepreneurs have responsibilities to commercialize science and technology for public benefit, and the public itself is often called on for the responsible support of science and technology. Consumers are admonished to be responsible users of technology. Yet the abstract noun *responsibility* is no more than 300 years old and has emerged to cultural and ethical prominence in association with modern science and technology from diverse legal, social, professional, religious, and philosophical perspectives.

Legal Responsibility

The legal term for responsibility is *liability*. Law makes explicit certain customary understandings of liability in two areas: criminal law and civil law. Criminal law deals with those offenses prosecuted and punished by the state. Civil law includes breaches of explicit or implicit contract in which injured parties may sue for compensation or damages.

Criminal liability was originally construed to follow simply from a transgression of the external forum of the law—doing something the law proscribes or not doing something it prescribes. But as it developed in Europe under the influence of a Christian theology of sin, which stresses the importance of inner consent, criminal liability was modified to include appreciation of the internal forum of intent. The result is a distinction between unintended transgressions such as accidental homicide and intentional acts such as first-degree murder; punishments for the former are less severe than for the latter.

In contrast to the historical development of restrictions on criminal liability, civil liability has expanded in scope through delimitations on the requirements for intentionality. Civil liability can be incurred by contract or it can be what is called "strict liability." In the case of explicit or implicit contract, intentional fault or negligence (a kind of failure of intention) must be proved. In the case of strict liability there need be no fault or negligence per se.

The concept of strict or no-faulty liability as a special kind of tort for which the civil law provides redress developed in parallel with modern industrial technology. In premodern Roman law, for instance, an individual could sue for damages only when losses resulted from intentional interference with person or property, or negligence. By contrast, in the English common law case of *Rylands v. Fletcher*, decided on appeal by the House of Lords in 1868, Thomas Fletcher was held liable for damages caused by his industrial undertakings despite their unintentional and nonnegligent character. Fletcher, a mill owner, had constructed a water reservoir to support his mills. Water from the reservoir inadvertently leaked through an abandoned mine shaft to flood John Rylands's adjacent mine. Although he admitted Fletcher did not and perhaps could not have known about the abandoned mine shaft, Rylands sued for damages. The eventual ruling in his favor argued that the building of a dam, which raised the water above its "natural condition," in itself posed a hazard for which Fletcher must accept responsibility.

In the early twenty-first century, the most common kinds of civil liability are just such no-fault or prima facie liabilities related to "nonnatural" industrial workplaces and consumer products in which activities or artifacts in themselves, independent of intent, pose special hazards. In the United States one of the key cases establishing this principle was that of *Greenman v. Yuba Power Products, Inc.*, decided on appeal by the California Supreme Court in 1963. In the words of Chief Justice Roger Traynor, in support of the majority:

> A manufacturer is strictly liable in tort when an article he places on the market ... proves to have a defect that causes injury to a human being.... The purpose of such liability is to insure that the costs of injuries resulting from defective products are borne by the manufacturers ... rather than by the injured persons who are powerless to protect themselves.

Religious Responsibility

The term *responsibility* derives from the Latin *respondēre*, meaning "to promise in return" or "to answer." As such it readily applies to what is perhaps the primordial experience of the Judeo-Christian-Islamic tradition: a call from God that human beings accept or reject. Given this reference—together with its regular embodiment in the "responsorials" of liturgical practice—it is remarkable that the term did not, until the twentieth century, play any serious role in European religious-ethical traditions.

The discovery and development of religious responsibility has again paralleled rising appreciation of the ethical issues emerging from science and technology. It is in opposition, for instance, to notions of secularization and control over nature that the Protestant theologian Karl Barth (1886–1968) distinguished between worldly and transcendent relationships. God is the wholly other, the one who cannot be reached by scientific knowledge. There is thus a radical difference between the human attempt to reach God (which Barth calls religion) and the human response to God's divine revelation (a response Barth identifies as faith). In his *Church Dogmatics* (1932) Barth goes so far as to identify goodness with responsibility in the sense of responding to God.

Catholic theologians have been no less ready to make responsibility central to ethics. For the Canadian Jesuit Bernard Lonergan (1904–1984), "Be responsible" is a transcendental precept coordinate with duties to "Be attentive," "Be intelligent," and "Be reasonable." Responsibility also plays a prominent role in the documents of Vatican II. At one point, after referencing the achievements of science and technology, *Gaudium et Spes* (1965) adds that, "With an increase in human powers comes a broadening of responsibility of the part of individuals and communities" (no. 34). Later, this same document on the church in the modern world suggests that, "We are witnesses of the birth of a new humanism, one in which man is defined first of all by his responsibility toward his brothers and toward history" (no. 55).

The most sustained effort to articulate a Christian ethics of responsibility is, however, that of H. Richard Niebuhr's *The Responsible Self* (1963). In this work Niebuhr contrasts the Christian anthropology of the human-as-answerer to the secular anthropologies of human-as-maker and human-as-citizen. For human-as-maker, moral action is essentially consequentialist and technological. For human-as-citizen, morality takes on a distinctly deontological character. With human-as-answerer, the tension between consequentialism and deontology is bridged by responsiveness to a complex reality, by an interpretation of the nature of this reality—and by an attempt to fit in, to act in harmony with what is already going on. "What is implicit in the idea of responsibility is the image of man-the-answerer, man engaged in dialogue, man acting in response to action upon him" (p. 56). Niebuhr's ethics of responsibility is what might now be called an ecological ethics.

Responsibility in Philosophy

The turn to responsibility in philosophy, like that in theology, exhibits two faces: first, a reaction to the challenge posed by the dominance of scientific and technological ways of thinking; and second, an attempt to take into account the rich and problematic complexity of technological practice. The first is prominent in Anglo-American analysis discourse, the second in European phenomenological traditions of thought.

According to Richard McKeon (1957), interest in the concept of responsibility can be traced to diverse philosophical backgrounds, one of which is the Greek analysis of causality (or imputability) and punishment (or accountability) for actions. As McKeon initially notes: "Whereas the modern formulation of the problem [of responsibility] begins with a conception of cause derived from the natural sciences and raises questions concerning the causality of moral agents, the Greek word for cause, *aitia* (like the Latin word *causa*), began

as a legal term and was then extended to include natural motions" (pp. 8–9). But it was in efforts to defend moral agency against threats from various forms of scientific materialism that the term became prevalent in analytic philosophy. For instance, H. L. A. Hart's distinctions between four kinds of responsibility—role, causal, liability, and capacity—(Hart 1968) are all related to issues of accountability as they arise in a legal framework, where they can help articulate a theory of punishment to meet the challenges posed by modern psychology.

McKeon's general thesis is that the term *responsibility* appeared in late-eighteenth and early-nineteenth-century moral and political discourse—as an abstract noun derived from the adjective *responsible*—in coordination with the expansion of democracy. But there are also numerous historical connections between the rise of democracy and the development of modern technology. On the theoretical level, the possessive individualism of *homo faber*, developed by Thomas Hobbes and John Locke, prepared the way for democracy and the new industrial order. On the practical level, democratic equality and technology clearly feed off one another.

But the connection goes deeper. According to McKeon, responsibility was introduced into the political context because of the breakdown of the old social order based on hierarchy and duty, and the inability of a new one to function based strictly on equality and self-interest. Whereas the former was no longer supported by the scientific worldview, the latter led to the worst exploitative excesses of the Industrial Revolution. To address this crisis there developed the ideal of relationship, in which individuals not only pursued their own self-interest but also tried to recognize and take into account the interests of others.

Something similar was called for by industrial technology. Good artisans, who dutifully followed the ancient craft traditions, were no longer enough, yet neither should they just be turned loose to invent as they pleased. Thomas Edison, after creating a vote register machine for a legislature, in which he subsequently discovered the legislature had no interest, resolved never again to invent simply what he thought the world needed without first consulting the world about what it wanted. The new artisan must learn to respond to a variety of factors—the material world, the economy, consumer demand, and more. This is what turns good artisans into responsible inventors and engineers. As their technological powers increase, so will their need to respond to an increasing spectrum of factors, to take more into account. Carl Mitcham (1994) has described this as a duty *plus respicere*, from the Latin to include more in one's circumspection.

Another argument to this effect is provided by John Ladd (1981) who, in considering the situation of physicians, argues that the expansion of biomedical technology has increased the private practitioner's dependence on technical services and undermined professional autonomy. Moral problems concerning physicians and society can no longer rest on an ethics of roles but involve the ethics of power, "the ethical side of [which] is responsibility" (p. 42).

The metaphysical elaboration of responsibility has taken place primarily in European philosophy. Lucien Levy-Bruhl's treatise titled *The Idea of Responsibility* (1884) is its starting point. After sketching a history of the idea from antiquity to the late nineteenth century, Levy-Bruhl argues surprise that a concept so basic to morality and ethical theory had not previously been subject to systematic investigation, especially since it is also manifested in a variety of ways across the whole spectrum of reality. There is responsibility or responsiveness at the level of physical matter, as atoms and molecules interact or respond to each other. Living organisms are further characterized by a distinctive kind of interaction or responsiveness to their environments and each other.

Extending this metaphysical interpretation Hans Jonas (1984), another philosopher in the European tradition, explored implications for science and technology. Responsibility is not a central category in previous ethical theory, Jonas argued, because of the narrow compass in premodern scientific knowledge and technological power. "The fact is that the *concept* of responsibility nowhere plays a conspicuous role in the moral systems of the past or in the philosophical theories of ethics." The reason is that "responsibility ... is a function of power and knowledge," which "were formerly so limited" that consequences at any distance "had to be left to fate and the constancy of the natural order, and all attention focused on doing right what had to be done now" (p. 123).

> All this has decisively changed. Modern technology has introduced actions of such novel scale, objects, and consequences that the framework of former ethics can no longer contain them.... No previous ethics had to consider the global condition of human life and the far-off future, even existence, of the race. These now being an issue demands ... a new conception of duties and rights, for which previous ethics and metaphysics provide not even the principles, let alone a ready doctrine. (pp. 6 and 8)

The new principle thus made necessary by technological power is responsibility, and especially a responsibility toward the future.

What for Jonas functions as a deontological principle, Caroline Whitbeck (1998) has argued may also name a virtue. When children are described as reaching "an age of responsibility," this indicates that they are able to "exercise judgment and care to achieve or maintain a desirable state of affairs" (p. 37). Acquiring the ability to exercise such judgment is to become responsible. At the same time, the term *responsibility* continues to name distributed obligations to practice such a virtue derived either from interpersonal relationships or from special knowledge and powers. "Since few relationships and knowledge are shared by everyone, most moral responsibilities are special moral responsibilities, that is, they belong to some people and not others" (p. 39).

Consideration of the special responsibilities that belong to scientists and engineers has been a major theme in advancing discussions of science, technology, and ethics. Although overlapping, these two discussions have nevertheless mostly taken place among different professional groups.

Scientific Responsibility

Efforts to define the social responsibility of scientists have involved an refinement of the representative Enlightenment view that science has the best handle on truth and is thus essentially and under all conditions beneficial to society. From such a perspective, the primary responsibility for scientists is thus to pursue and extend their disciplines.

Historically this responsibility found expression in Isaac Newton's hope for science as theological insight, Voltaire's belief in its absolute utility, and Benedict de Spinoza's thought that in science one possesses something pure, unselfish, self-sufficient, and blessed. A classic manifestation is the great French *Encyclopédie* (1751–1772), which sought "to collect all the knowledge that now lies scattered over the face of the earth, to make known its general structure to the men among whom we live, and to transmit it to those who will come after us." Such a project, wrote Denis Diderot, demands "intellectual courage."

The questioning of this tradition has roots in the Romantic critique of scientific epistemology and industrial practice, but did not receive a serious hearing among scientists themselves until after World War II. Since then one may distinguish three phases.

PHASE ONE: RECOGNIZING RESPONSIBILITIES. In December 1945 the first issue of the *Bulletin of the Atomic Scientists* led off with a statement of the goals of the newly formed Federation of Atomic (later American) Scientists. Members should "clarify . . . the . . . responsibilities of scientists in regard to the problems brought about by the release of nuclear energy" and "educate the public [about] the scientific, technological, and social problems arising from the release of nuclear energy." Previously scientists would have described their responsibilities as restricted to doing good science, not falsifying experiments, and cooperating with other scientists. Now, because of the potentially disastrous implications of at least one branch of science, scientists felt their responsibilities enlarge. They were called on to take into account more than the procedures of science; they must respond to an expanded situation.

The primary way that atomic scientists responded over the next decade to the new situation created by scientific weapons technology was to work for placing nuclear research under civilian control in the United States and to further subordinate national to international control. They did not, however, oppose the unprecedented growth of science. As Edward Teller wrote in 1947, the responsibility of the atomic scientists was not just to educate the public and help it establish a civilian control that would "not place unnecessary restrictions on the scientist," it was also to continue to pursue scientific progress. "Our responsibility," in Teller's words, "is [also] to continue to work for the successful and rapid development of atomic energy" (p. 355).

PHASE TWO: QUESTIONING RESPONSIBILITY. During the mid-1960s and early 1970s, a second-stage questioning of scientific responsibility emerged. Initially this questioning arose in response to the growing recognition of the problem of environmental pollution—a phenomenon that cannot be imagined as alleviated by simple demilitarization of science or increases in democratic control. Some of the worst environmental problems are caused precisely by democratic availability and use—as with pollution from automobiles, agricultural chemicals, and aerosol sprays, not to mention the mounting burden of consumer waste disposal. Rachel Carson's *Silent Spring* (1962) was an early statement of the problem that called for an internal transformation of science itself. But an equally focal experience during this second-stage movement toward an internal restructuring of science was the Asilomar Conference of 1975, which addressed the dangers of recombinant DNA research.

After Asilomar, the dangers of recombinant DNA research turned out to be not as immediate or as great as feared, and some members of the scientific community became resentful of post-Asilomar agitation—although others actually argued for even more stringent guidelines than those proposed (Sinsheimer 1976,

1978). Increased possible consequences nevertheless again broadened the scope of what could be debated as the proper responsibility of scientists. Robert L. Sinsheimer, for instance, himself a respected biological researcher and chancellor of the University of California, Santa Cruz, argued that modern science was based on two faiths. One is "a faith in the resilience of our social institutions . . . to adapt the knowledge gained by science . . . to the benefit of man and society more than the detriment"—a faith that "is increasingly strained by the acceleration of technical change and the magnitude of the powers deployed" (Sinsheimer 1978, p. 24). But even more telling is

> a faith in the resilience, even in the benevolence, of Nature as we have probed it, dissected it, rearranged its components in novel configurations, bent its forms, and diverted its forces to human purpose. The faith that our scientific probing and our technological ventures will not displace some key element of our protective environment, and thereby collapse our ecological niche. A faith that Nature does not set booby traps for unwary species. (Sinsheimer 1978, p. 23)

This new argument was commensurate with the development of what Jerome R. Ravetz (1971) saw as the replacement of "academic science" by "critical science"—which is in turn related to what others have termed public interest science. Or as William W. Lowrance (1985) argued, beyond responsibility in the first-stage sense, there is a need to incorporate in science itself what he referred to as principles of "stewardship."

PHASE THREE: REEMPHASIZING ETHICS. The attempt to transform science from within was overtaken in the mid-1980s by a new external criticism not of scientific products (knowledge) but of scientific processes (methods). A number of high-profile cases of scientific misconduct raised questions about whether public investments in science were being wisely spent. Were scientists simply abusing a public trust? Moreover, some economists began to question whether, even insofar as scientists did not abuse the public trust, but followed ethical research practices—which was surely mostly the case—scientific research was as much of a stimulus to economic progress as had been thought.

The upshot was that the scientific community undertook a self-examination of its ethics and its efficiency. Efforts to increase ethics education, or education in what became known as the responsible conduct of research, became required parts of science education programs, especially in the biomedical sciences at the graduate level. And increased efficiency in grant administration and management became issues for critical assessment. Since the 1990s scientists have increasingly been understood to possess social responsibilities that include the promotion of ethics and efficiency in the processes of doing science.

At the same time, scientists have also attempted to reemphasize the importance of science to national health care, the economy, environmental management, and defense. In the face of the AIDS epidemic, biomedical research presents itself as the only answer. Computers and biotechnologies are offered as gateways to new international competitive advantage and the creation of whole new sectors of jobs. Global climate change, it is argued, can be adequately assessed only by means of computer models and the science of complexity. Finally, especially since 9/11, new claims have been made for science as a means to develop protections against the dangers of international terrorism. The social responsibility of science is defended as the ethically guided production of knowledge that addresses a broad portfolio of social needs: the promotion of health, the creation of jobs, the protection of the environment, and the defending of Western civilization.

Engineering Responsibility

Applied science professionals such as technologists and engineers are more subject than scientists to both external (legal, political, or economic) and internal (ethical) regulation. Indeed, engineers have since the early twentieth century attempted to formulate explicit principles of professional responsibility—precisely because of the technological powers they wield. Historically, similar discussions did not originate among scientists until the second half of the twentieth century, and scientific organizations remain in the early twenty-first century less likely to have formal codes of conduct than engineering associations.

Engineering associations aspire to the formulation of codes of conduct similar to those found in medicine or law. But unlike medicine, which is ordered toward health, or law, the end of which is justice, it is less obvious precisely what constitutes the engineering ideal that could serve as the basis for a distinctive internalist ethics of responsibility. The original engineer (Latin *ingeniator*) was the builder and operator of battering rams, catapults, and other "engines of war." Engineering was originally military engineering. As such, the power of engineers, no matter how great, was significantly less than the organized strength of the army as a whole. Moreover, as with all other soldiers, their behavior was guided primarily by their obligations to obey hierarchical authority.

The eighteenth-century emergence of civil engineering in the design of public works such as roads, water supply and sanitation systems, lighthouses, and other nonmilitary infrastructures did not initially alter this situation. Civil engineers were only small contributors to larger processes. But as technological powers in the hands of engineers began to enlarge, and the number of engineers increased, tensions mounted between subordinate engineers and their superiors. The manifestation of this tension is what Edwin T. Layton Jr. (1971) called the "revolt of the engineers," which occurred during the late nineteenth and early twentieth centuries. It is in association with this revolt and its aftermath that *responsibility* enters the engineering ethics vocabulary.

One influential if failed effort at formulating engineering responsibility led to what was known as the technocracy movement and its idea that engineers more than politicians should wield political power. Henry Goslee Prout, a former military engineer who had become general manager of the Union Switch and Signal Company, speaking before the Cornell Association of Civil Engineers in 1906, described the profession in just such leadership terms: "The engineers more than all other men, will guide humanity forward.... On the engineers ... rests a responsibility such as men have never before been called upon to face" (quoted in Akin 1977, p. 8). At the height of this dream of expanded engineering responsibility, Herbert Hoover became the first civil engineer to be elected president of the United States, and an explicit technocracy movement fielded its own candidates for elective office. The ideology of technocracy sought to make engineering efficiency an ideal analogous to medical health and legal justice.

During World War II a different shift took place in the engineering conception of responsibility: not from company and client loyalty to technocratic efficiency but from private to public loyalty. A chastened version of responsibility nevertheless emphasized the potential for opposition between social and corporate interests. Having failed in trying to be responsible for everything, engineers came to debate the scope of more limited responsibilities—to themselves, to employers, and to the public. The need for this debate is still clearly dictated by the powers at their command and the problems such powers pose, even though it is not obvious that engineering entails responsibilities of any specific character.

With engineering under attack as a cause of environmental pollution, for the design of defective consumer goods, and as too willing to feed at the trough of the defense contract, one American engineer writing in the mid-1970s summed up the situation as follows. He first admitted that,

> Unlike scientists, who can claim to escape responsibility because the end results of their basic research can not be easily predicted, the purposes of engineering are usually highly visible. Because engineers have been claiming full credit for the achievements of technology for many years, it is natural that the public should now blame engineers for the newly perceived aberrations of technology. (Collins 1973, p. 448)

In other words, engineers had oversold their responsibilities and were being justly criticized. The responsibilities of engineers are in fact quite limited. They have no general responsibilities, only specific or special ones:

> There are three ways in which the special responsibility of engineers for the uses and effects of technology may be exercised. The first is as individuals in the daily practice of their work. The second is as a group through the technical societies. The third is to bring a special competence to the public debate on the threatening problems arising from destructive uses of technology. (Collins 1973, p. 449)

This debate, formalized in various technology assessment methodologies and governmental agencies, can be read as a means of subordinating engineers to the larger social order. In comparing responsibility in engineering with responsibility in science, it may thus appear that there has been more of a contraction than an expansion. Yet the issue of responsibility has so intensified that engineers now consciously debate the scope of their responsibilities in relationship to issues not previously acknowledged.

Too Much Responsibility?

One common worry about certain technologies is that they undermine human responsibility. For instance, reliance on computers in medical diagnostic processes or strategic missile defense systems transfers some decision making responsibilities from human beings to computers. But the same computer systems that assume practical responsibility for diagnosis or defense call for the exercise of a higher ideal of responsibility in their design and deployment. It is precisely because modern technology calls for so much responsibility at the ideal level that observers can be so sensitive to the issue at the practical level. It is not at all clear, for instance, that computers have in any way deprived human beings of

responsibilities they formerly had. What physicians of the early nineteenth-century would have been responsible for diagnosing and then treating the array of obscure diseases for which twenty-first-century physicians are held accountable? It is more likely that new technologies make possible certain responsibilities which they can also be configured to assist.

But this raises a question: Are the responsibilities thus called forth truly reasonable? From the perspective of prudence, one should not take on or give to another too much responsibility. To do so is to invite failure if not disaster. Although exact boundaries are not easy to determine in advance, once overstepped they are difficult to recover. In light of this principle of prudence, then, one must ask: Can the principle of responsibility, and those who are called to live up to it, really bear the added burden being placed on it and them by contemporary science and technology?

CARL MITCHAM

SEE ALSO *Christian Perspectives; Engineering Ethics; Responsible Conduct of Research.*

BIBLIOGRAPHY

Akin, William E. (1977). *Technocracy and the American Dream: The Technocrat Movement, 1900–1941.* Berkeley and Los Angeles: University of California Press.

Collins, Frank. (1973). "The Special Responsibility of Engineers." In "The Social Responsibility of Engineers," ed. Harold Fruchtbaum. Spec. issue, *Annals of the New York Academy of Sciences*, 196(10): 448–450.

Durbin, Paul T., ed. (1987). "Technology and Responsibility." In *Philosophy and Technology*, vol. 3. Boston: Dordrecht, Netherlands: D. Reidel. Seventeen papers from a conference; includes an annotated bibliography on the theme.

Greenman v. Yuba Power Products, Inc., 59 Cal. 2d 57 (1963).

Hart, H. L. A. (1968). *Punishment and Responsibility: Essays in the Philosophy of Law.* New York: Oxford University Press. See especially the "Postscript: Responsibility and Retribution."

Jasanoff, Sheila. (1995). *Science at the Bar: Law, Science, and Technology in America.* Cambridge, MA: Harvard University Press. Good on changes in the character of tort law in relation to technology.

Jonas, Hans. (1984). *The Imperative of Responsibility: In Search of an Ethics for the Technological Age*, trans. Hans Jonas and David Herr. Chicago: University of Chicago Press.

Ladd, John. (1981). "Physicians and Society: Tribulations of Power and Responsibility." In *The Law-Medicine Relation: A Philosophical Exploration*, ed. Stuart F. Spicker, Joseph M. Healey Jr., and H. Tristram Engelhardt Jr. Dordrecht, Netherlands: D. Reidel.

Layton, Edwin T., Jr. (1971). *The Revolt of the Engineers*. Cleveland: Press of Case Western Reserve University. Reprint, with new introduction, Baltimore: Johns Hopkins University Press, 1986.

Levy-Bruhl, Lucien. (1884). *L'idée de responsabilité* [The idea of responsibility]. Paris: Hachette.

Lowrance, William W. (1985). *Modern Science and Human Values*. New York: Oxford University Press.

McKeon, Richard. (1957). "The Development and the Significance of the Concept of Responsibility." *Revue Internationale de Philosophie* 11(1:39): 3–32.

Mitcham, Carl. (1987). "Responsibility and Technology: The Expanding Relationship." In *Technology and Responsibility*, ed. Paul T. Durbin, pp. 3–39. Dordrecht, Netherlands: D. Reidel. The present analysis is based on this earlier work.

Mitcham, Carl. (1994). "Engineering Design Research and Social." In Kristin Shrader-Frechette, *Ethics of Scientific Research*, pp. 153–168. Lanham, MD: Rowman and Littlefield.

Niebuhr, H. Richard. (1963). *The Responsible Self*. San Francisco: Harper and Row.

Ravetz, Jerome R. (1971). *Scientific Knowledge and Its Social Problems*. Oxford: Clarendon Press.

Rylands v. Fletcher, LR 3 HL 330 (1868).

Sinsheimer, Robert L. (1976). "Recombinant DNA—on Our Own." *BioScience* 26(10): 599.

Sinsheimer, Robert L. (1978). "The Presumptions of Science." *Daedalus* 107(2): 23–35.

Teller, Edward. (1947). "Atomic Scientists Have Two Responsibilities." *Bulletin of the Atomic Scientists* 3(12): 355–356.

U.S. Committee on Science, Engineering, and Public Policy. Panel on Scientific Responsibility and the Conduct of Research. (1992–1993). *Responsible Science: Ensuring the Integrity of the Research Process*. 2 vols. Washington, DC: National Academy Press.

Whitbeck, Caroline. (1998). *Ethics in Engineering Practice and Research*. Cambridge, UK: Cambridge University Press.

ANGLO-AMERICAN PERSPECTIVES

In the English language *responsibility* is generally defined as a quality or state of being answerable or accountable for acts or decisions. However, the term *responsibility* and its cognatex *responsible* are used in a variety of ways. H .L. A. Hart illustrated that variety with the following story of a drunken sea captain who lost his ship at sea.

> As captain of the ship, X was responsible for the safety of his passengers and crew. But on his last voyage he got drunk every night and was responsible for the loss of the ship with all aboard. It was rumoured that he was insane, but the doctors considered that he was responsible for his actions. Throughout the voyage he behaved quite irre-

sponsibly, and various incidents in his career showed that he was not a responsible person. He always maintained that the exceptional winter storms were responsible for the loss of the ship, but in the legal proceedings brought against him he was found criminally responsible for his negligent conduct, and in separate civil proceedings he was held legally responsible for the loss of life and property. He is still alive and he is morally responsible for the deaths of many women and children" (Hart 1968, p. 211).

Four Types of Responsibility

Hart uses this story to identify four different senses of responsibility: role responsibility, causal responsibility, liability responsibility, and capacity responsibility. Role responsibility refers to the duties and obligations a person has by virtue of occupying a role such as mother, doctor, or captain of a ship. When a person occupies a role, others expect certain kinds of behavior and hold that person accountable for failure to do what is expected. In this context individuals have duties to behave in certain ways that can be referred to as role responsibilities. Causal responsibility is attributed to things and events as well as persons. In the case of events one might say of the terrorist attack on September 11, 2001, that the event has been causally responsible for instilling fear in many U.S. citizens. In the case of persons a particular action by a person is specified as the cause of or the major causal contribution to an untoward event or occurrence. For example, a person's failure to stop at a stop sign may be said to be causally responsible for the ensuing accident. Causal responsibility may or may not be connected to blameworthiness. Thus, if the person failed to stop at the stop sign because she had a heart attack, she may not be blameworthy but her failure to stop is still causally responsible for the accident. Similarly, even if a person unknowingly or under coercion pressed a button that detonated a bomb, that person would be causally responsible for the resulting damage.

Liability responsibility often refers to legal liability and identifies the person or group that is expected to pay damages or make compensation or sometimes explain (give an account of what happened) in situations in which harm is done. Liability often but not always accompanies causal responsibility or blameworthiness. Strict liability refers to holding an individual liable—to pay damages, make compensation, or give an explanation of what happened—when that individual is not causally connected to the event and has done nothing wrong. An example would be holding a company liable for harm that resulted from a defect in one of its products despite the fact that the company did everything possible to make the product safe. Capacity responsibility refers to the capability (generally psychological) a person must possess to be considered morally responsible for his or her behavior. For example, if an individual lacked the ability to reason and to understand and control his or her behavior, it would be inappropriate to hold that person responsible for his or her actions.

In describing this fourfold distinction it is helpful to bring in the notion of blameworthiness. Being blameworthy or at fault is another sense of responsibility that depends on the other uses of that term. A person typically is considered blameworthy when (1) the person had capacity responsibility (that is, had the ability to understand and control his or her behavior); (2) the person did something he or she was not supposed to do (such as fail to perform a role-responsibility); and (3) the person's act or omission was causally responsible for an untoward event or harm. For example, a person would be blameworthy if while working as a night security person for a bank (and having the capacities of most human beings) he or she forgot to check to see if a door was properly locked and consequently allowed a burglar to get into the bank and steal money.

In addition to Hart's fourfold distinction and the concept of blameworthiness, moral philosophers have distinguished many different kinds of responsibility, including personal, collective, moral, legal, diminished, prospective, and retrospective responsibility. Thus, discussions of responsibility must attend carefully to the differing meanings of the term.

Analytic moral philosophers have focused largely on capacity responsibility and especially the connection between freedom and responsibility. For individuals to be responsible for their behavior, it would seem that they must be free to act as they do. If individual behavior were entirely determined, say, because it is predetermined by God or results from external causal forces such as genetics, upbringing, and circumstances, it would seem that individuals could not be held responsible for what they do: Their behavior is not in their control.

With this in mind, moral philosophers have focused on giving an account of human freedom without denying the various factors that influence human behavior. Often scholarship on this topic has focused on what it means to say that a person is free or "could have done otherwise."

By contrast, some philosophers have argued that ascriptions of responsibility should be seen as forward-looking (prospective) social practices. In this context human freedom is not a requirement. For example, ascriptions of responsibility can be understood to be mechanisms for exerting pressure on individuals to behave in certain ways. Society holds individuals responsible for their behavior to exert pressure on them to behave in socially desirable ways. When individuals behave in socially undesirable ways, society disapproves and tells them they are bad. Society uses the law to threaten and actually punish individuals when they engage in undesirable behavior. This is done to instill in individuals a sense of responsibility for their actions, a sense of responsibility that influences how they behave. Understanding responsibility in this way gives responsibility ascriptions a utilitarian and deterministic foundation. Responsibility ascriptions are utilitarian practices aimed at achieving good results. This account eliminates an element at the heart of notions of responsibility and at the core of the connection between freedom and being human: a sense that what it means to be human involves carrying the weight of responsibility for one's actions.

Responsibility in Science and Technology

A host of important responsibility issues arise in the fields of science, engineering, and technology. The issue that has received the most attention involves the responsibilities of scientists and engineers for the production of scientific knowledge and technological products. Because science and engineering give human beings enormous power for good and ill, questions about the responsibility of scientists and engineers, both individually and collectively, have always surrounded scientific and technological endeavors. The question became particularly prominent in the twentieth century with the creation and use of the first atomic bomb and later with the production of civilian nuclear power. The question persists in the early twenty-first century in regard to genetic engineering, surveillance technologies, cloning, and biological weapons. Are scientists and engineers considering the social and moral implications of what they are doing? Do they have a responsibility to stop what they are doing or to speak out when they think the risks of their work or that of their colleagues are too great?

Evidence of concern about the scientists' or engineers' responsibility for their work is seen, for example, in the ongoing fascination with Mary Shelley's *Frankenstein* (1818), a science fiction story in which a doctor-scientist uses scientific and technical prowess to bring a humanlike monster composed of separately acquired body parts to life. Doctor Frankenstein is horrified at the sight of his creation and immediately flees his laboratory; he does nothing until the beast begins to interfere in his life. Left to its own devices the beast wreaks havoc on the lives of Doctor Frankenstein and others.

The Frankenstein story is an indictment of those who fail to think about the implications of their attempts to create new knowledge, products, and techniques; it is an indictment of those who refuse to take responsibility for what they create. Whatever Mary Shelley's intentions were in writing *Frankenstein*, the story serves as a morality tale for a technoscientific world. Its relevance to a world in which biological weapons, clones, and powerful surveillance technologies have already been created is evident.

Failure and Disaster

The Frankenstein story suggests that scientists and engineers should consider the implications of their work before they do it and take responsibility for that work after it is done. More often than not responsibility issues arise after knowledge has been created and technological endeavors have been undertaken and some sort of failure subsequently leads to a disaster. Then attempts are made to trace back role responsibilities and identify who is to blame. For example, when the *Challenger* spaceship and more recently the *Columbia* crashed, public attention turned to figuring out what went wrong and who was responsible. Engineers, as well as managers, were put on the spot. Who made the decision to launch? Were there not signs that a problem existed? Who had failed to fulfill their responsibilities?

Similar questions arise for all technological failures, especially those which have catastrophic results, such as the Three Mile Island accident; the disaster at Bhopal, India; the DC10 airplane crash; and the Hyatt Regency hotel collapse. After September 11, 2001, questions were raised about the structural design of the World Trade Center as well as the failure of American intelligence organizations.

Although responsibility issues can and do arise independent of science and technology, the issues surrounding technological disasters seem particularly daunting because of their complexity. Modern technologies are so complex that the individuals involved in their development, production, distribution, and use often cannot understand fully the projects to which they are contributing. Because of that complexity there must be a division of labor, and this means that engineers and scientists often work on pieces of a larger project. This challenges traditional notions of responsibility, for how can individuals be responsible for what they are doing when they cannot fully comprehend what they are doing?

Information technology is a good example of this issue. Many computer programs consist of millions of lines of computer code. Can a single individual be responsible for all the lines of code in a program? No one can be expected to understand the entire program, and so how can particular individuals be held responsible for the program? Computer scientists develop testing procedures and standards for reliability, but there are limits to what they can be expected to do. Moreover, when projects are divided into parts, there is a danger of something falling into the cracks or of error being introduced when the parts are put together. The complexity of modern technologies poses daunting challenges both retrospectively in tracing back failure and prospectively in assigning responsibilities for large projects in a way that minimizes the likelihood of failure.

Many scientific and engineering professional associations acknowledge that their members have social and professional responsibilities both individually and collectively. Professional organizations are an important means of addressing some of those responsibilities. One method professional societies use is to adopt and promulgate codes of ethical and professional conduct.

DEBORAH G. JOHNSON

SEE ALSO *Engineering Ethics; Normal Accidents; Unintended Consequences.*

BIBLIOGRAPHY

Feinberg, Joel. (1970). *Doing and Deserving.* Princeton, N.J.: Princeton University Press.

Hart, Herbert Lionel Adolphus. (1968). "Chapter IX Postscript: Responsibility and Retribution." In *Punishment and Responsibility: Essays in the Philosophy of Law.* Oxford: Clarendon Press.

Johnson, Deborah G. (1989). "The Social/Professional Responsibility of Engineers." *Annals of The New York Academy of Sciences* 577: 106–114.

Shelley, Mary. (1995). *Frankenstein.* New York, Pocket Books. Originally published in 1818.

Zimmerman, Michael J. (2001). "Responsibility." In *Encyclopedia of Ethics*, 2nd edition, ed. L. C. Becker and C. B. Becker. New York: Routledge.

GERMAN PERSPECTIVES

In the German philosophical tradition the concept of responsibility (*Verantwortung*) has been accorded special and extensive treatment, especially in relation to science and technology. The following introduction to this tradition begins with a description of responsibility as a relational construct and then distinguishes three basic levels of responsibility: action responsibility, role responsibility, and universal moral responsibility.

Basic Concept

The German word *Verantwortung* derives from the Middle High German and originally meant simply "to answer," probably in response to an accusative question such as "Did you do X?" The concept of responsibility is thus evaluative and attributive as well as descriptive. A person can be *held* (to be) responsible, which introduces the normative or ethical dimension into human experience.

The concept of responsibility implies a multidimensional structure linked to assignment, attribution, and imputation, in ways that may be analyzed and interpreted with respect to the following model:

Someone S (the subject or bearer of responsibility, which can be a person or a corporation)

is responsible for A (actions, consequences, situations, tasks)

to O (addressees or "objects" of responsibility)

under the supervision or judgment of J (some judging or sanctioning agent)

in relation to N (a prescriptive or normative criterion of attribution)

and accountable within context C (a sphere or realm of human activity).

For example, a person (S) is responsible to other motorists and pedestrians (O) for stopping at traffic lights (A) under the supervision of the police or courts (J) in relation to the traffic laws (N) when driving an automobile (C). This makes responsibility a five- or six-place relation, although some of the relations may overlap. For instance, it is possible for an addressee (O) and supervisor (J) to be the same.

Following work in the development of attribution theory by the social psychologist Fritz Heider (1896–1988) and the social phenomenologist Alfred Schutz (1899–1959), it was the Polish logician I. M. Bochenski (1987 [1947]) who first defined responsibility in terms of the logic of relations. For Bochenski, however, responsibility was a two- or perhaps a three-place relation: Someone (S) is responsible for action (A) to another person (O).

As an attributive, relational construct, responsibility is also an interpretative concept with social func-

tions. It can be expressed as an attributive, relational norm (controlling expectations regarding action and behavior). Responsibility further implies that a person (S) must justify actions, action consequences, situations, tasks, and so forth (A) in front of an addressee (O) and before an agent (J) in respect to which the responsible party has obligations or duties in accordance with standards, criteria, or laws (N). Responsible parties are accountable for their own actions or under specified conditions for the actions of others. Parents, for example, are liable for certain behaviors of their children, and corporations for certain behaviors of their employees. (This tends to apply more to wrongdoings than to achievements.) The concept of responsibility thus structures social reality and social relations.

One may further differentiate between the typical bearers of responsibility in terms of active roles and observer roles. Specifically, one may impute or attribute a particular responsibility to oneself as an actor or to others from the multiple perspectives of a participant, observer, or scientist, in relation to general rules and norms. Particular cases of attribution instantiate general patterns of responsibility. The attribution of responsibility is an active process both in self-interpretation and in the interpretation of the actions of others. The concept of responsibility is thus implicated in self-understandings and projections of ideals for social order.

Types and Levels of Responsibility

Types of responsibility occur at three basic levels: individual actions, social roles, and universal moral principles. Such distinctions are justified by appeal to "ideal typ(ic)al" prevalence, similar but not identical to Max Weber's *Idealtypen* or ideal types. In what follows, diagrammatic schema are used to condense and illustrate hierarchical models of different types of responsibility, with different levels or strata referring to different dimensions of interpretation. The first diagram is more abstract and calls for more interpretative constructs, such as particular kinds of responsibilities, than the others.

QUALIFICATIONS. In general, the three levels are constituted by analytic and perspectival constructs that may overlap and all apply (although in different ways) to a single real case of responsibility. That is, concrete instances of responsibility attribution may be analyzed not only on a formal or abstract level (as illustrated in the first diagram) but also from a more concrete point of view (as with role or moral responsibility). Although usually any one analysis on a specific level is tied to a certain interpretation (e.g., some particular role), this does not preclude another interpretation (from, say, the moral point of view).

Within the different levels of these schematic constructs are further analytic constructs that are also able to be attributed to individuals or groups. Even in their more concrete forms, constructs are to be understood as analytic distinctions. That is, collective or group responsibility seldom precludes individual or personal responsibility, although collective responsibility cannot be reduced to or derived from individual or personal responsibility alone. The same applies to institutional responsibility. Moreover, there are conceptual connections or analytic relations between some juxtaposed or subordinated subtypes.

ACTION RESPONSIBILITY. The most obvious and general level of responsibility is that which involves being responsible for the results or consequences of one's own actions. This may be termed the prototyp(ic)al case of (causally oriented) action responsibility. A subject is held responsible for the outcomes of his or her actions in an instance for which he or she is accountable. An engineer designing a bridge or a dam is responsible to the supervisor, employer, client, and/or general public for his or her design in terms of technical correctness, safety, cost, feasibility, and more. A scientist is not responsible for the outcome of an experiment or research project but is responsible for the conduct of the research and the reporting of its results.

Frequently, accountability questions are raised in negative cases, when one or more of these criteria are not fulfilled. The breaking of a dam may be the result of such factors as honest mistakes in statics or dynamics analyses; careless, negligent, or even criminal misconduct; incompetence; and the use of substandard materials. The need to withdraw or revise technical reports in science may likewise be attributable to honest mistakes or malfeasance. In any particular case it is important to identify the particular negative action responsibility. Professional scientists and engineers have responsibilities to the public to ensure high standards in their work, to avoid risks of disasters insofar as this is compatible with reasonable costs, and to report results fully and completely without fabricating or falsifying data. The responsibility to avoid mistakes, failures, and poor quality products, processes, systems, and so on is part and parcel of action responsibility. Different types of action responsibility are shown in Figure 1.

The most commonly discussed cases of action responsibility are individual action responsibility. But if a group is acting collectively or if individuals participate in joint group action, then what may be called coresponsibility arises as a distinctive phenomenon. Coresponsibility is the sharing of responsibility by participating members in a group action. Responsibility for group actions is also sometimes called collective or group responsibility, and the circumstances in which this can be legitimately attributed to groups—especially large ones such as a nation-state or ethnic classification—are highly contentious. Mostly such attributions are rejected or justified only under very special cases on the grounds that groups should not be punished (or rewarded) for the actions of individuals. In practice, however, such punishments are quite common (as in warfare where they may be apologized for as "collateral damage").

ROLE RESPONSIBILITY. A second level of responsibility is constituted by role and universal moral responsibility. In accepting a role or fulfilling a task (e.g., by taking on a well-defined job) a role holder usually bears some responsibility for acceptable or optimal role fulfillment. Role responsibility is not opposed to or fundamentally different than individual action responsibility, but manifests action responsibility at a level other than that of human action as such. Indeed, as the examples already cited in discussing action responsibility indicate, most of these roles will entail individual action responsibilities, or can be thought of as constituting particular instances of individual action responsibility.

These roles or duties might be assigned in a formal way or be more or less informal. They can even be legally ascribed or at least legally relevant. Different types of roles and responsibilities, including legal responsibilities, are presented in diagrammatic form in Figure 2.

In corporate or institutional settings, role patterns include leadership responsibility (with respect to external and internal instances, addressees, and agents) as a special form of associated institutional role responsibility. In addition, there is the corporate responsibility of firms, corporations, or other social institutions such as government agencies and even nongovernmental organizations insofar as these have special tasks to perform or obligations to fulfill with respect to clients, the public, or members of the organization or corporation. This type of responsibility can also have a legal, moral, or neutral character, which may or may not coincide with group or institutional responsibility.

Other examples of role responsibility that deserve explicit mention include not only legal responsibility but also pedagogical responsibility, religious responsibility, political (citizen) responsibility, and more. In an

FIGURES 1–2

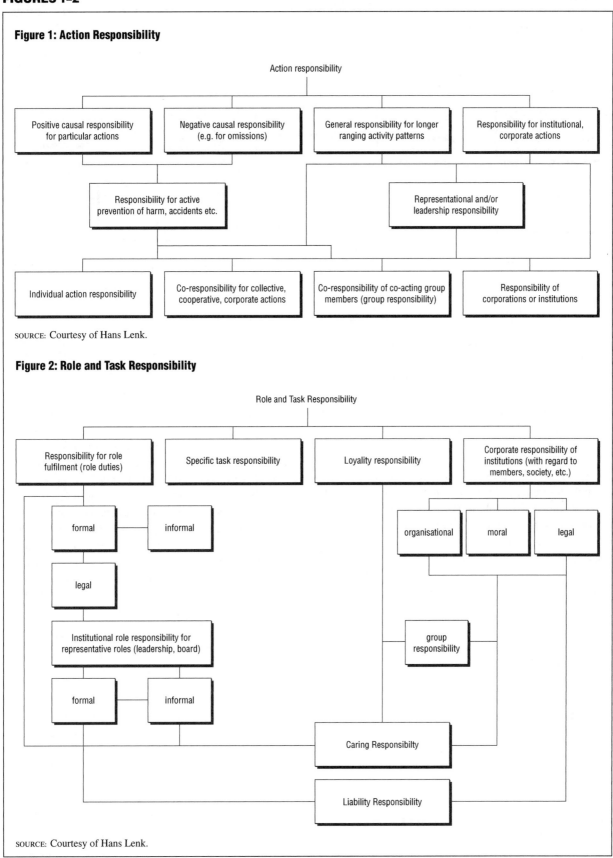

Figure 1: Action Responsibility

Action responsibility

Positive causal responsibility for particular actions

Negative causal responsibility (e.g. for omissions)

General responsibility for longer ranging activity patterns

Responsibility for institutional, corporate actions

Responsibility for active prevention of harm, accidents etc.

Representational and/or leadership responsibility

Individual action responsibility

Co-responsibility for collective, cooperative, corporate actions

Co-responsibility of co-acting group members (group responsibility)

Responsibility of corporations or institutions

SOURCE: Courtesy of Hans Lenk.

Figure 2: Role and Task Responsibility

Role and Task Responsibility

Responsibility for role fulfilment (role duties)

Specific task responsibility

Loyality responsibility

Corporate responsibility of institutions (with regard to members, society, etc.)

formal

informal

organisational

moral

legal

legal

group responsibility

Institutional role responsibility for representative roles (leadership, board)

formal

informal

Caring Responsibilty

Liability Responsibility

SOURCE: Courtesy of Hans Lenk.

FIGURE 3

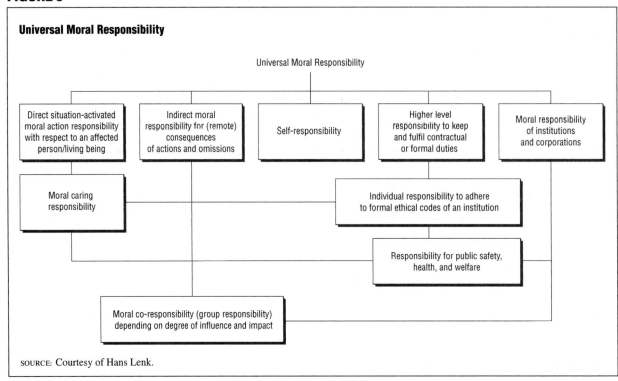

Universal Moral Responsibility

SOURCE: Courtesy of Hans Lenk.

advanced scientific and technological society one might also speak of consumer responsibility.

UNIVERSAL MORAL RESPONSIBILITY. Universal moral responsibility provides a different specification for the functioning of individual action responsibility than that associated with role responsibility. Not all action responsibility and role responsibility is specifically moral in character or moral to the same degree. To have a responsibility to be on time for an appointment because of a particular role has more an efficiency than a moral character; it is a responsibility that keeps some particular organizational system functioning more smoothly than would otherwise be the case.

Action responsibility and role responsibility take on a specifically moral character when an agent's actions and the results of those actions are directed toward persons or living beings (including even the agent) whose well-being is directly affected by the agent's activity. With regard to others such affects can be direct or indirect, can be defined by contractual or formal duties, and can inhere in institutions or corporations. By way of diagrammatic summary, see Figure 3.

For Hans Jonas (1984) universal moral responsibility can become pronounced with regard to the uses of technology that have the potential for environmental or human destruction such as nuclear weapons or genetic

engineering. Caring responsibility is not only role related (with different kinds of scientists or engineers exhibiting it in different degrees) but also general for those who inhabit a highly scientific and technological society—that is, those who promote and benefit from advanced scientific and technological activities. According to Jonas's argument, members of a scientific and technological society, by virtue of participating in such a society, and because of the tremendous potential for intentional and nonintentional destruction present in the society, become responsible for ensuring the well-being of all persons and other living beings affected by their specific actions in the form of a general and permanent obligation.

A few more restricted observations on various types of universal moral responsibility related especially to science and technology are as follows:

- The remote consequences of an agent's activity—possibly combined with the impacts of other people's commissions or omissions—may create an indirect moral (co-) responsibility. For instance, neglecting a safety check or wrongly certifying the airworthiness of an airplane could contribute to loss of life when coupled with a less than expert pilot or other crew member.

- Corporate moral responsibility frequently coincides with, but need not be identical to, the moral coresponsibility of members of a decision-making board. Therefore corporate moral responsibility is not to be analytically confused with the moral coresponsibility of group members partaking in a collective action or decision-making process. (Questions of responsibility distribution are increasingly important in assessing responsibility in the virtual environments created by computers and information systems, where teams of programmers have created web-based utilities in which people differentially interact to produce multiple types of products.)

- To abide by the ethics code of a professional society is a combination of indirect responsibilities. As such it is certainly a moral obligation. Thus beside immediate action- or impact-oriented responsibilities, scientists and especially engineers take on, through their professions, higher-level moral responsibilities to fulfill contractual or role duties and promises and to live up to the ethical standards of their professional organizations, not to infringe established laws, and more, inasmuch as the fulfillment of a task, contract, or role does not contradict another overriding moral norm or right. In engineering ethics codes the responsibility to protect public safety, health, and welfare has (since World War II) increasingly been considered paramount.

General Commentary

The previous review aims to summarize in somewhat schematic or outline form the consensus of an extended tradition of critical reflection on responsibility in the German philosophical traditions. These traditions run at least from Gottfried Wilhelm Leibniz (theodicy) through Immanuel Kant (categorical responsibility) and G. W. F. Hegel (idealist responsibility) and Karl Marx (economic responsibility) to the phenomenological tradition (Edmund Husserl through Martin Heidegger to Schutz) and critical social theory versus systems theory (Jürgen Habermas versus Niklas Luhmann). Since World War II, discussions within the Verein Deutscher Ingenieure (Society of German Engineers) have been especially concerned with conceptualizing responsibility in relation to science and technology. The 2002 "Fundamentals of Engineering Ethics" highlights the topic of responsibility in its first major paragraph. The most general discussion of responsibility in this context has occurred in the work of such philosophers as Karl Jaspers, Günther Anders, and Hans Jonas—drawing attention to new moral responsibilities engendered by nuclear weapons, environmental pollution, and genetic engineering. Hans Lenk and Matthias Maring have since

the 1980s worked to synthesize the many achievements within these traditions.

One of the important notes to emphasize about this schematic synthesis is that there exists a differentiated interplay among the identified levels and types of responsibilities, universal moral obligations being but one case. Moral responsibility may be activated by a special type of action and in connection with a special role, but its key characteristic is universality. Moral responsibility as such is not peculiar to a specific person or role but applies to everyone in a similar situation or role. Moral responsibility is nevertheless individualized in the sense that it cannot be delegated, substituted, displaced, replaced, or off-loaded by the respective person, corporation, or organization. Neither can it be diminished, divided, dissolved, or done away with by being shared by a number of people. Moral responsibility is both irreplaceable and unable to be diminished.

With regard to conflicts between different responsibilities or types, priority rules have been developed for adjudicating, regulating, or at least mitigating conflicts and for combining different responsibilities when they are present at the same time. In the last analysis, the presence of a situation- and context-dependent responsibility under the auspices of practical (concrete) humanity should, from the moral point of view, prevail or override any partial and nonmoral responsibility. That is, human rights trump role responsibility rights. One of the challenges of a technoscientific society is to explore ways in which such a priority can be operationalized in and through scientific and technological developments, not just among technical professionals but in society as a whole.

HANS LENK

SEE ALSO *Technology Assessment in Germany and Other European Countries.*

BIBLIOGRAPHY

Über einige strukturelle Probleme der Verantwortung." In *Über den Sinn des Lebens und über die Philosophie*, ed. J. M. Bochenski. Freiburg, Basel, Vienna: Herder.

French, Peter A. (1984). *Collective and Corporate Responsibility*. New York: Columbia University Press.

Jonas, Hans. (1984). *The Imperative of Responsibility: In Search of an Ethics for the Technological Age*, trans. Hans Jonas and David Herr. Chicago: University of Chicago Press.

Lenk, Hans. (1982). *Zur Sozialphilosophie der Technik* [Toward a social philosophy of technology]. Frankfurt am Main: Suhrkamp.

Lenk, Hans. (1987). "Ethikkodizes für Ingenieure" [Codes of ethics for engineers]. In *Technik und Ethik* [Technology and ethics], ed. Hans Lenk and Günter Ropohl. Stuttgart: Reclam.

Lenk, Hans. (1992). *Zwischen Wissenschaft und Ethik* [Between science and ethics]. Frankfurt am Main: Suhrkamp.

Lenk, Hans. (1994). *Macht und Machbarkeit der Technik* [Power and feasibility of technology]. Stuttgart: Reclam.

Lenk, Hans. (1998). *Konkrete Humanität: Vorlesungen über Verantwortung und Menschlichkeit* [Concrete humanity: Lecture on responsibility and humanity]. Frankfurt am Main: Suhrkamp.

Lenk, Hans. (2003). "Responsibility for Safety and Risk Minimization: Outline of an Attribution-Based Approach Regarding Modern Technological and Societal Systems." *Human Factors and Ergonomics in Manufacturing* 13(3): 203–222.

Lenk, Hans, and Matthias Maring. (2001). "Responsibility and Technology." In *Responsibility: The Many Faces of a Social Phenomenon*, ed. Ann Elisabeth Auhagen and Hans-Werner Bierhoff. London: Routledge.

Lenk, Hans, and Günter Ropohl, eds. (1987). *Technik und Ethik* [Technology and ethics]. Stuttgart: Reclam. 2nd edition, 1993.

Schutz, Alfred. (1967). *The Phenomenology of the Social World*, trans. George Walsh and Frederick Lehnert. Evanston, IL: Northwestern University Press.

Werhane, Patricia H. (1985). *Persons, Rights, and Corporations*. Englewood Cliffs, NJ: Prentice-Hall.

RESPONSIBLE CONDUCT OF RESEARCH

• • •

The *responsible conduct of research* (RCR) is one of two major components of research ethics. The essence of the concept is that RCR is central to the practice of science: " [T]he responsible conduct of research is not distinct from research; on the contrary, competency in research *encompasses* the responsible conduct of that research and the capacity for ethical decision making" (Institute of Medicine 2002, p. 9). The emphasis is on professional responsibilities and the extent to which the scientific research community and its members, as a profession, determine, recognize, and adhere to professional standards and values (Carr-Saunders and Wilson 1933). RCR assumes that: (1) there are identifiable, shared standards of practice and behavior that can and should be made explicit; (2) these standards are, consciously or unconsciously, acknowledged by members of the community; and (3) they are standards that research supervisors are expected to instill in trainees.

History

The term RCR is closely related to that of *research integrity*, which it tended to replace as a term of art in the 1990s. It is probably derived from the 1989 Institute of Medicine document *Responsible Conduct of Research in the Health Sciences* (1989) and the concept is further reinforced and reflected in the National Institutes of Health (NIH) requirement that pre- and post-doctoral trainees funded by the NIH receive some formal education in the proper conduct and reporting of research (NIH 1989). However its roots no doubt date from the 1980s when various professional scientific societies, including, for example, the American Chemical Society and Sigma Xi, developed and promulgated codes of conduct for their members (Sigma Xi 1984, Jorgensen 1995, Johnson 1999). In the 1990s the NIH requirement is credited with motivating the biomedical research community to develop educational programs to formalize and make explicit to trainees the expectations of the scientific community with regard to the procedures and processes involved in carrying out and publicizing the results of scientific investigation.

Although inherent in the notion of RCR is professional competence and integrity, education and training in RCR includes many other aspects of scientific research practice. Common usage of the term RCR is a bit of a misnomer because, in point of fact, it includes a wide range of elements beyond the conduct of research that are fundamental to the practice of scientific research. It encompasses not only the experimental process itself, but also closely associated processes such as the dissemination of research findings, the implications of competition among colleagues and its potential impact on the evaluation of research results, and the training of future scientists. As the leading agency emphasizing the importance of education in RCR, the Office of Research Integrity (ORI) in the Office of Public Health and Science in the U.S. Department of Health and Human Services (DHHS) has identified elements central to research practice and appropriate for explicit discussion in the context of RCR (Office of Research Integrity 2005). These topics are:

- data acquisition, management, sharing, and ownership;
- humane treatment of research subjects including both humans and laboratory or other non-human animals;
- allegations of research misconduct;
- recognition of, and management or elimination of, conflicts of interest and conflicts of commitment;

- the mentor/trainee relationship and associated responsibilities;

- publication practices and the responsibilities of authorship;

- the peer review process;

- the expectations of collaborators regarding the nature of collaborative research, appropriate recognition of contributions to the work, and the allocation of responsibility.

Assessment

RCR, in contrast to most formulations of scientific misconduct and integrity, does not solely nor primarily focus on the more egregious and unacceptable practices of fabrication, falsification, and plagiarism (FFP). Instead the notion of RCR implies that there are less responsible, as well as irresponsible, practices. Put another way, there are a range of research practices from the preferred, through accepted but discouraged to prohibited practices. Serious deviation from accepted practices in carrying out research, or in reporting the results of research, may be considered unacceptable by some members of the scientific research community.

A major emphasis of RCR is education in the form of making explicit for both trainees and peers what is often implicit in research practice. However debate continues over how best to assess the efficacy of that education. This stems in part from a lack of consensus on the extent to which the goals of RCR education should include not only explicit understanding of the standards and values of the community and the expectations of both colleagues and society regarding professional behavior, but also training in ethical decision making (Institute of Medicine 2002).

STEPHANIE J. BIRD

SEE ALSO Misconduct in Science.

BIBLIOGRAPHY

Carr-Saunders, Alexander P., and Wilson, P. A. (1933). The Professions. Oxford: Oxford University Press.

Institute of Medicine. (1989). Responsible Conduct of Research in the Health Sciences. Washington, DC: National Academies Press.

Institute of Medicine. (2002). Integrity in Scientific Research: Creating an Environment That Promotes Responsible Conduct. Washington, DC: National Academies Press.

Johnson, David. (1999). "From Denial to Action: Academic and Scientific Societies Grapple with Misconduct." In Perspectives on Scholarly Misconduct in the Sciences, ed. John M. Braxton. Columbus: Ohio State University Press.

Jorgensen, Andrew. (1995). "Survey Shows Policies on Ethical Issues Still Lacking Enforcement Mechanisms." Professional Ethics Report 8(1): 1, 6.

National Institutes of Health (NIH). (1989). "Reminder and Update: Requirement for Programs on the Responsible Conduct of Research in National Research Service Award Institutional Training Programs." NIH Guide for Grants and Contracts 18.

Sigma Xi, The Scientific Research Society. (1984). Honor in Science. Research Triangle Park, NC: Author. 2nd edition, revised and enlarged, 1986.

INTERNET RESOURCE

Office of Research Integrity. Available from http://ori.hhs.gov. Web site of the organization.

RESTORATION ECOLOGY

SEE Ecological Restoration.

RHETORIC OF SCIENCE AND TECHNOLOGY

• • •

Rhetorical inquiry is a multidisciplinary field of study devoted to the critical examination of discourse. Initiated in classical times, it cultivates an "ability, in each [particular] case, to see the available means of persuasion" (Aristotle 1991, p. 36). As an academic field, rhetoric of science and technology is the study of how scientists and non-scientists use arguments to advance claims about science and technology.

The idea that there is a *rhetoric* of science and technology may strike some as perverse and others as obvious. In popular parlance, the term rhetoric connotes something less than truthful, the ranting of politicians who evade substantive dialogue. When tied to science and technology, rhetoric can sound like a curse, staining the purity of certain knowledge and precise measurement with the mark of ideological bias and political maneuvering. But to those who study the rhetoric of science and technology, the term has no such connotation. Instead it is steeped in its ancient tradition and denotes the careful study of how texts are designed to seek the assent of an audience. When those texts are from the realm of science and technology, the means of persuasion utilized include such factors as appeals to disciplinary assumptions and values, the demonstration of methodological rigor, and the selection of language that suggest the neutral observation of nature.

Historical Development

The negative connotations attached to rhetoric are largely the result of a lengthy conflict with philosophy, in which the latter claimed the more valued side of oppositions between opinion and truth, form and content, passion and reason. Yet recent developments in philosophy and other fields recognize these dichotomies as problematic, resulting in a general resurgence of interest in the tradition of rhetorical inquiry, a tradition maintained by enclaves of scholars working mostly in departments of Speech Communication and English in the United States.

Developments in the philosophy, sociology, and history of science have also contributed to the rise of scholarship on the rhetoric of science and technology. Science studies scholars have shown that what one era recognizes as the truth of a scientific theory is seen by a later era as mere opinion, supplanted when an authorizing scientific community accepts a new truth claim. The fact that this transformation occurs by way of arguments addressed to a particular audience, that it often entails a significant shift in values and beliefs by people with an investment in the outcome of those arguments, and that it is frequently marked by controversy, makes rhetorical inquiry a natural approach to the study of such moments.

The idea that communication between scientists and the public might have a rhetorical dimension, or that new technologies may be promoted through rhetorical means, is rarely disputed. Thus the rhetorical examinations of these aspects of science and technology are likewise promising scholarly pursuits in an age when science and technology play such an important role in the development of public attitudes and policies.

The first hint that rhetorical inquiry might be applied to scientific discourse began appearing in the journals of rhetoricians in the 1970s. There were theoretical essays exploring the developments in philosophy and sociology of science that contributed to the possibility for a rhetoric of science (Weimer 1977; Overington 1977), research that began to examine the persuasive nature of specific scientific texts (Campbell 1975), and a general call for scholarship in this new area (Wander 1976). The birth of the field was announced when two books appeared almost simultaneously with nearly identical titles: Lawrence J. Prelli's *A Rhetoric of Science* (1989) and Alan G. Gross's *The Rhetoric of Science* (1990). Both fruitfully applied classical rhetorical concepts to the study of scientific truth claims.

In 1991 Randy Allen Harris wrote a thorough review of the nascent field, defining its relationship to other fields and organizing the scattered research into useful taxonomic categories. In 1993 the American Association for the Rhetoric of Science and Technology held its inaugural meeting at the National Communication Association convention, where it continues to meet annually. The field has continued to develop with the aid of such professional supports as the University of Iowa's Project on the Rhetoric of Inquiry, graduate programs specializing in the study of rhetoric in science and technology at the University of Pittsburgh and the University of Minnesota, and a series of books on the Rhetoric of the Human Sciences published by the University of Wisconsin Press. Research has generally grown along two paths: studies of the arguments made by scientists when they address other scientists, and scholarship that focuses on the relationship between science or technology and the public.

Internal Rhetorics of Science

The most heavily researched area in this growing field is the internal rhetoric of scientists, that is, the discourse scientists use when addressing other scientists, either within their own discipline or across disciplines. Because most people think the internal discourse of scientists is resistant to rhetorical scrutiny, scholars blazing the trail have focused on establishing that even the most specialized communication can be examined usefully through the lens of rhetorical analysis. The prototypical scientific research article has been the subject of much research. For example Watson and Crick's famous 1953 *Nature* report, "A Structure for Deoxyribose Nucleic Acid," has been examined in several unrelated studies that explain its persuasive design through rhetorical theories pertaining to voice, ethos, irony, kairos, stasis, and narrative (Bazerman 1988, Halloran 1984, Gross 1990, Miller 1992, Prelli 1989, Fisher 1994). An entire volume of essays has been written on the rhetoric of a single journal article by Stephen Jay Gould and Richard Lewontin (Selzer 1993).

More evidence that the research report was the primary focus for early rhetoricians of science is the fact that some of the first books in the field were devoted to illuminating writing practices in this genre. For example Charles Bazerman's *Shaping Written Knowledge* (1988) contrasts the scientific article with other forms of academic discourse and traces historical changes and disciplinary differences in the design of the experimental report. It shows how even scientists "use, transform, and invent tools and tricks of the symbolic trade" to shape claims so that they are judged novel and truthful by other scientists (p. 318). In *Writing Biology* (1990), Greg Myers looks at the review process to examine the way authors and editors, operating with different interests, negotiate the status of a scientific claim in a journal article. His book further traces the way two controversies are played out in scientific journals, where scientists interpret their own words and those of their opponents as freely and expertly as any debater in the public forum.

In addition to the rhetoric of the experimental article, landmark scientific monographs such as Newton's *Opticks* (1704) and Darwin's *On the Origin of Species* (1859) and have received sustained attention from scholars of rhetoric seeking to understand how scientists persuade their colleagues to accept radical new theories. The most successful scientists are often the ones who are also master rhetors, capable of adapting new ideas to the presuppositions of their audiences rather than making a frontal assault on a standard paradigm with the irresistible force of a revolutionary theory.

Rhetorical studies have done a particularly good job of showing how the style in which a scientific claim is communicated has an influence on how a scientific community thinks about that claim, and vice versa. Jeanne Fahnestock's careful account of rhetorical figures in science demonstrates that language does "much of our thinking for us, even in the sciences, and rather than being an unfortunate contamination, its influence has been productive historically, helping individual thinkers generate concepts and theories that can then be put to the test" (Fahnestock 1999, p. xi).

Because facilitating the growth of knowledge is the central activity of scientists, the way in which scientists use the tools of language and argument to advance knowledge claims has received the most attention from scholars of the rhetoric of science. Another internal rhetoric of science that receives less attention, either because it is considered less central or because its character is less contested and thus less shocking when discovered, is the way in which scientists persuade one another that a particular line of research holds future promise. Myers devotes a chapter of his book to the rhetoric of the grant proposal, a genre of scientific writing that must convince reviewers a research program deserves funding because of its potential interest to the scientific community and the professional ethos of the authors. Leah Ceccarelli (2001) examines motivational texts of science to show that scientists who employ a strategic ambiguity of language are better able to persuade colleagues from different disciplines to overcome

barriers separating their fields and engage in new interdisciplinary lines of research. These internal discourses of science that do not seek the assent of colleagues to a particular truth claim, but instead seek future action from fellow scientists, have been less studied by rhetoricians, but may be just as important to the ultimate development of science.

For the most part, research on internal rhetorics of science tends to be descriptive and explanatory in nature, uncovering the rhetorical practices at the heart of scientific activity. But some of it has an implicit prescriptive character, suggesting other resources of language and argument that scientists might use to shape science in more useful or ethical ways. In contrast research on external rhetorics of science and technology tends to be more explicit in its criticism of current communication practices and more direct in its recommendations for change.

External Rhetorics of Science and Technology

The ways in which scientists communicate with the public and the ways in which nonscientists communicate about scientific or technological issues are more obviously rhetorical in nature, and ripe for critical commentary. Popularization is one genre of scientific writing that is a natural subject for rhetorical analysis. By contrasting journal articles written for specialists with popularizations on the same topics, rhetorical inquiry has shown that popular accounts remove hedges and qualifications for scientific claims while emphasizing the uniqueness of observations (Fahnestock 1986). Because of these changes, the public may get a distorted view of the certainty and significance of a scientific knowledge claim, something that can be dangerous when the subject has important social implications. Rhetorical analysis contrasting internal rhetorics of science with popularizations has also demonstrated that while the former emphasize the activities of the scientists and the conceptual structure of the discipline in which they are working, the latter emphasize the activities of the objects being studied (Myers 1990). Again distortion may result, with public audiences developing an image of science as the unmediated observation of external nature, without the interference of scientists who employ theoretical apparatus or make methodological decisions.

Rhetorical inquiry has also brought critical attention to the situation in which an expert takes a new scientific theory away from its disciplinary origins and argues before public audiences, thus eluding accountability to the controls of a specialized scientific community (Lyne and Howe 1990). Popularization may be the genre of science writing that does the most to break down the barrier that exists between the two cultures of scientists and nonscientists, but its tendency to misrepresent science as a non-controversial activity of observation by disinterested individuals has ethical consequences, especially when the public is asked to make decisions about matters for which science and technology do not have indisputable answers.

Situations in which the public must act on technical matters despite a lack of scientific consensus have been the subject of several case studies in the rhetoric of science and technology. Examining cases as diverse as the recombinant DNA controversy of the 1970s (Gross 1990, Waddell 1990) and disputes over the accuracy of missile defense technology in the 1980s and 1990s (Mitchell 2000), rhetorical critics have analyzed debates about the public control of contemporary scientific and technological developments. Most have supported the findings of an early study of the discourse surrounding the Three Mile Island incident (Farrell and Goodnight 1981). When technical reason usurps the place of more appropriate modes of public deliberation about matters of social or political import, a crisis of communication is the result. In each case study, rhetorical patterns that promote democratic participation are endorsed as an alternative to the dysfunctional assumption that people can rely on science and technology to solve their most serious public problems.

Another type of scholarship on external rhetorics of science and technology takes a more historical approach, scrutinizing the documentary evidence surrounding a particular scientific field or technological development to uncover the specific discursive forms that reflect and shape public attitudes. The scope of such rhetorical histories can be broad, as it is in Celeste Condit's 1999 study of public debates about human heredity from 1900–1995, or narrow, as in Charles Bazerman's 1999 study of how Thomas Edison and the people around him represented light to the public from 1878–1882. In both cases though, the purpose of the rhetorical study is not to critique the oversimplification of popularizations, nor to valorize public deliberation over technological decision making, but to demonstrate the complicated ways in which science, technology, and culture interact in the public mind.

Conclusion

Although the rhetorical study of science and technology can be broadly divided into the examination of internal and external communication, there is work within the field that breaks out of this neat mold. For example the rhetoric of technology typically makes no distinction between internal and external genres, but examines both in the patterns of communication unique to "enterprises concerned with the development, production, and marketing of artifacts and practices" (Miller 1994, p. 92). There also are various fields of rhetorical study that intersect with the rhetoric of science and technology, but are not typically considered a part of it, such as the rhetorical study of medicine, mathematics, economics, or communication technologies.

Study of rhetoric in science and technology is an important but young field that sometimes suffers lack of confidence in communicating outside its peer group. A scan of citation practices in the literature demonstrates that most rhetoricians of science and technology are familiar with related research done in philosophy, history, and sociology of science, but the reverse is rarely true. Publishing mostly in journals read by other rhetoricians, or in books that are marketed to Speech Communication and English departments, they do little to communicate their findings to other science studies scholars or to scientists and the public. This is unfortunate, as the rhetorical critic's tools of close reading and argument analysis illuminate aspects of texts and debates that would benefit scholars in other fields. Perhaps with time, the rhetoric of science and technology will mature into a field that acts as a full and equal participant in the community of science studies scholars. At that point perhaps it will also do more to export its findings especially to scientists and citizens who must evaluate scientific discourse to make fully informed ethical decisions about science and technology.

LEAH CECCARELLI

SEE ALSO *Communication Ethics; Discourse Ethics; Knowledge; Science, Technology, and Literature.*

BIBLIOGRAPHY

Aristotle. (1991). *On Rhetoric: A Theory of Civic Discourse,* trans. George A. Kennedy. New York: Oxford University Press. Classical work on rhetoric that is still much used.

Bazerman, Charles. (1988). *Shaping Written Knowledge: The Genre and Activity of the Experimental Article in Science.* Madison: University of Wisconsin Press.

Bazerman, Charles. (1999). *The Languages of Edison's Light.* Cambridge, MA: The MIT Press.

Campbell, John Angus. (1975). "The Polemical Mr. Darwin." *Quarterly Journal of Speech* 61: 375–390. An early essay in the author's corpus on Darwin's Origin.

Ceccarelli, Leah. (2001). *Shaping Science with Rhetoric: The Cases of Dobzhansky, Schrödinger, and Wilson*. Chicago: University of Chicago Press.

Condit, Celeste Michelle. (1999). *The Meanings of the Gene: Public Debates about Human Heredity*. Madison: University of Wisconsin Press.

Fahnestock, Jeanne. (1986). "Accommodating Science: The Rhetorical Life of Scientific Facts." *Written Communication* 3: 275–296.

Fahnestock, Jeanne. (1999). *Rhetorical Figures in Science*. New York: Oxford University Press.

Farrell, Thomas, and G. Thomas Goodnight. (1981). "Accidental Rhetoric: The Root Metaphors of Three Mile Island." *Communication Monographs* 48: 271–300.

Fisher, Walter R. (1994). "Narrative Rationality and the Logic of Scientific Discourse." *Argumentation* 8: 21–32.

Gross, Alan. (1990). *The Rhetoric of Science*. Cambridge, MA: Harvard University Press. The second printing, 1996, has a new preface that reviews some work in the field.

Halloran, S. Michael. (1984). "The Birth of Molecular Biology: An Essay in the Rhetorical Criticism of Scientific Discourse." *Rhetoric Review* 3: 70–83.

Harris, R. Allen. (1991). "Rhetoric of Science." *College English* 53: 282–307.

Lyne, John, and Henry F. Howe. (1990). "The Rhetoric of Expertise: E. O. Wilson and Sociobiology." *Quarterly Journal of Speech* 76: 134–151.

Miller, Carolyn. (1992). "*Kairos* in the Rhetoric of Science." In *A Rhetoric of Doing: Essays on Written Discourse in Honor of James L. Kinneavy*, eds. Stephen P. Witte, Neil Nakadate, and Roger D. Cherry. Carbondale: Southern Illinois University Press.

Miller, Carolyn. (1994). "Opportunity, Opportunism, and Progress: *Kairos* in the Rhetoric of Technology." *Argumentation* 8: 81–96.

Mitchell, Gordon. (2000). *Strategic Deception: Rhetoric, Science, and Politics in Missile Defense Advocacy*. East Lansing: Michigan State University Press.

Myers, Greg. (1990). *Writing Biology: Texts in the Social Construction of Scientific Knowledge*. Madison: University of Wisconsin Press.

Overington, Michael A. (1977). "The Scientific Community as Audience: Toward a Rhetorical Analysis of Science." *Philosophy and Rhetoric* 10: 143–164.

Prelli, Lawrence. (1989). *A Rhetoric of Science: Inventing Scientific Discourse*. Columbia: University of South Carolina Press. Introduces readers to classical rhetorical theory, then adjusts that theory to apply it to science.

Selzer, Jack, ed. (1993). *Understanding Scientific Prose*. Madison: University of Wisconsin Press. Various critics apply different methods to analyze a single scientific essay.

Waddell, Craig. (1990). "The Role of *Pathos* in the Decision-Making Process: A Study in the Rhetoric of Science Policy." *Quarterly Journal of Speech* 76: 381–400.

Wander, Philip C. (1976). "The Rhetoric of Science." *Western Journal of Communication* 40: 226–235.

Weimer, Walter B. (1977). "Science as a Rhetorical Transaction: Toward a Nonjustificational Conception of Rhetoric." *Philosophy and Rhetoric* 10: 1–29.

RIGHTS AND REPRODUCTION

• • •

It is in the context of reproduction, interpreted broadly, that many of the issues concerning the ethics of science and technology have arisen. The birth of the first so-called "test tube" baby in 1978 set in motion an ongoing process of questioning interventions such as assisted reproduction, embryo research, and cloning.

The notion that human beings possess a legitimate inclination to conceive and bear children is part of traditional natural law teaching. For instance, Thomas Aquinas (1225–1274) argued that human beings share those inclinations "which nature has taught to all animals," such as sexual intercourse, education of offspring, and so forth (*Summa theologiae* I–II, Q. 94, a.2). As traditional natural law was transformed into modern natural right, and human sexuality became increasingly mediated by science and technology so as to become both more productive and subject to human control, the intersection of human rights and having children (termed variously both reproduction and procreation) became increasingly contentious.

One contention centers around what are termed *reproductive rights*, generally indicating women's right to control whether, when, and how they bear children. There is clearly an important gender dimension to the issues. The right to be free from interference such as sterilization, on the one hand, and the right to abortion, on the other, have been important historical landmarks in women's control over their fertility.

Historically, the content of the reproductive rights has gradually increased, however, beyond freedom from interference to include a right with a much wider scope, such as the right to positive assistance in reproduction (that is, the use of technology in the case of infertility); and also to choice of the *kind* of children one has (for example, sex selection and genetic factors). Reproductive rights in this sense remain hotly contested at least to some degree: While there is widespread acceptance of in vitro fertilization, some potential means of assisted reproduction continue to be regarded by many as unacceptable, such as reproductive cloning. The right to choose to avoid preventable genetic disorder in one's children is also regarded as problematic by those who find such choices expressive of intolerance towards difference.

Disagreements depend to a considerable extent on different views on what the fundamental basis of the right is—for example, on whether the right to reproduce is claimed a natural right, or as an aspect of autonomy— and on how these concepts themselves are understood. For example, it might appear strained to argue for a *natural* right to reproduce by *artificial* means, unless it is argued that the artificial is necessary in order to fulfill a natural purpose of human life. Should infertility be regarded as a disease that needs treatment, or just an unfortunate inability to satisfy one's wishes? Again, while on the one hand an autonomy argument might be deployed to suggest that the right to reproduce is an aspect of doing as one wants with one's body, on the other hand, in so far as reproduction has effects on others and requires the allocation of health care resources, it is difficult to see how the argument can, by itself, provide an argument for the cooperation of others. The welfare of future children is a consideration that may compete with that of the reproductive rights of adults.

RUTH CHADWICK

SEE ALSO *Assisted Reproduction Technology (ART)*; *Eugenics.*

RIGHTS THEORY

• • •

Rights are generally defined as justified claims for the protection of general interests. In this sense, human beings have been described as having rights to property, "to life, liberty, and the pursuit happiness" (United States Declaration of Independence, 1776), as "free and equal in rights" (Declaration of the Rights of Man and Citizen, 1789), and as having rights "to share in scientific advancement and its benefits" (Universal Declaration of Human Rights, 1948). More recently civil rights or liberties to freedom of speech and assembly have been complemented by proposals for social, economic, and welfare rights to minimum levels of shelter, food, and medical care. What was initially a quite limited relation of rights to science and technology, insofar as their advancement rested on the protection of intellectual property rights, has become increasingly a question of consumer rights to certain levels of material benefit and

safety related especially to technology. The assessment of such diverse claims nevertheless requires appreciation of the broader philosophical discussion of rights and various analytic distinctions introduced to clarify numerous complications.

Fundamental Distinctions

As initial observations have already indicated, the notion of rights has become deeply embedded in modern societies, but it has critics precisely because of its origin in particular socio-cultural contexts and because of its relationship to individualism. "Rights express the idea that respect for a given interest is to be understood from the point of view of the individual whose interest it is" (Waldron 1993, p. 576). While this statement arguably overlooks the fact that it is not only individuals but also groups that may be held to have rights, as seen in debates about rights of particular minorities, it soon becomes clear that this does not avoid questions of individual rights: Some of the most difficult issues with group rights concern relationships of the individual *to* the group.

The classic and most systematic attempt to delineate different kinds of rights was that of Wesley Hohfeld (1919), who identified a number of distinct categories. Some of the ways in which the term *rights* would be used, he argued, would be more accurately captured by the term *privileges*. These are to be contrasted with rights "in a strict sense," which Hohfeld categorized as claim-rights.

If a person X has a claim-right, in Hohfeld's sense, there must be at least one person who has a duty to X with regard to that claim. This is the thesis of correlativity of rights and duties: A claim can normally be met only through the efforts, or at least the non-interference, of others. This thesis has come to be regarded as definitive of rights.

To say that X has a privilege, however, has no such implication. A privilege is a liberty to do something, which may be either of a general or a special kind. In the general sense a privilege to act in a certain way is simply the absence of a duty to avoid doing it. No one is in a position to make a counter-claim against the person. In the special sense, however, a privilege is a liberty that is *exceptional*, that is, it is not enjoyed by other persons—for example, informed consent on the part of patients allows health care professionals certain liberties to do things to them which may be invasive, which would not be permissible in other circumstances. Hohfeld also distinguished claim-rights from other terms such as powers and immunities.

Questioning Correlativity

The thesis of the correlativity of rights and duties is problematic. First, different aspects of correlativity have been distinguished: the moral and the logical (Feinberg 1973). The moral correlativity thesis states that in order to have rights individuals must have and accept duties themselves. This is controversial because it would rule out the rights of persons with mental incapacity. Some would argue that all human beings have rights, pre- as well as postnatally, even if it is not possible to hold them to be subject to duties.

The logical correlativity thesis is concerned with what X's rights imply for others' duties. In terms of Hohfeld's claim-rights, which are *defined* in terms of the duties they imply for others, then questions arise about what those duties are, for where rights *do* imply duties, these may be of different kinds. For example, if X has a right to something, while it may not be the case that there is any person Y in particular who has a duty to do anything to help X to get that something, it may yet be appropriate to say that everyone has a duty not to *prevent* X from getting it.

More generally, however, the question has to be asked whether the correlativity thesis is true of all rights. The term *rights* has certain uses in political discourse, which go beyond claim-rights. This is described as the rhetorical or "manifesto" use. Thus Onora O'Neill writes: "A 'right to food' could be satisfied by earning enough money to buy food, by having enough land to grow it or by having friends and family with obligations to provide it; in each case there would be an entitlement to food ... But without one or other determinate institutional structure, these supposed economic rights amount to rhetoric rather than entitlement" (O'Neill 2000, p. 125). Such "rights" arguably do not imply duties on the part of anyone in particular.

Furthermore, even where rights imply duties, it does not appear to be the case that the converse applies: That where there are duties there are always corresponding rights. If one accepts that X has a right to the fulfilment of a promise, there must be someone who has a duty to fulfil it, and if Y has a duty to fulfil a promise, there must presumably be someone who has a right to have that duty fulfilled. Promising involves correlativity of this kind. Other duties, however, such as those involved in scientific inquiry, for example, are not of this type. Duties to pursue truth, avoid fraud, and publish results are arguably not best explained in terms of other people's rights, but arise from the nature and purpose of the activity itself.

Claim Elements

If one accepts that in addition to claim-rights in the strict sense there are also wider uses of the term, it is still possible nevertheless to regard rights as including a claim *element*. In order for a claim to be a right, however, it must be justified. The two elements, a claim and a justification, are common to both Hohfeldian claim-rights and manifesto rights. Different moral theories will attempt to justify rights in different ways, however, and it is the type of justification to which appeal is made that categorizes a right as of one sort or another. In the case of a claim-right to the fulfilment of a promise, that the promise was made, and thus a duty incurred, will form part of the justification. In the case of a manifesto right, however, the justification could be in terms of moral judgments about what should be the case, and may be based more on moral ideals of principles of justice than on duties.

There are at least two further distinctions it is important to consider in thinking about the claim element of a right: the contrast between *negative* and *positive* and the importance of rights of *voice* and rights of *exit*. The distinction between negative and positive rights depends on what they imply for others—either non-interference or positive action, respectively. The right to freedom of scientific inquiry, for example, might be construed negatively as a freedom right not to be prevented from pursuing a particular line of research. At the same time, it might be argued that freedom of scientific inquiry is meaningless unless research is funded, which might imply positive action on the part of others such as governments and research councils.

Claim-rights may change their content over time from negative to positive, depending on the social context. Thus at one time the right to reproduce was construed as a negative right to be free from interference: the right to choose whether or not to reproduce. Over time, however, it has been argued to include not only the right to decide on the number and spacing of one's children, but also, by some, the type of children one has. This has led to arguments about the extent of the different duties for others, including the provision of contraception advice, assisted reproduction where necessary, and sex selection. Again, X's right to life implies a duty on everyone not to kill X, but might also require, in a certain circumstance, that a bystander who has the ability to save X from drowning has a duty to do so.

The distinction between *rights of voice* and *rights of exit* (see Hirschman 1970) is particularly prominent in discussions of group rights. Rights of exit include the right not to participate—that is, the right to choose not to accept traditional practices of the group, such as practicing a certain religion. There is a view that the individual's right to exit from a group is essential if groups are to claim rights. A formal right of exit, however, may be insufficient to protect some oppressed group members, such as in the case of women traumatized by domestic violence. Rights of voice, as the name implies, involve the ability to participate in decision-making and to express one's preferences in, for example, political decision-making. The relationship between the two is complex: Arguably individuals should not need to exit if they have a right to exercise their voice within the group so that things can be changed from within.

The debate about rights of voice and rights of exit demonstrates the close association of rights talk with liberalism. Historically rights emerged in the context of liberalism, being concerned with essential freedoms and limiting government power, but there is an issue concerning the extent to which they should be limited to freedoms *to* do certain things, such as freedom of speech and movement, or whether they also embrace freedoms *from* such conditions as poverty. The distinction between negative and positive rights, describable as a distinction between freedom rights (liberal, freedom *to* rights) and rights of recipience or welfare rights, reflects underlying differences in political philosophy and justification.

Natural Rights

In moral and political argument, rights are used sometimes as starting points, sometimes as conclusions. A prominent example of the use of rights as starting points is to be found in Robert Nozick's *Anarchy, State and Utopia* (1974), the first sentence of which states: "Individuals have rights, and there are things no person or group may do to them (without violating their rights)." Nozick sees himself as operating in the tradition of the seventeenth-century philosopher John Locke (1632–1704), arguing that human beings have certain "natural" rights.

The notion depends on state of nature theory and natural law. The idea of a state of nature is a hypothetical state external to society, in which human individuals are unaffected by social conditioning, and which operates as a device for critical reflection on existing societies. The laws of different societies assign to their citizens or subjects different rights and duties. But beyond this, it is argued, there are natural laws and natural rights, which provide a point from which to criticize the laws in any particular society (such as laws that allow for institutions such as slavery). Locke argued that in a state of nature there would be a natural law that "no-one should harm another in his life, liberty or property" (*Two Treatises of Government,* ed. Peter Laslett, 2nd edition, Cambridge University Press, 1967).

The idea of natural rights has been heavily criticized, most notably by Jeremy Bentham (1748–1832) who described it as "nonsense upon stilts." Bentham argued: "From real laws come real rights; from imaginary laws come imaginary rights." ("Anarchical Fallacies" in *The Works of Jeremy Bentham,* Vol. II, ed. J. Bowring; Edinburgh: William Tait 1843). The doctrine of natural law confuses the questions of what the law is and what the law ought to be. While one can criticize the law from a moral point of view, in order to do this one needs a perspective such as that of utilitarianism, not the notion of natural law.

The idea of the state of nature has also been criticized as ahistorical by Marxist and feminist critics. The objection is that there is no universal human nature, no pre-social state of nature. What people are like, as well as their values and expectations, are the products of the society in which they live. There is a strand in natural law thinking that natural rights should be *evident* to everyone. But even those philosophers who employ the notion of a state of nature differ over how it is to be understood, and there is further disagreement over what rights there are. Property, for example, is high on the list of Locke and Nozick, but it is by no means evident to all that it is a *natural* right. From an opposing point of view the so-called "natural" right to property is a historically conditioned expression of the interest of those who have it. Rights are seen as institutionalizing certain interests at the expense of others. The debate about property rights has been particularly pertinent in science and technology, in the context of intellectual property and patenting, for example in relation to the human genome. The distinction between what is discovered and what is invented relies on a notion of what exists by nature, but controversy continues over what can legitimately be patented.

Human Rights

Nevertheless the idea that there are universal and timeless rights grounded in enduring features of human nature has persisted. The United Nations Declaration of Human Rights (1948) and the European Convention on Human Rights (1950) are expressions of this idea, although dispute has raged over how many of the rights contained in these documents are real rather than manifesto rights.

Despite traditional criticisms of natural rights, Tom Campbell (1983) has argued that socialists need not object to the notion of human rights as protectors of fundamental human interests, if this notion is divorced from the ahistorical concept of a state of nature and from the traditional view about what rights human beings have. On this view, the problem with the tradition of liberal western democracy in which the notion of natural rights flourished has been the concentration of thinkers in that tradition on "freedom" rights at the expense of "welfare" rights. To focus on freedom rights can seem callous when people's basic food needs are not being met.

The objection to this from those who favor freedom rights is that welfare rights cost money, and therefore are not always feasible. In order to count as a genuine human right, any given right must be "practicable, universal, and paramount." Consider again the example of the right to reproduce. If this is understood as the right to be free from interference, then it might appear to cost nothing. If it is interpreted as a right to in vitro fertilization (IVF), however, the costs could spiral out of control. Nevertheless, the so-called freedom rights also cost money and it might be better to think in terms of basic rights rather than accepting the negative-positive distinction (see Shue 1980). The right to freedom from interference in one's private life, for example, might require the provision of some machinery of justice, including a police force.

Thus the idea of natural rights as starting points runs into difficulties, while the notion of human rights has become a site of political struggle between competing political ideologies. So on what basis can an argument for rights be put forward? It is possible to put forward arguments on utilitarian grounds, giving reasons why people should be free to do certain things or why they should receive particular goods and services: in other words, that they should have rights to do x and y or to receive p and q because to do so leads to good consequence. In this sense the term *right* is quite vulnerable to being trumped by other considerations, as this way of reasoning does not regard rights as attaching to individuals in quite the same way as in the natural rights tradition: as integral to what is understood by a human being.

Rights as Conclusions

Ronald Dworkin (1977) has argued that rights themselves should be regarded as trumps over some background justification for political decisions that state a goal (such as one based on utilitarian reasoning) for the community as a whole. An example would be that if someone has a right to publish pornography, this means that it is for some reason wrong for officials to act in vio-

lation of that right, even if they (correctly) believe that the community as a whole would be better off if they did.

Dworkin argues for a "rights-based morality" in contrast to one based on either duties or goals. His arguments started with the claim that government must treat those whom it governs with equal concern and respect. He identified his aim as that of examining how far a theory of rights can be constructed from the abstract idea that government must treat people as equals. It was Dworkin's contention that utilitarianism does not do this. Despite its claim that "each counts for one and no one for more than one," he argued that utilitarianism is corrupted by external preferences, where external preferences are preferences we have regarding other people. An example might be that people who are homophobic do not only have a preference regarding their own sexuality but also have an external preference that others should not be free to embrace homosexuality. If the majority shared these external preferences the minority could experience discrimination and hardship. In the context of science and technology, some people object so strongly to possibilities such as human reproductive cloning that they not only wish not to engage in it themselves, but want it to be universally prohibited, although others argue that it could be contemplated as an application of the individual's right to reproduce.

Therefore in a society where the background justification is utilitarian, rights are needed to act as trumps over the outcome of utilitarian calculations. It is important to note that Dworkin does not want to exclude all external preferences (for example, charitable ones), but only those that fail to treat human beings with equal concern and respect. Thus he argues for basing political morality around a fundamental right to equal concern and respect.

Objections to Rights-Based Morality

Rights-based morality nevertheless overlooks crucial features of the moral landscape (see, for example, O'Neill 2000). Rights are adversarial, and may be useful when opposing oppressive governments—perhaps particularly in drawing attention to the plight of particular groups—but apart from such situations it may be more appropriate to look to another framework, such as that of duties. This way of looking at things, drawing more on the thought of Immanuel Kant (1724–1804), directs attention to what people ought to *do* rather than what they ought to get. Duties "formulate the requirements to which Declarations of Rights merely gesture," but rights have acquired popularity, argues O'Neill, because they appear to offer something to everyone (O'Neill 2000) without focusing on the associated and varied costs. While rights-talk is pervasive, it is important always to be alert to the question of justification of any particular rights claim.

As should be clear from the present discussion, although rights are easily asserted with regard to many aspects of science and technology, the full legitimization of such assertions is much more difficult. It may be that individuals have rights to intellectual property in particular forms of scientific inquiry, and that consumers have rights to be protected from invasions of privacy by means of surveillance technologies. It may be that individuals have a right to exit certain aspects of scientific and technological development, and that different publics have the right to a voice in the governance of science. However, for what are often no more than manifesto rights to become fully warranted claims will in many instances require further reflective consideration than has to date been achieved.

RUTH CHADWICK

SEE ALSO *Human Rights; Right to Die; Right to Life; Rights and Reproduction.*

BIBLIOGRAPHY

Campbell, Tom. (1983). *The Left and Rights: A Conceptual Analysis of the Idea of Socialist Rights.* London: Routledge & Kegan Paul. An examination of the extent to which socialist rights, as opposed to liberal rights, can be upheld.

Dworkin, Ronald. (1977). *Taking Rights Seriously.* London: Duckworth. Argues for the priority of the right to equal concern and respect.

Feinberg. Joel. (1973). *Social Philosophy.* Englewood Cliffs, NJ: Prentice-Hall. Contains a very useful introduction to the concept of rights.

Hirschman, Albert O. (1970). *Exit, Voice and Loyalty: Response to Decline in Firms, Organizations, and States.* Cambridge, MA: Harvard University Press. Argues for the distinction between rights of voice and rights of exit.

Hohfeld, Wesley N. (1919). *Fundamental Legal Conceptions as Applied in Judicial Reasoning.* New Haven, CT: Yale University Press. A classic categorization of rights.

Nozick, Robert. (1974). *Anarchy, State, and Utopia.* Oxford: Blackwell. Takes rights as starting points.

O'Neill, Onora. (2000). *Bounds of Justice.* Cambridge, UK: Cambridge University Press. Argues for treating principles of obligation as more basic than rights.

Shue, Henry. (1980). *Basic Rights: Subsistence, Affluence, and U.S. Foreign Policy.* Princeton, NJ: Princeton University Press. Argues for "basic" rights rather than prioritizing "freedom" or "welfare" rights.

Waldron, Jeremy, ed. (1984). *Theories of Rights.* Oxford: Oxford University Press. A collection of essays on topics such as natural rights. Includes an essay by Ronald Dworkin on rights as trumps.

Waldron, Jeremy. (1993). "Rights." In *A Companion to Contemporary Political Philosophy*, ed. Robert E. Goodin and Philip Pettit. Oxford: Blackwell. An overview.

RIGHT TO DIE

• • •

For literally hundreds of thousands of years, human beings recognized death as inevitable feature of the human condition—to be avoided when possible, but ultimately accepted as necessity. Indeed, one of the distinctive features of moral reflection involved considerations of how properly to approach death. The idea that one had a right to die rather than a necessity to accept death with grace would have been inconceivable. In the last half of the twentieth century, however, advances in scientific medicine and technology fundamentally altered the traditional framework of reflection on death. As it becomes increasingly possible to prolong death and to extend the life span, it becomes necessary not simply to accept death but to consider a possible right and in some cases even responsibility to die.

The claim that the individual has a right to die presupposes not only advances in science and technology however but also individualism: It requires a view about the individual having control over his or her own life. This is by no means a claim that has won universal support. In some societies the individual life has been regarded as belonging to the king or ruler who could command its sacrifice; another view is that it is for a divine being either to bestow or to take away life.

Even within an individualistic framework, however, the basis and limits of a right to die are not always clear. First, as rights are commonly supposed to impose duties on others, a right to die may require others at least to refrain from interfering, and possibly also to provide assistance, so speaking of a right to die may have a number of meanings, such as death through assisted suicide; rejection of treatment, food, and hydration; and euthanasia. Common to all these may be an argument that the individuals should be free, where possible, to choose the timing and manner of their death. They differ in their implications for other people involved (that is, what exactly other people have to do or refrain from doing in order to allow the individual to exercise their right).

Arguments for a Right to Die

To be free to choose to die, when life has ceased to hold any attraction or meaning, might be supported on the basis of a respect for autonomy. The simplest case appears to be that of the individual who both wants, and has the means, to commit suicide. The implications for third parties in this case are simply to refrain from interference. The German philosopher Immanuel Kant (1724–1804), however, argued that the rational agent could not consistently will to end one's own life, for this would mean that the same will that naturally wanted to extend life at the same time wanted to end it, and this involved a contradiction. He also argued that to commit suicide involved failing to treat rationality in one's own person as an end. There are clearly difficulties with this Kantian argument, because it fails to recognize any circumstances in which ending one's own life would be a rational course of action. It is significant, however, that mental health legislation has commonly regarded suicidal impulses as evidence of lack of rationality.

If autonomy is interpreted in the manner of the English philosopher John Stuart Mill (1806–1873), then there are strong grounds supportive of a right to die. Mill argued that when people interfere on paternalistic grounds to prevent persons from harming themselves, it is likely they will interfere wrongly, because individuals are in general the best judges of what is in their own interests. This suggests that an individual may very well be in a good position to know when life no longer has any meaning or value for the individual whose life it is. The onlooker may try to engage in rational argument, but should not forcibly interfere.

The issues become more complicated when the duties that the right implies for others involve positive assistance, such as in assisted suicide, which is distinguished from voluntary euthanasia on the grounds that the patient remains the agent. Apart from the legal requirements that may apply in different jurisdictions, there are questions about what obligations, if any, there are to assist. This is particularly problematic where individuals who wish to die do not have the means or ability to take their own life—for example, when they are incapacitated to the extent that they cannot help themselves. The assistance required may be providing the means, such as administering a drug, or withdrawing of food and fluids. There are issues here, also, about *who* is being asked to provide assistance—whether it is a friend, or family member, or a health professional. There are special questions about professional roles and the extent to which the obligations of professionals differ from those of others. Withdrawal of food and fluids may be regarded not as treatment, which an individual is entitled to refuse, but as a basic human right that is inalienable, or as basic care that should never be withdrawn.

Several key cases have addressed the issues of the right to die in the absence of the capacity for autonomous decision-making. The 1976 Karen Quinlan case (In re Quinlan, 70 NJ 10, 355 A.2d 647 [1976]) in the United States decided that Quinlan's right to privacy supported her right to be removed from a ventilator. In the United Kingdom, the 1993 Tony Bland case (Airedale NHS Trust *v.* Bland [1998] HL) concerned a patient who had been in a persistent vegetative state for more than three years when the hospital sought a declaration that it could lawfully withdraw all forms of life support. The case went up to the House of Lords, who argued not that it was in Bland's best interests to die but that it was not in his best interests to prolong his life in those circumstances, and that it was lawful to withdraw feeding.

In 2005 the case of Terri Schiavo, a Florida woman in a persistent vegetative state for more than a decade, became a cause célèbre because of basic disagreements between her husband and her parents over whether a feeding tube should be removed. For the husband, consistently supported by the courts, this would allow her to die in accord with previously expressed desires not to become dependent on extraordinary technological means. For the parents and many religious supporters, this was tantamount to murder.

The sense that individualism, and individual choice, need to be mediated by a sense of natural limits, also has the potential to facilitate acceptance of a responsibility to die, especially at a time of ever increasing possibilities for intervention.

RUTH CHADWICK

SEE ALSO *Death and Dying; Euthanasia; Euthanasia in the Netherlands; Persistent Vegetative State.*

BIBLIOGRAPHY

Brock, Dan W. (1993). *Life and Death: Philosophical Essays in Biomedical Ethics.* Cambridge: University Press. Includes essays on life and death decisions in the clinic, and on the value of prolonging life.

Callahan, Daniel. (1987). *Setting Limits: Medical Goals in an Aging Society.* New York: Simon and Schuster. An argument against prolonging life at all costs.

Regan, Tom, ed. (1980). *Matters of Life and Death: New Introductory Essays in Moral Philosophy.* New York: Random House. Includes discussions of suicide.

Singer, Peter. (1996). *Rethinking Life and Death: The Collapse of Our Traditional Ethics.* New York: St. Martin's Griffin. Includes a discussion of the Bland case.

RIGHT TO LIFE

• • •

What was once considered fate or a gift, that is, human life, is increasingly thought of as subject to manipulation or control by means of scientific research and biomedical technology. The ability to regulate fertility and pregnancy on the basis of knowledge and desire, along with psychological studies of child development and the potentials of genetic engineering—not to mention the potential of nuclear weapons and other runaway technologies to destroy all life on the Earth—have conspired to promote consideration of possible rights to existence of those forms of life that have become increasingly subject to the unintended impacts or conscious manipulation of others.

The Right to Life: The Narrow Sense

When the right to life is spoken of, it is normally human life that is meant, although there are arguments for extending the scope of the right to other life forms. To restrict the right to the human species may attract the charge of speciesism. For present purposes, however, the discussion will be confined to humans.

The force of the right to life, insofar as it imposes obligations on others, is normally to stress the wrongness of killing, rather than a positive right to be brought into existence. This is because it is difficult if not impossible to identify someone who would be wronged by not being brought into existence. There is also controversy over whether someone can be wronged *through* being brought into existence, as in the debate about "wrongful life." It does not follow, however, that the right to life has no implications for positive aid. There are differences of opinion about the extent to which a right to life could impose obligations to *save* another's life.

One of the most difficult problems facing a right to life, however, concerns the definition of human life—especially with regard to its beginning and ending. There is disagreement both about when life begins and about when it ends. Some would say that the question of when life begins is not the right question, because life is ongoing. The germ cells are alive and life continues from generation to generation. What is normally meant, however, is the life of an identifiable individual—but even this is not clear-cut, with some putting more emphasis on the concept of the person than on that of human being. There is a similar issue at the end of life—whether what is important is the death of the organism or the end of everything one recognizes as personhood.

Arguments for a Right to Life

Not all arguments for the view that killing a human being is wrong use the terminology of rights. It might depend upon a view about the sanctity of human life. The doctrine of the sanctity of human life might be religiously informed: Life is a gift from God and therefore sacred. It is a way of expressing the view that life has intrinsic value—it is valuable in itself, from beginning to end, and it is wrong to destroy it.

Those who support the sanctity of life doctrine typically also take a conservative view about the beginning and ending of life, the presumption being that there is something of intrinsic value from the moment of conception, and that while there is life there is a being worthy of respect at the end of life. On the sanctity of human life view there can be no trade-offs. In other words, it would not be permissible to kill one innocent person in order to raise the quality of life of others, or even because of the opinion that a person's quality of life is no longer worth living. Critics point out that this has implications for social policy. How can there be justification for taking money away from life-saving enterprises and giving it to those that can at best only improve quality of life for some people? Upholders of the sanctity of life doctrine here fall back on a distinction between negative and positive, holding that the doctrine imposes the obligation not to kill, but not necessarily to save at all costs.

Ronald Dworkin has stressed the importance of distinguishing the sanctity of life view from the view that the individual is a person with rights and interests. According to the doctrine of the sanctity of life, life has intrinsic value even if it is not in a person's own interests to continue living, and even if the focus of discussion is not a person with interests of its own. Thus the sanctity of life doctrine provides an objection to abortion even if is not presumed that the fetus is a person.

Arguments that do depend on rights, however, have to face the problem that different rights of different individuals may conflict, and it is not always clear how they are to be balanced against each other. Utilitarianism offers a way in which to balance the interests of different persons. It is sometimes criticized for being willing to sacrifice one life to save more, because the individual life is not regarded as sacred. Although killing is directly wrong, it is not *absolutely* wrong on this view. For a utilitarian, killing is wrong because of its consequences, both for the person concerned and for third parties. It is wrong to the extent that it prevents happiness, destroys a "worthwhile life," or creates misery. The person killed loses the chance of any future happiness. Third parties may suffer side effects such as distress at the loss of the person and fear for their own fate if the protection against killing is weakened.

A potential killer may nevertheless judge that the person in question does not have a life worth living. While side effects provide some protection against someone carrying out this sort of calculation, there is still a problem in hypothetical situations where adverse side effects can be ruled out. A further argument is that if someone wants to go on living, that is evidence that they have a life that is worthwhile.

If what is valued is the *amount* of happiness or worthwhile life, rather than the intrinsic value of the individual life, then in some circumstances this can be maximized by killing one person to save five. In many cases, again, this objection can be met by pointing to the undesirable side effects of a policy that is willing to sacrifice individuals. At the same time, because utilitarians see consequences as more important than the means of arriving at those consequences, they are less impressed by the distinction between killing and failing to save. Failing to help a person when help is available can be just as bad.

Hard Cases

For some, the right to life is inalienable—it cannot be given up. Others take the view that it can be forfeited; for example, by murderers, so that capital punishment becomes a justifiable form of killing. The greatest controversy, however, occurs over the issues of euthanasia, embryo experimentation, and abortion. In the latter two cases the disagreement is not so much over the right to life *per se* as over the status of the embryo and fetus. What some regard as the possessor of rights, others regard as a collection of cells and the issue has to be resolved by social decision-making, such as laws permitting embryo experimentation for a certain limited time.

Broader Views

A wider interpretation of the right to life could embrace notions of the right to survival of the human species overall. Concerns about environmental degradation and human conflict have led to calls for a balance between the quality of the environment and the sanctity of the dollar, rather than a focus on quality of life and sanctity of life in medical interventions. Such a global bioethics stresses the importance of acceptable survival for the human species. Others go beyond the survival of the human species, expanding the circle of morality to include other species, and respect for all life.

RUTH CHADWICK

SEE ALSO *Abortion.*

BIBLIOGRAPHY

Dworkin, Ronald. (1993). *Life's Dominion: An Argument about Abortion, Euthanasia, and Individual Freedom.* New York: Knopf. Argues that liberals and conservatives have misunderstood each other's positions. Attempts to clarify what it means to say that life is intrinsically valuable.

Glover, Jonathan. (1977). *Causing Death and Saving Lives.* Harmondsworth: Penguin. A classic study of the arguments relating to killing and letting die, including the sanctity of life, autonomy and rights.

Harris, John. (1985). *The Value of Life.* London: Routledge and Kegan Paul. Presents an argument about what it means to value life, with reference to a number of specific areas.

Potter, Van Rensselaer (1988). *Global Bioethics.* East Lansing: Michigan State University Press. Offers an account of bioethics that includes biology and humanistic knowledge to promote acceptable survival.

Singer, Peter (1981). *The Expanding Circle: Ethics and Sociobiology.* Oxford: Clarendon Press. A discussion of expansion of ethical concern.

RIGHT TO REPRODUCE

SEE *Rights and Reproduction.*

RISK AND EMOTION

• • •

Technologies, particularly if they are new, often give rise to emotional reactions that are based on perceived risks. Recent examples of such technological risks involve cloning and genetically modified food; the use of nuclear energy continues to spark heated and emotional debates. Empirical research has shown that people rely on emotions in making judgments about what constitutes an acceptable risk (Slovic 1999). However, this does not answer the question of whether judgments that are based on emotions can provide a better understanding of the moral acceptability of risks than do judgments that do not take the emotions into consideration. Many scientists dismiss the emotions of the public as a sign of irrationality. Should engineers, scientists, and policy makers involved in developing risk regulation take the emotions of the public seriously?

Emotions and Moral Judgments

There are two major traditions in modern moral theory that deal with the role of emotions, going back to the Enlightenment thinkers David Hume (1711–1776) and

Immanuel Kant (1724–1804). For the Scottish philosopher Hume ethics is based not on reason but on the emotions, particularly the sentiment of benevolence, which reason assists in achieving its goals. In opposition to that view the German philosopher Kant maintained that ethics depends on the rational determination of human conduct, with the emotions tending to function as distractions. In neither case, however, are the emotions understood to function in a cognitive manner to reveal something about the world. They are either the noncognitive source of moral value or a noncognitive distraction from moral rationality.

A quite different minority tradition in moral theory, however, grants the emotions cognitive value. This line of thought goes back to Aristotle (1925) who argued that through emotions we perceive morally salient features of concrete situations. In Hume's time the economist Adam Smith (1723–1790) suggested in *Theory of the Moral Sentiments* (1759) that emotional sympathies for others through imaginative identification with their pleasures and pains can provide knowledge about how other people experience the world. For Max Scheler the emotions are the motivators of decent behavior; they reveal the basic moral facts of life (Scheler 1913–1916).

In the 1970s such theories of the cognitive power of the emotions were given new support by developments in neurobiology, psychology, and the philosophy of the emotions. For scholars as diverse as Ronald De Sousa (1987), Robert Solomon (1993), Antonio Damasio (1994), and Martha Nussbaum (2001) emotions and cognitions are not mutually exclusive. Rather, to have moral knowledge, it is necessary to experience certain emotional states.

To be able to have moral knowledge, a person has to know or be able to imagine how it feels to be in a certain situation and to be treated by others in certain ways as well as how it feels when one is humiliated and hurt or cherished and embraced. These emotions are fundamental features of human life that point to what morality is really about. It is not possible to understand moral life without knowing these emotions and without having the ability to feel sympathy and compassion for others. Hence, only beings with the ability to have emotions can make justified moral judgments. The moral point of view implies that people can feel with others or at least imagine what their emotions might be like and that people care about morally important aspects of the lives of others (Schopenhauer 1969, Scheler 1970).

Emotions and Judging the Acceptability of Risks

A cognitive theory of emotions provides new insights about emotions toward acceptable risks. With the traditional picture one would have to choose between the horns of the Hume-Kant dilemma: either take emotions seriously but forfeit claims to rationality or emphasize rationality at the expense of the emotions. With a cognitive theory of emotions, however, one can argue for taking emotions seriously in order to achieve a more comprehensive rationality, particularly with respect to the moral acceptability of technological risks.

As an example, if people are forced against their will to do something they consider dangerous, this is most likely to result in emotions of anger or frustration. However, that is a completely reasonable response. A prima facie injustice has been done to them, and only if they can be persuaded that there are good reasons why they should undergo this specific risk will their anger subside. In contrast, if no good explanation can be given, they will remain upset. In fact, one might find a person irrational who would not get upset by such an injustice. One would judge a person confused who said, "I know company X is not respecting my rights by building this chemical plant in my neighborhood without informing me or asking my consent, and I think it is not fair, but I don't care." A moral judgment that does not lead to an appropriate emotion is seriously flawed.

Some cognitive theories of emotions would take this analysis even further and claim that without certain feelings or emotions a person is unable to have appropriate moral judgments (e.g., De Sousa 1987, Solomon 1993, Damasio 1994, Nussbaum 2001). When people fail to become outraged in response to abridgments of their autonomy, they may not fully grasp the injustice being done to them.

Moreover, people find it morally reasonable not only for the victim of an injustice to be outraged but also for witnesses to be affected in the same way. People even expect that those who inflict an injustice on others should be forced to reassess their actions if they truly care about those they harm. When such agents are unmoved by feelings of sympathy, they are thought of as hard-hearted and egoistical. Emotions thus help assess not only one's own situation but that of others as well as one's own actions in relation to others. In such ways emotions may lead to fairer social arrangements concerning technological risks.

Evaluation of Emotions Concerning Risks

The idea that emotions are useful pathways to moral knowledge concerning risks does not entail the idea that emotions are infallible as normative guides. Emotions also can be wrongheaded or misguided. Emotions can help people focus on certain salient aspects, but they also can lead people astray. Engineers may be enthusiastic about their products and overlook certain risks. The public may be ill informed and thus focus only on risks and overlook certain benefits. Both parties may be biased, and their emotions may reinforce those biases.

In such situations followers of Hume might claim that emotions should rule. Followers of Kant, by contrast, might argue that emotions should be set aside in favor of purely rational analysis. Those who adopt a cognitive theory of the emotions would defend the emotions as a potential source of new knowledge. Not only can reason be brought to bear in a critical manner on the emotions, the emotions may be used as a basis for critical assessments of reason. Indeed, the emotions themselves may be played off against each other in pursuit of mutual emotional assessment. One example would be the development of affective appreciation through sympathy with opposing perspectives. Engineers might try to make an emotional identification with the perspectives of the public, and vice versa, and those who benefit from technology might try to appreciate the perspectives of those who incur its costs. Without emotions being brought into the mix, well-founded judgments about the moral acceptability of technological risks are unlikely.

S A B I N E   R O E S E R

SEE ALSO *Emotion; Emotional Intelligence; Hume, David; Kant, Immanuel; Risk; Risk Assessment; Risk Perception; Risk Society.*

BIBLIOGRAPHY

Aristotle. (1925). *Nichomachean Ethics*, trans. W. D. Ross. London: Oxford University Press. A classic source of emotional cognitivism in ethics.

Damasio, Antonio. (1994). *Descartes' Error: Emotion, Reason, and the Human Brain.* New York: Putnam. Readable classic in recent neuropsychology, arguing that without emotions, people cannot be practically rational.

De Sousa, Ronald. (1987). *The Rationality of Emotion.* Cambridge, MA: MIT Press.

Nussbaum, Martha. (2001). *Upheavals of Thought.* Cambridge, UK: Cambridge University Press. A strong defense of cognitivism in the philosophy of emotions.

Scheler, Max. (1913–1916). *Der Formalismus in der Ethik und die Materiale Wertethik. Formalism in Ethics and Non-Formal Ethics of Values: A New Attempt toward a Foundation of an Ethical Personalism* (1973), trans. Manfred S. Frings and Roger L. Funk. Evanston, IL: Northwestern University Press.

Scheler, Max. (1970). *The Nature of Sympathy,* trans. Peter Heath. New York: Archon Books. A philosopher from the continental phenomenological tradition argues that people acquire moral knowledge through sympathy.

Schopenhauer, Arthur. (1969). *The World as Will and Representation,* trans. E. F. J. Pane. New York: Dover. Originally published as Die Welt Als Wille und Vorstellung (1819). Defends the importance of sympathy in ethics.

Slovic, Paul. (1999). "Trust, Emotion, Sex, Politics, and Science: Surveying the Risk-Assessment Battlefield." *Risk Analysis* 19: 689–701. Psychologist who studies the role of the emotions in assessing risk.

Smith, Adam. (1976). *Theory of Moral Sentiments.* Oxford: Oxford University Press. Defends the importance of moral emotions.

Solomon, Robert. (1993). *The Passions: Emotions and the Meaning of Life.* Indianapolis: Hacket. Solomon started a renewed interest in emotional congitivism in philosophy in the 1970s.

RISK AND SAFETY: OVERVIEW

• • •

Risk and safety are polyvalent concepts with numerous and overlapping ethical complexities in relation to science and technology. As such they are dealt with in a number of different entries.

In technical terms, scientific phenomena may exhibit certainty, risk, or uncertainty. Situations of *certainty* have a probability of 1. For example, all things being equal, it is certain (probability = 1) that water freezes when cooled below 0° Celsius. Cases of *risk* have some numerical probability between 0 and 1, based on a known or assumed model of what causes the outcome under study. For instance, the risk of tossing "heads" on a fair coin has a probability of 0.5, because the model is known. In risk assessment, the risk of something is typically defined as the average annual estimated probability of causing a fatality. Cases of *uncertainty* cannot be defined *a priori* in terms of probabilities, because of inadequate knowledge. To assign legitimate or scientifically valid probabilities, one needs experimental or frequency/statistical data; an example is data on automobile accidents for drivers of a given age. In many cases of uncertainty there may simply be no adequate data.

Despite the name, "risk assessors" typically do not assess cases of risk (with known or well-established probability between 0 and 1), but situations of great uncertainty. When people have "risk" knowledge, they do not need risk assessment. In part because they address uncertainty in extremely complex situations, risk assessments usually err between four to six orders of magnitude

(Shrader-Frechette 1991). That is, fatalities predicted by risk assessments typically are (later proved to be) wrong by factors of 10,000 to 1,000,000. Most predictions are too low and exhibit an "overconfidence bias" in favor of some technology (Kahneman and Tversky 2000; Kahneman, Slovic, and Tversky 1982).

It is against this technical background that the following entries on risk need to be read: "Risk Assessment," "Risk Ethics," and "Risk Perception." Other entries—such as "Risk and Emotion" and "Risk Society"—make an effort to move beyond the more strictly technical understanding of risk.

There is no technical concept of safety analogous to that of risk. Nevertheless, according to an influential analysis by William W. Lowrance, safety can be defined in terms of risk: "A thing is safe if its risks are judged to be acceptable" (1976, p. 8). Mike W. Martin and Roland Schinzinger pointed out as early as 1983 that this definition needs a qualifier: The judgment of acceptability needs to be done with adequate knowledge. Free consent is not enough; it must be free and informed.

Langdon Winner, however, has gone further and warned against defining safety in terms of risk. According to Winner, traditional efforts to promote safety had a clear goal of eliminating certain "workplace dangers" or "health hazards." But when the promotion of safety involves assessing risks in terms of their acceptability, the goal fades into "studying, weighing, comparing, and judging circumstances about which no simple consensus is available" (1986, p. 143). It is against this critical background that the articles on "Safety Engineering: Historical Emergence," "Safety Engineering: Practices," and "Safety Factors" need to be considered.

There are also a number of articles that are related to the concepts of risk and safety. Among these it is useful to mention "Exposure Limits" and "Hazards." Even more specific topics include "Radiation" and "Regulatory Toxicology."

CARL MITCHAM

BIBLIOGRAPHY

Kahneman, Daniel, and Amos Tversky, eds. (2000). *Choices, Values, and Frames*. New York: Russell Sage Foundation; Cambridge, UK: Cambridge University Press.

Kahneman, Daniel; Paul Slovic; and Amos Tversky, eds. (1982). *Judgment under Uncertainty: Heuristics and Biases*. Cambridge, UK: Cambridge University Press.

Lowrance, William W. (1976). *Of Acceptable Risk: Science and the Determination of Safety*. Los Altos, CA: William Kaufmann.

Martin, Mike W., and Roland Schinzinger. (1983). *Ethics in Engineering*. New York: McGraw-Hill. 4th edition, 2005.

National Research Council. Committee on Risk Characterization. (1996). *Understanding Risk: Informing Decisions in a Democratic Society*. Washington, DC: National Academy Press.

Resnik, Michael D. (1987). *Choices: An Introduction to Decision Theory*. Minneapolis: University of Minnesota Press. A good introduction to distinctions between certainty, risk, and uncertainty.

Shrader-Frechette, Kristin S. (1991). *Risk and Rationality: Philosophical Foundations for Populist Reforms*. Berkeley: University of California Press.

Winner, Langdon. (1986). *The Whale and the Reactor: A Search for Limits in an Age of High Technology*. Chicago: University of Chicago Press.

RISK ASSESSMENT

• • •

Many decisions involve an intuitive assessment of risk; this subjective risk assessment is usually called *risk perception*. Risk assessment is also a formalized approach to evaluating risk, often defined as a function of the probability and magnitude of loss or harm from an event. Risk assessment is often thought of as ethically obligatory, but since it can be done in more than one way, it is itself subject to ethical assessment.

Risks are routinely assessed formally for a wide variety of human endeavors, from drinking tap water to operating nuclear power plants; for natural hazards such as earthquakes, hurricanes, and floods; and for the human use of and exposure to chemicals and other substances such as arsenic or phthalates. Risks may also be defined and assessed in terms of specific harms or losses to people, for example a person's lifetime risk of dying of heart disease, or aquatic ecosystem risks from anthropogenic eutrophication (that is, being overburdened with nutrients as a result of human action). While failing to assess risk can lead to Faustian bargains with the future, risk assessments for public policy can be risky in themselves, as illustrated by the effects of transnational debates about risk assessments of genetically modified organisms, vaccines, and terrorism.

Methods

As described in *Risk Assessment in the Federal Government* (known as the "Red Book," 1983), risk assessment consists of four steps: hazard identification, dose-response assessment, exposure assessment, and risk characterization. More broadly, risk assessment entails identifying and characterizing an underlying hazard—including its sources, pathways, effects for given exposures, and mitigating factors, and estimating the associated contingent probabilities. In effect, formal risk assessment requirements are intended to insure that human and even ecological health is considered in decisions with other primary objectives.

For example, product risk assessment may be required by law, as in the case of new pharmaceuticals in the United States. In the United States, the Food and Drug Administration (FDA) assesses the adequacy of new drug risk assessments, including how they are conducted. The FDA also determines what constitutes permissible risk for a licensed product. As risk assessments are generally conducted in the service of specific risk management objectives, the two are mutually dependent (Committee on Risk Assessment of Hazardous Air Pollutants 1994). In some venues separation of risk assessment and risk management is considered critical to protect the science of risk assessment from contamination by management or political pressures. However, many formal risk assessment processes now include participation by multiple stakeholders to deliberate about risk management objectives and values, in addition to experts' technical analyses (Stern and Fineburg 1996).

Human health risk assessments are of necessity carried out at a population or group level—that is, for a statistical person rather than an identified individual. They are based on extrapolations from animal studies; on experimental tests of human product use, which usually involve relatively small samples; or on epidemiological studies, which rely on statistical controls. Recent developments in risk assessment have included the ability to tailor risk assessment results interactively for subpopulations, as is illustrated by online risk calculators that determine an individual's risk based on a few personal characteristics. But individual differences can make the population health risk assessments applied in policy decisions more or less applicable, for which reason minority populations may be poorly served by general risk assessments. An example in point is airbags in cars, which when designed to optimally protect average adults may harm or kill children.

Environmental risk assessment, as required for example in environmental impact statements, has focused largely on risks to human health and the economy, but increasingly addresses ecological endpoints. Because selection of assessment endpoints can determine the structure and outcome of decisions, it is inherently controversial. Assessing risks from ozone only in terms of economic loss from damage to automobile tires paints a very different picture of the size of the risks than if the assessment also takes into account acute respiratory or cardiovascular events triggered by exposure to ozone, or possible ecological effects of ozone, such as reduced growth rates and plant deformation.

Basic Issues

By focusing on probabilistic loss, risk assessment frames management choices in terms of threat reduction and loss avoidance. Common criticisms of risk assessments have included that they are based on an overly narrow conceptualization of benefits, or that the dimensions of harm included are insufficient or inappropriate. It is difficult to incorporate into a risk assessment even proxy measures for intangibles—such as quality of life—or other poorly defined or understood endpoints. In part to take into account uncertainties, risk assessments are sometimes designed to produce estimates of risk that err on the high side, for example by using upper bounds of estimated risks, rather than averages. Those risk assessment procedures that have been codified by government entities incorporate scientific procedures, including requirements for representative empirical data, statistical analyses, and quality control in the form of peer review. Some also include ethical requirements, such as human subjects review, or the participation of parties who may have a substantive interest in the value at risk.

Four issues are key to risk assessment as currently practiced. The first is what is valued, how and by whom it is valued, and the distributive implications thereof. The selection of assessment endpoints can have far-from-obvious implications, as the airbag example illustrates. Assessing values remains a methodological and ethical challenge (Fischhoff 1991, Slovic 1995).

The second is the treatment and interpretation of uncertainty—both uncertainty stemming from limits to what is known, and uncertainty stemming from inherent variability (see Morgan and Henrion 1990). Especially in the case of extremely rare and catastrophic events, the selection of a distribution function or simulation procedure with which to analyze uncertainties can influence the outcome of the assessment considerably. Similarly, choosing how to represent the results of the risk assessment and the uncertainty therein can influence how recipients interpret and use the assessment.

The third key issue is the substitutability implied or assumed by risk assessment, as it often requires comparative values. As has been illustrated in discussions of protected values and irreversible effects, in reality trade-offs are sometimes impossible or unethical.

Fourth is that technically competent risk assessment requires significant resources, is both analytically and data-intensive, and can be difficult to interpret. Risk assessments that are carried out for new drugs, for example, require expertise in toxicology and epidemiology and investments in large studies, which still may not be large enough to discover devastating rare or long-term adverse effects.

Risk assessments may produce risk characterizations that are not readily used to compare or prioritize risks. For example, ecological risk assessments may conclude simply that a specific species is at some risk of extinction, while a human health risk assessment may produce an estimated probability of a specific health endpoint within a given timeframe, for example a five percent probability of being diagnosed with breast cancer within five years. Comparing the two is difficult.

For this reason it is desirable that risk assessment outcomes be translatable to a common measure, such as an abstract measure of utility, or monetary value. Summary endpoints like the probability of human mortality or morbidity, or economic loss, can be presented in a common metric that facilitates at least some comparisons, such as disability adjusted life years, or monetary value. But choice of a common metric itself can be problematic, both because individuals may not agree on the equivalence of different forms of bodily injury or harm and because not all endpoints can be equally well represented by all measures. In addition, some measures, such as dollars, carry their own meaning, which may or may not facilitate the risk assessment depending on how that meaning is construed.

However, no single metric or endpoint necessarily constrains environmental, technological, or human health risk assessments. Although many risk assessors with economic training might prefer to use dollars as a summary endpoint, doing so is not a requirement of risk assessment, but a methodological choice with ethical implications. The identification and definition of possible endpoints to consider, the valuation of these, and the estimation of their contingent probabilities all entail some degree of judgment and choice.

ANN BOSTROM

SEE ALSO *Risk; Risk and Emotion; Risk-Cost-Benefit Analysis; Risk Ethics; Risk Perception; Risk Society.*

BIBLIOGRAPHY

Committee on the Institutional Means for the Assessment of Risks to Public Health, Commission on Life Sciences, National Research Council. (1983). *Risk Assessment in the Federal Government: Managing the Process.* Washington, DC: National Academy Press. Known as the "Red Book," this is probably the most cited publication on risk assessment.

Committee on Risk Assessment of Hazardous Air Pollutants, Board on Environmental Studies and Toxicology, Commission on Life Sciences, National Research Council. (1994). *Science and Judgment in Risk Assessment.* Washington DC: National Academy Press. An update on the "Red Book."

Fischhoff, B. (1991). "Value Elicitation: Is There Anything in There?" *American Psychologist* 46(8): 835–847. Summarizes the literature on value elicitation succinctly and with eloquence.

Morgan, M. Granger, and Max Henrion, with Mitchell Small. (1990). *Uncertainty: A Guide to Dealing with Uncertainty in Quantitative Risk and Policy Analysis.* Cambridge, UK: Cambridge University Press. A key publication on uncertainty.

Slovic, P. (1995). "The Construction of Preference." *American Psychologist* 50(5): 364–371. A must-read for anyone interested in values or preferences.

Stern, Paul C., and Harvey V. Fineberg, eds. (1996). *Understanding Risk: Informing Decisions in a Democratic Society.* Washington, DC: National Academy Press. Articulates the importance of deliberation as well as analysis in risk assessment.

UK Royal Society. (1992). *Risk Analysis, Perception, and Management: Report of a Royal Society Study Group.* London: The Royal Society. An important summary of EU perspectives on risk assessment and related issues.

RISK ETHICS

• • •

Risk ethics is an emerging branch of philosophy that investigates the moral aspects of risk and uncertainty. Although one originating motivation in the pursuit of science and technology was an effort to reduce risk and uncertainty present in the natural world, it has been increasingly appreciated that the scientific and technological world presents its own constructed risks. Recognizing that one form of risk (natural) is overcome only at the cost of another form of risk (involved with science or technology) has stimulated critical reflection on risk in ways that did not occur in the absence of technological risk.

A Brief Introduction to Risk Concepts

Risk has vernacular and technical meanings. In everyday language a risk is simply a danger. But in relation to science and technology, risk is often defined as the probability of some harm. The probability of a benefit is often called a chance. According to another common definition, risk is identified with the value obtained by multiplying the probability of some harm or injury by its magnitude. With any attempt to spell out the details of how this might be done, however, problems arise since it is not clear that there is a single measure for all harms or injuries. Attempts have been made to measure all health effects in terms of quality-adjusted life years (Nord 1999). Risk-benefit analysis goes one step further and measures all harms in monetary terms (Viscusi 1992). However, as several critics have pointed out, such unified approaches depend on controversial value assumptions and may be difficult to defend from an ethical point of view (Shrader-Frechette 1992).

Independent of methodological issues, however, are the assumptions of traditional moral philosophy, which has focused on situations in which the morally relevant properties of human actions are both well-determined and knowable. In contrast, moral problems in real life often involve risk and uncertainty. According to common moral intuitions it is unacceptable to drive a vehicle in such a way that the probability is 1 in 10 that one runs over a pedestrian, but acceptable if this probability is 1 in 1 billion. (Otherwise one could not drive at all.) It is far from clear how standard moral theories can account for the difference and explain where the line should be drawn.

Utilitarianism

In utilitarian ethics, all moral appraisals are reducible to assignments of utility, a (numerical) measure of moral value. Furthermore, the utility of human actions is assumed to depend exclusively on their consequences. According to utilitarianism one should always choose the alternative that has the highest utility, that is, the best consequences.

One utilitarian approach to risk is *actualism*, according to which the moral value of a risky situation is equal to the utility of the outcome that actually materializes. For example, suppose that an engineer decides not to reinforce a bridge in advance of it being subject to an exceptionally heavy load, although there is a 50 percent risk that the bridge will collapse under such use. If all goes well and the bridge carries the load, then according to the actualist standpoint what the engineer did was right. But examples such as this show that actualism cannot provide meaningful action guidance. Even if actualism is accepted as a method for retrospective

moral assessment, another theory is needed to guide decision-making about the future.

One such theory is *expected utility maximization,* which has become the standard utilitarian approach to risk. According to this theory, the utility of the prospect that an outcome may occur is obtained by multiplying the utility of the outcome itself by its probability. Then, the action with the highest probability-weighted value should be chosen. According to this rule, an action with the probability 1 in 10 to kill a person is five times worse than an action with the probability 1 in 50 of the same outcome. This method for weighing potential outcomes is routinely used in risk analysis.

In intuitive arguments about risk, it is common to give the avoidance of very large disasters, such as a nuclear accident costing thousands of human lives, a higher priority than is warranted by probability-weighted utility calculations. For instance, people clearly worry more about the possibility of airplane crashes (low-probability but high-cost events) than automobile accident deaths (which are higher-probability but lower-cost events). Expected utility maximization disallows such cautious decision-making. Proponents of precautionary decision-making may see this as a disadvantage of utility maximization, whereas others may see it as a useful protection against costly over-cautiousness.

Just like other forms of utilitarianism, expected utility maximization is strictly impersonal. Persons have no role in the ethical calculus other than as bearers of utilities whose values are independent of those who carry them. Therefore, a disadvantage affecting one person can always be justified by a sufficiently large advantage to some other person. No moral distinction is made between the act of exposing oneself to a serious danger in order to gain some advantage and the act of exposing someone else to the same danger for the same purpose. This is a problematic feature of utilitarian theory in general that is often aggravated in problems involving risk.

Duty- and Rights-Based Theories

A moral theory that is based on duties (rather than on the consequences of actions) is called deontological or duty-based. A moral theory in which rights have the corresponding role is called rights-based.

Robert Nozick formulated the problem for rights-based theories in dealing with risks in this way: "Imposing how slight a probability of a harm that violates someone's rights also violates his rights?" (Nozick 1974, p. 7). Similarly, one may ask the following question about deontological theories: "How large must the probability be that one's action will in fact violate a duty for that action to be prohibited?"

One possible answer to these questions is to prescribe that a (rights- or duty-based) prohibition to bring about a certain outcome implies a prohibition to cause an increase in the probability of that outcome (even if the increase is very small). But such a far-reaching extension of rights and duties is socially untenable. Human society would be impossible if people were not allowed to perform actions such as car driving that involve a small risk of developing into a violation of some prohibition.

It seems clear that rights and prohibitions may lose their force when probabilities are sufficiently small. The most obvious way to account for this is to assign to each duty or right a probability limit below which it is not valid. However, no credible way to derive such a limit has been proposed. It is also implausible to draw the line between acceptable and unacceptable probabilities of harm with no regard to the benefits involved. (In contrast, such weighing against benefits is easily accounted for in utilitarian theories.)

Contract Theories

According to contract theories, the moral principles that rule humans' dealings with each other derive from a contract between all members of society. The social contract prohibits certain actions, such as actions that lead to the death of another person. Under what conditions should it also prohibit actions with a low but nonzero probability of leading to the death of another person? The most obvious response to this question is to extend the criterion that contract theory offers for the determinate case, namely consent among all those involved, to cases involving risk and uncertainty. This can be done in two ways because consent, as conceived in contract theories, can be either actual or hypothetical.

According to the criterion of actual consent, all members of society would have a veto over actions that expose them to risks. This would make it virtually impossible, for example, to site industries that are socially necessary but give rise to emissions that may disturb those living nearby. With a rule of actual consent, a small number of nonconsenting persons would be able to create a society of stalemates, to the detriment of everyone else. Therefore, actual consent is not a realistic criterion in a complex society in which everyone performs actions with marginal effects on the lives of many others.

Contract theory has a long tradition of operating with the hypothetical consent that is presumed to be given by every hypothetical participant in an ideal deci-sion situation such as described in John Rawls's "original position." Unfortunately, none of the ideal situations constructed by contract theorists seems to have made the moral appraisal of risk and uncertainty easier or less dependent on controversial values than the corresponding appraisals in the real world.

Widening the Issue

Many discussions of risk have been limited by an implicit assumption that excludes important ethical aspects. It is assumed that once we have moral appraisals of actions with determinate outcomes, we can more or less automatically derive moral appraisals of actions whose outcomes are "probabilistic mixtures" of such determinate outcomes. Suppose, for instance, that moral considerations have led us to attach well-determined values to two outcomes X and Y. Then we are supposed to have the means needed to derive the values of mixed options such as 70 percent chance of X and 30 percent chance of Y. The crucial assumption is that the probabilities and values of nonprobabilistic alternatives completely determine the values of probabilistic alternatives.

In real life, however, there are always other factors in addition to probabilities and utilities that properly influence our moral appraisals of an uncertain or risky situation. We need to know not only the values and probabilities of potential outcomes, but also who exposes whom to risk and with what intentions, the extent to which the exposed person was informed, whether or not the person consented, and more.

Perhaps the most important foundational problem in risk ethics is the conflict between two principles that both have intuitive appeal. They can be called the collectivist and the individualist principles in risk ethics (Hansson 2004). According to the collectivist principle of risk ethics, exposure of a person to a risk is acceptable if and only if this exposure is outweighed by a greater benefit either for that person or others. According to the individualist principle, exposure of a person to a risk is acceptable if and only if this exposure is outweighed by a greater benefit for that person only.

The collectivist principle dominates traditional risk analysis, but if carried to extremes it will lead to neglect of individual rights. The individualist principle is equally problematic, because it allows minorities to prevent social progress. It is a major challenge for risk ethics to find a reasonable and principled compromise between these two extreme positions.

SVEN OVE HANSSON

SEE ALSO *Risk Assessment; Risk Perception.*

BIBLIOGRAPHY

Hansson, Sven Ove. (2003). "Ethical Criteria of Risk Acceptance." *Erkenntnis* 59(3): 291–309. Discusses how risks can be dealt with in different moral theories.

Hansson, Sven Ove. (2004). "Weighing Risks and Benefits." *Topoi* 23: 145–152. Discusses different ways to weigh risks against benefits.

Hansson, Sven Ove, and Martin Peterson. (2001). "Rights, Risks, and Residual Obligations." *Risk, Decision, and Policy* 6(3): 157–166. Discusses the obligations that follow from exposing the public to risk.

Nord, E. (1999). *Cost-Value Analysis in Health Care: Making Sense Out of QALYs.* Cambridge, UK: Cambridge University Press. One unified measure of human injuries.

Nozick, Robert. (1974). *Anarchy, State, and Utopia.* New York: Basic Books. Discusses rights-based approaches to risk.

Shrader-Frechette, Kristin. (1992). "Science, Democracy, and Public Policy." *Critical Review* 6(2–3): 255–264. A critical appraisal of cost-benefit analysis.

Thomson, Judith. (1985). "Imposing Risk." In *To Breathe Freely: Risk, Consent, and Air*, ed. Mary Gibson. Totowa, NJ: Rowman and Allanheld. Person-related aspects of risk exposure.

Viscusi, K. (1992). *Fatal Tradeoffs: Public and Private Responsibilities for Risk.* New York: Oxford University Press. The author is a leading proponent of risk-benefit analysis.

RISK PERCEPTION

• • •

Risk perception has been defined variously as perceived or subjective probability estimates of death, other judgments of probable harm or loss, psychological states such as fear or traumatic stress, beliefs about causal processes resulting in harm or loss—that is, mental models of hazardous processes, or attitudes toward the activity, event, product, or substance in question. Risk perception, in which risk is assessed subjectively, often without formal decomposition into probability and harm, is frequently treated as folk or lay *risk assessment.*

When elicited as subjective probability or frequency of mortality, risk perceptions can agree or disagree with actuarial information, where such exists, and can in some instances be validated or invalidated by science. Comparisons of lay and expert risk perceptions, together with research on the effects of risk communication, illustrate that expertise and information can have large effects on risk perceptions. Such comparisons have been used to make the ethical claim that non-experts are irrational when they fear risks that experts deem

acceptable, such as risks from genetically modified organisms. Shrader-Frechette points out that those framing risk questions control the answers, and suggests that to deal with the great uncertainties surrounding, for example, ecological risks, the burden of proof should fall on those proposing that a risk is acceptable. Shrader-Frechette also proposes a three-category framework for risk, as an alternative applying the effect-no effect (or acceptable-unacceptable) dichotomized view of science to risks. In her view, serious risks for which the complexities and uncertainties are so great that we lack sufficient information to make a decision fall into a third category (e.g., Shrader-Frechette, 1994). However, as intuitive statisticians, both experts and non-experts are subject to predictable judgmental biases (Fischhoff, Bostrom, and Jacobs Quadrel 2002; Gilovich, Griffin, and Kahneman 2001; Kahneman, Slovic, and Tversky 1982). Personal experiences also affect risk perceptions, though if not repeated their effects may disappear over time. That communities enact policies to reduce their seismic risks following large earthquakes and resist or ignore them at other times testifies to this, as do differences between life scientists and other scientists in their risk perceptions.

Schools of Thought

Risk perception research since the 1970s has been characterized by several schools of thought, each of which is associated with particular disciplinary backgrounds and methodological predilections. Psychometric research and cultural theory are among the most widely acknowledged.

Psychometric research on risk perception proceeded by analogy with measurements of physical perceptions—such as light, weight, or heat—in attempting to establish reliable, validated psychological scales for perceived risk. By eliciting people's judgments on dimensions such as dread, familiarity, catastrophic potential, and control, researchers were able to predict, to some extent, risk acceptance judgments. This research produced a risk factor space, the two dimensions of which were how familiar, controllable, and understood risks are, and how much people dread them, including judgments of catastrophic potential. For example, the risks from nuclear power are typically perceived as highly unknown and dreaded, landing in the upper right quadrant of those two dimensions, where as the risks from bicycles are perceived as known and are not dreaded, putting them in the lower left quadrant. This vein of research is best characterized in works by Paul Slovic, Baruch Fischhoff, Sarah Lichtenstein, and colleagues (Slovic 2000).

Cultural theory stems from anthropologist Mary Douglas's writings on risk and culture. Among the best-known tests of cultural theory are those that employ grid/group theory, in which it has been shown that people's attitudes toward risks are a product of their degree of individualism, egalitarianism, and hierarchy or collectivism. Related research on worldviews posits that risk perceptions are a function of attitudes toward science and technology in particular, but also other attitudes.

Another approach is to treat risk perception as an instance of information processing. Information processing is cognitive, social, and affective (Damasio 1994). Cognitive processes such as categorization, similarity judgments, and inference from mental models are, from an information processing perspective, all components of risk perception. Recent research shows that there is a strong relationship between affect and perceived risk. There is a commonly observed inverse relationship between perceived risk and perceived benefit. Under time pressure, which limits analytic thought and increases reliance on affect, this inverse relationship strengthens (Finucane, Alhakami, Slovic, and Johnson 2000). Further, introducing information that changes one's affective evaluation of an item, for example information that associates nuclear power with clean air and pastoral scenes, can systematically change both the related risk and benefit judgments.

People seem prone to using an "affect heuristic" that improves judgmental efficiency by deriving both risk and benefit evaluations from a common source: affective reactions to the stimulus item. The mechanisms for these effects may be hardwired in our brains, in the amygdala, through which all thought passes. Animal studies suggest that the amygdala coordinates multiple fear systems, and that fear is a potent determinant of memory, learning, and salience.

Ethical Issues

People's behavior depends on their risk perceptions. Given this dependency, whose risk perceptions should prevail to determine societal priorities is often contested. Further, technical risk assessments generally apply to a statistical person or to a population, and so are not directly applicable to an individual or that individual's perceptions of his or her own risk. Therein lies the central ethical dilemma posed by risk perceptions, exacerbated by their variability and vulnerability to judgmental biases.

In addition, overarching ethical principles conflict with manipulations of risk perceptions that may, at face value, seem in the public interest. Principles such as those in the U.S. Bill of Rights are vulnerable to perceived needs precipitated by risk perceptions. As the U.S. Public Law 107–56 (commonly known as the U.S. Patriot Act, 2001) and the U.K. Anti-Terrorism, Crime, and Security Act (also 2001) illustrate, it is easy to delimit transparency of government, judicial checks on legislative and executive branches, and civil liberties and equal treatment of citizens under the guise of reducing risks, even without evidence that the measures enacted will actually reduce risks.

The literature on risk perception across different domains of science and technology is daunting. Health, environmental, and technological risk perception, and to some extent hazard perception, are largely separate bodies of research. Health risk perception research is rooted primarily in social psychology, and has been dominated by the health belief model, the theory of reasoned action, and variants thereon. This research is influenced by the extended parallel process model, which predicts that people who believe something poses a serious risk to them personally will engage in fear control rather than risk control if they do not believe that they can control the risk effectively (Witte 1992). Environmental and technological risk perception research has drawn more broadly on social and cognitive sciences, including the theories and models cited above. Methods have varied from informal and sometimes misleading reliance on casual observations, such as of focus groups, to carefully designed and implemented surveys and experiments. Anthropology and ethnographic methods of studying risk perceptions have grown in importance, as practitioners have recognized their value in improving the design of risk interventions, as well as providing a fuller account of how people perceive risk.

Spatial and temporal dimensions of risk perceptions remain to be fully explored, and will likely provide further insights into risk behaviors.

ANN BOSTROM

SEE ALSO *Risk; Risk and Emotion; Risk-Cost-Benefit Analysis; Risk Ethics; Risk Perception; Risk Society.*

BIBLIOGRAPHY

Damasio, Antonio R. (1994). *Descartes' Error: Emotion, Reason and the Human Brain.* New York: Putnam.

Douglas, Mary, and Aaron Wildavsky. (1983). *Risk and Culture: An Essay on the Selection of Technical and Environmental Dangers.* Berkeley: University of California Press.

Finucane, M.L.; A. Alhakami; P. Slovic; and S. M. Johnson. (2000). "The Affect Heuristic in Judgments of Risks and Benefits." *Journal of Behavioral Decision Making* 13: 1–17.

Fischhoff, B.; A. Bostrom; and M. Jacobs Quadrel. (2002). "Risk Perception and Communication." In *Oxford Textbook of Public Health*, 4th edition, ed. Roger Detels, et al. Oxford, UK: Oxford University Press.

Gilovich, Thomas; Dale Griffin; and Daniel Kahneman, eds. (2001). *Heuristics and Biases: The Psychology of Intuitive Judgment*. Cambridge, UK: Cambridge University Press.

Kahneman, Daniel; Paul Slovic; and Amos Tversky, eds. (1982). *Judgment Under Uncertainty: Heuristics and Biases*. Cambridge, UK: Cambridge University Press.

Shrader-Frechette, K. (1994). "Science, Environmental Risk Assessment, and the Frame Problem." *BioScience* 44(8), 548–552.

Slovic, Paul. (2000). *The Perception of Risk*. London: Earthscan.

Witte, Kim. (1992). "Putting the Fear Back into Fear Appeals: The Extended Parallel Process Model." *Communication Monographs* 59: 329–349.

RISK SOCIETY

• • •

The concept of risk, long associated with the language of maritime trade and insurance, has become a key term for characterizing contemporary Western societies. Important early contributions to the development of this analysis were the work of Patrick Lagadec (1981), who coined the term *risk civilization*, and that of Mary Douglas and Aaron Wildavsky (1982). However, Ulrich Beck's *Risk Society* (1992), originally published in German in 1986, was the decisive contribution to a new theory of society. Beck's conceptualization has inspired research that focuses on the implications of science and technology for the social and natural environment and on the increasing use of risk analysis in discussions of public policies related to science and technology, and which involve ethical questions.

Reflexive Modernity

Beck's theory represents a continuation of the German tradition of an ethical questioning of modernity, including science and technology, that runs from Max Weber (1864–1929) through Jürgen Habermas (b. 1929). In contrast to postmodern theories that present late twentieth-century social transformations as going beyond modernism, Beck argues that modernity is going through an unintended and unseen phase that is forcing it to confront the premises and limits of its own model. Modernization has become, in his words, "reflexive." The concept of reflexive modernization, which was introduced by Beck and developed in a subsequent work with Anthony Giddens and Scott Lash (Beck, Giddens, and Lash 1994), propounds a "radicalization" of modernity in which the dynamics of individualization, globalization, gender revolution, underemployment, and global risks undermine the foundations of classical industrial modernity and make old concepts obsolete. The internal dynamism of modernity brings it up against the previously unknown possibility of global self-destruction as a result of the risks generated by certain technologies.

Beck thus depicts the risk society as coextensive with reflexive modernity. In the same way that "simple modernity" produced goods and services that presented challenges involving just distribution, reflexive modernity is producing risks that must be distributed justly.

An Expanded Concept of Risk

Many theoretical works in other disciplines had previously analyzed the risk concept, although more narrowly: economics, behavioral theory (in particular decision making and game theory), anthropology, and technology assessment.

In economics, where the concept has always been fundamental, prevailing interpretations make a clear distinction between risk and uncertainty. Whereas risk can be assessed and calculated in terms of its numerical probabilities, uncertainty cannot be treated in that manner. Introduced at the beginning of the twentieth century by Frank Knight (1885–1972) and John Maynard Keynes (1883–1946), this distinction made possible the recognition of the ontologically contingent nature of economic behavior and its aggregate outcomes. An economic agent cannot avoid wide margins of uncertainty or eliminate it by means of the application of more information or scientific knowledge.

The anthropological work of Douglas and Wildavsky (1982) diverges from this classical approach in emphasizing the subjective aspect of risk and the ways in which risk is assessed and perceived by individuals. Their work helped significantly to shift attention away from a probabilistic approach to the cultural framework of risk perception. Variations in the understandings and perceptions of risk in different societies demonstrate the cultural relativism involved in judgments of risk.

Beck's main contribution was to build risk systematically into a theory of modern society and its dilemmas. Risk is seen as a defining feature of society itself, forming the dark side of industrial successes, technical and scientific progress, and economic growth. It has stimulated changes in social relations, family structure, political and cultural organization, and even the self.

Unlike the threats of early industrialization, the risks of "late modernity" (nuclear, chemical, genetic, ecological, etc.) are generated by techno-economic decisions and considerations of utility. The novel aspect of contemporary risk society is that people's decisions as a civilization lead to problems and dangers that radically contradict the established language of control and conventional techniques of calculation. Current risks are not socially, spatially, or temporally demarcated; there are no clear-cut solutions; and it is difficult to trace responsibility or assess compensation for those who are affected. In addition, human perception fails to notice many of the risks: they become visible only through scientific interpretation (as in the case of stratospheric ozone depletion), which in turn increases dependence on experts.

Beck focuses above all on environmental and health risks, especially genetic technology. He later extended the concept of risk to global financial crises and transnational terrorist networks (Beck 2002). Bringing together such disparate phenomena enables him to identify relevant trends in modern societies but has the drawback of implying a less fragmented world than that which Beck perceives.

Niklas Luhmann (1993 [1991]) has enriched "risk society" analysis with his theory of autopoietic systems. Here risk is a specific form of dealing with the future that has to be decided in the context of probability and improbability. The uncertain and unforeseeable nature of the future arises not only from complexity and people's cognitive limitations but also from the decision-making process itself. There is a long hiatus between when a decision is made and when its consequences are felt, with random factors affecting them. To talk of risks is to see future losses as the consequence of a decision that has been made. For Luhmann this is where "risk" differs from "danger," with danger being attributable to external causes and corresponding to those "affected" by decisions. Although the distinction is slight because "one person's risk is another person's danger," it points to the key issue of acceptance of risk decisions.

Developments and Implications

Beck's message on the relationship between science, technology, politics, and ethics in late modernity is that our language does not inform future generations of the dangers people create when they use certain technologies. As it develops technologically, society encounters the difference between two worlds: the language of quantifiable risk, in which people think and act, and that of nonquantifiable insecurity, which people also are

creating. As risks become more complex and the need for precise calculations increases, there is growing doubt about the ability of science to control and foresee those risks. This situation has shaken the belief that technological and social progress go together and has forced science to acknowledge both its collateral effects and its inherent epistemological limitations. The concept of "world risk society" (Beck 1999) draws attention precisely to the limited controllability of globalized and artificially produced risks.

In these circumstances human responsibility for technological advancement is an ethical issue that is both relevant and complex. For Beck the processes and techniques of risk management block out responsibility. Modern society operates as a "laboratory" in which no one in particular must answer for the negative effects of technological experimentation. The institutions of modern society recognize the existence of risk but permit an "organized irresponsibility" (Beck 1995 [1988]). Pollution, along with its increasingly global impact in the form of climate change, graphically illustrates this paradox. The greater the environmental degradation is, the more laws and environmental regulations there are, but at the same time no institution seems to be specifically responsible.

Technologically induced risks lead to calls for the demonopolization of scientific expertise, its subjection to social scrutiny, and extension of democratic accountability to science, technology, economics, and government. For this to be achieved politics must "(re)-invent" itself and focus on issues previously regarded as apolitical. What once was the exclusive province of science has become the subject of intense political debate, as in the case of biotechnology. In this context individual citizens, movements, and interest groups participate and influence political decisions in the field that Beck describes as "sub-politics," which is located beyond the formal representative institutions of the political system.

Because the concept of risk is probabilistic in nature, it tends to deny inherent uncertainties and place greater emphasis on scientific control over randomness, contingencies, and chance. In the vast literature on risk there are authors who argue, however, that the language of uncertainty would be more appropriate for a better understanding of the current world, full of indeterminacies and contingencies, whether inherent in the world or epistemic. Underlying this argument would be lack of knowledge of the statistical probability of many of the possible outcomes, public distrust of the estimates produced by experts, potential margins of error, and the random unpredictability of nature and human behavior

(Martins 1998). This approach has affinities with the work of authors who underline the ontological nature of uncertainty that is inherent in the natural and social worlds and focus on "ignorance," "catastrophes," and "accidents" (see, for example, Perrow 1984). It differs from the work of those who stress above all the social perception of risks (such as Douglas and Wildavsky 1982).

Beck often is said to alternate between the realist and the constructivist approaches and to absorb uncertainty into the general category of risk. However, he cannot be said to limit risk to the perceptual aspect or to avoid a strong emphasis on uncertainty. There are several studies of practical situations in which risk is not limited to perceptions, such as the subpolitics of medicine. At the same time, in light of the emphasis Beck places on deregulation, uncertainty, and contingency, his "risk society" cannot properly be understood according to the probability model. In introducing the notions of "unintended consequences and unawareness" into his theory of reflexive modernity instead of emphasizing the "knowledge," as Giddens and Lash do, Beck recognizes that there are areas of unknowability, contingency, and ignorance. For this reason his theoretical approach lends itself to multiple interpretations that lie between the concepts of risk and uncertainty.

These issues are relevant because a decision based on risk or uncertainty is not neutral in its political consequences. Risk is associated with prevention, whereas uncertainty is associated with precaution (Godard et al. 2002). Risk may lead to a process of risk-mitigating negotiation and agreement, whereas uncertainty may lead to risk-avoiding prudence. The possibility of rejecting certain techno-economic decisions and actions has provoked a lively ongoing debate about the advisability of the "precautionary principle" at a time of rapid technological change.

HELENA MATEUS JERÓNIMO

SEE ALSO *Risk and Emotion; Risk Assessment; Risk-Cost-Benefit Analysis.*

BIBLIOGRAPHY

Beck, Ulrich. (1992 [1986]). *Risk Society: Towards a New Modernity*, trans. Mark Ritter. London: Sage.

Beck, Ulrich. (1995 [1988]). *Ecological Politics in an Age of Risk*, trans. Amos Weisz. Cambridge, UK: Polity Press.

Beck, Ulrich. (1999). *World Risk Society*. Cambridge, UK: Polity Press.

Beck, Ulrich. (2002). "The Terrorist Threat: World Risk Society Revisited." *Theory, Culture & Society* 19(4): 39–55.

Beck, Ulrich; Anthony Giddens; and Scott Lash. (1994). *Reflexive Modernization: Politics, Tradition and Aesthetics in the Modern Social Order*. Cambridge, UK: Polity Press.

Douglas, Mary, and Aaron Wildavsky. (1982). *Risk and Culture: An Essay on the Selection of Technical and Environmental Dangers*. Berkeley and London: University of California Press.

Godard, Olivier; Claude Henry; Patrick Lagadec; and Erwann Michel-Kerjan. (2002). *Traité des Nouveaux Risques: Précaution, Crise, Assurance*. Paris: Gallimard.

Keynes, John Maynard. (1921). *A Treatise of Probability*. London: Macmillan.

Knight, Frank H. (1921). *Risk, Uncertainty and Profit*. Boston: Houghton Mifflin.

Lagadec, Patrick. (1981). *La Civilisation du Risque: Catastrophes Technologiques et Responsabilité Sociale*. Paris: Seuil.

Luhmann, Niklas. (1993 [1991]). *Risk: A Sociological Theory*, trans. Rhodes Barrett. Berlin and New York: Walter de Gruyter.

Martins, Hermínio. (1998). "Risco, Incerteza e Escatologia: Reflexões sobre o *Experimentum Mundi* Tecnológico em Curso" [Risk, uncertainty and eschatology: Reflections on the ongoing experimentum mundi]. *Episteme* 2: 41–75.

Perrow, Charles. (1984). *Normal Accidents: Living with High-Risk Technologies*. New York: Basic Books.

ROADS AND HIGHWAYS

• • •

Roads and highways have been principal means by which entire economies and societies have emerged and grown over time. They have contributed positively to the spread of ideas, cultures, languages, inventions, goods, and services. Disease, enslavement, tribute, and warfare have also spread through networks of roads and highways to devastate entire peoples and areas and immeasurably alter the course of history.

Early Roads and Highways

The first roads dating back to the dawn of civilizations (c. 3000 B.C.E.) were little more than dirt paths worn down by frequent travel from one location to another via wheeled vehicles. Rivers were the main highways of this time period, as goods and people moved up and down their courses and any city that desired to rise to importance was located on a river. Yet within a period of a few hundreds of years, roads became a commonplace and began to reshape the geopolitical history of entire regions. Even during this period, rivers continued to be the most economical way to transport large quantities of goods, with roads being used to link river trade to cities and towns throughout entire regions.

The earliest roads were designed to bear the weight of wheeled traffic including carts, large wagons, and swift chariots. Within cities the main thoroughfares were paved and varied in width from two to ten meters. A series of "narrow streets" connected to these "broad streets" within cities and enabled populations within them to increase substantially in size. The Neo-Babylonians and Assyrians constructed royal roads that linked major cities across their empires. The Persians took over many of the practices of the Assyrians and maintained excellent royal roads, some as long as 2,670 kilometers (1,650 miles). These roads featured "excellent inns," as noted by the Greek historian Herodotus (c. 485–c. 425 B.C.E.), as well as special parks so that the king or his senior administrators could take their rest in leisure when traveling across the vast reaches of the Persian Empire. Similarly, in ancient Egypt roads were constructed both within large cities and linking cities and regions of Egypt and her territories to one another. The typical Egyptian road was about five meters in width. Outside of cities, most roads in the ancient Near East were unpaved but had been carefully prepared and leveled, and were regularly maintained. The ancient Greeks did not favor roads and built only a skeleton of dirt roads from one region to another until the time of Alexander the Great (356–323 B.C.E.), who saw the need for better roads linking the rapidly expanding segments of his empire.

The Romans, likely expanding upon earlier techniques of the Etruscans, took road building to new heights of engineering excellence, constructing two, four, six, and eight lane highways connecting all key parts of the Empire. Roads themselves became a symbol of the might of Rome and the certainty that if they were needed, a Roman army would arrive swiftly to deal with any sociopolitical unrest or the incursion of enemies from outside its borders. Roman surveyors determined the optimum location and direction of roads, favoring straight traces whenever possible. Roman engineers constructed roads that would last for centuries through careful attention to the underlying base materials, superb drainage to keep water away from the road and its foundation, the careful use of stones and cement, and regular repair. Many miles of these Roman roads survive throughout the former Empire and quite a few modern roads follow the exact course as their Roman predecessors. While originally designed for military purposes, the roads became the means by which Roman ideas, life, and culture spread across the Empire. All roads carried mileage markers, always delineated in terms of their distance from the imperial city of Rome, a reminder to all of the might and power of the Empire. By the time of Diocletian (245–c. 313), there were 372 main roads throughout the Empire,

covering a distance of some 85,000 km (nearly 53,000 miles). The Romans went well beyond any of their predecessors in the extent and interconnectivity of the system of secondary and primary roads they created and maintained across the Empire. In Roman Britain alone, more than 9,656 km (6,000 miles) of roads were constructed and maintained. Bridges and tunnels, milestones to enable travelers to instantly know their location, wooden signposts, and many other "modern" features of roads and highways were common throughout the Empire. The roads created ideal conditions for the growth of a postal service for government use and also a private postal service employed by wealthy citizens. A series of posts were set up so that couriers only had to move from one posting station to another—a design that would later be used by the famous but short-lived Pony Express in the American West.

Roads were not a distinctly Roman and European phenomena. The Qin and Han dynasties of China created a highly integrated network of roads, mainly for military use, in the second century B.C.E. The first Qin Emperor, Qin Shihuangdi (c. 259–210 B.C.E.) constructed 7,000 km (4,350 miles) of roads radiating out from his capital city of Xianyang in northern China. One hundred years later, there were more than 35,000 km (21,750 miles) of roads in northern China serving an empire of some 4 million square kilometers (1.5 million square miles). Similarly, the Incas created an empire running from Ecuador to central Chile and held it together via a network of more than 10,000 roads built across some of the most difficult mountain terrain in the world. Remnants of these Incan royal highways still exist in the early twenty-first century, and many modern roads follow the traces of these roads as a continuing tribute to the foresight and skills of these early highway engineers in South America.

The road systems developed by the Romans throughout the western Empire declined considerably after the fall of Rome, while those in the East continued to be maintained to a reasonable degree both under the Eastern Roman emperors and their Muslim conquerors. Many medieval roads in Europe declined to little more than dirt roads and were subject to flash floods and steady deterioration.

Modern Roads and Highways

Roads in the West began to be vigorously revived in the seventeenth century with the introduction of street lighting, ferry services, and emerging regulations from local, regional, and national governments. Central governments began to assume more direct responsibility and control for roads and centralized planning and maintenance became common, supported by general tax revenues.

Pierre Trésaguet (1716–1796), director of the École des Ponts et Chaussées in Paris in the mid-eighteenth century, had studied long and well the achievements of the Romans. His department had responsibility for some 40,000 km (25,000 miles) of roads throughout France, many built in the exact traces of earlier Roman roads. Trésaguet ensured that exacting road preparation methods were employed, following earlier Roman techniques, and the road system throughout France improved dramatically under his tenure. Two Scottish engineers made similar improvements throughout Britain in the early nineteenth century. Thomas Telford (1757–1834) built an exquisite model road between London and Holyhead demonstrating the superiority of preparing a very solid and carefully constructed roadbed before providing surfacing materials. While expensive to build, it was vastly superior to other roads. John McAdam (1756–1836), his fellow Scot, pioneered the use of natural materials as the base of a roadbed and developed methods to highly compact these materials to provide the same type of firmness that Telford achieved, only with much lower production costs. The surface material used on his road and the entire type of road took its name after him—macadam. By the late nineteenth century, the use of asphalt and portland cement (first used in Scotland in 1865) also became common and the maintenance required for roads became much less labor intensive.

Roads, including early toll roads such as the Lancaster Turnpike in Pennsylvania where travelers had to pay a fee to enter and/or exit the road, were a principal means of commerce in colonial America, and traces to the American West eventually were turned into roads that enabled white settlers to push rapidly westward in search of new lands and opportunities.

The advent of trains and railroads in North America, Europe, and elsewhere provided new opportunities to create many smaller secondary roads that linked many smaller towns and farming areas to commercial nodes. Consequently, these roads became the means by which goods and services circulated far more widely than was economically feasible before with attendant mobility of goods, people, and ideas. The combination of railroads and roads during the Civil War, for example, enabled large and rapid movements of troops and influenced the outcome of many a Civil War battle.

The nineteenth century saw the introduction of steam-powered equipment to construct roads, with the most important invention being the steamroller of Louis

Lemoine and Amedee Jean Ballaison. These steamrollers quickly found their way to India and other nations far from Europe and the United States. New gasoline powered vehicles provided even more powerful machines to build roads and also led to more plentiful traffic for roads, resulting in yet further expansion of networks of roads across nations. By the nineteenth century it was common for city roads to be made of portland cement, and bitumen (pitch) or concrete used for cross country routes. Rural roads in the hinterlands continued to consist of dirt and packed gravel.

The first multi-lane, limited access highway in North America was constructed during 1917–1925 as the Bronx River Parkway, a New York thoroughfare still in use in the twenty-first century. The first bona fide superhighway in the United States was the 160-mile Pennsylvania Turnpike from Middlesex to Pittsburgh that opened in 1940 and quickly outdistanced expectations as 2.4 million vehicles used it annually within the first few years. Adolf Hitler (1889–1945) and Benito Mussolini (1883–1945) were aficionados of superhighways, and under their direction, massive superhighways were constructed in Italy and Germany in the 1930s that enabled the rapid movement of troops. President Franklin Delano Roosevelt appointed a National Interregional Highway Commission in 1941 and a Federal Aid Highway Act was approved in 1944 that authorized $1.5 billion for interstate highway construction. By the time of the Eisenhower administration, the federal highway legislation resulted in the construction of more than 64,000 km (40,000 miles) of highways running across the United States in both north and south and east and west orientations. Many states, such as New York, Pennsylvania, Ohio, and Illinois, also built their own extensive toll roads that connected in networks running particularly throughout the northeast. In the early twenty-first century similar highway systems can be found throughout the world, and the proportional number of miles of such highways within a nation serves as a rough gauge of its economic status in the world. These massive networks of superhighways and their linked secondary roads enabled the massive growth of suburbs and attendant suburban "flight," substantially altering the tax base and quality of life of central cities–a situation readily observed in places such as Atlanta, Boston, Chicago, London, Los Angeles, Paris, and Philadelphia.

Highway Engineering and Ethical Issues

Highway planning in the early 2000s is a complex branch of civil engineering that is designed to move goods and people efficiently, effectively, and safely across large distances. It includes attention to forecasting demand, acquiring land from various parties, designing roads and arteries that make for safe and aesthetically pleasing experiences for highway users, moderating costs, and providing for long-term maintenance and expansion when needed. Traffic volume is generally measured in terms of annual average daily traffic, which allows for derivation of a figure that avoids the inevitable peaks and troughs of traffic flow in any given day, week, or month. An entire route is divided into zones and then estimates are made about travel between zones and the amount of travel that will be undertaken by different modes of transport (for example, trucks, cars, buses). A maximum theoretical traffic flow rate is calculated using reasonable parameters of environmental, highway, and traffic conditions. A further factor taken into account in planning is what level of service the road will need to bear that will be acceptable to its users. Travel is an inherently subjective experience, and planners attempt to find an acceptable level of service (LOS), avoiding the extremes of very good (index A) and very poor (index F).

A number of additional factors need to be considered. All human technological applications have environmental effects. Highways directly affect matters such as noise pollution from horns, tires on road surfaces, engines, the speed of traffic, and shock effects from heavy loads on road surfaces; air pollution due to carbon monoxide, nitrogen oxides, volatile hydrocarbons, sulfur oxides, and particulate matter from exhaust fumes as well as evaporation from road surfaces; water pollution due to runoff that picks up oils, trash, and other materials from road surfaces; and environmental effects from the initial siting of the highway and its continued maintenance. These latter effects can include changing migration patterns and habitats of birds, mammals, amphibians, fish, and other creatures, as well as increased road kills (which number substantially more than one million mammals per year in America alone). Sometimes road kills result in the total extinction of a species or a severe threatening of its existence, such as with the Florida panthers.

Highway design includes attention to both aesthetics and safety issues. Each highway has to surmount certain physical challenges that the land presents, and decisions have to be made about how much to use the natural features of the land in construction or to substantially alter them. Modern highways attempt to utilize natural materials and natural roadbeds as much as possible, because it is far cheaper than completely excavating and hauling away such materials and replacing

them with others. Sometimes the natural material base is not conducive to the type of heavy travel a particular road will be required to bear and then such steps have to taken.

A much larger portion of land is required than just that needed for the roadway itself. Most highways require a median that is almost equal in size to the width of the one or more lanes on one side of a divided highway. Then the outer edge of the driving lane requires a shoulder so that vehicles have a space to move off the road safely when they encounter vehicular or other problems. A drainage ditch is usually found outside the shoulder to handle runoff from the driving lanes, which are sloped in such a way that water runs off the highway quickly. The ditch also serves as the means to handle runoff from surrounding land on either side of the road cut to keep water off the road surface and prevent erosion from undermining the pavement or roadbed.

Pavement materials for roads and highways have to meet technical standards in order to be used. All materials must be sufficiently strong and durable to meet the required criteria that planners have established for that particular type of road. A typical highway is a composite of many different types of materials that are laid down in a carefully defined sequence and constantly checked to verify that they meet required specifications. Materials include sand, gravel, crushed rock, portland cement, asphaltic cement, lime, and, increasingly frequently, recycled materials such as crushed glass, scraps from old roadways, and pulverized tires.

Road geometry takes account of the steepness of curves, the slope of hills and valleys (road grades), passing maneuvers on varied terrain, and the need to maximize clear lines of sight. This is further complicated by situations where two highways meet one another, where a whole series of considerations must be addressed to plan and construct effective intersections and interchanges that enable a smooth and safe flow of traffic.

The actual siting of highways is always a complex decision that involves balancing factors such as travel time, vehicle operation cost, accessibility, environmental effects, societal acceptability, safety, total cost of construction, and viable alternative routes. Increasingly, local, state, and federal governments in many countries have to use the concept of eminent domain to assert their primary claim over land held by owners reluctant to relinquish their claims, frequently because they are opposed to the siting of the highway through their property. Government agencies generally are required by law to provide a fair-market value price to the owners.

The impact of interstate highways on commerce, migration, immigration, and employment growth has been the subject of much study. The overall findings indicate that, in general, counties or administrative units that reside alongside interstate highways see an increase in net immigration, employment growth, and commercial activity, while counties that have been bypassed by the interstate suffer net migration, a loss of employment over time, and declining commercial activity. The large amounts of particulate matter generated from major roadways has been identified as a source of chronic exposure that produces negative health effects within communities, especially in children and adults suffering from various respiratory preconditions.

Roads and highways also have to be managed by agencies to ensure that traffic flow is maintained at a reasonable level and that users of the roadway obey traffic laws that are designed to maintain such flows. Traffic signals of many different varieties have been developed, and a set of international standards have been developed for signs so that drivers can travel virtually around the globe and know what they are supposed to do in particular situations. Toll booths, highway exit and entry, emergency breakdown services, quick response to traffic accidents, enforcing traffic laws, and many other facets of roads and highways are generally under-appreciated by users but essential to maintaining a working system of roads and highways. Driver error, including falling asleep at the wheel, is by far the most common source of traffic accidents and deaths and injuries to drivers, pedestrians, and wildlife.

Future Developments

Computerization is the next major innovation in roads and highways, and virtually every industrialized nation has a wide range of current applications in the area of intelligent transportation systems (ITS). These include automated toll booths where vehicles with appropriate stickers on their vehicles can pass through the booth and automatically be billed for their trip rather than having to stop and manually deliver money or tokens to a human or automated operator. Many interstates or roads in heavily congested areas of the world use computers to regulate entry into the highway as traffic lights and barriers allow only one vehicle at a time onto the highway such that mergers happen more seamlessly and the flow of traffic on the road is not impeded by entering traffic. Many cities have sophisticated computer systems that regulate traffic signals across the city with the timing of signals changing throughout the day to accommodate the daily ebb and flow of traffic to and from major

zones within the city. Cameras placed in strategic positions in cities and mobile camera units elsewhere increasingly document speeding vehicles with attendant tickets being subsequently issued to the offenders. Global positioning technology makes it feasible to track vehicles anywhere in the world, and many large transport companies already utilize this technology to keep track of their vehicles both on the road and also across railroad systems in seamless global transportation networks that enable managers to ensure that their products arrive at required destinations in a timely manner and in good condition.

ITS planners have created plans for intermodal transport systems that utilize advanced telecommunications and computer systems to move goods across entire continents through underground tunnels or highways dedicated solely to the movement of freight. These intelligent systems would only require human operators on points of entry or exit within the system, and once on the network, goods could be accelerated greatly in their passage to desired destinations. Similar designs exist for automobiles of the future that would go on "autopilot" once the human operator had placed the vehicle on the superhighway. Computers would then guide the vehicle to the required exit point and then the human operator would take over control functions to move the vehicle safely off the superhighway. Such a system would alleviate the traffic jams so familiar to major interstate highway systems during peak flow times and enable resources to be used more efficiently.

The widespread use of ITS raises a host of ethical issues, many not particularly unique to these applications but part of a broad set of issues common to technological innovations. Increasingly the operators of these systems would have knowledge of one's whereabouts and be able to track the movement of a single individual across a city, state, or even potentially around the globe as these various systems come online and interconnect both operationally and informationally. Technical managers would also be able to shape human perceptions and experiences of reality by varying conditions on these systems—for example, deciding that today's optimal travel time from point A to point B will be 25.8 minutes, and programming the system to deliver these results. It should be noted, however, that highway engineers have always shaped human perceptions of the surrounding environment and influenced ways of life going back to where the first roads were constructed (all artifacts have politics, as Langdon Winner has argued), how structures actually are designed (for example, low bridges on the Wantagh Parkway in New York designed by Robert Moses (1888–1981) specifically to keep buses off the parkway), and via the distinct sociotechnical roles that engineers play in public policy making.

DENNIS W. CHEEK

SEE ALSO *Networks*.

BIBLIOGRAPHY

Adkins, Lesley, and Roy A. Adkins. (1994). *Handbook to Life in Ancient Rome*. New York: Facts on File. A one volume reliable reference that includes numerous references to Roman roads.

Astour, Michael C. (1995). "Overland Trade Routes in Ancient Western Asia." In *Civilizations of the Ancient Near East*, ed. Jack M. Sasson. New York: Scribners. A definitive treatise summarizing current historical and archaeological information on this topic.

Carlson, Daniel. (1995). *At Road's End: Transportation and Land Use Choices for Communities*. Washington, DC: Island Press. A study of the complex set of factors that should be considered in planning and siting of highways.

Chen, Wai-Fah, and J. Y. Richard Liew, eds. (2003). *The Civil Engineering Handbook*, 2nd edition. Boca Raton, FL: CRC Press. A massive standard reference work that provides copious details about the construction of highways and other matters in civil engineering.

Dorsey, David A. (1997). "Roads." In Volume 4 of *The Oxford Encyclopedia of Archaeology in the Near East*, ed. Eric M. Meyers. New York: Oxford University Press. A summary article by a noted expert on roads in the ancient world and particularly the Middle East.

Garrison, William L., and Jerry D. Ward. (2000). *Tomorrow's Transportation: Changing Cities, Economies, and Lives*. Boston, MA: Artech House. Explores key options and trade-offs concerning future transportation needs, challenges, and opportunities.

Lewis, Tom. (1997). *Divided Highways: Building the Interstate Highways, Transforming American Life*. New York: Viking. A journalistic depiction of the ways in which interstate highways transform culture with a focus on the situation in the United States.

Motavalli, Jim. (2001). *Breaking Gridlock: Moving Toward Transportation that Works*. San Francisco: Sierra Club Books. Argues for a more environment-friendly approach to transportation systems while avoiding simplistic solutions.

Rose, Mark H., and Bruce E. Seely. (1990). "Getting the Interstate System Built: Road Engineers and the Implementation of Public Policy, 1955–1985." *Journal of Policy History* 2(1): 23–55. An important study of a neglected topic: How do engineers interact with policymakers to help shape public decisions about transportation systems?

Sachs, Wolfgang. (1992). *For Love of the Automobile: Looking Back into the History of Our Desires*, trans. Don Reneau. Berkeley: University of California Press. A provocative, wide-ranging, sociological study of the ways in which the automobile shapes desires and the ways in which automobiles are designed to appeal to basic human desires.

Sussman, Joseph. (2000). *Introduction to Transportation Systems*. Boston: Artech House. An expert summary of all major issues related to transportation systems that grew out of a course the author has taught on this topic for many years at the Massachusetts Institute of Technology.

Vance, James E., Jr. (1990). *Capturing the Horizon: The Historical Geography of Transportation since the Sixteenth Century*. Baltimore, MD: Johns Hopkins University Press. The definitive one-volume history in English of transportation in modern times.

Winner, Langdon (1986). *The Whale and the Reactor: A Search for Limits in an Age of High Technology*. Chicago: University Of Chicago Press. A series of important essays by a noted political scientist about technology; see especially chapter two, "Do Artifacts Have Politics?" for a discussion of Robert Moses and the Wantagh Parkway.

ROBOTS AND ROBOTICS

• • •

Robots are programmable machines capable of moving around in and interacting with their physical environment. The word *robot* was popularized by Karel Capek (1890–1938) in his play *R.U.R.*, where he used it to refer to a race of manufactured humanoid slaves; robots are machines that can do the work of humans. It is debatable whether merely remote-controlled devices should count as robots, although many devices popularly thought of as robots are of this nature. Similarly, computer programs such as virtual "autonomous agents" and web "bots" are not, strictly speaking, robots as they lack the ability to manipulate the physical world.

The term *robotics* was coined by Isaac Asimov and refers to the study and use of robots. Research into robotics began in the 1940s, alongside research into cybernetics and computers. The first commercial robots were produced for industrial applications in manufacturing in the 1960s. As computing technology began to improve rapidly in the 1980s and 1990s, a number of writers such as Hans Moravec (1998) and Ray Kurzweil (1992) made arguably exaggerated claims on behalf of robots, suggesting that they would soon possess consciousness and intelligence. Major limitations on the tasks that can be performed by robots—especially in real environments—remain, largely due to a lack of success in reproducing "intelligence" and robust locomotive and sensory systems. The vast majority of existing robots

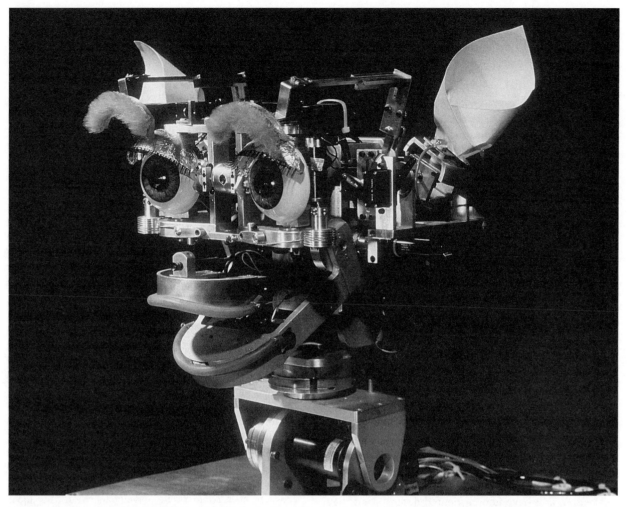

Kismet, a robot created by Dr. Cynthia Breazeal at MIT. She developed Kismet for her doctoral research in expressive social exchange between humans and humanoid robots. (© *Rick Friedman/Corbis.*)

are industrial robots, which perform a limited range of repetitive tasks in a controlled environment.

The ethical, political, and legal issues surrounding robots can be roughly grouped into two categories: those that are raised by existing technologies and a more speculative set that would arise if genuinely "intelligent" or conscious robots were to become a reality.

Existing technologies largely raise questions relating to their social impact (Weiner 1961). The main impact of robotics thus far has been to displace persons from jobs in manufacturing industries. It might be argued that by replacing workers in industries where jobs tended to be both highly paid and skilled, robots have had a negative impact on human happiness. Alternatively, it might be argued that robots have contributed to human happiness by eliminating the necessity of repetitive and occasionally dangerous work. The economies of scale and other increases in efficiency that robotics have made possible would also need to be taken into account in this calculation. Access to robots could conceivably become a source of inequality in a society where robots play a significant role.

Another area where it seems likely that robots will have dramatic social impacts is warfare. A number of types of remote control and semi-autonomous devices are already deployed by militaries around the world. It seems likely that fully autonomous robots will play a role in wars conducted by industrialized nations in the future.

The use of robots in military contexts raises many difficult ethical and legal issues. They offer to reduce casualties amongst friendly combatants, but in doing so may decrease the threshold of war. "Smart weapons" may allow commanders to attack military targets with greater precision and thus lower the risk of civilian casualties in war. However, the possession of such weapons by one side only may increase the likelihood and extent of asymmetrical warfare and consequently of increased civilian casualties. There are also ethical and legal questions surrounding the allocation of responsibility for deaths caused when such weapons go astray, resulting in attacks on targets that are not legitimate under the rules of war.

More prosaically, a number of quite advanced robots are now manufactured as entertainment devices and "robot pets." The development of robot toys suggests that there is a need to scrutinize the educative and communicative functions of these robots. There are also questions surrounding the ethics of human/robot interactions. Are robots appropriate objects of emotional attitudes? If not, then designing robots to encourage such investment may be wrong.

A much larger, more complex, but also speculative, set of issues would arise if robots were to achieve any degree of consciousness, or genuine intelligence.

At what point would such creations deserve moral concern? What rights should they have? While these questions are regularly raised by writers in the area, little serious philosophical work has been done on these subjects, perhaps reflecting a lack of faith that the technology will become a reality.

Yet much contemporary moral theory, which grounds moral status in the capacities of individuals, suggests that sentient robots would be deserving of the same moral regard as other sentient creatures. If robots can feel pain, then humans will have obligations to avoid causing them pain. If they become self-conscious, can reason, and have future-oriented desires, then they will be worthy of the same moral regard and respect as human persons. This suggests that it would be entirely appropriate to feel grief stricken by the "death" of a robot, to feel remorse for killing a robot, and even sometimes to choose to save the life of a robot over that of a human being.

This last scenario might serve as a test of the moral status of robots. Humans will know that robots are moral persons when they feel that the choice between the survival of a robot and of a person is a genuine moral dilemma. This might be called the "Turing Triage Test," after Alan Turing's famous test for when a machine can be said to think. If this test is a valid one, it suggests that what is required for robots to become persons may include the ability to express subtle and complex emotional states through their bodily appearance.

As well as the question of how people should treat robots, there is also the question of how robots are expected to treat people. What ethical precepts should they be designed to obey? Isaac Asimov's " three laws of robotics" are a famous attempt to answer some of these questions. Yet, as Asimov's stories demonstrate, much more will need to be done before humans become confident that intelligent robots could safely take their place alongside humanity. These questions would become especially urgent if artificially intelligent robots might be capable of reproducing themselves and thereby pose a threat to the human species. If robotics researchers *are* on the verge of creating entities that will be more intelligent than humans and that may compete with humanity for dominance over the planet, then this is a momentous decision, which should only be made after extensive public deliberation.

ROBERT SPARROW

SEE ALSO *Androids; Artificial Intelligence; Artificial Morality; Asimov, Isaac; Robot Toys; Turing Tests.*

BIBLIOGRAPHY

Asimov, Isaac. (1950). *I, Robot.* New York: Gnome Press. A collection of science fiction short stories that did much to popularize the idea of robots.

Brooks, Rodney Allen. (2003). *Robot: The Future of Flesh and Machines.* London: Allen Lane. A popular account of the history and probable future of robots and robotics by a leading robotics researcher.

Kurzweil, Ray. (1992). *The Age of Intelligent Machines.* Cambridge, MA: MIT Press. A history of research into "artificial intelligence" alongside an extremely speculative discussion of possible future developments in this area, including the ethical and social issues that they may raise.

Menzel, Peter, and Faith D'Aluisio. (2000). *Robo Sapiens: Evolution of a New Species.* Cambridge, MA: MIT Press. A survey of the state of robotics research at the beginning of the twenty first century in the form of a photo-essay and a set of interviews with researchers from around the world.

Moravec, Hans. (1998). *Robot: Mere Machine to Transcendent Mind.* New York: Oxford University Press. An extremely speculative discussion of the possible future of robotics, which argues that machines will be more intelligent than humans by the year 2050.

Turing, Alan. (1950). "Computing Machinery and Intelligence." *Mind* 59(236): 433–460. An influential paper in which Turing sets out his famous "imitation game" as a means of determining when a machine can properly be said to think.

Wiener, Norbert. (1961). *Cybernetics: Or Control and Communication in the Animal and the Machine,* 2nd edition. Cambridge, MA: MIT Press. An important work that sets out the theoretical basis for the discipline of *cybernetics*—and therefore for robotics. It also includes discussion of the possible social impact of the development of robots.

ROBOT TOYS

• • •

Robots combine sensors, computation, and motors to interact intelligently with their environment. Robot *toys* need to be so cheap and robust that they can be used as playthings. While there is a long history of toys that look like robots, only recently has the cost of computation dropped sufficiently to allow the sale of truly functional robotic toys. This entry focuses on three examples of this new genre of toy that should be of interest from the ethics perspective: Lego MindStorms robot construction sets and Furby interactive robotic pet by Tiger Toys, and Sony Aibo robot dog.

Lego and Furby: Some Contrasts

These two very different kinds of robotic toys were both introduced in 1998, had a large impact, and contrast in several interesting ways. Lego MindStorms and Furby represent two types of toys that Gary Cross (1997) finds typical of twentieth century U.S. toy production: the educational and the novelty toy. Lego MindStorms Robotics Invention System extended the Lego Technic construction system to include a programmable computer controller brick (the RCX), sensors and motors, and computer interface and programming environment. Lavish documentation and support (reflecting a long nurturing by educators) allowed users to build a variety of working robots, ranging from traditional light-guided rovers to static room alarms. Although MindStorms was expensive, included more than 700 pieces, and required considerable assembly and a personal computer, it was nevertheless an immediate success with both children and adults. It became widely used in schools and colleges and has remained in production for a number of years.

By contrast Furby was a plush but inexpensive, stand-alone, interactive toy. Multiple sensors (light, touch, sound, infra-red) drove a single motor, which, via a series of ingenious cams, controlled several motions of the ears, eyes, eyelids, mouth, and rear body (Pesce 2000). Enormously popular in its first season, with long lines at toy stores and price premiums featured on TV news, more than 12 million Furbys were sold in one year. Yet just as quickly the fad passed and in the early twenty-first century Furbys are no longer produced.

Robotic toys fall into two groups: the programmable and the pre-programmed. MindStorms takes programmability to the limit: One can choose which of several general purpose programming languages to use. The Furby was pre-programmed.

Another contrast is in terms of transparency and openness. MindStorms was released as a normal, closed (although very well documented) product. That is, one could run its code but not change it except in predefined ways. After a brief struggle with fans and hackers, Lego agreed to release the technical specifications and allow programming access to the RCX's ROMs. As a result MindStorms became an extensible open-source system for constructing robots. Indeed it has become a platform for a large variety of languages and operating systems. By contrast Furby remained a closed system. It was pre-programmed and an epoxy blob hid its computational abilities and electronics. Moreover its capacities were not documented but shrouded in rumour and advertising hype, so it was difficult to know what the toy could actually do. Could Furbys really learn?

Ethics

Interactive robotic toys raise special issues for ethics. First, robot toys face some special ethical requirements. As robots they interact with children in the real world, so they must be safe. Contrast virtual robot-building software such as the early Apple computer game RoboWar. Virtual battle robots can *fire projectiles* at each other in their on-screen arena without endangering people. Real robot toys are different: As programmed robots, they are capable of initiating unexpected actions; as toys they cannot be cordoned off from human contact in the way that real factory robots typically are.

Second, more subtly, robot toys face design challenges to keep contact with the real world fun and educational. The environment is a great teacher, providing feedback on feasible design for free. But the price can be costly; think of testing whether a Furby can swim or a Lego robot can navigate in sand. The ideal of a platform is helpful here (Danielson 1999). For example MindStorms pushes most electrical considerations down into the platform it provides. The connectors allow polarity to be reversed, but otherwise the user need not be aware of the electrical properties of the sensors and motors.

Third, interactive robotic toys may even change moral categories. Surprisingly Sherry Turkle has found that children categorize their Furbys in a new way: "Children describe these new toys as *sort of alive* because of the quality of their emotional attachments to the Furbies and because of their fantasies about the idea that the Furby might be emotionally attached to them" (Turkle 2000). These children appear to be assigning interactive toys to a third class, between the animate and the inanimate, because of how they interact with them. In a related development, robotic toy pets have been found useful in rehabilitation in Japan (Goodale 2001). These preliminary research results suggest that human relations with emotionally evocative and involving robotic companions will be ethically complex.

Aibo

The third example, Sony's Aibo robotic dog, raises some additional contrasts and ethical issues. Aibo was introduced in 1999 in the United States and Japan. Although very expensive, it sold out in Japan "in just 20 minutes" (Yoshida 2001). Aibo has never sold very well outside of Japan. This difference points to Japan's distinctive history and culture with respect to robots in general and robotic toys in particular. While Aibo's price and sophistication place it with the Lego system, there was an ethically interesting contrast: When Aibo owners hacked its software in order to personalize and extend its capabilities, Sony reacted to block them and protect its intellectual property. Lego, in contrast, opened MindStorms by publishing its source code. Third, Aibo's advanced capabilities allow it to function as a pet much better than the much simpler Furby. Aibos' cognitive and moral status is thus much more ambiguous (see Turkle 1995, chap 3). On one side, the animal rights organization People for the Ethical Treatment of Animals (PETA) claims "the turn toward having robotic animals in place of real animals is a step in the right direction" (MacDonald 2004). But research on actual attitudes towards Aibo find that owners "rarely attributed moral standing" (Peter Kahn, Friedman, and Hagman 2002).

Future Developments

Robotic toys will become ever more sophisticated interactively. Furby, for instance, gave rise to the more capable and expensive Aibo. Robotic toys may thus be a mechanism for increasing the pace of ethically challenging technological change. The toy industry is well known for driving down costs, in order to sell large volume blockbusters. (Furby was brought to market in less than a year and at less than one-half the expected price point.)

In the wake of Furby, there thus exists an increasing number of young new users of a technology, acquired over a short time, along with the design and industrial capacity to make more of the next version very quickly.

MIT roboticist Rodney Brooks, for example, has predicted that the first robots to establish a wide household presence will be robotic toys. This is a recipe for rapid technological and attitude change and little time for ethical reflection.

PETER DANIELSON

SEE ALSO Education; Entertainment; Popular Culture; Robots and Robotics; Safety Engineering: Practices.

BIBLIOGRAPHY

Cross, Gary. (1997). *Kids' Stuff: Toys and the Changing World of American Childhood.* Cambridge, Mass: Harvard University Press.

Danielson, Peter. (1999). "Robots for the Rest of Us or for the 'Best' of Us." *Ethics and Information Technology* 1:77–83.

Goodale, Carol. (2001). "Researchers Study How Robots Interact with Children." *IEEE Spectrum*, November 2001.

MacDonald, G. Jeffrey. (2004). "If you Kick a Robotic Dog, Is It Wrong?" *Christian Science Monitor*, Feb. 05 2004.

Pesce, Mark. (2000). *The Playful World : How Technology Is Transforming Our Imagination.* New York: Ballantine Books.

Kahn, Peter, Jr., Batya Friedman, and Jennifer Hagman. (2002). "I care about him as a pal: Conceptions of Robotic Pets in Online AIBO Discussion Forums." Paper read at CHI 2002, at Minneapolis, Minnesota.

Turkle, Sherry. (1995). Life on the Screen: Identity in the Age of the Internet. New York: Simon and Schuster.

Yoshida, Noriyuki. (2001). "Robot People and Robo Cats and Dogs." *Look Japan*, October 2001.

INTERNET RESOURCE

Turkle, Sherry. (2000). "A New Kind of Object: From Rorschach to Relationship." Available from http://www.edge.org/3rd_culture/story/101.html.

ROTBLAT, JOSEPH

• • •

The physicist Sir Joseph Rotblat (b. 1908), born in Warsaw, Poland, on November 4, was a member of the Manhattan Project, which developed the atomic bomb in the United States. In November 1944, when it became clear that Nazi Germany would not be able to develop a bomb and affect the outcome of World War II, he became the only scientist working on the weapon who

Joseph Rotblat, b. 1908. Rotblat is a Polish physicist who received the Nobel Peace Prize in 1995 in conjunction with the Pugwash Conferences on Science and World Affairs, for their efforts towards nuclear disarmament. (*Hulton Archive/Getty Images*)

resigned prior to its being used against Japan. This principled stand, that the benefits of nuclear power should only be used for peaceful purposes, has been a hallmark of Rotblat's career and was instrumental in his sharing the 1995 Nobel Peace Prize with the Pugwash Conferences on Science and World Affairs, the organization he helped found in 1957 to work for the complete elimination of nuclear weapons.

After earning his doctorate in physics from the University of Warsaw in 1937, Rotblat moved to the United Kingdom in 1939 where he worked with James Chadwick at the University of Liverpool on the feasibility of atomic fission.. Having lost his family in his native Warsaw when the Nazis invaded Poland in September 1939, Rotblat soon moved with other émigré scientists to Los Alamos, New Mexico, to contribute to the Manhattan Project. Following his resignation from the project, he moved back to the United Kingdom where he took up positions as Director of Research in Nuclear Physics at the University of Liverpool (1945–1949) and then as Professor of Physics at the University of London (1950–1976), specializing in the medical applications of nuclear radiation.

From his early years working with Chadwick to his association with Bertrand Russell and Albert Einstein as a signatory of the famous 1955 Russell-Einstein Manifesto, which called on scientists to work for the abolition of warfare and nuclear weapons, Rotblat has dedicated his professional and personal life to exposing the fallacy of nuclear deterrence and arguing for the immorality and illegality of nuclear weapons. Because of the role of scientists in creating first the atomic and then the hydrogen bombs, Rotblat believed scientists had both moral and professional duties to ensure that such weapons would not be used against humanity. From the first Pugwash Conferences meeting held in Pugwash, Nova Scotia, in July 1957, to the 2003 Pugwash annual conference that returned to Nova Scotia, he worked tirelessly in calling upon the global scientific community to maximize only the beneficial applications of science and technology.

In his final speech as President of Pugwash in 1997, Rotblat reiterated the principle that led to his resignation from the Manhattan Project in 1944: "Many scientists are still not willing to face reality. Many discourage or actively hamper young scientists from being concerned with the social impact of science ... Scientists have to realize that what we are doing has an impact ... on the whole destiny of humankind" (Rotblat 1997, pp. 248–249). Still active in Pugwash and in the movement to eliminate nuclear weapons in his nineties, Rotblat has been a source of inspiration for several generations of scientists around the world with his fundamental belief in the promise of science and technology to improve the human condition and eliminate war as a social institution.

JEFFREY BOUTWELL

SEE ALSO *Atomic Bomb; Pugwash Conferences.*

BIBLIOGRAPHY

Hinde, Robert, and Joseph Rotblat. (2003). *War No More: Eliminating Conflict in the Nuclear Age.* London: Pluto Press.

Rotblat, Sir Joseph. (1967). *Pugwash: A History of the Conferences on Science and World Affairs.* Prague: Czechoslovak Academy Of Sciences.

Rotblat, Sir Joseph. (1972). *Scientists in the Quest for Peace: A History of the Pugwash Conferences.* Cambridge, MA: MIT Press, 1972.

Rotblat, Sir Joseph, ed. (1998). *Nuclear Weapons: The Road to Zero* Boulder, CO: Westview Press.

ROUSSEAU, JEAN-JACQUES

● ● ●

Jean-Jacques Rousseau (1712– 1778), who was born in Geneva on June 28 and died on July 2 in Paris, was a self-taught genius who became the leading critic of the Enlightenment vision of an essential harmony between science and society, technology and ethics. As a mid-century member of a circle of intellectuals working on the *Encyclopédie*, a comprehensive attempt to synthesize scientific knowledge and technological skills for social utility, Rousseau's questioning nevertheless had the effect of contributing to the French Revolution and extending modernity.

Brilliant, intellectually disciplined, independent minded, and well-educated, Rousseau arrived in Paris in 1741 and proceeded to impress and become friends with some of the notable Enlightenment intellectuals, especially Denis Diderot (1713–1784) and Jean le Rond d'Alembert (1717–1783). Yet his independent free-thinking temperament found outlet in two prize-winning essays that attacked modern science, technology, enlightenment, and early modern political philosophy as undermining virtue and happiness: *The Discourse on the Arts and Sciences* (1750) subsequently called *The First Discourse*; and *The Discourse on the Origin and Foundation of Inequality Among Men* (1753), subsequently called *The Second Discourse*.

The First Discourse waged war against the modern project as a dangerous dream, corrupt and corrupting in its origin, means, ends, and consequences. The essential features of the dream are fundamental yet simple: The universe is matter in motion, neutral, even hostile to humankind: It was neither created by God for, nor naturally ordered to, human good. Yet knowledge of a certain kind is possible (mathematical physics) and can constitute power over nature, render it predictable and hence controllable for human ends. The pursuit of human good, in turn, is to be guided by calculative, rational, enlightened self-interest ultimately oriented to peace, health, material prosperity, comfort, and bodily pleasure. The climactic scene is to be life in healthful longevity and pleasurable prosperity. There looms on the horizon the specter of universal gratification, even if by means of the scientific manipulation of human nature itself.

The core of Rousseau's response is that because scientific knowledge can be useful, the talented few may seek it with different motives and purposes. Some will be moved by pride, seeking honor, glory, and even tyranny. Others are ultimately moved by fear, especially of death as well as of pain and suffering. Yet

Jean-Jacques Rousseau, 1712–1778. The Swiss-born philosopher, author, political theorist, and composer ranks as one of the greatest figures of the French Enlightenment. (*AP/Wide World Photos.*)

desires for peaceful prosperity are but vain diversions from the hard facts of life, recognition of which is required for the possible achievement of true virtue and happiness.

The Second Discourse deepens the argument by suggesting that the root of the problem is reason itself. First, reason includes the human ability to compare oneself with others. This capacity makes possible pride, the love of self over all others. Thus reason contributes to the human selfishness that engenders tyranny. Second, reason can also construct ideas, even of time, and hence of the future. This ability of reason brings the idea of one's ultimate future to mind—that is, death and its terrors—and hence breeds the fear of death. Whereas reason had been previously considered natural to human beings and good, Rousseau argues that in some way it is neither.

Rousseau's argument rests on a reinterpretation of human history. Whereas Aristotle (384 B.C.E.–322 B.C.E.), for instance, considered human history to be cyclical, believers in the Bible saw history as providentially headed toward the end-time, and the moderns argued for history as human progress, Rousseau proposed that human history is in large measure decay from the natural goodness of an early time. From Rousseau's perspective, reason itself is an accidental, artificial acquisition that separates humans from our natural goodness, so that nurture becomes opposed to nature.

In this way Rousseau raised the question, Why reason or science? After all, he claimed, the purpose of science cannot be known by science. Neither can science answer the most important questions—Is life good and What is the good life?

Rousseau's own answer to this fundamental question may be sketched as follows: Tyranny not death is the greatest preventable evil; hence issues of justice and political philosophy are more important than science. Additionally, human sociability, virtue, and happiness are rooted less in reason than in the passions, particularly sentiments such as love, beauty, romance, and pity or sympathy and compassion. Hence, Rousseau's novels and memoirs such as *Julie, Or, The New Heloise* (1761) and *Emile: Or, On Education* (1762) contain striking portraits of the loving, romantic couple; the joys of family life; the sense of community in the tribe or nation; as well as the pleasing sentiment associated with life itself.

As fundamental and coherent as Rousseau's attack on and attention to science and enlightenment may be, he was—and remains—a paradoxical, if not contradictory, teacher. Alongside attacks on reason are to be found high praise of Isaac Newton (1642–1727), René Descartes (1596–1650), and especially Francis Bacon (1561–1626) as the preceptors of the human race. Socrates (c. 470 B.C.E.–399 B.C.E.) (or Plato [428 B.C.E.– 347 B.C.E.]) is his self-proclaimed master, as a genius moved by pure not vain curiosity. Moreover, Rousseau did not live the life he taught as good. He philosophized while directing others to find happiness in noble sentiments.

Perhaps these tensions may be explained by Rousseau's vision of the human as a complex being oriented to conflicting goods: the goods of the body and of the soul, of the community and the individual, of life and truth, and, moreover, of the good of the few, theoretical pursuits, and the good of all others, practical pursuits, of theory and practice. The least one can conclude is that perhaps Rousseau took his stand as a middle-man, as the in-between being, as philosopher also concerned with the happiness of humankind, and, as such, forged his own place among the future teachers of the human race. Certainly many of the questions he raised have subsequently become themes in on-going discussions of science, technology, and ethics, even when they are not always explicitly referenced to Rousseau.

LEONARD R. SORENSON

SEE ALSO *Education; French Persepctives; Nature; Social Contract.*

BIBLIOGRAPHY

Masters, Roger D. (1968). *The Political Philosophy of Rousseau.* Princeton, NJ: Princeton University Press. A comprehensive account of Rousseau's complexity through the lens of his political philosophy.

Masters, Roger, and Christopher Kelly, eds. (1990–present). *The Collected Writings of Rousseau.* Hanover, NH: University Press of New England.

Melzer, Arthur M. (1990). *The Natural Goodness of Man: On the System of Rousseau's Thought.* Chicago: University of Chicago Press. Argues for the unity of Rousseau's thought in his notion of natural goodness.

Scott, John T. (1992). "The Theodicy of the Second Discourse: The Pure State of Nature and Rousseau's Political Thought." *American Political Science Review* 81: 696–711.

Sorenson, Leonard R. (1990). "Natural Inequality and Rousseau's Political Philosophy in his *Discourse on Inequality.*" *The Western Political Quarterly* 763–788. Proposes that Rousseau cannot be understood without a full account of his vision of philosophy.

Sorenson, Leonard R. (1992–1993). "Rousseau's Socratism." *Interpretation* 20: 135–155.

Strauss, Leo. (1947). *Natural Right and History.* Chicago: University of Chicago Press. Explains the practical intent and consequences of Rousseau's teaching.

Strauss, Leo. (1947). "On the Intention of Rousseau." *Social Research* 14: 455–487. Explores the theoretical intention of Rousseau.

Velkley, Richard, L. (2002). *Being After Rousseau: Philosophy and Culture in Question.* Chicago: University of Chicago Press. A fine account of Rousseau's influence on Kant and Heidegger.

ROYAL COMMISSIONS

• • •

Royal Commissions, or commissions of inquiry, are part of the executive arm of some Commonwealth governments that are rooted in the British parliamentary system. Their main function is to inform the government and often to deal with broad topics of social, cultural, or economic importance. The reports of Royal Commissions, whether interim or final, are tabled before a nation's parliament and regularly released as parliamentary papers.

Formation and Composition

In the United Kingdom a Royal Commission consists of three or more (usually five) Commissioners, including the Lord Chancellor, who are privy counselors appointed by letters patent to perform certain functions on the queen's behalf (United Kingdom Parliament 2003). Canadian or Australian counterparts sometimes produce minority reports that are more significant than the majority findings (Canadian Press Newswire 1996).

The 1868 Inquiries Act in Canada initiated a process by which Royal Commissions could be appointed by the cabinet to carry out full and impartial investigations of specific national problems. The terms of reference for the commission and the powers and names of the commissioners are stated officially in an order-in-council. The findings are reported to the cabinet and the prime minister for appropriate action. The names of commissions usually refer to the chair or commissioners. An example is the Royal Commission on National Development in the Arts, Letters, and Sciences, which was named the Massey Commission after Vincent Massey, who chaired it from 1949 to 1950 ("Index to Federal Royal Commissions" 2003).

Australia and New Zealand have implemented Royal Commissions as a means to find out facts. As in all other jurisdictions Royal Commissions in those countries are given special powers to compel the attendance of witnesses, compel the production of documents, and give special privilege to persons who give evidence before the commission so that they cannot be prosecuted or subjected to subsequent legal actions (Fitzsimmons 2003).

Scientific, Technological, and Ethical Issues

Royal Commissions have been used frequently to deal with significant scientific, technological, and ethical issues. New Zealand established the Royal Commission on Genetic Modification to develop suggestions for a new regulatory structure for its agri-food (agribusiness) sector. That commission looked for possible strategies for co-managing the range of interested parties involving new corporatist and managerial dimensions of food governance (Le Heron 2003). The Australian Aboriginal Deaths in Custody Commission, which sat from 1987 to 1991, made 339 recommendations in an attempt to prevent more deaths (Fitzsimmons 2003).

Canada's 1989 Royal Commission on New Reproductive Technologies was established to act as the official forum for public deliberation on a complex issue. According to Francesca Scala (2002), the commission showed great promise for defining questions of infertility treatment and related scientific research questions and matters of public concern. Scala argues, however, that the commission's stance in favor of reproductive technologies resulted from the government's capitulation to the powerful interests of the biomedical industry.

Controversies

At their best Royal Commissions are seen as independent bodies that allow for significant public input. They are, however, not without controversy and often are used by governments to gain breathing room on controversial issues, with costs running into the tens of millions of dollars and reports that take years to produce, with no obligation on the part of the government to act on those recommendations.

The Royal Commission on New Reproductive Technologies was launched in 1989 and released its final report in 1993. It received advice from 40,000 individuals and organizations with an interest in the matter (Wood 2002). After expenditures of more than $30 million the bottom line recommendation was that Canada needed laws to govern reproductive and genetic technologies (RGT). As a result the federal government placed a moratorium on nine controversial issues, including sex selection, human embryo cloning, and the buying and selling of eggs, sperm, and embryos. The resulting introduction of Bill C-47 died on the on the order table when the 1997 election was called. The second attempt to create RGT laws, Bill C-247, failed during its second reading in the Canadian parliament.

The Massey Commission in Canada submitted 146 recommendations under eight headings. As a result of those recommendations a federal scientific research policy was created, the National Library (now Library and Archives Canada) was created, actions were taken to create the Natural Sciences and Engineering Research Council and the Social Sciences and Humanities Research Council, and additional resources were provided to support universities as well as students. The impact of the commission's recommendations continues to affect research communities across Canada more than fifty years after the publication of its report.

The Royal Society of New Zealand considered the Royal Commission on Genetic Modification to be part of an effort "promoting excellence in science and technology" (Royal Society of New Zealand). The commission provided a forum for the submission of reports from a diverse range of sources that included the Maori

Congress, Friends of the Earth, New Zealand Biotechnology Association, Human Genetic Society, Grocery Marketers Association, Quakers, Anglicans, DuPont, CarterHolt, and Greenpeace.

Despite criticism regarding costs, political diversion, and lack of direct influence on final decisions, Royal Commissions often provide vital material for long-range policy decisions and are valuable as vehicles for consciousness-raising (O'Malley 2002). Ted Hodgetts, a retired political science professor who worked on Royal Commissions, stated that it sometimes takes years to measure a commission's value, particularly if a commission deals with longer-term arrangements. However, through the process of osmosis and seepage, the ideas enter the general discourse.

Royal Commissions maintain an arm's-length distance from the government of the day and provide impartiality and great inclusively of ideas, especially for ideas and opinions that do not correspond to the dominant political ideology. They generally avoid getting bogged down in party politics, as occurred with the hearings dealing with former U.S. President Bill Clinton's involvement in the Whitewater land deal and the raid on the Branch Davidian compound in Waco, Texas (Canadian Press Newswire 1996). Their usefulness in dealing with complex societal, scientific, technological, and ethical issues probably will continue far into the future.

PETER LÉVESQUE

SEE ALSO *Bioethics Committees and Commissions; Enquete Commissions.*

BIBLIOGRAPHY

Harrison, Barry. (2002). "A Comparative Examination and Assessment of National Study Commissions (Canada, United States)." Ph.D. diss., West Virginia University, Morgantown, WV.

Le Heron, Richard (2003). "Creating Food Futures: Reflections on Food Governance Issues in New Zealand's Agri-Food Sector." *Journal of Rural Studies* 19(1): 111–125.

Scala, Francesca. (2002). "Experts, Non-Experts, and Policy Discourse: A Case Study of the Royal Commission on New Reproductive Technologies." Ph.D. diss. Carleton University, Ontario, Canada.

INTERNET RESOURCES

Canadian Press Newswire. (1996). "Litigation Slows Inquiries, Raising Questions about Their Effectiveness: Instead of Giving Answers, Some Royal Commissions Create Questions." Available at http://www.dialogweb.com/servlet/logon?Mode=1

Fitzsimmons, Hamish. (2003). "Royal Commission Effectiveness Questioned." ABC Online. Available at http://www.abc.net.au.

"Index to Federal Royal Commissions." National Library of Canada. Available at http://www.nlc-bnc.ca/7/6/g6-120-e.html.

O'Malley, Martin. (2002). "An Inquiry into Inquiries." CBC News Online. Available at http://www.cbc.ca/news/indepth/background/royal_commissions.html.

Royal Society of New Zealand. (2003). "The Royal Commission on Genetic Modification." Available at http://www.rsnz.govt.nz/.

United Kingdom Parliament. (2003). "Royal Commissions." Available at http://www.parliament.the-stationary-office.co.uk/pa/ld.htm.

Wood, Owen. (2002). "Laws: Reproductive Technologies." CBC News Online. Available at http://www.cbc.ca/news/indepth/background.html.

ROYAL SOCIETY

• • •

Dating itself from 1660, the Royal Society of London originated with informal gatherings that began fifteen years earlier and then received its Royal Charter in 1662 as one of the first institutions devoted to the advancement of science. It has been the model for many scientific organizations formed since, not only in the United Kingdom but throughout the world. An independent charitable organization whose members have been selected for their eminence in the fields of science, technology, or medicine since the middle of the eighteenth century, the Royal Society was historically influential in establishing the processes of science and the scientific method as we understand them today.

Historical Impact

From the earliest days of the Society, religious or political affiliation was not a membership criterion. In principle, anyone could be a member; there was even a membership category for foreign nationals. In practice, however, the difficulties of travel kept many potential members from joining a group that met weekly in London, and membership fees were steep enough to exclude many others. In addition, lack of government financing spurred the Society to seek members from the upper social strata who presumably would be generous with their support. This may have inhibited lower-ranked individuals from joining a group that set a high social tone (Hunter 1982). Moreover, it has been suggested that the evolving criteria used for establishing scientific credibility deliberately excluded women and people of color (Harraway 1997). It was not until 1945 that the first woman was elected to the Fellowship. It was not until the tail end of the twentieth century that programs addressing diversity issues were put in place.

Henry Oldenburg (1615–1677), a man of German birth, was the first secretary of the Society (from 1660 to 1677), and as such became responsible for soliciting reports from around the world for publication in the *Philosophical Transactions of the Royal Society,* the oldest science journal still in publication. He was also instrumental in devising methods to secure works against plagiarism, a common problem of the day. These processes were precursors of contemporary notions of peer review and the credit due the first to publish a result. Moreover, in assessing the credibility of reports received, the Royal Society played a central role in establishing scientific norms for impartiality and absence of bias.

The inductive method as expressed by Francis Bacon (1561–1626) was the source of inspiration for many early members of the Society, including Robert Boyle (1627–1691) and Sir Isaac Newton (1642–1727). Adherents to this method proceed by gathering facts through experimentation and observation and then using such collected facts to infer general relationships. Boyle, one of the founding members, was instrumental in defining the experimental method, developing procedures for conducting, validating, documenting, and interpreting experiments. Newton served from 1703 to 1727 as the twelfth president of the Royal Society, the first scientist to hold the title.

Given the lack of external funding and the consequent need to solicit membership from the aristocracy, it was not until the 1800s that membership became the province of professional scientists. During this timeframe the government increasingly looked to the Royal Society for advice on matters of science and technology—a relationship that continues into the twenty-first century. The Royal Society also became increasingly successful in gaining government support for scientific expeditions, particularly to the Arctic and Antarctic. In mid-century, the government initiated a yearly science research grant program, the funds of which were administered by the Royal Society.

This century also saw increasingly successful efforts by the Royal Society to influence the legislative process. One notable example was an effort to modify the proposed language of the Cruelty to Animals Act of 1876, which would have eliminated experiments using animals not directly related to "saving or prolonging human life, or alleviating human suffering." The bill in its original form would have absolutely prohibited the use of dogs or cats in research. As passed, the prohibition against experimentation on cats and dogs was removed and restrictions generally loosened, though a license and inspection process was put in place (Hall 1984).

Recent Impact

At the beginning of the twenty-first century, the goals of the Society are to "push back the frontiers of knowledge and to improve the quality of life in Britain and globally" (Royal Society 2005). The Society continues to publish the *Philosophical Transactions* as well as other peer-reviewed science publications, and rewards achievement through induction of new Fellows and by bestowing medals and other awards to deserving individuals. The Society also acts as the United Kingdom's Academy of Science, providing scientific advice on science policy issues such as funding, and on public policy issues with a scientific or technical component such as cloning. It further represents UK science internationally. The Society continues to act as a funding agency, providing grant support to researchers as well as resources for science and math teachers.

ETHICS OF SCIENCE. The Royal Society does not have a written ethics policy, though the "quality of life" clause in the Society's mission statement could be taken for the beginnings of one. The statutes of the society allow for expelling a Fellow for conduct injurious to the character or interests of the Society.

During his 2004 Anniversary Address to the Society, Lord Robert May, its president, addressed the work the Society had done over the previous year in assessing scientific rules of conduct, specifically in regards to biological research. Among a variety of other issues, May noted his concerns about the peer review process, the unwillingness of some to consider other scientific views, and publication policies.

SCIENCE IN SOCIETY. In 1985, the Royal Society published a report on the public understanding of society that took the view that the general public did not know enough about science to make informed decisions and that more education was needed to correct this. However, given the negative reaction to the handling of science issues since then, including public concerns about genetically modified foods, the Society's approach to policy issues that affect the public has changed.

One outcome of this change was the establishment of a Science in Society program. This program has several components, one of which, the Dialogue initiative, is set up as a series of workshops between scientists and people of all walks of life. The purpose of these workshops is to develop consensus recommendations on topics of science or technology. The Royal Society carries these recommendations forward to the appropriate policy makers. Recent topics included trust in science, genetic testing, and cybertrust and information security.

Another component of the Science in Society program is a scheme whereby individual Members of Parliament (MPs) and a scientist from their district are paired up and allowed to experience each other's world. The scientists are briefed on the workings of government and accompany their MP during their daily activities. The MPs reciprocate by spending time in the scientist's laboratory. The aim is to both establish mutual understanding as well as to develop relationships.

SCIENCE POLICY. Each year, the Society provides reports on a wide variety of policy issues. In early 2005, the major policy topics included animals in research, bioweapons, climate change, the military use of depleted uranium, the environment, stem cells and cloning, nanoscience and nanotechnology, infectious diseases in livestock, humans in research, and genetically modified plants.

Increasingly these reports include sections summarizing societal concerns and the various ethical viewpoints held by stakeholders. Generally these reports do not choose a specific ethical standpoint; leaving that to society and the legislative process, but there are exceptions. For example, the 2003 report *Measuring Biodiversity for Conservation* takes the view that as a minimum "each generation should pass on a set of opportunities no less than what itself inherited."

RUTH DUERR

SEE ALSO *American Association for the Advancement of Science; National Academies; Newton, Issac.*

BIBLIOGRAPHY

Atkinson, Dwight. (1999). *Scientific Discourse in Sociohistorical Context: The Philosophical Transactions of the Royal Society of London, 1675–1975.* Mahwah, NJ: Lawrence Erlbaum. A rhetorical and multidimensional analysis of the contents of the Philosophical Transactions and how that has changed over the life of the Royal Society.

Hall, Marie. (1984). *All Scientists Now: The Royal Society in the Nineteenth Century.* Cambridge, UK: Cambridge University Press. Recounts the history of the Royal Society during the period where it transitioned from a society consisting largely of wealthy amateurs to its modern form where membership is selected based on scientific distinction.

Haraway, Donna J. (1997). *Modest-Witness@Second-Millennium.FemaleMan-Meets-OncoMouse: Feminism and Technoscience.* New York: Routledge.

Hunter, Michael. (1982). *The Royal Society and its Fellows 1660–1700: The Morphology of an Early Scientific Institution.* Bucks, UK: The British Society for the History of Science. A scholarly analysis of the membership and activities of the Royal Society during its formative years.

Hunter, Michael. (1989). *Establishing the New Science: The Experience of the Early Royal Society.* Woodbridge, UK: Boydell. A series of essays on the activities and problems faced by the Royal Society during its early years.

Purver, Margery. (1967). *The Royal Society: Concept and Creation.* London: Routledge and Kegan Paul. A scholarly assessment of the activities that led to the formation of the Royal Society.

Royal Society of London. (1985). *The Public Understanding of Science.* London: Author.

Royal Society of London. (2003). *Measuring Biodiversity for Conservation.* London: Author.

INTERNET RESOURCE

Royal Society of London. (2005). Available from http://www.royalsociety.org/. A good overview of the Royal Society, its history, and current activities.

RUSSELL, BERTRAND

• • •

Bertrand Arthur William Russell (1872–1970) was a British philosopher, logician, mathematician, and essayist as well as a champion of humanitarian ideals and influential critic of nuclear weapons. Best known as one of the founders of analytic philosophy, Russell was born into an aristocratic family in Trelleck, Monmouthshire, Wales, on May 18. In 1890, he entered Trinity College, Cambridge, where he later held a professorship until he was dismissed in 1916 for writing pacifist propaganda and leading anti-war protests. Russell then traveled, lectured, and continued to write both philosophical treatises and social and moral essays. He rejoined the faculty at Trinity College in 1944 and received the Nobel Prize in Literature and the British Order of Merit in 1950. After World War II, he became a leading figure in the effort to control nuclear weapons proliferation. Russell died at Penrhyndeudraeth, Wales, on February 2.

Logic, Mathematics, and Philosophy

Through his early examination of the philosophy of G. W. Leibniz, Russell became convinced that logical analysis is the most important method for philosophical investigation. So motivated, he set about the tasks of

Bertrand Russell, 1872–1970. The Welsh mathematician, philosopher, and social reformer made original and decisive contributions to logic and mathematics and wrote with distinction in all fields of philosophy. (*The Library of Congress.*)

making logic a more robust and powerful field and clearing away conceptual difficulties that had impeded its progress. One such difficulty was posed by a paradox that Russell himself discovered in 1901: The set of all sets that are not members of themselves is a member of itself if and only if it is not a member of itself. Russell's Paradox undermined naïve set theory, which served as the foundation of mathematics. Russell's own solution to the paradox was his theory of types of sets, which led to the foundation of modern axiomatic set theory. In his seminal work, *Principia Mathematica* (1910–1913), written jointly with Alfred North Whitehead (1861–1947), he attempted to derive all of mathematics from a restricted set of logical axioms. Although undermined by Gödel's proof that some propositions in any axiomatic system of suitable complexity remain undecidable, the formal system was a major intellectual achievement.

Along with G. E. Moore (1873–1958), Russell is credited with founding analytic philosophy, which rejected idealism and what is regarded as meaningless or incoherent philosophy in favor of clear and precise propositions. For Russell the application of analytic meth-

ods to traditional philosophical problems could resolve long-standing disputes. For example, in "On the Relations of Universals and Particulars" (1911) he claimed that logical arguments could resolve the ancient problem of universals. Among his most important contributions to the philosophy of language is his "theory of descriptions" expounded in "On Denoting" (1905). Russell was also a teacher of Ludwig Wittgenstein (1889–1951), the founder of that version of analytic philosophy known as linguistic philosophy, who later eclipsed his mentor in terms of philosophical importance. Karl Popper (1902–1994) and W. V. Quine (1908–2000) were also heavily influenced by Russell, and in fact Popper once referred to him as "the greatest philosopher since Kant" (1976, p. 109).

Science and Technology in Society

In his autobiography (1967–1969), Russell divulged that he was moved by a profound sympathy for the suffering of humankind. This motivated him to write about political and moral issues and to practice social activism. His ethical writings include *Why I Am Not a Christian* (1927) and *Marriage and Morals* (1929), both of which aroused popular antipathy. In fact, he lost a lectureship at City College in New York in 1940 because he was deemed "morally unfit" to teach. Russell's experiments in social and political activism included peace protests during World War I (for which he served six months in jail), three unsuccessful campaigns for a seat in Parliament, and founding and operating an experimental school from the late 1920s to the early 1930s. He also served as president of the International War Crimes Tribunal in 1967, which investigated the conduct of the United States during the Vietnam War.

Russell's views about the role of science in society are outlined in such works as Icarus, or the Future of Science (1924), in which he fears "that science will be used to promote the power of dominant groups, rather than to make men happy. Icarus, having been taught to fly by his father Daedalus, was destroyed by his rashness. I fear that the same fate may overtake the populations whom modern men of science have taught to fly" (p. 1).

In *The Impact of Science on Society* (1951) Russell discussed the potential for science to be utilized for mass psychological propaganda, and he made an unsettling observation about the potential for biological warfare to limit human population growth. In a 1958 essay, "The Divorce between Science and 'Culture'," he argued that governments and citizens must have better science education in order to avoid the potential disasters presented by modern science and technology.

Although he maintained a general optimism about science, including some controversial applications, Russell was concerned about a cultural lag in which human knowledge was expanding more quickly than the ability to utilize it wisely. Nowhere was this concern more evident than in his efforts to fight nuclear weapons and their international proliferation. The opening lines of "The Bomb and Civilization" (1945) expressed both his faith in science and his panic about how science can be easily misused: "It is impossible to imagine a more dramatic and horrifying combination of scientific triumph with political and moral failure than has been shown to the world in the destruction of Hiroshima." It should be noted, however, that while the United States still had a monopoly on nuclear arms, Russell advocated a preemptive war against Stalin, whom he argued was as evil as Hitler (Johnson 1989).

In 1954 Russell delivered his "Man's Peril" broadcast on the BBC, condemning the hydrogen bomb test at Bikini Atoll. The following year Russell and Albert Einstein issued the Russell-Einstein Manifesto, which called for a conference of scientists to discuss "what steps can be taken to prevent a military contest of which the issue must be disastrous to all parties?" This manifesto stimulated the first Pugwash Conference on Science and World Affairs in 1957.

In 1958, Russell became the founding president of the Campaign for Nuclear Disarmament (CND), which promoted nonviolent demonstrations to eradicate nuclear weapons and other weapons of mass destruction. In 1961 (at age 89), he was imprisoned for one week in connection with anti-nuclear protests. Two years later, he established the Bertrand Russell Peace Foundation, to promote his vision of peace, human rights, and social justice. Russell's last essay, "1967," took up the imminent doom presented by nuclear weapons in the scenario of obstinate sovereign states and argued that the only solution is to realize that "peace is the paramount interest of everybody."

CARL MITCHAM
VOLKER FRIEDRICH

SEE ALSO *Atomic Bomb; Haldane, J. B. S.; Pugwash Conferences.*

BIBLIOGRAPHY

Johnson, Paul. (1989). *Intellectuals.* New York: Harper and Row. Exposes the hypocrisies in the private lives of several intellectuals. The pertinent section on Russell is pp. 204–207.

Popper, Karl. (1976). *Unended Quest: An Intellectual Autobiography,* rev. edition. LaSalle, IL: Open Court.

Russell, Bertrand. (1903). *Principles of Mathematics.* London: Cambridge University Press.

Russell, Bertrand. (1905). "On Denoting." In *Mind,* Vol. 14. Reprinted in Bertrand Russell, 1973, *Essays in Analysis* (London: Allen and Unwin).

Russell, Bertrand. (1911). "On the Relations of Universals and Particulars." *Proceedings of the Aristotelian Society* 12: 1–24. Reprinted in Bertrand Russell, 1956, *Logic and Knowledge* (London: Allen and Unwin).

Russell, Bertrand. (1924). *Icarus, or the Future of Science.* New York: E.P. Dutton.

Russell, Bertrand. (1927). *Why I Am Not a Christian.* London: Watts.

Russell, Bertrand. (1929). *Marriage and Morals.* London: George Allen and Unwin.

Russell, Bertrand. (1951). *The Impact of Science on Society.* New York: Columbia University Press.

Russell, Bertrand. (1967, 1968, 1969). *The Autobiography of Bertrand Russell,* 3 vols. London: George Allen and Unwin.

Whitehead, Alfred, and Bertrand Russell. (1910–1913). *Principia Mathematica,* 3 vols. Cambridge: Cambridge University Press.

INTERNET RESOURCES

Russell, Bertrand. (1945). "The Bomb and Civilization." *Glasgow Forward* 39(33): 1, 3. Available from: http://www.humanities.mcmaster.ca/~russell/brbomb.htm#table

Russell, Bertrand. (1967). "1967," Available from: http://www.humanities.mcmaster.ca/~russell/bressay.htm

RUSSIAN PERSPECTIVES

• • •

Russian perspectives on science, technology, and ethics come from two sources: those outside and those inside Russia. Because of the historical impact of the Communist Revolution of 1917, the absorption of Russia into the Soviet Union (1922–1991) for much of the twentieth century, the role of Marxism as the official Soviet ideology, and a strong expatriate intellectual community, scholars outside Russia have created a substantial body of literature analyzing Russian-Soviet-Marxist-Communist perspectives on science and technology, including much related to ethics. While referencing some of this literature, the present entry nevertheless emphasizes discussions as they have developed within Russia itself.

Russian discussions of ethics in relation to science and technology have exhibited both strong positivist commitments to scientific and technological progress and equally vigorous criticisms of science and technology as destructive of traditional Russian values. A brief introduction to these discussions, emphasizing technology, may be divided into three periods: pre-Soviet, Soviet, and post-Soviet. The post-Soviet period has revived and extended some perspectives prominent during the pre-Soviet period.

Originating Discussions

Pre-Soviet Russian history may be divided from the point of view of the scientific and technological progress into three major periods. The first runs from the invasion of the legendary Scandinavian warrior Rurik in the 800s through Mongol (or Tartar) invasions in the 1200s to the rise of Ivan the Terrible in the 1500s and then to the beginning of the Romanov reign in the 1600s. The second takes place during the reign of Peter the Great (1682–1725). In his lifetime, two special schools for training engineers were established, the Engineering School in 1700, and the Mathematical-Navigation School in 1701. Peter the Great introduced engineering training into the Naval Academy, regimental schools, and even religious colleges. He founded the St. Petersburg Academy of Sciences in 1724. As the great modernizer of Russia, it was Peter who brought modern science and technology into the motherland, and thus it was during this second period that discussions relevant to science, technology, and ethics increasingly came to the fore.

The third period begins from the foundation of the first high engineering schools and runs to the Communist Revolution (1917). In 1809, the Institute of the Corps of Engineers of Rail Transport was set up in Russia for theoretical training for engineers and higher technological education. At that time, many vocational and secondary technical schools had already been transformed into higher technical schools and institutes. The Technological Institute in St. Petersburg, for example, had been created in 1862 as a school for foremen from the lower social strata, such as peasants and artisans. In Moscow, a Higher Technical School was established in 1868 following the reorganization of a vocational school (dating from 1830). These new higher educational establishments concentrated on the theoretical side of their curricula (Gorokhov 1998).

One of the most important contributors to such discussions was the Russian engineer Peter K. Engelmeyer (1855–1942). Engelmeyer's positivism is evident in the following words: "Our nineteenth, technological century is . . . the century of unprecedented conquest of the forces of nature. Technology has conquered for us space and time, matter and power, being the power itself that irrepressibly turns the wheel of progress" (Engelmeyer 1898, p. 6). For Engelmeyer the technological worldview dominated the nineteenth century because of an inward tendency of European culture to address real problems with real power. The genius of humanity over the previous two centuries had constructed a human-made microcosm within the larger natural one, making it possible for human beings to satisfy their physical needs to an extent previously unknown. Because of this Engelmeyer saw engineers as the leaders or technological elite in society, and argued for a new system of engineering education to promote the realization of this ideal. The emergence of technocracy in the twentieth century revealed how "efficient" such societal management can be. But it was difficult to anticipate the unintended consequences of this boundless scientific and technological progress, especially in the military sphere.

During this same period Russia was also home to an opposed school of religious and cultural criticism of technology. Sergei N. Bulgakov (1871–1944), in an article titled "The Main Problems of the Theory of Progress," published in 1902, emphasized that in the twentieth century technological change was becoming a kind of theology. By means of modern technology all people of the future were supposed to be happy, proud, and free. To bring happiness to as many people as possible was taking the form of a super modern religion in which society equipped with technological knowledge played the role of God. But according to Bulgakov such technological optimism, which tries to create a material heaven on Earth and even obtain cosmic power, inevitably leads to immoral practices. Technology begins to dominate human beings rather than serve them, making them not happy but miserable. The state, having become the patron of science and technology, inevitably begins to demand that science and technology serve economic and military ends.

During the Soviet Period

In the seven decades from the Communist Revolution to the collapse of the Soviet Union, science and technology were treated in two different ways. On the one hand, they were given unquestioned ideological support; socialism itself was said to be scientific and to provide the strongest support for technology. On the other, political interference in both science and technology compromised their autonomy and efficiency.

The common view in the West that this was simply a corruption of science and technology has been challenged by, for instance, Nikolai Krementsov (1997). Krementsov distinguishes the period of the initial Stalinization of science (1929–1939), its achievements during World War II and up to Joseph Stalin's death (1940–1953), and the post-Stalin consolidation. For Krementsov, Soviet science was "big science" that, as in the United States, it involved a convergence of party-state agencies and the scientific community. Its dramatic achievements—from the atomic and hydrogen bombs (1949 and 1953) to *Sputnik I* (1957)—should not be overlooked. Even in areas of health and medicine Soviet science realized important human benefits. As Vadim J. Birstein (2001) and others have documented, however, science was also used to experiment on human beings; like scientific experimentation that amounts to torture anywhere, this presents a major challenge to the ethics of the scientific community.

> Yet from the beginning of the 1930s, the general ideological atmosphere in the Soviet Union radically changed; from now on the only way to create the new human being was to be sought not in biological, but in social changes. ... Meanwhile a lot of medical research done in the Soviet Union sometimes posed ethical and legal problems. The first attempt on the part of the authorities to regulate medical research took place in 1936. Narkomzdrav [the name of the Ministry of Health at that time] of the Russian Federation issued regulations determining the conditions of testing new medical devices and methods, which could be dangerous to the health and life of patients. ... These rather progressive regulations, however, were issued at the same time, when in the depths of the KGB, the secret "Laboratory X" worked on the creation and testing of toxic substances. ... There are some indications that the laboratory tried to create toxins which could be impossible to detect after victim's death; these substances were tested on prisoners." (Yudin 2004)

During the post-Stalin era impressive attempts were made to adopt cybernetics in order to deal with the emerging problems of a command model of science and technology policy. Additionally, the theory of a new Scientific Technology Revolution (STR) that integrated science and technology anticipated by decades Western European notions of technoscience—and sought to maintain a close link between technoscience and social values.

Among the most insightful non-Russian scholars of Russian science and technology in relation to questions of ethics and politics is Loren R. Graham. In *What Have We Learned about Science and Technology from the Russian Experience?* (1998), he summarizes a life of research on this topic. Although he admits that this short book is more about science and technology than Russia, it nevertheless draws useful conclusions about science and technology in Russia. According to Graham,

> The enormous Soviet scientific establishment, the world's largest, performed rather well in many areas, provided for the nation's military strength, and supplied most of the needs of heavy industry. But it did not do so well in terms of intellectual breakthroughs or outstanding achievements.... Political freedom may not be as necessary for the development of natural science as many of its advocates have claimed, but a combination of political freedom and generous financial support *are* necessary for the most creative achievements. One of the tragedies of Russian history is that science there has never enjoyed both financial support and political freedom, either under the Soviet system or today, although, in chronological sequence, it had first the one and then the other. (pp. 132–133)

Another tragedy, however, is the degree to which despite all the rhetoric about their socialist-humanist character under Communism, from the 1930s through the 1980s Soviet science and technology was also deeply antihuman and destructive of the environment.

Post-Soviet Discussions

One major reason for the collapse of the Soviet Union was its failures in regard to the development of an ethics of science and technology that was anything more than their simple promotion for political purposes. The ideology that science and technology might perfect the future of humanity makes no difference to the happiness of the present generation. Indeed, the contemporary squandering of natural resources and contamination of the environment are sacrifices of the future as well as the present, and call for the response of a new ethics (Danilov-Danilian 1999). It is just such a felt need to rethink the uses of science and technology that has led to a reconsideration of the ideas of some of those who were driven out of Russia by the Soviet regime.

One of these thinkers whose ideas have been resurrected is Nikolai Berdyaev (1874–1948). From the 1930s Berdyaev argued that the domination of technology would destroy the person and lead inevitably to dehumanization. To struggle against the hegemony of technology was thus necessary to save humanity. Once everything can be transformed or constructed then this power will be applied even to the human psyche. This precisely was embodied in the unprecedented program for the remolding of the people from the capitalist past in the forge of socialist reconstruction (Gorokhov 1992).

For Berdyaev technology is dehumanizing because it opposes the humanistic ideals of Renaissance culture. But Renaissance ideals also place human beings in an antagonistic relationship with the environment. The main contradiction of contemporary technological civilization is that modern technology creates unprecedented opportunities for human beings to invent needs and wants, which are then satisfied by destroying the natural world. Berdyaev sees the basic problem as a split between indifferent and apocalyptic attitudes toward technology. The former interprets technology as a personal matter of inventors and engineers, and assumes no responsibility for the results of human activity. The latter interprets technology as anathema, the triumph of the Antichrist. But neither response is satisfactory. One contemporary alternative has been the Russian "cosmicism" (Stepin 2002), which "opposes physicalist thinking in order to develop ideas of unity between human beings and the cosmos," both in religious and natural scientific terms.

Along with the work of Berdyaev, the thought of Bulgakov has also once again become important in Russia. Although he was educated initially as an economist with Marxist sympathies, Bulgakov's studies of agrarian life led him to criticize Marxist proposals for the centralization of agriculture. Then in the early 1900s, after a religious crisis, he rejected Marxism completely in favor of a "sophiological" interpretation of Russian orthodoxy and undertook studies for the priesthood. After teaching political economy and theology at a university in the Crimea, in 1922 he was exiled from Russia and eventually took part in establishing the Institute of Orthodox Theology (St. Sergius Theological Institute) in Paris, where he remained until his death.

For Bulgakov human beings must accept their own nature as well as the natural environment as given. To reject either nature is to invite disaster, personal or environmental. To live with the impression of their ever-increasing power may open boundless vistas for "cultural creativity," but it also places humans in increasing danger. The way out of the antagonism between economic activity based on scientific research into the mechanisms of nature and nature itself is the gradual "digestion" of the human-made back into the natural. Bulgakov's philosophy stimulates discussions of low-waste and environmentally friendly technology, as has indeed been the case in Russia during the early 2000s—although against the background of the triumphant march of technological civilization, such an appeal remains the voice of one crying in the wilderness. Yet contemporary efforts to develop a theory of sustainable development correlates to a great extent with the ideas of Bulgakov.

In post-Soviet Russia it is thus common to argue that there are limits to scientific and technological progress. It is not possible to realize, implement, or produce only what is planned, designed, and projected in scientific forecasts; not all the negative effects of the technological activity can be accurately projected. It is only possible to foresee certain risks with new scientific technologies. But this requires the development of moral responsibility in science and professional ethics in engineering. Yet the invention of nuclear weapons and other large-scale technologies has also revealed the limits of individual ethical responsibility for those operating in sociotechnical systems (*Inshenernaja etika* 1998). In biotechnology and genetic engineering there is also a need to develop a scientific and engineering ethics that would guide natural scientific and engineering research (Frolov and Yudin 1989).

An increasing interest in environmental ethics has thus become a significant part of Russian discussions. No longer can humans trust in the power of nature to take care of itself.

> The natural mechanisms are not sufficient at present to preserve the biosphere. New methods for regulations, based on the understanding of natural processes and to some degree also the management of such processes, are required. Anthropogenic regulation can forestall natural cataclysms and decrease the speed of dangerous processes. We must choose between immediate profit and long-term revenues in the usage of natural resources. (Marfenin 2000, p. 8)

In Russia there is concern that when human beings are too eager to dominate nature with science and technology, they may destroy nature and, at the same time, their ongoing economic growth. When humans threaten the biosphere as a whole they also threaten human society. The alternative is a new paradigm in science and technology based on an equal partnership between humans and the environment (Danilov-Danilian and Losev 2000).

Such critical reflections point toward the need for ethical assessments of science and technology. In the words of Stepin again:

> Scientific cognition and technological activity . . . involve a wide range of possible development trajectories . . . and are always faced with the problem of choosing a certain scenario out of the variety of possible scenarios of development. And

the landmarks for this choice are not only knowledge but also the moral principles that ban the methods of experiment and transformation that are dangerous for people. More and more often contemporary complex research programs and technological projects require the social expertise that includes some ethical components.... Human society must find the way-out of the global crises, but to do this we shall have to come through an epoch of spiritual transformation and elaboration of a new system of values. (Stepin 1988, pp. 19–20)

Concern for the practical elaboration of a new paradigm of scientific and technological development, one that does not separate theory and practice nor ethical responsibilities and scientific-technological power, that respects both society and nature, thus animates current Russian perspectives on science, technology, and ethics.

VITALY GOROKHOV

SEE ALSO *Berdyaev, Nikolai; Communism; Lysenko Affair; Marxism; Sakharov, Andrei; Tolstoy, Leo.*

BIBLIOGRAPHY

Bailes, Kendall E. (1978). *Technology and Society under Lenin and Stalin: Origins of the Soviet Technical Intelligentsia, 1917–1941.* Princeton, NJ: Princeton University Press. One of the most comprehensive and insightful externalist studies.

Berdyaev, Nikolai. (1949). *Der Mensch und der Technik.* Berlin: Cornelsen. English version of one essay in this book, "Man and Machine," included in *Philosophy and Technology,* ed. Carl Mitcham and Robert Mackey (New York: Free Press, 1972).

Birstein, Vadim J. (2001). *The Perversion of Knowledge: The True Story of Soviet Science.* Boulder, CO: Westview Press.

Bulgakov, Sergei N. (1990 [1912]). *Filosofija Khoziaistva.* Moscow: Nauka. Translation by Catherine Evtuhov as *Philosophy of Economy: The World as Household* (New Haven, CT: Yale University Press, 2000).

Danilov-Danilian, Viktor I. (1999). "Novaja etika i ekologicheskij vysov" [New ethics and environmental challenge]. In his *Ustoichivoie razvitie i problemy ekologicheskoj politiki* [Sustainable development and problems of environmental policy]. *ECOS: Federalnyj vestnik ekologicheskogo prava* [Federal journal of environmental law], no. 5: 75–91.

Danilov-Danilian, Viktor I., and Kim S. Losev. (2000). *Ekologicheskij vyzov i ustojchivoje razvitije* [Ecological challenge and sustainable development]. Moscow: Progress-Traditzija.

Engelmeyer, Peter K. (1898). *Tekhnicheskij itog 19 veka* [The technological outcome of the nineteenth century]. Moscow: Tip. Kaznacheeva.

Frolov, Ivan T., and Boris G. Yudin. (1989). *The Ethics of Science: Issues and Controversies,* trans. Lilia Nakhapetyan and Valentin Parnakh. Moscow: Progress Publishers.

Gorokhov, Vitaly G. (1992). "Politics, Progress, and Engineering: Technical Professionals in Russia." In *Democracy in a Technological Society,* ed. Langdon Winner. Dordrecht, Netherlands: Kluwer Academic.

Gorokhov, Vitaly G. (1997). *Peter Klimentievich Engelmeyer: Ingener mekhanik i filosopf tekhniki (1855–1941)* [P. K. Engelmeyer: Mechanical engineer and philosopher of technology (1855–1941)]. Moscow: Nauka.

Gorokhov, Vitaly G. (1998). "Technological Enlightenment in Russia in the 19th and Early 20th Century and the Problems of Advancement in the Philosophy of Technology." In *Techne* 3(2).

Gorokhov, Vitaly G. (2001). *Technikphilosophie und Technikfolgenforschung in Russland.* Bad Neuenahr-Ahrweiler, Germany: Europäische Akademie zur Erforschung von Folgen wissenschaftlich-technischer Entwicklungen.

Gorokhov, Vitaly G. (2003). "Philosophie der Technik von P. K. Engelmeyer als technischer Optimismus." In *Jahrbuch des Deutsch-Russisches Kollegs 2001–2002,* ed. Vitaly G. Gorokhov. Aachen: Shaker Verlag.

Graham, Loren R. (1972). *Science and Philosophy in the Soviet Union.* New York: Knopf. Sees dialectical materialism as a major intellectual achievement that serves as a unifying theme across Soviet physics, cosmology, genetics, biology, chemistry, cybernetics, physiology, and psychology. See also Graham's *Science, Philosophy, and Human Behavior in the Soviet Union* (New York: Columbia University Press, 1987) and *Science in Russia and the Soviet Union* (Cambridge, UK: Cambridge University Press, 1993).

Graham, Loren R. (1993). *The Ghost of the Executed Engineer: Technology and the Fall of the Soviet Union.* Cambridge, MA: Harvard University Press.

Graham, Loren R. (1998). *What Have We Learned about Science and Technology from the Russian Experience?* Stanford, CA: Stanford University Press.

Inshenernaja etika v Rossii i USA: Istorija i sozialno politicheskij kontekst [Engineering ethics: History, context, and significance. Abstracts. American-Russian Workshop]. (1997). Moscow: Akademija menegementa innovatyij.

Inshenernaja etika v siseme obrayovanija [Designing engineering ethics education in Russia. Abstracts. American-Russian Workshop]. (1998). Moscow: Akademija menegementa innovatyij.

Krementsov, Nikolai. (1997). *Stalinist Science.* Princeton, NJ: Princeton University Press. Balanced account by a Russian historian of science, with some emphasis on the case of T. D. Lysenko.

Marfenin, Nikolai N. (2000). "Ecology and Humanism." In *State of Russia in the Surrounding World, 2000: The Analytical Series,* abstracts, ed. Nikolai N. Marfenin. Moscow: IIUEPS Press, 8–9.

Medvedev, Zhores A. (1978). *Soviet Science.* New York: Norton. A short historical overview by an insider arguing the moral of a need to protect science from possible misuse.

Stepin, Viatcheslav S. (1998). "Ustojchivoje razvitije i problema tzennostej" [Sustainable development and the problem of values]. In *Tekhnika, obshestvo, okrushajushjaja Sreda: Materialy meshdunarodnoj nauchnoj konferentyii (18–19.06.1998, Moskva)* [Technology, society, and environment: Proceedings of an International Conference, 18–19 June 1998, Moscow], ed. Vitaly G. Gorokhov. Moscow: RAN, MNEPU.

Stepin, Viatcheslav S. (2002). *Civilization, Science, Culture.* Karlsruhe: German Russian College, University of Karlsruhe; Moscow: State University of Humanities, Institute of Philosophy of the Russian Academy of Sciences.

Stepin, Viatcheslav S. (2005). *Theoretical Knowledge.* Synthese Library, Vol. 326. Heidelberg: Springer.

Yudin, Boris. (2004). "Medical Ethics in Russia, History, Contemporary Period." In *Encyclopedia of Bioethics*, 3rd edition. New York: Macmillan Reference USA.

Yudin, Boris. (2004). "Human Experimentation in Russia/the Soviet Union." In *Twentieth Century Ethics of Human Subjects Research*, eds. V. Roelcke and G. Maio. Stuttgart: Franz Steiner Verlag.